Communications
in Computer and Information Science 771

Commenced Publication in 2007
Founding and Former Series Editors:
Alfredo Cuzzocrea, Xiaoyong Du, Orhun Kara, Ting Liu, Dominik Ślęzak,
and Xiaokang Yang

More information about this series at http://www.springer.com/series/7899

Jinfeng Yang · Qinghua Hu
Ming-Ming Cheng · Liang Wang
Qingshan Liu · Xiang Bai
Deyu Meng (Eds.)

Computer Vision

Second CCF Chinese Conference, CCCV 2017
Tianjin, China, October 11–14, 2017
Proceedings, Part I

 Springer

Editors
Jinfeng Yang
Civil Aviation University of China
Tianjin
China

Qinghua Hu
Tianjin University
Tianjin
China

Ming-Ming Cheng
Nankai University
Tianjin
China

Liang Wang
Institute of Automation
Chinese Academy of Sciences
Beijing
China

Qingshan Liu
Nanjing University of Information
    Science and Technology
Nanjing
China

Xiang Bai
Huazhong University of Science
    and Technology
Wuhan
China

Deyu Meng
Xi'an Jiaotong University
Xi'an
China

ISSN 1865-0929   ISSN 1865-0937   (electronic)
Communications in Computer and Information Science
ISBN 978-981-10-7298-7   ISBN 978-981-10-7299-4   (eBook)
https://doi.org/10.1007/978-981-10-7299-4

Library of Congress Control Number: 2017960863

This Springer imprint is published by Springer Nature
The registered company is Springer Nature Singapore Pte Ltd.
The registered company address is: 152 Beach Road, #21-01/04 Gateway East, Singapore 189721, Singapore

Preface

Welcome to the proceedings of the Second Chinese Conference on Computer Vision (CCCV 2017) held in Tianjin!

CCCV, hosted by the China Computer Federation (CCF) and co-organized by Computer Vision Committee of CCF, is a national conference in the field of computer vision. It aims at providing an interactive communication platform for students, faculties, and researchers from industry. It promotes not only academic exchange, but also communication between academia and industry. CCCV is one of the most important local academic activities in computer vision. In order to keep track of the frontier of academic trends and share the latest research achievements, innovative ideas, and scientific methods in the field of computer vision, international and local leading experts and professors are invited to deliver keynote speeches, introducing the latest theories and methods in the field of computer vision.

The CCCV 2017 received 465 full submissions. Each submission was reviewed by at least two reviewers selected from the Program Committee and other qualified researchers. Based on the reviewers' reports, 174 papers were finally accepted for presentation at the conference. The acceptance rate is 37.4%. The proceedings of the CCCV 2017 are published by Springer.

We are grateful to the keynote speakers, Prof. Katsushi Ikeuchi from University of Tokyo, Prof. Sven Dickinson from University of Toronto, Prof. Gérard Medioni from University of Southern California, Prof. Marc Pollefeys from ETH Zurich, Prof. Xilin Chen from Institute of Computing Technology, Chinese Academy of Science.

Thanks go to the authors of all submitted papers, the Program Committee members and the reviewers, and the Organizing Committee. Without their contributions, this conference would not be a success. Special thanks go to all of the sponsors and the organizers of the five special forums; their support made the conference a success. We are also grateful to Springer for publishing the proceedings and especially to Ms. Celine (Lanlan) Chang of Springer Asia for her efforts in coordinating the publication.

We hope you find CCCV 2017 an enjoyable and fruitful conference.

Tieniu Tan
Hongbin Zha
Jinfeng Yang
Qinhua Hu
Ming-Ming Cheng
Liang Wang
Qingshan Liu

Organization

CCCV 2017 (CCF Chinese Conference on Computer Vision 2017) was hosted by the China Computer Federation (CCF) and co-organized by the Civil Aviation University of China, Tianjin University, Nankai University, and CCF Technical Committee on Computer Vision.

Organizing Committee

International Advisory Board

Bill Freeman	Massachusetts Institute of Technology, USA
Gerard Medioni	University of Southern California, USA
Ramin Zabih	Cornell University, USA
David Forsyth	University of Illinois at Urbana-Champaign, USA
Jitendra Malik	University of California, Berkeley, USA
Fei-Fei Li	Stanford University, USA
Ikeuchi Katsushi	The University of Tokyo, Japan
Song-chun Zhu	University of California, Los Angeles, USA
Long Quan	The Hong Kong University of Science and Technology, SAR China

Steering Committee

Tieniu Tan	Institute of Automation, Chinese Academy of Sciences, China
Hongbin Zha	Peking University, China
Xinlin Chen	Institute of Computing Technology, Chinese Academy of Sciences, China
Jianhuang Lai	Sun Yat-Sen University, China
Dewen Hu	National University of Defense Technology, China
Shengyong Chen	Tianjin University of Technology, China
Yanning Zhang	Northwestern Polytechnical University, China
Tao Wang	IQIYI Inc.

General Chairs

Tieniu Tan	Institute of Automation, Chinese Academy of Sciences, China
Hongbin Zha	Peking University, China
Jinfeng Yang	Civil Aviation University of China

Program Chairs

Qinghua Hu   Tianjin University, China
Ming-Ming Cheng   Nankai University, China
Liang Wang   Institute of Automation, Chinese Academy of Sciences,
                            China
Qingshan Liu   Nanjing University of Information Science
                            and Technology, China

Organizing Chairs

Jufeng Yang   Nankai University, China
Pengfei Zhu   Tianjin University, China

Publicity Chairs

Qiguang Miao   Xidian University, China
Wei Jia   Hefei University of Technology, China

International Liaison Chairs

Andy Yu   ShanghaiTech University, China
Zhanyu Ma   Beijing University of Posts and Telecommunications,
                            China

Publication Chairs

Xiang Bai   Huazhong University of Science and Technology,
                            China
Deyu Meng   Xi'an Jiaotong University, China

Tutorial Chairs

Zhouchen Lin   Peking University, China
Xin Geng   Southeast University, China

Workshop Chairs

Zhaoxiang Zhang   Institute of Automation, Chinese Academy of Sciences,
                            China
Rongrong Ji   Xiamen University, China

Sponsorship Chairs

Yongzhen Huang   Watrix Technology
Huchan Lu   Dalian University of Technology, China

Demo Chairs

Liang Lin   Sun Yat-Sen University, China
Chao Xu   Tianjin University, China

Competition Chairs

Wangmeng Zuo	Harbin Institute of Technology, China
Jiwen Lu	Tsinghua University, China

Website Chairs

Changqing Zhang	Tianjin University, China
Shaocheng Han	Civil Aviation University of China

Finance Chairs

Guimin Jia	Civil Aviation University of China
Ruiping Wang	Institute of Computing Technology, Chinese Academy of Sciences, China

Coordination Chairs

Lifang Wu	Beijing University of Technology, China
Shiying Li	Hunan University, China

Program Committee

Haizhou Ai	Tsinghua University, China
Xiang Bai	Huazhong University of Science and Technology, China
Yinghao Cai	Chinese Academy of Sciences, China
Xiaochun Cao	Chinese Academy of Sciences, China
Hui Ceng	University of Science and Technology Beijing, China
Hongbin Cha	Peking University, China
Zhengjun Cha	University of Science and Technology of China, China
Hongxin Chen	HISCENE
Shengyong Chen	Zhejiang University of Technology, China
Songcan Chen	Nanjing University of Aeronautics and Astronautics, China
Xilin Chen	Chinese Academy of Science, China
Yongquan Chen	HuanJing Information Technology Inc., China
Hong Cheng	University of Electronic Science and Technology of China
Jian Cheng	Chinese Academy of Science, China
Mingming Cheng	Nankai University, China
Jun Chu	Nanchang Hangkong University, China
Yang Cong	Chinese Academy of Science, China
Chaoran Cui	Shandong University of Finance and Economics, China
Cheng Deng	Xidian University, China
Weihong Deng	Beijing University of Posts and Telecommunications, China
Xiaoming Deng	Chinese Academy of Science, China

Jing Dong	Chinese Academy of Science, China
Junyu Dong	Ocean University of China
Weisheng Dong	Xidian University, China
Fuqing Duan	Beijing Normal University, China
Bin Fan	Chinese Academy of Science, China
Xin Fan	Dalian University of Technology, China
Yuchun Fang	Shanghai University, China
Jufu Feng	Peking University, China
Jianjiang Feng	Tsinghua University, China
Xianping Fu	Dalian Maritime University, China
Shenghua Gao	ShanghaiTech University, China
Yong Gao	JieShang Visual technology Inc., China
Shiming Ge	Chinese Academy of Science, China
Xin Geng	Southeast University, China
Guanghua Gu	Yanshan University, China
Jie Gui	Chinese Academy of Science, China
Yulan Guo	National University of Defense Technology, China
Zhenhua Guo	Tsinghua University, China
Aili Han	Shandong University, China
Junwei Han	Northwestern Polytechnical University, China
Huiguang He	Chinese Academy of Science, China
Lianghua He	Tongji University, China
Ran He	Chinese Academy of Science, China
Yutao Hou	NVIDIA Semiconductor Technical Services Inc.
Zhiqiang Hou	Air Force Engineering University
Haifeng Hu	Sun Yat-sen University, China
Hua Huang	Beijing Institute of Technology, China
Di Huang	Beijing University of Aeronautics and Astronautics
Kaiqi Huang	Chinese Academy of Science, China
Qingming Huang	University of Chinese Academy of Sciences
Yongzhen Huang	Chinese Academy of Science, China
Yanli Ji	University of Electronic Science and Technology of China
Rongrong Ji	Xiamen University, China
Wei Jia	Chinese Academy of Science, China
Tong Jia	Northeastern University, China
Yunde Jia	Beijing Institute of Technology, China
Muwei Jian	Ocean University of China
Yugang Jiang	Fudan University, China
Shuqiang Jiang	Chinese Academy of Science, China
Cheng Jin	Fudan University, China
Lianwen Jin	South China University of Technology, China
Xiaoyuan Jing	Wuhan University, China
Xiangwei Kong	Dalian University of Technology, China
Jianhuang Lai	Sun Yat-sen University, China
Zhen Lei	Chinese Academy of Science

Hua Li	Chinese Academy of Science, China
Jian Li	National University of Defense Technology
Jia Li	Beijing University of Aeronautics and Astronautics
Teng Li	Anhui University, China
Xi Li	Zhejiang University, China
Ce Li	Lanzhou University of Technology, China
Chaofeng Li	Jiangnan University, China
Chunguang Li	Beijing University of Posts and Telecommunications, China
Chunming Li	University of Electronic Science and Technology of China
Jianguo Li	Intel
Jinping Li	University of Jinan, China
Lingling Li	Zhengzhou Institute of Aeronautical Industry Management, China
Peihua Li	Dalian University of Technology, China
Shiying Li	Hunan University, China
Yongjie Li	University of Electronic Science and Technology of China, China
Zechao Li	Nanjing University of Science and Technology, China
Ziyang Li	Chinese Academy of Science, China
Zongmin Li	China University of Petroleum, China
Shengcai Liao	Chinese Academy of Science, China
Liang Lin	Sun Yat-sen University, China
Weiyao Lin	Shanghai Jiao Tong University, China
Zhouchen Lin	Peking University, China
Chenglin Liu	Chinese Academy of Science, China
Haibo Liu	Harbin Engineering University, China
Heng Liu	Anhui University of Technology, China
Hong Liu	Peking University, China
Huaping Liu	Tsinghua University, China
Jing Liu	Chinese Academy of Science, China
Li Liu	National University of Defense Technology, China
Qingshan Liu	Nanjing University of Information Science and Technology, China
Shigang Liu	Shanxi Normal University, China
Weifeng Liu	China University of Petroleum, China
Yiguang Liu	Sichuan University, China
Yue Liu	Beijing Institute of Technology, China
Guoliang Lu	Shandong University, China
Huchuan Lu	Dalian University of Technology
Xiaoqiang Lu	Xi'an Institute of Optics and Precision Mechanics, Chinese Academy of Sciences, China
Jiwen Lu	Tsinghua University, China
Feng Lu	Beijing University of Aeronautics and Astronautics, China

Yao Lu	Beijing Institute of Technology, China
Bin Luo	Anhui University, China
Ke Lv	University of Chinese Academy of Sciences, China
Bingpeng Ma	University of Chinese Academy of Sciences, China
Huimin Ma	Tsinghua University, China
Wei Ma	Beijing University of Technology, China
Zhanyu Ma	Beijing University of Posts and Telecommunications, China
Lin Mei	Third Institute of Public Security
Deyu Meng	Xi'an Jiaotong University, China
Qiguang Miao	Xidian University, China
Rongrong Ni	Beijing Jiaotong University, China
Xiushan Nie	Shandong University of Finance and Economics
Jifeng Ning	Northwest A&F University, China
Jianquan Ouyang	Xiangtan University, China
Yuxin Peng	Peking University, China
Yu Qiao	Shenzhen Institutes of Advanced Technology, Chinese Academy of Sciences, China
Lei Qin	Chinese Academy of Sciences, China
Pinle Qin	North University of China
Jianhua Qiu	Beijing Wisdom Eye Technology, China
Chuanxian Ren	Sun Yat-sen University, China
Qiuqi Ruan	Beijing Jiaotong University, China
Nong Sang	Huazhong University of Science and Technology, China
Shiguang Shan	Chinese Academy of Science, China
Shuhan Shen	Chinese Academy of Science, China
Jianbing Shen	Beijing Institute of Technology, China
Linlin Shen	Shenzhen University, China
Peiyi Shen	Xidian University, China
Wei Shen	Shanghai University, China
Mingli Song	Zhejiang University, China
Fei Su	Beijing University of Posts and Telecommunications, China
Hang Su	Tsinghua University, China
Dongmei Sun	Beijing Jiaotong University, China
Jian Sun	Xi'an Jiaotong University, China
Taizhe Tan	Guangdong University of Technology, China
Tieniu Tan	Chinese Academy of Science, China
Xiaoyang Tan	Nanjing University of Aeronautics and Astronautics, China
Jin Tang	Anhui University, China
Jinhui Tang	Nanjing University of Science and Technology, China
Yandong Tang	Chinese Academy of Science, China
Zengfu Wang	Chinese Academy of Science, China
Hanzi Wang	Xiamen University, China

Hanli Wang	Tongji University, China
Hongyuan Wang	Changzhou University, China
Hongpeng Wang	Harbin Institute of Technology, China
Jinjia Wang	Yanshan University, China
Jingdong Wang	Microsoft Research
Liang Wang	Chinese Academy of Science, China
Qi Wang	Northwestern Polytechnical University
Qing Wang	Northwestern Polytechnical University
RuiPing Wang	Chinese Academy of Science, China
Shengke Wang	Ocean University of China
Shiquan Wang	Philips Research Institute of China
Sujing Wang	Chinese Academy of Science, China
Tao Wang	IQIYI Inc.
Wei Wang	Chinese Academy of Science, China
Yuanquan Wang	Hebei University of Technology, China
Yuehuan Wang	Huazhong University of Science and Technology, China
Yunhong Wang	Beijing University of Aeronautics and Astronautics, China
Shikui Wei	Beijing Jiaotong University, China
Gongjian Wen	National University of Defense Technology, China
Xiangqian Wu	Harbin Institute of Technology, China
Lifang Wu	Beijing University of Technology, China
Jianxin Wu	Nanjing University, China
Jun Wu	Chongqing Kaiser Technology, China
Yadong Wu	Southwest University of Science and Technology, China
Yihong Wu	Chinese Academy of Sciences, China
Yongxian Wu	South China University of Technology, China
Guisong Xia	Wuhan University, China
Shiming Xiang	Chinese Academy of Science, China
Xiaohua Xie	Sun Yat-sen University, China
Junliang Xing	Chinese Academy of Sciences, China
Hongkai Xiong	Shanghai Jiao Tong University, China
Chao Xu	Tianjin University, China
Mingliang Xu	Zhengzhou University, China
Yong Xu	Harbin Institute of Technology, China
Yong Xu	South China University of Technology, China
Xinshun Xu	Shandong University, China
Feng Xue	HeFei University of Technology, China
Jianru Xue	Xi'an Jiaotong University, China
Yan Yan	Xiamen University, China
Dongming Yan	Chinese Academy of Sciences, China
Junchi Yan	IBM China Research Institute, China
Bo Yan	Fudan University, China
Chenhui Yang	Xiamen University, China

Dong Yang	Beijing Wisdom Eye Technology Inc., China
Gongping Yang	Shandong University, China
Jian Yang	Beijing Institute of Technology, China
Jie Yang	Shanghai Jiaotong University, China
Jinfeng Yang	Civil Aviation University of China
Jufeng Yang	Nankai University, China
Lu Yang	University of Electronic Science and Technology of China
Wankou Yang	Southeast University, China
Jian Yao	Wuhan University, China
Mao Ye	University of Electronic Science and Technology of China
Xucheng Yin	University of Science and Technology Beijing, China
Yilong Yin	Shandong University, China
Xianghua Ying	Peking University, China
Xingang You	Beijing Institute of Electronics Technology and Application, China
Xinge You	Huazhong University of Science and Technology, China
Jian Yu	Beijing Jiaotong University, China
Shiqi Yu	Shenzhen University, China
Xiaoyi Yu	Peking University, China
Ye Yu	HeFei University of Technology, China
Zhiwen Yu	South China University of Technology, China
Jun Yu	Hanzhou Electronic Science and Technology University, China
Jingyi Yu	ShanghaiTech University, China
Xiaotong Yuan	Nanjing University of Information Science and Technology, China
Di Zang	Tongji University, China
Honggang Zhang	Beijing University of Posts and Telecommunications, China
Junping Zhang	Fudan University, China
Junge Zhang	Chinese Academy of Sciences, China
Lin Zhang	Tongji University, China
Shihui Zhang	Yanshan University, China
Wei Zhang	Fudan University, China
Wei Zhang	Shandong University, China
Wenqiang Zhang	Fudan University, China
Xiaoyu Zhang	Chinese Academy of Sciences, China
Yan Zhang	Nanjing University, China
Yanning Zhang	Northwestern Polytechnical University, China
Yifan Zhang	Chinese Academy of Sciences, China
Yimin Zhang	Intel China Research Center, China
Yongdong Zhang	Chinese Academy of Science, China
Yunzhou Zhang	Northeastern University

Zhang Zhang	Chinese Academy of Science, China
Zhaoxiang Zhang	Chinese Academy of Science, China
Guofeng Zhang	Zhejiang University, China
Yao Zhao	Beijing Jiaotong University, China
Cairong Zhao	Tongji University, China
Yang Zhao	HeFei University of Technology, China
Yuqian Zhao	Central South University
Weishi Zheng	Sun Yat-sen University, China
Wenming Zheng	Southeast University
Bineng Zhong	Huaqiao University, China
Dexing Zhong	Xi'an Jiaotong University, China
Hanning Zhou	Beijing Gourd Software Technology Inc., China
Rigui Zhou	Shanghai Maritime University, China
Zhenfeng Zhu	Beijing Jiaotong University
Liansheng Zhuang	University of Science and Technology of China
Beiji Zou	Central South University
Wangmeng Zuo	Harbin Institute of Technology, China

Sponsoring Institutions

WATRIX TECHNOLOGY

EXTREME VISION
极视角

AN
中科智谷
上海中科智谷人工智能工业研究院
Shanghai Academy Of Artificial Intelligence

SEGWAY
ROBOTICS

PERCIPIO.XYZ

Contents – Part I

Deep Neural Network

Biological Vision Inspired Visual Method

An Evolution Perception Mechanism for Organizing and Displaying 3D Shape Collections

Lingling Zi[(✉)], Xin Cong, and Yanfei Peng

School of Electronic and Information Engineering,
Liaoning Technical University, Huludao 125105, China
lingling19812004@126.com

Abstract. How to display the best view of shape collections to facilitate non-profession users is becoming a challenging problem in the field of computer vision. To solve this problem, an evolution perception mechanism for organizing and displaying shape collections (EPM) is proposed. On the one hand, the evolution tree based on Quartet analysis is constructed to effectively organize shapes from a global perspective. On the other hand, the shape snapshot method based on the view features of shapes is presented for each single shape in the evolution tree, so as to show the details of the shape from a local perspective. Moreover, the interactive display methods using the evolution trees are provided to guide users to effectively explore the shape collections. Experimental results show that EPM could provide the effective displaying ways that are close to the user's viewing.

Keywords: Evolution perception · Shape organization
Evolution tree · Shape displaying · Quartet analysis

1 Introduction

At present, the designing and construction process of the 3D shape models on behalf of 3D printing and customer culture is very cumbersome, which reduces the ordinary users' enthusiasm on using the 3D shapes [1–5]. Therefore, the methods of 3D shape display with convenience and efficiency are very important to promote the popularization and universalization of 3D shapes. 3D shapes are displayed and recognized through the optimal view of shapes. The optimal view could visualize the structural features of shapes and effectively describe the inherent feature details of 3D shapes [6–8]. For the single 3D shape, human beings reasonably choose the view position to get the optimal view [9,10]. But for 3D shape collections, it is not easy to select the optimal views [11,12]. Therefore, how to simulate the human's perception to automatically determine the best position of view is a hot issue in the field of computer vision.

In recent years, researchers have tried to quantify the visual habit of human observation objects through using the geometric information and semantic information of 3D shapes, and put forward many the methods to compute the optimal

© Springer Nature Singapore Pte Ltd. 2017
J. Yang et al. (Eds.): CCCV 2017, Part I, CCIS 771, pp. 3–13, 2017.
https://doi.org/10.1007/978-981-10-7299-4_1

view. These methods are divided into two categories. One is to define view-related functions to obtain the optimal view [13,14]. Due to the differences between the structure and geometric characteristics of 3D shapes, it is difficult to find appropriate definitions to apply all the shapes. The other is to use semantic information of shapes to obtain the differentiated view [15–18]. The obtained views vary with the classification of shape collections. So, this kind of method is more suitable for browsing shape collections. However, how to efficiently display the optimal views and how to describe the details of the shapes are still problems to be solved. For the first question, a qualitative analysis method based on Quartet is introduced and the shape organization method is proposed to realize the effective organization of 3D collections. For the second question, the single shape display method is presented to extract the optimal view of 3D shapes.

Biological genes, individuals and populations are the important factors in the biological evolution theory and the nature is to change the genes to achieve optimization [19,20]. Inspired by the theory of the biological evolution, we introduce them into the field of 3D shape modeling. Specially, the three-dimensional shape corresponds to the biological individual, the shape part corresponds to the biological gene and the shape collection corresponds to the biological population. The task of shape organizing and display could be described as follows. The intentions of user demands are regarded as the driving force. Under the action of driving force, the existing shape individuals are preformed self-adaptive adjusting to reform shape population. Specifically, the shape collections with the same semantics are taken as the initial shape population. Then shape evolutionary mechanism are performed to form new ways of display and we repeat this process until the user demands are satisfied.

In this paper, we propose an evolution perception mechanism (EPM) to organize and display 3D shape collections, which is shown in Fig. 1. The core idea of EPM is to use evolution concepts to achieve the task of shape organizing and displaying orienting user preferences, so as to resolve the problem of displaying the best view of shape collections. The main contributions of EPM are as follows.

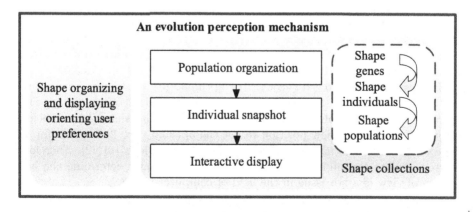

Fig. 1. The proposed method

(1) We describe shape collections as shape populations containing shape individuals. And the evolution tree based on quartet analysis is constructed to realize the global organization of shape populations.

(2) The individual snapshot method based on the view features of shapes is presented to show the details of the single shape and it mainly contains preprocessing, view sampling, saliency feature extraction and snapshot calculation.

(3) The interactive display approaches using the evolution trees are proposed to further facilitate users to effectively explore 3D shapes and they are applied to construct 3D shape system.

The rest of the paper is structured as follows. Section 2 presents the framework of EPM. Section 3 illustrates the implementation of EPM, including population organization, individual snapshot and interactive display. Section 4 presents experimental work to demonstrate our approach. The last section concludes the paper.

2 The Overview of EPM

The aim of EPM is not only displaying the overall effect of shape collections, but also reflecting the features of each single shape. To achieve this task, we adopt the concepts from theory of evolution. The shape collections with the same semantic class are regard as the shape populations. So, the goal of EPM is changed to display effectively the shape population. Figure 2 shows the framework of EPM and it contains three steps.

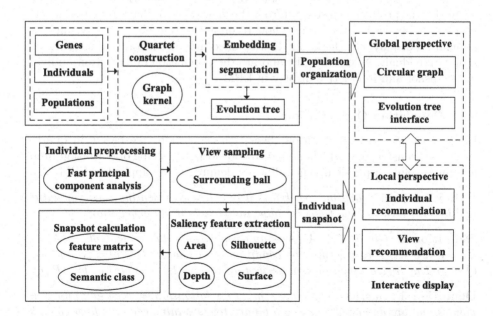

Fig. 2. The framework of EPM

(1) Population organization. We adopt the quartet analysis approach to organize shape population. The Quartet sequences are computed according to the geometric feature and semantic relationship between the individual genes. On this basis, the shape evolution tree could be constructed to effectively display the population.

(2) Individual snapshot. For each individual in the population, we use SVM classifier based on the features of visual saliency to determine the optimal view. This process mainly includes four parts: individual preprocessing, view sampling, saliency feature extraction and snapshot calculation.

(3) Interactive display. Based on the above methods, we design two interactive display approaches to browse and view 3D shapes. One is relocation display of shape population. The users could be free to choose any individual in the evolution tree. At the same time, the population could be reorganized to provide intuitive display. The other is individual recommendations and view recommendations. Based on the individuals chosen by users, multiple shapes and views are given in the form of individual snapshots.

3   The Details of EPM

Shape populations contain many shape individuals and shape individuals need to be mapped to two-dimensional views, which could be viewed and recognized by the users. The displaying effects of different views of the same shape individual are different and the best view could maximize the structural features of shape individuals to effectively characterize intrinsic features. Therefore, how to find the best view to achieve effective display of the shape population is a challenging problem. The problem is discussed from the view point of shape populations and shape individuals. For the former, how to organize properly to achieve efficient population display. For the latter, how to maximize structural characteristics of shape individuals. Next, we will elaborate the population organization, individual snapshot and interactive display in EPM.

3.1   Population Organization

Inspired by the idea of evolution theory, the evolution tree based on quartet analysis is constructed, which is used to transform the shape organization problem into evolution problem. The following definitions are applied in EPM.

Definition 1. *A evolution tree ET is a root tree representing each population and the leaf nodes of the tree are shape individuals in the population.*

Definition 2. *Given $A' \in A$ (A is any population), the individuals in A' are divided two set, P and Q, and the condition of division is $P \cup Q = A'$ and $P \cap Q = \phi$. The division relation is described an edge in ET, denoted as b= (P, Q).*

Definition 3. *Quartets reflect the semantic relationship between any four individuals and are described by using a binary tree spanning edge b, which could be denoted as $QU_b = \{uv|wg\}$. In which $u, v \in P$ and $w, g \in Q$.*

We adopt the semantic calculations method to compute Quartet sequences. Specifically, the semantic similarity scores are calculated for each individual. The higher the values of score, the closer the two individuals are. We use graph kernel GK to evaluate semantic similarity between individuals. Given graph $G(V, E)$ denotes gene relations of individuals, V denotes the nodes reflecting shape genes, and E denotes the different context relations among the genes. Combining geometric features of genes with context relations between genes, GK is computed by using Formula 1.

$$GK^l = \begin{cases} GK_N(n_A, n_B) \times \sum\limits_{\substack{n_s \in AN_{Gu}(n_A) \\ n_{s'} \in AN_{Gv}(n_B)}} GK_E(b, b') GK^{l-1}(G_u, G_v, n_s, n_{s'}) , l > 0 \\ GK_N(n_A, n_B) , l = 0 \end{cases} \quad (1)$$

In (1), l denotes the length of graph path, Gu and Gv are the relation graph for any individual Su and Sv, respectively. n_A and n_B are genes of the corresponding individual. $AN(x)$ denotes all adjacent nodes for x in G, b and b' are the edges which connect between m_A and n_s, n_B and n_s. GK_N denotes the node kernel, which reflects the similarity geometric characteristics of the two genes. GK_E denotes the edge kernel, if there is the relation between e_u and e_v, $GK_E(e_p, e_q)$ is 1, otherwise is 0.

The process of constructing ET contains two steps, embedding and segmentation. First, the points representing the individual are embedded in the matrix space according to semantic similarity scores. So, the problem of the embedding of the representative shape points is transformed into the semi-definite programming problem of the maximum cut, that is, satisfying the minimum value of Formula 2.

$$\sum_{1 \leq j \leq k} L(u_j, v_j, w_j, g_j) - \sum_{1 \leq j \leq k} Q(u_j, v_j, w_j, g_j), \ s.t. <p_i, p_i> = 1 (1 \leq i \leq m)$$

$$(2)$$

In which (u_j, v_j, w_j, g_j) represents any quartets in the k sequence. $L(a, b, c, d) = (ang(v_a, v_b) + ang(v_c, v_d))/2$, $Q(a, b, c, d) = (ang(v_a, v_c) + ang(v_a, v_d) + ang(v_b, v_c) + ang(v_b, v_d))/4$. On this basis, the recursive partitioning of shape population are performed to create ET. p_i represents any shape point in a matrix space.

3.2   Individual Snapshot

Individual snapshot method concentrates on finding good views of a single shape and it concludes four steps: individual preprocessing, view sampling, saliency feature extraction and snapshot calculation.

The first step is individual preprocessing and its task is to complete the correction and normalization of 3D shapes. The principal directions of the shapes are calculated by the fast principal component analysis technique and they are used for the direction of each coordinate axis of the shapes, so that the individuals in the population are in a consistent attitude. Then the distance between each vertex and centroid of the shapes are normalized, and the three-dimensional shapes with attitude correction and normalization are obtained.

The second step is view sampling and its task is to acquire the different two-dimensional views relating to the viewpoint position. Since the viewpoint can be continuously changed in the three-dimensional space, it is necessary to select the discrimination viewpoint to reduce the sample space. We adopt the surrounding ball to select the views and construct the sphere for each shape individual. Then uniform sampling is performed along the surface of the sphere to determine the positions of evenly distributed viewpoints. Based on the obtained positions, two-dimensional views of shape individuals in the different viewpoints are competed.

We extract the important features for the obtained views and mainly consist four parts: area property R_1, silhouette property R_2, depth property R_3 and surface curvature property R_4. The overall feature for each view are described by the feature function $F = \sum w_i R_i$, where w_i ($i = 1, 2, 3, 4$) is the property weighting factor. The distance function between any two views is European distance of the corresponding property feature, as shown in Formula 3. In which $\|\cdot\|$ is the European norm.

$$dist(v_i, v_j) = dist(F_i, F_j) = ||F_i - F_j|| \tag{3}$$

The last step is snapshot calculation. We change the problem of the snapshot calculation to the problem of classification evaluation, and find the view with the best classification performance among the individuals in the population as the optimal view. To measure the classification ability of each view, we compute the representative view according to the training set, use these views to construct the feature matrix and the semantic class labels, so as to train the classifier. In practices, the SVM classifier is adopted.

3.3   Interactive Display

Based on the above methods, the goal of the interactive display method is to provide the friendly means of browsing 3D shapes from the global and local aspects.

From a global perspective, we display the evolution trees of each population for users, which also serve as an overview of the overall shapes. The users are free to choose any individual in the evolution tree. Once the individual is selected, the remaining shapes are positioned according to semantic similarity scores to form a circular graph, whose center is the selected individual. This way could give an intuitive understanding, and show the contrast the currently selected shape with the other shapes in the circular graph. At the same time, at any time the users could return to the evolution tree interface to select a different shape, which is as the starting point for new shape exploration.

From a local perspective, we provide individual recommendation function and view recommendation function. For the former, we give multiple shapes relating the individuals chosen by users. For the latter, we give the views for recommended individual order by the computing of important features. That is, if the display view is not satisfied by users, other views could be provided in turn. In this way, we could shorten the time for observing the 3D shapes.

The proposed method could be applied to construct three-dimensional shape system and this system mainly contains three parts, which are demonstrated as follows.

(1) Semantic annotation. It mainly includes shape collection, shape preprocess, population labeling and gene function labeling. The way of shape collection is directional information acquisition and it provides choices of the resource sources for users. In view of the uncertainty of the orientation of 3D shapes, they are demanded to be preprocessed, including correcting and normalizing, so as to obtain the shapes with attitude correction and normalization.

(2) Resource management. It is responsible for creating a three-dimensional shape gallery and generating the index files to speed up the search. The shapes with semantic annotation are stored in the corresponding shape individual library, so as to complete the organization and storage of resources. And at the same time the label documents corresponding to shape resources are established.

(3) Shape display. It is responsible for organizing and display shape collections that conforms to the user's modeling needs. In the form of interface display, two friendly means are provided until shape population satisfying the users are presented. Also, many functions would be presented to users, such as individual recommendation, view recommendation and evolution trees.

4   Experimental Results and Discussion

We used experimental data from Princeton University database [21] and the COSEG dataset [22], including ants (the number of individuals was 30), chairs (the number of shape individuals was 50), goblets (the number of shape individuals was 25), vases (the number of shape individuals was 60), birds (the number of individuals was 26) and lambs (the number of shape individuals was 30). Then we validate the effectiveness of the proposed method from three aspects.

We compare EPM with the method of area projection (APM) to evaluate the effects of single shape individual. Considering the perception of users for the quality of the results, we select 20 shape individuals from each shape population and randomly sort the obtained results by using the methods of EPM and APM. Then 20 users are invited to vote the results and Fig. 3 shows the comparison results between the six shape populations. It can be seen that the values corresponding to the proposed method are always higher than those by the comparison method, indicating that the users could observe more richer shape information, and obtain better approximation the visual habits of user observation.

We use the view selection error (VSE) measurement to evaluate the consistency between the obtained view and the observing viewpoint of the users. The smaller the values of VSE method, the better the algorithm is to consistent with

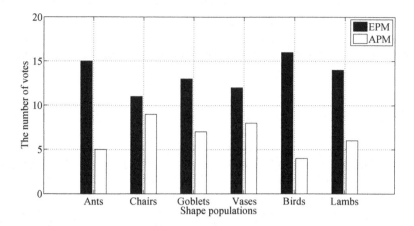

Fig. 3. Comparison of the vote results for EPM and APM methods

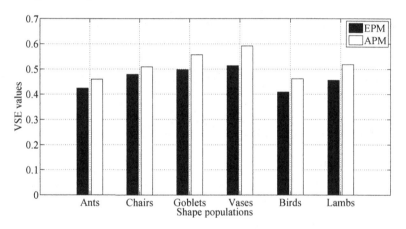

Fig. 4. Comparison of VSE values for EPM and APM methods

users' observing. We recorded the average VSE values corresponding to each shape population and Fig. 4 shows the comparison results for the six shape populations by using EPM and APM methods. From the figure, we can see that the values using the EPM method are always lower than the values using APM. The reason is that we take advantage of the important view feature, which would be helpful to observe viewpoint of 3D shapes.

Considering that the best view could maximize the structural features of 3D shapes, it is reasonable to check retrieval performance to reflect the quality of the best displaying view, so as to evaluate displaying of 3D shape collections. So, we apply EPM in the field of 3D shape retrieval. We random select five shape individuals from each shape population and record the precision and recall values. Figures 5 and 6 show respectively the average precision and the average recall

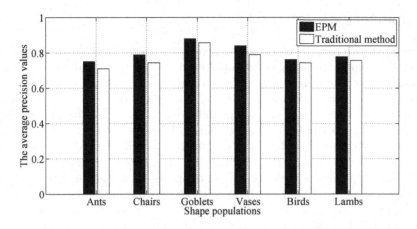

Fig. 5. The average precision results

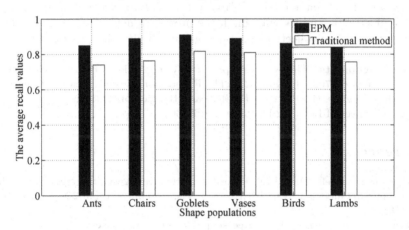

Fig. 6. The average recall results

using EPM and the traditional method. It can be seen that the former results are always higher than the latter, which indicates that the application of EPM could get better retrieval performance. Owing to the similar shape individuals in the evolution tree, we display them as the results of query extension, so as to close to the users' query intention.

In a word, we evaluate the EPM from the above the experiments and the proposed method provides the effective displaying ways that are close to the user's viewing. However, EPM is valid for man-made shapes with less complex structural features and it needs to introduce more factors in the evolution theory to ensure the completion of complex shapes. In addition, the problem of shape displaying is a subjective subject and it still lacks objective evaluation for characterizing people's subjective feelings.

5   Conclusions

In this paper, we propose an evolution perception mechanism for organizing and presenting shape collections and it could both display the overall effect of shape collections and reflect the features of each single shape. The framework of EPM is proposed by using evolutionary theory and shape collections are described as shape populations containing shape individuals. To realize the global organization of shape populations, shape evolution tree based on quartet analysis is constructed. To show the details of the single shape individual, the individual snapshot method based on the view features of shapes is presented and it contains individual preprocessing, view sampling, saliency feature extraction and snapshot calculation. Based on above methods, the interactive display using the evolution trees are proposed to further explore 3D shapes. Experimental results show that EPM could maximize the visualization of shape collections and get the higher exploration experience of users. Moreover, EPM could achieve good performance in 3D shape retrieval. However, evolution theory has extraordinary depth and solidity, and we will enrich the proposed method to improve the displaying of 3D shape collections containing the complex shapes in future.

Acknowledgements. This work is supported by the National Natural Science Foundation of China (61702241, 61602227); The Foundation of the Education Department of Liaoning Province (L2015225, LJYL019) and the Doctoral Starting up Foundation of Science Project of Liaoning Province (201601365).

References

1. Jackson, B., Keefe, D.F.: Lift-Off: using reference imagery and freehand sketching to create 3D models in VR. IEEE Trans. Vis. Comput. Graph. **22**(4), 1442–14511 (2016)
2. Tan, G., Zhu, X., Liu, X.: A free shape 3d modeling system for creative design based on modified catmull-clark subdivision. Multimed. Tools Appl. **76**(5), 1–18 (2017)
3. Spagnuolo, M.: Shape 4.0: 3D shape modeling and processing using semantics. IEEE Comput. Graph. Appl. **36**(1), 92–96 (2016)
4. Kalogerakis, E., Chaudhuri, S., Koller, D., Koltun, V.: A probabilistic model for component-based shape synthesis. TOG **31**(4), 55 (2012)
5. Han, Z., Liu, Z., Han, J., Bu, S.: 3D shape creation by style transfer. Vis. Comput. **31**(9), 1147–1161 (2015)
6. Huang, S.S., Shamir, A., Shen, C.H., Zhang, H., Sheffer, A., Hu, S.M., Cohen-Or, D.: Qualitative organization of collections of shapes via quartet analysis. TOG **32**(4), 71 (2013)
7. Chen, X., Xu, W.W., Yeung, S.K., Zhou, K.: View-aware image object compositing and synthesis from multiple sources. J. Comput. Sci. Technol. **31**(3), 463–478 (2016)
8. Che, L., Kang, F.: Viewpoint scoring approach and its application to locating canonical viewpoint for 3D visualization. In: Zhang, L., Song, X., Wu, Y. (eds.) AsiaSim/SCS AutumnSim-2016. CCIS, vol. 644, pp. 625–633. Springer, Singapore (2016). https://doi.org/10.1007/978-981-10-2666-9_63

9. Bircher, A., Kamel, M., Alexis, K., Oleynikova, H.: Receding horizon "Next-Best-View" planner for 3D exploration. In: IEEE International Conference on Robotics and Automation, pp. 1462–1468 (2016)
10. Schreck, T.: What features can tell us about shape. IEEE Comput. Graph. Appl. **38**(3), 82–97 (2017)
11. Kim, V.G., Li, W., Mitra, N.J., DiVerdi, S., Funkhouser, T.: Exploring collections of 3D models using fuzzy correspondences. TOG **31**(4), 54 (2012)
12. Secord, A., Lu, J., Finkelstein, A., Singh, M., Nealen, A.: Perceptual models of viewpoint preference. TOG **30**(5), 1–12 (2011)
13. Yamauchi, H., Saleem, W., Yoshizawa, S., Karni, Z., Belyaev, A., Seidel, H.P.: Towards stable and salient multi-view representation of 3D shapes. In: IEEE International Conference on Shape Modeling and Applications, p. 40 (2006)
14. Wang, W., Gao, T.: Constructing canonical regions for fast and effective view selection. In: IEEE CVPR, pp. 4114–4122 (2016)
15. Laga, H.: Semantics-driven approach for automatic selection of best views of 3D shapes. In: Eurographics Workshop on 3D Object Retrieval, pp. 15–22 (2010)
16. Qi, J., Tejedor, J.: Deep multi-view representation learning for multi-modal features of the schizophrenia and schizo-affective disorder. In: International Conference on Acoustics, Speech and Signal Processing, pp. 116–118 (2016)
17. Lin, H., Gao, J., Zhou, Y., Zhang, C.: Semantic decomposition and reconstruction of residential scenes from LiDAR data. TOG **32**(4), 66 (2013)
18. Valentin, J., Vineet, V., Cheng, M.M., Kim, D., Shotton, J., Kohli, P.: SemanticPaint: interactive 3d labeling and learning at your fingertips. TOG **34**(5), 154 (2015)
19. Young, M.: The Technical Writer's Handbook. University Science, Mill Valley (1989)
20. Wang, R., Pujos, C.: Emergence of diversity in a biological evolution model. J. Phys. Conference Ser. **604**, 012019 (2015)
21. Chen, X., Golovinskiy, A., Funkhouser, T.: A benchmark for 3D mesh segmentation. TOG **28**(3), 341–352 (2009)
22. COS. The Shape Coseg Dataset. http://web.siat.ac.cn/yunhai/ssl/ssd.htm

C-CNN: Cascaded Convolutional Neural Network for Small Deformable and Low Contrast Object Localization

Xiaojun Wu[1,2(✉)], Xiaohao Chen[1], and Jinghui Zhou[1]

[1] School of Mechanical Engineering and Automation, Shenzhen Graduate School,
Harbin Institute of Technology, Shenzhen 518055, Guangdong, China
`wuxj@hit.edu.cn`
[2] Shenzhen Key Laboratory for Advanced Motion Control and Modern
Automation Equipment, Shenzhen 518055, Guangdong, China
`http://smea.hitsz.edu.cn/Userfiles/wuxiaojun/`

Abstract. Traditionally, the normalized cross correlation (NCC) based or shape based template matching methods are utilized in machine vision to locate an object for a robot pick and place or other automatic equipment. For stability, well-designed LED lighting must be mounted to uniform and stabilize lighting condition. Even so, these algorithms are not robust to detect the small, blurred, or large deformed target in industrial environment. In this paper, we propose a convolutional neural network (CNN) based object localization method, called C-CNN: cascaded convolutional neural network, to overcome the disadvantages of the conventional methods. Our C-CNN method first applies a shallow CNN densely scanning the whole image, most of the background regions are rejected by the network. Then two CNNs are adopted to further evaluate the passed windows and the windows around. A relatively deep model net-4 is applied to adjust the passed windows at last and the adjusted windows are regarded as final positions. The experimental results show that our method can achieve real time detection at the rate of 14FPS and be robust with a small size of training data. The detection accuracy is much higher than traditional methods and state-of-the-art methods.

Keywords: Deep learning · Convolution neural network
Small deformable and low contrast detection

1 Introduction

Object localization is an important task in the process of industrial automatic production, for example, pick and place of an industrial robot, position localization in surface mount technology (SMT) etc. Template matching method is

X. Wu—This work was supported by the Basic Research Key Project of Shenzhen Science and Technology Plan (No. JCYJ20150928162432701, JCYJ2017041315253 5587, JCYJ20170307151848226).

J. Yang et al. (Eds.): CCCV 2017, Part I, CCIS 771, pp. 14–24, 2017.
https://doi.org/10.1007/978-981-10-7299-4_2

mostly adopted for this task in machine vision applications. NCC based template matching applies normalized cross correlation as the features [1], shape based template matching adopts edge feature for object localization [2]. In many cases, this kind of methods performs well, but these methods have drawbacks in that they are less robust against geometrical distortion including occlusion, deformation, illumination distortion, motion blur or extreme low contrast.

In recent years, CNN based object detection methods have achieved state-of-the-art result on classification tasks [3–5] and object detection tasks [6–8]. Current CNN based object detection methods mainly focus on general object detection in nature sense. As objects have different scales and different aspect ratios, sliding window based method [9–12] will lead to a very large computation complexity, so region proposal based and regression based methods are adopted. This kind of method was first proposed by Girshick et al. [6]. For accelerating, Fast-RCNN [13] by Girshick and Faster-RCNN [8] by Shaoqing Ren were proposed. Recently, regression models based methods became a new research hotspot [14, 15] and more rapid methods, like YOLO [16] and single shot multibox detector (SSD) [17], are proposed.

There are 3 main challenges for applying current CNN techniques to object localization in the industrial environments, (1) it's unrealistic to label a lot of training data for a single scene in industry; (2) the method should be fast enough to cope with the large-capacity production line; (3) it should be robust to deal with the variety of products. Current state-of-the-art methods, like SSD [7] and Faster-RCNN [8], are rapid, but these methods require a huge amount of training data and training time, which is not suitable for practical applications in industry.

In the industrial automatic inspections, objects usually have a fixed scale and aspect ratio, so we only need to scan the whole image by only one fixed window, which makes the computation complexity acceptable. What's more, this kind of methods have less false negative. But we find if only adoption a signal model, there are many false alarms in the localization results.

Above all, we design a cascaded convolutional neural network (C-CNN) based method for object localization in the industrial environment. The proposed C-CNN can achieve a rapid localization speed and it is robust even we only use a small size of training data. Our method runs 14 FPS on GTX970. In the following section, we present the overall framework of proposed method and the details, and then we introduce our experiments and compare our method to traditional template matching and current state-of-the-art CNN based methods.

2   Cascaded CNN Detector

For object localization in industry, our object detector is shown in Fig. 1. Given a testing image, we first resize the image to a small scale and use net-16 to densely scan the whole image to reject most of the background windows. Then two networks, net-32-1 and net-32-2, further reject the remained background windows. The passed windows are accepted as the rough detection results. We apply a

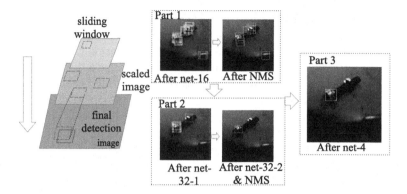

Fig. 1. Testing pipeline of our method. First we detect in the scaled image and adjust in the original image finally.

relatively deep model net-4 to adjust the passed windows. Non-maximum suppression [17] (NMS) is adopted to eliminate highly overlapped detection windows at the end of part 1 and part 2, seeing in Fig. 1.

2.1   The net-16

Training of the net-16. The structure of net-16 is shown as Fig. 2, which is a binary CNN classifier for classifying objects and backgrounds. We select Rectified Linear Units (ReLU) as our active function [18]. ReLU has been widely used in many work [3], and has been proved that it can improve expression ability of network and speed up convergence. Softmax loss function is adopted as our cost function; we also apply weight decay for avoiding overfitting.

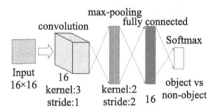

Fig. 2. CNN structures of net-16.

We crop the object patches as the positive examples, and other regions have an intersection over union (IOU) less than 40% with objects are regarded as negative examples. The numbers of negative examples are much more than positive examples. So we adopt rotating, Gaussian blur and Gaussian noise for data augmentation. We keep the positive and negative examples have a ratio 1 : 4 while training.

Testing of the net-16. We adapt fully convolutional network instead of densely scanning the whole image, which can eliminate redundancy computation and achieve same results. When testing, the fully connected layer of net-16 is converted to a convolutional layer with 7×7 kernel size.

For a $W \times H$ testing image with $a \times a$, we first resize the image to $w \times h$. Here, $w = 16/W$, $h = 16/H$. Then input the resized image into the converted net-16 and obtain a map of confidence. Every confidence refers to one window on the original image, and the windows with a confidence less than threshold $T_1 = 0.9$ will be rejected.

2.2   The net-32

The net-32 is divided into two sub-network net-32-1 and net-32-2. Both networks are binary classifiers.

Training of the net-32. Through our experiments, using only one network cannot reject fault detection, so we design a deep network net-32-1 and a shallow network net-32-2. The two network structures are shown as Fig. 3. Deep network structure will help to extract more semantic information, and the shallow structure can retain more details [19].

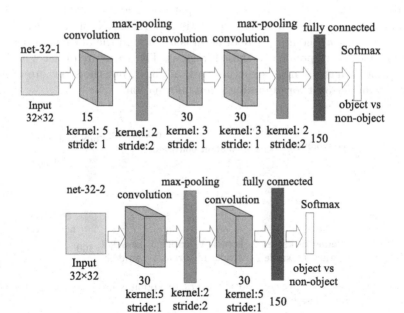

Fig. 3. CNN structures of the net-32-1 and net-32-2.

For the training of these two networks, we apply the trained net-16 to carry out hard negative mining. We use net-16 to scan the images, the windows which have confidence higher than T_1 and have an IOU less than 0.5 with positive windows will be regard as negative examples.

Considering the remaining windows of net-16 will be not accurate enough. We randomly sample windows which have same size with objects and have more than 70% IOU with ground truth as positive examples. We also employ data augmentation strategy in this part.

Testing of the net-32. We evaluate the passed windows of net-16 and the 8 windows around them by net-32. Given a passed window (x, y, a, a) centering at (x, y) of (a, a) size and the size of original image is $W \times H$. Then our evaluated windows are (x', y', a, a), where $x' = x \pm (rS_x)$, $y' = y \pm (rS_y)$. Here, $r \in \{-0.75, 0, 0.75\}$, $S_x = 2W/16$ and $S_y = 2H/16$. The windows corresponded to a confidence higher than threshold $T_2 = 0.85$ are regarded as detection windows.

2.3   The net-4

The net-16 and net-32 are not accurate enough, so we train another CNN, called net-4, to adjust the detecting windows. The net-4 is a 9-class classifier for the object pattern and its eight surrounding patterns.

Training of the net-4. As object region patterns and its surrounding patterns are similar to each other, we design a relatively large CNN for this task. For an object of size $a \times a$, the input size of net-4 is $a \times a$, and the outputs of net-4 are the confidence corresponded to these 9 regions. The structure of net-4 is shown as Fig. 4. net-4 has a similar structure with net-32-1, but the structure of net-4 is wider than net-32-1 for obtaining more information.

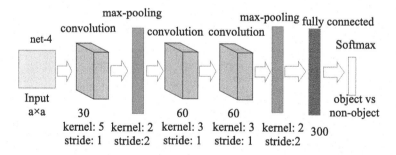

Fig. 4. net-4 consists of three convolution layers, two pooling layers, one fully connected layer and a softmax classifier.

For an object window (x, y, a, a), we crop the object windows' patches and 8 surrounding windows' patches as our training data. These nine windows can be expressed as

$$(x + r_x a, y + r_y a, a, a), \tag{1}$$

where $r_x \in \{-0.15, 0, 0.15\}$ and $r_y \in \{-0.15, 0, 0.15\}$. We also apply some data augmentation measures. It should be noted that the rotated patterns should be cropped from the rotated images instead of rotating the cropped patterns directly.

Testing of net-4. The net-4 accepts the passed windows as input and distinguishes which pattern of this window should be, then adjusts the center coordinate of the detection windows. The adjusting process is shown as Fig. 5.

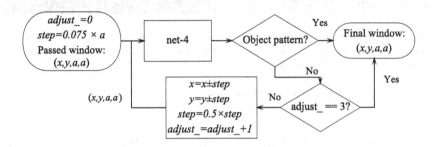

Fig. 5. The adjusting pipeline for a passed window.

3   Experiments and Analysis

In this section, we evaluate the performance of the proposed approach and compare our method to other methods. We adopt precision rate (P_{rate}) and recall rate (R_{rate}) to measure the performance of these algorithms. P_{rate} and R_{rate} are defined as Eq. (2).

$$\begin{cases} P_{rate} = \frac{TP}{TP+FP} \\ R_{rate} = \frac{TP}{TP+FN} \end{cases} \tag{2}$$

where TP is true positive; FP is false positive; FN is false negative.

3.1   Experiments Results

For evaluating our proposed algorithm, we test our method in different image sets. Each of image sets are compose of 100–200 images with resolution of 640 × 480. Figure 6 demonstrates some sample images from the testing image data sets. We can see that there are lots of interferences and the backgrounds are also very complex, and the targets are very small, blur and deformed, which make the detection task very difficult.

We only picked out and annotate about 20–30 images of them as the training data and the testing result is shown as Table 1. We find our method is robust although the training set is only composed of 20–30 images. All of P_{rate} and R_{rates} are higher than 97% in our experiments.

Fig. 6. Some sample images from testing image data sets, the top left called semi images, the top right called flower images, the bottom left called candy images, and the bottom right called screw images.

Table 1. Results of our proposed method

		Object	Background	P_{rate}	R_{rate}
Semi images	Positive	369	12	96.85%	98.66%
	Negative	5	0		
Flower images	Positive	208	0	100%	97.1%
	Negative	6	0		
Candy images	Positive	98	2	98.0%	98%
	Negative	2	0		
Screw images	Positive	279	3	98.93%	99.64%
	Negative	1	0		

3.2 Comparisons Results and Analysis

We compare our method with the shape based template matching in Mvtech Halcon [21] (from a Germany based machine vision company) and SSD [7] method. For template matching algorithm, only one template image is needed. For SSD [7], We also pick out about 20–30 images as training data. We compare these three methods in different image sets. The results are similar, so we only show the result of candy images in Table 2. In the candy images, the candy wrapping paper is reflecting, deformable, distorted and low contrast with the background. The result is shown as Fig. 7.

The results show that our method has a better performance than the template matching and SSD, the yellow box in Fig. 7. When the objects have large various visual changes and the shape features of objects are not obvious, the template matching nearly fails to detect the objects. Although we adopt 4-CNN models in our method, we apply multi-scale detecting and fully convolutional network to accelerate the algorithm. As we adopt the CNN in our detector, our method is easy to be parallelized on GPU. When using a moderate GPU card, GTX970, our

Table 2. Results of template matching and SSD

		Object	Background	P_{rate}	R_{rate}
Template matching	Positive	11	89	11%	11%
	Negative	89	0		
SSD	Positive	91	0	100%	91%
	Negative	9	0		

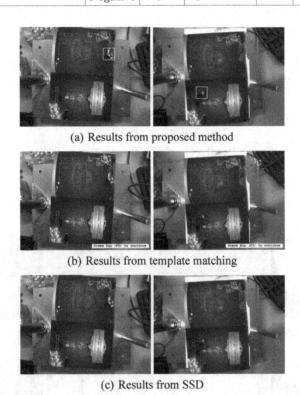

(a) Results from proposed method

(b) Results from template matching

(c) Results from SSD

Fig. 7. Object localization results of our proposed method (top), template matching (middle) and SSD (bottom). (Color figure online)

method can achieve about 14 FPS which is comparable to traditional template matching methods. The runtime comparison illustrates in Table 3.

We also compare our method with a commercial software ViDi Suite [22], a deep learning based industrial image analysis software, developed by a Swiss software firm to solve industrial vision challenges. We also provide 20–30 images as training data for ViDi. The result is shown in Table 4 and Fig. 8.

As the results shown in Table 4 and Fig. 8, the ViDi has a higher P_{rate}. But our method can obtain a better recall rate in three of these four image sets. Particularly, ViDi only has 75% R_{rate} in Candy images. As the background is

Table 3. Runtime comparison

Methods	Runtime for one image
Template matching (on CPU)	43 ms
SSD (on GTX970)	45 ms
Proposed method (on CPU)	633 ms
Proposed method (on GTX970)	72 ms

Table 4. Results of ViDi object localization method

		Object	Background	P_{rate}	R_{rate}
Semi images	Positive	369	0	100%	98.66%
	Negative	5	0		
Flower images	Positive	203	0	100%	95.75%
	Negative	25	0		
Candy images	Positive	75	0	100%	75%
	Negative	25	0		
Screw images	Positive	257	0	100%	91.79%
	Negative	23	0		

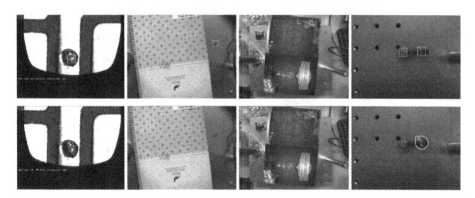

Fig. 8. The first row shows the results of proposed method and the second shows the localization results of ViDi.

usually stationary in industrial machine vision applications, the object localization is different from the wild object detection. There is not any publicly available dataset to compare.

4   Conclusion

Object localization is an important task in the industrial machine vision applications. Traditional template matching methods will completely fail to detect objects in some extreme cases and current CNN based methods are focused on general object detection in nature sense. We propose a cascaded CNN detector, C-CNN, specifically for object detection in industrial sense. The C-CNN method is proved to be robust through our experiments and can locate the objects in extremely poor quality images. It can outperform the traditional methods and the state-of-the-art methods with small number of the training images. Furthermore, the real time performance of our method is achieved on a moderate GPU. It can be utilized in practical machine vision systems.

References

1. ShinIchi, S.: Simple low-dimensional features approximating NCC-based image matching. Pattern Recognit. Lett. **32**(14), 1902–1911 (2014)
2. Hou, Q.Y., Lu, L.H., Bian, C.J., Zhang, W.: Template matching and registration based on edge feature. In: Photonics Asia International Society for Optics and Photonics, pp. 1429–1435 (2013)
3. Alex, K., Sutskever, I., Hinton, G.E.: ImageNet classification with deep convolutional neural networks. In: Neural Information Processing Systems (NIPS), pp. 1097–1105 (2012)
4. Szegedy, C., Liu, W., Jia, Y.Q., Sermanet, P., Reed, S., Anguelov, D., Erhan, D., Vanhoucke, V., Rabinovich, A.: Going deeper with convolutions. In: IEEE Conference on Computer Vision and Pattern Recognition (CVPR), pp. 1–9 (2015)
5. He, K.M., Zhang, X.Y., Ren, S.Q., Sun, J.: Deep residual learning for image recognition. arXiv preprint arXiv:1512.03385 (2015)
6. Ross, G., et al.: Rich feature hierarchies for accurate object detection and semantic segmentation. In: IEEE Conference on Computer Vision and Pattern Recognition (CVPR), pp. 580–587 (2014)
7. Liu, W., Anguelov, D., Erhan, D., Szegedy, C., Reed, S., Fu, C.-Y., Berg, A.C.: SSD: single shot multibox detector. arXiv preprint arXiv:1512.02325 (2015)
8. Ren, S., He, K., Girshick, R., Sun, J.: Faster R-CNN: towards real-time object detection with region proposal networks. IEEE Trans. Pattern Anal. Mach. Intell. **PP**(99), 1 (2016)
9. Sermanet, P., Eigen, D., Zhang, X., Mathieu, M., Fergus, R., Yann, L.C.: Overfeat: integrated recognition, localization and detection using convolutional networks. arXiv preprint arXiv:1312.6229 (2013)
10. Li, H.X., Lin, Z., Shen, X.H., Brandt, J., Hua, G.: A convolutional neural network cascade for face detection. In: IEEE Conference on Computer Vision and Pattern Recognition (CVPR), pp. 5325–5334 (2015)
11. Farfade, S.S., Saberian, M.J., Li, L.J.: Multi-view face detection using deep convolutional neural networks. In: International Conference on Multimedia Retrieval ACM, pp. 224–229 (2015)
12. Chen, X.Y., Xiang, S.M., Liu, C.L., Pan, C.-H.: Vehicle detection in satellite images by parallel deep convolutional neural networks. In: Asian Conference on Pattern Recognition (IAPR), pp. 181–185 (2013)
13. Girshick, R.: Fast R-CNN. In: IEEE International Conference on Computer Vision (ICCV), pp. 1440–1448 (2015)
14. Szegedy, C., Toshev, A., Erhan, D.: Deep neural networks for object detection. In: Neural Information Processing Systems (NIPS), pp. 2553–2561 (2013)

15. Sun, Y., Wang, X., Tang, X.: Deep convolutional network cascade for facial point detection. In: IEEE Conference on Computer Vision and Pattern Recognition (CVPR), pp. 3476–3483 (2013)
16. Redmon, J., Divvala, S., Girshick, R., Farhadi, A.: You only look once: unified, real-time object detection. arXiv preprint arXiv:1506.02640 (2015)
17. Adam, H., Hradi, M., Zemk, P.: EnMS: early non-maxima suppression. Pattern Anal. Appl. **15**(2), 121–132 (2012)
18. Glorot, X., Bordes, A., Bengio, Y.: Deep sparse rectifier neural networks. Aistats **15**(106), 275–283 (2011)
19. Wanli, O., et al.: Deepid-net: multi-stage and deformable deep convolutional neural networks for object detection. arXiv preprint arXiv:1409.3505 (2014)
20. Lecun, Y., Bottou, L., Bengio, Y., Haffner, P.: Gradient-based learning applied to document recognition. Proc. IEEE **86**(11), 2278–2324 (1998)
21. Mvtech Halcon. http://www.mvtec.com/products/halcon/
22. Vidi suite. https://www.vidi-systems.com/

Skeleton-Based 3D Tracking of Multiple Fish From Two Orthogonal Views

Zhiming Qian[1(✉)], Meiling Shi[2], Meijiao Wang[1], and Tianrui Cun[1]

[1] Chuxiong Normal University, Chuxiong 675000, China
qzhiming@126.com, {meijiaow,cuntianrui}@cxtc.edu.cn
[2] Qujing Normal University, Qujing 655011, China
mlshi@mail.ynu.edu.cn

Abstract. This paper proposes a skeleton-based method for tracking multiple fish in 3D space. First, skeleton analysis is performed to simplify object into feature point representation according to shape characteristics of fish. Next, based on the obtained feature points, object association and matching are achieved to acquire the motion trajectories of fish in 3D space. This process relies on top-view tracking that is supplemented by side-view detection. While fully exploiting the shape and motion information of fish, the proposed method is able to solve the problems of frequent occlusions that occur during the tracking process and has good tracking performance.

Keywords: 3D tracking · Fish tracking · Feature point
Stereo matching

1 Introduction

Animal behavior analysis has been a major research area in the natural sciences. As an important member of aquatic animals, fish occupy an essential position in the animal population. Fish ethology is of great significance to understanding animal behavior evolution and fisheries development, and thus is an active branch of study of animal behavior analysis [2,3,5].

Fish behavior is typically recorded as videos in the laboratory environment. Quantitative analysis of fish behavior using video images cannot be performed without acquiring the fish trajectory from the images. Traditionally, these trajectories would be obtained manually from each image frame. This method is inefficient and inaccurate. Due to developments in computer science, computer vision technology provides an effective approach for acquisition and quantitative representation of fish behavior.

Depending on the environment, video-based fish tracking can be classified into either 2D or 3D tracking. In 2D tracking, fish are confined to a container with shallow water. Thus, fish motion can be approximated as planar motion. Despite its ability to analyze fish behavior, this type of tracking can only be performed in 2D space and is unable to completely represent fish behavior. 3D

© Springer Nature Singapore Pte Ltd. 2017
J. Yang et al. (Eds.): CCCV 2017, Part I, CCIS 771, pp. 25–36, 2017.
https://doi.org/10.1007/978-981-10-7299-4_3

tracking has drawn a lot of attention from academia because it is capable of providing fish trajectories that represent true fish behavior by simulating the motion pattern of fish in a real-world environment.

Many researchers had proposed some effective methods for 3D fish tracking. Zhu et al. [12] designed a catadioptric stereo-vision system for obtaining the motion trajectories of several fish by using a video camera and two planar mirrors. Nimkerdphol and Nakagawa [6] proposed a method of the 3D coordinates of zebrafish (*Danio rerio*) by using two regular digital video camcorders placed nonplanar to an object. Butail and Paley [1] proposed a 3D tracking system for tracking the full-body trajectories of schooling fish using multiple cameras. Maaswinkel et al. [4] presented an automatic video tracking system which can record the swimming trajectories of single and groups of zebrafish by using a mirror system. Voesenek et al. [10] presented a method that can track the fishs body position, curvature and direction in 3D space from multiple viewpoints.

Although existing tracking methods are able to obtain 3D motion trajectories of fish in certain environmental scenarios, there are some problems that are not completely solved, such as large appearance changes of objects in different views and frequent occlusions among different objects. Therefore, it is still a challenge to track multiple fish in 3D space. In this paper, a skeleton-based 3D tracking method is proposed to acquire motion trajectories of multiple fish from two orthogonal views. The contributions of this paper are as follows:

(1) Propose an effective strategy to reconstruct 3D trajectories of multiple fish by using a combination of top-view tracking and side-view detection.
(2) Propose a feature point model for representing objects in different views, which enhances detection performance and reduces tracking difficulty.
(3) Propose a novel stereo matching method that combines the epipolar constraint and motion consistency constraint in order to reduce matching ambiguity.

2   The Proposed Method

Two synchronization cameras are used to record videos of objects motion in the rectangular container from the top and side view vertical to the water surface. First, the main skeletons for moving regions are extracted and the feature points that are most expressed in the skeletons are obtained. Next, feature points in neighboring frames are associated in the top view and 2D tracking results are acquired. Finally, 3D motion trajectories are reconstructed by matching top-view tracking with features points in the side view. Figure 1 shows the flow chart of the proposed tracking method.

2.1   Object Detection

Fish assumes a strip structure in two view images. Inspired by this observation, a skeleton analysis is performed to simplify objects from the 2D region to the 1D curve. Next, the feature points which consist of the skeleton endpoints and motion direction are obtained to represent the object.

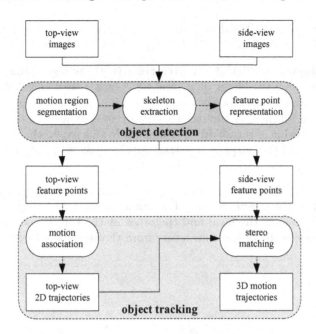

Fig. 1. Flow chart of the proposed method.

Moving Region Segmentation. In the laboratory environment, the video images usually consist of moving objects and nearly static background, and each object stays only for a short time in an area. Thus, moving regions can be segmented from the difference between the current frame and a background image.

$$R_t = \{(x,y) \in I| \quad | \underset{(I_1,\ldots,I_n)}{median}(x,y) - I_t(x,y)| > T_g\} \ . \tag{1}$$

where $I_t(x,y)$ denotes the t-th frame image, and $median(x,y)$ represents the background image which consists of the median image of the first n frames in each view. In order to minimize interference from the moving region's coarse edge during subsequent extraction of the skeleton, we first perform a morphologic operation on the moving region, fill in the holes within the region, and delete small interfering regions. The moving region is then smoothed using median filter.

Skeleton Extraction. The skeleton of each moving region is extracted via the augmented fast marching method (AFMM) [9]. The AFMM is a fast and efficient skeleton extraction method. First, an arrival time U is set for each point at the edge of the region, and then the value of U for the entire region is obtained by iteration of the fast marching method. Based on the distribution of U, the skeleton points can be defined as:

$$s = \{(i,j)|\max(|u_x|,|u_y|) > T_u\}$$
$$u_x = U(i+1,j) - U(i,j), \quad u_y = U(i,j+1) - U(i,j) \quad . \tag{2}$$

This equation shows that, for a given point (i,j), we regard this point as the skeleton point when the larger of u_x and u_y is larger than threshold T_u, where u_x and u_y denote the difference of U between this point and its neighbor in the x and y directions. In the skeleton extraction process, the skeleton represents more detail when the value of threshold T_u is smaller. In this paper, the threshold is set to a large value for the purpose of extracting the main skeleton and eliminating interference from small branches in the skeleton.

Feature Point Representation. The main skeleton of fish usually has two endpoints only, one at the head and the other at the tail. Based on this observation, we simplify the object's structure from the skeleton curve to the feature point.

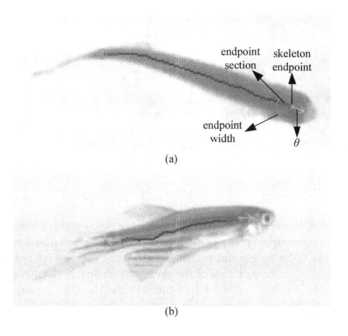

Fig. 2. Feature points in different views. (a) Top view. (b) Side view.

Assume the skeleton endpoint is se, where an endpoint segment $es = \{(x_i, y_i)|i = 1, \ldots, n\}$ consists of n skeleton points closest to se. The direction of the endpoint segment can be calculated based on the least squares method.

$$\theta = \arctan\left(\frac{\sum x_i \sum y_i - n \sum x_i y_i}{(\sum x_i)^2 - n \sum x_i^2}\right) \quad . \tag{3}$$

The position (p) of skeleton endpoint se and the direction (θ) are combined into a feature point $F(p, \theta)$, where the feature point is used to represent the object, as shown in Fig. 2. This representation has the following advantages: (1) Less data: only one point is necessary to effectively represent the object, thus improving the tracking efficiency, (2) Direction information: reduces the difficulty of stereo matching and effectively improves the accuracy of data association, and (3) Occlusion handling: even if an occlusion exists between objects, the occluded objects can still be represented effectively (Fig. 3), thereby greatly improving the ability of occlusion tracking.

The fish body in the top view becomes thinner from head to tail. Based on the characteristic, we draw a circle with each feature point in the top view as the center, and the shortest distance between the center and the edge is the radius r. The fish body width located at this point can be approximated as $2r$. Let w_h and w_t denote the width of the feature points in the head and tail of the fish body, respectively. Based on the shape of the fish body in the top view, it is observed that $w_h > w_t$, as shown in Fig. 2(a). Therefore, the tail feature point can be eliminated by comparing the width. The position and direction of the object is represented by the head feature point.

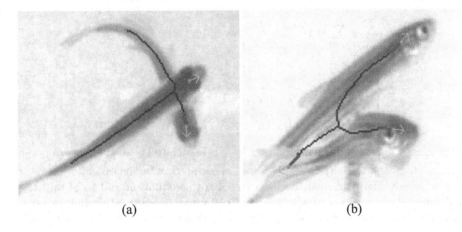

(a)   (b)

Fig. 3. Feature point representation when an occlusion exists between objects. (a) Top view. (b) Side view.

The shape variation of fish in the side view is complicated and it is challenging to recognize the head feature point from all feature points. Hence, all feature points are first used to represent the object, and an effort is made during the stereo matching process to determine the head feature point.

2.2   Objects Tracking

Top-View Tracking. The object's motion in the top view has the following characteristics: (1) The shape changes are relatively small compared to the side view, and (2) The motion states of the objects has good consistency between neighboring frames, which can be expressed as: the changes of position and direction are small for the same object, and the changes are large for different objects. According to these cues, we construct an association cost function for the objects based on the method proposed in [8].

Assume $F_{i,t}^{top} = (p_{i,t}^{top}, \theta_{i,t}^{top})$ is the head feature point of an arbitrary object i_t in the top view in frame t, $p_{i,t}^{top}$ and $\theta_{i,t}^{top}$ denote the position and direction of the feature point, respectively. The association cost function can be defined as:

$$cv(F_{j,t-1}^{top}, F_{i,t}^{top}) = \omega \left(\frac{pc(p_{j,t-1}^{top}, p_{i,t}^{top})}{pc_{\max}} \right) + (1 - \omega) \left(\frac{dc(\theta_{j,t-1}^{top}, \theta_{i,t}^{top})}{dc_{\max}} \right). \qquad (4)$$

where pc_{max} and dc_{max} denote the maximum motion distance and the maximum deflection angle of the object in neighboring frames, respectively. $pc(p_{j,t-1}^{top}, p_{i,t}^{top})$ and $dc(\theta_{j,t-1}^{top}, \theta_{i,t}^{top})$ represent the position change and direction change between $F_{j,t-1}^{top}$ and $F_{i,t}^{top}$, respectively. The association model can be defined as:

$$z = \sum_{j_{t-1}=1}^{m} \sum_{i_t=1}^{n} cv(F_{j,t-1}^{top}, F_{i,t}^{top}) x_{j_{t-1}i_t}$$

$$s.t. \begin{cases} \sum_{j_{t-1}=1}^{m} x_{j_{t-1}i_t} = 1 & (i_t = 1, \ldots, n) \\ \sum_{i_t=1}^{n} x_{j_{t-1}i_t} = 1 & (j_{t-1} = 1, \ldots, m) \\ x_{j_{t-1}i_t} = 1 \text{ or } 0 & (j_{t-1} = 1, \ldots, m; i_t = 1, \ldots, n) \end{cases} \qquad (5)$$

where $x_{j_{t-1}i_t}$ means the feature point $F_{j,t-1}^{top}$ is associated with the feature point $F_{i,t}^{top}$; $x = 0$ means the object $F_{j,t-1}^{top}$ is not associated with the object $F_{i,t}^{top}$.

The greedy algorithm can be used to solve this equation and to obtain the local optimal association of all objects. To improve performance, if the inter-object distance is larger than pc_{max}, the association is abandoned. Figure 4 shows the association process of feature points.

Stereo Matching. The purpose of stereo matching is to determine the third coordinate for each feature point in 2D trajectories of top-view tracking. In many binocular vision-based stereo matching methods, the number of matched objects is typically reduced first via the epipolar constraint. Next, the object is confirmed based on appearance similarity. However, this strategy is not feasible in the proposed method because the large angle between cameras causes the shape of the object to vary greatly between any two views, making it impossible to match objects based on appearance similarity. In order to address this problem, we match feature points from the top view to side view subject to the epipolar constraint and motion consistency constraint.

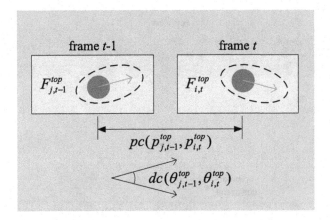

Fig. 4. Illustration of motion association in the top view.

(1) Epipolar constraint

Assume $F_{q,t}^{side} = (p_{q,t}^{side}, \theta_{q,t}^{side})$ is the feature point of an arbitrary object q_t in the side view in frame t, and $l_{i,t}^{side}$ is the corresponding epipolar line of $F_{i,t}^{top}$ in the side view. The association probability is determined by the Euclidean distance from $p_{q,t}^{side}$ to $l_{i,t}^{side}$ and the result of the epipolar constraint can be expressed as:

$$ec(F_{i,t}^{top}, F_{q,t}^{side}) = \begin{cases} 1, & d(p_{q,t}^{side}, l_{i,t}^{side}) < T_e \\ 0, & otherwise \end{cases}. \tag{6}$$

where T_e denotes the maximum matching distance under the epipolar constraint. If there is only one object under the epipolar constraint, the matching is performed successfully. If there is more than one object, we use motion consistency constraint to reduce matching ambiguity.

(2) Motion consistency constraint

If there are k feature points of possible matching $(F_{1,t}^{side}, \ldots, F_{k,t}^{side})$ for $F_{i,t}^{top}$ in the side view subjected to the epipolar constraint. According to Eq. (4), motion consistency constraint can be expressed as:

$$mc(F_{i,t}^{top}) = \arg \min_q cv(F_{i,t-1}^{side}, F_{q,t}^{side}) \quad (q = 1, \ldots, k). \tag{7}$$

where $F_{i,t-1}^{side}$ denotes the matching of the feature point $F_{i,t-1}^{top}$ in frame $t - 1$. This equation indicates that if the feature point of it achieves maximum motion consistency with that of the previous matching object in the side view, then matching is performed successfully. Figure 5 shows an example of stereo matching.

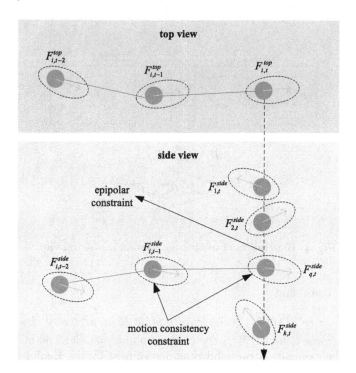

Fig. 5. Example of stereo matching. The dashed arrow indicates the epipolar line. An object in the top view can find k candidates on corresponding epipolar line at frame t. The matching object is determined by the epipolar constraint and motion consistency constraint.

3   Experimental Results and Discussion

3.1   Data Sets

In order to evaluate the proposed method, we choose zebrafish as the tracking object. The length of zebrafish is 1–3 cm. They were placed in a $20 \times 20 \times 20$ cm container with a water depth of about 18 cm. Two Flare 4M180-CL synchronized cameras were installed in the top-view and side-view directions. Video images were recorded at a rate of 90 fps and with a resolution of 2048×2040 pixels. According to the difference in the number of fish, we divide the test data into two sets: D1 for 10 fish and D2 for 15 fish. Each data set contains 1,000 consecutive frames.

3.2   Parameters Setting

In order to set the detection and tracking parameters, we select 300 frames from each view as parameter samples. The tracking results obtained under different parameters are compared with the ground-truth generated by visual examination to determine the optimal setting. The final setting of parameters is shown in Table 1.

Table 1. Parameters setting in the test process.

View	Detection		Tracking			
	T_g	T_u	ω	pc_{max}	dc_{max}	T_e
Top view	30	50	0.6	40	180	–
Side view	40	70	0.6	40	180	20

3.3   Evaluation Metrics

We analyze the tracking results with the following four performance metrics.

Trajectory completeness (TC): Defined as the ratio of the number of tracked objects to the number of ground-truth objects. The higher the value of this metric, the more complete the tracked trajectories.

Trajectory fragmentations (TF): Defined as the ratio of the number of obtained trajectories to the number of actual trajectories. The larger the value of this metric, the more frequently the trajectories are broken.

Mostly tracked trajectories (MT): Defined as the number of tracked trajectories which are correctly tracked for more than 80% of ground-truth trajectories. The larger the value of this metric, the greater the number of effective tracked trajectories.

Identity switches (IDS): Defined as the number of object identity switches. The smaller the value of this metric, the stronger the ability of the method to handle occlusions.

3.4   Results and Discussion

In order to evaluate the performance of the proposed method more effectively, we compare it with other two state-of-the-art methods, namely Wang et al.'s [11] and idTracker [7]. Experimental results are shown in Table 2.

From TC, it can be observed that the proposed method is able to produce very complete tracking results (over 94%) in the two tests and only 5.2% of the tracking information is lost at most. It can also be seen from TF and IDS that the proposed method can process occlusion effectively, because it is able to track the object accurately before and after occlusion occurs. There are at most twelve trajectory fragmentations and fourteen identity switches in the two tests, which means that track continuity and reliability are guaranteed. Finally, MT indicates that the proposed method can successfully obtain most of the objects trajectories. Only three trajectories are shorter than 80% of the lengths of ground-truth trajectories due to association and stereo matching errors. These results fully demonstrate feasibility and effectiveness of the proposed method. Acquired trajectories using the proposed method are shown in Fig. 6.

Table 2. Tracking performance on different test sets.

Method	Test Set	TC	TF	MT	IDS
Wang *et al.*	D1 (10 fish)	0.817	6.8	2	26
	D2 (15 fish)	0.764	11.6	3	54
idTracker	D1 (10 fish)	0.715	13.4	1	82
	D2 (15 fish)	0.536	28.3	0	142
Proposed	D1 (10 fish)	**0.969**	**1.4**	**9**	**6**
	D2 (15 fish)	**0.948**	**1.8**	**13**	**14**

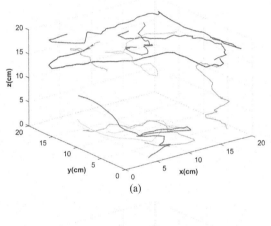

(a)

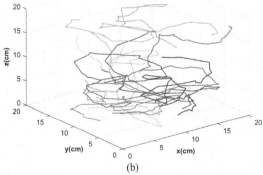

(b)

Fig. 6. Tracking results on different groups. (a) 10 fish. (b) 15 fish.

Wang et al. first detected objects in the top view and side views by using the eye-focused Gaussian mixture model and the eye-focused Gabor model, respectively. Then, tracking results were obtained by associating top-view tracking and side-view detection. Their method performs well under good lighting conditions. However, in the experiments, detection performance of their method is poor because of normal lighting conditions. This leads to reduced tracking performance. Compared with the method proposed by Wang et al., the detection performance of the proposed method is more stable, and can adapt to a variety of lighting conditions. In addition, the proposed method uses both position and direction information to associate and match objects, which can improve tracking performance.

idTracker is one of the best methods for multi-fish tracking. It obtains motion trajectories by performing feature matching on all objects across image frames based on appearance analysis. In the experiments, we track objects in images from the top view and side view using idTracker at the same time. Then, tracking results from the two directions are fused to acquire trajectories in 3D space. Experimental results show that idTracker cannot track the object effectively due to object appearance variations. While the object is swimming in 3D space, its appearance in the image varies with the focal length of the camera. Especially in the side-view direction, the shooting distance causes variation in object appearance, and variation of the object in the swimming direction also leads to considerable changes in its appearance across neighboring frames. As we know, idTrackers tracking performance is heavily dependent on its remarkable appearance recognition ability. Instability of object appearance in the experiment thus makes it more challenging to match appearance features. And idTracker produces many mismatches due to this reason. The experiments also showed that, due to the complexity in the object's appearance, the motion-based method is superior to the appearance-based method for 3D tracking.

4   Conclusion

A skeleton-based multiple fish 3D tracking method is proposed in this paper. The contributions of this paper are as follows. First, a feature point model is constructed based on the shape characteristics of fish to represent objects in multi-view images. The constructed model enhances detection performance and reduces tracking difficulty. Second, motion consistency constraint is performed to eliminate uncertainty in stereo matching and improve the accuracy of object tracking. Third, the proposed method provides an effective strategy to address object occlusions by using top-view tracking, where object shapes change slightly, supplemented by side-view detection.

Acknowledgments. The research work presented in this paper was supported by National Natural Science Foundation of China (No. 61363023), and the Program for Innovative Research Team (in Science and Technology) in University of Yunnan Province.

References

1. Butail, S., Paley, D.A.: 3d reconstruction of the fast-start swimming kinematics of densely schooling fish. J. R. Soc. Interface **9**(66), 77–88 (2012)
2. Delcourt, J., Denoël, M., Ylieff, M., Poncin, P.: Video multitracking of fish behaviour: a synthesis and future perspectives. Fish Fisheries **14**(2), 186–204 (2013)
3. Dell, A.I., Bender, J.A., Branson, K., Couzin, I.D., et al.: Automated image-based tracking and its application in ecology. Trends Ecol. Evol. **29**(7), 417–428 (2014)
4. Maaswinkel, H., Zhu, L., Weng, W.: Using an automated 3d-tracking system to record individual and shoals of adult zebrafish. J. Vis. Exp. **82**(82), e50681 (2013)
5. Martineau, P.R., Mourrain, P.: Tracking zebrafish larvae in group-status and perspectives. Methods **62**(3), 292–303 (2013)
6. Nimkerdphol, K., Nakagawa, M.: Effect of sodium hypochlorite on zebrafish swimming behavior estimated by fractal dimension analysis. J. Biosci. Bioeng. **105**(5), 486–492 (2008)
7. Pérez-Escudero, A., Vicente-Page, J., Hinz, R.C., Arganda, S., de Polavieja, G.G.: idtracker: tracking individuals in a group by automatic identification of unmarked animals. Nature Methods **11**(7), 743–748 (2014)
8. Qian, Z.M., Wang, S.H., Cheng, X.E., Chen, Y.Q.: An effective and robust method for tracking multiple fish in video image based on fish head detection. BMC Bioinform. **17**(1), 251–263 (2016)
9. Telea, A., Wijk, J.J.V.: An augmented fast marching method for computing skeletons and centerlines. In: Vissym Proceedings of the Symposium on Data Visualisation, pp. 251–259 (2002)
10. Voesenek, C.J., Pieters, R.P.M., Leeuwen, J.L.V.: Automated reconstruction of three-dimensional fish motion, forces, and torques. PloS One **11**(1), e0146682 (2016)
11. Wang, S.H., Liu, X., Liu, Y., Zhao, J., Chen, Y.Q.: 3d tracking swimming fish school using a master view tracking first strategy. In: IEEE International Conference on BIBM, pp. 516–519 (2016)
12. Zhu, L., Weng, W.: Catadioptric stereo-vision system for the real-time monitoring of 3d behavior in aquatic animals. Physiol. Behav. **91**(1), 106–119 (2007)

A Retinal Adaptation Model
for HDR Image Compression

Xuan Pu, Kaifu Yang, and Yongjie Li[✉]

Key Laboratory for Neuroinformation of Ministry of Education,
University of Electronic Science and Technology of China,
Chengdu 610054, China
{yangkf,liyj}@uestc.edu.cn

Abstract. High dynamic range (HDR) images are usually used to capture more information of natural scenes, because the light intensity of real world scenes commonly varies in a very large range. Humans visual system is able to perceive this huge range of intensity benefiting from the visual adaptation mechanisms. In this paper, we propose a new visual adaptation model based on the cone- and rod-adaptation mechanisms in the retina. The input HDR scene is first processed in two separated channels (i.e., cone and rod channels) with different adaptation parameters. Then, a simple receptive field model is followed to enhance the local contrast of the visual scene and improve the visibility of details. Finally, the compressed HDR image is obtained by recovering the fused luminance distribution to the RGB color space. Experimental results suggest that the proposed retinal adaptation model can effectively compress the dynamic range of HDR images and preserve local details well.

Keywords: Visual adaptation · High dynamic range
Image enhancement · Human visual system

1 Introduction

The range of the light intensity in natural scenes of the real world is so vast that the illumination intensity of sunlight can be 100 million times higher than that of starlight. Therefore, high dynamic range (HDR) images are usually used to capture more information of natural scenes. However, most display devices available to us have a limited dynamic range [21]. HDR tone mapping or dynamic range compression methods are usually required to reproduce the HDR images matching the dynamic range of the standard display devices, so that the details in both the dark and bright areas are as faithfully visible as possible.

In recent years, many HDR image compression methods have been proposed. Roughly, the existing methods can be classified into two categories: global operators and local operators. Global operators apply the same transformation to each pixel, without considering the spatial position. For example, Tumbling and

© Springer Nature Singapore Pte Ltd. 2017
J. Yang et al. (Eds.): CCCV 2017, Part I, CCIS 771, pp. 37–47, 2017.
https://doi.org/10.1007/978-981-10-7299-4_4

Rushmerier proposed a tone mapping operator in 1993 [25], which uses a single spatially invariant level for the scene and another adaptation level for the display. Although it can preserve the brightness value, but lost the visibility of high dynamic range scenes [12]. Instead of brightness, Ward [26] and Ferwerda et al. [8] aimed at preserving the contrast. They used a scaling factor to transform real-world luminance values to displayable range. They believe that a Just Noticeable Difference (JND) in the real world can be mapped as the JND on the display device. They built their global models based on the threshold versus intensity (TVI) function. Different from Ward's contrast-based scale factor that only considers the photopic lighting conditions, Ferwerda et al. added a scotopic component. These models can preserve the contrast well but may lose visibility of the very high and very low intensity regions [12]. In the year of 2000, Pattanaik et al. [20] proposed a new time-dependent tone reproduction operator following the framework proposed by Tumblin and Rushmeier. In short, although these global models are low computational cost, they are difficult to preserve the details with high contrast scenes.

In contrast, local operators adapt the mapping functions to the statistics and contexts of local pixels. In comparison to the global operators, local operators perform better on preserving details. However, a fundamental problem with local operators is the halos artifacts appearing around the high-contrast edges. For example, Pattanaik et al. [19] described a comprehensive computational model of human visual system adaptation and spatial vision for realistic tone reproduction. This model is able to display HDR scenes on conventional display devices, but the dynamic range compression is performed by applying different gain-control factors to each band-pass, which causes strong halo effects. To solve this problem, Durand et al. [6] presented a method with edge-preserving filter (bilateral filer) to integrate local intensities for avoiding halo artifacts.

In 2004, Ledda et al. [13] proposed a local model based on the work of Pattanaik et al. [20]. The adaptation part of that method tries to model the mechanisms of human visual adaptation, with the help of a bilateral filer to avoid halo artifacts. Along this line, there are also some other works that attempt to build bio-inspired methods for HDR image rendering and dynamic range compression [9,14,28]. Biologically, human visual system can respond to huge luminance range, depending on the ability of visual adaptation, mainly the dark adaptation and light adaptation. Dark adaptation refers to how the visual system recovers its sensitivity when going from a bright environment to a dark one, while light adaptation indicates the processing that visual system recovers its sensitivity when we go from a dark environment to a bright one. In the retina, cone photoreceptors respond to higher light levels while rods are highly sensitive to the light of the dark and dim condition. Therefore, the light and dark adaptations is related to the switch of cones and rods.

In this paper, we propose a new visual adaptation model to compress the dynamic range of HDR images according to the mechanisms of cones and rods in the retina. We design two separated channels (cone and rod channels) with different adaptation properties to compress the dynamic range of light and dark information in the HDR images. The simple receptive field model is further used to enhance the local contrast for improving the visibility of details. Finally, the compressed HDR image is obtained by recovering the fused luminance to the RGB color space. Experimental results suggest that the proposed retinal adaptation model can effectively compress the dynamic range of HDR image and preserve the scene details well.

2   The Proposed Model

In this study, we proposed a new dynamic range compression method aiming at modeling the mechanisms of visual adaptation. The input image is first separated into two luminance distribution maps for the rod and cone channels. Then the responses of rods and cones are obtained with the Naka-Rushton adaptive model [16] according to the local luminance of the scene. A local enhancement operator based on the receptive field of ganglion cells is then used to enhance the local contrast in both the cone and rod channels. Finally, cone and rod channels are adaptively fused into a unified luminance map with compressed dynamic range and enhanced local contrast. The flowchart of dynamic range compression in the proposed model is shown in Fig. 1. Finally, the luminance information is recovered to the RGB space to obtain the color images.

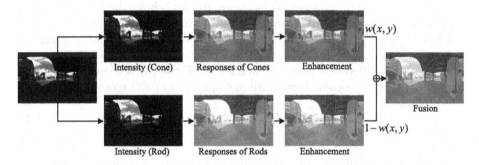

Fig. 1. The flowchart of dynamic range compression in the proposed model. The spatially varying weighting $w(x, y)$ for fusion is computed with Eq. (10).

2.1   Response of Photoreceptors

According to the visual adaptation mechanisms, cone and rod photoreceptors vary their response range dynamically to better adapt the available luminance of the environment. The model proposed by Naka and Rushton [16] has been widely used to describe the responses of cones [2,11] and rods [4]. It indicates that the response of a photoreceptor at any adaptation level can be simply described as $R = I^n/(I^n + \sigma^n)$, where I is the light intensity and σ is the semi-saturation parameter determined by the adaptation level, n is a sensitivity-control exponent which is normally between the range of 0.7–1.0 [3,17]. Thus, the dark and light adaptations can be explained by changing the value of σ at varying luminance levels. For example, on a sunny day, we cannot see well at the beginning when we

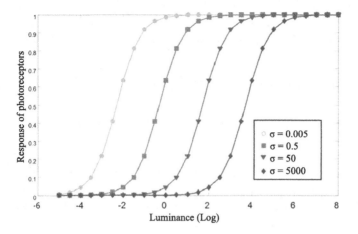

Fig. 2. Response to luminance with different adaptation levels.

enter a dark room. It is just because the value of σ is originally set at high adaptation level while I is of low intensity, which makes the response R almost zero. But after tens of minutes, visual sensitivity is restored after σ changes into a smaller adaptation level and the response R increases. Figure 2 shows the response curves changing with different adaptation levels by varying the value of σ.

HDR image compression mainly refers to compress the dynamic range of luminance information. In this work, we first convert the input HDR image (RGB) into two different luminance channels for further processing by rods and cones [19], i.e., the photopic luminance of cones (L_{cone}) and scotopic luminance of rods (L_{rod}):

$$L_{cone} = 0.25 \cdot I_{in}^R + 0.67 \cdot I_{in}^G + 0.065 \cdot I_{in}^B \tag{1}$$

$$L_{rod} = 0.702 \cdot X + 1.039 \cdot Y + 0.433 \cdot Z \tag{2}$$

where, I_{in}^R, I_{in}^G and I_{in}^B are respectively the R, G and B components of the given HDR image, while X, Y and Z are the three components when transforming the original RGB space to XYZ color space. Thus, we can obtain the cone and rod luminance responses referring to Naka-Rushton equation [16] according to

$$R_{cone}(x, y) = \frac{L_{cone}^n(x, y)}{L_{cone}^n(x, y) + \sigma_{cone}^n(x, y)} \tag{3}$$

$$R_{rod}(x, y) = \frac{L_{rod}^n(x, y)}{L_{rod}^n(x, y) + \sigma_{rod}^n(x, y)} \tag{4}$$

Following the previous work [27], there is a empirical relation between the adaptation parameter σ and visual intensity levels as

$$\sigma_{cone} = L_{cone}^\alpha \cdot \beta \tag{5}$$

$$\sigma_{rod} = L_{rod}^\alpha \cdot \beta \tag{6}$$

In this paper, we experimentally set the value of α as 0.69 for both the cone and rod channels, but β is a constant that differs for the rods and cones, i.e., $\beta = 4$ for the cone channel and $\beta = 2$ for the rod channel.

2.2 Local Enhancement and Fusion

The responses of cones and rods pass through the retina which contains the bipolar cells, horizontal cells, retinal ganglion cells, etc. In this paper, we just focus on the role of local enhancement by the retinal ganglion cells. We use a difference-of-Gaussians (DOG) model [23] to simulate the receptive filed of ganglion cells for enhancing the local contrast of the visual scene. Thus the output of ganglion cells (G_{cone} and G_{rod}) can be computed as:

$$G_{cone}(x,y) = R_{cone}(x,y) * \left(g(x,y;\sigma_{rf}^c) - k \cdot g(x,y;\sigma_{rf}^s)\right) \qquad (7)$$

$$G_{rod}(x,y) = R_{rod}(x,y) * \left(g(x,y;\sigma_{rf}^c) - k \cdot g(x,y;\sigma_{rf}^s)\right) \qquad (8)$$

$$g(x,y;\sigma_{rf}) = \frac{1}{2\pi\sigma_{rf}^2} \exp(-\frac{x^2+y^2}{2\sigma_{rf}^2}) \qquad (9)$$

where $*$ denotes the convolution operator, σ_{rf}^c and σ_{rf}^s are respectively the standard deviations of Gaussian shaped receptive field center and its surround, which are experimentally set to be 0.5 and 2.0 in this work. k denotes the sensitivity of the inhibitory annular surround.

To combine the ganglion cell's outputs along the cone and rod channels, we use a sigmoid function to describe the weighting of cone and rod systems when fusing the two signals.

$$w(x,y) = \frac{1}{0.2 + L_{cone}(x,y)^{-0.1}} \qquad (10)$$

where the fixed parameters (i.e., 0.2 and -0.1) are experimentally set.

From Eq. (10), the value of $w(x,y)$ is increased with the increasing of luminance, which basically matches the light and dark adaptation mechanisms. In the photopic condition, the cone system makes more contribution, while the rod system is more sensitive in the scotopic range. Thus, the fused luminance response is given by

$$L_{out}(x,y) = w(x,y) \cdot G_{cone}(x,y) + (1 - w(x,y)) \cdot G_{rod}(x,y) \qquad (11)$$

2.3 Recovering of the Color Image

To recombine the luminance information into a color image and assure color remains stable, we keep the ratio between the color channels constant before and after compression [10]. Considering that for some input images with high saturation, the output images may appears over saturation when using this simple rule, we add an exponent s to control the saturation of the output image [21], which is described as

$$I_{out}^c(x,y) = L_{out}(x,y) \left(\frac{I_{in}^c(x,y)}{L_{in}(x,y)} \right)^s, \quad c \in \{R,G,B\} \tag{12}$$

where $I_{out}^c(x,y), c \in \{R,G,B\}$ are the RGB channels of the output image, $L_{in}(x,y)$ is replaced using $L_{cone}(x,y)$. The exponent s is given as a parameter between 0 and 1, and we set $s = 0.8$ for the most scenes.

3 Experimental Results

In this section, we evaluated the performance of the proposed method by comparing our method with some existing algorithms. We considered two representative local algorithms including the methods proposed by Durand and Dorsey [18] and Meylan et al. [15], and one typical global algorithm proposed by Pattanaik et al. [20]. The model of Durand et al. uses a bilateral filter to integrate the local intensity. The algorithm proposed by Meylan et al. is an adaptive Retinex model. The method of Pattanaik et al. is aimed at simulating certain mechanisms of human visual system, but with a global operation.

Experimental results are shown in Figs. 3, 4, 5, 6 and 7. In Fig. 3, we tested the methods with an indoor HDR scene named as "office". Figure 3(a)–(d) lists

(a) original image   (b) our operator   (c) Durand and Dorsey[6]   (d) Meylan *et al.*[15]

(e) original scenes at different exposure levels

Fig. 3. Comparison of the results with local algorithms on the indoor "office" image.

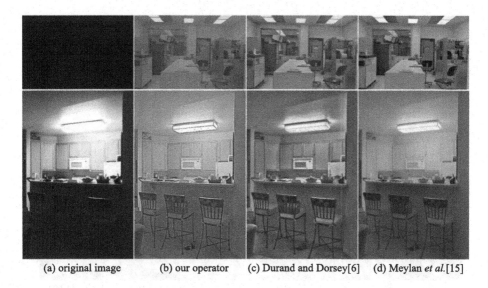

(a) original image   (b) our operator   (c) Durand and Dorsey[6]   (d) Meylan *et al.*[15]

Fig. 4. Comparing the results with local algorithms (inside scenes).

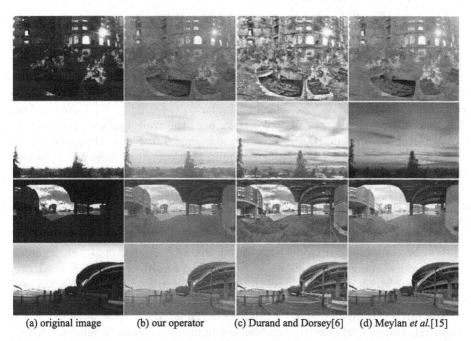

(a) original image   (b) our operator   (c) Durand and Dorsey[6]   (d) Meylan *et al.*[15]

Fig. 5. Comparison of the results with local algorithms on four outdoor scenes.

the results of the whole image (the first row), the local details in the dark area (the second row), and the local details in the bright area (the third row). For the dark regions, our and Meylan's results are better than that of the Durand's

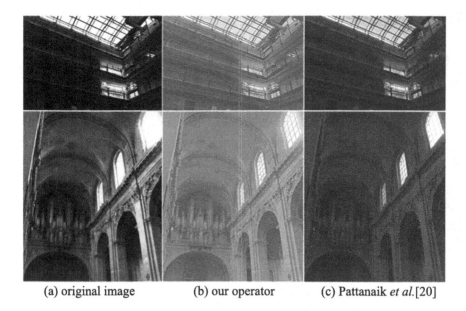

(a) original image   (b) our operator   (c) Pattanaik *et al.*[20]

Fig. 6. Comparing the results with a global algorithm.

method. For the bright region, our method obtains good luminance compression and preserves more details. The last row (Fig. 3(e)) shows different exposure levels of the HDR image, it clearly shows that our model and Durand's method still keep the wall a little blue which is similar to the original HDR image, but the Meylan's result has serious color cast, e.g., the wall appears even a little yellow. Other two inside HDR scenes are also shown in Fig. 4. Durand's model has serious contrast reversals artifacts around high contrast edges. For instance, unlocked lamps on the ceiling in top image of Fig. 4, their edges are out of shape.

We also selected four outdoor HDR scenes to give more comparisons. The results are shown in Fig. 5. In some situations, Durand's algorithm could lose some information, while our method preserves more clear details in both the dark and light situations with less color cast.

We also compared our model with the method of Pattanaik et al. [20], a global tone reproduction operator (shown in Fig. 6). The results of Pattanaik et al. are cited from pfstools (http://pfstools.sourceforge.net/tmo_gallery/). We can see that Pattanaik's results lose much color information and have lower contrast.

We further compared our method with Ashikhmin's tone mapping algorithm [1], Drago's adaptive logarithmic mapping [5], Durand's bilateral filtering [6], Fattal's gradient domain compression [7], Reinhard's photographic tone reproduction [22], Tumblin's fovea interactive method [24] and Ward's contrast-based operator [26]. The results of these methods are downloaded from the Max Planck Institut Informatik (MPII) (http://resources.mpi-inf.mpg.de/tmo/NewExperiment/TmoOverview.html). Two examples of Napa Valley and Hotel Room scenes in MPII dataset are shown in Fig. 7. We calculated the entropy

Fig. 7. comparison on Napa Valley and Hotel Room from MPII. In every scene from top to bottom and left to right there are the original HDR image, our results, Ashikhmin [1], Drago [5], Durand [6], Fattal [7], Reinhard [22], Drago's Retinex, Tumblin [24] and Ward [26] respectively.

Table 1. Quantitative comparison on images from MPII with entropy as metric.

	Napa Valley	Hotel Room	Atrium Morning	Atrium Night	Memorial
Ashikmin [1]	7.416	6.966	7.778	7.972	7.769
Drago [5]	7.381	6.756	7.421	7.907	7.753
Durand [6]	7.182	6.695	7.046	7.424	7.291
Fattal [7]	7.600	7.445	8.031	7.455	8.504
Reinhard [22]	7.232	6.817	7.375	7.860	7.837
Tumblin [24]	7.016	6.391	6.995	7.491	7.668
Ward [26]	7.416	6.694	7.251	7.915	7.783
Proposed	**7.941**	**7.267**	**7.830**	**8.309**	**8.030**

values as quantitative metric. The entropy of our method and the seven compared methods on five scenes are listed in Table 1. A higher entropy score means the richer information contained in an image, indicating the better performance of the method.

4   Conclusion

In this paper, we proposed a visual adaptation model based on the retinal adaptation mechanisms to compress the dynamic range of HDR images. The model has considered the sensitivity changes based on the light and dark adaptation mechanisms and the receptive field property of retinal ganglion cells. We have compared our results with several typical compression operators, on both indoor and outdoor scenes. In general, the proposed model can efficiently compress the dynamic range and enhance the local contrast. Considering that some high level information in the visual system can usually serve as an important guidance to improve the low level visual processing, in the future work we will consider to add the global visual effect when compressing the dynamic range.

Acknowledgments. This work was supported by the Major State Basic Research Program under Grant 2013CB329401, and the Natural Science Foundations of China under Grant 61375115, 91420105. The work was also supported by the Fundamental Research Funds for the Central Universities under Grant ZYGX2016KYQD139.

References

1. Ashikhmin, M.: A tone mapping algorithm for high contrast images. In: 13th Eurographics Workshop on Rendering, Pisa, Italy, 26–28 June 2002, p. 145. Association for Computing Machinery (2002)
2. Baylor, D., Fuortes, M.: Electrical responses of single cones in the retina of the turtle. J. Physiol. **207**(1), 77–92 (1970)
3. Boynton, R.M., Whitten, D.N.: Visual adaptation in monkey cones: recordings of late receptor potentials. Science **170**(3965), 1423–1426 (1970)
4. Dowling, J.E., Ripps, H.: Adaptation in skate photoreceptors. J. Gen. Physiol. **60**(6), 698–719 (1972)
5. Drago, F., Myszkowski, K., Annen, T., Chiba, N.: Adaptive logarithmic mapping for displaying high contrast scenes. Comput. Graph. Forum **22**, 419–426 (2003). Wiley Online Library
6. Durand, F., Dorsey, J.: Fast bilateral filtering for the display of high-dynamic-range images. ACM Trans. Graph. (TOG) **21**, 257–266 (2002). ACM
7. Fattal, R., Lischinski, D., Werman, M.: Gradient domain high dynamic range compression. ACM Trans. Graph. (TOG) **21**, 249–256 (2002). ACM
8. Ferwerda, J.A., Pattanaik, S.N., Shirley, P., Greenberg, D.P.: A model of visual adaptation for realistic image synthesis. In: Proceedings of the 23rd Annual Conference on Computer Graphics and Interactive Techniques, pp. 249–258. ACM (1996)
9. Gao, S., Han, W., Ren, Y., Li, Y.: High dynamic range image rendering with a luminance-chromaticity independent model. In: He, X., Gao, X., Zhang, Y., Zhou, Z.-H., Liu, Z.-Y., Fu, B., Hu, F., Zhang, Z. (eds.) IScIDE 2015. LNCS, vol. 9242, pp. 220–230. Springer, Cham (2015). https://doi.org/10.1007/978-3-319-23989-7_23

10. Hall, R.: Illumination and Color in Computer Generated Imagery. Springer, New York (2012)
11. Hiroto, I., Hirano, M., Tomita, H.: Electromyographic investigation of human vocal cord paralysis. Ann. Otol. Rhinol. Laryngol. **77**(2), 296–304 (1968)
12. Larson, G.W., Rushmeier, H., Piatko, C.: A visibility matching tone reproduction operator for high dynamic range scenes. IEEE Trans. Vis. Comput. Graph. **3**(4), 291–306 (1997)
13. Ledda, P., Santos, L.P., Chalmers, A.: A local model of eye adaptation for high dynamic range images. In: Proceedings of the 3rd International Conference on Computer Graphics, Virtual Reality, Visualisation and Interaction in Africa, pp. 151–160. ACM (2004)
14. Li, Y., Pu, X., Li, H., Li, C.: A retinal adaptation model for HDR image compression. In: Perception, vol. 45, pp. 45–45. Sage Publications Ltd., London (2016)
15. Meylan, L., Susstrunk, S.: High dynamic range image rendering with a retinex-based adaptive filter. IEEE Trans. Image Process. **15**(9), 2820–2830 (2006)
16. Naka, K., Rushton, W.: S-potentials from colour units in the retina of fish (cyprinidae). J. Physiol. **185**(3), 536–555 (1966)
17. Normann, R.A., Werblin, F.S.: Control of retinal sensitivity. J. Gen. Physiol. **63**(1), 37–61 (1974)
18. Paris, S., Durand, F.: A fast approximation of the bilateral filter using a signal processing approach. In: Leonardis, A., Bischof, H., Pinz, A. (eds.) ECCV 2006. LNCS, vol. 3954, pp. 568–580. Springer, Heidelberg (2006). https://doi.org/10.1007/11744085_44
19. Pattanaik, S.N., Ferwerda, J.A., Fairchild, M.D., Greenberg, D.P.: A multiscale model of adaptation and spatial vision for realistic image display. In: Proceedings of the 25th Annual Conference on Computer Graphics and Interactive Techniques, pp. 287–298. ACM (1998)
20. Pattanaik, S.N., Tumblin, J., Yee, H., Greenberg, D.P.: Time-dependent visual adaptation for fast realistic image display. In: Proceedings of the 27th Annual Conference on Computer Graphics and Interactive Techniques, pp. 47–54. ACM Press/Addison-Wesley Publishing Co. (2000)
21. Reinhard, E., Heidrich, W., Debevec, P., Pattanaik, S., Ward, G., Myszkowski, K.: High Dynamic Range Imaging: Acquisition, Display, and Image-Based Lighting. Morgan Kaufmann, Burlington (2010)
22. Reinhard, E., Stark, M., Shirley, P., Ferwerda, J.: Photographic tone reproduction for digital images. ACM Trans. Graph. (TOG) **21**(3), 267–276 (2002)
23. Rodieck, R.W., Stone, J.: Response of cat retinal ganglion cells to moving visual patterns. J. Neurophysiol. **28**(5), 819–832 (1965)
24. Tumblin, J., Hodgins, J.K., Guenter, B.K.: Two methods for display of high contrast images. ACM Trans. Graph. (TOG) **18**(1), 56–94 (1999)
25. Tumblin, J., Rushmeier, H.: Tone reproduction for realistic images. IEEE Comput. Graph. Appl. **13**(6), 42–48 (1993)
26. Ward, G.: A contrast-based scalefactor for luminance display. In: Graphics Gems IV, pp. 415–421. Academic Press Professional Inc., San Diego (1994)
27. Xie, Z., Stockham, T.G.: Toward the unification of three visual laws and two visual models in brightness perception. IEEE Trans. Syst. Man Cybern. **19**(2), 379–387 (1989)
28. Zhang, X.-S., Li, Y.-J.: A retina inspired model for high dynamic range image rendering. In: Liu, C.-L., Hussain, A., Luo, B., Tan, K.C., Zeng, Y., Zhang, Z. (eds.) BICS 2016. LNCS, vol. 10023, pp. 68–79. Springer, Cham (2016). https://doi.org/10.1007/978-3-319-49685-6_7

A Novel Orientation Algorithm
for a Bio-inspired Polarized Light Compass

Guoliang Han, Xiaoping Hu, Xiaofeng He$^{(\boxtimes)}$, Junxiang Lian,
Lilian Zhang, and Yujie Wang

College of Mechatronics and Automation, National University of Defense Technology,
Changsha 410073, Hunan, China
hanguoliang09@nudt.edu.cn, hexf_lit@126.com

Abstract. Many animals, such as honey bees and tarantulas, can navigate with polarized light, the key of which is to extract compass information from skylight polarization pattern. Many groups have conducted research on bionic polarization sensors and obtained good orientation results in the open-sky circumstances. However, if the sky is obscured by leaves or buildings, the skylight polarization pattern will be greatly affected. This paper presents an unsupervised method for polarization navigation when the sky is partly blocked. First of all, we introduce the core components of polarized light compass and the measurement method of skylight polarization pattern. Then, an unsupervised method is used to extract the sky region according to the single scattering Rayleigh model. Finally, we calculate solar meridian vector and heading angle using pixels of sky region. Results show that polarization navigation, featuring high anti-interference and no accumulation error, is suitable for outdoor autonomous navigation.

Keywords: Polarized light compass · Bio-inspired navigation
Skylight polarization pattern

1 Introduction

Navigation is essential for a variety of human activities. Current navigation approaches are mostly depended on global navigation satellite system (GNSS). However, the signals of GNSS are easy to be disturbed, and possibly unavailable in many terrestrial environments such as indoors, near tall buildings. The remarkable navigation ability that many animals possess provide good reference for us. Animals, such as sand ants, honey bees, tarantulas, and part of birds, can derive compass information from the pattern of polarized light that appears due to scattering of sunlight in sky. This compass information help them reach their destination accurately in spite of long-distance navigation.

Many research groups have done a plenty of work on skylight polarization measurements and corresponding navigation methods [1,2]. At present, the measurements of polarized light can be divided into two categories: single-source

© Springer Nature Singapore Pte Ltd. 2017
J. Yang et al. (Eds.): CCCV 2017, Part I, CCIS 771, pp. 48–59, 2017.
https://doi.org/10.1007/978-981-10-7299-4_5

based and image based. Point source navigation sensors use multiple photodiodes to extract polarization information of incident light from certain directions. Lambrinos et al. constructed a polarization compass using photosensitive diodes and successfully employed it for robot navigation [3]. Chu and co-workers enhanced the polarization sensitivity mechanisms and proposed an angle estimation algorithm [4,5]. However, the point-source based sensors are extremely susceptible to environmental interference. In order to minimize the effect of noisy and improve the robustness, image-based polarization measurements are exploited by researchers. Pust and Shaw built an imaging Stokes-vector polarimeter to study the skylight polarization pattern [6]. Lu et al. used Hough transform to extract solar meridian from points having a polarization angle of nearly 90 in polarization imaging [7]. Based on single scattering Rayleigh model, Wang et al. designed a solar meridian detection algorithm taking into account polarization information of all the pixels [8]. However, all these polarization sensors are working in the open-sky circumstances, the polarization pattern will be greatly affected when the sky is obscured by leaves or buildings.

In this paper, we proposes an unsupervised method for polarization imaging navigation when the sky is partly blocked. For the first time, we extracted the sky region from angle of polarization pattern based on single scattering Rayleigh model. After extracting the sky region, we calculated the solar meridian using pixels of sky region. The proposed method were tested in outdoor experiments. The rest of this paper is organized as follows: Sect. 2 introduces the polarized light compass and orientation algorithm. Section 3 tests proposed method in both turntable experiment and dynamic vehicle experiment, then presents the results and discussion. Finally, conclusions are drawn in Sect. 4.

2   Materials and Methods

2.1   System Description

The designed PL-compass (Polarized Light Compass) is shown in Fig. 1(a). The PL-compass is composed of 4 measurement units and the four units are arranged in square layout to ensure the maximum overlap of the image. Each measurement unit contains one CCD camera (GC1031CP, Smartek), one linear polarizer and one wide-angle lens (F1.4~F16, focal length 3.5 mm), as shown in Fig. 1(b). The resolution of each camera is 1034×778, and its angle of view is $77° \times 57.7°$. The linear polarizers are fixed between the camera and wide-angle lens. Theoretically, four linear polarizers are arranged in $0°, 45°, 90°$ and $135°$ with respect to reference axis, the arrangement of polarizers are shown in Fig. 1(c). Wang et al. [9] have proved that using this installation mode could minimize the effect of noise when computing the polarization pattern. In order to guarantee the synchronization of exposure time for all the four cameras, cameras are triggered by external falling edges.

We calibrated the intrinsic and extrinsic parameters of four cameras using the Camera Calibration Toolbox for Matlab. After that, the gain factor of each camera and installation error of polarizers are calibrated with a high-precision turntable and a standard light source. For detailed description of calibration process, one can reference our previous work [10].

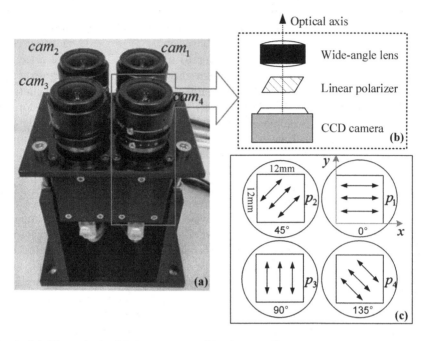

Fig. 1. (a) The polarized light compass. (b) The installing structure of each measurement unit. (c) The arrangement of the four linear polarizers.

2.2   Observation of Sky Polarization Angle Pattern

For an incident light from particular direction, the responses of four cameras can be described as:

$$f_j = \frac{1}{2}K_j I[1 + d cos(2\phi - 2\beta_j)] \tag{1}$$

where $j = 1, 2, 3, 4$ denotes the number of cameras, K_j denotes the gain factor of the jth camera, d and ϕ denote the polarization degree and polarization angle of the incident light respectively, β_j denotes the orientation of the polarizer in j^{th} camera.

Equation (1) can be rewritten as:

$$2f_j/K_j = I d \cos 2\phi \cos 2\beta_j + I d \sin 2\phi \sin 2\beta_j + I, j = 1, 2, 3, 4 \tag{2}$$

Define

$$D = \begin{bmatrix} \cos 2\beta_1 \sin 2\beta_1 \ 1 \\ \cos 2\beta_2 \sin 2\beta_2 \ 1 \\ \cos 2\beta_3 \sin 2\beta_3 \ 1 \\ \cos 2\beta_4 \sin 2\beta_4 \ 1 \end{bmatrix}, X = \begin{bmatrix} x_1 \\ x_2 \\ x_3 \end{bmatrix} = \begin{bmatrix} I d \cos 2\phi \\ I d \sin 2\phi \\ I \end{bmatrix}, F = \begin{bmatrix} 2f_1/K_1 \\ 2f_2/K_2 \\ 2f_3/K_3 \\ 2f_4/K_4 \end{bmatrix} \tag{3}$$

Then Eq. (2) can be described in matrix form:

$$F = DX \tag{4}$$

According to the output information of CCD cameras, Eq. (4) can be solved by least square estimation.

$$\widehat{X} = (D^T D)^{-1} D^T F \tag{5}$$

The angle of polarization for incident light is given by:

$$\phi = \frac{1}{2} \arctan\left(\frac{x_2}{x_1}\right) \tag{6}$$

2.3 Extraction of the Sky Region

In clear sky, the scattering particles are mainly composed of atmospheric molecules, the size of which is much smaller than the wavelength of incident light. The actual measurements of polarization angle pattern agree well with the single scattering Rayleigh model [9]. However, if the sky is obscured by leaves or buildings, the measurements of corresponding pixels will have large deviations from the real value. In order to solve this problem, we establish a criteria to extract the pixels of sky region according to the single scattering Rayleigh model.

The right-hand Cartesian coordinate frames shown in Fig. 2 are defined as below:

Navigation coordinate frame ($ONED$): The N axis is the geographic north, E is geographical eastward, D is to the vertical downward direction, E axis and D axis are not shown in the diagram.

Camera coordinate frame ($OX_lY_lZ_l$): Select the 1st camera as the reference camera, O is the center of image, X_l and Y_l are parallel to the row and column axis of CCD sensor respectively. When the sensor is leveled, the axis Z_l points to the zenith direction.

Incident rays coordinate frame ($O_iX_iY_iZ_i$): The axis Z_i points to the observation direction, and the X_i axis is in the plane OPP' and perpendicular to the observation direction (the axis Y_i is not shown).

O represents the position of observer, S is the position of the sun on the celestial sphere. The zenith angle and azimuth angle of the sun in the navigation coordinate frame is γ_S and φ_S respectively. P is the observation point on the celestial sphere. The zenith angle and azimuth angle of the observation point P in the camera coordinate frame are γ and α respectively. The solar azimuth angle in camera coordinate frame is α_S, the polarization angle of scattered light is ϕ.

According to the astronomical ephemeris, after inputting the time and location information, solar zenith angle γ_S and azimuth angle φ_S can be obtained.

Estimation of the Solar Meridian Vector
In camera coordinate frame, vector $\overrightarrow{OS}$ and $\overrightarrow{OP}$ are represented as below:

$$\begin{aligned}
\overrightarrow{OS_l} &= [\sin\gamma_S \cos\alpha_S \ \sin\gamma_S \sin\alpha_S \ \cos\gamma_S]^T \\
\overrightarrow{OP_l} &= [\sin\gamma \cos\alpha \ \sin\gamma \sin\alpha \ \cos\gamma]^T
\end{aligned} \tag{7}$$

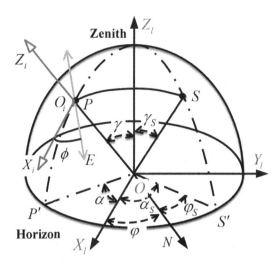

Fig. 2. Description of the single scattering Rayleigh model.

We can use single scattering Rayleigh theory to describe the scattering process of sun light, where the E-vector of scattered light e is perpendicular to the scattering plane, the expression of e in camera coordinate frame is given by:

$$e^l = \overrightarrow{OS_l} \times \overrightarrow{OP_l} = \begin{bmatrix} sin\alpha_S \sin\gamma_S cos\gamma - \sin\alpha \cos\gamma_S \sin\gamma \\ \cos\alpha \cos\gamma_S \sin\gamma - \cos\alpha_S \sin\gamma_S cos\gamma \\ sin(\alpha - \alpha_S) \sin\gamma_S \sin\gamma \end{bmatrix} \qquad (8)$$

The transformation matrix C_l^i from Camera coordinate frame l to incident rays coordinate frame i can be calculated as follows:

$$\mathbf{C}_l^i = \begin{bmatrix} \cos\gamma & 0 & -\sin\gamma \\ 0 & 1 & 0 \\ \sin\gamma & 0 & \cos\gamma \end{bmatrix} \begin{bmatrix} \cos\alpha & \sin\alpha & 0 \\ -\sin\alpha & \cos\alpha & 0 \\ 0 & 0 & 1 \end{bmatrix} \qquad (9)$$

The expression of $\overrightarrow{PE}$ in incident rays coordinate frame is given by:

$$\overrightarrow{PE_i} = C_l^i e^l = \begin{bmatrix} -\sin(\alpha - \alpha_S) \sin\gamma_S \\ \cos\gamma_S \sin\gamma - \sin\gamma_S \cos\gamma \cos(\alpha - \alpha_S) \\ 0 \end{bmatrix} = \begin{bmatrix} \cos\phi \\ \sin\phi \\ 0 \end{bmatrix} \qquad (10)$$

The last equation can be obtained directly from the incident rays coordinate frame, then the relation between ϕ and α_S can be obtained from Eq. (10):

$$\cos\gamma \cos(\alpha - \alpha_S) - \sin(\alpha - \alpha_S) \tan\phi = \cot\gamma_S \sin\gamma \qquad (11)$$

Assuming $A = \cos\gamma, B = \tan\phi$, according to angle sum and difference identities, we can obtain the following equation

$$\sin(\alpha - \alpha_S + \beta) = \frac{\cot\gamma_S \sin\gamma}{\sqrt{A^2 + B^2}} \qquad (12)$$

where $\beta = \arctan \frac{A}{B}$. Then the solar azimuth angle α_S can be obtained by:

$$\alpha_S = \alpha + \beta - \arcsin(\frac{\cot \gamma_S \sin \gamma}{\sqrt{A^2 + B^2}}) or \alpha_S = \alpha + \beta + \arcsin(\frac{\cot \gamma_S \sin \gamma}{\sqrt{A^2 + B^2}}) - \pi \quad (13)$$

The zenith angle γ and azimuth angle α of the observation point P in the camera coordinate frame l can be calculated by the following formula:

$$\begin{aligned} \gamma &= \arctan(\frac{\sqrt{(x_p - x_c)^2 + (y_p - y_c)^2}}{f_c}) \\ \alpha &= \arctan(\frac{y_p - y_c}{x_p - x_c}) \end{aligned} \quad (14)$$

where f_c denotes the focal length of the camera, (x_c, y_c) is the image principal point, each pixel (x_p, y_p) in the image is corresponding to an observation point P in the sky.

We calculate the meridian direction α_S of all pixels, then histogram statistic method is used to analyze its probability distribution. The peak of the histogram distribution is the estimated solar meridian direction $\hat{\alpha}_S$.

The estimation of solar meridian vector in camera coordinate system is given by:

$$\hat{s} = \left[\sin(\gamma_S)cos(\hat{\alpha}_S) \ \sin(\gamma_S)\sin(\hat{\alpha}_S) \ \cos(\gamma_S) \right] \quad (15)$$

Sky Area Detection

Defining E-vector matrix $E = \left[e_1 \cdots e_N \right]$ for all the pixels, where N is the total number of pixels in the image. The solar meridian vector is perpendicular to the E-vector, so the criteria to extract effective pixels can be empirically established as following:

$$\left| E^T \hat{s} \right| < 0.1 \quad (16)$$

According to Eq. (16), we can extract the pixels of sky region and obtain a subset of E-vector matrix E_u. If the number of effective pixels is less than 20% of N, we will reject this frame data.

2.4   Computation of the Azimuth Angle

After extracting the sky region, the azimuth angle is accurately computed using pixels of sky region. In the actual measurement, the optimal estimation of the sun vector s can be expressed as the following optimization problem:

$$\min_s(s^T E_u E_u^T s), s.t. s^T s = 1 \quad (17)$$

Equation (17) can be solved by Lagrange multiplier method. Define the following Lagrange cost:

$$L(s) = s^T E_u E_u^T s - \lambda(s^T s - 1) \quad (18)$$

where λ is called the Lagrange multiplier. Taking the derivatives of L(s) with respect to s and set it to zero, we have:

$$E_u E_u^T s = \lambda s \quad (19)$$

Equation (19) indicates that the optimal s is an eigenvector of $E_u E_u{}^T$ with eigenvalue λ. Substituting (19) into (18), we obtain

$$L(s) = \lambda \tag{20}$$

The eigenvector of $E_u E_u{}^T$ corresponding to its smallest eigenvalue is the optimal estimation of the sun vector. We find the eigenvectors and eigenvalues of $E_u E_u{}^T$ by finding the Singular Value Decomposition (SVD) of E_u.

Assuming s^* is the eigenvector of $E_u E_u{}^T$ corresponding to its smallest eigenvalue, then the optimal solar meridian direction α_S^* can be calculated by:

$$\alpha_S^* = arc\tan(\frac{s_2^*}{s_1^*}) \tag{21}$$

where s_1^* and s_2^* are the first and second elements of s^*, respectively.

The heading angle calculated by polarization pattern has the 180° ambiguity. We compute the mean brightness of left side and right side for original image to avoid this ambiguity, because the pixels near the sun are more likely to have higher brightness. The heading angle in navigation coordinate frame is calculated according to the solar azimuth φ_S and the optimal solar meridian direction α_S^*, which is given by:

$$\varphi = \varphi_S - \alpha_S^* \tag{22}$$

3   Experimental Results

3.1   Turntable Experiment

The orientation experiment was carried out in December 16, 2015 in Changsha. On this day, the weather was sunny and the sky was clear. In order to verify the heading accuracy of proposed method, the PL-compass was horizontally fixed on the precise angle-dividing table, which has a total of 391 minimal rotating steps with each step representing 0.9207°. In the experiment, the turntable started from 0°, and then steadily increased to 359.08° at a step of 27.6215° (30 minimal steps). After reaching 359.08°, the turntable rotated back to 0° at a step of −27.6215°. The experiment was conducted from 16:40 to 16:43 when the zenith angle of the sun was about 80.4°.

Figure 3 shows the angle pattern of sky polarization measured in the orientation experiment. The straight line in the graph represents the projection of solar meridian in the camera coordinate frame. It can be seen clearly that the angle pattern of sky polarization is approximately centrosymmetric relative to the image center, and the polarization angle of the solar meridian is positive or negative 90°. These results are consistent with the Rayleigh scattering model, that is, the E-vector of the scattered light is perpendicular to the scattering plane. Comparing the results of the first two frame, we can see that the polarization angle of the sky rotates with the rotation of the turntable. The turntable rotates from 0° to 359.08° and then back to the 0°, as shown in the horizontal

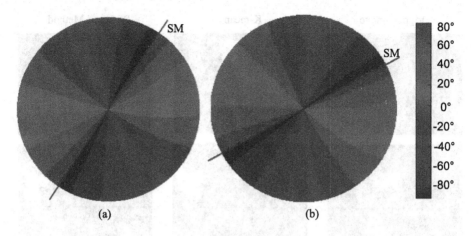

Fig. 3. The first two measurements of sky polarization pattern.

axis in Fig. 4. Set the initial heading angle of the system as reference, the orientation error is defined as the difference between the output of the PL-compass and the rotation angle of the turntable. The results of the orientation experiment is shown in Fig. 4, in which the maximum heading error is about 0.8° and the standard deviation of the error is about 0.32°.

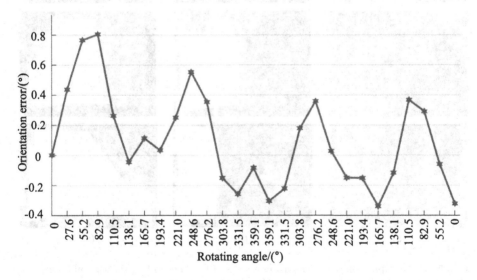

Fig. 4. Orientation error in turntable experiment.

3.2 Dynamic Vehicle Experiment

In order to verify the heading accuracy of the PL-compass in dynamic case, the PL-compass was fixed on the roof of a vehicle. The experiment was conducted in the campus of National University of Defense Technology (NUDT)

Original image   K-means   Proposed Method

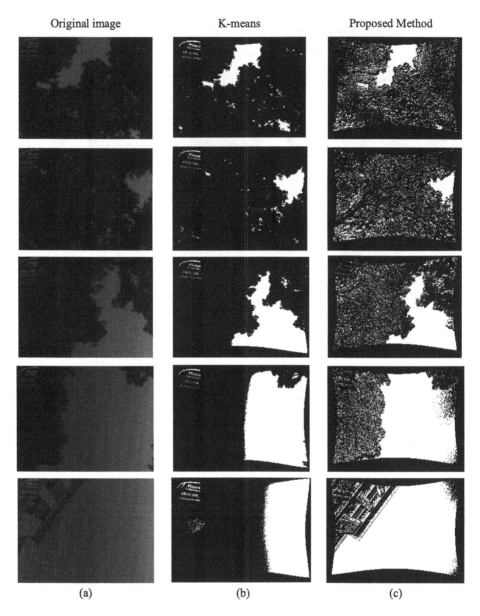

(a)   (b)   (c)

Fig. 5. The comparison of classification results of two methods. (a) Original image. (b) Classification results of K-means. (c) Classification results of proposed method.

in December 7, 2016 from 16:59 to 17:10. During the experiment, the vehicle drove along the road of campus. Sometimes the sky above the vehicle was partly blocked by tree leaves and buildings, these frames could be used to test the robustness and accuracy of proposed method. The benchmark of this experiment was GNSS/INS integrated navigation system, which was equipped on the

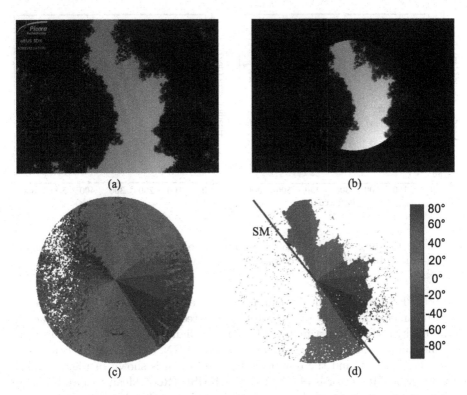

Fig. 6. (a) Original image. (b) Selection of region of interest. (c) Angle of polarization pattern. (d) Extracted sky region and estimated solar meridian.

vehicle. The heading accuracy of GNSS/INS is better than 0.05°. The sampling frequency of GNSS/INS and PL-compass were 100 Hz and 1 Hz respectively. The weather was fine for orientation of PL-compass. In this paper, we assumed the horizontal angle of PL-compass in dynamic vehicle experiment was equal to 0°.

The traditional unsupervised classification method is cluster analysis. In order to test the performance of our proposed algorithm, the classification results of proposed algorithm are compared with those of K-means algorithm, and the results are shown in Fig. 5. Although some areas are obscured by leaves or buildings, our sensor can extract sky polarization information from these regions, which is considered to be sky in our algorithm. The classification results for the first three images in Fig. 5 are basically the same, while the classification results for the last two pictures are largely different. This is because the brightness range of the sky is relatively large, under this situation, the proposed algorithm can extract the sky area more correctly than that by the K-means algorithm. The computation time of proposed algorithm for an image (778×1034) is about 0.11 s in Matlab, which is only about a quarter of that in K-means algorithm.

Figure 6(a) is the original image collected by PL-compass. After calibration, a circular mask was used to select a region of interest, the result is shown in

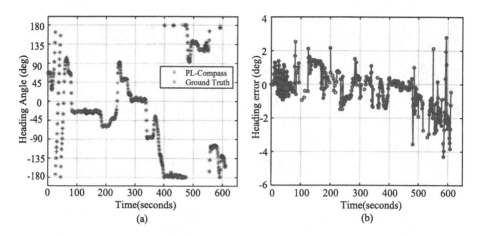

Fig. 7. (a) Heading angle of GNSS/INS and PL-compass in dynamic vehicle experiment. (b) Heading error of PL-compass compared with ground truth.

Fig. 6(b). Figure 6(c) is the measurement of the polarization angle pattern. By using our method, the sky region can be extracted unsupervised. The extraction result of sky region and estimation of solar meridian (SM) are shown in Fig. 6(d).

The comparison of heading angle obtained by GNSS/INS and PL-compass is shown in Fig. 7(a), and the difference between them is shown in Fig. 7(b). The maximum heading error is about 4.3°, the RMSE (Root Mean Square Error) of estimated heading angle is about 1.15°. It is obviously that the orientation errors of PL-compass in the last 100 s was larger than those in the first 500 s. This is mainly because the vehicle was going downhill at that time and the horizontal angle was largely deviated from 0°.

4   Conclusion

In this paper, a novel PL-Compass is proposed. The PL-compass mainly consists of 4 measurement units. Each measurement unit contains one CCD camera, one linear polarizer and one wide-angle lens. The skylight polarization pattern can be measured in realtime. An unsupervised method is introduced to extract the sky region. After extracting the sky region, we calculate azimuth angle in navigation coordinate frame using pixels of sky region. In turntable experiment, the measured skylight polarization pattern agrees well with the Rayleigh scattering model. The maximum error is about 0.8° and the standard deviation of the error is about 0.32°. In dynamic vehicle experiment, the maximum heading error is about 4.3°, the RMSE of estimated heading angle is about 1.15°. Experiment results show that the algorithm proposed in this paper can be used for polarization imaging navigation when the sky is partly obscured.

For future work, we will combine the PL-compass and inertial navigation system to reduce the effect of deviation of horizontal angle in dynamic vehicle experiment.

Acknowledgments. This work was supported by the National Natural Science Foundation of China (grants 61503403, 61573371 and 61773394).

References

1. Chahl, J., Mizutani, A.: Biomimetic attitude and orientation sensors. IEEE Sens. J. **12**(2), 289–297 (2011)
2. Sarkar, M., Bello, D.S.S., Hoof, C.V., Theuwissen, A.J.P.: Biologically inspired cmos image sensor for fast motion and polarization detection. IEEE Sens. J. **13**(3), 1065–1073 (2013)
3. Lambrinos, D., Kobayashi, H., Pfeifer, R., Maris, M., Labhart, T., Wehner, R.: An autonomous agent navigating with a polarized light compass. Adapt. Behav. **6**(1), 131–161 (1997)
4. Zhao, K.C., Chu, J.K., Zhang, Q., Wang, T.C.: A novel polarization angle sensor and error compensation algorithm for navigation. J. Astronaut. **30**(2), 503–509 (2009)
5. Chu, J.K., Zhao, K.C., Zhang, Q., Wang, T.C.: Construction and performance test of a novel polarization sensor for navigation. Sens. Actuators Phys. **148**(1), 75–82 (2008)
6. Pust, N.J., Shaw, J.A.: Dual-field imaging polarimeter using liquid crystal variable retarders. Appl. Opt. **45**(22), 5470–5478 (2006)
7. Lu, H., Zhao, K., You, Z., Wang, X.: Design and verification of an orientation algorithm based on polarization imaging. Qinghua Daxue Xuebao J. Tsinghua Univ. **54**(11), 1492–1496 (2014)
8. Wang, Y., Hu, X., Lian, J., Zhang, L., He, X.: Bionic orientation and visual enhancement with a novel polarization camera. IEEE Sens. J. **17**(5), 1316–1324 (2017)
9. Wang, Y., Hu, X., Lian, J., Zhang, L., Xian, Z., Ma, T.: Design of a device for sky light polarization measurements. Sensors **14**(8), 14916–14931 (2014)
10. Fan, C., Hu, X., Lian, J., Zhang, L., He, X.: Design and calibration of a novel camera-based bio-inspired polarization navigation sensor. IEEE Sens. J. **16**(10), 3640–3648 (2016)

Biomedical Image Analysis

A Brain MR Images Segmentation and Bias Correction Model Based on Students t-Mixture Model

Yunjie Chen$^{(\boxtimes)}$, Qing Xu, and Shenghua Gu

College of Mathematics and Statistics,
Nanjing University of Information Science and Technology, Nanjing 210044, China
priestcyj@nuist.edu.cn

Abstract. Accurate segmentation for magnetic resonance images is an essential step in quantitative brain image analysis. However, due to the existence of bias field and noise, many segmentation methods are hard to find accurate results. Finite mixture model is one of the wildly used methods for MR image segmentation; however, it is sensitive to noise and cannot deal with images with intensity inhomogeneity. In order to reduce the effect of noise, we introduce a robust Markov Random Field by incorporating new spatial information which is constructed based on posterior probabilities and prior probabilities. The bias field is modeled as a linear combination of a set of orthogonal basis functions and coupled into the model and makes the method can estimate the bias field meanwhile segmenting images. Our statistical results on both synthetic and clinical images show that the proposed method can obtain more accurate results.

Keywords: Students t-distribution · Gibbs random field
Magnetic resonance image · Intensity inhomogeneity

1 Introduction

Brain disease is one of the main diseases threatening human health. Using brain imaging to examine the qualitative and quantitative analysis of brain function is helpful for the effective diagnosis of brain disease [1]. Brain MR images have high contrast among different soft tissues and high spatial resolution across the entire field of view, which makes them easier to be analyzed than other medical images. Unfortunately, due to the imaging mechanism and the complexity of organ and tissue structure, MR images will appear gray inconsistency, weak boundary and other phenomena. Most segmentation methods are hindered by various imaging artifacts such as noise and intensity inhomogeneities [2].

Although researchers have proposed many methodologies [3–5] for segmentation, there remain some challenges which come from low contrast, intensity inhomogeneity and noise perturbation. In practical applications especially in the field of medical image segmentation, increasingly stringent requirements on

© Springer Nature Singapore Pte Ltd. 2017
J. Yang et al. (Eds.): CCCV 2017, Part I, CCIS 771, pp. 63–76, 2017.
https://doi.org/10.1007/978-981-10-7299-4_6

the accuracy of the segmentation results make it urgent to find more robust approaches. Finite mixture model (FMM) is one of the most widely used clustering models in the field of image segmentation. Clustering is the process of arranging data into groups that having similar characteristics. When the conditional probability in FMM is chosen as a Gaussian distribution, we call the model as the Gaussian mixture model (GMM) [6,7]. GMM can deal with the multivariate data problem effectively and flexibly because its small number of parameters can be estimated simply by adopting the expectation maximization (EM) algorithm. However, GMM wont get an accurate segmentation result when the image is influenced by heavy noise and have some outliers. In order to deal with such problems, some scholars have proposed to increase the number of mixture components of GMM to capture the heavy tails of the distribution which is at the expense of increasing the complexity of the model.

Students t-mixture model (SMM) proposed by Peel and McLachlan [8] is used as an alternative to GMM. Having an additional parameter called the degrees of freedom; SMM conforms to the elliptical symmetry which is wider than the circular symmetry of GMM. Hence, SMM is more robust than GMM in the presence of heavy outliers. However, SMM and GMM assume that each pixel in the image is independent and neither of them uses the prior information of the pixels in their neighborhood. Thus, GMM and SMM are still sensitive to noise.

In order to reduce the effect of noise, the mixture models with Markov random field (MRF) have been widely used for pixel labeling [9–11] where the prior distribution for each pixel corresponding to each label and depends on its neighboring pixels. Further, some models [12–17] were proposed to model the joint of the pixels in order to incorporate the spatial smoothness constraints between neighboring pixels. Among these improvements, EM algorithm was proposed to estimate parameters in the model by Diplaros et al. [14]. On the assumption that the hidden class labels of pixels are generated by prior distributions which share similar parameters for neighboring pixels, a pseudo-likelihood is introduced to couple neighboring prior information based on the entropy of Kullback-Leibler (KL) divergence [16]. Furthermore, Nikou et al. [15] proposed a family of smoothness priors based on the MRF which guarantee different smoothness strength for each cluster. This model also captures information in different spatial directions and obtained closed form update equations of the solutions by EM algorithm. Although these models reduce the influence of noise on the segmentation to some extent, most of the improvements are at the expense of increasing the model parameters and the complexity of the model.

Motivated by the aforementioned observation, in this paper, we propose a new method of spatial constraint construction. Spatially variant finite mixture model in [12] constructed the spatial constraint of the model only by using the priors of the current pixel and its neighborhood. On the basis of SVFMM, we firstly choose Students t-distribution instead of Gaussian distribution to deal with the situation of outliers. In order to improve the robustness of the model to noise, we construct a new penalty term for the model which incorporates with the posterior distributions and priors information. In the tth iteration of the EM algorithm, the prior distribution obtained in the M step only use the result of the *(t-1) th* iteration. Through the improvement of our model, the priors obtained in the *tth* iteration uses the posterior of the E step in the same iterative process which can correct the possible errors in the algorithm. However, in practice, there will be a phenomenon of intensity inhomogeneity which affects the segmentation accuracy. The bias field is mainly related to the reflection frequency and static noise field, Jungke et al. [18] clearly stated that the main challenges of medical image segmentation is for the bias field rather than the noise. In this paper, we use the basic function to fit the bias field to improve the robustness of the model.

2   Backgrounds

2.1   Students t-Mixture Model (SMM)

The finite mixture model is one of the most widely used clustering models for image segmentation. Let y^i denotes the observation at *ith* pixel of an image $(i = 1, ..., N)$ and x^i denotes the class label of the pixel. The density function for a $K-$component finite mixture model is

$$f(y|x; \theta_1, \theta_2, ..., \theta_k; \pi_1, \pi_2, ..., \pi_k) = \sum_{j=1}^{n} \pi_j \psi_j(y|x; \theta_j). \qquad (1)$$

For every *ith* pixel belongs to *jth* class, $p(x_i = j) = pi_j$ represents the prior distribution which satisfies $0 < pi_j < 1$ and $\sum_{j=1}^{n} \pi_j = 1$ s selected as a Gaussian distribution, (1) represents the Gaussian Mixture Model (GMM).

It has been proved that the Student's t-distribution has a heavier tail than the Gauss distribution, and the finite mixture model of the Student's t-distribution provides a more robust way than GMM. The density function of the Students t-distribution [19] is

$$t(y_i|\theta_j) = t(y_i|\mu_j, \Sigma_j, v_j)$$
$$= \frac{\Gamma(\frac{v_j}{2} + \frac{p}{2})}{\Gamma\frac{v_j}{2}} \frac{|\Sigma_j|^{-\frac{1}{2}}}{(\pi v_j)^{\frac{p}{2}}} \times [1 + \frac{(y_j - \mu_j)^T \Sigma_j^{-1}(y_j - \mu_j)}{v_j}]^{-\frac{v_j + p}{2}}, \qquad (2)$$

where μ_j, Σ_j and v_j refer to the mean, covariance and degrees of freedom of multivariate Students t–distribution.

It has been proved that Although FMM is widely used in image segmentation, it is assumed that each pixel is independent; no spatial information about the pixels is used. In order to overcome the drawback of GMM, the Spatially Variant Finite Mixture Model (SVFMM) [12] is proposed.

2.2   Spatially Variant Finite Mixture Model (SVFMM)

The SVFMM [13] method considers *a maximum a posteriori* (MAP) approach and introduces the spatial information based on the following Gibbs function [12, 20, 21]

$$P(\Pi) = \frac{1}{Z} exp(-U(\Pi)),$$

where

$$U(\Pi) = \beta \sum_{i=1}^{N} V_{N_i}(\Pi).$$ (3)

The Z is a normalizing constant and $U(\Pi)$ is an energy function. The function V_{N_i} stands for the clique potential associated with the neighborhood N_i and can be computed as follows

$$V_{N_i}(\Pi) = \sum_{m \in N_i} g(u_{i,m}),$$ (4)

where $u_{i,m}$ specifies the distance between two label vectors π^i and π^m which can be describe as

$$u_{i,m} = |\pi^i - \pi^m| = \sum_{j=1}^{K} (\pi_j^i - \pi_j^m)^2.$$ (5)

Function $g(u)$ in (4) should be monotonically increasing and nonnegative and one choice [21] can be

$$g(u) = (1 + u^{-1})^{-1}.$$ (6)

In recent years, SVFMM has gained a lot of research and improvement. One route in these studies is to construct a space constraint based on the Markov random field (MRF). Most of the improvement of [12,16,22] has better segmentation, but usually with the increase of the model parameters.

3 Proposed Method

We choose the Students t-distribution instead of Gaussian distribution to improve the robustness of the model to outliers. And in order to make full use of the spatial information between pixels to improve the accuracy of the model, we combine Students t-distribution and SVFMM. A new way of spatial constraint addition based on SVFMM was proposed, which can improve the accuracy of segmentation and control model complexity without increasing the number of parameters. The new distance in clique potential (4) is

$$u_{i,m} = \sum_{j=1}^{K} (\pi_j^i - \pi_j^m)^2 + (z_j^i - z_j^m)^2.$$ (7)

The z_j^i is the condition expectation values of the hidden variables which is often computed in the E-step of EM algorithm. Distance function 4 describes the difference between the prior probability of each pixel and its neighborhood, the value of the potential function 3 is large when the difference is obvious. We improve the distance to consider the latest posterior probability of each iteration, which can correct the probable error in the previous iteration. After adding the improved spatial information, the energy function is

$$f(y|\Theta) = \sum_{j=1}^{K} \pi_j t_j(y|\theta_j) \cdot p(\Pi)$$

$$= \sum_{j=1}^{K} \pi_j \frac{\Gamma(\frac{v_j}{2} + \frac{p}{2})|\Sigma_j|^{-\frac{1}{2}}}{\Gamma\frac{v_j}{2} \ (\pi v_j)^{\frac{p}{2}}} \times [1 + \frac{(y_j - \mu_j)^T \Sigma_j^{-1}(y_j - \mu_j)}{v_j}]^{-\frac{v_j+p}{2}} \cdot p(\Pi), \quad (8)$$

where

$$p(\Pi) = \frac{1}{z} exp(-\beta \sum_{i=1}^{N} \sum_{m \in N_i} (1 + (\sum_{j=1}^{K} (\pi_j^i - \pi_j^m)^2 + (z_j^i - z_j^m)^2)^{-1})^{-1}). \quad (9)$$

In fact, a major problem for segmentation of MR images is the intensity inhomogeneity due to the bias field. Under the influence of bias field, the image formation can be given by

$$I = I_0 \times B + N, \quad (10)$$

where I is the observed image, I_0 is the true image, B is the bias field and N is the noise.

We improve the spatial constraint method and use the base function to fit the bias field and coupled to the model. Our model is given as follows

$$f(y|\Theta) = \sum_{j=1}^{K} \pi_j t_j(y|\theta_j) \cdot p(\Pi)$$

$$= \sum_{j=1}^{K} \pi_j \frac{\Gamma(\frac{v_j}{2} + \frac{p}{2})|\Sigma_j|^{-\frac{1}{2}}}{\Gamma\frac{v_j}{2} \ (\pi v_j)^{\frac{p}{2}}} \times [1 + \frac{(y_j - B\mu_j)^T \Sigma_j^{-1}(y_j - B\mu_j)}{v_j}]^{-\frac{v_j+p}{2}} p(\Pi), \quad (11)$$

where

$$p(\Pi) = \frac{1}{z} exp(-\beta \sum_{i=1}^{N} \sum_{m \in N_i} (1 + (\sum_{j=1}^{K} (\pi_j^i - \pi_j^m)^2 + (z_j^i - z_j^m)^2)^{-1})^{-1}). \quad (12)$$

4 Parameter Learning

In this section we use the EM algorithm to estimate the parameters of the model. The log-likelihood function of our model is written in the form

$$L(\Pi, \Theta|Y) = \sum_{i=1}^{N} log\{\sum_{j=1}^{K} \pi_j t_j(y|\theta_j)\}$$

$$- logZ - \beta \sum_{i=1m \in N_i}^{N} \sum (1 + (\sum_{j=1}^{K} (\pi_j^i - \pi_j^m)^2 + (z_j^i - z_j^m)^2)^{-1})^{-1}). \quad (13)$$

Apply of the complete data condition in [12] and the Jensens inequality, we can get

$$J(\Pi, \Theta|Y) = \sum_{i=1}^{N} \sum_{j=1}^{K} z_j^i \{log\pi_j + log t_j(y|\theta_j)\}$$

$$- logZ - \beta \sum_{i=1}^{N} \sum_{m \in N_i} (1 + (\sum_{j=1}^{K} (\pi_j^i - \pi_j^m)^2 + (z_j^i - z_j^m)^2)^{-1})^{-1}). \quad (14)$$

Similar to the MRF-based methods [12,13], Z and in (14) are set to one. The new objective function is given by

$$J(\Pi, \Theta | Y) = \sum_{i=1}^{N} \sum_{j=1}^{K} z_j^i \{ log\pi_j + logt_j(y|\theta_j) \}$$

$$- \sum_{i=1}^{N} \sum_{m \in N_i} (1 + (\sum_{j=1}^{K} (\pi_j^i - \pi_j^m)^2 + (z_j^i - z_j^m)^2)^{-1})^{-1}). \quad (15)$$

With help of the scaling factor, Students t-distribution can be represented as a mixture of Gaussians, with the same mean and covariance as follows

$$t_j(y^i|\theta_j) = N(y^i|\mu_j, \frac{\Sigma_j}{u_j^i}), \quad (16)$$

where

$$N(y^i|\mu_j, \frac{\Sigma_j}{u_j^i}) = \frac{|\Sigma_j|^{-\frac{1}{2}}(u_j^i)^{\frac{p}{2}}}{(2\pi)^{\frac{p}{2}}} exp\{-\frac{1}{2}u_j^i(y^i - \mu_j)^T \Sigma_j^{-1}(y^i - \mu_j)\} \quad (17)$$

and

$$u_j^i \sim gamma(\frac{\upsilon_j}{2}, \frac{\upsilon_j}{2}) \quad (18)$$

According to [8], the complete-data log likelihood can be factored into three parts and can be given as

$$logL_c(\Theta) = logL_1c(\pi) + logL_2c(\upsilon) + logL_3c(\theta), \quad (19)$$

where

$$logL_1c(\pi) = \sum_{i=1}^{N} \sum_{j=1}^{K} z_j^i log\pi_j - \sum_{i=1}^{N} \sum_{m \in N_i} (1 + (\sum_{j=1}^{K} (\pi_j^i - \pi_j^m)^2 + (z_j^i - z_j^m)^2)^{-1})^{-1}), \quad (20)$$

$$logL_2c(\upsilon) = \sum_{i=1}^{N} \sum_{j=1}^{K} z_j^i \{ -log\Gamma(\frac{\upsilon_j}{2}) + \frac{1}{2}\upsilon_j log(\frac{\upsilon_j}{2}) + \frac{1}{2}\upsilon_j(logu_j^i - u_j^i) - logu_j^i \}, \quad (21)$$

$$logL_3c(\theta) = \sum_{i=1}^{N} \sum_{j=1}^{K} z_j^i \{ -\frac{1}{2}plog(2\pi) - \frac{1}{2}log|\Sigma_j| - \frac{1}{2}u_j^i(y^i - B_i\mu_j)^T \Sigma_j^{-1}(y^i - B_i\mu_j) \}. \quad (22)$$

We calculate the conditional expectation values of the hidden variables in the (k+1)th iteration E-step of EM algorithm as follows

$$E(z_j^i|y^i) = z_j^{i(k+1)} = \frac{\pi_j^{i(k)} t_j(y^i|\theta_j^{(k)})}{\sum_{l=1}^{K} \pi_l^{i(k)} t_l(y^i|\theta_l^{(k)})}, \quad (23)$$

$$E(u_j^i|y^i) = u_j^{i(k+1)} = \frac{v_j^{(k)} + p}{v_j^{(k)} + (y^i - \mu_j^{(k)})^T \Sigma_j^{(k)-1}(y^i - \mu_j^{(k)})}, \tag{24}$$

and

$$E(logu_j^i|y^i) = logu_j^{i(k)} - log(\frac{v_j^{(k)} + p}{2}) + \psi(\frac{v_j^{(k)} + p}{2}). \tag{25}$$

The M-step is calculated by maximizing the expectation of the complete data log likelihood form which can be given as

$$Q(\Theta;\Theta^{(k)}) = Q_1(\pi;\Theta^{(k)}) + Q_2(v;\Theta^{(k)}) + Q_3(\theta;\Theta^{(k)}), \tag{26}$$

where

$$Q_1(\pi;\Theta^{(k)}) = \sum_{i=1}^{N}\sum_{j=1}^{K} z_j^{i(k)} log\pi_j - \sum_{i=1}^{N}\sum_{m\in N_i} (1+(\sum_{j=1}^{K}(\pi_j^i - \pi_j^m)^2 + (z_j^i - z_j^m)^2)^{-1})^{-1}, \tag{27}$$

$$Q_2(v;\Theta^{(k)}) = \sum_{i=1}^{N}\sum_{j=1}^{K} z_j^{i(k)}\{-log\Gamma(\frac{v}{2}) + \frac{1}{2}v_j log(\frac{v_j}{2})$$
$$+ \frac{1}{2}v_j(logu_j^{i(k)} - u_j^{i(k)}) + \psi(\frac{v_j^{(k)} + p}{2}) - log(\frac{v_j^{(k)} + p}{2})\}, \tag{28}$$

and

$$Q_3(\theta;\Theta^{(k)}) = \sum_{i=1}^{N}\sum_{j=1}^{K} z_j^{i(k)}\{-\frac{1}{2}plog(2\pi) - \frac{1}{2}log|\Sigma_j|$$
$$+ \frac{1}{2}logu_j^{i(k)} - \frac{1}{2}u_j^{i(k)}(y^i - B_i\mu_j)^T \Sigma_j^{-1}(y^i - B_i\mu_j)\}. \tag{29}$$

The expression of the μ_j and Σ_j can be obtained by maximizing $Q_3(\theta;\Theta^{(k)})$ (Eq. 29),which yields

$$\frac{\partial Q_3}{\partial \mu_j^{(k+1)}} = \sum_{i=1}^{N} z_j^{i(k)} u_j^{i(k)} \Sigma_j^{-1}(y^i - B_i^{(k)}\mu_j^{(k+1)}) \tag{30}$$

and

$$\frac{\partial Q_3}{\partial \Sigma_j^{-(k+1)}} = \frac{1}{2}\sum_{i=1}^{N} z_j^{i(k)}\{\Sigma_j^{-(k+1)} - u_j^{i(k)}(y^i - B_i^{(k)}\mu_j^{(k+1)})^T(y^i - B_i^{(k)}\mu_j^{(k+1)})\}. \tag{31}$$

Let $\frac{\partial Q_3}{\partial \mu_j^{(k+1)}} = 0$ and $\frac{\partial Q_3}{\partial \Sigma_j^{-(k+1)}} = 0$ yields

$$\mu_j^{(k+1)} = \frac{\sum_{i=1}^{N} z_j^{i(k)} u_j^{i(k)} y^i}{\sum_{i=1}^{N} z_j^{i(k)} u_j^{i(k)}} \tag{32}$$

and

$$\Sigma_j^{(k+1)} = \frac{\sum_{i=1}^N z_j^{i(k)} u_j^{i(k)} (y^i - B_i^{(k)} \mu_j^{(k+1)})^T (y^i - B_i^{(k)} \mu_j^{(k+1)})}{\sum_{i=1}^N z_j^{i(k)}} \qquad (33)$$

Setting the partial derivative $Q_2(v; \Theta^{(k)})$(Eq. 28) with respect to v_j, we have

$$\frac{\partial Q_2}{\partial v_j^{(k+1)}} = \frac{1}{2} \sum_{(i=1)}^N z_j^{i(k)} \{ log(\frac{v_j}{2}) - \psi(\frac{v_j}{2}) + 1 + logi_j^{i(k)} - u_j^{i(k)}$$

$$+ \psi(\frac{v_j^{(k)} + p}{2}) - log(\frac{v_j^{(k)} + p}{2}) \}. \qquad (34)$$

Then setting (34) equal to zero, it means that $v_j^{(k)}$ needs to be iteratively computed to solve the following equations:

$$log(\frac{v_j}{2}) - \psi(\frac{v_j}{2}) + 1 + \frac{\sum_{i=1}^N z_j^{i(k)} \{ logu_j^{i(k)} - u_j^{i(k)} \}}{\sum_{i=1}^N z_j^{i(k)}} + \psi(\frac{v_j^{(k)} + p}{2}) - log(\frac{v_j^{(k)} + p}{2}) = 0. \qquad (35)$$

In order to maximize $Q_1(\pi; \Theta^{(k)})$(Eq. 27) with respect π_j^i we set the derivative equal to zero and obtain the following quadratic [15]

$$4[\sum_{m \in N_i} \dot{g}(u_{i,m})](\pi_j^{i(k+1)})^2 - 4[\sum_{m \in N_i} \dot{g}(u_{i,m})](\pi_j^{i(k+1)}) - z_j^{i(k)} = 0, \qquad (36)$$

where $\dot{g}(u_{i,m})$ indicates the derivative of $g(Eq.(6))$. N_i in the above equation is the neighborhood of the ith pixel whose label vectors π^m has not been updated in kth step. The two roots of (36) are

$$\pi_j^{i(k+1)} = \frac{[\sum_{m \in N_i} \dot{g}(u_{i,m})\pi_j^m] \pm \sqrt{[\sum_{m \in N_i} \dot{g}(u_{i,m})\pi_j^m]^2 + z_j^{i(k+1)}[\sum_{m \in N_i} \dot{g}(u_{i,m})]}}{2[\sum_{m \in N_i} \dot{g}(u_{i,m})]} \qquad (37)$$

Many scholars assumed that the Bias field B is smooth and slowly varying in small local region [2] and some of them [23,24] model B to be a linear combination of L smooth basis functions $S_1, ..., S_l$:

$$B(y) = \sum_{l=1}^L q_l s_l(y), \qquad (38)$$

where q_l is the combination coefficient s_l satisfies:

$$\int_\Omega s_i(x) s_j(x) dx = \delta_{ij}. \qquad (39)$$

$\delta_{ij} = 1$ for $i = j$ and $\delta_{ij} = 0$ for $i \neq j$. In Eq.(36) the bias field B can be written as:

$$B_i = Q^T S_i \qquad (40)$$

and $Q_3(\theta; \Theta^{(k)})$(Eq. 29) can be written as:

$$Q_3(\theta; \Theta^{(k)}) = \sum_{i=1}^{N} \sum_{j=1}^{K} z_j^{i(k)} \{ -\frac{1}{2} plog(2\pi) - \frac{1}{2} log|\Sigma_j|$$

$$+ \frac{1}{2} logu_j^{i(k)} - \frac{1}{2} u_j^{i(k)} (y^i - Q^T S_i \mu_j)^T \Sigma_j^{-1} (y^i - Q^T S_i \mu_j) \} \quad (41)$$

Setting the partial derivative of (41) with respect to Q equals zero, we have

$$\frac{\partial Q_3(\theta; \Theta^{(k)})}{\partial Q} = \partial \{ \sum_{i=1}^{N} \sum_{j=1}^{K} z_j^{i(k)} \{ -\frac{1}{2} plog(2\pi) - \frac{1}{2} log|\Sigma_j| + \frac{1}{2} logu_j^{i(k)}$$

$$- \frac{1}{2} u_j^{i(k)} (y^i - Q^T S_i \mu_j)^T \Sigma_j^{-1} (y^i - Q^T S_i \mu_j) \} \} / \partial Q$$

$$\Longrightarrow \sum_{i=1}^{N} S_i S_i^T \sum_{j=1}^{K} z_j^i u_j^i \mu_j^T \Sigma_j^{-1} \mu_j Q = \sum_{i=1}^{N} S_i \sum_{j=1}^{K} z_j^i u_j^i y^i \Sigma_j^{-1} \mu_j$$

$$\Longrightarrow Q = (\sum_{i=1}^{N} S_i S_i^T \sum_{j=1}^{K} z_j^i u_j^i \mu_j^T \Sigma_j^{-1} \mu_j)^{-1} \sum_{i=1}^{N} S_i \sum_{j=1}^{K} z_j^i u_j^i y^i \Sigma_j^{-1} \mu_j. \quad (42)$$

We summary the computation process of our algorithm as follows:

Step.1 Initialize the parameters of the model and select 3×3 square as the neighborhood window.

Step.2 E-step: Calculate z_j^i, u_j^i and $logu_j^i$ by using Eqs.(23)–(25), respectively.

Step.3 M-step: Calculate $\mu_j, \Sigma_j, \upsilon_j, \pi_j$ and Q by using Eqs.(32)–(33), (35), (37) and (42) respectively.

Step.4 Go to Step.2 until the object function converges.

5   Experiment Results and Discussion

In this section, we experimentally evaluate our proposed method in a set of MR images. The clinical images are generated from Internet Brain Segmentation Repository (IBSR, http://www.cma.mgh.harvard.edu/ibsr/). We also evaluate GMM [6], SVFMM [12], SMM [8] and SMM-SVFMM for comparison. Our experiments have been developed in Matlab R2012a and are carried out with clinical brain MR images. All methods are initialized by using K-means method with 100 iterations.

Firstly, we compare our model with other methods in order to show its robustness to noise. Figure 1(a) shows the initial image with Gaussian noise (with 30% noise level). Figure 1(b) is the standard segmentation results. Figure 1 (c)-(f) show the segmentation result for GMM, SVFMM, SMM and SMM-SVFMM respectively. Figure 1(g) shows the corresponding segmentation results of our method. It can be seen from the results that our model is more robust to noise by using the new model Gibbs theory. In addition, Fig. 1(h)-(m) are the details of Fig. 1(b)-(g) which shows that our method can better preserve the details of the initial image compared with other methods. By comparing Fig. 1(l) and Fig. 1(m), this paper presents a new method of spatial constraint which is superior to the SVFMM model.

In order to quantitatively analyze the segmentation effect of various segmentation methods, we use Js values [25] as a metric to evaluate their performance.

$$Js(S_1, S_2) = \frac{S_1 \bigcap S_2}{S_1 \bigcup S_2} \tag{43}$$

S_1 in (43) is the segmentation result and S_2 is the ground truth. The value of Js should be higher when the result is more accurate. Under different noise conditions, the average results are listed in Table 1. The values show that our method has a more accurate segmentation results than GMM, SVFMM, SMM and SMM-SVFMM under different noise environment.

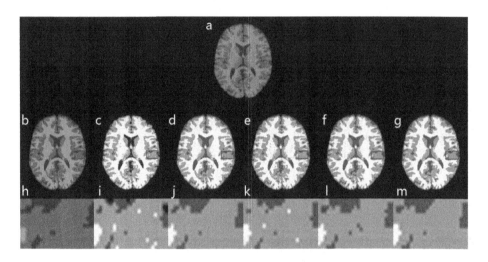

Fig. 1. (a) Segmentation image with 30% noise level. (b) Original four-class image. (h) Figure (b) details. (c) GMM segmentation result. (i) Figure(c) details. (d) SVFMM segmentation result. (j) Figure(d) details. (e) SMM segmentation result. (k) Figure(e) details. (f) SMM-SVFMM segmentation result. (l) Figure(f) details. (g) Segmentation result of proposed method. (m) Figure(g) details.

Furthermore, in order to show the advantage of our model when the image effected by intensity inhomogeneity, we compare our method with GMM, SVFMM, SMM and SMM-SVFMM. Figure 2(a) shows the image affected by noise and bias field (with 3% Gaussian noise and 40% bias field). Figure 2(b) shows the initial without noise and been influenced by gray inhomogeneity. Figure 2(c)-(f) show the segmentation result for GMM, SVFMM, SMM and

Table 1. Comparison of segmentation accuracy of different models under different noise conditions

Method	Var : 0.02 (MCR%)	Var : 0.03 (MCR%)	Var : 0.04 (MCR%)	Var : 0.06 (MCR%)
GMM	70.22	67.89	66.43	64.84
SVFMM	78.35	75.37	73.50	71.13
SMM	76.74	73.46	72.09	70.42
SMM-SVFMM	79.19	76.31	74.52	72.50
Proposed	79.72	77.20	75.91	73.88

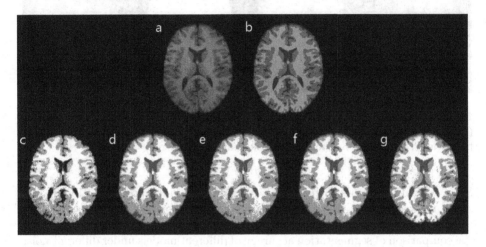

Fig. 2. (a) Corrupted by 3%Gaussian noise and 80% bias field. (b) Original four-class image. (c) GMM (d) SVFMM (e) SMM (f) SMM-SVFMM (g) proposed method.

Table 2. Comparison of segmentation accuracy of different models under different noise and bias field conditions

Percent of bias field	GMM	SVFMM	SMM	SMM − SVFMM	Propsed
10%	93.89	97.47	97.01	97.52	97.58
20%	93.67	97.26	96.85	97.32	97.55
30%	90.35	94.53	94.15	94.60	94.68
40%	92.94	96.50	96.27	96.52	97.51
60%	91.29	95.07	95.11	95.07	97.36
80%	88.41	93.37	93.75	93.37	97.28
100%	86.12	91.85	92.51	91.85	93.09

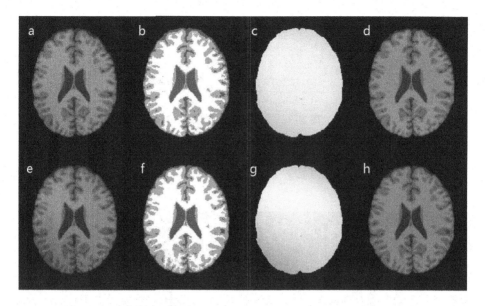

Fig. 3. (a) Corrupted by 3%Gaussian noise and 30% bias field. (b) Proposed method (c) bias field fitted (d) corrected image (e) Corrupted by 3% Gaussian noise and 80% bias field. (f) Proposed method (g) bias field fitted (h) corrected image.

SMM-SVFMM which still have noise and intensity inhomogeneity. Figure 2(g) shows the result of our model. The results comparison indicates that our model is more robust to noise and intensity inhomogeneity than other models. Table 2 is the comparison of segmentation accuracy of different models under different noise and bias field conditions. The comparison shows that our method has the highest accuracy under the different degrees of intensity inhomogeneity. Figure 2(a) and (e) are the image to be segmentation with 3% Gaussian noise and were affected by 30% and 80% percent of bias field. Figure 3(b) and (f) are segmentation result by our model which shows that our method is robust to every degree of intensity inhomogeneity.

6   Conclusion

In this paper, we proposed a new model based on SVFMM by use the Students t-distribution instead of the Gaussian distribution and introduce a new robust model Gibbs theory. Furthermore, the base function is used to fit the bias field and coupled to the Markov Random Field in order to enhance the robustness of the intensity inhomogeneity. Moreover, we simultaneously use the EM algorithm for parameters estimation. From the experimental results, we can conclude that the proposed model is more robust to noise and bias fields than other methods. When the neighborhood radius is 5, the accuracy of the model is higher than 3.

Acknowledgement. The authors would like to thank the anonymous reviewers for their valuable comments and suggestions to improve the quality of the paper. This work was supported in part by the National Nature Science Foundation of China 61672291.

References

1. Boesen, K., Rehm, K., Schaper, K., Stoltzner, S., Woods, R., et al.: Quantitative comparison of four brain extraction algorithms. Neuroimage **22**(3), 1255–1261 (2004)
2. Chen, Y., Zhang, H., Zheng, Y., Jeon, B., Wu, Q.M.J.: An improved anisotropic hierarchical fuzzy c-means method based on multivariate student t-distribution for brain MRI segmentation. Pattern Recognit. **60**(C), 778–792 (2016)
3. Mclachlan, G.J., Peel, D.: Advanced Algorithmic Approaches to Medical Image Segmentation. Springer, Heidelberg (2006). https://doi.org/10.1007/978-0-85729-333-6
4. Permuter, H., Francos, J., Jermyn, I.: A study of Gaussian mixture models of color and texture features for image classification and segmentation. Pattern Recognit. **39**(4), 695–706 (2006)
5. Ma, Z., Tavares, J.M.R.S.: A review of algorithms for medical image segmentation and their applications to the female pelvic cavity. Comput. Methods Biomech. Biomed. Eng. **13**(2), 235 (2010)
6. Titterington, D.M., Smith, A.F.M., Makov, U.E.: Statistical Analysis of Finite Mixture Distributions. Wiley, Hoboken (1985)
7. Jain, A.K., Duin, R.P.W., Mao, J.: Statistical pattern recognition: a review. IEEE Trans. Pattern Anal. Mach. Intell. **22**(1), 4–37 (2000)
8. Peel, D., Mclachlan, G.J.: Robust mixture modelling using the t distribution. Stat. Comput. **10**, 335–344 (2000). Kluwer Academic Publishers
9. Forbes, F., Peyrard, N.: Hidden Markov random field model selection criteria based on mean field-like approximations. IEEE Trans. Pattern Anal. Mach. Intell. **25**(9), 1089–1101 (2003)
10. Celeux, G., Forbes, F., Peyrard, N.: Em procedures using mean field-like approximations for Markov model-based image segmentation. Pattern Recognit. **36**(1), 131–144 (2003)
11. Nguyen, T.M., Wu, Q.M.J.: Fast and robust spatially constrained Gaussian mixture model for image segmentation. IEEE Trans. Circuits Syst. Video Technol. **23**(4), 621–635 (2013)
12. Sanjay-Gopal, S., Hebert, T.J.: Bayesian pixel classification using spatially variant finite mixtures and the generalized EM algorithm. IEEE Trans. Image Process. **7**(7), 1014–1028 (1998). A Publication of the IEEE Signal Processing Society
13. Blekas, K., Likas, A., Galatsanos, N.P., Lagaris, I.E.: A spatially constrained mixture model for image segmentation. IEEE Trans. Neural Netw. **16**(2), 494–498 (2005)
14. Diplaros, A., Vlassis, N., Gevers, T.: A spatially constrained generative model and an em algorithm for image segmentation. IEEE Trans. Neural Netw. **18**(3), 798–808 (2007)
15. Nikou, C., Galatsanos, N.P., Likas, A.C.: A class-adaptive spatially variant mixture model for image segmentation. IEEE Trans. Image Process. **16**(4), 1121–3 (2007). A Publication of the IEEE Signal Processing Society

16. Sfikas, G., Nikou, C., Galatsanos, N.: Edge preserving spatially varying mixtures for image segmentation. In: 2008 IEEE Conference on Computer Vision and Pattern Recognition, CVPR 2008, pp. 1–7. IEEE (2008)
17. Nikou, C., Likas, A.C., Galatsanos, N.P.: A Bayesian framework for image segmentation with spatially varying mixtures. IEEE Trans. Image Process. **19**(9), 2278–2289 (2010)
18. Jungke, M., Seelen, W.V., Bielke, G., Meindl, S., Grigat, M., Pfannenstiel, P.: A system for the diagnostic use of tissue characterizing parameters in N M R tomography. In: de Graaf, C.N., Viergever, M.A. (eds.) Information Processing in Medical Imaging, pp. 471–481. Springer, Boston (1988). https://doi.org/10.1007/978-1-4615-7263-3_31
19. Bishop, C.M.: Pattern Recognition and Machine Learning. Information Science and Statistics. Springer, New York (2006)
20. Zhang, Y., Brady, M., Smith, S.: Segmentation of brain MR images through a hidden Markov random field model and the expectation-maximization algorithm. IEEE Trans. Med. Imaging **20**(1), 45–57 (2001)
21. Green, P.J.: Bayesian reconstructions from emission tomography data using a modified EM algorithm. IEEE Trans. Med. Imaging **9**(1), 84 (1990)
22. Bouguila, N., Ziou, D.: A Powreful finite mixture model based on the generalized Dirichlet distribution: unsupervised learning and applications. In: International Conference on Pattern Recognition, vol. 1, pp. 280–283. IEEE Computer Society (2004)
23. Ji, Z., Liu, J., Cao, G., Sun, Q., Chen, Q.: Robust spatially constrained fuzzy cmeans algorithm for brain MR image segmentation. Pattern Recognit. **47**(7), 2454–2466 (2014)
24. Li, C., Gore, J.C., Davatzikos, C.: Multiplicative intrinsic component optimization (MICO) for MRI bias field estimation and tissue segmentation. Magn. Reson. Imaging **32**(7), 913–923 (2014)
25. Zhang, H., Wu, Q.M., Nguyen, T.M.: Incorporating mean template into finite mixture model for image segmentation. IEEE Trans. Neural Netw. Learn. Syst. **24**(2), 328–335 (2013)

Define Interior Structure for Better Liver Segmentation Based on CT Images

Xiaoyu Zhang[1(✉)], Yixiong Zheng[2], and Bin Zheng[3]

[1] State Key Lab of CAD&CG, Zhejiang University, Hangzhou, China
xiaoyuzhang@zju.edu.cn
[2] The 2nd Affiliated Hospital, Zhejiang University, Hangzhou, China
zyx_xxn@126.com
[3] Surgical Simulation Research Lab, University of Alberta, Edmonton, Canada
bin.zheng@ualberta.ca

Abstract. Liver Segmentation has important application for preoperative planning and intraoperative guiding. In this paper we introduce a new approach by defining the interior structure (hepatic veins) before segmenting the liver from nearby organs. We assume that cells of the liver should lay within a certain distance of the hepatic veins. Therefore, a clear segmentation on hepatic veins will facilitate our segmentation on liver voxel. We build a probabilistic model which adopts four main features of the liver cells based on this idea and implement it on the open source platform 3DMed. We also test the accuracy of this method with four groups of CT data. The results are similar when compared to human experts.

Keywords: Liver · Probabilistic and statistical methods · Vessels
X-ray imaging and computed tomography

1 Introduction

In clinic practice, CT scan is one of the most popular methods to obtain digital images of human livers. In the last few decades, computer scientists have made great efforts to separate the liver from other organs in the abdominal CT. For example, the Grand Challenge of MICCAI 2007 calls for many liver segmentation methods, the summary of which can be found in a 2009 publication of Heimann *et al.* These methods could be divided into three categories, including *gray level based* method, *shape based* method and *pattern based* method [14].

The intensity of CT image is used for segmentation in the gray level based method. Several classic algorithms like region growing, level set and graph cut can fit into this category. The region growing method, first introduced by Pohle *et al.* [19], is widely employed to segment brain tissues [1], thoracic aorta [15] and abdominal organs [23]. In MICCAI 2007, both Rusko *et al.* [20] and Beck and Aurich [2] employ a 3D region growing tool to segment the liver. This method is further improved by Yuan *et al.* [24] and Elomorsy [8] through adding pre- and

© Springer Nature Singapore Pte Ltd. 2017
J. Yang et al. (Eds.): CCCV 2017, Part I, CCIS 771, pp. 77–88, 2017.
https://doi.org/10.1007/978-981-10-7299-4_7

post-processing steps to handle the leakage problem of region growing method. It is also used as an initialization step for more complex pipelines presented by Goryawala *et al.* [9] and Chu *et al.* [5]. In this paper, we also use the region growing method to find the rough area of the liver.

The two most frequently used models for the shape based method are statistical shape model (SSM) [6] and probabilistic atlases (PA) [10,17]. However, the newly developed Sparse Shape Composition (SSC) model challenges these two methods by claiming that there is not an accurate "mean shape" for all livers [22]. It decomposes the liver to multiple regions and thus increases the flexibility of the shape by taking account of shape within each region.

The multiple texture features of the liver in CT are taken into account in the pattern based method. For example, Christ *et al.* [4] and Ben-Cohen *et al.* [3] all train a fully convolutional neural network (FCNN) to achieve the semantic liver segmentation from the CT images, while Danciu *et al.* [7], Luo *et al.* [13] and Zhang *et al.* [25] use SVM to train the classifier.

In this study, we take a different approach by taking the internal structure inside the liver as the reference for liver segmentation. Specifically, we segment vessel system in the liver, then use it to identify the component of liver. In our algorithm, we choose the hepatic vein system from the three tube systems inside the liver [21], because it displays the most obvious intensity contrast against cellular tissue of the liver. We also choose the CT images obtained at portal venous phase during which the agent highlights the hepatic veins most significantly.

In brief, we (1) segment hepatic veins from abdominal CT; (2) use the tree-structure of vein to define which mess around them belongs to the liver; (3) apply isolated point cleaning algorithm to selected liver voxels to further define liver boundary; (4) compare resultant liver boundary to a gold standard, i.e. the manual defined liver boundary by an experimenter on each slide of CT images.

2   Method

In our approach, we construct a probabilistic model to segment liver. There are four probabilities corresponding to four different features of the liver voxel in our probabilistic model, including the intensity, location, distance to hepatic veins and connectivity of the candidate voxels. The "probability" here is a generalized name for the four measurements we use for these features. It could be binary or continuous function.

2.1 P_1: Liver Intensity Range Probability

P_1 is to search for voxels in a certain intensity range defined by μ_l (center) and ρ_l (range). This idea is valid because the liver occupies a continuous and relatively concentrated area in the histogram of abdominal CT. To define the intensity range of liver, we set a threshold manually according to the real-time visual feedback of our interface. When the voxel intensity $I(\boldsymbol{x})$ is beyond this threshold, the value of P_1 will be set to 0 directly. When $I(x)$ falls inside the pre-defined intensity threshold, the P_1 will be calculated according to their similarity to the center of the threshold (Formula (1)).

$$P_1(x) = \begin{cases} 1 - \frac{|I(x-\mu_l)|}{\rho_l} & |I(x) - \mu_l| \leq \rho_l \\ 0 & |I(x) - \mu_l| > \rho_l \end{cases} \tag{1}$$

Before applying the intensity-based filter, we are also aware of noises in the original CT images. So we apply the anisotropic diffusion filter (Perona-Malik [18]) to remove small gratitude change caused by noise as well as keep the large gratitude change caused by intensity difference near liver edge. Figure 1 shows the improved outcome of P_1 when the images are smoothed by the anisotropic diffusion filter first. Noises inside the liver are significantly removed (Fig. 1(d)).

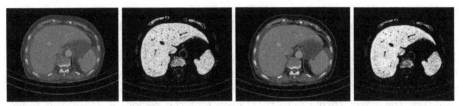

(a) Original Image   (b) P_1 Result for (a)   (c) Smoothed Image   (d) P_1 Result for (c)

Fig. 1. Noises in P_1 result are significantly eliminated after the anisotropic diffusion filter is applied.

2.2 P_2: Liver Location Probability

The liver, for the majority of human, is located at the same position of the body [12]. This feature is used to define location probability P_2. The goal of P_2 is to produce a rough area of liver. We employ the region growing algorithm and morphological operations to achieve this goal.

In particular, we put a seed inside the liver on an arbitrary slice. Then we apply the region growing algorithm with a small similarity threshold to expand the region from the seed point. This is to prevent the region from growing to other neighboring organs. As a result, holes appear as shown in Fig. 2(b). So we further apply the morphological dilation to fill in the holes (Fig. 2(c)). The radius of the dilation is kept small to better preserve the outline of the liver. Small radius of the dilation leaves some holes not completely filled. So we finally apply the morphological closing and all of the holes can be perfectly filled (Fig. 2(d)).

2.3 P_3: Hepatic Vein Neighborhood Probability

The hepatic veins inside the liver are ideal reference for defining the boundaries of the liver. The hepatic cells need to be nourished by blood from haptic veins that

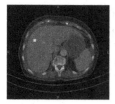 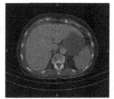 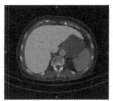 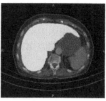

(a) Input Image and One Seed Point   (b) Result of Region Growing   (c) Result of Morphological Dilation   (d) Result of Morphological Closing

Fig. 2. Region growing algorithm and morphological operations are employed to get a rough area of the liver.

reach every corner of the liver. Meanwhile, the hepatic veins could only supply blood to cells located within a certain distance to them. This is the anatomical foundation for P_3.

For vessel extraction, we apply a threshold filter within the region of P_2 instead of applying the region growing algorithm to the whole volume. Because the former is not affected by the connectivity of the vessel structure in CT images and is able to detect all the candidate vessel points within the reference region (Fig. 2). To formulate this method, we define the hepatic vein possibility P_{3v1}. For the voxel with intensity $I(\boldsymbol{x})$ out of the region of P_2 or the range $(\mu_v \pm \rho_v)$, we set P_{3v1} to 0. And for voxels that are within the region of P_2 and the range $(\mu_v \pm \rho_v)$, we calculate the P_{3v1} according to their intensity similarity to a predefined center μ_v.

$$
P_{3v1}(\boldsymbol{x}) = \begin{cases} 1 - \frac{|I(\boldsymbol{x}-\mu_v)|}{\rho_v} & |I(\boldsymbol{x}) - \mu_v| \leq \rho_v \ \land \ P_2(\boldsymbol{x}) > 0 \\ 0 & |I(\boldsymbol{x}) - \mu_v| > \rho_v \ \lor \ P_2(\boldsymbol{x}) \leq 0 \end{cases} \tag{2}
$$

Wide coverage of the second method also brings new problems. The most critical one is that those isolated points with similar intensity to hepatic veins may be faultily selected, which decreases the accuracy of P_3 and further reduces the efficiency of the algorithm. We develop an algorithm to prevent these isolated points from being wrongly added to the point cloud of hepatic veins. In this algorithm, the 26-neighbourhood of every candidate vessel points is checked and only those that have a certain number of vessel points in their neighborhood could be preserved. A more detailed description of this process is shown in Algorithm 1.

For the hepatic vein neighborhood probability P_3, we calculate it based on the point cloud of hepatic veins. We find the n-nearest neighbors of each voxel from the point cloud and record the distance between each pair. Then the probability of current voxel $\boldsymbol{x}$ as a near neighbor of hepatic veins can be measured by the ratio of the hepatic vein possibility $P_{3v1}(\boldsymbol{x})$ and corresponding distance $distance(\boldsymbol{x}, \boldsymbol{x_i})$ between $\boldsymbol{x}$ and $\boldsymbol{x_i}$. As we consider n neighbors of each voxel, the final possibility is an average of n ratios generated by previously described method. Algorithm 2 shows details of this process.

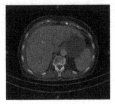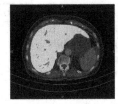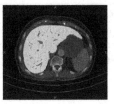

(a) Original Image   (b) Vessel Extracted   (c) Vessel Extracted
                            by Region Growing   by Threshold

Fig. 3. The coverage of threshold based vessel extraction method (red area in (c)) is wider than that of region growing based method (blue area in (b)). (Color figure online).

Algorithm 1. Clean Isolated Points

Input: point cloud of candidate hepatic veins PC_{can},
      radius r_N of the checking neighborhood,
      minimum number n_N of vessel points in a qualified point's neighborhood
Output: cleaned point cloud of hepatic veins PC_{clean}
1: **for each** point p in PC_{can} **do**
2:    $n_{neighbor} \leftarrow 0$;
3:    **for each** neighboring point p_n that $\|p_n, p\| \leq r_N$ **do**
4:       **if** $(p_n \in PC_{can})$ **then**
5:          $n_{neighbor} + +$;
6:       **end if**
7:       **if** $(n_{neighbor} \geq n_N)$ **then**
8:          **add** p_n **to** PC_{clean}
9:          break;
10:      **end if**
11:   **end for**
12: **end for**

$$P_3(\boldsymbol{x}) = \Phi(P_{3v1}(\boldsymbol{x}), d) \sim \frac{1}{n} \sum_{\boldsymbol{x} \in N} \frac{P_{3v1}(\boldsymbol{x_i})}{distance(\boldsymbol{x}, \boldsymbol{x_i})} \qquad (3)$$

The running on P_3 for the whole volume is a time-consuming process. So we construct a k-d tree for the point cloud of hepatic veins to reduce the algorithm complexity. In this way it turns the time complexity of n-nearest neighbor searching from $O(n)$ to $O(logn)$. And we just run P_3 for voxels with non-zero P_1 value, because it is a necessary but not sufficient condition for the liver voxel.

In brief, there are two key steps for running the hepatic vein neighborhood probability P_3: (1) choose candidate hepatic vein points and assign hepatic vein probability P_{3v1}; (2) calculate the final P_3 value based on P_{3v1} and P_2. Algorithm 2 shows details of this process.

Algorithm 2. Calculate Hepatic Vein Neighborhood Probability P_3

Input: original image I, liver location probability P_2
Output: hepatic vein neighborhood probability P_3
1: **for each** voxel x in liver image I **do**
2:    **if** $(\mathbf{abs}(I(x) - \mu_v) < p_v)$ **then**
3:        $P_{3v1}[x] \leftarrow 1 - \mathbf{abs}(I(x) - \mu_v)/p_v$;
4:    **else**
5:        $P_{3v1}[x] \leftarrow 0$;
6:    **end if**
7:    **if** $(P_2[x])$ **then**
8:        $P_{3v}[x] \leftarrow P_{3v1}[x]$;
9:        **add** x **to** PC_{can};
10:    **else**
11:        $P_{3v}[x] \leftarrow 0$;
12:    **end if**
13:    $PC_{clean} \leftarrow \mathbf{CleanIsolatedPoints}(PC_{can})$;
14: **end for**
15: **Normalize**(P_{3v});
16: **for each** voxel x in liver image I **do**
17:    $X_{neighbor} \leftarrow \mathbf{FindNNeareastPoint}(x, PC_{clean}, n)$;
18:    **for each** voxel x_i in $X_{neighbor}$ **do**
19:        $P_3[x] += P_{3v}[x_i]/\mathbf{Distance}(x, x_i)$;
20:    **end for**
21:    $P_3[x] /= n$;
22: **end for**
23: **Normalize**(P_3);

2.4 P_4: Liver Voxel Neighborhood Probability

Since the human liver is a solid and interconnected entity, the goal of P_4 is to punish voxels that are not close to other liver voxels. Strictly speaking, we are not calculating a new probability. We just apply the cleansing algorithm previously used for P_3 to further remove isolated points.

We first construct a point cloud with all voxels with their P_3 bigger than a predefined lower bound ρ_{low}. Then we apply the cleansing algorithm (Algorithm 1) to find isolated points inside the point cloud. Finally, we get the value of P_4 for each voxel according to Formula (4). For isolated voxels, we don't remove it directly but subtract a constant κ from its P_3 value.

$$P_4(x) = \begin{cases} P_3(x) - \kappa & P_3(x) \geq \rho_{low} \wedge \; x \; is \; isolated \\ P_3(x) & P_3(x) \geq \rho_{low} \wedge \; x \; is \; not \; isolated \\ 0 & P_3(x) < \rho_{low} \end{cases} \quad (4)$$

2.5 Final Probability

The final probability is calculated for every voxel as the product of P_1 and P_4. If the probability is larger than a predefined constant η, the corresponding voxel is considered as a liver voxel and gets the mask value *true*. Otherwise, it gets the mask value *false*. The output of the final result is a binary image which stores the region of segmented liver.

3   Experiment and Results

3.1   Experimental Setup and Evaluation Measures

We set up a control experiment to evaluate the accuracy of our algorithms for liver segmentation. The experiment environment was built on the open source medical image processing platform 3DMed developed by the Chinese Academy of Science [16]. We develop a plugin in it to test our algorithm interactively.

The evaluation is measured by variables used in Grand Challenge of MACCAI 2007. We compare the segmentation outcomes with expert-generated reference and rate them according to five measures: **Volumetric Overlap Error, Relative Volume Difference, Average Symmetric Surface Distance (ASD), Root Mean Square Symmetric Surface Distance (RMSD)** and **Maximum Symmetric Surface Distance (Hausdorff distance) (MSD)** [11]. Besides, we add the accuracy measure as shown in Formula (5). It is to show the difference between the algorithm segmented result A and the manual segmented result B more clearly.

$$Accuracy = 100(|A \cap B|/|B|) \qquad (5)$$

3.2   Results

Vessel Extraction. The accuracy of vessel extraction is important because it serves as the foundation for other algorithms in this pipeline. Our algorithm extract liver vessels based on the intensity and location. We compared our vessel extraction outcomes to the result from the dynamic region growing tool provided by Mimics 17.0 on the same dataset. Figure 4 shows that our outcome (Fig. 4 (a)) is more accurate than the Mimics output (Fig. 4(b)) in terms of the ability to eliminate unnecessary part connected to hepatic veins. And in our method, much thinner branch of the hepatic veins could be identified and displayed to surgeons (Fig. 4(a)).

Liver Segmentation. In fact, we produce two segmentation results in our pipeline for each dataset. The first one is the rough liver area reported simply by the location probability P_2. We call it "location based method". The second one is the final segmentation result jointly determined by four probabilities mentioned above. We call it "vessel distance based method". Segmentation outcomes of these two models were then compared to default outcome of Mimics 17.0 (Table 1).

The location based method produces better segmentation result than the vessel distance based method, and both of them produce better segmentation

(a) Our Method   (b) Dynamic   Region
                        Growth in Mimics 17.0

Fig. 4. The vessel structure extracted by our method (a) is better than that extracted by the dynamic region growing tool provided in Mimics 17.0.

Table 1. A Comparison of segmentation result

Method	Accuracy [%]	Overlap error [%]	Volume diff. [%]	Average dis. [mm]	RMS dis. [mm]	Max. dis. [mm]
Location Based Method	**94.2 ± 1.5**	**16.2 ± 5.3**	**8.3 ± 6.3**	**1.2 ± 0.2**	4.9 ± 3.3	**54.1 ± 19**
Vessel Distance Based Method	87.3 ± 6.8	30.422.9	9 ± 31	10.9 ± 15	**4 ± 2.2**	134.3 ± 109
Mimics 17.0	58 ± 13.7	50.4 ± 14	−21.4 ± 21	5.5 ± 3.6	9.1 ± 5.4	64.4 ± 38

Note: Number in bold is the best result for each measure

result than Mimics 17.0. The outcome of location based method shows significantly higher accuracy rate, lower overlap error and surface distance compared to the Mimics output. The vessel distance based method is not that superior, but it still produces higher accuracy as well as lower overlap error, volume difference (absolute value), RMS distance and Max. Distance than the Mimics output. We notice that the vessel distance based method can achieve shortest RMS distance compared to the location based method and Mimics. This indicates that the segmentation result of vessel distance based method is good at finding all the candidate voxels of the liver, but it may introduce mistakes in a more detailed view.

Result Comparison. Four sets of CT images are obtained and manually segmented by a human expert. The expert's segmentation serves as the ground truth. Outputs from Mimics, location based and vessel distance based algorithms are compared to the ground truth in terms of 3D volume (Fig. 5).

By inspecting results in Fig. 5, we can find that the location based method has the ability to eliminate other tissues that are connected to or have similar intensity to the liver. The vessel distance based method, however, shows lower capacity to remove those neighboring points whose intensity lies in the same threshold as liver. The result produced by the "dynamic region growing" tool of Mimics presents an unstable capacity to segment the liver. For some part of the liver, Mimics is able to produce clear and smooth boundary that is comparable to the location based method. But for other parts, leaking to other neighboring organs such as the spleen and kidney is likely.

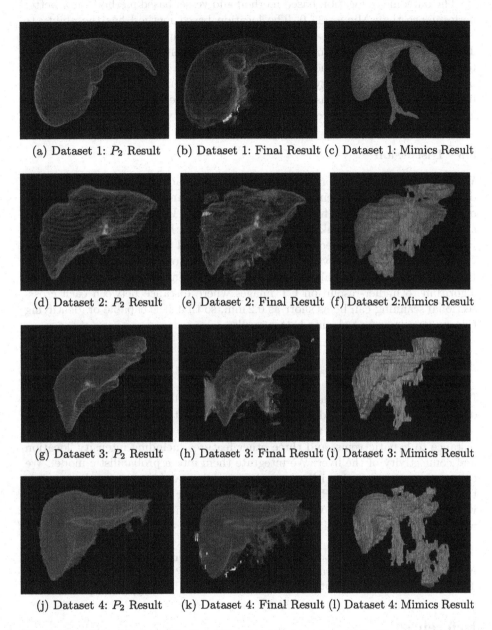

(a) Dataset 1: P_2 Result   (b) Dataset 1: Final Result (c) Dataset 1: Mimics Result

(d) Dataset 2: P_2 Result   (e) Dataset 2: Final Result (f) Dataset 2:Mimics Result

(g) Dataset 3: P_2 Result   (h) Dataset 3: Final Result (i) Dataset 3: Mimics Result

(j) Dataset 4: P_2 Result   (k) Dataset 4: Final Result (l) Dataset 4: Mimics Result

Fig. 5. The first column is P_2 Result by location based method. The second row is final result by vessel distance based method. The thrid column is the Mimics result by the tools from Mimics 17.0. The 3D visualization of the segmentation results shows that the location based method produces the most accurate result among the three segmentation strategies.

The outcome of location based method and vessel based method show better performance than Mimics 17.0. The location based method has the ability to eliminate other tissues that are connected to or have similar intensity to the liver. The vessel distance based method can achieve shortest RMS distance compared to the location based method and Mimics. However, the vessel distance based method shows lower capacity to remove those neighboring points whose intensity lies in the same threshold as liver. The result produced by the "dynamic region growing" tool of Mimics presents an unstable capacity to segment the liver.

3.3 Discussion

We believe that the major reason for the unsatisfactory results of the vessel distance based liver segmentation approach is our limited ability to perform an optimal vessel extraction from current CT images. In the cases we study, the cross sectional scanning taken by CT is at 0.5–1 mm horizontally and 1.3–5 mm vertically, which is not precise enough to show small blood vessel like capillary. Capillaries are the kind of vessel we expect to use for our distance based method. Unfortunately, it is not available with current CT images. We plan to apply this algorithm to higher-resolution CT images where distance between each cross-sectional scanning can be as short as 0.2 mm, so that it is capable of identifying thinner vessels and producing better result.

4 Conclusion

In this paper we propose an interior structure based liver segmentation strategy on CT images. The main idea is to extract the hepatic vein structure and then use it as a reference to find the boundary of the liver. Our pipeline considers the distance to hepatic veins, and three other features including intensity, location and connectivity of the liver. We integrate them into a probabilistic model. We test our segmentation algorithm on real clinical data and the result is better than the output from Mimics. However, it is not superior to location based method when we perform the test with conventional CT scan. We hope our algorithm will yield a better outcome when applied to the high-resolution data.

Acknowledgements. We would like to thank the anonymous reviewers for their helpful comments. This project was supported by the Major Program from the Science and Technology Research Foundation of Zhejiang Province, China (No. 2014C04008-1).

References

1. Agrawal, S., Panda, R., Dora, L.: A study on fuzzy clustering for magnetic resonance brain image segmentation using soft computing approaches. Appl. Soft Comput. **24**, 522–533 (2014)
2. Beck, A., Aurich, V.: Hepatux-a semiautomatic liver segmentation system. In: 3D Segmentation in the Clinic: A Grand Challenge, pp. 225–233 (2007)

3. Ben-Cohen, A., Diamant, I., Klang, E., Amitai, M., Greenspan, H.: Fully convolutional network for liver segmentation and lesions detection. In: Carneiro, G., et al. (eds.) LABELS/DLMIA -2016. LNCS, vol. 10008, pp. 77–85. Springer, Cham (2016). https://doi.org/10.1007/978-3-319-46976-8_9
4. Christ, P.F., et al.: Automatic liver and lesion segmentation in CT using cascaded fully convolutional neural networks and 3D conditional random fields. In: Ourselin, S., Joskowicz, L., Sabuncu, M.R., Unal, G., Wells, W. (eds.) MICCAI 2016. LNCS, vol. 9901, pp. 415–423. Springer, Cham (2016). https://doi.org/10.1007/978-3-319-46723-8_48
5. Chu, C., et al.: Multi-organ segmentation based on spatially-divided probabilistic atlas from 3D Abdominal CT images. In: Mori, K., Sakuma, I., Sato, Y., Barillot, C., Navab, N. (eds.) MICCAI 2013. LNCS, vol. 8150, pp. 165–172. Springer, Heidelberg (2013). https://doi.org/10.1007/978-3-642-40763-5_21
6. Cootes, T.F., Hill, A., Taylor, C.J., Haslam, J.: The use of active shape models for locating structures in medical images. In: Barrett, H.H., Gmitro, A.F. (eds.) IPMI 1993. LNCS, vol. 687, pp. 33–47. Springer, Heidelberg (1993). https://doi.org/10.1007/BFb0013779
7. Danciu, M., Gordan, M., Florea, C., Vlaicu, A.: 3D DCT supervised segmentation applied on liver volumes. In: 2012 35th International Conference on Telecommunications and Signal Processing (TSP), pp. 779–783. IEEE (2012)
8. Elmorsy, S.A., Abdou, M.A., Hassan, Y.F., Elsayed, A.: K3. A region growing liver segmentation method with advanced morphological enhancement. In: 2015 32nd National Radio Science Conference (NRSC), pp. 418–425. IEEE (2015)
9. Goryawala, M., Gulec, S., Bhatt, R., McGoron, A.J., Adjouadi, M.: A low-interaction automatic 3D liver segmentation method using computed tomography for selective internal radiation therapy. BioMed Res. Int. **2014**, Article ID 198015, 12 p. (2014). https://doi.org/10.1155/2014/198015
10. Heimann, T., Meinzer, H.P.: Statistical shape models for 3D medical image segmentation: a review. Med. Image Anal. **13**(4), 543–563 (2009)
11. Heimann, T., Van Ginneken, B., Styner, M.A., Arzhaeva, Y., Aurich, V., Bauer, C., Beck, A., Becker, C., Beichel, R., Bekes, G., et al.: Comparison and evaluation of methods for liver segmentation from CT datasets. IEEE Trans. Med. Imaging **28**(8), 1251–1265 (2009)
12. Kirbas, C., Quek, F.: A review of vessel extraction techniques and algorithms. ACM Comput. Surv. (CSUR) **36**(2), 81–121 (2004)
13. Luo, S., Hu, Q., He, X., Li, J., Jin, J.S., Park, M.: Automatic liver parenchyma segmentation from abdominal CT images using support vector machines. In: ICME International Conference on Complex Medical Engineering, CME 2009, pp. 1–5. IEEE (2009)
14. Luo, S., Li, X., Li, J.: Review on the methods of automatic liver segmentation from abdominal images. J. Comput. Commun. **2**(02), 1 (2014)
15. Martínez-Mera, J.A., Tahoces, P.G., Carreira, J.M., Suárez-Cuenca, J.J., Souto, M.: A hybrid method based on level set and 3D region growing for segmentation of the thoracic aorta. Comput. Aided Surg. **18**(5–6), 109–117 (2013)
16. Mitk (Medical Imaging Tookit) (2014). http://www.mitk.net/index.html
17. Park, H., Bland, P.H., Meyer, C.R.: Construction of an abdominal probabilistic atlas and its application in segmentation. IEEE Trans. Med. Imaging **22**(4), 483–492 (2003)
18. Perona, P., Malik, J.: Scale-space and edge detection using anisotropic diffusion. IEEE Trans. Pattern Anal. Mach. Intell. **12**(7), 629–639 (1990)

19. Pohle, R., Toennies, K.D.: Segmentation of medical images using adaptive region growing. In: Medical Imaging 2001, pp. 1337–1346. International Society for Optics and Photonics (2001)
20. Rusko, L., Bekes, G., Nemeth, G., Fidrich, M.: Fully automatic liver segmentation for contrast-enhanced CT images. In: MICCAI Wshp. 3D Segmentation in the Clinic: A Grand Challenge, vol. 2, no 7 (2007)
21. Schiff, E.R., Sorrell, M.F., Maddrey, W.C.: Schiff's Diseases of the Liver, vol. 1. Lippincott Williams & Wilkins, Philadelphia (2007)
22. Shi, C., Cheng, Y., Liu, F., Wang, Y., Bai, J., Tamura, S.: A hierarchical local region-based sparse shape composition for liver segmentation in CT scans. Pattern Recogn. **50**, 88–106 (2016)
23. Yang, X., Le Minh, H., Cheng, K.T.T., Sung, K.H., Liu, W.: Renal compartment segmentation in DCE-MRI images. Med. Image Anal. **32**, 269–280 (2016)
24. Yuan, R., Luo, M., Wang, S., Wang, L., Xie, Q.: A method for automatic liver segmentation from multi-phase contrast-enhanced CT images. In: SPIE Medical Imaging, p. 90353H. International Society for Optics and Photonics (2014)
25. Zhang, X., Tian, J., Xiang, D., Li, X., Deng, K.: Interactive liver tumor segmentation from CT scans using support vector classification with watershed. In: Engineering in Medicine and Biology Society, EMBC, 2011 Annual International Conference of the IEEE, pp. 6005–6008. IEEE (2011)

Computer Vision Applications

GPU Accelerated Image Matching
with Cascade Hashing

Tao Xu[1], Kun Sun[1], and Wenbing Tao[1,2(✉)]

[1] National Key Laboratory of Science and Technology on Multi-spectral Information
Processing, School of Automation, Huazhong University of Science and Technology,
Wuhan, Hubei, People's Republic of China
{xutao15,sunkun,wenbingtao}@hust.edu.cn

[2] Hubei Key Laboratory of Medical Information Analysis and Tumor Diagnosis
and Treatment, School of Biomedical Engineering,
South-Central University for Nationalities, Wuhan, Hubei, People's Republic of China

Abstract. SIFT feature is widely used in image matching. However,
matching massive images is time consuming because SIFT feature is a
high dimensional vector. In this paper, we proposed a GPU accelerated
image matching method with improved Cascade Hashing. Firstly, we
propose a disk-memory-GPU data exchange strategy and optimize the
load order of data, so that the proposed method can deal with big data.
Then, we parallelize the Cascade Hashing method on GPU. An improved
parallel reduction and an improved parallel hashing ranking are proposed
to fulfill this task. Finally, extensive experiments are carried out to show
that our image matching is about 20 times faster than the SiftGPU,
nearly one hundred times faster than the CPU Cascade Hashing, and
hundreds of times faster than the CPU Kd-Tree based matching.

Keywords: Image matching · SIFT · Cascade Hashing · GPU

1 Introduction

Image feature matching is to find the correspondence between images. SIFT
feature [11] is widely used since it has strong anti-interference ability and high
matching accuracy. However, each SIFT feature is represented by a high 128
dimensional vector and the time complexity of brute search is $O\left(N^2\right)$, where N
is the average number of feature points in the image. In order to accelerate the
speed of image matching, a series of methods based on search tree are proposed,
such as Kd-Tree and M-Tree [14]. The time complexity of Kd-Tree is $O\left(N \log N\right)$
[7], which is much faster than brute search.

In 2014, Cheng *et al.* proposed a fast image feature matching method based
on Cascade Hashing in 3D reconstruction [5]. The Cascade Hashing (CasHash)
uses the Locality Sensitive Hashing [4] (LSH) to quickly determine and reduce
a candidate point set, so CasHash avoids fully calculating Euclidean distance
between all possible points. As is reported, CasHash is hundreds of times faster

© Springer Nature Singapore Pte Ltd. 2017
J. Yang et al. (Eds.): CCCV 2017, Part I, CCIS 771, pp. 91–101, 2017.
https://doi.org/10.1007/978-981-10-7299-4_8

than brute force matching. Although CasHash is very fast when matching a pair of images, its efficiency drops quickly when the number of image pairs to be matched is extremely large. This is a common case in Structure from Motion [3,8] on very large unordered image sets, which requires exhaustive matching between images to infer scene overlap. With the rapid development of hardware, GPU card is becoming more and more popular. It has been used for high-performance and massively parallel computation. A single GPU card can achieve similar or better performance than a cluster of CPUs, while the cost and land occupation is much smaller [10]. In this paper, we want to make full use of such property of the GPU to further accelerate the matching procedure.

The flow chart of our proposed approach is shown in Fig. 1. CasHash is parallelized on GPU by using CUDA [16]. To achieve this goal, some further improvements and optimizations are made. Firstly, we propose a disk-memory-GPU data exchange strategy and optimize the load order for massive images. Then we use LSH to determine and reduce a candidate point set. After this we use improved parallel hashing ranking to select a few nearest neighbors. Finally, we use improved parallel reduction to calculate Euclidean distance between the nearest neighbors and copy output asynchronously.

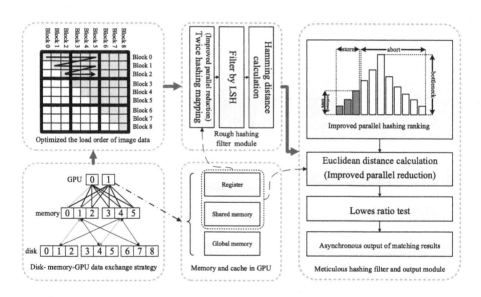

Fig. 1. The flow chart of our proposed approach.

2   Review of Cascade Hashing

CasHash [5] uses two layers of LSH for efficient nearest neighbor search. LSH [4] is a kind of Approximate Nearest Neighbor (ANN) method, whose average time complexity is $O(1)$. Through hashing mapping, the neighbors can be assigned to the same bucket, that is, the same hashing code. It consists of the following steps.

Step 1. Hashing lookup with multiple tables. All feature points are mapped into m-bit short hashing codes by LSH with different random hashing functions, returning a series of hashing codes of each image. Every image establishes multiple hashing lookup tables. Each query point in image I treats points with the same hashing code in image J as its candidate points.

Step 2. Hashing remapping. All the candidate points are mapped into n-bit $(n > m)$ hashing codes with LSH. Then the Hamming distances between the query point in image I and its candidate points on image J are computed.

Step 3. Hashing ranking. The Hamming distance after hashing remapping ranges from 0 to n. CasHash sort the points by their hamming distances to the query point in ascend order. The top k nearest neighbors are selected as the final candidates.

Step 4. Euclidean distance calculation. Finally, the Euclidean distances of the SIFT feature descriptors between the query point in image I and its k nearest neighbors on image J are computed. Then the point with the smallest distance which also passes Lowe's ratio test [11] is selected as the matching point.

3   Cascade Hashing Matching Based on GPU

In this paper, we propose the disk-memory-GPU data exchange strategy, the improved parallel reduction and the improved parallel hashing ranking methods, which are used to modify and parallelize CasHash [5] on GPU.

3.1   The Disk-Memory-GPU Data Exchange Strategy

Since the data can not be fully loaded into the memory or the GPU at once, we organize them into blocks and groups. We use 100 images as a data block, and bundle 10 data blocks(1000 images) as a data group. We construct a three level Disk-Memory-GPU data exchange strategy. First, several data groups are loaded from the disk to the memory. Then several data blocks are copied from the memory to the GPU. Once a data block or group is no longer needed, it is replaced by a newly loaded data block or group.

An example of the disk-memory-GPU data exchange strategy is given in Fig. 2. The image data is divided into 9 data blocks and 3 data groups. Each small square represents a data block, and each rectangle consisting of three small squares represents a data group. The red line indicates that the first and second groups are loaded from disk to memory, and data block 0 and 1 are loaded from memory to GPU. The black line indicates other possible load tasks. In Fig. 2 for example, two data groups can be loaded mostly from disk to memory and two data blocks can be loaded mostly from memory to GPU. However, if the load task order is not optimal, some data blocks will be re-loaded many times after they are released, which may lead to heavy I/O cost.

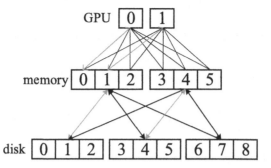

Fig. 2. The diagram of the disk-memory-GPU data exchange strategy. The red line indicates which group and block are loaded now, and the black line indicates other possible load tasks. (Color figure online)

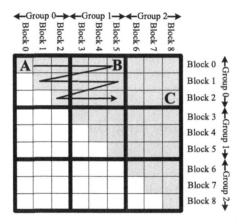

Fig. 3. The optimized load order in the diagram of the disk-memory-GPU data exchange strategy

We propose an optimized load task order which is shown in Fig. 3. As many data groups as possible are loaded from disk to memory. In Fig. 2 for example, data group 0 and 1 are loaded. From small number data groups, data blocks in data group i are matched with themselves, and then with data blocks in group $i+1, i+2, \ldots$. When the data group i is no longer needed, it can be free from memory. As many data blocks as possible are loaded from memory to GPU. In Fig. 2 for example, data block 0 and 1 are loaded. From small number data blocks, images in data block j are matched with themselves, and then with images in group $j+1, j+2, \ldots$. When the data block j is no longer needed in the data groups loaded, it can be free from GPU. The data blocks are loaded in row-major order in the data groups loaded and each cell in the matrix is a load task. For example, the load task of the cell A in Fig. 3 is to load group 0 from hard disk to memory, and then load block 0 from memory to GPU. After the load task of the cell B, data block 0 is no longer needed in the data groups loaded, it can be free from GPU. After the load task of the cell C, data group 0 is no longer needed, it can be free from memory.

3.2 Fast Computation of Hashing Code on GPU

In [4], Charikar defined the hashing function by using inner product similarity. In step 1 and step 2 of CasHash, when calculating the hashing code, we need to compute the inner product of the high dimensional vector. Let d be the dimension of the query vector, so the CUDA kernel function of the hashing map is $Hashing\_kernel <<< N * n, d >>> (PR, PQ)$, where PR and PQ are the pointers of random vectors r_k and SIFT feature descriptors in the GPU global memory, respectively. Since the SIFT feature point is 128 dimensional, $d = 128$. N is the number of points, and n is the length of hashing code. The kernel function totally calls $N * n$ blocks and each block calls d threads. The inner product vector needs to do d accumulation. Since the GPU global memory access is relatively slow and GPU needs to access GPU global memory repeatedly in the accumulation step, we use the GPU shared memory to reduce access delay [9]. The shared memory is a kind of GPU cache. The products are stored in the shared memory temporarily when doing accumulation, thus the arithmetic speed is improved.

There are 128 threads participating in operation at the same time in each block. Each thread reads a component of r from PR and a component of x from PQ, and then multiplies them on shared memory. Data access in shared memory is very fast, but using serial accumulation in each block will activate only one thread, which will reduce arithmetic efficiency. The way to deal with vector accumulation in general parallel computing is parallel reduction [15], which is also calculated in the shared memory.

The classic GPU parallel reduction is showed in Fig. 4. The accumulation of each block added in the shared memory makes full use of threads. For example, in the first round, the accumulation of 128 elements calls 64 threads, and then obtains 64 results. Consequently, the numbers of elements and threads are reduced by half after every round of operation until the last round, which uses one thread to add two elements. Accumulation is a process of reducing the number of threads used. The algorithm greatly reduces the idle threads in the accumulation process.

The access time for GPU global memory, GPU shared memory and GPU registers are hundreds of operational cycles, 10 operational cycles and a single operation cycle [6], respectively. So if registers instead of shared memory is used in the parallel reduction, it can further reduce the delay. However, all the registers could only be accessed by a single thread. If we use the registers in the early stage of parallel reduction, most threads will be idle. In this paper, we improve the parallel reduction by using both the shared memory in the early stage and the registers in the final stage to achieve faster speed.

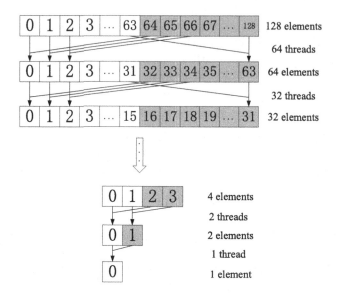

Fig. 4. The diagram of general parallel reduction on GPU

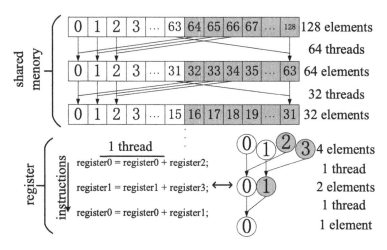

Fig. 5. The diagram of improved parallel reduction on GPU

The proposed method is shown in Fig. 5, the registers are used instead of the shared memory in the last N_r rounds of the reduction. Since the number of threads is small in the last several rounds, using registers will not waste too many threads but save much shared memory access time. In Fig. 5, in the last two rounds, the classic method of parallel reduction requires to access shared memory 2 times, need to do addition 2 times. So it needs about 22 cycles to complete the operation of the last two calculations. And after the use of registers, need to access registers 3 times, need to do addition 3 times. So it needs about 6 cycles to complete the operation of the last two calculations. Thus the proposed method greatly improves the speed of operation. In Sect. 4, we show the relationship between the N_r rounds registers involved in and the time of image matching.

3.3 Improved Parallel Hashing Ranking

In step 3 of CasHash, we propose an improved parallel hashing ranking method. Each GPU block ranks the Hamming distance between a query points and its candidate points in an image, and the process of constructing a index table in a GPU block is shown in Fig. 6. As the histogram showed, the horizontal axis is Hamming distance and the vertical axis is the number of points that fall in the bucket. On the one hand, because the candidate points have passed through rough filtering, the points are the neighbors of the query point. So few Hamming distances are large. On the other hand, there are 2^{128} kinds of different hashing codes, so few Hamming distances are small. Therefore in experiments, the candidate points are concentrated in the middle of the hashing bucket. If two candidate points have the same Hamming distance, there will be conflict when storing them into the same hashing bucket. In order to avoid the error caused by the conflict, it is necessary to use the atomic operation of CUDA. The atomic operation can make the conflict become serial task [17], which will reduce the efficiency of operation. and the time of conflict in the middle of the hashing bucket is most. The efficiency of parallel computation follows the bucket principle, so the computation time depends on the points number in the middle of the hashing bucket.

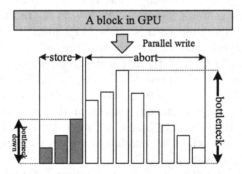

Fig. 6. The diagram of the bottleneck of parallel hashing ranking. The bucket number of the grey buckets is smaller than the threshold, and the bucket number of the white buckets is bigger than the threshold.

However, we only need the first k nearest neighbors, and k is less than the number of hashing buckets, so most of the k nearest neighbors is in the first few buckets. In Fig. 6, we set a threshold in parallel hashing ranking. The candidate points whose Hamming distance to the query greater than the threshold are discarded and not stored in the bucket and the candidate points whose Hamming distance to the query smaller than the threshold τ are stored in the bucket. Therefore the threshold τ will reduce the computational bottleneck and improve efficiency. But if the threshold is too small, we may filter the real matching point which will reduce matching accuracy. In this paper, the influence of the threshold on the matching accuracy and matching time is illustrated by experiments in Sect. 4.

3.4   Improved Parallel Euclidean Distance Calculation

In step 4 of CasHash, calculating the Euclidean distances still spends a lot of time in CPU, and the Euclidean distance formula is $\sum_{i=0}^{d-1} \left(x_i^1 - x_i^2 \right)^2 = \Delta x \cdot \Delta x$. Therefore, the Euclidean distance is essentially a vector inner product, which is the same as that in the Sect. 3.2. Since CPU and GPU are different devices, the operations of CPU and GPU are asynchronous [12]. When doing image matching on GPU, CPU can copy the output from GPU memory to host memory and then to hard disk. This method hides the output time, and improves the overall operation efficiency.

4   Experiments and Results

In this paper, we extracted SIFT features from the tested image data sets and perform exhaustive image matching. That is to say, when the image number is N, the number of matching pairs is $N(N-1)/2$. In the experiment, the proposed method is compared with Kd-Tree [14], SiftGPU [2] and the original CasHash [5]. The experiments of Kd-Tree and the original CasHash are tested on E5-2630 v3@2.4GHz CPU without parallel computation. SiftGPU and the proposed method are tested on one GTX Titan X GPU. All the experiments are running in Ubuntu 14.04 operating system. Table 1 shows the comparison of four methods above. The experiments show that the speedup of our method is more obvious when the number of images is larger, such as the "pozzoveggiani" data set. This is because when the number of images increases, the scale of the problem increases quadratically.

Figure 7 shows the matching time on the "erpbero" dataset when using registers in different last N_r rounds. When N_r increases, the matching time will first decrease and then rise after N_r is larger than a certain value. This is because that using registers in the last few rounds will reduce memory access time, but if registers are used too early in the parallel reduction a lot of threads will be idle. We empirically set $N_r = 3$ to achieve the best performance in our experiments.

Figure 8(a) shows the number of matching points and matching time on the "erpbero" dataset for different τ. Figure 8(b) gives the matching accuracy for different τ on the public Oxford dataset [13], which offers the Homography transformation as the ground truth. If the match pairs satisfy $x_1 - Hx_2 < \epsilon$, where H is the given Homography matrix between image pairs and we set $\epsilon = 6$, we assume that the two keypoints compose a accurate matching pairs. In Fig. 8, when the threshold τ is about 40, the matching points don't decrease and we can guarantee high matching accuracy and less matching time.

Table 1. Results on matching experiments. In (d), Data-Rome16K [1] (15178 images) 1.152e8 pairs 7891 mean points; Data-Dubrovnik6K [1] (6044 images) 1.826e7 pairs 7438 mean points.

(a) Data-pozzoveggiani (54 images) 1431 pairs 8306 mean points

Method	time(s)	speed(pairs/s)	speedup
Kd-Tree	698.921	2.05	1.00×
CasHash	183.218	7.81	3.81×
SiftGPU	30.103	47.54	23.22×
Ours	2.496	573.32	280.02×

(b) Data-erpbero (259 images) 33411 pairs 8641 mean points

Method	time(s)	speed(pairs/s)	speedup
Kd-Tree	1.479e4	2.26	1.00×
CasHash	1461.525	22.86	10.12×
SiftGPU	752.164	44.42	19.66×
Ours	34.394	971.42	429.91×

(c) Data-Aos_Hus (811 images) 328455 pairs 7768 mean points

Method	time(s)	speed(pairs/s)	speedup
Kd-Tree	1.456e5	2.26	1.00×
CasHash	2.800e4	11.73	5.20×
SiftGPU	6971.441	47.11	20.89×
Ours	292.541	1122.77	497.81×

(d) Our Method on Some Large Data Set.

Data Set	time(s)	speed(pairs/s)
Dubrovnik6K	1.054e4	1093
Rome16K	1.565e5	1167

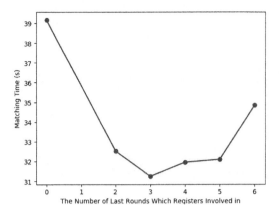

Fig. 7. Results on the improved parallel reduction. The horizontal axis is the last N_r rounds registers involved in.

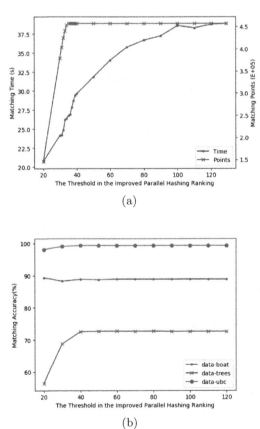

(a)

(b)

Fig. 8. The influence of the threshold τ. The influence of the threshold τ on the matching points and matching time is showed in (a). The influence of the threshold τ on the matching accuracy is showed in (b).

5   Conclusions

In this paper, we proposed a GPU accelerated image matching algorithm with CasHash for fast feature matching of massive images. We proposed an improved parallel reduction and an improved parallel hashing ranking to optimize and parallelize the Cascade Hashing. For massive images, we proposed a disk-memory-GPU data exchange strategy and we optimize the load order of image data.

Acknowledgements. This work is supported by the National Natural Science Foundation of China (Grant 61772213 and 61371140), also in part by Grants 2015CFA062, 2015BAA133 and 2017010201010121.

References

1. Cornell 3D Location Recognition Datasets. http://www.cs.cornell.edu/projects/p2f/
2. SiftGPU. http://cs.unc.edu/ccwu/siftgpu/
3. Agarwal, S., Furukawa, Y., Snavely, N., Simon, I.: Building Rome in a day. Commun. ACM **54**, 105–112 (2011)
4. Charikar, M.: Similarity estimation techniques from rounding algorithms. In: Proceedings of 34th Annual ACM Symposium (2002)
5. Cheng, J., Leng, C., Wu, J., Cui, H., Lu, H.: Fast and accurate image matching with cascade hashing for 3D reconstruction. In: Proceedings of IEEE Computer Society Conference on Computer Vision and Pattern Recognition, pp. 1–8. IEEE, June 2014
6. Cook, S.: CUDA programming: a developer's guide to parallel computing with GPUs (2012)
7. Cormen, T.H. (ed.): Introduction to Algorithms. MIT Press, Cambridge (2009)
8. Crandall, D., Owen, A., Snavely, N., Huttenlocher, D.: Discrete continuous optimization for large-scale structure from motion. In: Computer Vision and Pattern Recognition, pp. 3001–3008 (2011)
9. Hong, S., Kim, H.: An analytical model for a GPU architecture with memory-level and thread-level parallelism awareness. In: ACM SIGARCH Computer Architecture News, vol. 37, no. 3, p. 152 (2009)
10. Lee, A., Yau, C., Giles, M.B., Doucet, A., Holmes, C.C.: On the utility of graphics cards to perform massively parallel simulation of advanced Monte Carlo methods. J. Comput. Graph. Stat. **19**(4), 769–789 (2010)
11. Lowe, D.G.: Distinctive image features from scale-invariant keypoints. Int. J. Comput. Vis. **60**(2), 91–110 (2004)
12. Micikevicius, P.: 3D finite difference computation on GPUs using CUDA. In: Proceedings of 2nd Workshop on General Purpose (2009)
13. Mikolajczyk, K., Schmid, C.: A performance evaluation of local descriptors. IEEE Trans. Pattern Anal. Mach. Intell. **27**(10), 1615–1630 (2005)
14. Muja, M., Lowe, D.: Fast approximate nearest neighbors with automatic algorithm configuration. VISAPP (1) (2009)
15. Nickolls, J., Buck, I., Garland, M., Skadron, K.: Scalable parallel programming with CUDA. Queue **6**, 40–53 (2008)
16. Nvidia, C.: Programming Guide (2010)
17. Sanders, J., Kandrot, E.: CUDA by Example: An Introduction to General-Purpose GPU Programming, Portable Documents. Addison-Wesley, Boston (2010)

Improved Single Image Dehazing with Heterogeneous Atmospheric Light Estimation

Yi Lai[1,2,3(✉)] and Ying Liu[1,2]

[1] School of Telecommunication and Information Engineering,
Xian University of Posts and Telecommunications,
Xi'an 710121, People's Republic of China
laiyi0614@163.com
[2] Key Laboratory of Electronic Information Application Technology for Scene
Investigation, Ministry of Public Security, Xian 710121, People's Republic of China
[3] Key Laboratory of Spectral Imaging Technology of Chinese Academy of Sciences,
Xi'an 710119, People's Republic of China

Abstract. Images captured in foggy or hazy weather conditions are
often degraded by the scattering of atmospheric particles, which seriously reduces the performance of outdoor computer vision processing
systems. Single image haze removal algorithm has been considered to be
an efficient dehazing method in recent years. The key to this type of approach is the estimation of atmospheric light. In this paper, an improved
single image dehazing algorithm with heterogeneous atmospheric light
estimation is presented to enhance the quality of hazy images. First, the
heterogeneous atmospheric light is calculated with max-pooling. Second,
a haze-free image can be recovered with the estimated atmospheric light
based on dark channel prior. The experimental results on a variety of
hazy images demonstrate that the addressed method outperforms state-
of-the-art approaches through the assessment of dehazing effect and algorithm cost.

Keywords: Dehazing · Heterogeneous atmospheric light estimation
Max-pooling

1 Introduction

Outdoor images captured under foggy weather condition are often seriously
degraded by the turbid medium (e.g., dust, water-droplets) in the atmosphere.
And the distant objects in the degraded images lose the color fidelity and become
blurred with their surrounding, as demonstrated in Fig. 1(a). This problem seriously reduces the performance of outdoor computer vision processing systems.
Thus, effective and robust dehazing (or haze removal) methods are strongly desired
in both computational photography and computer vision applications [1,2], such
as outdoor surveillance, intelligent driving, and satellite remote sensing system.

© Springer Nature Singapore Pte Ltd. 2017
J. Yang et al. (Eds.): CCCV 2017, Part I, CCIS 771, pp. 102–112, 2017.
https://doi.org/10.1007/978-981-10-7299-4_9

(a)   (b)

(c)   (d)

Fig. 1. Image dehazing by presented approach. (a) Input hazy image. (b) Dehazed image. (c) Estimated heterogeneous atmospheric light. (d) Estimated medium transmission.

Available methods for image dehazing are broadly classified as image enhancement based approaches and image restoration based schemes [3]. The approaches based on image enhancement employ image processing algorithms to restore the dehazed image, enhance the visibility of a haze image, and highlight the valuable information in a scene. The advantage of this kind of method is that it is very convenient with the commonly used image enhancement algorithms. However, the recovered image usually suffers from significant halos and distorted colors. In order to recover high-quality dehazed images, the schemes based on image restoration build a atmospheric scattering model based on the scattering effect of atmospheric aerosol particles on the light. This is a special kind of haze removal method based on the physical mechanism of image degradation, thus the restored image is more photo-realistic, and the details information are also preserved well.

Early haze removal methods based on image restoration [4–7] usually require multiple images or additional information to be available. Though these methods can ameliorate the visibility of hazy images, they cannot be employed in applications where additional information or multiple images are not obtainable. Therefore, single image haze removal has been a hot spot of research given its wider application range [8]. Recently, a lot of progresses have been made since the introduction of single image haze removal. These image dehazing algorithms [1,2,9–14] remove the haze under certain priors or assumptions, such as dark channel prior (DCP) [2], and then recover the haze-free image with a haze model. And the advantage of them is to demand only a single input image.

Image dehazing mainly includes two tasks, estimating the atmospheric light and calculating the medium transmission, as shown in Fig. 1(c) and (d), respectively. Inaccurate estimation of the atmospheric light can incur unwanted color shift. As a result, the haze-free image cannot be obtained visually pleasing. Therefore, the estimation of the atmospheric light is the key technique for image dehazing. The homogeneous atmospheric light estimation considers the light both absorbed and scattered by turbid mediums, so it is widely applied in image dehazing. In [10], the brightest pixel of the entire hazy image was regarded as the atmospheric light. And in [2], the top 0.1% brightest pixels in the dark channel were first picked up, and then the one with the highest intensity was selected as the estimation of the atmospheric light. [12] filtered each color channel of an input image using a minimum filter, and then the maximum value of each color channel was estimated as the atmospheric light.

These methods mentioned above simply approximate the atmospheric light under an assumption that the whole scene has the same value for the atmospheric light. Namely, the atmospheric light within an entire image is homogeneous. This assumption leads to that many state-of-the-art algorithms for dehazing get much darker dehazed images, especially objects with dark color in scenes. In fact, the atmospheric light is not a valid one when there are multiple point light sources such as the sun, street lights, or vehicle headlights at night. Hence, these existing approaches often fail by mistakenly taking white objects (e.g., clouds) as the atmospheric light. Meanwhile, the amount of scattering depends on the distances of the scene points from the camera, and the degradation varies spatially [2].

In order to efficiently and robustly remove haze from a single input image, an improved dehazing scheme based on the heterogeneous atmospheric light estimation is presented in this paper. The main works are as follows. Firstly, we think that an entire image has not uniform value for the atmospheric light. In other words, the atmospheric light within an image is heterogeneous. And then, the heterogeneous atmospheric light can be obtained with max-pooling. Finally, combining the estimated heterogeneous atmospheric light and a dark channel prior dehazing method, we can recover a haze-free image with faithful colors and fine image details. As a result of our experiments, the haze-free image restored by the developed scheme achieves more satisfying image quality at an acceptable algorithm cost compared with state-of-the-art approaches.

The rest of this paper is organized as follows. The atmospheric scattering model is treated in Sect. 2. The improvements for dehazing is addressed in Sect. 3. Experimental results are shown in Sect. 4. Finally, we conclude the whole paper in Sect. 5.

2   Atmospheric Scattering Model

The atmospheric scattering model (ASM) is the basic model of image dehazing [1]. To address the formation of a hazy image, the atmospheric scattering model is first presented by McCartney in [15], and further developed by Narasimhan and Nayar [4,5] later. When the atmospheric light is homogenous, the model can be defined as follows:

$$I(x) = J(x)t(x) + A(1 - t(x)) \tag{1}$$

where x is the position of the pixel within the image, $I(x)$ is the observed intensity representing the hazy image, $J(x)$ is the scene radiance describing the haze-free image. A denotes the atmospheric light, and $t(x)$ indicates the medium transmission representing the portion of the light that is not scattered and reaches the camera. The goal of image dehazing is to estimate A and $t(x)$, and then recover $J(x)$ from $I(x)$ based on (1).

3   Improvements for Dehazing

To create a photo-realistic haze-free image, two improvements to the DCP approach [14] are given. One is to produce the heterogeneous atmospheric light estimation, which accurately describes how the amount of scattering from one or multiple point light sources is spatial-variant. The other is to decide how a dehazed image is recovered efficiently. Therefore, the two improvements for the addressed dehazing method are discussed in this section.

3.1   Heterogeneous Atmospheric Light Estimation

The estimation of the atmospheric light is a fundamental and challenging problem for the haze removal [16]. Under the assumption of homogeneous atmospheric light, the darker regions in these images become very dark after haze removal and some important details in the images are often lost. Therefore, the assumption of homogeneous atmospheric light does not hold in most cases. Besides, the occlusion of big objects to the light source and different albedos or absorptivities of different objects in a scene can incur the heterogeneous atmospheric light. Based on these observations, we assume that the atmospheric light is heterogeneous in whole scene. As a result, we present a novel solution which is useful for estimating heterogeneous atmospheric light from a single hazy image directly.

According to (1), the heterogeneous ASM is formulated by:

$$I(x) = J(x)t(x) + A(x)(1 - t(x)) \tag{2}$$

where $A(x)$ is the heterogeneous atmospheric light, and is dependent on the position x in a scene. The difference between (2) and (1) is that $A(x)$ is a variable, not a constant. Actually, (1) can be regarded as a special case of (2) when the $A(x)$ is a constant.

Additionally, in the ideal case, the medium transmission $t(x)$ could be zero as the scenery objects that appear in the image could be very far from the camera, and we have:

$$I(x) = A(x), \ t(x) \to 0 \tag{3}$$

Equation (3) shows that the intensity of the pixel can represent the value of the atmospheric light $A(x)$ when $t(x)$ tends to zero.

Therefore, $A(x)$ can be calculated by $I(x)$ approximately when the medium transmission $t(x)$ tends to be very small. When $t(x)$ becomes very small, all of the light radiated from objects in a scene are almost scattered and cannot reach the camera. Hence $I(x)$ becomes nearly haze-opaque. Under this circumstance, $I(x)$ can be taken as the approximation of the heterogeneous atmospheric light $A(x)$.

Inspired by this idea, we propose a novel method for the estimation of heterogeneous atmospheric light. Firstly, in order to make the haze in the input image become white color, the white balance operation is conducted on the input hazy image $I(x)$. Secondly, the max-pooling [17] is performed on the output image of white balance to obtain a small heterogeneous atmospheric light. Here the size of the window for max-pooling is N × N. And then, a guided image filtering [14] strategy is employed to suppress the block artifacts incurred by the max-pooling. Finally, according to the size of original hazy image, a bicubic interpolation way is used to enlarge the small heterogeneous atmospheric light, and then the heterogeneous atmospheric light $A(x)$ can be obtained, as shown in Fig. 1(c).

3.2 Dehazing with Heterogeneous Atmospheric Light Estimation

Image dehazing is essentially under-constrained problem if the input is a single haze image. In order to handling this problem, the explore for additional priors or constraints are generally required. Dark channel prior (DCP) for single image dehazing [2] is a better solution compared with most haze removal methods. Therefore, we estimate the medium transmission $t(x)$ based on (1). Under the DCP assumption, referring to (1), we easily derive the medium transmission $t(x)$ as follows:

$$t(x) = 1 - \min_{y \in \Omega(x)} \left(\min_{c \in \{R,G,B\}} \frac{I^c(y)}{A^c} \right) \tag{4}$$

where I^c is a color channel of I, $\Omega(x)$ is a local patch center at x, $\min_{c \in \{R,G,B\}}$ is conducted on each pixel, and $\min_{y \in \Omega(x)}$ is a minimum filter. Hence we can directly obtain the estimation of $t(x)$. To refine the medium transmission $t(x)$, a guided image filtering is used to smooth the image.

The aim of image dehazing is to recover the scene radiance $J(x)$. After finishing the estimation of the heterogeneous atmospheric light $A(x)$ and the medium transmission $t(x)$, referring to (2), the scene radiance can be restored by:

$$J(x) = \frac{I(x) - A(x)}{\max(t(x), \varepsilon)} + A(x) \tag{5}$$

where ε is a small constant (typically 0.1) for avoiding division by zero. One haze-free image recovered with the developed scheme is illustrated in Fig. 1(b).

4   Experimental Results and Analysis

To confirm the performance of our proposed dehazing method, several experiments are conducted for a set of real-world hazy images. The tests are implemented with the MatlabR2014a and performed on an Intel Core4 3.1 GHz computer that has 8 GB memory. We compare the qualitative results, quantitative ones, and computing time complexity of the developed algorithm with state-of-the-art algorithms, including Tarel's [11], He's [14], and Meng's methods [12].

Figures 2, 3, 4, 5, and 6 demonstrate the qualitative comparison of different results on some real-world images, including 'Road', 'Swans', 'People', 'Manhattan 2', and 'Yosemite 1'. As depicted in Figs. 2(e), 3(e), 4(e), 5(e), and 6(e), the haze-free images recovered by the developed algorithm are much visually pleasing than the other three methods. However, there are several noticeable artifacts in the outputs using the other approaches. As shown in Figs. 2(b), 3(b), 4(b), 5(b), and 6(b), the restored images by Tarel's method often suffer from distorted colors and significant halos. And the recovered images by He's algorithm are too dark as depicted in Figs. 2(c), 3(c), 4(c), 5(c), and 6(c). While Meng's scheme often tends to produce some geometric distortion and some hazes still remain in the restored images as illustrated in Figs. 2(d), 3(d), 4(d), 5(d), and 6(d).

To compare with the other three approaches quantitatively, we use the visible edge gradient [18] metrics. The metrics include the percentage of new visible edges e, the normalized gradient mean of visible edges r, and the percentage of saturated black and white pixels σ. The value of e evaluates the ability of the method to restore edges which were not visible in the hazy image $I(x)$ but are in the restored image $J(x)$. The value of $\bar{r}$ expresses the quality of the contrast restoration. The value of σ measures contrast by detecting the number of saturated black and white pixels.

The metrics are separately defined as follows:

$$e = \frac{n_r - n_0}{n_0} \tag{6}$$

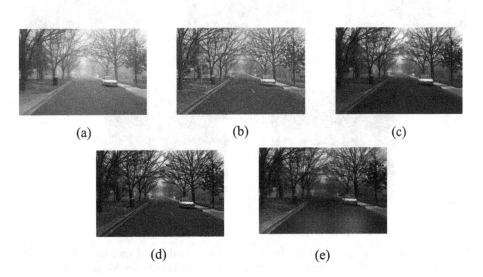

(a)   (b)   (c)

(d)   (e)

Fig. 2. Qualitative comparison of different results on real-world images of 'Road' (a) Input hazy images. (b) Tarel's results [11]. (c) He's results [14]. (d) Meng's results [12]. (e) Proposed method's results.

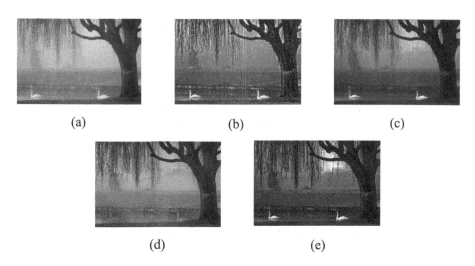

(a)   (b)   (c)

(d)   (e)

Fig. 3. Qualitative comparison of different results on real-world images of 'Swans' (a) Input hazy images. (b) Tarel's results [11]. (c) He's results [14]. (d) Meng's results [12]. (e) Proposed method's results.

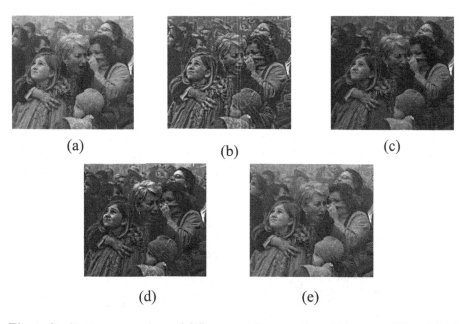

(a)   (b)   (c)

(d)   (e)

Fig. 4. Qualitative comparison of different results on real-world images of 'People' (a) Input hazy images. (b) Tarel's results [11]. (c) He's results [14]. (d) Meng's results [12]. (e) Proposed method's results.

Fig. 5. Qualitative comparison of different results on real-world images of 'Manhattan 2' (a) Input hazy images. (b) Tarel's results [11]. (c) He's results [14]. (d) Meng's results [12]. (e) Proposed method's results.

$$\bar{r} = \exp\left(\frac{1}{n_r} \sum_{P_i \in \psi_r} \log r_i\right) \tag{7}$$

$$\sigma = \frac{n_s}{w \times h} \tag{8}$$

where n_0 and n_r denote the cardinal numbers of the set of visible edges in the hazy image $I(x)$, respectively in the contrast restored image $J(x)$. ψ_r represents the visible edges set in $J(x)$, P_i is the pixel of the visible edge, and r_i indicates the Sobel gradient ratio between P_i and $I(x)$. n_s is the number of the saturated black and white pixels, w and h are separately the width and the height of $I(x)$.

Tables 1 are some results of quantitative measurements. The notations employed to describe the metrics are the same as that used in [17]. Obviously the proposed method has shown better performance in most cases. It is reasonable that the presented scheme can properly describe that the amount of scattering from one or multiple point light sources is spatial-variant. Therefore the heterogeneous atmospheric light can be estimate accurately, and the haze-free image of high quality with faithful colors and fine edge details can be obtained.

To test the complexity of the discussed approaches, the average computing time for the algorithms is evaluated for the tested datasets. The computing time complexity is shown in Table 2. Because the He's method has the better use of stronger priors or assumptions for image haze removal, it has the lowest computation complexity. Meng's scheme involves time consuming for iterative operations to calculate the medium transmission map. Due to applying standard median filter to obtain a satisfactory dehazed image quality, Tarel's method has the highest computation complexity. The proposed algorithm has to employ extra operations to obtain the heterogeneous atmospheric light map, and hence demands more run-time. Comparing with the other three methods, the computing time complexity of the treated way is moderate.

According to these results, the developed methods can work efficiently with only minor increase of the computational complexity.

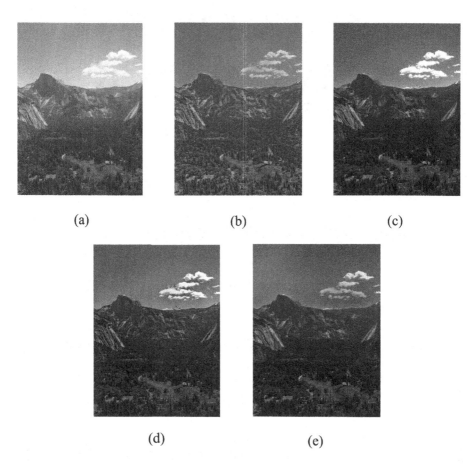

Fig. 6. Qualitative comparison of different results on real-world images of 'Yosemite 1' (a) Input hazy images. (b) Tarel's results [11]. (c) He's results [14]. (d) Meng's results [12]. (e) Proposed method's results.

Table 1. Quantitative comparisons

Input	Metric	Tarel [11]	He [14]	Meng [12]	Proposed
Road	e	0.7092	0.2847	0.5916	**0.8285**
	$\bar{r}$	2.1068	1.5855	2.1972	**2.4803**
	σ	0	0	0.0142	0.0042
Swans	e	0.7344	0.2993	0.5119	**0.8072**
	$\bar{r}$	3.0743	1.2414	2.1858	**3.4336**
	σ	0.0004	0	0.0085	0
People	e	0.5022	0.2358	0.3463	**0.6258**
	$\bar{r}$	2.1850	0.9225	1.4441	**3.0584**
	σ	0.0004	0.0064	0.0692	0

Table 2. Computing time

Dehazing method	Average computing time (s)
Tarel [10]	13.16
He [13]	1.05
Meng [11]	5.74
Proposed method	**1.23**

5 Conclusions

In this paper, an improved single image dehazing scheme based on heterogeneous atmospheric light estimation is developed to get a better visual quality of haze-free images. The improvements include two things. One is the heterogeneous atmospheric light can be estimated by max-pooling. The other is a haze-free image is recovered with the obtained atmospheric light. Because the derived atmospheric light can be adjusted adaptively according to the input image brightness, some encouraged results are obtained. Experiment results confirm the efficiency of the presented algorithms with only minor increase of the computing time complexity. In the future, more efficient medium transmission estimation is a very important research topic.

Acknowledgement. This work was supported in part by the National Natural Science Foundation of China (41504115, 61202183), Strengthening Police science and technology Foundation from Ministry of Public Security (2016GABJC51), Shaanxi Provincial Education Department (16JK1691), the Open Projects Program of National Laboratory of Pattern Recognition (201700013), and by the Open Research Fund of Key Laboratory of Spectral Imaging TechnologyChinese Academy of Sciences (LSIT201709D).

References

1. Cai, B.L., Xu, X.M., Jia, K., Qing, C.M.: Dehazenet: an end-to-end system for single image haze removal. IEEE Trans. Image Process. **25**(11), 5187–5198 (2016)
2. He, K., Sun, J., Tang, X.: Single image haze removal using dark channel prior. IEEE Trans. Pattern Anal. Mach. Intell. **33**(12), 2341–2353 (2011)
3. Wang, W.C., Yuan, X.H., Wu, X.J., Liu, Y.L.: Fast image dehazing method based on linear transformation. IEEE Trans. Multimedia **19**(6), 1142–1155 (2017)
4. Narasimhan, S.G., Nayar, S.K.: Contrast restoration of weather degraded images. IEEE Trans. Pattern Anal. Mach. Intell. **25**(6), 713–724 (2003)
5. Narasimhan, S.G., Nayar, S.K.: Vision and the atmosphere. Int. J. Comput. Vis. **48**(3), 233–254 (2002)
6. Kopf, J., et al.: Deep photo: model-based photograph enhancement and viewing. ACM Trans. Graph. **27**(5), 116 (2008)
7. Shwartz, S., Namer, E., Schechner, Y.Y.: Blind haze separation. In: IEEE Computer Society Conference on Computer Vision and Pattern Recognition (CVPR 2006), New York, USA, pp. 1984–1991, June 2006
8. Tang, K., Yang, J.C., Wang, J.: Investigating haze-relevant features in a learning framework for image dehazing. In: IEEE Computer Society Conference on Computer Vision and Pattern Recognition (CVPR 2014), Columbus, USA, pp. 2995–3002, June 2014
9. Fattal, R.: Single image dehazing. ACM Trans. Graph. **27**(3), 72 (2008)
10. Tan, R.T.: Visibility in bad weather from a single image. In: IEEE Computer Society Conference on Computer Vision and Pattern Recognition (CVPR 2008), Anchorage, USA, pp. 1–8, June 2008
11. Tarel, J.P., Hautire, N.: Fast visibility restoration from a single color or gray level image. In: IEEE International Conference on Computer Vision (ICCV 2009), Kyoto, Japan, pp. 2201–2208, September/October 2009
12. Meng, G.F., Wang, Y., Duan, J., Xiang, S., Pan, C.: Efficient image dehazing with boundary constraint and contextual regularization. In: IEEE International Conference on Computer Vision (ICCV 2013), Sydney, Australia, pp. 617–624, December 2013
13. Zhu, Q.S., Mai, J.M., Shao, L.: A fast single image haze removal algorithm using color attenuation prior. IEEE Trans. Circ. Syst. Video Technol. **24**(11), 3522–3533 (2015)
14. He, K., Sun, J., Tang, X.: Guided image filtering. IEEE Trans. Pattern Anal. Mach. Intell. **35**(6), 1397–1409 (2013)
15. McCartney, E.J.: Optics of the Atmosphere: Scattering by Molecules and Particles. Wiley, New York (1976)
16. Feng, C., Zhuo, S.J., Zhang, X.P., Shen, L.: Near-infrared guided color image dehazing. In: IEEE International Conference on Image Processing (ICIP 2013), Melbourne, Australia, pp. 2363–2367, Septembr 2013
17. Murray, N., Perronnin, F.: Generalized max pooling. In: IEEE Computer Society Conference on Computer Vision and Pattern Recognition (CVPR 2014), Columbus, USA, pp. 2473–2480, June 2014
18. Hautire, N., Tarel, J.P., Aubert, D.: Blind contrast enhancement assessment by gradient ratioing at visible edges. Image Anal. Stereol. J. **27**(2), 87–95 (2008)

Spatiogram and Fuzzy Logic Based Multi-modality Fusion Tracking with Online Update

Canlong Zhang[1,2]($\boxtimes$), Zhixin Li[1,2], Zhiwen Wang[3], and Ting Han[1]

[1] Guangxi Key Lab of Multi-source Information Mining and Security,
Guangxi Normal University, Guilin, China
zcltyp@163.com
[2] Guangxi Experiment Center of Information Science,
Guilin University of Electronic Technology, Guilin, China
[3] College of Computer Science and Communication Engineering,
Guangxi University of Science and Technology, Liuzhou, China

Abstract. Multi-modality fusion tracking is an interesting but challenging task. Many previous works just consider the fusion of different features from identical spectral image or identical features from different spectral images alone, which makes them be quite distinct from each other and be difficult to be integrated naturally. In this study, we propose an unified tracking framework to naturally integrate multiple different modalities via innovative use of spatiogram and fuzzy logic. Specifically, each modal target and its candidate are first represented by second-order spatiogram and their similarity is measured. Next, a novel objective function is built by integrating all modal similarities, and then a joint target center-shift formula is gained by performing mathematical operation on the objective function. Finally, the optimal target location is gained recursively by applying the mean shift procedure. Besides, a model update scheme via particle filter is developed to capture the appearance variations. Our framework allows the modalities to be original pixels or other extracted features from single image or different spectral images, and provides the flexibility to arbitrarily add or remove modality. Tracking results on the combination of infrared gray-HOG and visible gray-LBP clearly demonstrate the excellence of the proposed tracker.

Keywords: Spatiogram · Fuzzy logic · Particle filter · Mean shift

1 Introduction

Target tracking is a major foundation in visual surveillance, human-machine interface, vehicle navigation, video scene analysis, etc. Numerous methods have been reported [1], and these methods can be classified into two categories: using single-modality [2–5] and using multi-modality [6–10]. By leveraging the combined benefit of using different modalities while compensating for failure in individual modality, multi-modality system offers considerable advantage over single-modality system.

© Springer Nature Singapore Pte Ltd. 2017
J. Yang et al. (Eds.): CCCV 2017, Part I, CCIS 771, pp. 113–127, 2017.
https://doi.org/10.1007/978-981-10-7299-4_10

The fusion of different spectral images is significant in multi-modality tracking, because each spectral image provides disparate, yet complementary information about a scene. It offers an improved operational robustness, because distinct physical sensing principles compensate for particular perception shortcomings. Besides the fusion of multi-spectral images, the combination of multiple features from identical spectral image is also widely researched. The multi-feature fusion can better represent the target than single feature under complex dynamic circumstance by taking advantage of the complementarity of different features. However, above two types of multi-modality tracking method act in their own way, and few method can integrate them in a natural mean. In this paper, we propose a unified framework to cope with the problem by employing spatial histogram representation and mean shift search. Being fast and robust, the Mean Shift Tracker (MST) [5] has been used widely and many extensions and variants were proposed [11–14]. Instead of completely ignoring the spatial structure of the object features in original MST, Birchfield et al. [11] introduced a spatial histogram (abbreviated to *spatiogram*) that was formed by weighting each bin of histogram with the mean and covariance of the locations of the pixels that contribute to that bin. The spatiogram-based method not only improves the accuracy of the target appearance model but also preserves the rapidity of mean shift tracking, so we choose it to build our framework. The main contributions of our work are as follows:

- Establish a unified framework of multi-modal target tracking based on joint spatiogram representation, and the framework is flexible enough to handle any number of modalities;
- Design a fast multi-input multi-output fuzzy system to adaptively adjust the weight of each modality for evaluating the target state pre frame;
- Develop a particle filter based model update scheme to keep track of the most representative reference spatiogram throughout the tracking procedure;

The rest of paper is organized as follows. In Sect. 2, we first introduce the related work. Next, the spatiograms and their similarity is sketched in Sect. 3. After that, respectively detail the proposed tracking algorithm and fuzzy logic based weight adjust method in Sect. 4. Our model update scheme based particle filter is described in Sect. 5. Experimental results on the method are reported in Sect. 6. We conclude this paper in Sect. 7.

2   Related Work

The cooperation of visible and infrared imagery is the most active topic in multi-spectral combination. Infrared sensor can detect relative difference in the amount of thermal energy emitted from objects, and is independent of illumination, making it more effective than visible camera under poor lighting condition, but infrared imagery can't provide color and texture features. Visible sensor is oblivious to temperature difference, but is more effective than infrared sensor when objects are at thermal crossover, provided that the scene is well illuminated and the objects have color signatures different from the background. Therefore, by combining their data, a tracker can perform better than one that uses only single sensor. There are many methods for tracking color-infrared target. Reference [6] proposed a framework by fusing the outputs of multiple spatiogram trackers. Reference [7] presented a multi-cue mean-shift tracking approach based on fuzzified region dynamic image fusion. Reference [8] proposed a tracking approach for visible-infrared target using tracking-before-fusion, in which the visible and infrared targets were tracked individually, and their results were fused. However, the final result may be very bad when any one of their tracking results is poor. Reference [9] proposed a compressive spatial-temporal Kalman fusion tracking algorithm that allowed independent features to be integrated, whose fusion model was formulated with both spatial and temporal fusion coefficients. However, its Bayesian classification requires a large number of samples to gain enough accuracy, which leads to large computational burden. Reference [10] proposed an infrared-visible target tracking method using joint sparse representation, in which only one target template was required for infrared or visible image. The method need solve only one ℓ_1 norm minimization problem, so its time cost is dramatically decreased. However, it is the single target template that leads to the loss of diversity of target templates, thus reducing the robustness of tracker.

In recent years, many multi-feature fusion tracking algorithms have been reported. Reference [15] proposed a fusion tracking method that employed local steering kernel descriptor and color histogram to represent the target object. Reference [16] proposed a new joint sparse representation model for robust feature-level fusion tracking, which dynamically prevented unreliable feature being fused by taking advantage of the sparse representation. Reference [17] developed a particle filter tracking algorithm based on the fusion of color histogram and edge orientation histogram. Reference [18] located the target through independent multi-feature fusion and region-based temporal difference model under particle filter framework. Reference [19] proposed a multi-feature fusion tracking framework by employing hashing method to fuse different features to generate compact binary feature. Reference [20] presented a novel online object tracking algorithm by using multi-feature channels with adaptive weights.

3   The Spatiograms and Their Similarity

Spatiogram [11] is generalization of histogram that includes potentially higher order moments. Conventional histogram is a zeroth-order spatiogram, while second-order spatiogram contains spatial mean and covariance for each histogram bin.

Denote by $\hat{h} = \{\hat{p}_u, \hat{\boldsymbol{\mu}}_u, \widehat{\Sigma}_u\}_{u=1}^m$ the m-bin reference second-order spatiogram of the target. In each frame of the image sequence, the center point z of the image region in which the spatiogram $h(z) = \{p_u(z), \boldsymbol{\mu}_u(z), \Sigma_u(z)\}_{u=1}^m$ is closest to $\hat{h}$ is sought. Let $\{\boldsymbol{x}_i\}_{i=1}^n$ denote the (normalized) coordinates of the pixels in a candidate region centered at z, then the feature probability of the uth bin is

$$p_u(z) = \sum_{i=1}^{n} k(\|x_i - z\|^2)\delta[b(x_i) - u], \tag{1}$$

where $b(x)$ is the bin number $(1, \cdots, m)$ associated with the feature at location x, δ is the Kronecker delta function, and $k(x)$ is a kernel profile that assigns smaller weights to pixels farther from the circle center. Note that the kernel is normalized such that its profile satisfies $\sum_{i=1}^{n} k(\|x_i - z\|^2) = 1$. The coordinate mean $\mu_u(z)$ and covariance $\Sigma_u(z)$ of pixels belong to the bin u are

$$\mu_u(z) = \frac{1}{\sum_{k=1}^{n} \delta[b(x_k) - u]} \sum_{i=1}^{n} (x_i - z)\delta[b(x_i) - u], \tag{2}$$

$$\Sigma_u(z) = \frac{1}{\sum_{k=1}^{n} \delta[b(x_k) - u] - 1} \sum_{i=1}^{n} (x_i - \mu_u(z))(x_i - \mu_u(z))^{\mathrm{T}}\delta[b(x_i) - u], \tag{3}$$

Similarly, the calculation of reference spatiogram $\hat{h}$ can be regarded as a special case of $h(z)$ that $z = 0$.

The similarity between the spatiograms is measured by

$$\rho(z) \triangleq \rho[h(z), \hat{h}] = \sum_{u=1}^{m} \psi_u(z)\sqrt{p_u(z)\hat{p}_u}, \tag{4}$$

where $\sqrt{p_u(z)\hat{p}_u}$ is used to measure the similarly between features of candidate region and target image, while

$$\psi_u(z) = \frac{4|\Sigma_u(z)\widehat{\Sigma}_u|^{\frac{1}{4}}}{|\widetilde{\Sigma}_u|^{\frac{1}{2}}} \exp\left\{-\frac{1}{2}(\mu_u(z) - \hat{\mu}_u)^{\mathrm{T}}(\widetilde{\Sigma}_u(z))^{-1}(\mu_u(z) - \hat{\mu}_u)\right\} \tag{5}$$

is used to measure the similarity in spatial arrangement between these features, and $\widetilde{\Sigma}_u(z) = 2(\Sigma_u(z) + \widehat{\Sigma}_u)$.

4   Proposed Multi-modality Tracker

Assume there are N different modalities (or N reference spatiograms), then it is a fact that each candidate state should correspond to N image patches with different modalities. That is to say, at any time instant t, all well registered N image patches with different modalities should share same dynamic state (including location, size and shape), because they merely use different modalities to describe identical real object.

4.1   Joint Spatiogram Representation

Denote by $\hat{h}_j = \{\hat{p}_u^j, \hat{\mu}_u^j, \widehat{\Sigma}_u^j\}_{u=1}^{m}$ the jth modality reference spatiogram. Given a candidate state centered at z, then there are N candidate spatiograms with different modalities, and we denote by $h_j = \{p_u^j(z), \mu_u^j(z), \Sigma_u^j(z)\}_{u=1}^{m}$ the jth modality candidate spatiogram. The similarity between the jth modality candidate spatiogram and its reference spatiogram is measured by

$$\rho_j(z) = \sum_{u=1}^{m} \psi_u^j(z)\sqrt{p_u^j(z)\hat{p}_u^j} \tag{6}$$

where

$$\psi_u^j(z) = \frac{4|\Sigma_u^j(z)\widehat{\Sigma}_u^j|^{\frac{1}{4}}}{|\widetilde{\Sigma}_u^j|^{\frac{1}{2}}} \exp\left\{-\frac{1}{2}(\mu_u^j(z) - \hat{\mu}_u^j)^{\mathrm{T}}(\widetilde{\Sigma}_u^j(z))^{-1}(\mu_u^j(z) - \hat{\mu}_u^j)\right\}, \tag{7}$$

and $\widetilde{\Sigma}_u^j(z) = 2(\Sigma_u^j(z) + \widehat{\Sigma}_u^j)$.

For the multi-modality fusion tracking, whether a candidate state should be accepted or not is decided by joint similarity of all modality candidates and their corresponding targets, so we define the joint similarity (or object function) as

$$\rho(z) \triangleq \sum_{j=1}^{N} \alpha_j \rho_j(z), \tag{8}$$

where $0 \leq \alpha_j \leq 1$ is the weight that reflects the reliability of the jth modality to evaluating the target state in current frame, and $\sum_{j=1}^{N} \alpha_j = 1$.

4.2   Target Localization

The goal of target localization is to estimate the target translation $\hat{z}$ that maximizes the joint similarity in (8). Denote by $\hat{z}_0$ the estimated target location in the previous frame. Approximating the joint similarity (8) in the current frame by its first-order Taylor expansion around the values $p_u^j(\hat{z}_0)$ and $\mu_u^j(\hat{z}_0)$ results in

$$\rho(z) \approx \sum_{j=1}^{N}\sum_{u=1}^{m} \alpha_j \psi_u^j(\hat{z}_0)\sqrt{p_u^j(\hat{z}_0)\hat{p}_u^j}((\widetilde{\Sigma}_u^j(\hat{z}_0))^{-1}(\hat{\mu}_u^j - \mu_u^j(\hat{z}_0))\mu_u^j(z) + \cdots$$

$$\sum_{j=1}^{N}\sum_{u=1}^{m} \frac{\alpha_j}{2}\psi_u^j(\hat{z}_0)\sqrt{\hat{p}_u^j/p_u^j(\hat{z}_0)}p_u^j(z) + C, \tag{9}$$

where

$$p_u^j(z) = \sum_{i=1}^{n} k(\|x_i^j - z\|^2)\delta[b(x_i^j) - u], \tag{10}$$

$$\mu_u^j(z) = \frac{1}{\sum_{k=1}^{n}\delta[b(x_k^j) - u]} \sum_{i=1}^{n}(x_i^j - z)\delta[b(x_i^j) - u], \tag{11}$$

and C is independent of z. Taking the derivative of (9) with respect to z yields

$$\frac{\partial\rho(z)}{\partial z} = \sum_{j=1}^{N}\sum_{i=1}^{n} w_i^j k'(\|z - x_i^j\|^2)(z - x_i^j) - \sum_{j=1}^{N}\sum_{u=1}^{m} w_u^j, \tag{12}$$

where

$$
\left.
\begin{aligned}
w_i^j &= \sum_{u=1}^{m} \frac{\alpha_j}{2} \psi_u^j(\hat{\boldsymbol{z}}_0)\sqrt{\hat{p}_u^j/p_u^j(\hat{\boldsymbol{z}}_0)}\delta[b(\boldsymbol{x}_i^j) - u] \\
\boldsymbol{w}_u^j &= \alpha_j \psi_u^j(\hat{\boldsymbol{z}}_0)\sqrt{p_u^j(\hat{\boldsymbol{z}}_0)\hat{p}_u^j}((\widetilde{\boldsymbol{\Sigma}}_u^j(\hat{\boldsymbol{z}}_0))^{-1}(\hat{\boldsymbol{\mu}}_u^j - \boldsymbol{\mu}_u^j(\hat{\boldsymbol{z}}_0)))
\end{aligned}
\right\}
\tag{13}
$$

Set $\frac{\partial \rho(\boldsymbol{z})}{\partial \boldsymbol{z}} = 0$ and solve for $\boldsymbol{z}$:

$$
\hat{\boldsymbol{z}} = \frac{\sum\limits_{j=1}^{N}\sum\limits_{i=1}^{n} w_i^j g(\|\boldsymbol{z}_0 - \boldsymbol{x}_i^j\|^2)\boldsymbol{x}_i^j - \sum\limits_{j=1}^{N}\sum\limits_{u=1}^{m} \boldsymbol{w}_u^j}{\sum\limits_{j=1}^{N}\sum\limits_{i=1}^{n} w_i^j g(\|\boldsymbol{z}_0 - \boldsymbol{x}_i^j\|^2)},
\tag{14}
$$

where $g(x) = -k'(x)$. If we use the Epanechnikov profile [5] then the derivative of the kernel is constant, thus the iteration (14) is reduced to

$$
\hat{\boldsymbol{z}} = \left(\sum_{j=1}^{N}\sum_{i=1}^{n} w_i^j \boldsymbol{x}_i^j - \sum_{j=1}^{N}\sum_{u=1}^{m} \boldsymbol{w}_u^j\right) \Big/ \sum_{j=1}^{N}\sum_{i=1}^{n} w_i^j,
\tag{15}
$$

To enable the physical meaning of (15) to evident, we rearrange the numerator to obtain

$$
\hat{\boldsymbol{z}} = \sum_{j=1}^{N}\sum_{i=1}^{n}(w_i^j \boldsymbol{x}_i^j - \boldsymbol{v}_i^j) \Big/ \sum_{j=1}^{N}\sum_{i=1}^{n} w_i^j,
\tag{16}
$$

where $\boldsymbol{v}_i^j = \sum_{u=1}^{m}(\boldsymbol{w}_u^j \delta[b(\boldsymbol{x}_i^j) - u]/\sum_{k=1}^{n}\delta[b(\boldsymbol{x}_k^j) - u])$. It is easy to see from (16) that each pixel in each modality candidate patch casts a vote proportional to $\|w_i^j \boldsymbol{x}_i^j - \boldsymbol{v}_i^j\|$ in the direction of $\boldsymbol{x}_i^j - \boldsymbol{v}_i^j/w_i^j$ for the Mean-Shift offset toward or away from $\hat{\boldsymbol{z}}$, which also shows that the optimal location of the target is determined by all modalities.

4.3   Adjusting Weights Using Fuzzy Logic

The weights in most algorithms are assumed to be unchanged during the tracking, but the fact is that the importance (reliability) of each modality changes over time. Usually, the target doesn't change drastically between consecutive frames, but has always a few difference, so we can only fuzzily rather than determinately estimate the change. In this paper, we use the similarity between the tracking result and the reference model to measure the change, and employ the fuzzy logic method to calculate the weights. Specifically, the inputs of fuzzy system are the reliabilities

$$
e_j(\hat{\boldsymbol{z}}) = \frac{1}{\sqrt{2\pi}\delta} \exp\left(-\frac{1 - \rho_j(\hat{\boldsymbol{z}})}{2\delta^2}\right)
\tag{17}
$$

index $j = 1,\ldots,N$, where $\rho_j(\hat{\boldsymbol{z}})$ is the similarity between the tracking result and the reference model in current frame. Notice that the smaller the similarity, the lower the reliability. The outputs of fuzzy system are the weights α_j index $j = 1, \ldots, N$ in the next frame.

In this study, we apply the singleton fuzzification, product inference, and centroid defuzzification to build the fuzzy system [21]. Each input variable e_j is fuzzified with five linguistic variables, labeled SR, S, M, B, BR, partitioned on the interval [0, 1], and each output variable α_j is also fuzzified with nine linguistic variables, labeled ST, VS, SR, S, M, B, BR, VB, BT, partitioned on the interval [0, 1], where ST stands for smallest, VS for very smaller, SR for smaller, S for small, M for middle, B for big, BR for bigger, VB for very bigger, BT for biggest. The membership functions of e_j and α_j are all Gaussian function, as shown in Fig. 1(a) and (b).

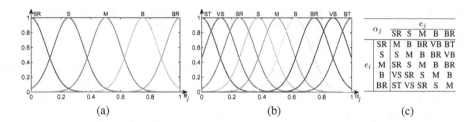

(a)   (b)   (c)

Fig. 1. (a) Membership function for e_j (b) membership function for α_j (c) the control rule bases for a dual-input and single-output fuzzy system.

Our fuzzy system is a typical multi-input multi-output system with N inputs and N outputs, and if directly using the IF-THEN rule of N inputs inferring one output then resulting in $N \times 5^N$ rules, which is time-consuming. Therefore, we convert the fuzzy system into several dual-input and single-output (DISO) fuzzy systems, and define the fuzzy rule of each DISO system as

$$\text{IF } e_j \text{ is } A_1^k \text{ and } e_i \text{ is } A_2^k \text{ THEN } \alpha_j^i \text{ is } B^k, k = 1, \ldots, K \qquad (18)$$

where A_1^k and A_2^k belong to {SR, S, M, B, BR}, B^k belongs to {ST, VS, SR, S, M, B, BR, VB, BT}, α_j^i $(i \neq j)$ is the weight of modality j relative to modality i, and K is the total number of rules. The control rule bases originating from (18) is shown in Fig. 1(c), apparently, $K = 25$ and $\alpha_i^j = 1 - \alpha_j^i$. After performing the fuzzy control procedure to acquire all α_j^i, the α_j is calculated by

$$\alpha_j = (\alpha_j^1 + \cdots + \alpha_j^{(j-1)} + \alpha_j^{(j+1)} + \cdots + \alpha_j^N)/(N-1) \qquad (19)$$

Finally, all α_j are normalized such that they satisfy $\sum_j \alpha_j = 1$. Since the derivation of each α_j needs $N-1$ noninteractive DISO fuzzy systems, the total number of DISO systems is $N(N-1)/2$, thus the total fuzzy rules can be reduced to $25 \times N(N-1)/2$.

Algorithm 1 summarizes the proposed Multi-modality Joint Spatiogram Tracking (MJST) procedure, which is implemented in MATLAB+MEX. The model update will be detailed in the next section.

Algorithm 1. Multi-modality fusion tracking via spatiogram and fuzzy logic

Input: each modal target image $\{I(\boldsymbol{x}_i^j, 1)\}_{i=1}^n$ centered at $\boldsymbol{z}_1$ in the first frame, the corresponding reference spatiogram $\{\hat{p}_u^j(1), \hat{\boldsymbol{\mu}}_u^j(1), \widehat{\boldsymbol{\Sigma}}_u^j(1)\}_{u=1}^m$, its weight α_j, and γ;

1: **for** $t = 2, \ldots, T$ **do**

2:    Compute the current candidate spatiogram $\{p_u^j(\boldsymbol{z}_{t-1}), \boldsymbol{\mu}_u^j(\boldsymbol{z}_{t-1}), \boldsymbol{\Sigma}_u^j(\boldsymbol{z}_{t-1})\}_{u=1}^m$ by (1) to (3) and its similarity $\rho_j(\boldsymbol{z}_{t-1})$ with its reference spatiogram by (6) to (7), as well as their $\rho(\boldsymbol{z}_{t-1})$ by (8);

3:    Compute each weights w_i^j and $\boldsymbol{w}_u^j$ by (13), and find new candidate location $\boldsymbol{z}_t$ by (15);

4:    Compute the new candidate spatiogram $\{p_u^j(\boldsymbol{z}_t), \boldsymbol{\mu}_u^j(\boldsymbol{z}_t), \boldsymbol{\Sigma}_u^j(\boldsymbol{z}_t)\}_{u=1}^m$ by (1) to (3) and its similarity $\rho_j(\boldsymbol{z}_t)$ with its reference spatiogram by (6) to (7), and their $\rho(\boldsymbol{z}_t)$ by (8);

5:    If $\rho(\boldsymbol{z}_t) < \rho(\boldsymbol{z}_{t-1})$, then set $\boldsymbol{z}_t \leftarrow \frac{1}{2}(\boldsymbol{z}_{t-1} + \boldsymbol{z}_t)$ and go to Step 4;

6:    If $\boldsymbol{z}_t = \boldsymbol{z}_{t-1}$ or the number of iterations reached M_{max}, then go to Step 7. Otherwise, set $\boldsymbol{z}_{t-1} \leftarrow \boldsymbol{z}_t$ and $\rho(\boldsymbol{z}_{t-1}) \leftarrow \rho(\boldsymbol{z}_t)$, and go to Step 3;

7:    Compute all reliability $e_j(\boldsymbol{z}_t)$ by (17) and adjust all weight α_j using fuzzy logic method;

8:    For each modality, establish the particle filter with the state variable $\{I(\boldsymbol{x}_i^j, t-1)\}_{i=1}^n$ and its current observation $\boldsymbol{h}_j(\boldsymbol{z}_t)$ by (20) to (22), run the filter procedure to gain optimal state $\{\hat{I}(\boldsymbol{x}_i^j, t)\}_{i=1}^n$, and obtain the new reference spatiogram $\{\hat{p}_u^j(t), \hat{\boldsymbol{\mu}}_u^j(t), \widehat{\boldsymbol{\Sigma}}_u^j(t)\}_{u=1}^m$ according to (23);

9: **end for**

5   Model Update via Particle Filter

In practical tracking scenarios, the target model should be updated so that it can capture the appearance variations due to illumination or pose changes. Some self-learning methods use the weighted sum of the current tracking result and current model to get new model, which is easy to realize but also prone to drift from the target because of the accumulation of errors. Other semi-supervised learning methods are suggested to avoid the drift problem, but the classifier-based tracking cannot be applicable to our case. Motivated by the Ref. [12], we propose a model update mechanism based on particle filter, and its details are introduced as follows.

5.1   Spatiogram Filtering

In most of tracking methods, the particle filter keeps tracking of object changes in position and velocity, not the changes of object appearance. We use particle filter for filtering object spatiogram so as to obtain the optimal estimate of the target model.

In frame t, let $I(\boldsymbol{x}_i, t)$ be the feature value (e.g., intensity, gradient, texture, etc.) at the normalized coordinate $\boldsymbol{x}_i$ in object image, and $\boldsymbol{h}(\{I(\boldsymbol{x}_i, t)\}_{i=1}^n)$ be the corresponding spatiogram that is obtained by performing (1) to (3) on $\{I(\boldsymbol{x}_i, t)\}_{i=1}^n$. The object appearance variations in essence are the value changes of all $I(\boldsymbol{x}_i, t)$. Since the change of each $I(\boldsymbol{x}_i, t)$ is independent, we can filter each of them independently. Thus, the state prediction equation of each $I(\boldsymbol{x}_i, t)$ is given by

$$I(\boldsymbol{x}_i, t) = I(\boldsymbol{x}_i, t-1) + \omega_i(t-1), \qquad i = 1, 2, \cdots, n, \tag{20}$$

where $I(\boldsymbol{x}_i, t)$ is called the state of pixel $\boldsymbol{x}_i$ in the frame t, and $I(\boldsymbol{x}_i, t-1)$ is its counterpart in the frame $t-1$. $\omega_i(t-1)$ specifies the state noise owing to the object appearance variation, which is assumed to be Gaussian, and furthermore, to have the same variance σ_ω^2 for all $\boldsymbol{x}_i$.

To obtain measurements of the filters, the previous reference spatiogram $\boldsymbol{h}(\{I(\boldsymbol{x}_i, t-1)\}_{i=1}^n)$ is matched with the current frame, and yielding a new spatiogram $\boldsymbol{h}(\boldsymbol{z}_t)$ at the convergence position $\boldsymbol{z}_t$ provided by the mean-shift tracking algorithm. Then, $\boldsymbol{h}(\boldsymbol{z}_t)$ is used as measurement for $\boldsymbol{h}(\{I(\boldsymbol{x}_i, t)\}_{i=1}^n)$, and the observation equation is

$$\boldsymbol{h}(\boldsymbol{z}_t) = \boldsymbol{h}(\{I(\boldsymbol{x}_i, t)\}_{i=1}^n) + \boldsymbol{v}(t), \tag{21}$$

where $\boldsymbol{v}(t)$ models the noise in the image signal. Since each bin in the spatiogram is independent, the (21) can be rewritten as

$$h_u(\boldsymbol{z}_t) = h_u(t) + \boldsymbol{v}_u(t), \qquad u = 1, 2, \cdots, m, \tag{22}$$

where $h_u(\boldsymbol{z}_t) = [p_u(\boldsymbol{z}_t), \boldsymbol{\mu}_u(\boldsymbol{z}_t), \boldsymbol{\Sigma}_u(\boldsymbol{z}_t)]^{\mathrm{T}}$, $\boldsymbol{h}(\boldsymbol{z}_t) = \{h_u(\boldsymbol{z}_t)\}_{u=1}^m$, $h_u(t) = [\hat{p}_u(t), \hat{\boldsymbol{\mu}}_u(t), \widehat{\boldsymbol{\Sigma}}_u(t)]^{\mathrm{T}}$, $\boldsymbol{h}(\{I(\boldsymbol{x}_i, t)\}_{i=1}^n) = \{h_u(t)\}_{u=1}^m$, and $\boldsymbol{v}_u(t)$ is the observation noise, which is assumed to be Gaussian, and furthermore, to have the same variance σ_v^2 for all u. Because the $\boldsymbol{\Sigma}_u$ is 2-D diagonal matrix, we only update its two diagonal entries, thus the h_u is a 5-D vector.

We call $\boldsymbol{h}(\{I(\boldsymbol{x}_i, t-1)\}_{i=1}^n)$ current model, and $\boldsymbol{h}(\boldsymbol{z}_t)$ observation model. Particle filter then yields a tradeoff between the two models and provides us an optimal estimate of the object model that is called candidate model, denoted by $\boldsymbol{h}(\{\hat{I}(\boldsymbol{x}_i, t)\}_{i=1}^n)$, where $\{\hat{I}(\boldsymbol{x}_i, t)\}_{i=1}^n$ is the optimal particle. In implement, the $\{I(\boldsymbol{x}_i, t)\}_{i=1}^n$ is treated as a particle whose likelihood weight is calculated by $\frac{1}{\sqrt{2\pi}\sigma_v} \exp\{-\frac{1-\rho[\boldsymbol{h}(\boldsymbol{z}_t), \boldsymbol{h}(\{I(\boldsymbol{x}_i, t)\}_{i=1}^n)]}{\sigma_v^2}\}$ incorporating with (4), and the $\{\hat{I}(\boldsymbol{x}_i, t)\}_{i=1}^n$ is determined by taking the weighted mean of all particles.

5.2 Update Criterion

It is not suitable for us to accept the candidate model all the time. We should try to find a robust criterion to decide whether the candidate mode $\boldsymbol{h}(\{\hat{I}(\boldsymbol{x}_i, t)\}_{i=1}^n)$ should be accepted because over-update could make tracker sensitive to outliers like occlusions or dramatic appearance changes.

We take the similarity $\rho[\boldsymbol{h}(\boldsymbol{z}_t), \boldsymbol{h}(\{I(\boldsymbol{x}_i, t-1)\}_{i=1}^n)]$ between the observation model and the current model as the criterion. If the similarity is smaller than the threshold γ, which implies that the object encounters dramatic appearance changes, then we reject $\boldsymbol{h}(\{\hat{I}(\boldsymbol{x}_i, t)\}_{i=1}^n)$ and remain using current model, otherwise accept it. The final model update formula is as follows

$$\hat{\boldsymbol{h}}(t) = \begin{cases} \boldsymbol{h}(\{I(\boldsymbol{x}_i, t-1)\}_{i=1}^n) & \rho[\boldsymbol{h}(\boldsymbol{z}_t), \boldsymbol{h}(\{I(\boldsymbol{x}_i, t-1)\}_{i=1}^n)] < \gamma \\ \boldsymbol{h}(\{\hat{I}(\boldsymbol{x}_i, t)\}_{i=1}^n) & \rho[\boldsymbol{h}(\boldsymbol{z}_t), \boldsymbol{h}(\{I(\boldsymbol{x}_i, t-1)\}_{i=1}^n)] \geq \gamma \end{cases} \tag{23}$$

Once $\boldsymbol{h}(\{\hat{I}(\boldsymbol{x}_i, t)\}_{i=1}^n)$ is accepted, the $\{\hat{I}(\boldsymbol{x}_i, t)\}_{i=1}^n$ is also saved.

6   Experiments

To verify the flexibility, accuracy and efficiency of the proposed tracker, we tested six registered infrared-visible sequences which involve general difficulties like night, shade, cluster, crossover and occlusion. We also compared the proposed tracker to the state-of-the-art methods such as ℓ_1 tracker (L1T) [4], joint sparse representation tracker (JSRT) [10], and fuzzified region dynamic fusion tracker (FRD) [7]. All tracking results are obtained by running these trackers on an Intel Dual-Core 2.6 GHz CPU with 8 GB RAM, using the same initial positions for fair comparison. The L1T can tackle only one modality at a time.

Set $M_{max} = 20$, $\gamma = 0.4$, and the number of particles 50 for model update. We use the combination of infrared gray, infrared HOG (histogram of oriented gradient), visible gray, and visible LBP (local binary pattern), so $N = 4$.

6.1   Combination of Infrared Gray-HOG and Visible Gray-LBP

The LBP [22] is very effective to describe the image texture features, and has advantages such as fast computation and rotation invariance, which facilitates the wide usage in the fields of texture analysis, image retrieval, object tracking, etc. In LBP, each pixel is assigned a texture value, which can be naturally combined with the gray of the pixel to represent targets. The LBP operator labels the pixel in an image by thresholding its neighborhood with the center value and considering the result as a binary number.

The HOG technique [23] counts occurrences of gradient orientation in localized portions of an image, and is very effective to describe the object shape feature. HOG is invariant to geometric and photometric transformations, so is used widely in object detection. See [23] for the extraction steps of HOG. To enable HOG to naturally combine with the gray of image, we employ the visualization method suggested in [24].

Generally, the visible image contains more texture features, whereas shape feature is more obvious in the infrared image, so we extract LBP feature from visible image and HOG feature from infrared image. Figure 2 shows the extraction of LBP and HOG features from visible and infrared images respectively.

Figure 3 shows the screenshots of some sampled tracking results on video1-video6.

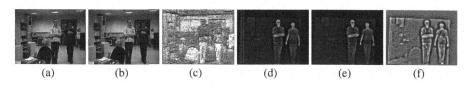

(a)   (b)   (c)   (d)   (e)   (f)

Fig. 2. Extracting LBP and HOG features from visual and infrared images respectively, (a) visible image, (b) visible gray, (c) visible LBP, (d) infrared image, (e) infrared gray, (f) infrared HOG.

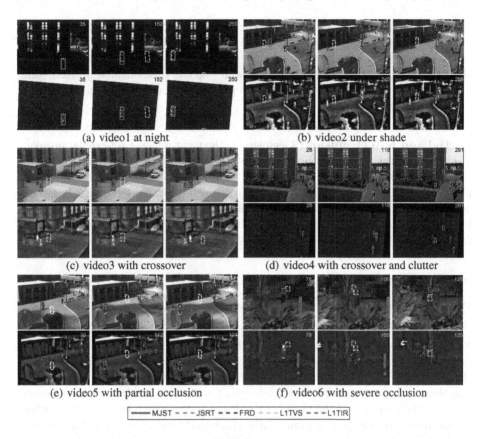

(a) video1 at night   (b) video2 under shade

(c) video3 with crossover   (d) video4 with crossover and clutter

(e) video5 with partial occlusion   (f) video6 with severe occlusion

MJST — — — JSRT ▪ — ▪ FRD ▪ ▪ ▪ L1TVS — ▪ — L1TIR

Fig. 3. Screenshots of some sampled tracking results, where L1TIR represents using L1T to track infrared target, and L1TVS represents using L1T to track visible target.

Night and Shade: At night without light, the target in visible image is usually obscure (see Fig. 3(a)), so the tracker that only using visible camera often fails to hold the target. However, by using or cooperating with infrared camera, our tracker can hold well the target, because the infrared sensor is independent of illumination. In Fig. 3(b), the man in black is walking into a shade area, and the contrast between him and his background is very low in visible image but no for infrared image, which is similar to walking at night, so our tracker can hold the target well. It is notable that our tracker successfully overcome the occlusion of lamppost (see frame 245 and 288 in Fig. 3(b)), which is mainly attribute to the use of online model update.

Crossover and Clutter: The tracker is easily attracted by the distracter when target is at thermal crossover or in cluttered background. For example, the FRD is attracted by the right man when two men are at thermal crossover (see frame 78 in Fig. 3(c)), and FRD and L1TVS are attracted by the cluttered background (see frame 118 and 291 in Fig. 3(d)). The distracter presents similar appearance as the target, so can give good match to the target, which makes it be difficult to discriminate such a distracter only using single source or matching score. However, our tracker can stick with the target all the time because of using joint compressive representation of infrared and visual modalities, thus can always capture the target well.

Partial and Full Occlusion: Occlusion is the most general yet crucial problem in object tracking, and it is classified into partial and full. For partial occlusion, our tracker can successfully overcome it (see frame 245 in Fig. 3(b), frame 112 in Fig. 3(e), and frame 100 in Fig. 3(f)), because both it use model update to handle occlusion. For full occlusion, as showed in Figs. 3(f) and 4(f), almost all trackers lost the targets, which is because that the targets have completely disappeared from our sight.

6.2   Quantitative Comparison

We use five criteria, i.e., center offset error $\varrho = \sqrt{(x_G - x_T)^2 + (y_G - y_T)^2}$, average center offset error $\bar{\varrho} = \frac{1}{q} \sum_i \varrho_i$, overlap ratio $\epsilon = \frac{area(R_G \bigcap R_T)}{area(R_G \bigcup R_T)}$, average overlap ratio $\bar{\epsilon} = \frac{1}{q} \sum_i \epsilon_i$, and success rate $sr = \frac{1}{q} \sum_i f(\epsilon_i - 0.5)$, where (x_G, y_G, R_G) is the center and region of target given by manual, (x_T, y_T, R_T) is the result given by the tacker, and q is the number of frames. The $f(x)$ is step function, if $x \geq 0$, then $f(x) = 1$, else $f(x) = 0$. The overlap ratio is used to evaluate the accuracy of the tracking result pre frame, and whose value is 1 when the estimated region overlaps fully with the ground truth, thereby obtaining best tracking result. An ideal tracker should have less center offset error, and high overlap ratio and success rate.

The L1T can track only one of visible and infrared at a time, for easy to compare with other fusion trackers, we integrate the tracking results of L1TVS and L1TIR as following: (1) the time consumer of L1T is equal to the time sum of L1TVS and L1TIR, and (2) the target state of L1T is equal to the weight average of the target states of L1TVS and L1TIR, where the weight is induced by

Table 1. Quantitative comparison of MJST, JSRT, FRD and L1T.

	MJST			JSRT			FRD			L1T		
	$\bar{\varrho}$	$\bar{\epsilon}$	sr	$\bar{\varrho}$	$\bar{\epsilon}$	sr	$\bar{\varrho}$	$\bar{\epsilon}$	sr	$\bar{\varrho}$	$\bar{\epsilon}$	sr
video1	4.28	0.81	1.00	83.3	0.19	0.21	6.97	0.75	0.93	36.4	0.32	0.36
video2	2.31	0.77	0.98	69.5	0.07	0.07	5.33	0.64	0.77	35.3	0.07	0.06
video3	1.47	0.84	1.00	17.6	0.46	0.51	2.75	0.84	0.95	2.64	0.78	0.90
video4	5.44	0.75	0.93	53.5	0.30	0.38	20.9	0.27	0.36	7.46	0.70	0.87
video5	19.1	0.65	0.78	79.6	0.12	0.15	24.0	0.42	0.59	20.6	0.56	0.68
video6	9.79	0.62	0.73	16.7	0.43	0.51	3.63	0.78	0.85	23.5	0.53	0.61
averag	7.07	0.74	0.90	53.4	0.26	0.31	10.6	0.62	0.74	21.0	0.49	0.58

their observation likelihood in a mean square error way. The quantitative results are summarized in Table 1, where red font indicates the best performance while the blue font indicates the second best ones. The detail of overlap ratio pre frame is shown in Fig. 4. It can be observed that, compared with other trackers, our tracker obtains the highest tracking accuracy and success rate.

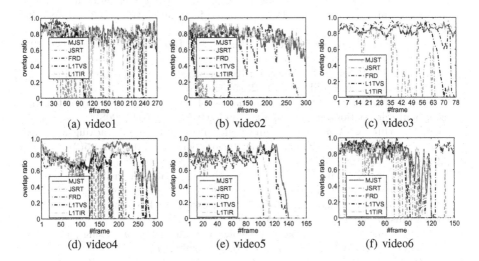

Fig. 4. The overlap ratio between tracked region and manually marked ground truth

7 Conclusions

In this paper we had proposed a unified multi-modality tracking framework by cooperating fuzzy logic. In the framework, different types of modalities can be arbitrary added, removed and naturally integrated through joint spatiogram representation. In addition, the target model update strategy enables the tracker promptly respond to the variations of appearance. Experiment results on six datasets demonstrate that our approach performs best. Our current tracker is not quite suitable for tracking full occlusion targets.

Acknowledgments. This work was supported by the National Natural Science Foundation of China (Grant Nos. 61365009, 61663004, 61462008), the Natural Science Foundation of Guangxi Province of China (Grant Nos. 2014GXNSFAA118368, 2016GXNS-FAA380146, 2013GXNSFAA019336), the Doctoral Research Foundation of Guangxi Normal University, and Guangxi Collaborative Innovation Center of Multisource Information Integration and Intelligent Processing. The authors would like to thank the anonymous reviewers for their constructive comments.

References

1. Wu, Y., Lim, J., Yang, M.H.: Online object tracking: a benchmark. In: IEEE Conference on Computer Vision & Pattern Recognition, pp. 2411–2418 (2013)
2. Zhang, C.L., Tang, Y.P., Li, Z.X., et al.: Dual-kernel tracking approach based on second-order spatiogram. J. Electron. Inf. Technol. **37**(7), 1660–1666 (2015)
3. Henriques, J.F., Caseiro, R., Martins, P., et al.: High-speed tracking with kernelized correlation filters. IEEE Trans. Pattern Anal. Mach. Intell. (PAMI) **37**(3), 583–596 (2014)
4. Mei, X., Ling, H.: Robust visual tracking using l1 minimization. In: The 12th International Conference on Computer Vision, pp. 1436–1443 (2009)
5. Comaniciu, D., Ramesh, V., Meer, P.: Kernel-based object tracking. IEEE Trans. Pattern Anal. Mach. Intell. (PAMI) **25**(5), 564–577 (2003)
6. Ciarno, C., Noel, E.O., Alan, S.: Thermo-visual feature fusion for object tracking using multiple spatiogram. Mach. Vis. Appl. **19**(5), 483–494 (2008)
7. Xiao, G., Yun, X., Wu, J.M.: A multi-cue mean-shift target tracking approach based on fuzzified region dynamic image fusion. Sci. China Inf. Sci. **55**(3), 577–589 (2012)
8. Xiao, G., Yun, X., Wu, J.M.: A new tracking approach for visible and infrared sequences based on tracking-before-fusion. Int. J. Dyn. Control **4**(1), 40–51 (2016)
9. Yun, X., Jing, Z.L., Xiao, G., et al.: A compressive tracking based on time-space Kalman fusion model. Sci. China: Inf. Sci. **1**, 1–15 (2016)
10. Liu, H.P., Sun, F.C.: Fusion tracking in color and infrared images using joint sparse representation. Sci. China: Inf. Sci. **55**(3), 590–599 (2012)
11. Birchfield S.T.: Spatiograms versus histograms for region-based tracking. In: Proceedings of IEEE Conference on Computer Vision and Pattern Recognition, SanDiego, California, USA, pp. 1158–1163 (2005)
12. Peng, N.S., Yang, J., et al.: Model update mechanism for mean-shift tracking. J. Syst. Eng. Electron. **16**(1), 52–57 (2005)
13. Leichter, I.: Mean shift trackers with cross-bin metrics. IEEE Trans. Pattern Anal. Mach. Intell. **34**(4), 695–706 (2012)
14. Tomas, V., Jana, N., Jiri, M.: Robust scale-adaptive mean-shift for tracking. Pattern Recogn. Lett. **49**(1), 250–258 (2014)
15. Zoidi, O., Tefas, A., Pitas, I.: Visual object tracking based on local steering kernels and color histograms. IEEE Trans. Circ. Syst. Video Technol. **23**(5), 870–882 (2013)
16. Lan, X., Ma, A.J., Yuen P.C.: Multi-cue visual tracking using robust feature-level fusion based on joint sparse representation. In: IEEE Conference on Computer Vision and Pattern Recognition (CVPR), pp. 1194–1201 (2014)
17. Xu, F., Zhao, L.: A particle filter tracking algorithm based on adaptive feature fusion strategy. In: Proceedings of IEEE World Congress on Intelligent Control and Automation (WCICA), pp. 4612–4616 (2012)
18. Lu, X., Song, L., Yu, S., et al.: Object contour tracking using multi-feature fusion based particle filter. In: Proceedings of IEEE Conference on Industrial Electronics and Applications (ICIEA), pp. 237–242 (2012)
19. Ma, C., Liu, C.C., Peng, F.R., et al.: Multi-feature hashing tracking. Pattern Recogn. Lett. **69**(1), 62–71 (2016)
20. Jiang, H.L., Li, J.H., Wang, D., et al.: Multi-feature tracking via adaptive weights. Neurocomputing **207**, 189–201 (2016)

21. Ross, T.J.: Fuzzy Logic with Engineering Applications, 3rd edn. McGraw-Hill Inc., New York (1995)
22. Timo, O., Matti, P., Topi, M.: Multiresolution gray-scale and rotation invariant texture classification with local binary patterns. IEEE Trans. Pattern Anal. Mach. Intell. **24**(7), 971–987 (2002)
23. Navneet, D., Bill, T.: Histograms of oriented gradients for human detection. In: IEEE Computer Society Conference on Computer Vision & Pattern Recognition, vol. 1, no. 12, pp. 886–893 (2005)
24. Vondrick, C., Khosla, A., Pirsiavash, H., et al.: Visualizing object detection features. Int. J. Comput. Vis. **119**(2), 145–158 (2016)

A Novel Real-Time Tracking Algorithm Using Spatio-Temporal Context and Color Histogram

Yong Wu[1], Zemin Cai[1(✉)], Jianhuang Lai[2], and Jingwen Yan[1]

[1] Department of Electronic Engineering, College of Engineering,
Shantou University, Shantou, China
{15ywu1,zmcai,jwyan}@stu.edu.cn

[2] Key Laboratory of Machine Intelligent and Advanced Computing,
Sun Yat-sen University, Ministry of Education, Guangzhou, China
stsljh@mail.sysu.edu.cn

Abstract. Spatio-temporal context (STC) is one of the most important features in describing the motion in videos. STC-based tracking algorithm achieved good performance on real-time tracking. However, it is very difficult to perform precise tracking in complex situations like heavy occlusion, illumination changes, and pose variation. In this paper, we propose a real-time tracking method which is robust to target variation during tracking single-object. Experiments on some challenging sequences highlights a significant improvement of tracking accuracy over the state-of-the-art methods.

Keywords: Spatio-temporal context · Color histogram · Deformation

1 Introduction

Target tracking is one of the basic problems in computer vision and a great number of methods [1–7, 10] have been proposed to address this problem. Among the existing methods, single-object tracking has been extensively and actively studied in recent years because of numerous important applications such as military, entertainment, and security fields. There are many single-object trackers like STC [1], TLD [2], KCF [3], and DSST [4], which have good performance.

Generally, the state-of-the-art algorithms [5–7, 10] use model adaptation to employ information presented in later frames, however the tracker need to update the model using previous frames. The great weakness is that small errors may accumulate in and cause model drift. Finally, the trackers will miss the target while there is a drastic change in appearance.

Spatio-temporal context (STC) is one of the most important information of the moving object in videos and is useful while tracking this target. The STC-based tracking algorithm proposed in [1] achieved good performance on real-time tracking. The tracker runs at 350 FPS at an i7 machine. However, it will always fail while there is large change in appearance such as heavy occlusion, illumination changes, and pose variation, as show in Fig. 1. Because the model we

© Springer Nature Singapore Pte Ltd. 2017
J. Yang et al. (Eds.): CCCV 2017, Part I, CCIS 771, pp. 128–144, 2017.
https://doi.org/10.1007/978-981-10-7299-4_11

learn that needs to depend strongly on tracked object's spatial layout. However, color histogram has robustness to variation in shape.

In this paper, we propose a novel and fast tracking algorithm, concerning both the deformations and the color changes of the tracking target. Combination of two kinds of cost function from different models in a dense translation search significantly improves the adaptability to the change in surroundings and tracking stability. Meanwhile, the efficiency is acceptable.

2   Related Work

Tracking by Detection. A lot of previous works have addressed the problem of objects tracking. Motion segmentation [8,9,14,19,20,22,30] plays a very important role in some early works. However, motion segmentation goes wrong if camera is static, and these traditional algorithms can't distinguish the object category. Hence, tracking by detection framework recently becomes more and more popular along with the improvement of object detection. The framework can overcome the weakness of motion segmentation mentioned above. The major process for tracking by detection framework includes global data association and SMC. The global data association depends on image information from the following frames and history images. For example, Nevatia [13] used a human detector to suit greedy data with a Bayesian framework. STC is also a tracking algorithm with Bayesian framework. Huang *et al.* [11] continued to develop this method by a model of 3 levels. After that, Li *et al.* [31] extended Huang's approach, and they choose some important features with nonparametric models to train data. And afterwards, like Leibe *et al.* and Zhang *et al.* kept it going. Eventually, the tracking algorithms are improved via these global data association. It cannot be online tracking for moving targets with this global data association. SMC can achieve a goal that is real time tracking because the algorithm needn't future frames. They only need to apply some effective strategies of estimation from previous images to gain right associations. The more features obtained, the more precision achieved. Particle filter is proposed as a classical SMC. However, sometimes, detection will be false, and multiple calibration cameras [15] and 3-D depth estimation is necessary. Andriluka *et al.* proposed a novel human detection to conquer the pedestrians of tracking. Kuo *et al.* [21] designed an algorithm to learn appearance model from every interaction target. Many particle filter frameworks do not detect of final sparse to obtain enough robustness. These methods aforementioned that can locate people by a nice pedestrian detector, then estimated the object position by a particle filter.

Reducing Model Drift. Grabner [40] proposed a semi-supervised online algorithm based on online Boosting which updates classifiers via combining label samples and unlabel ones for online tracking. To some extent, semi-supervised can reduce model drift. Meanwhile, boosting classifier is able to adapt changing of appearance after obtaining context prior. Babenko *et al.* presented an online multiple instance learning (MIL) tracking algorithm in [32], by updating

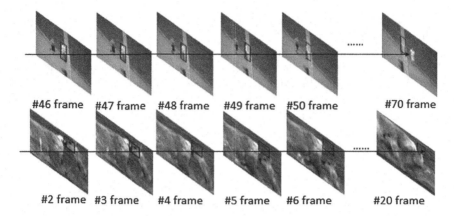

Fig. 1. Failure cases of STC tracker proposed in [1]. Top (occlusion): tracking single-object (a white T-shirt woman) (Jogging), using a red rectangle; top left: in #46 frame STC tracker can track the woman; but in #50 frame the tracking model drifts and failed (#70); Bottom (pose variation): STC tracker tracks a deer, but the tracker missed the target due to deformation and pose variation.

classifier from positive samples that are comprised current tracking target surrounding area of interest. It employed gradient boosting based on probability model and jointing multiple instance learning framework. The algorithm aforementioned can reduce loss due to false tracking and adds a probability of samples belong to target. Wang [28] pointed a concept of target's purity that means target reduce interfere because of background by using online appearance model to handle noise. The appearance model is constructed by Bayesian classifier and the positive sample obtained from the area of surrounding target. And negative samples obtained from the area of surrounding target. STC also has the problem of model drift because the tracker can't be always correct.

Long Term Tracking. Kalal proposed a long-term tracking algorithm named Tracking-Learning-Detection (TLD) [2,16,23,27] by combining online detection and tracking. The tracker can re-detection the object when missed target appeared. The tracker and detector can run in parallel of TLD algorithm. And the TLD obtains final result by confidence depending on detector and tracker. The tracker applies a median optical flow [38] of consistency forward-backward estimator. And detector update by employing a random ferns classifier with P-N learning algorithm. The TLD selects unlabel samples from the previous moving trajectory.

Robustness to Deformation. Spatio-temporal context (STC) is the very important feature for measuring the deformation of moving objects in videos. STC-based tracking algorithm has a number of advantages, especially in time-save running. However, the tracker is easy to fail when shape deformation occurs. Color histogram (the LBP also has the same property) has robustness to deformation because color histogram feature has nothing to do with the position of pixels. However, when it comes to tracking mission, color histogram is not sufficient when it suffers similarity between the moving target and background (for example, the target is red and background also is red). Hence, we consider combining the STC feature with color histogram to develop a more robust algorithm for moving object tracking. Color histogram feature is intuitional in computer vision applications and is easy to extract. Simultaneously, it is very effective in object tracking tasks [12]. The difficulty in pursuing deformation features is how to learn a suitable model for deformation. There is a better way to fix tracking deformation problem even the model is hard to build. PixelTrack [25] presented an accumulate voting strategy instead of deformable model, although it hasn't been demonstrated yet in real scene.

3   The Proposed

In the current frame t, object location is given via rectangle R_t, where (Eq. 1) tracking window R_t is chosen from a set S_t to maximize the C function.

$$R_t = \arg \max_{R \in S_t} C(M(t), P_t). \tag{1}$$

The function M is a cost value of our algorithm. C represents transformation between confidence value and parameters P in order to obtain optimum rectangular window. Where

$$P_t = \arg \min_{p \in \Omega} L(p, X_t) + \alpha R_l(p). \tag{2}$$

The loss function $L(x)$ needs to be small and is related to previous images and the object's location X_t, and $X_t = (r_i, R_i)_{i=1}^t$ is a regularization term with weight α to prevent over-fitting. In order to combine spatio-temporal context feature with color histogram which are extracted from two different models, the tracking problem need to be figured out in a different way. For dealing with the problems, making it easier, we address a confidence function of a linear combination of STC (M_2) and histogram scores (M_1).

$$C_{t+1}(x) = \gamma_1 M_1(x) + \gamma_2 M_2(x) \tag{3}$$

Depending on 3-channel feature image (i.e. RGB feature) φ_r, histogram score can be calculated by

$$M_1(\varphi, h) = h^T (\frac{1}{|\triangle|} \sum_{e \in \triangle} \varphi [e]). \tag{4}$$

where e is a location averaged from the image patches of possible regions. And h is a histogram weight vector with 3-scale. $\triangle \in \mathbb{Z}^2$ is of finite grid. We transformed $h^T \varphi [e]$ into $\varsigma_{(h,\varphi)} [e]$ and then obtained

$$M_1(\varphi, h) = \frac{1}{|\triangle|} \sum_{e \in \triangle} \varsigma(h, \varphi) [e]. \tag{5}$$

After color histogram feature calculation, spatio-temporal context (STC) is also computed efficiently via a nice scheme for real-time tracking. In the current frame, given the target location X^* (the first frame object location is given), the object location likelihood can be computed:

$$
\begin{aligned}
M(x) &= P(X \mid o) \\
&= \sum_{c(z) \in X^c} P(X, c(z) \mid o) \\
&= \sum_{c(z) \in X^c} P(X \mid (c(z), o)) P(c(z) \mid o).
\end{aligned}
\tag{6}
$$

where X^c (context feature) is setted $X^c = c(z) = (I(z), z) \mid z \in \Omega_c(X^*)$, and I(z) is defined as image intensity. $\Omega_c(X^*)$ is the neighborhood of the location. The spatial relationship between the target location and context information is connected by conditional probability $P(X, c(z) \mid o)$, then we deploy the total probability formula to transform it which turned into appearance of the local information context prior probability $P(c(z) \mid o)$ and spatial information $P(X \mid c(z), o)$.

3.1 Loss Function

In the current frame, the loss function is defined by summation of previous images' losses with a weighted ω_t,

$$
L(P, x_n) = \sum_{t=1}^{n} \omega_t l(x_t, R_t, P).
\tag{7}
$$

where loss function $l(x, R, P)$ is defined as

$$
l(x, R, P) = \min \|R - H\|_2^2.
\tag{8}
$$

R is optimum rectangle window, and $H = \arg \max_{R \in S_t} (M, P))$. We consider to apply Struck algorithm with SVM [18] to optimize Eq. (7). However, it will be expensive to deal with the problem using Struck algorithm. Eventually, we solve the equation and have

$$
h_t = \arg \min_{h} (L_c(h, X_t) + \frac{1}{2} \|h\|^2).
\tag{9}
$$

Now, there are two score functions M_1, M_2, and we have two parameters γ_1, γ_2, where $\gamma_1 + \gamma_2 = 1$. The specific value can be achieved through many trials. Under ideal conditions, the exactly moving target will make the function $C(x) = 1$. Corresponding, a higher probability with combination of confidence map and histogram score is indispensable.

3.2 STC Tracking Model

Spatial Context Model. The conditional probability $P(X \mid c(z), o)$ in Eq. (6) can be denoted as

$$
P(X \mid c(z), o) = f^{sc}(X - z)
\tag{10}
$$

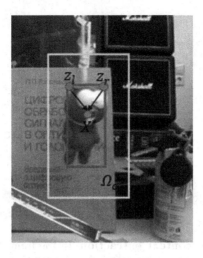

Fig. 2. Lemming model

where $f^{sc}(x)$ is a function. Independent variable X indicates the object location (the center of the target rectangle window). And its surrounding location is marked as z. Hence the function $f^{sc}(x)$ denotes a spatial relationship between an object and its spatial information. Noticed that $f^{sc}(x)$ is not a function of radially symmetric (as shown in Fig. 2) (i.e., $f^{sc}(X - z) \neq f^{sc}(|X - z|)$), this method can fix ambiguities at a certain extent. Normally, a tracker tracks a moving object (the upper left) in Fig. 2 using its appearance (z_l), and will make a mistake by using z_r because of their similar spatial relationship. Based on these analyses, we combined this spatial relationship and the color histogram features to develop a new tracking scheme. Even the circumstance of Fig. 2 was appeared in the tracking video, we can distinguish it.

Context Prior Model. The context prior probability function of Eq. (6) is defined as

$$P(c(z) \,|\, o) = I(z)(z - X^*). \tag{11}$$

where $I(x)$ is the image intensity (gray value). And we thought the image feature we extract is too simple. Firstly, we prepare to use LBP feature, but we discover it reduce speed after experiment. And $\omega(x)$ is a weighted function

$$\omega(x) = ae^{-\frac{|X|^2}{\sigma^2}} \tag{12}$$

a is a constant which is restricted set [0,1] of $P(c(z) \,|\, o)$, and σ is a parameter of scale. The Eq. (12), looks like a kernel density estimate function, we are inspired by this. The closer the context location z is to the target X^* (the rectangle center of tracking), the more important it is to tracking target of next frame, meantime, it gets higher weight (i.e., the closer to center of object is, the more important it is).

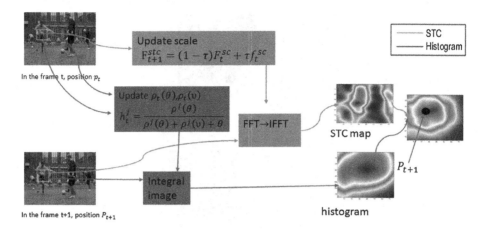

Fig. 3. Spatio-temporal context extraction. In the frame t, the target (red ball) was obtained (indicated using a red rectangle). Frame t was used as a pre-image. Then spatial weighted function (Eq. (12)) can achieve features for searching in a region (yellow rectangle) from the context prior model (Eq. (11)). And Eq. (14) can be transformed to the frequency domain for fast tracking. Finally, inverse transformation was conducted and the tracking target was estimated through updating the STC model. **Color histogram computation**. In the frame t, $\rho_t(\nu)$ and $\rho_t(\theta)$ was updated using Eq. (23) and color histogram h_t can be calculated using ridge regression. The moving targets new location was estimated again through the histogram results. However, the tracking locations have large errors either using STC map or histogram because of similarity between the background and the foreground (a red ball). Targets location has a significant correction via combining these two kinds of information. It demonstrated that our tracker has a better performance. (Color figure online)

Confidence Map Model. The confidence map of a target is defined as

$$M(X) = P(X \mid o) = be^{-\left|\frac{X-X^*}{\alpha}\right|^{\beta}}. \tag{13}$$

where b is a limit coefficient, α is a scale parameter, β is a shape parameter of spatio-temporal context. The target location ambiguity is a classical problem for visual tracking. We have learned from many examples that the closer to tracking object, the larger probability of ambiguity appears. As the STC, when is large, the confidence map value is away to the center of the tracking target. When β is too small, the fewer positions are learned. Finally, we chosen a balanced value $\beta = 1$.

Learning Spatial Context Model. Depending on the previous functions (Eqs. 11 and 13), we can have a new function $M(x)$ as

$$\begin{aligned}
M(X) &= be^{-\left|\frac{X-X^*}{\alpha}\right|^{\beta}} \\
&= \sum_{z \in \Omega_c} (X^*) f^{sc}(X-z) I(z) \omega(z-X^*) \\
&= f^{sc}(X) \otimes (I(X)\omega(X-X^*)).
\end{aligned} \tag{14}$$

Symbol $\otimes$ is a convolution operator. The function (14) can be transformed to the frequency domain for fast tracking, using the Fast Fourier Transform (FFT).

$$f^{sc} = \mathbf{F}^{-1}(\frac{\mathbf{F}(be^{-\left|\frac{X-X^*}{\alpha}\right|^{\beta}})}{\mathbf{F}(I(X)\omega(X-X^*))}).\qquad(15)$$

Essentially, the tracking problem is a fast version of detection, while the first frame is given or employing the state-of-the-art detector [17]. In our experiment, the first location of moving object was giving previously. In the frame t, a $f_t^{sc}(X)$ from Eq. 15 is obtained by learning. Then the STC model can be updated to $\mathbf{F}_{t+1}^{stc}$ using $f_t^{sc}(X)$. Next, our algorithm estimate the targets new location in the next frame (t+1). In (t+1) frame, we extract useful features using spatio-temporal context and color histogram and scan the local location in Ω_c. The targets new location X_{t+1}^* is calculated via maximizing confidence map function value given by

$$X_{t+1}^* = \arg\max_{X \in \Omega_c}(X_t^*)M_{t+1}(X).\qquad(16)$$

where

$$M_{t+1} = \mathbf{F}^{-1}(\mathbf{F}(\mathbf{F}_{t+1}^{stc}(X)) \circ \mathbf{F}(I_{t+1}(X)\omega(X-X^*))).\qquad(17)$$

and $\circ$ is the Hadamard product.

The STC model can be updated by using

$$F_{t+1}^{stc}(X) = (1-\tau)F_t^{sc} + \tau f_t^{sc}.\qquad(18)$$

Update of Scale. While tracking moving objects in video, the new location is obtained by maximizing the confidence map introduced in Eq. (17), STC tracking model is updated simultaneously. Hence, the parameters like σ and ω should be updated to fit into new model. Parameters can be updated based on Algorithm 1 introduced as follow:

Algorithm 1. *scale updating on a sequence with K frames*
for k=1 to K do,
1: calculate $M_t(X_t^*)$, and $M_{t-1}(X_{t-1}^*)$
2: $M_t(X_t^*)$ divide by $M_{t-1}(X_{t-1}^*)$, and let $s_t' \leftarrow \sqrt{\frac{M_t(X_t^*)}{M_{t-1}(X_{t-1}^*)}}$
3: Let $S_t \leftarrow \frac{1}{n}\sum_{i=1}^{n} s_{t-i}'$
4: update s by using $S_{t+1} = (1-\varepsilon)S_t + \varepsilon\bar{S}_t$
5: $\sigma_{t+1} \leftarrow s_t\sigma_t$
end for

$M_t(X_t^*)$ is the confidence map function, s_t' is the scale of estimated confidence between two adjacent frames. In order to reduce the influence of noise, s_{t+1} is obtained through filtering from n consecutive images.

3.3   The Color Histogram Score Function

Transform of Function. The color histogram score can be calculated easily and fast in the searching rectangle. Local binary pattern is an alternative feature to achieve similar tracking results, however, it will increase the complexity of computation significantly. The loss function of current frame can be determined by

$$l_{hist}(x, R, h) = \sum_{(q,y) \in \nabla} \left\| h^T (\sum_{e \in \triangle} \varphi_{M_1(x,q)}[e]) - y \right\|^2. \tag{19}$$

∇ is a set of pairs (q, y), where q is the selected rectangle and y defines the corresponding object. φ is the transformation which requires a lot of computational resource for the 3-channel color histogram feature. Eventually we propose another way to define $\varphi(x) = e_{k(x)}$, where e_i is a Dirac gamma matrix. Then, we use PLT method [29] to calculate $h^T \varphi[e] = h^{k[e]}$. Linear regression was employed to feature extraction for foreground and background regions $\vartheta \in \mathbb{Z}^2$ and $\nu \in \mathbb{Z}^2$ respectively. Then the loss function can be divided into two parts,

$$l_{hist}(x, R, h) = \frac{1}{|\vartheta|} \sum_{e \in \vartheta} (h^T \varphi[e] - 1)^2 + \frac{1}{|\nu|} \sum_{c \in \nu} (h^T \varphi[e])^2. \tag{20}$$

Furthermore, we can turn it into

$$l_{hist}(x, R, h) = \sum_{j=1}^{S} [\frac{N^j(\vartheta)}{|\vartheta|} (h^j - 1)^2 + \frac{N^j(\nu)}{|\nu|} (h^j)^2]. \tag{21}$$

where $N^j(\vartheta) = |(e \in \vartheta, k[e] = j)|$ is the number of pixels in the region ϑ, and $N^j(\nu) = |(e \in \nu, k[e] = j)|$. Ridge regression can be used to solve this equation and we have

$$h_t^j = \frac{\rho^j(\vartheta)}{\rho^j(\vartheta) + \rho^j(\nu) + \vartheta}. \tag{22}$$

Update of Parameters. Inspired by the introduction in [34], parameters $\rho_t(\vartheta)$ and $\rho_t(\nu)$ in Eq. (22) can be determined by

$$\begin{cases} \rho_t(\vartheta) = (1 - \eta)\rho_{t-1}(\vartheta) + \eta \rho_t^{'}(\vartheta) \\ \rho_t(\nu) = (1 - \eta)\rho_{t-1}(\nu) + \eta \rho_t^{'}(\nu) \end{cases}. \tag{23}$$

For a S-channel feature image, $\rho_t(G)$ is the vector of ρ_t^j, where j = 1S. And η is 0.04, while the nbins of color histograms is $32 \times 32 \times 32$.

Integral Color Histogram. The histogram score function can be calculated using integral image, as show in Fig. 3. We choose multiple possible targets from many rectangular regions. The histograms are extracted from each patch. And we compare the histograms which are calculated from these regions. At the point (x, y) of the image, the integral image holds the sum of all the pixels included the rectangular region defined by the top-left corner of the image and

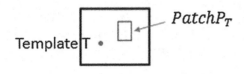

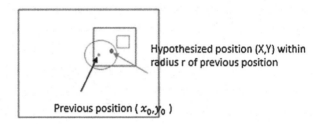

Fig. 4. Template patch P_T and the corresponding image patch P_I, for a hypothesized position

the point (x, y). We can achieve the statistical features from all the pixels of the rectangular regions, including the 4 points of the corners. We construct every bin of the histogram from the integral image by counting the number of pixels which fall into the bin. After that, we can obtain the number of pixels in the region we are interested. Finally, the color histogram of the rectangle window was achieved. Once we get the histogram data, we can compute the color histogram easily, and we evaluate a hypothesis on the current object's position.

Given a target X from the previous frame, we want to locate it in the current frame. X can be represented by a template image T. Our mission is to determine the position and the scale of X in the current frame. When we start tracking, based on the prior knowledge about position and scale from the previous, we can search the new position in the neighborhood.

We assume that the target location is (x_0, y_0) in the previous frame, and the searching radius is r. The rectangular patch is defined as $P_T = (dx, dy, x_h, y_w)$ in the template, as show in Fig. 4. Its center is substituted (dx, dy) by the template center. And x_h, y_w are half height and width respectively. The object's position is (x, y) in the current frame. Then the image patch P_T owns a corresponding patch P_I, and its center is $(x + dx, y + dy)$, and x_h, y_w are half height and width. The correspondence is shown in Fig. 4.

3.4 The Combination

Finally, the object is tracked via a confidence map and histogram score function. In the frame $t + 1$, the center of the rectangular window is determined by Eq. (24) from the previous frame.

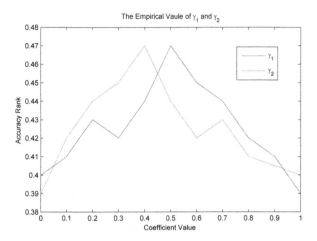

Fig. 5. The empirical value of γ_1 and γ_2

$$C_{t+1}(x) = \gamma_1 M_1(x) + \gamma_2 M_2(x)$$
$$= \gamma_1 \frac{1}{|\triangle|} \sum_{e \in \triangle} \varsigma(h, \varphi)[e] + \gamma_2 \boldsymbol{F}^{-1}((\boldsymbol{F}_{t+1}^{stc}(x)) \circ \boldsymbol{F}(I_{t+1}(x)\omega(x - X^*))). \tag{24}$$

where γ_1 and γ_2 are the combination coefficients. We recommend empirical values of γ_1 and γ_2 from lots of experiments, as shown in Fig. 5. In our experiments, we choose $\gamma_1 = \gamma_2 = 0.5$. The complete tracking algorithm is defined as follows in Algorithm 2.

Algorithm 2. *The proposed tracking method*
1: receive histogram score $M_1(x)$
2: develop confidence map $M_2(x) \leftarrow P(X|o)$
3: update parameters using Eq. (23)
4: calculate spatio-temporal context using Eq. (18)
5: estimate the target's position from the combined equation Eq. (24)
6: end

3.5   Search Paradigm

When the moving object is located in the current frame, our tracker need to estimate the target's new location in the next frame. In consideration of the camera's high frame rate and the movement distance between adjacent frames, we just need to search in a small region around the pre-existing tracking window instead of the whole frame. In order to have a fair comparison in tracking speed, according to [1], we double the size of the rectangular window as a search region in the current frame. And our tracker should update some parameters. Meanwhile, the scale needs to be changed by STC scenario (Algorithm 1). Lots of information about previous frames is useful in updating these parameters.

Actually, considering both the deformation area and texture preservation, our tracker combines the STC feature with color histogram to estimate new central point of the tracking window in the next frame. Spatio-temporal context is an important feature in describing the deformation. STC-based tracker has good performance in moving object tracking, especially in tracking speed. However, STC-based tracker always fails when there is severe occlusion or appearance change. Color histogram can make up these disadvantages because it is irrelevant to the deformation of the moving object. Similarly, histogram calculation is fast and our tracker can achieve object tracking in time.

4   Evaluation

We evaluate our algorithm using 11 video sequences (and those videos are with complex and challenged scenes) to challenge heavy occlusion, severe deformation and motion blur with two representative state-of-the-art trackers (TLD, STC and). For gaining a fair evaluation, we run each algorithm on a same i5 machine. We analyze success rate (SR) (%), running time, failures, and average frame per second (FPS). There is a standard to evaluate our algorithm, except that, we also have rules to judge our tracker from running time and robustness. The tracking miss occurs when the tracking area (like a head of a person) drifts 50%.

4.1   Experiment Results Analysis

Our tracking algorithm is based on spatio-temporal context (STC), so we compare it with a STC-based tracker [1] and a pretty tracking algorithm TLD [2]. Locally orderless tracking (LOT) [10] and LIT [39] are also compared because of their good performance in tracking. The SR and FPS are calculated and analyzed. Each sequence of the dataset has labeled ground truth. Success rate is defined as $socre^k = \frac{area(ROI_c \cap ROI_g)}{area(ROI_c \cup ROI_g)}$, where ROI_g is a ground-truth bounding box and ROI_c is an estimated bounding box by a tracker. Our algorithm will catch the target when $socre^k > 0.5$ in the k frame. Without doubt, Table 1 indicates that our tracker has a better performance in accuracy of tracking. Although the calculating speed is lower than STC method, the efficiency of our algorithm is satisfactory. One other thing to note is that the color bins used in histogram calculation is $32 \times 32 \times 32$. The efficiency can be improved while we use $16 \times 16 \times 16$. STC-based tracking method achieved 210 FPS in average. However, the accuracy is decreased significantly while facing heavy occlusion and pose variation situations, as shown in Fig. 1. TLD method is also a nice tracking algorithm though it is not so well if it suffers complex situations. Its average success rate (SR) is about 62.8%, lower than our tracking algorithm.

4.2   Robustness Analysis

Tracking problem is never an easy work, and there exist many factors like illumination, occlusion, deformation and abrupt motion, which will influence the

Table 1. Success rate (SR), Frame per second (FPS), and the number of failure (failures). The total number we evaluated is 3532. The results in boldface indicate the best.

Sequence	STC	TLD	Our tracker	LOT	LIT
Ironman	20.0	**30.0**	20.0	6.0	6.0
Matrix	10.0	12.0	**34.0**	0	5
Ball1	0	9.5	**100**	61.0	19.0
David3	**100**	80.0	**100**	98.0	79.4
Deer	11.3	10.0	**100**	9.9	11.3
Football	27.6	13.8	74.6	16.6	13.8
Jumping	**100**	9.6	31.9	**100**	**100**
Skiing	12.3	10.0	**14.8**	7.4	12.3
FaceOcc2	**100**	**100**	**100**	**100**	**100**
Lemming	**90**	83.0	72.6	52.4	44.9
David	**100**	**100**	90	20	26.0
Failures	1100	1314	**800**	1597	1679
Average SR	68.9	62.8	**77.3**	54.8	52.5
Average FPS	**210**	90	40	0.5	0.7

accuracy or even cause tracking failure. Our proposed tracker runs at 40 FPS on an i5 machine with 8GB RAM. Relatively, STC had strong robustness, in which is proved by running FaceOcc2 sequence that is partially occluded. However, STC is unworkable if it meets heavy occlusion (see #61 of the Jogging sequence in Fig. 1). The algorithm we proposed can track the target well in this Jogging sequence. The Deer sequence shown in Fig. 6 includes a moving object with high speed. We received a bad feedback when apply the STC algorithm to track the deer (see #30 and #60 of Deer sequence in Fig. 6).

From Table 1, We can see that our tracker achieves satisfactory performance. Comparing with TLD and STC method, our tracker requires more processor time. However, our tracking result is more encouraging and it can achieve object tracking in real time. LOT [10] and LIT [39] tracking methods receive acceptable performance in average SR. However, they are outside the range of real-time trackers.

More specifically, our proposed tracking framework can track moving object even the target has high speed and severe pose variation. We found out that STC tracker proposed in [1] almost fail in situations when abrupt motion or similarity between background and foreground (area of tracking) exists. For example, the ball1 sequence shown in Fig. 6 contains a moving object (red ball). And the texture of its background (the red wall) is similar with this target. The STC method cant locate the ball fundamentally (see #5, #30, and #60 frame in Fig. 6). The proposed tracking algorithm shows a pretty performance that the moving object is tracked correctly in each frame. Our tracker is more robust than

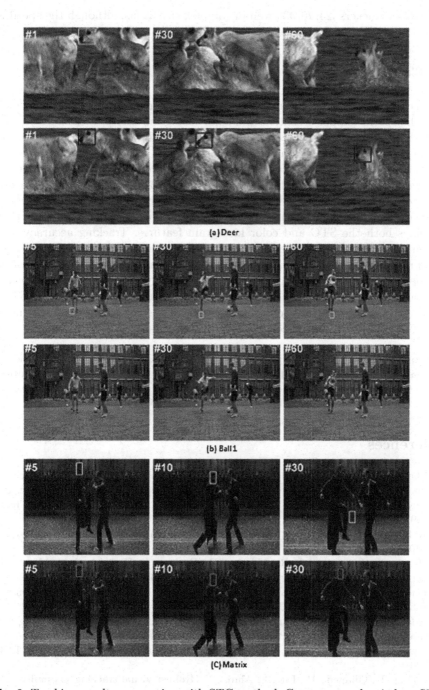

Fig. 6. Tracking results, comparing with STC method. Green rectangle window: STC method. Red rectangle window: our tracker (Color figure online)

some other trackers [24,26,33] even in complex situations, although the speed is lower than STC-based method in [1] and TLD in [2].

5   Conclusion

In this paper, we present a novel real time tracking method, which concerns both the spatio-temporal context and color histogram information. STC framework shows a nice performance to illumination and blur of motions. Particularly, tracking with STC features has a high speed which achieved 210 PFS in an i5 machine. However, it is hard to adapt to the situations of deformation and heavy occlusion. On the other hand, histogram features have nothing to do with the pixels' spatial relationship in the frame, which means it isn't sensitive to moving object's deformation. Hence, we proposed a simple tracking framework that combines both the STC and color histogram features. Tracking accuracy and computational complexity are the two main challenges in object tracking issue. Our proposed tracker earns a significant performance, even though it suffers complex situations like shape deformation and heavy occlusion. We also notice that our tracker is still a real-time algorithm although it is slower than STC-based methods. Finally, we would like to say that the color histogram scheme is a very useful information which can't be ignored.

Acknowledgments. This project was supported by the National Natural Science Foundation of China (61203249, 61471228), Guangdong Natural Science Foundation (2015A030313439) and Foundation of Key Laboratory of Machine Intelligence and Advanced Computing of the Ministry of Education (MSC-201601A).

References

1. Zhang, K., Zhang, L., Liu, Q., Zhang, D., Yang, M.-H.: Fast visual tracking via dense spatio-temporal context learning. In: Fleet, D., Pajdla, T., Schiele, B., Tuytelaars, T. (eds.) ECCV 2014. LNCS, vol. 8693, pp. 127–141. Springer, Cham (2014). https://doi.org/10.1007/978-3-319-10602-1_9
2. Kalal, Z., Mates, J., Mikolajczyk, K.: Tracking-learning-detection. TPAMI **34**(7), 1409–1422 (2012)
3. Henriques, J.F., Caseiro, R., Martins, P., Batista, J.: High-speed tracking with kernelized correlation filters. TPAMI **37**(3), 583–596 (2015)
4. Danelljan, M., Häger, G., Khan, F.S., Felsberg, M.: Accurate scale estimation for robust visual tracking. In: BMVC (2014)
5. Adam, A., Rivlin, E., Shimshoni, I.: Robust fragments-based tracking using the integral histogram. In: CVPR, pp. 798–805 (2006)
6. Kwon, J., Lee, K.M.: Visual tracking decomposition. In: CVPR, pp. 1269–1276 (2010)
7. Zhang, T., Ghanem, B., Liu, S., Ahuja, N.: Robust visual tracking via multi-task sparse learning. In: CVPR, pp. 2042–2049 (2012)
8. Qu, W., Schonfeld, D.: Real-time decentralized articulated motion analysis and object tracking from videos. TIP **16**(8), 2129–2137 (2007)

9. Isard, M., MacCormick, J.: BraMBLe: a Bayesian multiple-blob tracker. In: ICCV, pp. 34–41 (2001)
10. Oron, S., Bar-Hillel, A., Levi, D., Avidan, S.: Locally orderless tracking. In: CVPR, pp. 1940–1947 (2012)
11. Huang, C., Wu, B., Nevatia, R.: Robust object tracking by hierarchical association of detection responses. In: Forsyth, D., Torr, P., Zisserman, A. (eds.) ECCV 2008. LNCS, vol. 5303, pp. 788–801. Springer, Heidelberg (2008). https://doi.org/10.1007/978-3-540-88688-4_58
12. Possegger, H., Mauthner, T., Bischof, H.: In defense of color-based model-free tracking. In: CVPR, pp. 2113–2120 (2015)
13. Wu, B., Nevatia, R.: Detection and tracking of multiple, partially occluded humans by Bayesian combination of edgelet based part detectors. IJCV **75**, 247–266 (2007)
14. Poznyak, A.S., Yu, W., Sánchez, E.N., Përez, J.P.: Nonlinear adaptive trajectory tracking using dynamic neural networks. TNN **10**(6), 1402–1411 (1999)
15. Kim, K., Davis, L.S.: Multi-camera tracking and segmentation of occluded people on ground plane using search-guided particle filtering. In: Leonardis, A., Bischof, H., Pinz, A. (eds.) ECCV 2006. LNCS, vol. 3953, pp. 98–109. Springer, Heidelberg (2006). https://doi.org/10.1007/11744078_8
16. Kalal, Z., Matas, J., Mikolajczyk, K.: Online learning of robust object detectors during unstable tracking. In: ICCV, pp. 1417–1424 (2009)
17. Dollar, P., Appel, R., Belongie, S., Perona, P.: Fast feature pyramids for object detection. TPAMI **36**(8), 1532–1545 (2014)
18. Solera, F., Calderara, S., Cucchiara, R.: Learning to divide and conquer for online multi-target tracking. In: ICCV, pp. 4373–4381 (2015)
19. Zhao, T., Nevatia, R.: Tracking multiple humans in crowded environment. In: CVPR, pp. 406–413 (2004)
20. Yang, F., Paindavoine, M.: Implementation of an RBF neural network on embedded systems: real-time face tracking and identity verification. TNN **14**(5), 1162–1175 (2003)
21. Kuo, C.H., Huang, C., Nevatia, R.: Multi-target tracking by on-line learned discriminative appearance models. In: CVPR, pp. 685–692 (2010)
22. Wang, H., Suter, D., Schindler, K., Shen, C.: Adaptive object tracking based on an effective appearance filter. TPAMI **29**(9), 1661–1667 (2007)
23. Kalal, Z., Matas, J., Mikolajczyk, K.: P-N learning: bootstrapping binary classifiers by structural constraints. In: ICCV, pp. 49–56 (2010)
24. Collins, R.T.: Mean-shift blob tracking through scale space. In: CVPR, pp. 234–240 (2006)
25. Duffner, S., Garcia, C.: PixelTrack: a fast-adaptive algorithm for tracking non-rigid objects. In: ICCV, pp. 2480–2487 (2013)
26. Zhang, T., Ghanem, B., Liu, S., Ahuja, N.: Robust visual tracking via multi-task sparse learning. In: CVPR, pp. 2042–2049 (2012)
27. Kalal, Z., Mikolajczyk, K., Matas, J.: Face-TLD: tracking-learning-detection applied to faces. In: ICIP, pp. 3789–3792 (2010)
28. Wang, P., Qiao, H.: Online appearance model learning and generation for adaptive visual tracking. Trans. Circ. Syst. Video Technol. **21**(2), 156–169 (2011)
29. Kristan, M., Pflugfelder, R., Leonardis, A., Matas, J., Porikli, F., Cehovin, L., Nebehay, G., Fernandez, G., Vojir, T., Gatt, A., et al.: The visual object tracking VOT2013 challenge results. In: ICCVW, pp. 98–111 (2013)

30. Song, X., Cui, J., Zha, H., Zhao, H.: Vision-based multiple interacting targets tracking via on-line supervised learning. In: Forsyth, D., Torr, P., Zisserman, A. (eds.) ECCV 2008. LNCS, vol. 5304, pp. 642–655. Springer, Heidelberg (2008). https://doi.org/10.1007/978-3-540-88690-7_48
31. Li, Y., Huang, C., Nevatia, R.: Learning to associate: HybridBoosted multi-target tracker for crowded scene. In: CVPR, pp. 2953–2960 (2009)
32. Viola, P., Platt, J.C., Zhang, C.: Multiple instance boosting for object detection. In: Advances in Neural Information Processing Systems, pp. 1417–1424 (2005)
33. Dinh, T.B., Vo, N., Medioni, G.: Context tracker: exploring supporters and distracters in unconstrained environments. In: CVPR, pp. 1177–1184 (2011)
34. Bertinetto, L., Valmadre, J., Golodetz, S.: Staple: complementary learners for real-time tracking. In: ICCV, pp. 1401–1409 (2016)
35. Hare, S., Saffari, A., Torr, P.H.: Struck: structured output tracking with kernels. In: ICCV, pp. 263–270 (2011)
36. Zhang, K., Zhang, L., Yang, M.-H.: Real-time compressive tracking. In: Fitzgibbon, A., Lazebnik, S., Perona, P., Sato, Y., Schmid, C. (eds.) ECCV 2012. LNCS, vol. 7574, pp. 864–877. Springer, Heidelberg (2012). https://doi.org/10.1007/978-3-642-33712-3_62
37. Ross, D., Yuan, J., Yang, M.-H.: Incremental learning for robust visual tracking. IJCV **77**, 125–141 (2008)
38. Kalal, Z., K. Mikolajcayk, K., Matas, J.: Forward-backward error: automatic detection of tracking. In: ICPR, pp. 2756–2759 (2010)
39. Mei, X., Ling, H.: Robust visual tracking and vehicle classification via sparse representation. PAMI **33**(11), 2259–2272 (2011)
40. Grabner, H., Leistner, C., Bischof, H.: Semi-supervised on-line boosting for robust tracking. In: Forsyth, D., Torr, P., Zisserman, A. (eds.) ECCV 2008. LNCS, vol. 5302, pp. 234–247. Springer, Heidelberg (2008). https://doi.org/10.1007/978-3-540-88682-2_19

A New Visibility Model for Surface Reconstruction

Yang Zhou[1,2], Shuhan Shen[1,2(✉)], and Zhanyi Hu[1,2]

[1] NLPR, Institute of Automation, Chinese Academy of Sciences,
Beijing 100190, People's Republic of China
{yang.zhou,shshen,huzy}@nlpr.ia.ac.cn
[2] University of Chinese Academy of Sciences,
Beijing 100049, People's Republic of China

Abstract. In this paper, we propose a new visibility model for scene reconstruction. To yield out surface meshes with enough scene details, we introduce a new visibility model, which keeps the relaxed visibility constraints and takes the distribution of points into consideration, thus it is efficient for preserving details. To strengthen the robustness to noise, a new likelihood energy term is introduced to the binary labeling problem of Delaunay tetrahedra, and its implementation is presented. The experimental results show that our method performs well in terms of detail preservation.

Keywords: Surface reconstruction · Visibility · Scene detail

1 Introduction

Scene reconstruction from images is a fundamental problem in Computer Vision. It has many practical applications in entertainment industry, robotics, cultural heritage digitalization and geographic systems. For decades, it has been studied due to its low cost of data acquisition and various usages. In recent years, researchers have made tremendous progress in this field. For small objects under controlled conditions, current scene reconstruction methods could achieve great results [7,16]. However, when it comes to large scale scenes with multi-scale objects, they would have some problems with the completeness and accuracy, especially when dealing with scene fine details [18].

Scene details like small scale objects and object edges are essential part of scene surfaces. However, preserving scene details in multi-scale scenes has been a difficult problem. The existing surface reconstruction methods either ignore the scene details or rely on further refinement to restore them. This is because firstly, compared with noise, the supportive points in such areas are sparse, making it difficult to distinguish true surface points from false ones; and secondly, the models and associated parameters employed in existing methods are not particularly adept for large scale ranges, where scene details are usually compromised for overall accuracy and completeness. While the above first case seems unsolvable

© Springer Nature Singapore Pte Ltd. 2017
J. Yang et al. (Eds.): CCCV 2017, Part I, CCIS 771, pp. 145–156, 2017.
https://doi.org/10.1007/978-981-10-7299-4_12

due to the lack of sufficient information, we only focus on the second case in this work. In particular, we propose a new visibility model along with a new energy formulation for surface reconstruction. Our method follows the idea in [6,11], with the visibility model and energy formulation replaced with ours. To make our method adept for large scale ranges, we propose a new visibility model with adaptive ends of lines of sight; in the meantime, a new likelihood energy term is introduced to keep the proposed method robust for noise. Experimental result shows that our visibility model is efficient for preserving scene details.

2   Related Work

Recently, a variety of works have been done to advance 3D scene reconstruction. To deal with small objects, silhouette based methods [5,13] are proposed. The silhouettes provide proper bounds for the objects, which helps to reduce the computing cost and yield a good model for the scene. However, good silhouettes rely on effective image segmentation, which remains a difficult task so far. Volumetric methods like space carving [1,10,24], level sets [8,15] and volumetric graph cut [14,20,22] often yield good results for small objects. But the computational and memory costs increase rapidly as scene scale grows, which makes them unsuitable for large scale scene reconstruction.

For outdoor scenes, uncontrollable imaging conditions and multiple scale structure make it hard to reconstruct scene surfaces. A common process of large scale scene reconstruction is to generate the dense point cloud from calibrated images first, and then extract the scene surface. The dense point cloud can be generated by depth fusion method [2,17,19] or by feature expansion methods [3]. Once the dense point cloud is generated, scene surface can be reconstructed by Possion surface construction [9], by level-set like method [21] or by graph cut based methods [6,11,12]. Graph cut based methods are exploited most for its efficiency and ability of filtering noise. In these methods, Delaunay tetrahedra are generated from dense point cloud first, then a visibility model is applied to weight the facets in Delaunay tetrahedra, finally the binary labeling problem of tetrahedra is solved and the surface is extracted. The typical visibility model is introduced in [12]. The basic assumption of the visibility model in [12] is that the space between camera center and 3D point is free-space, while the space behind 3D point is full-space. Then this typical visibility model is promoted by a refined version in [11], namely soft-visibility, to cope with noisy point clouds.

In this paper, we follow the idea of [6,11] and exploit the visibility information in the binary labeling problem of Delaunay tetrahedra. The main contributions of our proposed method are the introductions of (a) a new visibility model and (b) a new energy term for efficient noise filtering. Experimental comparison results show that our method rivals the state-of-the-art methods [2,3,6,19] in terms of accuracy and completeness of scene reconstruction, but performs better in terms of detail preservation.

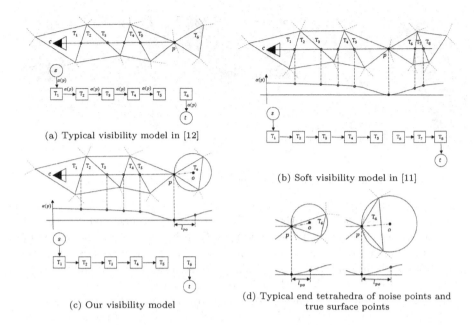

(a) Typical visibility model in [12]

(b) Soft visibility model in [11]

(c) Our visibility model

(d) Typical end tetrahedra of noise points and true surface points

Fig. 1. (a) Typical visibility model, (b) soft visibility model and (c) our visibility model in 2D. (d) From left to right: typical end tetrahedron (in 2D) of noise points and that of true surface points on densely sampled surfaces.

3   Visibility Models and Energies

The input of our method is a dense point cloud generated by MVS methods. Each point is attached with the visibility information recording that from which views the point is seen. Then, Delaunay tetrahedra are constructed, and the scene reconstruction problem is formulated as a binary labeling problem with a given visibility model. The total energy of the binary labeling problem is

$$E_{total} = E_{vis} + \lambda_{like}E_{like} + \lambda_{qual}E_{qual} \tag{1}$$

where λ_{like} and λ_{qual} are two constant balancing factors; E_{vis} and E_{like} are defined in Sects. 3.2 and 4.1; E_{qual} is the surface quality energy in [11]. By minimizing E_{total} with Maxflow/Mincut algorithm, the label assignment of Delaunay tetrahedra is yielded and a triangle mesh is extracted, which consists of triangles lying between tetrahedra with different labels. Further optional refinement can be applied as described in [23]. In the following subsections, we introduce the existing visibility models and our new one.

3.1   Existing Visibility Models

The typical visibility model [12] assumes that, for each line of sight v, the space between camera center c and point p is free-space, and that behind point p

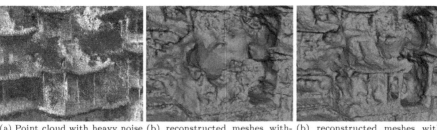

(a) Point cloud with heavy noise (b) reconstructed meshes with- (b) reconstructed meshes with
                                      out the likelihood energy   the likelihood energy

Fig. 2. Surface reconstruction without and with the likelihood energy.

is full-space. Thus, the tetrahedra $(T_1-T_5$ in Fig. 1(a)) intersected by segment (c, p) should be labeled as outside, and that $(T_6$ in Fig. 1(a)) right behind point p as inside. For a single line of sight, the above label assignment is desirable. While taking all lines of sight into consideration, some surface part might not be handled properly. This violates the label assignment principle described above. To quantize the conflicts, the facets intersected by lines of sight are given weights of $W(l_{T_i}, l_{T_j})$, and those for punishing bad label assignments of the first and the last tetrahedron are $D(l_{T_1})$ and $D(l_{T_{N_v+1}})$, respectively. Therefore, the visibility energy is the sum of the penalties of all the bad label assignments in all the lines of sight, as

$$E_{vis}^{typical} = \sum_{v \in \mathcal{V}} [D(l_{T_1}) + \sum_{i=1}^{N_v-1} W(l_{T_i}, l_{T_{i+1}}) + D(l_{T_{N_v+1}})] \qquad (2)$$

where if $l_{T_1} = 1$, $D(l_{T_1}) = \alpha(p)$; otherwise, $D(l_{T_1}) = 0$; if $l_{T_{N_v+1}} = 0$, $D(l_{T_{N_v+1}}) = \alpha(p)$; otherwise, $D(l_{T_{N_v+1}}) = 0$; if $l_{T_i} = 0 \wedge l_{T_j} = 1$, $W(l_{T_i}, l_{T_j}) = \alpha(p)$; otherwise, $W(l_{T_i}, l_{T_j}) = 0$; $\alpha(p)$ is the weight of point p, which is its photo-consistency score; $\mathcal{V}$ is the visibility information set; N_v is the number of tetrahedra intersected by the line of sight v, indexed from the camera center c to the point p; $N_v + 1$ denotes the tetrahedron behind the point p; l_T is the label of tetrahedron T, with 1 stands for inside and 0 for outside.

The typical visibility model described above is effective, but it has several flaws in dealing with dense and noisy point clouds. In such cases, the surfaces yielded by [12] tend to be overly complex, with bumps on and handles inside the model [11]. One possible solution is the soft visibility in [11], shown in Fig. 1(b). The basic assumptions of the two visibility models are similar. The differences are that in the soft visibility model, the edge weights are given a factor $(1 - e^{-d^2/2\sigma^2})$ according to the distance d between the point p and the intersecting point, and the end of the line of sight is shifted to a distance of $k\sigma$ along it.

3.2   Our Proposed Visibility Model

Though the soft visibility model is effective to filter noise points, it sometimes performs badly in preserving details, especially in a large scene containing small

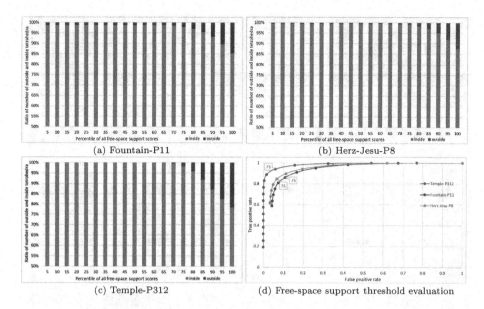

Fig. 3. This figure shows the ratios of number of outside and inside tetrahedra with different percentiles of all free-space support scores on (a) Fountain-P11 [18], (b) Herz-Jesu-P8 [18] and (c) Temple-P312 [16]. (d) The true positive rates and false positive rates of different free-space support threshold.

scale objects (see experimental results). According to our observations, this happens mainly because of the strong constraint imposed on the end of line of sight in the tetrahedron $k\sigma$ from the point p along the line of sight. It could be free-space even the point p is a true surface point in some cases.

To balance noise filtering and detail preserving, we propose a new visibility model shown in Fig. 1(c). In our visibility model, we keep the relaxed visibility constraints in the space between the camera center c and the point p, and set the end of line of sight in the tetrahedron right behind the point p. We observe that the end tetrahedra of noise points and true surface points have different shapes, as shown in Fig. 1(d). In addition, to determine the weight of the t-edge of the end tetrahedron, we compare the end tetrahedra of noisy points and true surface points on public datasets. Figure 1(d) shows a typical end tetrahedron of a noisy point and that of a true surface point on densely sampled surfaces in 2D space. Noise points tend to appear a bit away from true surface, which makes the end tetrahedra thin and long, and true surface points are often surrounded by other true surface points, which makes their end tetrahedra flat and wide. Based on these observations, we set a weight of $\alpha(p)(1 - e^{-kr^2/2\sigma^2})$ to the t-edge of the end tetrahedron, where r is the radius of the circumsphere of the end tetrahedron. Thus the corresponding visibility energy is formulated as

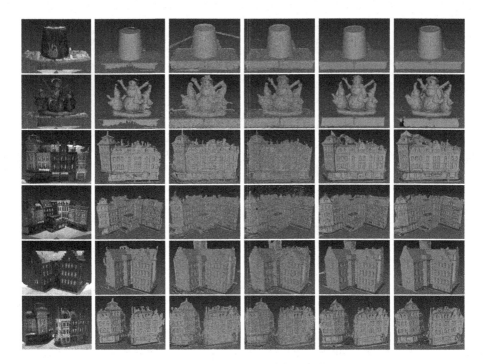

Fig. 4. Results of the five methods on MVS Data Set [7]. From left to right: the reference model, Tola et al. [19], Furukawa and Ponce [3], Campbell et al. [2], Jancosek and Pajdla [6] and ours.

$$E_{vis} = \sum_{v \in \mathcal{V}} [D(l_{T_1}) + \sum_{i=1}^{N_v - 1} W(l_{T_i}, l_{T_{i+1}}) + D(l_{T_{N_v+1}})] \tag{3}$$

where if $l_{T_1} = 1$, $D(l_{T_1}) = \alpha(p)$; otherwise, $D(l_{T_1}) = 0$; if $l_{T_{N_v+1}} = 0$, $D(l_{T_{N_v+1}}) = \alpha(p)(1 - e^{-kr^2/2\sigma^2})$; otherwise, $D(l_{T_{N_v+1}}) = 0$; if $l_{T_i} = 0 \wedge l_{T_j} = 1$, $W(l_{T_i}, l_{T_j}) = \alpha(p)(1 - e^{-d^2/2\sigma^2})$; otherwise, $W(l_{T_i}, l_{T_j}) = 0$. Although Eqs. 2 and 3 look identical, they differ from each other in term $W(l_{T_i}, l_{T_{i+1}})$ and $D(l_{T_{N_v+1}})$. One may refer to the text for details.

4   Likelihood Energy for Efficient Noise Filtering

In both the typical visibility model and our proposed one, the end of each line of sight is set in the tetrahedron right behind the point p. This practice could weaken the ability of noise filtering. When the surface is sampled very noisy, part of the surface fails to be reconstructed, as shown in Fig. 2(b). This is mainly due to the unbalanced links of s-edges and t-edges in the s-t graph. The s-edges are heavy-weighted and concentrative, while the t-edges are low-weighted and scattered. The noisier the point cloud is, the greater the gap between s-edges and t-edges will be.

Table 1. Performance of 5 methods on MVS data set [7]

Dataset	scan1				scan24				scan44			
	$\bar{A}$	$A_{0.5}$	$\bar{C}$	$C_{0.5}$	$\bar{A}$	$A_{0.5}$	$\bar{C}$	$C_{0.5}$	$\bar{A}$	$A_{0.5}$	$\bar{C}$	$C_{0.5}$
Tola [19]	**0.2449**	0.1918	0.3994	0.3213	**0.3937**	0.2656	0.8512	0.4340	**0.6410**	**0.3251**	0.7382	0.4160
Furu [3]	0.4533	0.2517	0.4105	0.3533	0.7455	0.3125	0.4828	0.3692	1.0951	0.4906	0.5264	0.3872
Camp [2]	0.4428	0.2550	0.3988	0.3601	3.4601	0.5227	0.4010	0.2954	1.2926	0.5641	**0.3807**	**0.2976**
Jan [6]	0.3595	0.2314	0.4246	0.3444	0.6054	**0.2032**	**0.3488**	**0.2574**	0.8098	0.36739	0.5652	0.3907
ours	0.3053	**0.1760**	**0.3062**	**0.2556**	0.7865	0.2354	0.4230	0.2874	1.2309	0.5410	0.6787	0.4577

$\bar{A}$: mean accuracy; $A_{0.5}$: median accuracy; $\bar{C}$: mean completeness; $C_{0.5}$: median completeness.

4.1   Likelihood Energy

To solve the problem, we introduce a likelihood energy E_{like} to the total energy of the binary labeling problem. E_{like} is defined as

$$E_{like} = \sum_{i=1}^{N} D(l_{T_i}), \text{ where } D(l_{T_i}) = \begin{cases} U_{out}(T_i) & \text{if } l_{T_i} = 1 \\ U_{in}(T_i) & \text{otherwise} \end{cases} \quad (4)$$

where N is the total number of Delaunay tetrahedra. E_{like} measures the penalties of a wrong label assignment. For each tetrahedron, a penalty is given for its wrong label, i.e. $U_{out}(T_i)$ or $U_{in}(T_i)$. To evaluate the likelihood of the label assignment of a tetrahedron, we employ the measure *free-space support* [6], which is used to measure the emptiness of a compact space. For each line of sight (c, p) that intersects tetrahedra T, it contributes to the emptiness of T_i, thus increasing the outside probability of T. The free-space support $f(T)$ of tetrahedron T is defined as

$$f(T) = \sum_{(c,p) \in S_T} \alpha(p), \text{ with } S_T = \{(c, p) \,|\, (c, p) \cap T \neq \emptyset\} \quad (5)$$

In order for free-space support $f(T)$ of tetrahedron T to adequately describe the likelihood energy, we set $U_{out}(T) = \lambda f(T)$ and $U_{in}(T) = \lambda(\beta - f(T))$, where $f(T)$ is the free-space support of tetrahedron T, λ and β are constants.

4.2   Implementation of the Likelihood Energy

For the likelihood term E_{like}, we link two edges for each vertex i in s-t graph. One is from source s to vertex i, with weight $U_{out}(T_i)$; the other is from vertex i to sink t, with weight $U_{in}(T_i)$. However, we observe that the tetrahedra with lower $f(T)$ are easily to be mislabeled, thus we only link the vertices whose correspondent tetrahedra have lower $f(T)$ than the 75th percentile of all $f(T)$s to sink t, and cross out the s-edges of all vertices.

Note that the free space support threshold of the 75th percentile is empirically set. We generate dense point clouds from 3 datasets, Fountain-P11 [18], Herz-Jesu-P8 [18] and Temple-P312 [16]. Then we label the tetrahedra with the method in [6], and take the result as quasi-truth. Finally, we evaluate the ratios

(a) Thin objects   (b) Object edges

Fig. 5. Comparison with the method of Jancosek and Pajdla [6]. The result of our method is on the left columns of (a) and (b) and that of [6] on the right columns of them.

of number of outside and inside tetrahedra in different proportions of all free-space support scores, as shown in Fig. 3. From Fig. 3(a)–(c) we can see that, generally tetrahedron with $f(T)$ lower than the 75th percentile of all $f(T)$s has a high probability to be truly inside. From Fig. 3(d), when the free-space support threshold is set as the 75th percentile of each dataset, both true positive rate and false positive rate are reasonable. Therefore, we set the free space support threshold as the 75th percentile of all free-space support scores. From Fig. 2, we can see that the likelihood energy is helpful for filtering noise.

5   Experimental Results

The input dense point cloud is generated with the open source library Open-MVG[1] and OpenMVS[2]. Sparse point cloud is generated by OpenMVG, then densified with OpenMVS. Delaunay tetrahedralization is computed using CGAL[3] library. Maxflow/Mincut algorithm [4] is used. Our method is tested on public benchmark MVS Data Set [7] and some of our own datasets.

Figure 4 shows the results on the MVS Data Set [7]. From left to right, the reference model, the results of Tola et al. [19], Furukawa and Ponce [3], Campbell et al. [2], Jancosek and Pajdla [6] and ours are shown. The final meshes given by Tola et al. [19], Furukawa and Ponce [3] and Campbell et al. [2] are generated by poisson surface reconstruction method [9] and trimmed, which are provided in

---

[1] http://imagine.enpc.fr/~moulonp/openMVG/.

[2] http://cdcseacave.github.io/openMVS/.

[3] http://www.cgal.org/.

(a) JinDing dataset

(b) NanChan Temple dataset

Fig. 6. Result of (a) JinDing dataset and (b) NanChan Temple dataset. In the first row of (a) and (b), example image(s) and dense point cloud are shown; in the second row of (a) and (b), overview result and details of our method are shown in left 2 columns and those of Jancosek and Pajdla [6] in right 2 columns.

MVS Data Set; the meshes of Jancosek and Pajdla [6] are generated by Open-MVS which contains an reimplementation of [6]. Table 1 shows the quantitative results of the five methods. From Fig. 4 and Table 1, though our method is not quantitatively the best, the meshes generated by our method preserves more details than the others. To better visualize the performances of Jancosek and Pajdla [6] and our method, we enlarge some parts in Fig. 4, and present in Fig. 5, with the result of our method on the left columns of (a) and (b) and that of [6] on the right columns of them. Figure 5(a) shows the performance of the 2 methods on thin objects. In some cases, [6] fails to reconstruct them completely; even the complete ones are less detailed than those by our method. In Fig. 5(b), our method shows a tremendous capability of preserving sharp object edges, compared with the result of [6]. From Fig. 5, we can notice that our method performs well in dealing with small scale objects and sharp edges, while [6] fails to reconstruct those parts completely and loses some details.

Figure 6 show the results of our method and Jancosek and Pajdla [6] on our own datasets, the JinDing and NanChan Temple datasets. The JinDing dataset consists of 192 images taken by a hand-held camera, with the resolution of 4368×2912. The NanChan Temple dataset contains 2006 images, with 1330 of them taken by a hand-held camera and the rest by an unmanned aerial vehicle (UAV). The ground images have a resolution of 5760×3840 and the aerial ones of 4912×3264. The dense point cloud contains $31M$ and $40M$ points for the two datasets respectively. Note that we downsample the images twice in NanChan Temple dataset so that the point cloud would not be too dense to be processed. In Fig. 6, example image(s), dense point cloud, result overview and details of our method and Jancosek and Pajdla [6] are presented. From the closer view of the result of our method, details like elephant's tusk and ears, cloth folds, beam edges and roof drainages are preserved. In the meantime, the two models of our method avoid over-complexity. The results show that our method performs better than [6] on our datasets.

6 Conclusions and Future Work

In this paper, we present a novel visibility model and a new energy term for surface reconstruction. Our method takes dense point clouds as input. By applying Delaunay tetrahedralization to the dense point cloud, we divide the scene space into cells, and formulate the scene reconstruction as a binary labeling problem to label each tetrahedron inside or outside. To yield out meshes with enough scene details, we introduce a new visibility model to filter out noise and select true surface points. The proposed visibility model keeps the relaxed visibility constraints and takes the distribution of points into consideration, thus it is efficient for preserving details. To strengthen the robustness to noise, a new likelihood energy term is introduced to the binary labeling problem. Experimental result shows that our method performs well in dealing with scene details like thin objects and sharp edges, which will help to attach semantic knowledge to the model. So, for the future work, we would like to investigate to segment the model and add semantic knowledge to each part of the scene.

Acknowledgments. This work was supported by the Natural Science Foundation of China under Grants 61333015, 61421004 and 61473292.

References

1. Broadhurst, A., Drummond, T.W., Cipolla, R.: A probabilistic framework for space carving. In: Proceedings of 8th IEEE International Conference on Computer Vision, ICCV 2001, vol. 1, pp. 388–393. IEEE (2001)
2. Campbell, N.D.F., Vogiatzis, G., Hernández, C., Cipolla, R.: Using multiple hypotheses to improve depth-maps for multi-view stereo. In: Forsyth, D., Torr, P., Zisserman, A. (eds.) ECCV 2008. LNCS, vol. 5302, pp. 766–779. Springer, Heidelberg (2008). https://doi.org/10.1007/978-3-540-88682-2_58
3. Furukawa, Y., Ponce, J.: Accurate, dense, and robust multiview stereopsis. IEEE Trans. Pattern Anal. Mach. Intell. **32**(8), 1362–1376 (2010)
4. Goldberg, A.V., Hed, S., Kaplan, H., Tarjan, R.E., Werneck, R.F.: Maximum flows by incremental breadth-first search. In: Demetrescu, C., Halldórsson, M.M. (eds.) ESA 2011. LNCS, vol. 6942, pp. 457–468. Springer, Heidelberg (2011). https://doi.org/10.1007/978-3-642-23719-5_39
5. Guan, L., Franco, J.S., Pollefeys, M.: 3D occlusion inference from silhouette cues. In: 2007 IEEE Conference on Computer Vision and Pattern Recognition, pp. 1–8. IEEE (2007)
6. Jancosek, M., Pajdla, T.: Exploiting visibility information in surface reconstruction to preserve weakly supported surfaces. In: International Scholarly Research Notices 2014 (2014)
7. Jensen, R., Dahl, A., Vogiatzis, G., Tola, E., Aanæs, H.: Large scale multi-view stereopsis evaluation. In: 2014 IEEE Conference on Computer Vision and Pattern Recognition, pp. 406–413. IEEE (2014)
8. Jin, H., Soatto, S., Yezzi, A.J.: Multi-view stereo reconstruction of dense shape and complex appearance. Int. J. Comput. Vis. **63**(3), 175–189 (2005)
9. Kazhdan, M., Bolitho, M., Hoppe, H.: Poisson surface reconstruction. In: Proceedings of 4th Eurographics Symposium on Geometry Processing, vol. 7 (2006)
10. Kutulakos, K.N., Seitz, S.M.: A theory of shape by space carving. Int. J. Comput. Vis. **38**(3), 199–218 (2000)
11. Labatut, P., Pons, J.P., Keriven, R.: Robust and efficient surface reconstruction from range data. In: Computer Graphics Forum, vol. 28, pp. 2275–2290. Wiley Online Library (2009)
12. Labatut, P., Pons, J.P., Keriven, R.: Efficient multi-view reconstruction of large-scale scenes using interest points, delaunay triangulation and graph cuts. In: 2007 IEEE 11th international conference on computer vision. pp. 1–8. IEEE (2007)
13. Laurentini, A.: The visual hull concept for silhouette-based image understanding. IEEE Trans. Pattern Anal. Mach. Intell. **16**(2), 150–162 (1994)
14. Lempitsky, V., Boykov, Y., Ivanov, D.: Oriented visibility for multiview reconstruction. In: Leonardis, A., Bischof, H., Pinz, A. (eds.) ECCV 2006. LNCS, vol. 3953, pp. 226–238. Springer, Heidelberg (2006). https://doi.org/10.1007/11744078_18
15. Pons, J.P., Keriven, R., Faugeras, O.: Multi-view stereo reconstruction and scene flow estimation with a global image-based matching score. Int. J. Comput. Vis. **72**(2), 179–193 (2007)
16. Seitz, S.M., Curless, B., Diebel, J., Scharstein, D., Szeliski, R.: A comparison and evaluation of multi-view stereo reconstruction algorithms. In: 2006 IEEE Computer Society Conference on Computer Vision and Pattern Recognition (CVPR 2006), vol. 1, pp. 519–528. IEEE (2006)
17. Shen, S.: Accurate multiple view 3D reconstruction using patch-based stereo for large-scale scenes. IEEE Trans. Image Process. **22**(5), 1901–1914 (2013)

18. Strecha, C., von Hansen, W., Van Gool, L., Fua, P., Thoennessen, U.: On benchmarking camera calibration and multi-view stereo for high resolution imagery. In: IEEE Conference on Computer Vision and Pattern Recognition, CVPR 2008, pp. 1–8. IEEE (2008)
19. Tola, E., Strecha, C., Fua, P.: Efficient large-scale multi-view stereo for ultra high-resolution image sets. Mach. Vis. Appl. **23**(5), 903–920 (2012)
20. Tran, S., Davis, L.: 3D surface reconstruction using graph cuts with surface constraints. In: Leonardis, A., Bischof, H., Pinz, A. (eds.) ECCV 2006. LNCS, vol. 3952, pp. 219–231. Springer, Heidelberg (2006). https://doi.org/10.1007/11744047_17
21. Ummenhofer, B., Brox, T.: Global, dense multiscale reconstruction for a billion points. In: Proceedings of the IEEE International Conference on Computer Vision, pp. 1341–1349 (2015)
22. Vogiatzis, G., Torr, P.H., Cipolla, R.: Multi-view stereo via volumetric graph-cuts. In: 2005 IEEE Computer Society Conference on Computer Vision and Pattern Recognition (CVPR 2005), vol. 2, pp. 391–398. IEEE (2005)
23. Vu, H.H., Labatut, P., Pons, J.P., Keriven, R.: High accuracy and visibility-consistent dense multiview stereo. IEEE Trans. Pattern Anal. Mach. Intell. **34**(5), 889–901 (2012)
24. Yang, R., Pollefeys, M., Welch, G.: Dealing with textureless regions and specular highlights-a progressive space carving scheme using a novel photo-consistency measure. In: ICCV, vol. 3, pp. 576–584 (2003)

Global-Local Feature Attention Network
with Reranking Strategy
for Image Caption Generation

Jie Wu[1], Siya Xie[1], Xinbao Shi[1], and Yaowen Chen[2(✉)]

[1] College of Engineering, Shantou University, No. 243 Daxue Road, Shantou, China
{13jwu1,14syxie,16xbshi}@stu.edu.com
[2] The Key Laboratory of Digital Signal and Image Processing of Guangdong,
Shantou University, No. 243 Daxue Road, Shantou, China
ywchen@stu.edu.com, ywchen@stu.edu.cn

Abstract. In this paper, a novel framework, named global-local feature
attention network with reranking strategy (GLAN-RS), is presented for
image captioning task. Rather than only adopt unitary visual information in the classical models, GLAN-RS explore attention mechanism to
capture local convolutional salient image maps. Furthermore, we adopt
reranking strategy to adjust the priority of the candidate captions and
select the best one. The proposed model is verified using the MSCOCO
benchmark dataset across seven standard evaluation metrics. Experimental results show that GLAN-RS significantly outperforms the state-
of-the-art approaches such as M-RNN, Google NIC etc., which gets an
improvement of 20% in terms of BLEU4 score and 13 points in terms of
CIDER score.

Keywords: Image caption · Global-local feature attention network
Reranking strategy · Candidate captions

1 Introduction

With the rapid development of the artificial intelligence in recent years, image
caption has been a hot spot in computer vision and natural language processing.
Image caption is a multidisciplinary subject involving signal processing, pattern
recognition, computer vision and cognitive sciences, which aims to describe the
content of the input image by learning dense mappings between images and
words. Image caption can be applied to the image retrieval, children's education
and the life support for visually impaired persons, which has a positive role in
promoting social life.

Due to the advances of deep neural network [1], some state-of-the-art models
have been presented for solving the challenges of generating image captions.
For example, Mao et al. [2] propose a multimodal recurrent neural network

---

J. Wu—Student Paper.

© Springer Nature Singapore Pte Ltd. 2017
J. Yang et al. (Eds.): CCCV 2017, Part I, CCIS 771, pp. 157–167, 2017.
https://doi.org/10.1007/978-981-10-7299-4_13

(MRNN) for sentence generation. Xu et al. [3] explore attention mechanisms that capture salient feature of the raw image to generate caption. However, they only adopt unitary image feature instead of feeding global and local visual information simultaneously. Furthermore, they are easy to generate caption irrelevant to image content.

In order to overcome the above limitations, we propose a new global-local feature attention network with reranking strategy (GLAN-RS) for addressing image-captioning task. As illustrated in Fig. 1, GLAN-RS consists of global-local feature attention network to exploit visual information and a reranking strategy to select the consensus caption in candidate captions.

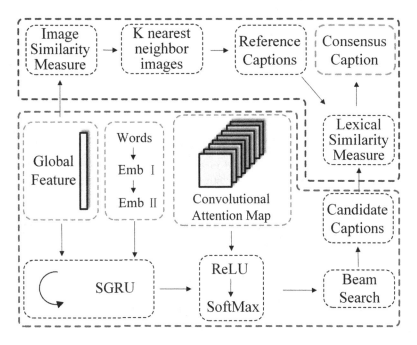

Fig. 1. The framework of GLAN-RS, which consists of global-local feature attention network (red dashed) and reranking strategy network (blue dashed). Emb denotes the dense word representation with two integrated layers. SGRU denotes the Stacked Gated Recurrent Unit. (Color figure online)

Our contributions are as follows:

Firstly, GLAN-RS combines global image feature and local convolutional attention maps for capture holistic and salient visual information simultaneously.

Secondly, we explore nearest neighbor approach to calculate the image similarity and obtain their reference captions. Thus we can determine the best candidate caption by finding the one with highest score respect to the reference captions.

Moreover, we validate the effectiveness of GLAN-RS comparing with the state-of-the-art approaches [2,4] consistently across seven evaluation metrics, which show that GLAN-RS get an improvement of 20% in terms of BLEU4 score and 13 points in terms of CIDER score.

2   Proposed Method

2.1   Encoder-Decoder Framework for Image Caption Generation

We adopt the popular encoder-decoder framework for image caption generation, where convolution neural network (CNN) [5,6] encodes the image into a fixed-dimension feature vector and then a stacked Gated Recurrent Unit (SGRU) [7] decodes the visual information into sematic information. Given an input image and its corresponding caption, the encoder-decoder model directly maximize the log-likelihood function of the following objective:

$$\theta^* = \arg\min_{\theta} \sum_{i=0}^{L} \log p(s_i | f_I, (\theta))$$ (1)

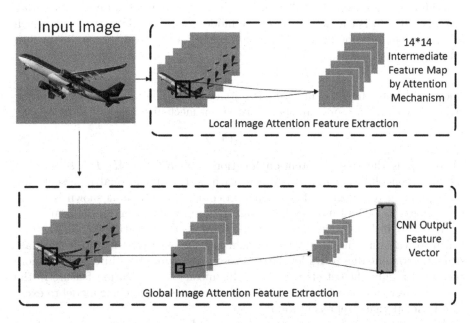

Fig. 2. The global image attention features and regional convolutional attention feature extraction process.

2.2   Global Image Attention Feature

Where f_I denotes the global image representation of the raw image, s_i denotes a sequence of words in a sentence of length L and is the parameters of the model. Images information should be encoded as fix-length vectors to feed into SGRU. CNN is used to extract the image feature map f_I from a raw image I to give an overview of the image content to the model:

$$f_I = CNN(I) \tag{2}$$

As is shown is Fig. 2, we capture the global image attention feature in the last fully-connected layer.

2.3   Local Convolutional Image Attention Feature

In the encoder-decoder framework with SGRU and local convolutional attention map, the conditional probability of the log-likelihood function is modeled as:

$$logp(s_t|s_{1:t-1}) = tanh(h_t, c_t) \tag{3}$$

where c_t is the attention context vector obtained in the third layer of the fifth convolutional layer. In this paper, we utilize the SGRU with two hidden layers to modulate the information flow inside the unit instead of applying a separate memory cells. h_t is the activation of the hidden state in SGRU at time t, which is a linear combination between previous activation h_{t-1} and the candidate activation h_t':

$$h_t = (1 - z_t)h_{t-1} + z_t h_t' \tag{4}$$

where z_t denotes how the unit updates its content. The method for computing attention context vector c_t in spatial attention mechanism is defined as:

$$c_t = f_{att}(h_t, I) \tag{5}$$

where f_{att} is the spatial attention function. $I \in R^{d \times k} = [I_1, I_k]$, $I_i \in R^d$ is the local convolutional image features, each is a d dimensional representation corresponding to a region of the image in $conv5, 3$ layer. As is shown is Fig. 2, we exploit the 14 * 14 intermediate feature map by attention mechanism.

We show in Fig. 3 raw image and visualize its CNN features in Fig. 4, from which we can find the fifth convolutional layer is effective to capture accurate semantic content due to its low spatial resolution while the third layer is more effective to highlight details precisely. This means the activation feature in the $conv5, 3$ layer can detect some pivotal image areas and guide our model to excavate regional visual representation.

We feed the spatial image features I_i and hidden state h_t through a hyperbolic tangent neural function followed by a softmax function to generate the attention distribution over the k regions of the image:

$$\alpha_t = \frac{exp(e_t)}{\sum_{t=1}^{k} exp(e_t)} \tag{6}$$

Fig. 3. The visualization of the raw image.

Fig. 4. The visualization of the partial activation feature in $conv5, 3$ layer.

$$e_t = tanh(w_i I_i + w_h h_t) \tag{7}$$

where w_i, w_h are the project parameters to be learnt. α_t is the spatial attention weight over features in I_i. The attention context vector c_t is computed bases on the attention distribution:

$$c_t = \sum_{t=1}^{k} \alpha_t I_i \tag{8}$$

We combine c_t and h_t to predict the next word as in Eq. (3). As is shown in Fig. 5, we design a bimodal layer to process the information from the local image attention feature and the activation of the current hidden layer of SGRU. The Cascaded semantic layer is explored to capture dense syntactic representation.

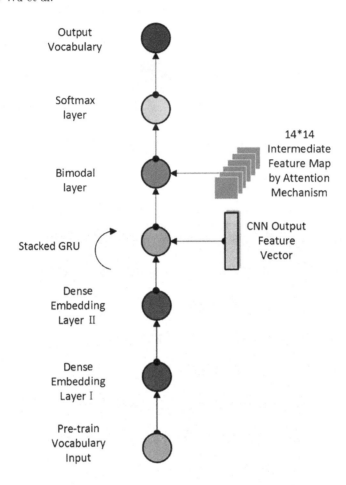

Global and Local Image Attention Model

Fig. 5. The flowchart of GLAN model. Modules of different colors represent layers of different functions. The input layer feeds the words one by one. (Color figure online)

2.4   Reranking Strategy

For a test image, we first utilize global-local feature attention network to create the hypothetical captions adopting Beam Search algorithm. Then we use nearest neighbor approach to find the similar images and their corresponding reference captions. In this paper, Euclidean distance measure $D_E(x_1, x_2)$ is adopted to calculate the feature similarities, which are defined by:

$$D_E(x_1, x_2) = \sqrt{\sum_{n=1}^{N} (x_{1k} - x_{2k})^2} \tag{9}$$

We define consensus caption C^* as the one that has the highest accumulated similarity with the reference captions. The consensus caption selective function is defined by [8]:

$$c^* = \arg\max_{c_1 \in H} \sum_{c_2 \in R} Sim(C_1, C_2) \tag{10}$$

where H is the set of hypothetical captions while R is the set of reference captions. $Sim(c_1, c_2)$ is the accumulated lexical similarity function between two captions c_1, c_2

$$Sim(C_1, C_2) = BPe^{\frac{1}{N}\sum_{n=1}^{N} log p_n} \tag{11}$$

where p_n is the modified n-gram precision and BP is a brevity penalty for the short sentences, which is computed by:

$$BP = \min(1, e^{1-\frac{b}{c}}) \tag{12}$$

where b is the length of the reference sentence while c is the length of the candidate sentence.

3   Experimental Results

We evaluate the performance of GLAN-RS using the MSCOCO dataset [9,10], which contains 82,783 images for training, 40,504 for validation, with 5 reference captions provided by human annotators [11]. We randomly sample 1000 images for testing. We utilize two classical CNN models, i.e., IV3 network [5] and VggNet [6] to en-code the visual information. Seven standard evaluation metrics are adopted for evaluations, including BLEU scores (i.e. B-1, B-2, B-3 and B-4) [12] and MSCOCO validate toolkit (ROUGE-L, CIDEr and METEOR) [13–15].

3.1   Datasets

Microsoft COCO Caption is a datasets released by Microsoft Corporation and contains almost 300,000 images with five reference sentences; The MS COCO Caption datasets is created for image caption, which not only consists of rich images and captions, but also provides the computing servers and code of evaluation merits. The MS COCO Caption datasets has become the first choice of researchers in recent year. Figure 6 illustrates some pictures in MSCOCO datasets.

3.2   Evaluation Metrics

In order to validate the quality of the caption generated by the model, we choose seven automatic evaluation merits that have a high correlation with human judgments. The higher the score of merits, the better effect of image caption.

Bleu [12] was designed for automated evaluation of statistical machine translation, which can be adopted to measure similarity of descriptions. Given the

Fig. 6. The illustration of pictures in MSCOCO datasets.

diversity of possible image descriptions, Bleu may penalize candidates which are arguably descriptive of image content. CIDEr [13] is a special automatic evaluation metrics for image caption, which measures the consistency of image caption by calculating TF-IDF weights of each n-tuple. ROUGE [14] is a set of evaluation metrics designed to evaluate text summarization algorithms. ROUGE-L uses a measure based on the Longest Common Subsequence. Meteor [15] is the harmonic mean of unigram precision and recall that allows for exact, synonym, and paraphrase matchings between candidates and references. The researchers can calculate precision, recall, and F-measure by Meteor.

3.3   Model Details

We trained all sets of weights using stochastic gradient descent with changed learning rate and no momentum. We set fix learning rate in initial iteration period and decline 15% every iteration period after initial iteration period

Dropout and ReLu function are adopted to avoid gradient vanishing or exploding. All weights are randomly initialized with the range of $(-1.0, 1.0)$. we use 1028 dimensions for dense embedding and the size of the SGRU is set to 2048. To infer the sentence given an input image, we use Beam Search algorithm, which keeps the best m generated captions. For convenience, we introduce the architectural details in the following:

GLAN+VGG uses VGGnet to extract the 4096-dimensions holistic visual embedding features while **GLAN+IV3** utilizes Inception V3 network to encode the 2048-dimensions holistic image representations. **GLAN+RS** applies the Euclidean distance function to re-rank the order of the caption. We denote **GLAN+RS** as our proposed model to compare with several state-of-the-art models.

3.4   Performance Comparison of GLAN-RS with the State-of-the-Art Models

In order to verify the superiority of our proposed model, we compare its performance with some state-of-the-art methods including: Google NIC [4], GLSTM [11], Attention [3] and MRNN [2].

Table 1. Comparison results on MSCOCO dataset by using VggNet to extract image features. (-) indicates an unknown metric.

Model	CIDEr	B-4	B-3	B-2	B-1	ROUGEL	METEOR
M-RNN [2]	-	25.0	35.0	49.0	67.0	-	-
GLSTM [16]	81.25	26.4	35.8	49.1	67.0	-	22.74
GLAN+VGG	89.2	29.2	39.3	53.0	70.0	51.3	23.3

Table 2. Comparison results on MSCOCO dataset by using IV3 network to extract image features. (-) indicates an unknown metric.

Model	CIDEr	B-4	B-3	B-2	B-1	ROUGEL	METEOR
Google NIC [4]	-	24.6	32.9	46.1	66.6	-	-
Hard-Attention [3]	-	25.0	35.7	50.4	71.8	-	23.04
Soft-Attention [3]	-	24.3	34.4	49.2	70.7	-	23.9
M-RNN [2]	86.4	27.0	36.4	50.0	68.0	50.0	23.1
GLAN+IV3	90.2	29.7	39.5	53.8	69.8	51.5	23.6
GLAN+RS	99.3	31.5	42.0	55.9	72.7	53.2	24.6

Tables 1 and 2 list the comparison results, where we respectively encode visual embedding features with VggNet and IV3 network. From the tables, we can obtain the follow conclusion:

1. GLAN performs best for almost all the listed evaluation criteria with using VggNet and IV3 network respectively to extract the visual embedding feature. This may due to the reason that our model employ global and local attention feature to learn deep image-to-word mapping, which is able to exploit semantic context to high quality image captions.
2. **GLAN+IV3** outperforms **GLAN+VGG**, which means robust image representation benefits the performance.
3. Attention mechanism is able to capture the latent visual representation to get higher Bleu1 score and generate high-level caption.
4. Reranking strategy makes it possible to select the caption with the highest accumulated similarity with the reference captions.
5. **GLAN+RS** has a remarkable improvement on the all evaluation criteria. **GLAN+RS** significantly outperforms the state-of-the-art approaches such as M-RNN, Google NIC etc., which gets an improvement of 20% in terms of BLEU4 score and 13 points in terms of CIDER score.

3.5   Generation Results

In order to qualitatively validate the effective of GLAN-RS, we select several images to generate captions by GLAN-RS model, which is shown in the Fig. 7.

A: dog jumping in the
air to catch a frisbee.

B: a man flying through the
air while riding a snowboard.

C: a small child holding a
donut with sprinkles.

D: a red fire hydrant sitting
in the middle of a forest

E: a horse pulling a
carriage down a road

F: an empty street at night
with a traffic light

Fig. 7. The illustration of the generated caption by GLAN-RS model.

As is shown in Fig. 7, our proposed model can detect the activity phrases such as "catch a frisbee", "ride a snowboard" and "pullig a carriage" in different instances, which indicates that GLAN-RS is able to capture the key action information in the image. our model notices the "red fire hydrant" and "donut with sprinkles" in instance (D) and (C), which validate that our model make it possible to exploit latent detail information. It's amazing to find that the phrase "at night" is generated for instance (F), which show that GLAN-RS benefits its distinctive encoder-decoder framework which combines the global and regional attention features to capture semantic-related visual concept representation.

4   Conclusion

In this paper, we propose a novel global-local feature attention network with re-ranking strategy (GLAN-RS) for image caption generation. We utilize global image feature and local convolutional attention maps for exploiting visual representation. The consensus caption is selected by nearest neighbor approach and reranking strategy. The experimental results on benchmark MSCOCO dataset demonstrate the superiority of GLAN-RS for generating high quality image captions.

Acknowledgements. We are grateful for the financial support from the Innovative Application and Re-search Project of Guangdong Province (No. 2016KZDXM013) and the Science & Technology Project of Shantou City (A201400150).

References

1. Krizhevsky, A., Sutskever, I., Hinton, G.: ImageNet classification with deep convolutional neural networks. In: NIPS (2012)
2. Mao, J., Xu, W., Yang, Y., Wang, J., Yuille, A.: Deep captioning with multimodal recurrent neural networks (mRNN). arXiv preprint arXiv:1412.6632 (2014)
3. Xu, K., Ba, J., Kiros, R., Cho, K., Courville, A.C., Salakhutdinov, R., Zemel, R.S., Bengio, Y.: Show, attend and tell: neural image caption generation with visual attention. CoRR, abs/1502.03044 (2015)
4. Vinyals, O., Toshev, A., Bengio, S., Erhan, D.: Show and tell: a neural image caption generator. In: CVPR (2015)
5. Szegedy, C., Liu, W., Jia, Y., Sermanet, P., Reed, S., Anguelov, D., Erhan, D., Vanhoucke, V., Rabinovich, A.: Going deeper with convolutions. arXiv preprint arXiv:1409.4842 (2014)
6. Simonyan, K., Zisserman, A.: Very deep convolutional networks for large-scale image recognition. CoRR, abs/1409.1556 (2014)
7. Chung, J., Gulcehre, C., Cho, K., Bengio, Y.: Empirical evaluation of gated recurrent neural networks on sequence modeling. CoRR, abs/1412.3555 (2014)
8. Devlin, J., Gupta, S., Girshick, R., Mitchell, M., Zitnick, C.L.: Exploring nearest neighbor approaches for image captioning. arXiv preprint arXiv:1505.04467 (2015)
9. Lin, T.-Y., Maire, M., Belongie, S., Hays, J., Perona, P., Ramanan, D., Dollár, P., Zitnick, C.L.: Microsoft COCO: common objects in context. In: Fleet, D., Pajdla, T., Schiele, B., Tuytelaars, T. (eds.) ECCV 2014. LNCS, vol. 8693, pp. 740–755. Springer, Cham (2014). https://doi.org/10.1007/978-3-319-10602-1_48
10. Chen, X., Fang, H., Lin, T.Y., Vedantam, R., et al.: Microsoft coco captions: data collection and evaluation server. arXiv:1504.00325 (2015)
11. Rashtchian, C., Young, P., Hodosh, M., et al.: Collecting image captions using Amazon's mechanical turk. In: NAACL Hlt 2010 Workshop on Creating Speech and Language Data with Amazon's Mechanical Turk, pp. 139–147 (2010)
12. Papineni, K., Roukos, S., Ward, T., Zhu, W.J.: BLEU: a method for automatic evaluation of machine translation. In: ACL, pp. 311–318 (2002)
13. Vedantam, R., Zitnick, C.L., Parikh, D.: Cider: consensus-based image description evaluation. In: CVPR (2015)
14. Lin, C.-Y.: Rouge: a package for automatic evaluation of summaries. In: Text Summarization Branches Out: Proceedings of the ACL-04 Workshop, vol. 8 (2004)
15. Banerjee, S., Lavie, A.: Meteor: an automatic metric for MT evaluation with improved correlation with human judgments. In: Proceedings of the ACL Workshop on Intrinsic and Extrinsic Evaluation Measures for Machine Translation and/or Summarization, vol. 29, pp. 65–72 (2005)
16. Jia, X., Gavves, E., Fernando, B., et al.: Guiding the long-short term memory model for image caption generation. In: ICCV (2015)

Dust Image Enhancement Algorithm Based on Color Transfer

Hao Liu, Ce Li$^{(\boxtimes)}$, Yuqi Wan, and Yachao Zhang

College of Electrical and Information Engineering,
Lanzhou University of Technology, Lanzhou 730050, China
xjtulice@gmail.com

Abstract. The increasing dust weather has seriously affected the quality of captured images. Therefore, research of the dust image quality enhancement has become an important hotspot in the field of computer vision. Compared with pictures which obtained in sunny day, dust images have some obvious problems such as low-quality in definition and light, hue yellowing and so on. Amid such questions, available methods cannot always prevail. Hence, this paper proposed a dust image enhancement algorithm based on color transfer. First of all, applying the scene gist feature to select a target image which has the highest similarity with input image in clear images database; then, the target image color information is passed to input image through the color transfer algorithm; finally, utilizing the contrast limited adaptive histgram equalization algorithm to restore the definition in dust image. The experimental results show that our algorithm can effectively enhance dust images quality in different degree, get more image details and resolve the definition and hue correction problems. We proposed a absolute color difference index to measure this method. The experiments demonstrate that the proposed algorithm outperforms the state-of-the-art models.

Keywords: Dust image · Color transfer · Image enhancement

1 Introduction

In recent years, more and more dust weather has seriously affected images quality. With the rapid development of artificial intelligence, the identification of dust images still constraint the increasing of self-driving cars. So how to enhance dust images is an important problem need to be solved.

Dust weather means wind raise a large number of dust into the air. Massive solid particles make the air turbid that reduced the visibility. Because of increasing dust particles in the air, the scattering effect of light is enhanced. So the color of dust images is yellowing, brightness and clarity is obviously reduced.

Currently, dust image enhancement algorithms can be divided into two types: one is improved defog algorithm [1–3] because dust images and fog images also caused by light scattering and reflection from various particles in the air [4–6]. These algorithms tend to ignore the difference between dust and fog that dust

© Springer Nature Singapore Pte Ltd. 2017
J. Yang et al. (Eds.): CCCV 2017, Part I, CCIS 771, pp. 168–179, 2017.
https://doi.org/10.1007/978-981-10-7299-4_14

is a solid particle with color. Hence, this kind of methods cannot resolve the problem of hue yellowing and low brightness in dust images. The other type is using different ways to work out different questions in dust images [7–9]. We can correct color cast, increase light and improve definition separately. These algorithms solved the basic problem of dust image and get a good result. But the implementation process of these methods always bring in a large number of parameters, thus the application range of dust image enhancement algorithms is limited.

Due to shortcomings of traditional algorithms, we proposed a dust image enhancement algorithm based on color transfer. Firstly, applying scene gist feature [10] to select the target image which has the highest similarity scene with the input image; then, the color transfer algorithm [11] is used to transfer the color information in target image to input image; finally, utilizing the contrast limited adaptive histgram equalization algorithm (CLAHE) [12] to restore the definition in dust image. We also design a index to test the effect of our result. Important intermediate results of this algorithm are show in Fig. 1.

(a)   (b)   (c)   (d)

Fig. 1. Important intermediate result images of this algorithm. (a) input image, (b) target image, (c) image after color transfer, (d) result image.

2   Related Works

Bertalmío [4] proposed a Retinex algorithm based on kernel function which has better reinforcing effect for dust images. However, this method cannot effectively solve the problem about hue yellowing in dust images. Yan [5] use partially overlapped sub-block histogram equalization to improve dust images. This algorithm effectively get a better contrast in the enhancement image. But this method result will bring hue distortion. Li [8] using dark channel prior proposed a single image restoration algorithm. This algorithm improves dust image clarity, but the brightness of result image is dim. Above algorithms are improving defog algorithm to enhance dust images. However, these ways only resolve the question of contrast in dust images, while ignoring the color correction of image enhancement.

In order to obtain better image enhancement effect on dust images, many scholars began to solve the hue yellowing, low brightness and low-quality definition separately. Jiang [7] proposed a dust image enhancement algorithm base on gray world and dark channel. This method effectively improve dust images contrast and clarity but it also has more serious color problems. Ning et al. [13] proposed a new algorithm based on hue adjustment and contrast enhancement. Though this method use Gauss model to adjust the color in dust images and singular value decomposition to enhance the dust image quality. It also bring too many parameters, which leads a bad effect on images that have heavy dust.

Some of the above mentioned algorithms just pay attention to the effect of dust image clarity enhancement, while ignoring color problems in dust image. Although other algorithms solve color and clarity questions at the same time. They bring many parameters in calculation process that have a bad influence on the applicability of this algorithm. Therefore, we proposed a dust image enhancement algorithm based on color transfer, which obtaining a better result and reducing the number of parameters.

3   Proposed Approach

3.1   Image Matching Based on Scene Gist Feature

How to choose a clearer image which has similar scene consistent with the input image scene is a key problem of the proposed algorithm. A good target image not only proved the effectiveness of scene matching algorithm, but also a similar scene can get better result in process of color transfer. In order to solve this problem, the scene gist feature [10] is used to select the target image. The image matching process based on the scene gist feature [10] is shown in Fig. 2.

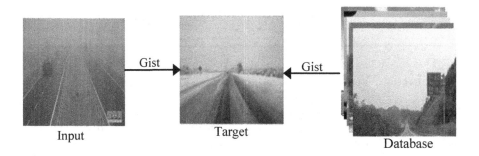

Fig. 2. Image matching based on scene gist

Scene gist [10] is a feature that is commonly used in scene classification. In our method, we use it to select target image have the same scene with input image. The feature can be used to determine the semantic attributes of the scene according to low-level features and rough spatial structure of the scene.

Firstly, extract gist feature from input image and database images. Because scene gist is extracted by multiple low-level feature in scene images (such as spatial frequency, color and texture informations) to create structure about whole scene. Before we select target image which has similar scene with dust image, we should remove the effect of color feature on image scenes. Therefore, when we extracting scene gist feature in image, the multi-scale and multi-direction gabor filter is adopted to extract the texture features in images. This way can reduce the influence of color features on matching process. Secondly, the filtered image is divided into $4 * 4$ blocks. The global feature information of the image is extracted from each block to get scene gist. Next, the euclidean distance about scene gist between input image and database images is calculated. Finally, the image has the smallest euclidean distance in database is the target image.

Target image is selected in a clear images database. So if the database has large number of clear, normal and different scene images are important. So far, we do not find any dust images matching database. According requirement above, we designed a clear images database. Using this database, our method get a good result. In this paper, database design process will be described in Sect. 4.1.

3.2 Color Transfer Model

In 2001, Reinhard et al. [11] proposed a method about color transfer between two different images. The method effectively passes the color of the target image to the input image in CIELab color space. In this way, we descend parameters number in processing procedure and the application scope of our algorithm is extended. This method is decomposed into three steps:

Step 1: In order to removing correlation between different color channels, we convert the target image from RGB color space to CIELab color space.

Step 2: We calculate the mean and variance of each color channel in the target image. Continue doing this work for input image. In this step, mean represents images detail features and variance represents the hue information of the image.

Step 3: According (1), we restore the input image;

$$I_k = \frac{\delta_t^k}{\delta_s^k}(S^k - mean(S^k)) + mean(T^k), \qquad k \in (l, a, b) \tag{1}$$

In this model, S is the input image, T is the target image, δ_s and δ_t is the variance of the input image and target image; $mean(S)$ and $mean(T)$ is the mean of the input image and target image; $k \in (l, a, b)$ represents three color channels. The main process are shown in Fig. 3.

From the second line in Fig. 3, we can see that the input image RGB is not uniform. After color transfer operation, the result image RGB channel are consistent distribution. It means no color cast in image. In summary, through color transfer we could effectively solve the problem about hue yellowing and low brightness in dust image.

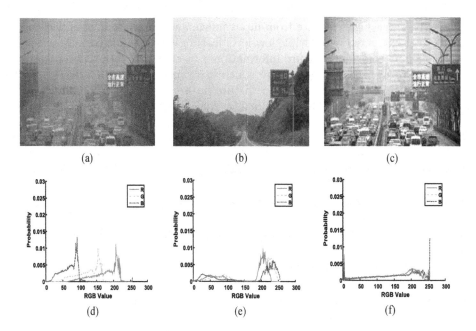

Fig. 3. Color transfer process. (a) input image, (b) target image, (c) image after color transfer. Second line images are the corresponding images of RGB. (Color figure online)

3.3   CLAHE Algorithm

We can find from Fig. 3 that the image after color transfer can solve the color problems in dust images. But problem still exist in result images about low-quality in definition. In order to solve this problem, we utilizing CLAHE algorithm [12] to enhance contrast of the image and finally get a clear result image. CLAHE algorithm [12] has two stages, first we calculate the local histogram of the image. Second, we change the image contrast by redistributing the brightness after set the contrast range. This algorithm can improve the local contrast of the image. Through this way, we can get more image details and raise the image definition.

4   Experimental Results and Analysis

4.1   Database

The purpose of this algorithm is to enhance the dust image to get a clear image. In order to ensure the practical application of our algorithm, we should restrict database images. In the database, images which have reflection of light, shadow, color cast and a large area have same color should remove to avoid the bad affect about dust image enhancement. Because the dust image mainly exist in cities, streets and other artificial scene images, we select these kinds of clear images to get a good match result.

According to above principles, we choose artificial scene images in Oliva and Torralba's database in MIT [14] and Li's database [9] first. Then, remove ineligible images. Finally, we get a clear images database which have 1898 images in total. The color image size in this database are $256 * 256$ (Fig. 4).

(a)   (b)

Fig. 4. Database image and ineligible image examples. (a) examples of database image, (b) examples of ineligible image. (Color figure online)

4.2   Subjective Experiment Results of Dust Images

All experiment results that we shown in this paper are obtained under the same computer environment (Intel (R), Core (TM), i5-4590CPU@3.30 GHz, 8.00 GB, RAM).

The subjective visual effects that we compared with the Retinex algorithm [4], Jiang algorithm [7], and original image, are shown in Fig. 5. It can be seen from this comparison about the dust image enhancement result, not only our method effectively resolve the question about color cast in dust images, but also retains more image details. It proves that our method is beneficial to dust image further processing. From the first image and the fourth image, compared with other two algorithms, it can be seen that our algorithm can improve the brightness of the image. At the same time, as shown in the eighth image, Retinex algorithm and Jiang algorithm dust image enhancement effect is not very ideal, but our algorithm get a better result on enhance severe dust images.

(a)   (b)   (c)   (d)

Fig. 5. Subjective results of dust images enhancement. (a) input image, (b) Retinex [4], (c) Jiang [7], (d) our method.

4.3 Objective Experimental Results of Dust Images

In this paper, we use three indexes that include local information entropy, average gradient and contrast to evaluate results of dust images enhancement. The specific data are shown in Tables 1, 2 and 3.

Table 1. Local entropy of our method compared with others.

Num	Input	Retinex [4]	Jiang [7]	Ours
1	2.6725	3.0455	3.3250	**3.9834**
2	1.8357	2.9464	2.8072	**3.6976**
3	2.7978	2.8455	3.4613	**4.1602**
4	3.6162	3.8168	3.7825	**4.5758**
5	2.8107	2.8780	3.2255	**4.0927**
6	3.1080	3.2605	3.8062	**4.1780**
7	2.8368	3.1701	3.2275	**3.9335**
8	2.3166	3.1621	3.0139	**3.9860**

Table 2. Average gradient of our method compared with others.

Num	Input	Retinex [4]	Jiang [7]	Ours
1	0.0190	0.0191	0.0203	**0.0327**
2	0.0049	0.0109	0.0078	**0.0189**
3	0.0216	0.0226	0.0244	**0.0353**
4	0.0314	0.0383	0.0434	**0.0629**
5	0.0271	0.0280	0.0297	**0.0449**
6	0.0497	0.0498	0.0552	**0.0754**
7	0.0159	0.0210	0.0203	**0.0257**
8	0.0060	0.0150	0.0080	**0.0319**

Table 3. Contrast of our method compared with others.

Num	Input	Retinex [4]	Jiang [7]	Ours
1	66.153	64.327	81.805	**151.238**
2	6.480	18.992	13.005	**59.721**
3	64.457	70.590	80.729	**169.661**
4	62.494	90.456	120.361	**240.995**
5	99.342	106.457	117.975	**266.771**
6	155.574	155.693	194.608	**358.645**
7	34.895	60.831	61.522	**95.477**
8	9.443	48.631	22.002	**236.468**

Image information entropy is usually used to measure the richness of image informations. The average gradient can be used to represent the relative clarity of images. The high contrast is very helpful for image clarity, detail performance and gray level representation. Therefore, these objective quality evaluation indexes can give a comprehensive evaluation to measure the quality of images.

As shown in Table 1, the local information entropy of the proposed algorithm is higher than other two algorithms, which shows that the proposed algorithm preserves more details in dust image. In Table 2, the average gradient of our algorithm is better than other algorithms, it represents our results are more clear than others. In contrast of Table 3, our method result is higher than other algorithms. It shows that our algorithm not only removes the hue deviation of dust images, but also improves the contrast of images greatly and keep images color undistorted. The results of our algorithm are more conducive to the subsequent processing of dust images.

4.4   Simulative Dust Image Experimental Results

In order to further verify the effectiveness of our algorithm, we find four clear images and then use photoshop software to add a mask to adjust image color, contrast and brightness. After these steps, we convert a clear image into a dust image. Then, dust images are processed by Retinex algorithm [4], Jiang algorithm [7], and our algorithm respectively. The final experimental results are shown in Fig. 6.

| (a) | (b) | (c) | (d) | (e) |

Fig. 6. Subjective results of simulative dust images enhancement. (a) original image, (b) simulative dust image, (c) Retinex [4], (d) Jiang [7], (e) our method. (Color figure online)

As shown in Fig. 6, the Retinex algorithm [4] results are slant in red, Jiang algorithm [7] overall slant in blue, while our algorithm does not exist these problems. To better describe the result about our method, we will evaluate the quality of result images and original images respectively.

In this part, we use an integrated local natural image quality evaluation (IL-NIQE) [15] was proposed by Zhang in 2015. This algorithm which calculate image informations like color, texture, contrast and so on in different databases to train an image quality evaluation model. We use IL-NIQE to evaluate the quality of the original images and result images are shown in Fig. 6 through IL-NIQE. The experimental results are show in Table 4. In this image evaluation model measure result, we can find our algorithm is significantly similar with original images than other algorithms. It also objectively confirmed the effectiveness of the proposed algorithm.

Table 4. IL-NIQE values of our method compared with others.

Num	Input	Retinex [4]	Jiang [7]	Ours
1	16.977	24.116	22.956	**22.165**
2	17.460	20.770	21.528	**18.759**
3	17.861	18.705	23.611	**18.228**
4	16.122	22.657	19.941	**16.301**

Although we verified the effectiveness of the proposed algorithm from this evaluation model. But this method requires a lot of training data to construction the evaluation model and also expend more time. Because of the restriction of the database, this method cannot generated objective evaluation results for any picture. Therefore, in order to get a simple and effective method to evaluation dust image enhancement result, we proposed a color similarity evaluation index named absolute color difference (ACD). First, we calculate the difference of two images between each color channel. Then, we get absolute value from these difference and finally calculating the mean of the absolute value. The calculation equation of the algorithm is shown in (2).

$$M = \frac{1}{3} \sum_{c=1}^{3} mean(|I_1^c - I_2^c|) \tag{2}$$

M is the color absolute difference, c represents R, G, B three channels of images, I_1 is the original image, I_2 is the result image. It can be seen from (2) that smaller absolute color difference represents smaller color difference between two images. It also means better effect of dust image enhancement.

To verifying the objective of the evaluation criteria, we use photoshop software to add a yellow mask to Fig. 7(a). The opacity of mask are 30%, 60%, and 90%, respectively. Then we use absolute color difference to calculate the color similarity between these pictures with original picture. The results of ACD are 4.89, 9.71 and 14.60, respectively. Hence, we can concluded that with the increase of mask opacity, the absolute color difference is improving, the greater difference with original color. We illustrate the absolute color difference can effectively evaluate image color similarity.

Fig. 7. Images with different levels of mask. (a) original image. (b) 30% opacity mask image. (c) 60% opacity mask image. (d) 90% opacity mask image. (Color figure online)

Table 5. ACD of our method compared with others.

Num	Retinex [4]	Jiang [7]	Ours
1	11.000	43.000	**2.000**
2	30.667	61.333	**28.667**
3	72.667	110.000	**31.667**
4	22.333	25.333	**3.000**

After verifying the objective validity of the absolute color difference on two images color similarity. We use this index assess images in Fig. 6. The experimental results are shown in Table 5. From the comparison of the absolute color difference between different algorithms, we can find that the algorithm is proposed in this paper is superior to Retinex algorithm [4] and Jiang algorithm [7]. These results further proves the superiority of our algorithm.

5   Conclusions

Aim at the problems of dust image about hue yellowing, low-quality brightness and definition, we put forward a dust image enhancement algorithm based on color transfer. Firstly, we through the scene gist feature select a target image in clear images database which has the highest similarity with input image. Then, we transfer the target image color information to the input image that solve color questions in dust image. Next, we use CLAHE algorithm to increase image definition. Finally, we achieve the purpose of image enhancement for dust. The subjective and objective experimental results show that the proposed algorithm is superior to traditional algorithms. In the future, we will study how to apply this algorithm in more vision tasks such as dust image text detection and recognition.

Acknowledgments. The paper was supported in part by the National Natural Science Foundation (NSFC) of China under Grant No. 61365003 and Gansu Province Basic Research Innovation Group Project No. 1506RJIA031.

References

1. Rahman, Z., Jobson, D.J., Woodell, G.A.: Multi-scale retinex for color image enhancement. In: IEEE International Conference on Image Processing, vol. 3, pp. 1003–1006 (1996)
2. Jobson, D.J., Rahman, Z., Woodell, G.A.: A multiscale retinex for bridging the gap between color images and the human observation of scenes. IEEE Trans. Image Process. **6**(7), 965–976 (1997)
3. He, K., Sun, J., Tang, X.: Single image haze removal using dark channel prior. IEEE Trans. Pattern Anal. Mach. Intell. **33**(12), 2341–2353 (2011)
4. Bertalmío, M., Caselles, V., Provenzi, E.: Issues about retinex theory and contrast enhancement. Int. J. Comput. Vis. **83**(1), 101–119 (2009)
5. Yan, T., Wang, L., Wang, J.: Video image enhancement method research in the dust environment. Laser J. **4**, 23–25 (2014)
6. Wu, Y.: Research on enhancement algorithms for image and video in dust environment. M.Sc. thesis, Inner Mongolia University (2016)
7. Jiang, H.: Research of image enhancement algorithms is extreme weather conditions. M.Sc. thesis, Shenyang University of Technology (2016)
8. Li, H., Zhong, M.: Restoration algorithm of dust weather images based on dark prior. J. Kashgar Univ. **37**(3), 42–45 (2016)
9. Li, M.: Research of scene classification algorithm based on hypercomplex DCT domain. M.Sc. thesis, Lanzhou University of Technology (2015)
10. Aude, O.: Gist of the scene. Neurobiol. Attention **696**(64), 251–258 (2005)
11. Reinhard, E., Adhikhmin, M., Gooch, B., Shirley, P.: Color transfer between images. IEEE Comput. Graph. Appl. **21**(5), 34–41 (2001)
12. Reza, A.M.: Realization of the contrast limited adaptive histogram equalization (CLAHE) for real-time image enhancement. J. VLSI Sig. Process. **38**(1), 35–44 (2004)
13. Ning, Z., Shanjun, M., Li, M.: Visibility restoration algorithm of dust-degraded images. J. Image Graph. **21**(12), 1585–1592 (2016)
14. Oliva, A., Torralba, A.: Modeling the shape of the scene: a holistic representation of the spatial envelope. Int. J. Comput. Vis. **42**(3), 145–175 (2001)
15. Zhang, L., Zhang, L., Bovik, A.C.: A feature-enriched completely blind image quality evaluator. IEEE Trans. Image Process. **24**(8), 2579–2591 (2015)

Chinese Sign Language Recognition with Sequence to Sequence Learning

Chensi Mao, Shiliang Huang, Xiaoxu Li, and Zhongfu Ye[✉]

Department of Electronic Engineering and Information Science,
National Engineering Laboratory for Speech and Language Information Processing,
University of Science and Technology of China, Hefei 230027, Anhui, China
{mcs111,huangsl8,lxx123}@mail.ustc.edu.cn, yezf@ustc.edu.cn

Abstract. In this paper, we formulate Chinese sign language recognition (SLR) as a sequence to sequence problem and propose an encoder-decoder based framework to handle it. The proposed framework is based on the convolutional neural network (CNN) and recurrent neural network (RNN) with long short-term memory (LSTM). Specifically, CNN is adopted to extract the spatial features of input frames. Two LSTM layers are cascaded to implement the structure of encoder-decoder. The encoder-decoder can not only learn the temporal information of the input features but also can learn the context model of sign language words. We feed the images sequences captured by Microsoft Kinect2.0 into the network to build an end-to-end model. Moreover, we also set up another model by using skeletal coordinates as the input of the encoder-decoder framework. In the recognition stage, a probability combination method is proposed to fuse these two models to get the final prediction. We validate our method on the self-build dataset and the experimental results demonstrate the effectiveness of the proposed method.

Keywords: Sign language recognition · Long short-term memory
Convolutional neural network · Trajectory

1 Introduction

Sign language is a kind of gestures expressed by the movements of hands and body to exchange information for deaf and mute people. Recently, sign language recognition (SLR) has drawn increasing attention in the computer vision community in order to provide an efficient mechanism to convenient the communication between the normal and hearing impaired people.

SLR has been widely studied over the years. Previous work on SLR can be divided into two categories: traditional methods and deep learning based methods. For traditional methods, hand-crafted features are usually adopted, such as Histogram of Oriented Gradient (HOG) [1] and Histogram Of Oriented Optical Flow (HOF) [2]. However, the hand-crafted features are usually too simple to represent the sign language and are difficult to design. Besides, Hidden

© Springer Nature Singapore Pte Ltd. 2017
J. Yang et al. (Eds.): CCCV 2017, Part I, CCIS 771, pp. 180–191, 2017.
https://doi.org/10.1007/978-981-10-7299-4_15

Markov Model (HMM) is a classical method to model the temporal information [3,4], but HMM is based on the assumption of Markov property , which is not always true for sign language in practice.

Recently, deep learning has achieved great success in computer vision. Convolutional neural network (CNN) and Recurrent Neural Network (RNN) are two typical deep learning architectures. CNN has a strong ability to learn effective representative of images, and has led to a series breakthroughs in many computer vision tasks, including image classification [5] and object detection [6]. RNN shows strong ability in modeling temporal information, so it's widely adopted in sequence modeling tasks such as speech recognition [7] and machine translation [8]. Inspired by the success of deep learning, researchers start to apply learning based methods to SLR. In these methods, the hand-crafted features are replaced with the learned features. For example, in [9], a multichannel CNN is developed to enhance the features for hand posture classification. Later on, LSTM is adopted in [10,11] to model the temporal information. All these methods regard SLR as a classification problem, while neglect the inherent context relationship between sign language words. As the developing of deep learning, encoder-decoder network has been proposed for many sequence tasks. In [12], Venugopalan *et al.* employed CNN for feature extraction and pour the feature into the encoder-decoder network to realize video caption. In [13], a unified network consisting of convolutional layers, recurrent layers, transcription layers was proposed for scene text recognition.

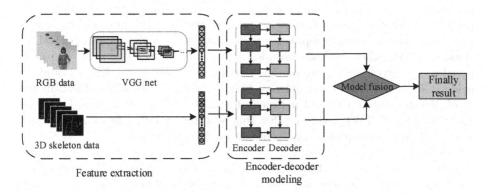

Fig. 1. The proposed sequence to sequence learning framework. This framework consists three modules, including feature extraction, encoder-decoder modeling and model fusion.

In this paper, we propose a sequence to sequence framework for SLR. The aim of SLR is actually to translate a gesture into its corresponding words, which is actually a sequence to sequence mapping problem similar to video caption. The overall framework is illustrated as Fig. 1. The proposed framework consists of three modules: feature extraction, encoder-decoder modeling and model fusion. The inputs of the framework are captured by Kinect2.0, including RGB image sequences and skeletal coordinates. Firstly, the image sequences are fed into the CNN to extract the spatial features. Then the spatial features are used as the input of the encoder-decoder for temporal modeling. The encoder-decoder module consists of two LSTM layers. The encoder LSTM layer is used to map the input features to a fixed-length vector, while the decoder LSTM is used to learn the context relationship of the sign language. Secondly, the skeletal coordinates are used as auxiliary features and are directly fed into another encoder-decoder module. Finally, in the model fusion module, a probability combination method is proposed to fuse the two models to get the final prediction.

The remainder of this paper is organized as follows. In Sect. 2, we will give an a brief review of the related work. The proposed sequence to sequence framework will be introduced in detail in Sect. 3. The experimental results are shown in Sect. 4, and Sect. 5 concludes the paper.

2   Related Work

The sign language recognition methods can be divided into two categories: traditional methods and deep learning based methods. Traditional SLR methods usually use hand-crafted features. In [1], HOG feature is adopted for hand shape representation. Several other researchers focus on trajectory feature to realize more robust SLR [14] or gesture recognition [15]. As to temporal modeling, HMM has been widely used. Starner et al. [16] utilized HMM to model American Sign Language acquired by tracking the user's unadorned hands using single camera. In [17], Gao et al. proposed a method using the self-organizing feature maps (SOFM) as different signers feature extractor to transform input signs into significant and low-dimensional representations that can be well modeled by HMM. Besides HMM, other methods are also explored. Sminchisescu et al. [18] proposed a framework for human motion recognition based on conditional random fields (CRFs) and maximum entropy Markov models (MEMMs). Lichtenauer et al. [19] presented a hybrid approach for SLR by using statistical Dynamic Time Warping (DTW), assuming that time warping and classification should be separated because of conflicting likelihood modeling demands.

As for deep learning based SLR method, Barros et al. [9] developed a multi-channel CNN to enhance the features for hand posture classification. Wu et al. [20] employed dynamic neural networks to process skeleton and depth and RGB (RGB-D) images of sign language data, and then combined the network with HMM to realize gesture recognition. Later on, Pigou et al. [11] proposed an end-to-end architecture incorporating temporal convolutions and bidirectional recurrence in order to acquire better temporal feature. More recently, Liu et al. [10] presented an end-to-end method for SLR based on LSTM, using trajectory acquired by using Microsoft Kinect as the input of the network.

3   Proposed Method

In this section, we will introduce the proposed sequence to sequence framework in detail. As shown in the Fig. 1, the whole framework can be divided into three modules: feature extraction, encoder-decoder modeling and model fusion. We will introduce the three modules in the following sections.

3.1   Feature Extraction

The sign language data used in this work is captured by Microsoft Kinect2.0, including image sequences and 3D skeletal points. The image sequences and skeletal coordinates are jointly used, considering that multi-modality of data is beneficial for recognition. As is shown in the Fig. 1, we utilize CNN to extract the feature of image sequences rather than designing hand-crafted feature. In this work, we use the pre-trained VGG-16 [21] as the feature extractor, which is an outstanding model trained on the ImageNet dataset [22]. The image sequences are fed into the VGG-16 and the fc7 layer units are selected as the image feature, which is a vector of 4096 dimensions.

Kinect2.0 has the skeletal tracking ability by recording space coordinates (x, y, z) of 25 skeleton points of human body, such as hands, wrists and elbows. The 25 coordinates are concatenated to a vector of 75 dimensions. Therefore, each frame of the image sequences corresponds to a string of space coordinates with 75 dimensions. We take this kind of data as the trajectory features considering that the movements of hands or body of the sign language can also play an important role in recognition.

3.2   Encoder-Decoder Modeling

The Background of LSTM. RNN are the most popular choice to solve the sequence to sequence mapping problems. RNN can model the temporal relationship within the input sequence by using the recurrent feedbacks. For the simple RNN, it may suffer from the gradient vanishing and exploding problem [23]. Therefore, some new architectures have been proposed to alleviate these problems. The long short-term memory (LSTM) [24] model is an enhanced RNN architecture to implement the recurrent feedbacks using various learnable gates and has been widely adopted in temporal modeling. Given the input x_t, the forget gate f_t, input gate i_t, output gate o_t, input modulation gate $\tilde{C}_t$ and memory cell C_t can be computed by the following equations :

$$f_t = sigm(W_{xf}x_t + W_{hi}h_{t-1} + b_f) \tag{1}$$

$$i_t = sigm(W_{xi}x_t + W_{hi}h_{t-1} + b_i) \tag{2}$$

$$\tilde{C}_t = tanh(W_{xc}x_t + W_{hc}h_{t-1} + b_c) \tag{3}$$

$$C_t = f_t \odot C_{t-1} + i_t \odot \tilde{C}_t \tag{4}$$

$$o_t = sigm(W_{xo}x_t + W_{ho}h_{t-1} + b_o) \tag{5}$$

$$h_t = o_t \odot tanh(C_t) \tag{6}$$

where $sigm$ is the sigmoid function, $tanh$ is the hyperbolic tangent non-linearity, $\odot$ stands for element-wise product with the gate value, and the weight matrices denoted by W_{ij} and biases b_j are the network parameters which should be optimized during the training process.

Encoder-Decoder Network for SLR. Typical methods regard SLR as a classification task, while in this paper we formulate SLR as a sequence to sequence problem, mainly based on two observations as follows:

(i) Some isolated Chinese sign language usually consists of several separated gestures and each gesture corresponds to some specific Chinese characters.
(ii) Other Chinese sign language cannot be further separated, but there still exists inherent context relationship among the Chinese characters.

Figure 2 provides two typical examples to explain this observation. Figure 2(a) represents the Chinese sign language word 'post office', which can be divided into two isolated gestures corresponding to two Chinese characters: 'post' and 'house', respectively. The first five frames correspond to 'post' and the rest five frames mean 'house'. Figure 2(b) illustrates the sign language word 'train', which consists of two characters in Chinese. Although the sign language word 'train' cannot be further divided into isolated gestures, there still exists context relationship between the two characters of 'train' in Chinese. Considering the two types of characteristics of Chinese sign language, we formulate Chinese SLR as a sequence to sequence problem. Inspired by the application of LSTM in video caption, we design an encoder-decoder network to handle it.

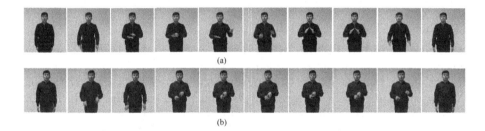

(a)

(b)

Fig. 2. Two typical Chinese sign language words widely used in the daily life. (a) 'post office', (b) 'train'.

The encoder-decoder network is shown in Fig. 3, which consists of two LSTMs (blue blocks and yellow blocks) and a softmax layer. The first LSTM (blue blocks) is acted as the encoder and the other LSTM (yellow blocks) is acted as the decoder. At the encoding stage, the LSTM is used to map the input feature $(x_1, x_2, ..., x_n)$ to a fixed-length vector which contains the temporal information. After all the input features of sign language sample are exhausted, there will be a mark to indicate the beginning of decoding. At the decoding stage, the output of the encoder is fed into the second LSTM to learn the context model of sign words. We establish a vocabulary by taking the Chinese characters as the basic elements. The final output is a string of Chinese characters from the vocabulary. The probability $p(z_t)$ of every character in the vocabulary at each time t will be estimated using a softmax layer:

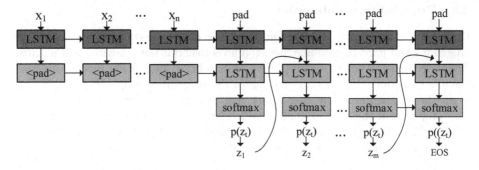

Fig. 3. The encoder-decoder framework. The top LSTM is the encoder mapping the inputs to a temporal fixed-length vector, and the bottom LSTM is the decoder which build a language model and output a sequence of characters. EOS is used to indicate the end of decoding. Zeros padding is used when there is no input at the time step. (Color figure online)

$$p(z_t|h_t) = \frac{exp(W_z h_t)}{\sum_{z \in V} exp(W_z h_t)} \tag{7}$$

where h_t is the output vector of decoder LSTM at time t, z_t is a basic element in the vocabulary, W_z is the parameters of this layer and V is the whole vocabulary. The final outputs can be estimated by the conditional probability $p(z_1, z_2, ..., z_m | x_1, x_2, ..., x_n)$, which can be computed by the following equation:

$$p(z_1, z_2, ..., z_m | x_1, x_2, ..., x_n) = \prod_{t=1}^{m} p(z_t | h_t, z_1, z_2, ..., z_{t-1}) \tag{8}$$

Image and Trajectory Sequence Modeling. In the feature extraction stage, CNN is utilized to extract the spatial features of image sequences. The output of CNN is a vector of 4096 dimensions. The 4096-dimensional vector is firstly embedded to 500 dimensions by a fully connected layer which will be trained jointly with the encoder-decoder network. The embedding operation is utilized to reduce the complexity of model. After embedding, the image feature is fed into the encoder-decoder network frame by frame. During the encoding process, the image feature sequence is mapped into a fixed-length representation. During the decoding process, the characters sequence $(z_1, z_2, ..., z_m)$ is generated. In this way, we can establish the images sequence model, termed as $model_{rgb}$.

As to the trajectory feature, we directly utilize the 75-dimensional coordinate vector as the input of another encoder-decoder. In this way, we can establish the trajectory model (denoted as $model_{tra}$). This model can be used as a complementary of images sequence model, because skeletal coordinates provide important motion information of the human body.

3.3 Probability Combination

Unlike data and feature level fusion where the data or features are usually het-erogenous and hard to combine, decision level fusion is more flexible and nature. Therefore, in this paper we propose a probability combination method to fuse the image model and trajectory model. In the recognition step, the probabilities of all the characters in the vocabulary will be computed for each model at the time step t. And the final prediction is determined by combining the two proba-bilities through the following equation, so the decoding output of time t will be the character which corresponds to the maximum decoding probability.

$$p(z_t = w) = \alpha \times p_{rgb}(z_t = w) + (1 - \alpha) \times p_{tra}(z_t = w) \tag{9}$$

where w is a certain character in the vocabulary, α is the weighting parameter which will be tune on testing dataset, and p_{rgb} and p_{tra} are the probabilities estimated by $model_{rgb}$ and $model_{tra}$, respectively.

4 Experiments

In this section, we conduct a series of experiments to evaluate the effectiveness of the proposed approach on our self-built dataset captured by the Kinect2.0. First, we will introduce the dataset and the detail of experiment settings. Then the experimental results and analysis are provided.

4.1 Dataset and Experiment Settings

The dataset adopted in the experiment consists of 90 Chinese sign language words widely used in our daily life, such as 'airplane' and 'train', unlike simple gesture, sign language is usually more complex which consists of several simple gestures. A sign language word sample usually consists of 80–120 frames. There are 100 samples for each word, and these samples are captured by Kinect2.0. Therefore, this dataset is composed of 9000 samples in total. For each sample, both RGB frames and skeletal coordinates are recorded. Specifically, the cap-tured RGB frame is of size 1920×1080, skeletal coordinates are (x, y, z) of 25 skeleton points of human body. In the experiment, we divide the dataset into two parts, 70% of the data is utilized as the training set and the rest as test set. We use the pre-trained VGG-16 as the spatial feature extractor of images. For an input image, we extract the $fc7$ layer with 4096 units in VGG net as the image feature and then embed it to a 500-dimensional vector. For the encoder-decoder, the number of units of each LSTM is set to be 1000. We use the deep learning framework Caffe [25] to train the models. The network parameters are optimized using stochastic gradient descent (SGD), and the cross-entropy loss is used as loss function.

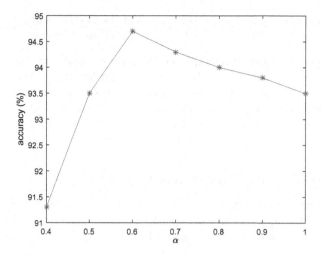

Fig. 4. Recognition accuracy with different α values

4.2   Results and Analysis

After the experiment setting and data preparation, we use the self-build dataset to evaluate the performance of our method. We firstly study the impact of different α values on the final recognition accuracy. We vary this parameter from 0.4 to 1, and the result is illustrated in Fig. 4. It can be seen that we can obtain the best recognition accuracy when α is set to be 0.6, which is also used in the following experiment.

For comparison, two typical methods [10, 26] are selected as the representative algorithms. The method in [26] is utilizing hand shape features (HOG features) and trajectory features to represent the sign videos, and using HMM to realize recogntion. We also implement the algorithm proposed in [10], where LSTM is employed to learn the temporal information of trajectory. The recognition accuracy of the above methods are shown in Table 1.

Table 1. Average accuracy of the proposed method

	Features	Accuracy
HMM [26]	HOG feature + trajectory	88.9%
LSTM_fc [10]	Normal skeleton joints	92.6%
$model_{rgb}$	Gesture images	93.5%
$model_{fusion}$	Gesture images + trajectory	94.7%

We can see from the Table 1 that our method outperforms the compared methods when only using the $model_{rgb}$. And it is noted that when fusing RGB model and trajectory model, the accuracy can be further improved by about 1.2%. From the experimental results, we can draw the following conclusions:

(i) The proposed sequence to sequence learning method has the capability to learn the inherent context relationship in sign language. And this method is more reliable for the Chinese sign language recognition task.
(ii) Since the trajectory data provide effective and precise position information by recoding the space coordinates, it can be used as an enhancement for recognition.

4.3   Results of Using Hand Region Images

We also conduct the experiment by using the hand region images instead of the full images as the input of the framework. The method in [27] are adopted to extract the hand region. The experimental results are shown in Table 2.

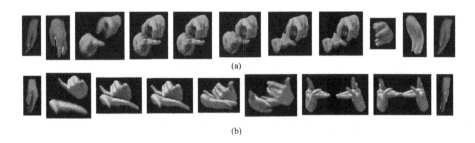

Fig. 5. Two Chinese sign language words after hand region extracting. (a) 'train', (b) 'berth'.

As can be seen from the Table 2, the recognition accuracy can be further improved by 3.1% compared with using the full gesture images. This is mainly due to the fact that the most meaningful movements center on the hand region which is shown in Fig. 5. Besides, there is only tiny difference between some movements of different gestures, in which case the body and clothes may interfere, so the full images features extracted by CNN cannot be discriminated effectively during recognition. Therefore, using the hand region instead of the full images can be more beneficial for the recognition accuracy.

Table 2. Results using hand region images and trajectory

	Features	Accurancy
$model_{rgb}$	Hand images	97.1%
$model_{fusion}$	Hand images + trajectory	97.8%

5   Conclusion

In this paper, a sequence to sequence framework based on CNN and LSTM is proposed for Chinese sign language recognition. The propose sequence to sequence framework consists of three modules: feature extraction, encoder-decoder modeling and model fusion. Specifically, we use CNN to extract image sequence features, an encoder-decoder network is utilized to build the end-to-end temporal model, and finally image model and trajectory model are fused to get recognition result. This framework can not only learn the spatial features of the input but also can learn the temporal information and context relationship in Chinese sign language. A series of experiments have been conducted, and the experimental results show that the proposed method outperforms the compared methods on our dataset. In the future, we will go on investigate how to further improve the recognition accuracy and apply this method to continuous sign language recognition.

Acknowledgments. This work is supported by the Fundamental Research Funds for the Central Universities (Grant no. WK2350000002).

References

1. Jangyodsuk, P., Conly, C., Athitsos, V.: Sign language recognition using dynamic time warping and hand shape distance based on histogram of oriented gradient features. In: Proceedings of the 7th International Conference on PErvasive Technologies Related to Assistive Environments, p. 50. ACM (2014)
2. Hernández-Vela, A., Bautista, M.Á., Perez-Sala, X., Ponce-López, V., Escalera, S., Baró, X., Pujol, O., Angulo, C.: Probability-based dynamic time warping and bag-of-visual-and-depth-words for human gesture recognition in RGB-D. Pattern Recogn. Lett. **50**, 112–121 (2014)
3. Grobel, K., Assan, M.: Isolated sign language recognition using hidden Markov models. In: 1997 IEEE International Conference on Systems, Man, and Cybernetics. Computational Cybernetics and Simulation, vol. 1, pp. 162–167. IEEE (1997)
4. Yang, R., Sarkar, S.: Gesture recognition using hidden Markov models from fragmented observations. In: 2006 IEEE Computer Society Conference on Computer Vision and Pattern Recognition (CVPR 2006), vol. 1, pp. 766–773. IEEE (2006)
5. Krizhevsky, A., Sutskever, I., Hinton, G.E.: Imagenet classification with deep convolutional neural networks. In: Advances in Neural Information Processing Systems, pp. 1097–1105 (2012)
6. Ouyang, W., Wang, X.: Joint deep learning for pedestrian detection. In: Proceedings of the IEEE International Conference on Computer Vision, pp. 2056–2063 (2013)
7. Graves, A., Jaitly, N.: Towards end-to-end speech recognition with recurrent neural networks. ICML, vol. 14, pp. 1764–1772 (2014)
8. Sutskever, I., Vinyals, O., Le, Q.V.: Sequence to sequence learning with neural networks. In: Advances in Neural Information Processing Systems, pp. 3104–3112 (2014)

9. Barros, P., Magg, S., Weber, C., Wermter, S.: A multichannel convolutional neural network for hand posture recognition. In: Wermter, S., Weber, C., Duch, W., Honkela, T., Koprinkova-Hristova, P., Magg, S., Palm, G., Villa, A.E.P. (eds.) ICANN 2014. LNCS, vol. 8681, pp. 403–410. Springer, Cham (2014). https://doi.org/10.1007/978-3-319-11179-7_51

10. Liu, T., Zhou, W., Li, H.: Sign language recognition with long short-term memory. In: 2016 IEEE International Conference on Image Processing (ICIP), pp. 2871–2875. IEEE (2016)

11. Pigou, L., Van Den Oord, A., Dieleman, S., Van Herreweghe, M., Dambre, J.: Beyond temporal pooling: recurrence and temporal convolutions for gesture recognition in video. arXiv preprint arXiv:1506.01911 (2015)

12. Venugopalan, S., Rohrbach, M., Donahue, J., Mooney, R., Darrell, T., Saenko, K.: Sequence to sequence-video to text. In: Proceedings of the IEEE International Conference on Computer Vision, pp. 4534–4542 (2015)

13. Shi, B., Bai, X., Yao, C.: An end-to-end trainable neural network for image-based sequence recognition and its application to scene text recognition. IEEE Trans. Pattern Anal. Mach. Intell. 39, 2298–2304 (2016)

14. Sun, C., Zhang, T., Bao, B.K., Xu, C.: Latent support vector machine for sign language recognition with kinect. In: 2013 20th IEEE International Conference on Image Processing (ICIP), pp. 4190–4194. IEEE (2013)

15. Gowayyed, M.A., Torki, M., Hussein, M.E., El-Saban, M.: Histogram of oriented displacements (HOD): describing trajectories of human joints for action recognition. In: IJCAI (2013)

16. Starner, T., Weaver, J., Pentland, A.: Real-time american sign language recognition using desk and wearable computer based video. IEEE Trans. Pattern Anal. Mach. Intell. 20(12), 1371–1375 (1998)

17. Gao, W., Fang, G., Zhao, D., Chen, Y.: A chinese sign language recognition system based on SOFM/SRN/HMM. Pattern Recogn. 37(12), 2389–2402 (2004)

18. Sminchisescu, C., Kanaujia, A., Metaxas, D.: Conditional models for contextual human motion recognition. Comput. Vis. Image Underst. 104(2), 210–220 (2006)

19. Lichtenauer, J.F., Hendriks, E.A., Reinders, M.J.T.: Sign language recognition by combining statistical dtw and independent classification. IEEE Trans. Pattern Anal. Mach. Intell. 30(11), 2040–2046 (2008)

20. Wu, D., Pigou, L., Kindermans, P.J., Le, N.D.H., Shao, L., Dambre, J., Odobez, J.M.: Deep dynamic neural networks for multimodal gesture segmentation and recognition. IEEE Trans. Pattern Anal. Mach. Intell. 38(8), 1583–1597 (2016)

21. Simonyan, K., Zisserman, A.: Very deep convolutional networks for large-scale image recognition. arXiv preprint arXiv:1409.1556 (2014)

22. Deng, J., Dong, W., Socher, R., Li, L.J., Li, K., Fei-Fei, L.: Imagenet: a large-scale hierarchical image database. In: IEEE Conference on Computer Vision and Pattern Recognition. CVPR 2009, pp. 248–255. IEEE (2009)

23. Bengio, Y., Simard, P., Frasconi, P.: Learning long-term dependencies with gradient descent is difficult. IEEE Trans. Neural Netw. 5(2), 157–166 (1994)

24. Hochreiter, S., Schmidhuber, J.: Long short-term memory. Neural Comput. 9(8), 1735–1780 (1997)

25. Jia, Y., Shelhamer, E., Donahue, J., Karayev, S., Long, J., Girshick, R., Guadarrama, S., Darrell, T.: Caffe: convolutional architecture for fast feature embedding. In: Proceedings of the 22nd ACM international conference on Multimedia, pp. 675–678. ACM (2014)

26. Zhang, J., Zhou, W., Xie, C., Pu, J., Li, H.: Chinese sign language recognition with adaptive HMM. In: 2016 IEEE International Conference on Multimedia and Expo (ICME), pp. 1–6. IEEE (2016)
27. Jiang, Y., Tao, J., Ye, W., Wang, W., Ye, Z.: An isolated sign language recognition system using RGB-D sensor with sparse coding. In: 2014 IEEE 17th International Conference on Computational Science and Engineering (CSE), pp. 21–26. IEEE (2014)

ROI Extraction Method of Infrared Thermal Image Based on GLCM Characteristic Imitate Gradient

Hui Shen, Li Zhu$^{(\boxtimes)}$, Xianggong Hong, and Weike Chang

College of Information Engineering, Nanchang University,
Nanchang 330031, Jiangxi, China
lizhu@ncu.edu.cn

Abstract. Automatic inspection of UAV (unmanned aerial vehicle) vision in photovoltaic power station is of great significance in effectively capturing high-definition images and quickly detecting the fault area, which can reduce the risk of false detection and lower the cost of manual operation. However, due to the complexity of photovoltaic power station environment, disturbance often occurrence on images, leading up to the misjudgment of the fault area. We proposed a method to extraction region based on the gray level co-occurrence matrix (GLCM) and textural features. The image extraction from target area can be achieved by extracting feature images, gradient imitate and region filling. This method effectively combines the textural features of images with edge features of the gradient images. A comparison is made between the algorithm promoted in this paper and the grab cut method, on the basis of the labeled image segmentation algorithm. It turns out that the mean precision and mean recall of the proposed imitation gradient image extraction method are higher than that of the grab cut algorithm, and the *Recall*value, F index and J index are better than the grab cut algorithm. A new algorithm is proposed construct filling model by using gradient image and a morsel of texture feature calculated by GLCM method. The advantages of the proposed algorithm are fewer interactive tags, fewer manual labels. Therefore, the image detection of fault area can be better realized.

Keywords: GLCM · Contrast · Entropy · Imitate gradient

1 Introduction

As a kind of green energy, photovoltaic power generation has a very promising development and significant economic benefits. How to detect the fault area of solar panels efficiently and accurately is of great significance for photovoltaic power plants. During recent years, with the application of thermal imaging technology, the fault detection of large area solar panels has become possible, the UAV equipped with infrared thermal imager inspects the solar panel group

© Springer Nature Singapore Pte Ltd. 2017
J. Yang et al. (Eds.): CCCV 2017, Part I, CCIS 771, pp. 192–205, 2017.
https://doi.org/10.1007/978-981-10-7299-4_16

overhead, getting infrared thermal imaging of the photovoltaic plate area. The scene of heat balance, transmission distance, atmospheric attenuation, instrument imaging system and other factors will result in low-resolution images for human beings eyes, advent of edge blur. The clarity is even lower than that of the visible light image. SNR (signal-to-noise ratio) is often low when the image component distribution is complex [6]. In the area extraction of an infrared thermal image, the interference caused by the environmental factors in the image composition should not be ignored. There are a variety of interference factors, among which the instrument box, the hot pipe connected with the lower part of the plate area, the bare steel pipe bracket of photovoltaic panel are the typical ones. At the same time, the influence of the weather plays a role in the infrared thermal imaging figure, for example, cloudy shooting thermal imaging makes a little temperature difference between the plate and the environment come into being.

Therefore, how to make better use of the only image information in the image, to make it better for segmentation and extraction, is the problem we should study. The extraction of the photovoltaic panel is the key step for the thermal images segmentation, whose result will directly affect the later fault detection. Therefore, it is of great importance to extract the solar panels correctly and remove the environmental interference. Infrared thermal image ROI extraction method can be divided into two types, the digital image processing method and statistical based method. From the image processing method perspective, it can be classified as methods from threshold [17], textural analysis [13] and energy-based method [2,7] to infrared background clutter suppression filter [14,15].

However, these methods have special requirements for the image, when the SNR of part of the region is lower, or when the gray levels gap between the target area and interference region is not so big, it often makes the hybridity results of target area and interference region mix [18]. Besides, a region is required to be extracted for further classification. Recently, image extraction has been developed based on statistical methods, such as fuzzy C-means (FCM) clustering image extraction method [10,12]. Document [16] applies optimized Pulse-Coupled Neural Network (PCNN) into the image segmentation. However, when it comes to practical applications, the statistical method needs to pre-mark the image data manually. Besides, the built model needs a great number of labor experiments and image data, and continuous analyses to increase the effectiveness of model implementation. The introduction of statistical methods to extract ROI of images is obviously very difficult and time-consuming even for the processing of a small amount of dataset.

Therefore, an infrared thermal image ROI extraction method based on GLCM feature [1,3] imitation gradient based on the characteristics of infrared image data sets is proposed in this paper. The gray level gap, edge change and the texture structure of the image are used to construct a feature model. Afterward, the feature model will be used to extract the image ROI. The method is used to test 17 photovoltaic panels thermal images for multiple times in this paper. The extracted mean precision and mean recall value are relatively high according to the results.

2   Method

2.1   The Proposed Method

In this paper, the algorithm is used to extract the region of the thermal image of the photovoltaic panel through three main steps. The first step is to extract the feature images, and then obtain the "contrast" and "entropy" feature images of the original image by calculating GLCM. Then obtain the gradient image by calculating original image after cross-over process. In the second step, the imitate gradient images are achieved by weighting the three feature images. The third step is to fill the imitate gradient image and complete the extraction of the ROI region.

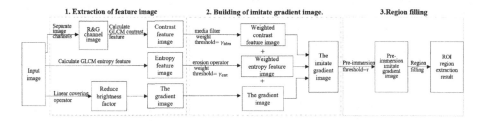

Fig. 1. The diagram of the proposed method is illustrated in Fig. 1.

2.2   Extraction of the Three Feature Image

Extraction of the Gradient Image. As the imaging range of the commercial hand-held image is small so that the local temperature is easy to become higher. This paper uses the method of weakening the influence of image brightness by linear covering operator and the covering operator is as follows:

$$f_2(x,y) = (1-\mu)f_0(x,y) + \mu f_1(x,y) \tag{1}$$

$(1-\mu)$ is the attenuation factor which weekend the impact of the original image $f_0(x,y)$. Supposing the all black image is $f_1(x,y)$, change the μ from 0 to 1, and then get the cross-dissolve image $f_2(x,y)$, which reduced the influence of the brightness factor. Change RGB mode to the HSI mode to get the I component image $I_2(x,y)$ of $f_2(x,y)$.

The method to calculate the difference between the horizontal and vertical component is used to solve the gradient. In this formula: $I_2(x,y)$ represents a pixel in the $f_2(x,y)$ and the gradient image $g(x,y)$ will be got after calculation as per the following method.

$$I_2(x,y) = 0.299 \times R_{f_2} + 0.587 \times G_{f_2} + 0.114 \times B_{f_2} \tag{2}$$

$$g(x,y) = |I_2(x,y) - I_2(x+1,y)| + |I_2(x,y) - I_2(x,y+1)| \tag{3}$$

Extraction of Contrast Feature Image. The experiment shows that the infrared thermal images of R channel mainly represents the contour feature and that of G channel represents the high temperature and low temperature features of the thermal images. Merger of the R and G channels images can concurrently represent the contour of and the region temperature of the image.

We combine the R and G channel images, and conduct lineal weighing to make up new components, and then change the merged two channels to be $I_3(x, y)$ by Eq. (26). Therein, $f_0(x, y)$ is the original input image. And R_{f_0}, G_{f_0} refer to the R channel image and G channel image. And $f_0(x, y)$ is the original input image.

$$I_3(x, y) = 0.299 \times R_{f_0} + 0.587 \times G_{f_0} \tag{4}$$

In the image, if the pixels which deviate from the diagonal have the larger value, namely, the brightness value of image varies quickly, the contrast will have a larger value, which reflects the clarity of the image and depth of texture. The deeper the texture is, the larger the contrast value will be [11]. In turn, it is smaller. In an $M \times N$ image, S_{xy} is set to represent the image window with distance of s from top to bottom which is in the center point (x, y). In the slipping pixel block area S_{xy}, $P(i, j)$ represents the probability that the point with the gray level i leaves away from a particular fixed position relation d (which is the number of the pixels with intervals in the space, recorded as g_{ij}, θ is direction to the point with the gray j level. The contrast value $con(x, y)$ of this point in the sliding pixel block area can be obtained in the fixed distance d and fixed direction θ.

$$P(i, j | d, \theta) = \frac{g_{ij}}{M \times N} \ (i, j \in S_{xy}) \tag{5}$$

$$con(x, y) = (i - j)^2 P^2(i, j | d, \theta) \tag{6}$$

The $con(x, y)$ of the four different directions is estimated and its mean value $\overline{con}(x, y)$ is estimated which can be regarded as the GLCM of the point S_{xy} in the region (x, y). We calculate $\overline{con}(x, y)$ of the mean contrast value of a pixel and integrate them into a matrix to get $c_1(x, y)$. The threshold K_0 is selected to perform banalization for $c_1(x, y)$ and then we can get the contrast feature image $c_2(x, y)$ of the image.

$$\overline{con}(x, y) = \frac{1}{4} \sum_\theta (i - j)^2 P^2(i, j | d, \theta) \ (i, j \in S_{xy}, \theta = 0°, 45°, 90°, 135°) \tag{7}$$

$$c_1(x, y) = \overline{con}(x, y) \tag{8}$$

$$c_2(x, y) = \begin{cases} 0, & otherwise \\ L_{\max}, & K_0 < \overline{con}(x, y) \end{cases} \tag{9}$$

Extraction of Entropy Feature Image. The entropy reflects the heterogeneity and complexity of the texture in the image, which is the measurement of the degree confusion in the image and the entropy is the measurement of the information possessed by the image. If the image has no any texture, and entropy will be close to 0. If the image has many fine textures, entropy value of the image will be very large; otherwise the image has less texture, the image entropy is small [13]. Similar to the process of estimation of the contrast feature, The representation form of the entropy feature is obtained as $ent(x, y)$ when the pixel in the region S_{xy}, where θ is the four directions. The mean value is estimated by calculating the entropy feature values $ent(x, y)$ in the four directions respectively to get a mean value $\overline{ent}(x, y)$. And the entropy feature image $e_1(x, y)$ of the whole image can be obtained after the completion of the image coverage of each small block.

$$ent(x, y) = -P(i, j|d, \theta) \log_c P(i, j|d, \theta) \tag{10}$$

$$\overline{ent}(x, y) = -\frac{1}{4} \sum_\theta [P(i, j|d, \theta) \log_c P(i, j|d, \theta)](\theta = 0°, 45°, 90°, 135°) \tag{11}$$

$$e_1(x, y) = \overline{ent}(x, y) \tag{12}$$

A threshold is selected to carry out the banalization for the entropy feature image $e_1(x, y)$, and the gray levels of the pixels being layered assign to 0 and $L_{\max}$ respectively. For the convenience of following superposition, we take the operation value of entropy and then get the entropy feature image $e_2(x, y)$.

$$e_2(x, y) = \begin{cases} 0, & otherwise \\ L_{\max}, & L_0 < e_1(x, y) < 0 \end{cases} \tag{13}$$

2.3   Building of Imitate Gradient Image

After processing contrast feature image by the 5×5 kernel size median filter, we get $c_3(x, y)$ as the result image by processing with median filter (Fig. 2).

$$c_3(x, y) = median[c_2(x, y)] \tag{14}$$

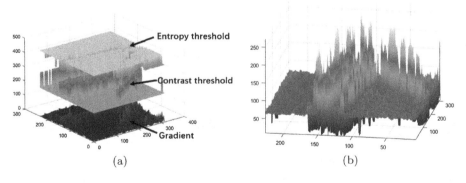

(a)   (b)

Fig. 2. (a) Three-layer feature image with linear superposition and sequence; (b) imitate gradient image.

The entropy feature image is carried out through adopting the morphology erosion [8], using ellipse operator and the kernel sized is as 3×3. The entropy feature image $e_3(x, y)$ which is acquired through adopting the morphological method. Use of the erosion method can decrease the disconnection of boundary of entropy feature image.

$$e_3(x, y) = e_2(x, y) \ominus B = \{ z \,|\, (B)_z \subseteq e_2(x, y) \} \tag{15}$$

After preprocessing the two feature images, then we need to layer marked weight of two images. The layering purpose is to build the gradient images similar with the shape of the ROI region. The entire imitate gradient image is divided into three regions, such as object region (ROI), environment areas and the boundary of environment and panel group. $g_{\max}(x, y)$ indicates the intensity of overall gray level transformation of the gradient image, namely, the degree of boundary's obviousness. It can be represented by formula (16).

$$g_{\max}(x, y) = \max\{g(x, y)\} \tag{16}$$

If the $g_{\max}(x, y)$ of image with the reduction of brightness is highly valued, it is bespoken that the boundary of the image is significantly obvious. To use the value containing $g_{\max}(x, y)$ to weigh two feature images (contrast feature image $c_3(x, y)$, and entropy feature image $e_2(x, y)$) is able to get the corresponding weighing coefficient γ and parameters α, β, and the formula goes as:

$$\gamma = \alpha \times g_{\max}(x, y) + \beta \tag{17}$$

Contrast feature image that is marked major outline, and finished the enhancement of major outline of the gradient image, and sets up barriers for environmental boundary and panel region. The parameters adopted in this experiment are $\alpha = -1, \beta = L_{\max}$ and the weighted value is γ_{idm}, balancing indistinct gradient image and clearly gradient image.

$$\gamma_{\mathrm{idm}} = L_{\max} - g_{\max}(x, y) \tag{18}$$

$$f_{idm}(x, y) = \begin{cases} \gamma_{\mathrm{idm}}, & c_3(x, y) = L_{\max} \\ 0, & otherwise \end{cases} \tag{19}$$

The negated entropy feature image is weighed by γ_{ent}. The parameters selected in the research are $\alpha = 0.6, \beta = 0$. Entropy feature is on the first floor in the simulation of the imitate gradient image.

$$\gamma_{ent} = 0.6 \times g_{\max}(x, y) \tag{20}$$

$$f_{ent}(x, y) = \begin{cases} \gamma_{ent}, & e_3(x, y) = L_{\max} \\ 0, & otherwise \end{cases} \tag{21}$$

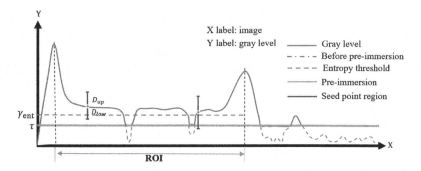

Fig. 3. The blue imaginary line is indicating the hole on entropy feature image. (Color figure online)

2.4   Region Filling

After using seed point filling the imitate gradient image by GLCM feature, still have miss segmentation due to the existence of holes and gray level rugged.

Pre-immersion of Gradient Image and Region Filling Process. The green line is sign for adopting the pre-immersion. The seed points are labelled by blue point, and the region of seed points are labelled by red line (Fig. 3).

For this reason, while selecting the seed point randomly, due to those disconnected regions, the region filling shall not be fulfilled in the targeted region. The pre-immersion is to eliminate the interference due to those disconnected regions, simplify the parameters and elevate the accuracy of region filling.

The immersion shall be carried out on imitate gradient in the lower level. The method adopted is to realize the reversed filling through the threshold segmentation. Such process is named as the pre-immersion, with the purpose of easily setting the parameters of negative difference. The negative difference is valued as τ, regulated among the interval of $(0 < \tau < \gamma_{ent})$, which means this value shall be not more than the filling value of the entropy feature image. The τ value shall firstly immerse the imitate gradient image. The image immersed result remarked as $f_{dst}(x, y)$.

$$f_{dst}(x, y) = \begin{cases} f(x, y), & otherwise \\ \tau, & f(x, y) \leq \tau \end{cases} \tag{22}$$

The region filling [9] is to combine the pixels or the subdomains to become the larger region in accordance with the formation standard if the pre-formation. The method initiates from a group of seed points. The up and low differences are adopted to control the gray scale scope for the seed point filling, and are expressed as $[D_{up}, D_{low}]$.

The seed point and the pixel of its connected region are expressed as (x, y), the difference of gray level is set within the range of $[D_{up}, D_{low}]$, the pixel of the connected region is labelled as 1, and the pixel not within such range is labelled as 0. The pixel labelled as 1 shall be added to the seed point set S, until the labelling is ended.

Our method fills the region through selecting the seed point, and such seed point can be selected manually, or can be acquired through fitting the center of maximized inscribed polygon based on the adoption of morphological method for the contract feature image. After the region filling, as the up difference is limited in scope, the holes shall be emerged internally. Hence, the extracted image from the overall region shall be more perfected through by hole filling.

3   Experiment

The image data collected from Jiangsu LINYANG Power Station, locating at the No. 666, LINYANG Road, Qidong Economic Development Area, Jiangsu Province 226200, China, with the coordinate position on the map of (121.639278, 31.817825). The thermal imaginary instrument adopted here is modeled as DM63 series thermal infrared image, which is researched and developed by Zhejiang Dali Technology Co., Ltd... The time of image acquisition was from Jul. to Aug. 2016. The image was shot in the morning, high noon and evening respectively. The weather while shooting the image was conditioned as sunny and cloudy weather. The pixel of the adopted thermal imager instrument is 320×240.

Our method compared with the "grab cut" method. And grab cut is a high-efficient segmentation algorithm based on foreground and background, which comprehensively uses two types of information of texture and boundary to perform the image segmentation. The algorithm of grab cut is set forth in the paper [2]. While in the method proposed in this paper acquiring the gradient image, the attenuation factor $1 - \mu$ is adopted as 0.4. The edge of the image has white laces because of the imaging. So three pixels should be excluded from the upper and lower boundary of the image when calculating the $g_{\max}(x, y)$. The images in the dataset are all sized as 320×240, in order to conveniently select the threshold value. So in the process of experiment, the $M \times N \times \overline{con}(x, y)$ and $M \times N \times \overline{ent}(x, y)$ is adopted to threshold the feature image use GLCM method (M and N refer to the number of pixel of the image from horizontal and vertical direction).

The parameters for obtainment of contrast feature image are set as follows:

As for S_{xy} the distance $s = 5$, the distance between pixels is $d = 4$, the gray level is set as $L = 32$, and the threshold values of contrast feature image are $K_0 = 50$, $L_{\max} = 255$. The parameters for obtainment of entropy feature image are set as follows: the separation distance between upper and lower boundaries is $s = 5$, the distance between pixels is $d = 4$, the gray level is set as $L = 16$, and the threshold values of entropy feature image are $L_0 = -64$, $L_{\max} = 255$. we choose seeds by manual in Experiment, the number of seeds there are less than three seed points by manual. In the course of region filling, there are less than 3 seed points by manual. In the course of prefill, the selected $\tau = 20$. In the step of region filling, the seed points are selected interactively, the negative difference value D_{low} is acquired as 20, and the positive difference value D_{up} is acquired as 15.

Grab cut method which is for comparison adopts two types of seed points One seed point is the foreground labelling, viz. the PV plane (Photovoltaic Panel) region, and the other is background labeling, viz. the non-PV region. The interactive operation is carried out twice.

The quantitative calculation method is adopted in the process of evaluation in order to evaluate the accuracy of different algorithms better. The image extracted by manual operation in region, as benchmark image, compares with the results extracted by two interactive methods (our method and the grab cut method). The 17 images are carried out the ROI extraction test, and each image operates 10 times. The selected evaluation standards include: precision, P, R, F [4], F_α (confidence $\alpha = 1$) the F measure is used to value between precision and recall, the results of F measure value more closely to 1, the more accurate are the results obtained by the algorithm [5], and Jaccard, J [5] for the evaluation. Being as a statistical magnitude, If J value is more approaching to 1, shows that the segmentation results more closely to the standard [4].

$$P = \frac{S_1}{S_2} \tag{23}$$

$$R = \frac{S_1}{S_3} \tag{24}$$

$$F_\alpha = \frac{(1+\alpha^2)PR}{(\alpha^2 P + R)}, (\alpha = 1) \tag{25}$$

$$J(S_1, S_3) = \frac{|S_1 \cap S_3|}{|S_1 \cap S_3|} \tag{26}$$

The set S_1 is perceived as the correct pixel point acquired through the algorithm (including the boundary point), the set S_2 refers to regional pixel points extracted by test method, and the set S_3 is the pixel set from the region extracted by manual operation. The operator $|\cdot|$ therein refers to the totality of pixel in the region.

4   Result

Figure 4 shows the comparison of the ROI extraction results between the algorithm presented in this paper and the "grab cut" method. The first set represents the original image, and the second set stands for the area extracted through the algorithm presented in the paper, while the third set shows the image area extracted through the comparison method of "grab cut". Five columns of images from left to right are featured by five typical forms respectively in accordance with the probabilities of occurrence. The images in the first column are most representative among the dataset, in which huge thermal differences between areas of the photovoltaic panel and external environment can be found and both of them possess abundant textures. Compared with the grab cut algorithm, the algorithm proposed in this paper can produce superior area extraction results.

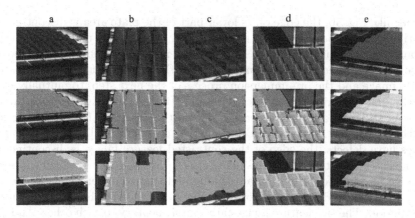

Fig. 4. Comparison of extraction results between the method in this paper and the grab cut method (Color figure online)

According to the images in the second column, it can be seen that when the target area is heavily blended with the environment, such as the pipe existing below the plate zone in the image, it is excluded from the target photovoltaic panel areas. This can easily cause the error of judgment for images at later phases (misjudgments of faults in non ROI). However, the method in this paper can accurately and integrally extract the ROI and demonstrate a better boundary integrity exacted from the areas. The images in the third column represent another sort of images in the dataset, which has a large proportion of target areas and abundant texture details. Under such circumstance, extracted area images could be fragmented when the threshold segmentation method is applied. Fortunately, the method this paper puts forward can achieve a relatively higher recall ratio with respect to the area extraction of images. The images in the fourth column refer to the target areas of images in the dataset, which possess complex textures and obvious color differences, leading up to the large information in the image contents. Although the algorithm in this paper has also accurately completed the target area extraction of the photovoltaic panel, there are still some drawbacks concerning the discrete boundaries. The fifth column demonstrates a relatively small thermal difference between ROI and non ROI in the image set of thermal imaging. In addition, the boundary is rather blurry. In the case when ROI areas possess fewer textures, the imaging weather photographed under the UAV inspection might be cloudy. The algorithm presented in this paper transcends the "grab cut" method in the boundary accuracy of area extraction, as well as the recall value.

Figure 5 shows the two discussions of the combination about feature image faced with two conditions. The first row corresponds to the original image, the second row corresponds to the contrasted feature image, the third row corresponds to the entropy feature image, and the fourth row corresponds to the regional extraction result. There are three major combinations: I. It can be seen from the Fig. 5b and f that the ridge of the contrasted feature image is discontinuous, like Fig. 5b, the edge of its lower part of the plate area is discontinuous, and the area Fig. 5c marked with the entropy feature can make up for these disconnections in the filling area. II. When the marked entropy feature image is sparse and scattered Fig. 5g, combining with the ridge built in the contrasted feature image and the pre-immersion process of the next step, to mutually complement and optimize the extracted result (Fig. 5h). The two feature images characteristics can complement each other on the defects of feature extraction, and they both can get more excellent extracted results. III. The two feature images have defects: the edge of the contrasted feature image is discontinuous, and the entropy feature image has many closed discrete regions, which will make the ridge incomplete during the area-filling and lead to under-segmentation (as shown in the second column of row d of Fig. 4). It needs to be improved in the latter period. The two feature images are complementary to each other. And also the automatic pre-immersion will improve the robustness of algorithm proposed in this paper.

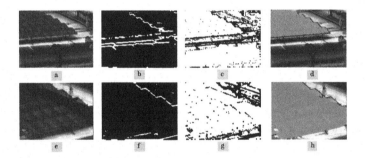

Fig. 5. Extraction of contrast feature image and entropy feature image

Figure 6 shows that the boundary of the ROI is continuously, region inside is plain and gray-level variance relatively small. While interference region has small but numerous holes, average gray-level of it is lower than object region and gray-level variance is larger. So the method builds a model that uses three different gray-level to make object regions, boundary and interference regions become layers. And its the base for the following region filling process.

Fig. 6. Imitated gradient image after weighting, (a) imitated gradient image of original image 7; (b) imitated gradient image of original Image 17

Table 1 is the comparison between our method and Grab Cut algorithm on ROI extraction of an infrared thermal image. The algorithm in this paper is better than the grab cut algorithm in the Recall value and J-index (contrasted with the standard value 1). There are a higher Precision P and a lower Recall value of Grab Cut method in the test of this dataset, which indicates that the grab cut method is under-segmentation which universally exists in this data set. Refer to the paper [19]. Use of Otsu method is a conventional method for regional extraction of photovoltaic panels in thermal imaging. We also used the

Table 1. Comparison of performance evaluation index between the method in this paper and grab cut

Image item	Performance Evaluation Index Comparison				
	Method	$\overline{P}$	$\overline{R}$	$\overline{F}$	$\overline{J}$
1	Grab cut	99.88%	63.20%	0.7582	0.6312
	Our method	99.23%	96.99%	0.9809	0.9626
2	Grab cut	99.58%	83.67%	0.9089	0.8337
	Our method	98.53%	98.30%	0.9841	0.9688
3	Grab cut	99.74%	91.38%	0.9529	0.9116
	Our method	99.94%	88.17%	0.9365	0.8812
4	Grab cut	77.37%	89.72%	0.8307	0.7106
	Our method	76.84%	89.89%	0.8276	0.7068
5	Grab cut	98.43%	87.92%	0.9254	0.8659
	Our method	79.82%	73.21%	0.7601	0.6189
6	Grab cut	95.74%	98.46%	0.9708	0.9432
	Our method	99.94%	79.66%	0.8859	0.7962
7	Grab cut	98.07%	75.46%	0.8475	0.7449
	Our method	95.87%	93.16%	0.9449	0.8956
8	Grab cut	94.99%	88.70%	0.9100	0.8390
	Our method	92.21%	91.82%	0.9198	0.8515
9	Grab cut	85.37%	87.67%	0.8599	0.7560
	Our method	96.09%	94.93%	0.9539	0.9130
10	Grab cut	72.94%	82.83%	0.7721	0.6351
	Our method	78.06%	77.55%	0.7754	0.6343
11	Grab cut	83.23%	90.54%	0.8541	0.7542
	Our method	99.27%	95.40%	0.9729	0.9473
12	Grab cut	99.52%	93.04%	0.9615	0.9261
	Our method	87.93%	99.65%	0.9342	0.8766
13	Grab cut	95.43%	67.13%	0.7750	0.6437
	Our method	79.77%	76.79%	0.7816	0.6431
14	Grab cut	90.00%	68.06%	0.7630	0.6394
	Our method	99.80%	96.24%	0.9799	0.9606
15	Grab cut	96.95%	72.03%	0.8245	0.7045
	Our method	99.97%	93.16%	0.9645	0.9313
16	Grab cut	87.92%	79.84%	0.8264	0.7172
	Our method	99.16%	95.10%	0.9709	0.9434
17	Grab cut	83.54%	96.40%	0.8871	0.8094
	Our method	99.51%	94.81%	0.9711	0.9438

Otsu method to test the data of this paper, the average Recall of 17 images we got is 69.02%, the average Precision is 93.16%, F-index is 0.7773, and J-index is 0.6670. Therefore, it is aware that the automatic image thresholding is ineffective for the regional extraction of this data set, and the Otsu method applies to the regional ex-traction of thermal imaging photovoltaic panels in desert areas. Therefore, the primary object of comparison in this paper is: grab cut method, which is also a regional extraction method based on seed.

Based on the calculation of the mean precision value, mean recall value, J index average and F test average, one-sided t-test is carried out on these performance indexes in paired, with results shown in Table 2.

Table 2. One-sided t-test on performance indexes of method here and *grab cut*

	P	*R*	*F*	*J*
p	$0.2911^{\#}$	0.0283^{*}	0.0229^{*}	0.0238^{*}

From the statistical result of the one-sided t-test, it can be seen that p is less than 0.05, which means it has statistical significance. For grab cut algorithm, it uses two types of seed points for testing the performance of non-ROI and ROI respectively, which thus increases the workload and calculation amount. Based on the performance index of the two algorithms in Table 1, it shows that the algorithm has a few of advantages over dividing the data set of the thermal imaging solar panels. The performance indexes R, F and J have statistical significance.

5   Conclusions

In order to enhance the accuracy in the fault extraction in the thermal imaging areas of photovoltaic panel, this paper puts forward a method to extract the regional thermal images based on the contrasted feature and entropy feature of GLCM algorithm. The novelty of this algorithm is that we combined the characteristics of the thermal image and built a model of the target area to carry out extraction of ROI area. The experiment results have demonstrated that the algorithm proposed in this paper is advantageous in area extraction performances, with a relatively high precision value and recall value, which can adapt with the area extraction of the infrared thermal image. Meanwhile, a comparison study has been conducted with the grab cut method. Less manual interference features in our method. Additionally, the operation and calculation of algorithm here are much easier. Further study will increase the accuracy and robustness of the algorithm under circumstances of a low signal-to-noise ratio and a complex interference composition.

Acknowledgment. This work is supported by the National Natural Science Foundation of China (No. 61463035), China Postdoctoral Science Foundation (2016M592117), Science and Technology Department of Jiangxi Province Science Foundation (20161BAB202045), Jiangxi Province Postdoctoral Research Excellent Foundation (2016KY01), Microsoft Azure Research Foundation (2017).

References

1. Haralick, R.M., Shanmugam, K.: Textural features for image classification. IEEE Trans. Syst. Man Cybern. **6**, 610–621 (1973)
2. Rother, C., Kolmogorov, V., Blake, A.: Grabcut: interactive foreground extraction using iterated graph cuts. In: ACM Transactions on Graphics (TOG), vol. 23, no. 3, pp. 309–314. ACM, August 2004
3. Shanmugan, K.S., Narayanan, V., Frost, V.S., Stiles, J.A., Holtzman, J.C.: Textural features for radar image analysis. IEEE Trans. Geosci. Remote Sens. **3**, 153–156 (1981)
4. Christensen, H.I., Phillips, P.J. (eds.): Empirical Evaluation Methods in Computer Vision, vol. 50. World Scientific, Singapore (2002)
5. Martin, D.R., Fowlkes, C.C., Malik, J.: Learning to detect natural image boundaries using local brightness, color, and texture cues. IEEE Trans. Pattern Anal. Mach. Intell. **26**(5), 530–549 (2004)
6. Xin Z. Research on the Visual Effect Enhancement Technology for Infrared Images. Tianjin University, pp. 40–42 (2009). (in Chinese)
7. Szelisk, R.: Computer Vision: Algorithms and Applications. Springer Science and Business Media, Berlin (2010). https://doi.org/10.1007/978-1-84882-935-0
8. Gonzalez, R.C., Woods, R.E.: Digital Image Processing, 3rd edn. Prentice-Hall, Inc., Upper Saddle River (2002)
9. Sarkar, S., Das, S.: Multilevel image thresholding based on 2D histogram and maximum Tsallis entropya differential evolution approach. IEEE Trans. Image Process. **22**(12), 4788–4797 (2013)
10. Nayak, J., Naik, B., Behera, H.S.: Fuzzy C-means (FCM) clustering algorithm: a decade review from 2000 to 2014. In: Jain, L.C., Behera, H.S., Mandal, J.K., Mohapatra, D.P. (eds.) Computational Intelligence in Data Mining - Volume 2. SIST, vol. 32, pp. 133–149. Springer, New Delhi (2015). https://doi.org/10.1007/978-81-322-2208-8_14
11. Mohanaiah, P., Sathyanarayana, P., Gurukumar, L.: Image texture feature extraction using GLCM approach (2014)
12. Saha, S., Alok, A.K., Ekbal, A.: Brain image segmentation using semi-supervised clustering. Expert Syst. Appl. **52**, 50–63 (2016)
13. Cremers, D., Rousson, M., Deriche, R.: A review of statistical approaches to level set segmentation: integrating color, texture, motion and shape. Int. J. Comput. Vis. **72**(2), 195–215 (2007)
14. Qi, S., Ma, J., Tao, C., Yang, C., Tian, J.: A robust directional saliency-based method for infrared small-target detection under various complex backgrounds. IEEE Geosci. Remote Sens. Lett. **10**(3), 495–499 (2013)
15. Bae, T.W., Zhang, F., Kweon, I.S.: Edge directional 2D LMS filter for infrared small target detection. Infrared Phys. Technol. **55**(1), 137–145 (2012)
16. Mohammed, M.M., Badr, A., Abdelhalim, M.B.: Image classification and retrieval using optimized pulse-coupled neural network. Expert Syst. Appl. **42**(11), 4927–4936 (2015)
17. Al-Amri, S.S., Kalyankar, N.V.: Image segmentation by using threshold techniques. arXiv preprint arXiv:1005.4020 (2010)
18. Zhang, S.F., Huang, X.H., Wang, M.: Algorithm of infrared background suppression and mall target detection. J. Image Graph. **21**(8), 1039–1047 (2016). (in Chinese)
19. Zhang, J., Jung, J., Sohn, G., et al.: Thermal infrared inspection of roof insulation using unmanned aerial vehicles. Int. Arch. Photogramm. Remote Sens. Spat. Inf. Sci. **40**(1), 381 (2015)

Learning Local Feature Descriptors with Quadruplet Ranking Loss

Dalong Zhang, Lei Zhao$^{(\boxtimes)}$, Duanqing Xu, and Dongming Lu

College of Computer Science and Technology, Zhejiang University,
Hangzhou 310027, China
{dalongz,cszhl,xdq,ldm}@zju.edu.cn

Abstract. In this work, we propose a novel deep convolutional neural network (CNN) with quadruplet ranking loss to learn local feature descriptors. The proposed model receives quadruplets of two corresponding patches and two non-corresponding patches, and then outputs features measured by L_2 norm with dimension (256d) close to that of SIFT, which thus can be easily applied to practical vision tasks by plug-in replacing SIFT-like features. Moreover, the proposed model mitigates the problem that the margin separating corresponding and non-corresponding pairs varies with samples caused by commonly used triplet loss, and improves the capacity of utilizing the limited training data. Experiments show that our model outperforms some state-of-the-art methods like TN-TG and PN-Net.

Keywords: Image feature · Patch matching · Feature extraction Convolutional neural network

1 Introduction

Feature representation for local image patches is the problem of mapping raw image patches to descriptors in certain feature spaces to discriminate the correspondence of patches, which is one of major research topics in various computer vision tasks, such as structure from motion [17], multi-view stereo reconstruction [18], image stitching and alignment [21], etc. Recently, the field has witnessed the trend that traditional hand-crafted image features [2,14], are evolving into current data-driven and learning-based image feature descriptors [3,20,22], especially CNN-based features [6,19,26].

In this work, we aim to exploit the deep CNNs [4,7,8,12] to learn good local feature descriptors which can be used as plug-in replacement for popular descriptors like SIFT [14]. To that end, the metric measurement and feature dimension are restricted: the feature dimension should be close to that of SIFT while the distance is measured by L_2 norm. Many efforts has been made to move in this direction, but Siamese based methods [6,19,26,27] may not make full use of the relation between patch pairs while commonly used triplet loss [1,13] may lead to the problem that the margin separating corresponding and non-corresponding

© Springer Nature Singapore Pte Ltd. 2017
J. Yang et al. (Eds.): CCCV 2017, Part I, CCIS 771, pp. 206–217, 2017.
https://doi.org/10.1007/978-981-10-7299-4_17

pairs varies with samples [23]. Learning a robust and discriminative descriptor that is competent for various patch-based computer vision tasks is still an open problem.

Inspired by the success of recent works [1,13,19,26], we propose a novel quadruplet CNN model to learn local feature descriptors. First, we construct our network with four branches sharing tied weights, receiving two corresponding patches and two non-corresponding patches simultaneously. Then, we minimize a quadruplet ranking loss to ensure that the distance between corresponding patches should be smaller than that between non-corresponding patches. At last, our model outputs a 256d feature descriptor measured by L_2 norm which can be easily used in a plug-in way to replace traditional descriptors. The proposed model enforces an interval between corresponding and non-corresponding patch pairs, which is helpful to distinguish the correspondence of them. Also we abrogate the 'anchor' used in triplet loss and design a online sampler to ensure any combination of positive pairs and negative pairs could contribute gradients in optimization process, which, as a result, pushes non-corresponding pairs far away from corresponding pairs without considering whether they refer to the same patch.

Overall, our main contributions are as follows: (1), we propose a novel quadruplet model to learn local feature descriptors and achieve better embedding results compared with some state-of-the-art methods; (2), we propose a way to improve the capacity of utilizing the limited data by online sampling and abrogating the 'anchor' when using quadruplet loss; (3), we apply our model to image matching task and it shows good generalization properties.

2   Related Work

Pretty a lot of works have been done to propose local feature descriptors and reached a practically applicable level of performance. Benefit from the growing availability of labeled data and increasing computational resources, data-driven methods for local descriptors have been a trend and already outperformed classic hand-crafted descriptors to some degree. Various techniques have been borrowed in carrying out such descriptors, including Linear Discriminant Analysis [3], boosting [22], convex optimization [20], and deep learning [10]. In this section, we will explore works based on deep CNNs, mainly organized by their architectures.

Siamese is one of the most frequently used architectures in learning local feature descriptors. Networks with Siamese architecture often has two branches sharing tied weights guided by pairwise loss. Works like [6,19,26] exploit this architecture to learn local feature representation. Simo-Serra et al. [19] trained a network outputting features measured by L_2 norm while [6,26] added a decision network to learn descriptors and metric simultaneously. Zagoruyko and Komodakis [26] also tried several variants of Siamese architecture like 2-channel model and pesudo-siam model and so on. Works with such architecture achieve significant improvement compared with their predecessor works [20,22]. But some works like [19] mainly focus on classifying corresponding pairs and non-corresponding pairs without ensuring reasonable variation between them, which may lead to sub-optimal problem [1] when used for different tasks.

Triplet is born with the advantage of ensuring an interval between corresponding pairs and non-corresponding pairs, and once applied to this field, it becomes one of the most successful architectures. Works like [1,13] achieve the state-of-the-art performance on the Multi-view Stereo Correspondence dataset (MVS) [3], proving the effectiveness of this architecture. Kumar et al. [13] tried to combine the Siamese and Triplet architecture with global loss, while [1] trained a quite simple network but achieved epic performance and extraction efficiency. However, triplets of training set is formed based on the 'anchor', which may vary according to samples. Unfortunately, the margin separating corresponding pairs and non-corresponding pairs is depending on the position of 'anchor' and thus also varies with samples. The margin varying problem prevents the triplet architecture from enlarging variation between corresponding pairs and non-corresponding pairs and increases the risk of overfitting.

Inspired by these works, we propose a novel quadruplet model to learn feature descriptors. The model increases the capacity of utilizing the limited training data and also mitigates the margin varying problem. Experiments show that our method leads to significant improvements in performance.

3   The Proposed Approach

The overall methodology of the proposed approach is utilizing the power of deep convolution neural network to extract features from image patches to discriminate the correspondence of patches. But to get good features, we not only need a powerful network, but also a good loss function to guide the network. In this section, we will first explain our loss function and then detail the network structure.

3.1   Loss Function

Our loss is mainly inspired by triplet loss [1,13], which is successful exploited in learning feature descriptors. We will analysis the triplet loss first and then introduce our quadruplet ranking loss.

Triplet Loss. Triplet loss is based on triplet (a, p, n), where a is the 'anchor' (reference patch) while p and n represent positive and negative patch refer to the anchor respectively. Such loss enforces that the distance between positive pairs should be smaller than that between negative pairs:

$$D(f(a), f(p)) + I < D(f(a), f(n)) \tag{1}$$

where $f(\cdot)$ is the neural network mapping the raw image to certain feature space and I is used to ensure an interval between positive pairs and negative pairs to

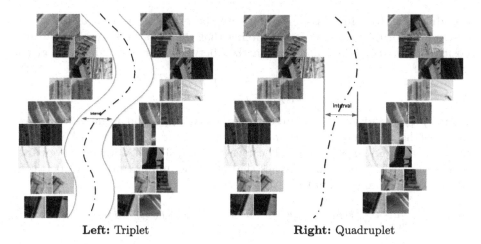

Left: Triplet   **Right:** Quadruplet

Fig. 1. Separating the positive and negative pairs with triplet and quadruplet loss. **Left:** Triplet loss enforces an interval between positive and negative pairs, but the margin varies with samples; **Right:** With quadruplet loss, the interval is ensured between any positive and negative pairs, thus even the minimum distance between them should be larger than the given interval, which works like support vectors in SVM [16].

separate them. $D(\cdot, \cdot)$ means a distance metric, and is often set to L_2 distance in this field. To satisfy the Eq. 1, triplet loss is formed as:

$$L(a, p, n) = max(0, I + \|f(a) - f(p)\|_2 - \|f(a) - f(n)\|_2) \qquad (2)$$

The loss is efficient to push negative pairs away from positive pairs as the optimization process won't stop until the distance I between positive and negative pairs is confirmed. The interval I increases the average distance between positive and negative 'class' and thus makes them more different from each other, which is helpful to discriminate them.

However, according to Eqs. 1 and 2, the loss will confirm a margin between positive and negative pairs with reference to the 'anchor', which means the margin is depending on the position of 'anchor'. During optimization process, although the training samples can be content to Eq. 1 and the interval between positive and negative pairs can be confirmed, the position of the interval (separating hyperplane or say margin) may be distorted with different samples (shown in Fig. 1). The varying margin increases the risk that the distance between some positive pairs is larger than that between negative pairs, which will damage the performance when discriminating the correspondence of patch pairs. Moreover, the margin varying problem prevents the triplet loss from enlarging variation between corresponding pairs and non-corresponding pairs, and thus becomes a bottleneck to improve performance. What's worse, if the margin is highly nonlinear, it may increase the risk of overfitting. Hard negative mining can be helpful, but it's unclear to define the hard negative samples and it's quite time-consuming to find the proper hard negative samples.

Quadruplet Ranking Loss. To mitigate the problem mentioned above, we design a quadruplet ranking loss enforcing that the distance between arbitrary positive pairs should be smaller than that between negative pairs with given interval I:

$$\begin{aligned} \|f(p_1) - f(p_2)\|_2 + I < \|f(n_1) - f(n_2)\|_2 \\ \text{s.t.} \quad \forall (p_1, p_2) \in P, \quad \forall (n_1, n_2) \in N \end{aligned} \tag{3}$$

where (p_1, p_2) and (n_1, n_2) represent the positive pair and negative pair and they together form a quadruplet (p_1, p_2, n_1, n_2). P and N mean the whole set of positive pairs and negative pairs. The constraint Eq. 3 pushes any negative pairs away from positive pairs and implies that even the minimum distance between them should be larger than the given interval I:

$$\max_{(p_1, p_2) \in P} \|f(p_1) - f(p_2)\|_2 + I < \min_{(n_1, n_2) \in N} \|f(n_1) - f(n_2)\|_2 \tag{4}$$

which is illustrated in Fig. 1. According to Eqs. 3 and 4, we design a quadruplet ranking loss:

$$L(p_1, p_2, n_1, n_2) = \max(0, I + \|f(p_1) - f(p_2)\|_2 - \|f(n_1) - f(n_2)\|_2) \tag{5}$$

We use SGD to minimize the loss Eq. 5, and the gradient can be derived as:

$$\begin{aligned} \frac{\partial L}{\partial f(p_1)} &= 2(f(p_1) - f(p_2)) \\ \frac{\partial L}{\partial f(p_2)} &= -2(f(p_1) - f(p_2)) \\ \frac{\partial L}{\partial f(n_1)} &= -2(f(n_1) - f(n_2)) \\ \frac{\partial L}{\partial f(n_2)} &= 2(f(n_1) - f(n_2)) \end{aligned} \tag{6}$$

The proposed quadruplet loss holds some appealing properties: (1), the margin now depends on the position of $\max_{(p_1, p_2) \in P} \|f(p_1) - f(p_2)\|_2$ rather than the 'anchor', thus the margin varying problem is mitigated to some degree (shown in Fig. 1). (2), the loss is flexible to handle any combinations of positive and negative pairs as we abrogate the 'anchor' when forming quadruplets. The loss is degraded to triplet loss if $p_1 = n_1$; otherwise, it represents the added constraint on distance between positive and negative pair with different 'anchor'. As a result, we can make full use of any combinations of positive and negative pairs to optimize the target function, which improves the capacity to utilize the limited training data. We also propose a online sampling method to provide sufficient quadruplets, which will introduce in the next section.

3.2   Network Structure

We build a quadruplet deep convolution network subject to Eq. 5 to learn local feature descriptors.

The overall network consists of four branches with tied weights as shown in Fig. 2. In consistent with recently proposed methods, our network receives quadruplets of gray-scale patches with size 64×64 as input (one patch each for the four branches). As the quadruplets is constructed with equal number of positive and negative patch pairs, we don't need to balance them to prevent the model from swing to one side during training. Moreover, each branch outputs a 256d vector measured by L_2 norm as learned feature descriptor, which is similar to the traditional local feature descriptors (such as SIFT). It's worth noting that, during inference, we just forward one branch to get the features as all these branches share the same weights.

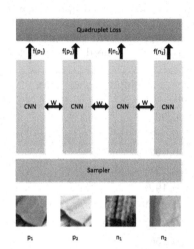 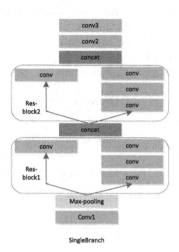

Fig. 2. Network Structure. The **left** shows the whole network with four branches sharing the same weights. The **right** part shows the structure of a single branch and the detail of layer parameters is reported in Table 1.

The single branch is designed referring to ResNet [7,8]. Each branch consists of two residual blocks along with other layers. We detail the structure of the single branch in Fig. 2 (the right part) and the parameters for each layer are listed in Table 1. Note that every convolution layer is followed with Batch Normalization layer [9] and ReLU activation [5] except for the last one, which are not shown in Fig. 2 and Table 1.

We also design a online sampler layer to enlarge training set. The sampler works as follows: (1), receive offline quadruplets $(p_{1,i}, p_{2,i}, n_{1,i}, n_{2,i})$ of training samples, where i implies the i-th quadruplet in current batch. (2), randomly select positive pair $(p_{1,m}, p_{2,m})$ and negative pair $(n_{1,k}, n_{2,k})$ from offline quadruplets to form a new quadruplet $(p_{1,m}, p_{2,m}, n_{1,k}, n_{2,k})$. (3), repeat the second step for certain times and add these new quadruplets to original quadruplets batch, then feed the expanded batch to the network. During training, we use the online sampler to double the offline quadruplets. The online sampler layer not only helps enlarge the training set, but also provides sufficient combinations of positive and negative pairs, which is helpful to discriminate the correspondence of patch pairs . Moreover, the sampling process is done within a batch, thus no extra patches from outside is needed.

We implement the work with caffe [11] and use SGD with learning rate 0.01 decreasing by a factor of 2 every 100,000 iterations, momentum 0.9 and weight decay 0.0001 to train our model. Margin I in Eq. 5 is set to 0.8 as $f(\cdot)$ is normalized. With NVIDIA GTX980Ti, it often takes about 2 hrs per epoch for training with $2M$ offline quadruplets (doubled by online sampling).

Table 1. The parameters for a single branch. Note that in this table Res-Block1 and Res-Block2 mean the two residual blocks in the single branch. The quadruple (f, s, p, o) means the parameters of the corresponding layer, mainly for convolution layer and pooling layer, representing filter size, stride size, padding size, and the num of the output feature maps, respectively. For pooling layer, there's no need to set the last parameter 'o', and we use '$-$' instead, like $(2, 2, 0, -)$.

Layer name	Output dimension	Detail	
Conv1	$96 \times 58 \times 58$	$(7, 1, 0, 96)$	
Max-Pooling	$96 \times 29 \times 29$	$(2, 2, 0, -)$	
Res-Block1	$192 \times 12 \times 12$	$(5, 1, 0, 192)$	
		$(3, 2, 0, 192)$	$(1, 3, 0, 192)$
		$(1, 1, 0, 192)$	
Res-Block2	$256 \times 6 \times 6$	$(5, 1, 0, 256)$	
		$(3, 1, 0, 256)$	$(1, 2, 0, 256)$
		$(1, 1, 0, 256)$	
Conv2	$128 \times 4 \times 4$	$(3, 1, 0, 128)$	
Conv3	$256 \times 1 \times 1$	$(4, 1, 0, 256)$	

4   Experiments

In this section, we will first introduce the dataset used for assessing our model, and then present the result with standard evaluation protocol. Also, we apply our model to image matching task.

4.1   Dataset and Evaluation Protocol

For training, we use the Multi-view Stereo Correspondence dataset (MVS) [3], which is collected by Winder et al. for learning descriptors. The dataset consists of three subsets, Liberty (LY), Notre Dame (ND), and Yosemite (YO), which were sampled from 3D recontructions of the statue of Liberty (New York), Notre Dame (Paris) and Half Dome (Yosemite). Patches were extracted around each interest point that generated by Difference of Gaussian detector (DoG) or Harris interest points detector. Each of the three subsets contains more than 450k grayscale image patches (64×64 pixels) with pre-generated label to form patch pairs, all with equal number of match and dismatch patch pairs. In our work, we use the DoG set and make data augmentation by rotating the pair of patches by 90, 180, and 270 degrees, and flipping the images horizontally and vertically.

To evaluate the performance of our model, we follow the evaluation protocol of [3]. We train the model on one subset and test on the other two subsets. And then We report the false positive rate at 95% recall (FPR95) on each of the six combinations of training and testing sets, as well as the mean across all combinations. Following [13,20,26], our test is done on the 100,000 patch-pair Liberty, Notre Dame, and Yosemite sets.

4.2 Results

Following the evaluation protocol mentioned above, we evaluate our model and some recently proposed methods with normalized SIFT as baseline. Meanwhile, we indicate the memory footprint for each of the descriptors by specifying their dimension, as summarized in Table 2. Moreover, we draw ROC curves to show the result of triplet and quadruplet with same network structure.

Table 2. The MVS matching result. Numbers are false positive rate at 95% recall, and the smaller the better. **Bold** numbers are the best results on the dataset.

Train	Dim.	ND	YO	LY	YO	LY	ND	Mean
Test		LY		ND		YO		
SIFT (no training) [14]	128d	29.84		22.53		27.29		26.55
Simonyan et al. (2014) [20] PR	<640d	16.56	17.23	9.88	9.49	11.89	11.11	12.69
Simonyan et al. (2014) [20] PR-proj	<80d	12.42	14.58	7.22	6.82	11.18	10.08	10.38
Zagoruyko et al. (2015) [26] siaml_2	192d	13.24	17.25	6.01	8.38	19.91	12.64	12.90
Zagoruyko et al. (2015) [26] 2streaml_2	256d	8.79	12.84	4.54	5.58	13.24	13.02	9.67
Simo-Serra et al. (2015) [19]	128d	8.50		4.59		14.59		9.22
Kumar et al. (2016) [13] TN-TG	256d	9.91	13.45	3.91	5.43	10.65	9.47	8.80
Balntas et al. (2016) [1] PN-Net	128d	8.27	9.76	3.81	4.45	9.55	7.74	7.26
Balntas et al. (2016) [1] PN-Net	256d	8.13	9.65	3.71	4.23	8.99	**7.21**	6.98
ours (Triplet ≈ 50 epoches)	128d	8.43	10.85	4.66	4.69	9.62	9.01	7.88
ours (Quadruplet ≈ 50 epoches)	128d	8.67	**9.58**	**3.61**	**3.87**	9.39	8.54	7.27
ours (Quadruplet)	256d	**6.87**	**8.63**	**2.77**	**3.16**	**8.30**	8.19	**6.32**

Our model show significant improvement in performance compared with the baseline and recently proposed works [1,13,19,20,26]. We reduce the average error rate from 26.5% (SIFT) to 6.32%, while win five out of all six combinations in terms of FPR95 compared with the state-of-the-art methods [1,13]. Our best model reduces the FPR95 by 2.48% and 0.66% compared with TN-TG [13] and PN-Net [1] separately. The comparison between our model and these previous works proves that it's powerful for our model to discriminate correspondence of patches.

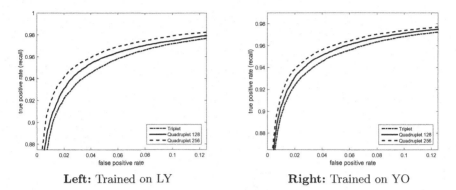

Left: Trained on LY   **Right:** Trained on YO

Fig. 3. ROC curves for model tested on ND subset. We draw the ROC curves of models with triplet (128d) and quadruplet (128d,256d). **Left:** Model trained on LY subset and tested on ND subset; **Right:** Model trained on YO subset and tested on ND subset.

We also train the model with triplet and quadruplet loss with same single branch and training epoches ($\approx$ 50 epoches), and results show that quadruplet reduce the average error rate by 0.6% compared with triplet. We draw ROC curves of models to show the performance of the triplet model and quadruplet model, which is illustrated in Fig. 3. The comparison is done firstly with same output feature dimension (128d), and then we add our best model with 256d feature into competition. We present the result that tested on ND subset for brief, and the ROC curves clearly show that the quadruplet model outperforms the triplet one. The comparison between the 128d and 256d model also shows that it's helpful to improve performance by enlarging the feature dimension with our model. Moreover, while [19,26] is based on siamese architecture and [1,13] utilize triplet architecture to learn feature descriptors, our quadruplet model achieve best result, which implies that it's potential to use the proposed quadruplet architecture to learn feature descriptors.

4.3   Application Examples

We also apply our model to practical image matching task. We extract the features from raw image as follows: (1), detect the interest points with DoG; (2), extract the patches according to the position, scale, and orientation of the interest points; (3), resize the patches to 64 × 64 and feed the patches to the network to get the final feature descriptors. These steps are very similar to that of the most frequently used descriptors — SIFT. It's also worth mentioning that, as our feature is measured by L_2 distance and the dimension (256d) is close to that of SIFT, many well studied techniques (KNN search et al.) can be retained, thus it's quite easy to apply our descriptor to practical vision tasks by plug-in replacing SIFT descriptors.

When compared with SIFT in image matching task, our model shows its advantages. Following [15], we compute the number of false match points (according to the fixed overlap error 50%) on a group of images with blur, light and viewpoint changes. As illustrated in Fig. 4, our descriptors are robust to changes and reduce the false match points by about 40% compared with SIFT. Note that, the images used in matching task is totally different from our training set, which proves the good generalization properties of our model.

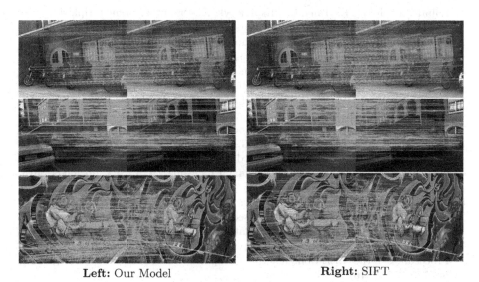

Left: Our Model   **Right:** SIFT

Fig. 4. Image matching examples. The left is our result while right SIFT's. **Top Row:** 'bikes' with blur. False matching points 58 vs. 112. **Middle Row:** 'leuven' with light change. False matching points 40 vs. 72. **Bottom Row:** 'graf' with viewpoint change. False matching points 59 vs. 82.

5   Conclusions

In this work, we proposed a quadruplet deep CNN model to learn local feature descriptors from raw data. Together with quadruplet ranking loss, we also design a sampler layer to make full use of limited training data. Our model achieves best results compared with SIFT and some state-of-the-art approaches, and shows good generalization properties when applied to practical image matching tasks. Moreover, our descriptor is able to replace traditional descriptors like SIFT in a plug-in way, thus many well studied techniques (KNN search et al.) can be retained. The main drawback of our method is that the process of training and inferring is time-consuming.

In future, we will simplify the network to accelerate the speed of feature extracting [1]. Moreover, we will try integrating our method with keypoint detection methods [14, 25] to build a unified end-to-end framework [24]. In this way, we can learn keypoint detection and feature representation at the same time, which is an interesting research direction to compete with current pipeline methods.

Acknowledgments. This work was financially supported by the Zhejiang Province Natural Science Foundation (Y16F020023).

References

1. Balntas, V., Johns, E., Tang, L., Mikolajczyk, K.: Pn-net: conjoined triple deep network for learning local image descriptors. arXiv preprint arXiv:1601.05030 (2016)
2. Bay, H., Ess, A., Tuytelaars, T., Van Gool, L.: Speeded-up robust features (SURF). Comput. Vis. Image Underst. **110**(3), 346–359 (2008)
3. Brown, M., Hua, G., Winder, S.: Discriminative learning of local image descriptors. IEEE Trans. Pattern Anal. Mach. Intell. **33**(1), 43–57 (2011)
4. Cun, Y.L., Boser, B., Denker, J.S., Howard, R.E., Habbard, W., Jackel, L.D., Henderson, D.: Handwritten digit recognition with a back-propagation network. In: Advances in Neural Information Processing Systems, pp. 396–404 (1990)
5. Dahl, G.E., Sainath, T.N., Hinton, G.E.: Improving deep neural networks for LVCSR using rectified linear units and dropout. In: 2013 IEEE International Conference on Acoustics, Speech and Signal Processing (ICASSP), pp. 8609–8613. IEEE (2013)
6. Han, X., Leung, T., Jia, Y., Sukthankar, R., Berg, A.C.: Matchnet: unifying feature and metric learning for patch-based matching. In: Proceedings of the IEEE Conference on Computer Vision and Pattern Recognition, pp. 3279–3286 (2015)
7. He, K., Zhang, X., Ren, S., Sun, J.: Deep residual learning for image recognition. In: Proceedings of the IEEE Conference on Computer Vision and Pattern Recognition, pp. 770–778 (2016)
8. He, K., Zhang, X., Ren, S., Sun, J.: Identity mappings in deep residual networks. In: Leibe, B., Matas, J., Sebe, N., Welling, M. (eds.) ECCV 2016. LNCS, vol. 9908, pp. 630–645. Springer, Cham (2016). https://doi.org/10.1007/978-3-319-46493-0_38
9. Ioffe, S., Szegedy, C.: Batch normalization: accelerating deep network training by reducing internal covariate shift. In: International Conference on Machine Learning, pp. 448–456 (2015)
10. Jahrer, M., Grabner, M., Bischof, H.: Learned local descriptors for recognition and matching. In: Computer Vision Winter Workshop, vol. 2 (2008)
11. Jia, Y., Shelhamer, E., Donahue, J., Karayev, S., Long, J.: Caffe: convolutional architecture for fast feature embedding. In: Eprint Arxiv, pp. 675–678 (2014)
12. Krizhevsky, A., Sutskever, I., Hinton, G.E.: Imagenet classification with deep convolutional neural networks. In: Neural Information Processing Systems, pp. 1097–1105 (2012)
13. Kumar, B., Carneiro, G., Reid, I., et al.: Learning local image descriptors with deep siamese and triplet convolutional networks by minimising global loss functions. In: Proceedings of the IEEE Conference on Computer Vision and Pattern Recognition, pp. 5385–5394 (2016)
14. Lowe, D.G.: Distinctive image features from scale-invariant keypoints. Int. J. Comput. Vis. **60**(2), 91–110 (2004)
15. Mikolajczyk, K., Schmid, C.: A performance evaluation of local descriptors. IEEE Trans. Pattern Anal. Mach. Intell. **27**(10), 1615 (2005)
16. Mohri, M., Rostamizadeh, A., Talwalkar, A.: Foundations of Machine Learning. MIT Press, Cambridge (2012)
17. Molton, N., Davison, A.J., Reid, I.: Locally planar patch features for real-time structure from motion. In: BMVC, pp. 1–10 (2004)

18. Seitz, S.M., Curless, B., Diebel, J., Scharstein, D., Szeliski, R.: A comparison and evaluation of multi-view stereo reconstruction algorithms. In: 2006 IEEE Computer Society Conference on Computer vision and pattern recognition, vol. 1, pp. 519–528. IEEE (2006)

19. Simo-Serra, E., Trulls, E., Ferraz, L., Kokkinos, I., Fua, P., Moreno-Noguer, F.: Discriminative learning of deep convolutional feature point descriptors. In: Proceedings of the IEEE International Conference on Computer Vision, pp. 118–126 (2015)

20. Simonyan, K., Vedaldi, A., Zisserman, A.: Learning local feature descriptors using convex optimisation. IEEE Trans. Pattern Anal. Mach. Intell. **36**(8), 1573–1585 (2014)

21. Szeliski, R.: Image alignment and stitching: a tutorial. Found. Trends® Comput. Graph. Vis. **2**(1), 1–104 (2006)

22. Trzcinski, T., Christoudias, M., Fua, P., Lepetit, V.: Boosting binary keypoint descriptors. In: Computer Vision and Pattern Recognition, pp. 2874–2881 (2013)

23. Ustinova, E., Lempitsky, V.: Learning deep embeddings with histogram loss. In: Neural Information Processing Systems, pp. 4170–4178 (2016)

24. Yi, K.M., Trulls, E., Lepetit, V., Fua, P.: LIFT: learned invariant feature transform. In: Leibe, B., Matas, J., Sebe, N., Welling, M. (eds.) ECCV 2016. LNCS, vol. 9910, pp. 467–483. Springer, Cham (2016). https://doi.org/10.1007/978-3-319-46466-4_28

25. Yi, K.M., Verdie, Y., Fua, P., Lepetit, V.: Learning to assign orientations to feature points. In: Computer Vision and Pattern Recognition, pp. 107–116 (2016)

26. Zagoruyko, S., Komodakis, N.: Learning to compare image patches via convolutional neural networks. In: Proceedings of the IEEE Conference on Computer Vision and Pattern Recognition, pp. 4353–4361 (2015)

27. Zbontar, J., LeCun, Y.: Computing the stereo matching cost with a convolutional neural network. In: Proceedings of the IEEE Conference on Computer Vision and Pattern Recognition, pp. 1592–1599 (2015)

Three-Dimensional Reconstruction of Ultrasound Images Based on Area Iteration

Biying Chen[1], Haijiang Zhu[1(✉)], Jinglin Zhou[1], Ping Yang[2], and Longbiao He[2]

[1] College of Information and Technology, Beijing University of Chemical Technology,
Beijing 100029, China
275185370@qq.com, zhuhj@mail.buct.edu.cn
[2] Division of Mechanics and Acoustic, National Institute of Metrology,
Beijing 100029, China

Abstract. Ultrasonic C scan technique has been extensively applied in nondestructive testing (NDT) in recent years. Aiming at ultrasonic C scanning original data, this paper presents a 3D reconstruction algorithm for ultrasonic C scanning image based on area iteration. Because the distance between two neighboring slices is small, the area chance between two neighboring images is also few. Therefore, we design an iterative rule on the area of the object to detect and extract the target contour of each image. After preprocessing of ultrasonic C image, we extract the contour of the object scanned by ultrasonic transducer. Then, we reconstruct the 3D ultrasonic C scanning image using marching cube algorithm. In experiments, we implement the proposed method for ultrasonic scanning data of tissue-mimicking phantom including a liver model, and we compare the proposed method with other methods. The results show that the proposed method can more accurately display the edge information of the scanned object, and improve the precision of region detection, especially suitable when the edge has the incomplete information.

Keywords: 3D reconstruction · Ultrasonic C scan · Speckle detection
Area iteration

1 Introduction

Although people discovered the fundamental physical phenomena of acoustic imaging and CT scanning in ancient times, the realization of efficient imaging device technology in the ultrasonic detection technology is hindered due to the lack of research on data acquisition and image processing in the long term [1]. With the development of science and technology, ultrasonic imaging instruments have made great progress in the past ten years. Around the field of rapid development, development of technology research on instruments, algorithms and bio-medical applications has increased exponentially. The performance of system function, image quality, imaging speed has made obvious progress. The

© Springer Nature Singapore Pte Ltd. 2017
J. Yang et al. (Eds.): CCCV 2017, Part I, CCIS 771, pp. 218–230, 2017.
https://doi.org/10.1007/978-981-10-7299-4_18

advancement of technology has also promoted the development of practical application. The objects of ultrasound scanning have almost related to the work piece, phantom, and all the organs of the body in the medical field.

With the development of 3D imaging by CT, PET and MRI, the application of ultrasound imaging has also been extended to 3D [2]. Compared with computed tomography (CT) imaging and magnetic resonance imaging (MRI), ultrasonic imaging has the advantages of being non-invasive, non-ionizing, portable and low-cost, less harmful than radiation with imaging methods, so it's more suitable for imaging in multi-field applications [3]. In recent years, with the development of 3D ultrasound imaging technology, it has been proved to become an increasingly important role in the practical imaging applications and reconstruction [4–6]. 3D ultrasound instruments are used by researchers and commercial companies as the operation of the device, and make the 3D ultrasound imaging technology integration to detection, inspection and reconstruction experiments [7–11]. It has been proved that ultrasound scanning equipment has been experimentally applied to the reduction of 3D ultrasound scanning phantom.

As 2D ultrasound imaging technology can only provide the scanning of tomographic image, which is unable to provide complete volume data of tissues and organs, the 3D ultrasound imaging technology has been paid attention in recent years [12]. Many 3D reconstruction algorithms focus on 2D ultrasound images. Generally, volume reconstruction methods can be classified into three categories according to their implementation [13,14], which are Voxel based method (VBM), Pixel based method (PBM) and Function based method (FBM). After the acquisition of 3D information, the reconstruction in the usual sense is to transform every pixel in the 2D image into a 3D coordinate system, which is called Pixel based method (PBM) [15]. Another simple method of 3D reconstruction and display is based on image features.

How to improve the robustness of the reconstruction algorithm is still a challenging problem in 3D ultrasound imaging, making it more accurate and efficient to meet the needs of clinical applications. At present, there are some special difficulties in the display of 3D ultrasound images [16]. Firstly, different with CT or MRI, brightness in ultrasound image does not have the meaning of density; ultrasound image reflects the changes of acoustic impedance in the ultrasonic propagation path of the human body. Therefore, the successful method of image processing in CT or MRI cannot be simply used in the processing of ultrasound images. Then, the quality of the original 3D data will directly affect the effect of image display. The signal-to-noise ratio is low Because of the inherent speckle noise in ultrasound images, which brings difficulties to the edge detection and segmentation. Last but not least, during the acquisition of 3D ultrasound image data, it is possible to have gaps in adjacent 2D planes. If we do not use methods such as spatial interpolation, the quality of the display will be directly affected by the gap.

In order to reduce the noise and interference in process, and improve the quality of reconstruction, the image processing of ultrasonic C scan imaging should be optimized. The main purpose of the improvement is [17]: reducing speckle noise, repairing the edge and improving the efficiency of time.

For these reasons, we present a 3D reconstruction algorithm for ultrasonic C scanning image based on area iteration. An iterative rule on the area of the object is designed to detect and extract the target contour of each layer image in the proposed method. The experimental result shows that the ambient noise and the artificial disturbance are obviously weakened, and the visualization performance is strong. In the processing of irregular image and the quality of the 3D imaging focusing on ultrasonic tissue-mimicking phantom, our method is superior to the existing method.

The rest of this paper is organized as follows: Sect. 2 presents the proposed method, which includes data collecting device and the steps of our method. The proposed method is verified by real data and images in Sect. 3 and conclusions are given in Sect. 4.

2   System and Methods

2.1   Ultrasonic C Scan System

Ultrasonic C scan technology is one of the key technologies to realize qualitative, directional, quantitative and nondestructive evaluation of defects. In this work, ultrasonic C scan equipment (C-SCAN-ARS), which is produced by Allrising Technology Co. Ltd. (Beijing), is used to collect 3D data. The device is shown in Fig. 1, which is controlled by the software of the ultrasonic C scan testing system. The system made accessible to researchers in the experiment, combining with scanning software UltrasonicImmersed, analysis software ImageView. The ultrasonic C scan system uses a computer to control ultrasonic transducer in liver mimicking phantom mold and scans it alternately, displaying the intensity of the reflected wave in the detection distance within a specific range (phantom internal) as the brightness change continuously. The internal cross section of the mold can be drawn.

Fig. 1. Ultrasonic C scanning equipment.

In the process of scanning phantom by ultrasonic C scan equipment, the problem of the edge loss was caused by the blurred ultrasound image and the loss of reflected wave, which should be considered and dealt with in the process of 3D imaging. However, many previous studies have focused on the image processing and reconstruction of regular objects, and it is difficult to realize the ultrasound image processing and reconstruction of irregular objects. In general, there are many factors that can affect the efficiency and quality of the ultrasound image acquisition, processing and final reconstruction. In general, the main factors reflect the following two aspects. One of the reasons is the irregular of phantom itself. There is a fluctuating, out of flatness and rough phenomenon, which has caused uneven signal intensity on the same surface. In the process of operation, the probe direction of ultrasonic C scan system remains vertical, usually forming a signal scattering phenomenon which can lead to decrease the received echo amplitude. On the other hand, image noise and interference of various types are included in scanning images, such as speckle noise, sound wave refraction, shadow interference and reverberation, which can lead to the decrease of image quality and the loss of the edge, making it difficult for the technical personnel to analyze and detect. Most of the noise interference comes from the interaction between the ultrasonic signal and the internal material.

2.2 The Proposed Method

The proposed method in this section is composed of several steps, and the flow chart is shown in Fig. 2. First, we take a series of 2D gray images from original C scanning data and preprocess each layer image, including image denoising and binarization. We then recognize all regions using 8-connected region detection method after filling empties and eliminating small holes for the binarized images. The next step is the largest area region extraction based iterative closest area algorithm. Finally, we reconstruct the 3D structure of the object using marching cube algorithm.

Firstly, we select mean filter and Gauss filter to ultrasonic C image because median filter can smooth pulse noise and protect the sharp edges of the image Gauss filter is suitable for Gauss noise. So the main methods adopted are median filter and Gauss filter method to eliminate the noise and filter.

Secondly, the image needs to be binarized to preserve the feature information of the image after image filtered. The OSTU threshold method is selected to process the smoothed image in our experiment. Here, the pixels are divided into two types: the target and the background. The result preserves the original edge features, and paves the way for the edge extraction and reconstruction subsequently.

The third step is the regions detection based 8-connected regions to each layer image. The connected area analysis has been more widely adopted in many applications of image processing and pattern recognition. This method can realize the coarse segmentation of the object identified, and the whole picture is decomposed into several connected sub-graphs to reduce the storage overhead of the system [18].

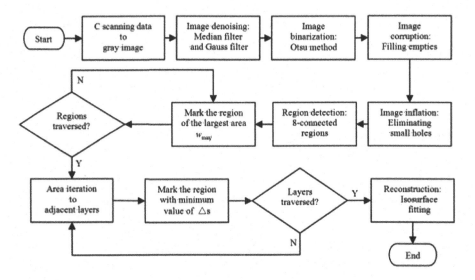

Fig. 2. The flow chart of the proposed approach.

After all the connected regions are recognized, we will extract the object region in each layer image. This paper proposes the object region detect method based iterative closest area algorithm.

Let two adjacent images be I_1 and I_2. The multiple regions are represented by $S = [s_1, s_2, \ldots s_i]$ in image I_1 and by $W = [w_1, w_2, \ldots w_i]$, where i and j are the number of the region detected. Figure 3 gives a diagram of two adjacent layer images.

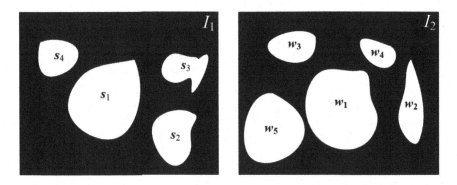

Fig. 3. The diagram of two adjacent layer images.

Suppose the number of ultrasonic C scanning layers is n. That is to say, we obtain n gray images. The steps of the iterative closest area algorithm are outlined as follows.

1. Selected random one binary image in the whole image set as image I_1. If the image is the top or the bottom, the image set is traversed down or up. Let the adjacent image be I_2.
2. Estimated the areas of all the detected regions in I_1, and marked the areas as s_1, s_2, s_3 and s_4. Likewise, the areas in I_2 are marked by as w_1, w_2, w_3, w_4 and w_5.
3. The center of each region is computed in the areas $s_i(i = 1, 2, 3, 4)$ and $w_j(j = 1, 2, 3, 4, 5)$. Let the center point be $p_i(i = 1, 2, 3, 4)$ and $q_j(j = 1, 2, 3, 4, 5)$.
4. As shown in Fig. 3, the maximum area s_i from the area set s_i is found in the image I_1 and the center point corresponding to s_i is labeled as p_1.
5. Compared the areas s_i in the image I_1 with t_j in the image I_2, and estimated the area difference Δs_i . Because the two images are neighbor, the difference between two detection regions is very little. Therefore, we can eliminate the tiny areas, even to a few noise points.
6. To determine the position of the detected region in the image I_2, we must match the center point p_1 of the maximum area s_1 in the image I_1 to that of the corresponding center point $q_j(j = 1, 2, 3, 4, 5)$ in the image I_2. We first define an objective function E

$$E = \| Hp_1 - q_j \|, j = (1, 2, 3, 4, 5) \tag{1}$$

where H is a transformation matrix.
7. The least squares method is used to solve the optimal solution subject $min(E)$, and the corresponding area region in the image I_2 can be obtained.
8. Taken the previous one (or next) layer of the image processed, and repeat the above step, and so on. Iterate over the whole 3D scanning data of the object to get the result of area iteration.

Finally, we use the different methods of 3D reconstruction. The methods of 3D reconstruction are mainly divided into two groups: Surface Rendering and Direct Volume Rendering. We choose algorithms from every groups and give the results of reconstruction and performance analysis.

3   Experiment and Result

3.1   Materials and Preparation

In order to use the 3D data of liver phantom by operating the ultrasonic C scan equipment, we made a phantom including a liver model using the ultrasonic tissue-mimicking materials. The size of the phantom is 180 mm * 100 mm * 80 mm. Figure 4(a) is a liver model, and Fig. 4(b) is an ultrasound phantom to simulate the human tissue. The ultrasonic probe in the experiment was selected of the 5 MHZ phantom of water immersion ultrasonic C scan, and the phantom of liver phantom will be immersed in water for ultrasonic C scan experiment, with the scanning depth according to the maximum value of reflected wave, about 450 mm.

Fig. 4. (a) Liver model, and (b) The phantom of ultrasonic tissue-mimicking materials.

3.2   Two-Dimensional Image Pre-processing

The first in this experiment was acquired the one-dimensional data by ultrasonic C scan, instead of selecting the image. The scanning range in the XOY plane is 190 mm * 140 mm. The scanning step is 1 mm and the sampling rate is up to 1000 MHZ.

This experiment utilized median filtering and Gauss filtering to smooth the original image (as shown in Fig. 5(a)). First, we used the median filter of a 3 * 3 template to smooth pulse noise. Next, the image is denoised by the Gauss filter, which the variance and the mean are 3 and 2, to suppress Gauss noise. The results of the smoothing process are shown in the Fig. 5(b) and (c). Compared with the original image, the number of noise points in the results is significantly reduced.

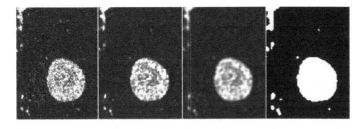

Fig. 5. (a) The original image, (b) The image used median filter, (c) The image used Gauss filter, and (d) A binary image by OSTU global threshold method.

In image binarization, OSTU threshold method is selected to process the smoothed image. The binarization image is shown in the Fig. 5(d). It can be seen that the result preserves the original edge features but a few small area regions are also detected. According to the characteristics of ultrasound images, the OSTU global threshold method is used to pre-segment for the image, then the 8-connected region detect algorithm is again employed to realize twice segmentation. The experimental result of single image shows that the improved algorithm has the improvement on the level of completeness in the region detection, but also has paid attention to the speed problem in detail. The algorithm effectively measures the number of pixels in each connected domain and eliminates the noise and interference, which is not affected by the shape of the region and improve the efficiency of edge detection.

The next step is to realize the largest area detection of the scanned object through the area iteration method. We presented the iteration result of the 8 layers image in the middle part of the scanned object. The results are displayed in Fig. 6. In this experiment, the edge of the extracted image is very clear. The area iterative method overcomes the shortcomings of time consuming and complexity, which provided a method of rapid and accurate evaluation in ultrasound scan results.

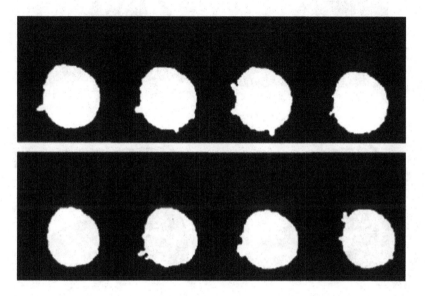

Fig. 6. Images of middle part using the proposed method.

3.3   Three-Dimensional Image Reconstruction

Prior to the reconstruction process, we use the two-dimensional data processing methods to ensure the useful data information can be retained and the useless data can be filtered out, which can improve the efficiency and the quality of reconstruction. In comparative experiments, we use the proposed method, the Gauss filtering and the edge detection to deal with two-dimensional data as a contrast . Then we choose three different methods to finish three-dimensional image reconstruction to verify the advantages of this proposed method. Through these different methods, different three-dimensional images can be reconstructed in different two-dimensional data processing methods.

Firstly, we choose the ISOSURFACE function in MATLAB for rendering 3D image of implicit function. In order to evaluate the reconstruction quality, we compared the reconstructed results of the Gauss filtering, the edge detection of canny, sobel and our improved algorithm. With using the method based on speckle detection and area iteration, the 3D reconstruction result of the experimental data is shown in the Fig. 7(a). The Gauss filtering reduction results are shown in Fig. 7(b). After the two simple and different kinds of edge extraction, the reduction results are shown as follows in Fig. 7(c) and (d).

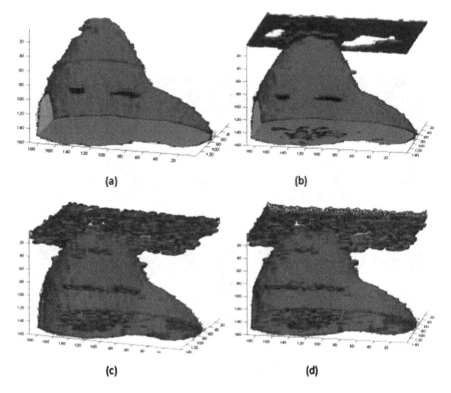

Fig. 7. Reconstruction results in contract: (a) The result of proposed method. (b) The result of Gauss filtering. (c) The result of Canny edge extraction. (d) The result of Sobel edge extraction.

Similarly, we choose a 3D volume reconstruction algorithm to restore the 3D image as the second validation. Before reconstruction can begin, in order to reduce the number of voxels to be processed in the reconstruction and to accelerate the speed of 3D reconstruction, three different methods should be used to preprocess the data set. With the useless voxel was reset to the background voxel, the 3D volume reconstruction can begin. The voxel is projected directly onto the plane and the number of projections of the data set should be calculated in this algorithm. With the algorithm, the reconstructed results of four smoothing methods can clearly generated in Fig. 8. We can use them to compare each performance on two-dimensional data filtering.

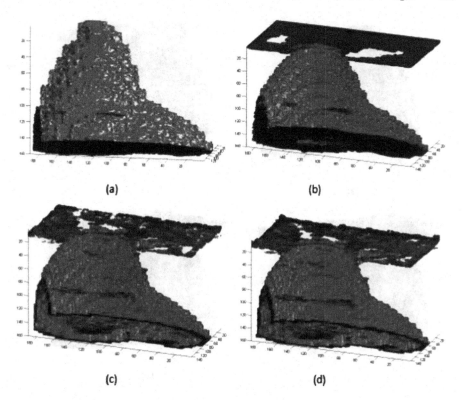

Fig. 8. Reconstruction results in contract: (a) The result of proposed method. (b) The result of Gauss filtering. (c) The result of Canny edge extraction. (d) The result of Sobel edge extraction.

Lastly, we choose the Marching Cubes algorithm to reconstruct 3D images. The algorithm is an important method for surface reconstruction of medical images and the key is to compute the isosurface. The results of reconstructed images based on different 2D images processing are shown in Fig. 9.

After comparing and verifying the results from these reconstructed methods above, we can come to the conclusion from four algorithms of 2D data processing, that is, when we use our improved algorithm to reconstruct the 3D phantom of the ultrasonic C scan image, the noise interference has reduced and the edge contour is clear, with the overall reconstruction quality much better than the other three methods. The three different reconstructed methods can prove the same conclusion without being accidental.

Concluded from the three image reconstruction, the scan data noise mainly concentrated in the top area of liver phantom, so we focus on the phantom quality of the top with the comparison of different treatment methods for scanning data after the reconstruction, so as to evaluate reconstruction method. According to the 3D phantom with the original data of the top 50 images from 1 to 50 for the top phantom reconstruction, it can be proved that the data area at the top has

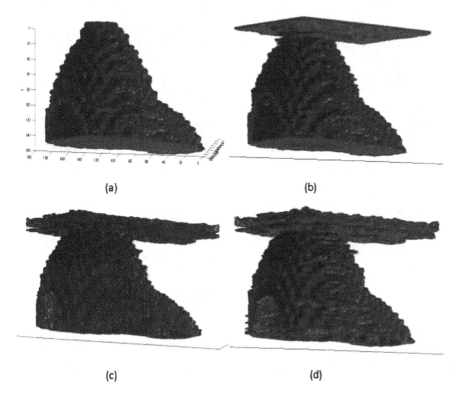

(a)   (b)

(c)   (d)

Fig. 9. Reconstruction results in contract: (a) The result of proposed method. (b) The result of Gauss filtering. (c) The result of Canny edge extraction. (d) The result of Sobel edge extraction.

aliasing noise seriously and original signal quality is poor. It is not ideal to the direct reduction of 3D phantom. By contrast, the methods of Gauss image filter and Canny edge extraction can weaken the noise interference to certain extents in Fig. 10(a) and (b), but they can't restore the missing edge and improve the quality of reconstruction. Figure 10(c) is the 3D phantom of the top with the use of 3D image reconstruction method based on speckle detection and area iteration.

Fig. 10. 3D reconstruction results of the top 50 images: (a) Reduction results of Gauss image filter. (b) Reduction results of Canny edge extraction. (c) Reduction results of our method.

The noise of the phantom is small, and the edge is clear and complete, with the phantom contour being true and accurate, which has optimized the results of image processing. At the same time, the computational time of this method is reasonable, because its image processing and iterative process are not complex, and there is no additional calculation and interference, so it is convenient to improve the quality and efficiency of the 3D reconstruction process.

4   Conclusion

Aiming at the irregular ultrasonic tissue-mimicking phantom of liver phantom, this paper has described the 3D reconstruction algorithm for ultrasonic C scanning image based on area iteration. The method calculated the contour area on the image with iterative technique according to the continuity of area, matching and extracting the contour area closest, which goes through the whole phantom data in turn and extract the entire contour, finally completing 3D reconstruction. The experimental results further verified the proposed method.

Acknowledgments. This work was supported in part by the National Natural Science Foundation of China under grant Nos. 61672084 and 61473025.

References

1. Sharmila, K., Sangeetha Lakshmi, G., Siva, S.A.: High-speed medical imaging in 3D ultrasound computer tomography. IRJET **2**(7), 346–350 (2015)
2. Prager, R.W., Ijaz, U.Z., Gee, A.H., Treece, G.M.: Three-dimensional ultrasound imaging. Proc. Inst. Mech. Eng. Part H J. Eng. Med. **224**(2), 193–223 (2010)
3. Jin, C.-Z., Nam, K.-H., Paeng, D.-G.: The spatio-temporal variation of rat carotid artery bifurcation by ultrasound imaging. In: Ultrasonics Symposium, pp. 1900–1903 (2014)
4. Riviere, C.N., Thakral, A., Iordachita, I.I., Mitroi, G., Stoianovici, D.: Predicting respiratory motion for active canceling during percutaneous needle insertion. In: Proceedings of the IEEE International Conference in Medicine and Biology Society, vol. 4, pp. 3477–3480 (2001)
5. Abolhassani, N., Patel, R.V.: Deflection of a flexible needle during insertion into soft tissue. In: Proceedings of the IEEE International Conference on Engineering in Medicine and Biology Society, vol. 1, no. 3, pp. 3858–3861 (2006)
6. Majewicz, A., Marra, S.P., van Vledder, M.G., Lin, M., Choti, M., Song, D.Y., et al.: Behavior of tip-steerable needles in ex vivo and in vivo tissue. IEEE Trans. Biomed. Eng. **59**(10), 2705–2715 (2012)
7. Abayazid, M., op den Buijs, J., de Korte, C.L., Misra, S.: Effect of skin thickness on target motion during needle insertion into soft-tissue phantoms. In: IEEE Ras and Embs International Conference on Biomedical Robotics and Biomechatronics, vol. 9, no. 4, pp. 755–760 (2012)
8. Hong, J., Dohi, T., Hashizume, M., Konishi, K., Hata, N.: An ultrasound-driven needle-insertion robot for percutaneous cholecystostomy. Phys. Med. Biol. **49**(3), 441–455 (2014)

9. Delgorge, C., Courreges, F., Bassit, L.A., Novales, C., Rosenberger, C., Smith-Guerin, N., et al.: A tele-operated mobile ultrasound scanner using a light-weight robot. IEEE Trans. Inf. Technol. Biomed. **9**(1), 50–58 (2005)
10. Onogi, S., Yoshida, T., Sugano, Y., Mochizuki, T., Masuda, K.: Robotic ultrasound guidance by B-scan plane positioning control. Procedia CIRP **5**(3), 100–103 (2013)
11. Kim, C., Chang, D., Petrisor, D., Chirikjian, G., Han, M., Stoianovici, D.: Ultrasound probe and needle-guide calibration for robotic ultrasound scanning and needle targeting. IEEE Trans. Biomed. Eng. **60**(6), 1728–1734 (2013)
12. Pierrot, F., Dombre, E., Dégoulange, E., Urbain, L., et al.: Hippocrate: a safe robot arm for medical applications with force feedback. Med. Image Anal. **3**(3), 285–300 (1999)
13. Solberg, O.V., Lindseth, F., Torp, H., Blake, R.E., Nagelhus Hernes, T.A.: Freehand 3D ultrasound reconstruction algorithms-a review. Ultrasound Med. Biol. **33**(7), 991–1009 (2007)
14. Barry, C.D., Gee, A.H., Berman, L.: Three-dimensional freehand ultrasound: image reconstruction and volume analysis. Ultrasound Med. Biol. **23**(8), 1209–1224 (1997)
15. Jensen, J.A., Nikolov, S.I., Gammelmark, K.L., Pedersen, M.H.: Synthetic aperture ultrasound imaging. Ultrasonics **1**(8), 5–15 (2006)
16. Ungi, T., Abolmaesumi, P., Jalal, R., Welch, M., Ayukawa, I., Nagpal, S., Lasso, A., Jaeger, M., Borschneck, P.D., Fichtinger, G., Mousavi, P.: Spinal needle navigation by tracked ultrasound snapshots. IEEE Trans. Biomed. Eng. **59**(10), 2766–2772 (2012)
17. Moon, H., Geonhwan, J., Park, S., Shin, H.: 3D freehand ultrasound reconstruction using a piecewise smooth Markov random field. Comput. Vis. Image Underst. **151**, 101–113 (2016)
18. Suzuki, K., Horiba, I., Sugie, N.: Fast connected-component labeling based on sequential local operations in the course of forward raster scans. In: International Conference on Pattern Recognition, vol. 2, no. 2, pp. 434–437 (2000)

A Generation Method of Insulator Region Proposals Based on Edge Boxes

Zhenbing Zhao[1]([⊠]), Lei Zhang[1], Yincheng Qi[1], and Yuying Shi[2]

[1] School of Electrical and Electronic Engineering,
North China Electric Power University, Baoding 071003, China
zhaozhenbing@ncepu.edu.cn
[2] School of Mathematics and Physics, North China Electric Power University,
Beijing 102206, China

Abstract. The generation of region proposals is the foundation of object detection. In the object detection task, the steady increase in complexity of classifiers may lead to improvement of detection quality, yet with the cost of increased computation time at the same time. One approach to overcome the tension between high detection quality and low computational complexity is through the use of "region proposals". High-quality insulator region proposals also play important roles in the detection of transmission line inspection images. This paper applies Edge Boxes to the localization of insulators in inspection images creatively, considering the characteristics of insulators' edge images, and combines these characteristics with Edge Boxes. As a result, more insulator region proposals are displayed. The experimental results show that, our method can effectively reduce the interference area, meanwhile, has high quality of region proposals with fast speed of calculation.

Keywords: Insulator · Inspection image · Object detection
Region proposal · Edge Boxes

1 Introduction

Ensure the reliability of the transmission line is an important part of the smart grid construction. Insulators are indispensable element on transmission lines with the dual function of electrical insulation and wire support. Besides, frequent faults of insulators may lead to large-scale blackout and huge losses [1,2]. One of the methods to improve the efficiency of insulator detection greatly is to process the transmission line inspection images by means of computer vision, and it can realize the request of intelligence and automation. Among them, the key and foundation of automatic detection is the localization of insulator in inspection images automatically [3].

Current trendy and top performing object detectors mostly employ region proposals to guide the detection and localization for objects [4,5]. In image challenges based on world famous data set such as PASCAL VOC [6] and ImageNet

© Springer Nature Singapore Pte Ltd. 2017
J. Yang et al. (Eds.): CCCV 2017, Part I, CCIS 771, pp. 231–241, 2017.
https://doi.org/10.1007/978-981-10-7299-4_19

[7], the object detectors which achieve outstanding performance and excellent effect all use the method of region proposals. For instance, the framework of R-CNN (Regions with CNN features) [8], which combined Selective Search [9] and CNN (Convolutional Neural Network) [10], has raised the detection rate from 35.1% to 53.7% in Pascal VOC Challenge. For an image, it first generates multiple region proposals and then, sets these proposals to fixed size to send to CNN for feature extraction and classification. Much follow-up work such as SPP-Net [11], Fast R-CNN [12] and Faster R-CNN [13] also involves the generation of region proposals, no matter in original images or feature maps obtained through deep neural network. Therefore, we settle down to finding a better generation method for region proposals, and it indeed has great significance in feature extraction and object localization.

In order to locate the insulators in transmission line inspection images automatically, and realize real-time detection and fault diagnosis on this basis, we choose the method of Edge Boxes [14], which has better performance in public data sets to be applied to transmission line inspection images that we obtained through professional equipment. However, the original intention of Edge Boxes is to detect the general objects in images, not specifically for insulators. If we expect to generate region proposals outstanding insulators only using Edge Boxes, the results are not satisfied. Therefore, in our method, we make full use of the prior characteristics of insulators and combine these characteristics with Edge Boxes. By processing the images with a series of operation, we finally generate the region proposals which contain more insulator parts. The experimental results show that our method can reduce the interference area effectively, and can ensure the follow-up phase of feature extraction to extract more pure insulator characteristics.

2 Related Work

2.1 Multiple-Scale Sliding Window

Inspired by the implementation of BING [15,16], multiple-scale Sliding Window is the first one widely used for generating region proposals. Many fixed size windows slide in equidistant step on the images which are transformed into different scales. Due to the search space is the whole image, the greatest advantage is that its miss rate is extremely low and it will not leave any proposals out. However, Multiple-Scale Sliding Window classifiers increase linearly with the number of windows tested, and while single-scale detection requires classifying around 10^4–10^5 windows per image, the number of windows grows by an order of magnitude for multi-scale detection. The huge search space and consumption of time further influenced the detection efficiency.

2.2 Selective Search

Instead of proposal generation method without any strategies like Multiple-Scale Sliding Window, Selective Search combines the strength of both exhaustive search [17] and segmentation. It uses the image structure to guide the sampling process like segmentation, meanwhile, it aims to capture all region proposals like exhaustive search. In Selective Search, a number of original areas are obtained by image segmentation, and then they are merged by strategy which based on color, texture and size. All object scales have to be taken into account. Compared with the traditional method with single strategy, Selective Search offered a variety of strategies, reduced the search space greatly and finally obtained more excellent results in object recognition. So far from Selective Search has been proposed, it has been widely used in many advanced object detection methods including R-CNN and Fast R-CNN. However, for the purpose of speeding up feature extraction, about 2000 region proposals generated from per image are still a stumbling block.

2.3   Edge Boxes

Edges provide a sparse but informative representation of an image. On the study of region proposals, Zitnick and Dollár [14] found a new way to generate object bounding box proposals which only use edges in an image. Edge Boxes took full advantage of the rich edge information in images, proposed a simple box objectness scoring method based on the number of edges that exist in the box and the number of contours that overlap the box's boundary. This novel method can reduce the number of generated region proposals effectively, meanwhile, the calculating speed and precision has improved greatly compared with Selective Search.

However, the assumption of object parameters such as shape and size are not suitable for insulators in Edge Boxes. There might be some omissions of insulators, or too much interference of other components. Therefore, improving the method of Edge Boxes to generate region proposals, which are more suitable for insulators, is an important content of this paper. We considering the characteristics of insulators' edge images, took a series of operations such as K-means clustering on CSS (Curvature Scale Space) points and circle on insulator subclass, combined insulators' characteristics with Edge Boxes to get better performance. We will describe our method in detail in Sect. 3.

3   Our Method

3.1   Framework

For the input inspection image, we first do some preprocessing including graying, threshold segmentation and remove of the redundancy small area. We extract the edge of images and the CSS points [18,19] in edge images. These CSS points are clustered into two subclasses by K-means. Then, we find CSS points which lie on suspected insulator subclass (the subclass that might be insulators) according to some certain rules, and use these points as the centers to form a set number of circles. This step can increase the number of edges that exist in the box completely which locates in the insulator. We put the images back to Edge Boxes scoring system, and now, the score of proposal box which contains insulator will increase, so as to make the output of the proposals contain more insulator subclasses. The framework of our method is shown in Fig. 1.

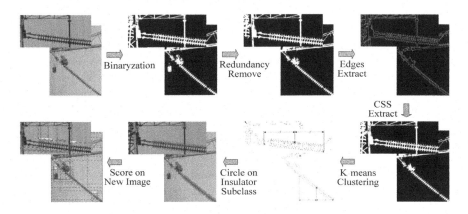

Fig. 1. Framework of our insulator region proposal generation method based on Edge Boxes (Color figure online)

3.2   Image Preprocessing

The process of graying and threshold segmentation towards insulator inspection images, transforming the original images into binary images, can realize the separation of foreground and background. Containing varieties of objects such as insulator strings, towers, wire and inspectors, it is also difficult to determine the position of insulators in foreground. Choosing the method of morphological filtering, we first operate the binary images with morphological erosion in order to separate objects at slender points and remove the noise of tiny areas. Then the operation of morphological dilation fills the internal holes and smoothes the larger objects' boundaries. These two morphological operations can remove most noise points, making the object edges smoother. As for the surviving small areas after filtering, we set a threshold to remove them. This step can eliminate the interference of impurity and improve the localization accuracy.

3.3   CSS Corners and K-Means Clustering

Contour curves [20] are extracted from the edge images. CSS corners are obtained as follows. Firstly, we calculate the curvature of each pixel point in contour curves under the high scale and choose the maximum curvature points as candidates for CSS corners. Secondly, if the curvature of one candidate point is greater than the preset threshold then mark this candidate point as the correct CSS corner. Finally, pinpoint all the correct CSS corners under the low scale.

Insulator string contains numbers of umbrella plates which have similarity in shape, meanwhile the curvature of each umbrella plate's edge is almost the same. This character makes the distribution of CSS corners extracted from insulators very uniform yet no evident regularity is found in other component such as tower, wire and inspectors. Figure 2 shows the distribution of CSS corners in insulator strings.

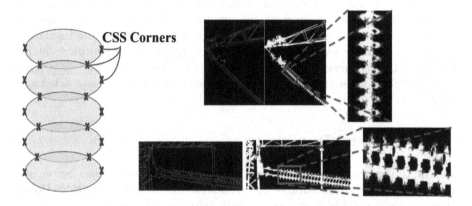

Fig. 2. CSS corners in insulator strings. We can clearly see that these points distribute neatly and evenly on the edge and the joint of umbrella plates

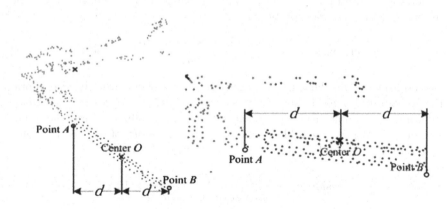

Fig. 3. Find suspected insulation subclasses by judging d (Color figure online)

We cluster all CSS corners into two subclasses through K-means. Between each subclass, we find point A to represent the point which has the smallest abscissa and point B to represent the point which has the biggest abscissa. O is the center of clustering and if A and B has similar distance d towards O, we deem it be the suspected insulation subclass. To make the distinction results more accurate, we further considered the ordinate. The blue points in Fig. 3 are the CSS corners we found in suspected insulation subclass.

3.4 Circle

In Edge Boxes, one observation is that the more contours wholly contained in a bounding box, the more likely it contains an object. For this grading rule, we believe that if we can increase the number of contours around suspected insulation subclasses, so that we can improve box score which contains insulators. To make things easier, we adopt the method of CIRCLE. To be specific, centered on the blue CSS corners we found in Sect. 3.3, we circle numbers of circles with minor radius and these circles can increase closed contours effectively. Schematic diagram can be seen in the last two steps in Fig. 1.

4   Experimental Result

Based on several experiments, we have verified that our method has better performance compared with Sliding Window, Selective Search and pure Edge Boxes. The comparison unfolds from three aspects: effectiveness, precision and speed.

4.1 Effectiveness and Precision

In this section, we put the ratio of region proposals which contain insulator as an evaluation criterion of our method, and it can reflect the effectiveness when generating insulator region proposals:

$$\text{effectiveness} = \frac{proposals\ contain\ insulators}{all\ proposals} \times 100\% \tag{1}$$

In addition to insulators, the proposals we generated usually also contain some background such as sky. Generally speaking, we hope that the area of insulator in the whole region proposal is as far as possible big and the area of background is as far as possible small. We define the index of precision to reflect the area ratio of groundtruth and proposals contain insulators:

$$\text{precision} = \frac{S_{groundtruth}}{S_{proposals\ contain\ insulators}} \times 100\% \tag{2}$$

Tables 1 and 2 give the effectiveness and precision of Sliding Window, Selective Search, Edge Boxes and our method (due to limited space, we present the experimental results on three images under each method). Table 1 shows the number of region proposals which contain insulators. Table 2 shows the ratio of all proposals when the precision reached 50%, 75% and 90% respectively. As we can see from the data in Tables 1 and 2, the number of proposals generated through pure Edge Boxes and our method has dropped a lot compared

Table 1. Effectiveness of different methods

Method	Threshold	All proposals	Which contain insulators/effectiveness
Sliding window	–	$\approx 10^4$	–
Selective search	$k = 200$	132	22/16.67%
	$k = 200$	178	52/29.21%
	$k = 200$	300	61/20.33%
Edge Boxes	score > 0.15	34	0/0.00%
	score > 0.24	5	1/20.00%
	score > 0.25	25	17/68.00%
Ours	score > 0.15	30	**7/23.33%**
	score > 0.24	44	35/**79.55%**
	score > 0.25	47	40/**85.11%**

Table 2. Precision of different methods

Method	Threshold	Precision $= 50\%$ number/ratio	Precision $= 75\%$ number/ratio	Precision $= 90\%$ number/ratio
Sliding window	–	–	–	–
Selective search	$k = 200$	6/4.54%	3/2.27%	0/0.00%
	$k = 200$	40/22.47%	14/7.87%	6/3.37%
	$k = 200$	58/19.33%	36/12.00%	20/6.67%
Edge Boxes	score > 0.15	0/0.00%	0/0.00%	0/0.00%
	score > 0.24	1/20.00%	1/20.00%	0/0.00%
	score > 0.25	17/68.00%	14/56.00%	10/40.00%
Ours	score > 0.15	7/**23.33%**	7/**23.33%**	3/**10.00%**
	score > 0.24	35/**79.55%**	35/**79.55%**	31/**70.45%**
	score > 0.25	40/**85.11%**	40/**85.11%**	38/**80.85%**

Table 3. Comparison of proposals generation speed

Method	Test image no. 1 (sec.)	Test image no. 2 (sec.)	Test image no. 3 (sec.)
Sliding window	$>>1$ min	$>>1$ min	$>>1$ min
Selective search	12.7626	16.8485	28.4467
Edge Boxes	0.5294	0.3952	0.4813
Ours	**0.5032**	**0.4027**	**0.4715**

with Sliding Window and Selective Search. Specially, when employing the pure Edge Boxes, few box contains insulators was detected, sometimes none. However, through employing the method we proposed in this paper, all indexes have greatly increased, no matter the quantity or the quality of region proposals. A more intuitive comparison can be seen in Fig. 6.

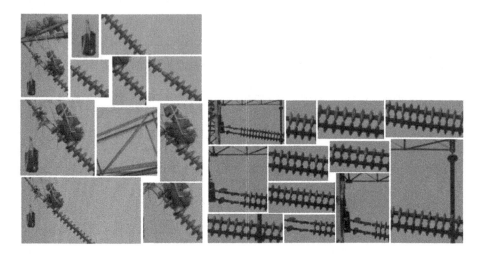

Fig. 4. Parts of the region proposals generated through our method

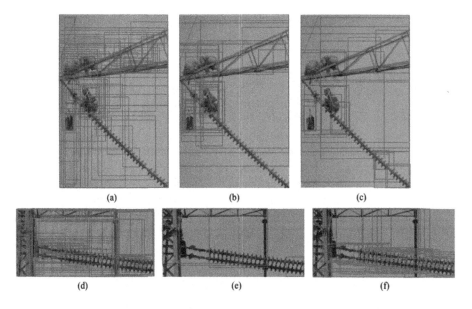

Fig. 5. Region proposals generated through different methods. (a) (d) Selective Search. (b) (e) pure Edge Boxes. (c) (f) Our method

Figure 4 shows parts of the generated region proposals through our method. Figure 5 shows the proposals generated through three different methods: (a) (d) Selective Search, (b) (e) pure Edge Boxes and (c) (f) the method proposed in this paper. Hundreds proposals are generated through Selective Search and these waste too much time. The number of proposals generated through pure Edge Boxes and our method has dropped a lot, meanwhile, proposals generated through our method are more concentrated around the insulators.

4.2   Speed

In the task of object detection, we hope it consumes less time in the period of generating region proposals. In order to accelerate the whole process of object detection, the number of proposals needs to be cut down and the generation speed needs to be expedited at the same time. Table 3 shows the time that four methods need (identical to Sect. 4.1, we show results on three images and the present time is the average of ten experiments).

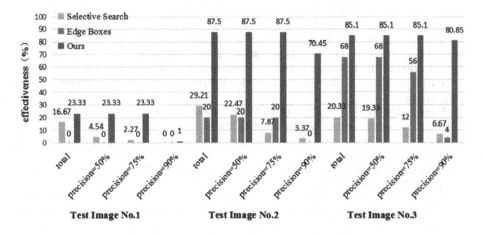

Fig. 6. Effectiveness and precision of different methods

Among them, Sliding Window took far more time than three others because of the search space lies on the whole image. Compared with Selective Search, pure Edge Boxes increased substantially in generating speed and probably only about 1.69% to 4.15% of it. In this paper, our method achieved a slight acceleration or flat, for instance, the speed increased by 4.95% in test image 1 and 2.04% in test image No. 3.

5   Conclusion

In this paper, in the process of locating insulators in transmission line inspection images, we overcome the shortcoming that the proposal areas failed to highlight the insulators, and propose a generation method of insulator region proposals based on Edge Boxes. We considering the characteristics of insulators' edge images, took a series of operations such as K-means clustering on CSS points and circle on insulator subclass, combined insulators' characteristics with Edge Boxes to get better performance. With this method we proposed, more insulator region proposals are displayed and less interference regions are presented. Furthermore, the experiment results showed that our method did well both in effectiveness and precision, and achieved fast computation speed at the same time.

Acknowledgments. This work was supported partially by National Natural Science Foundation of China (No. 61401154), Hebei Province Natural Science Foundation of China (No. F2016502101) and the Fundamental Research Funds for the Central Universities (No. 2015ZD20).

References

1. Wen, Hu.: Research of Electric Power Equipment Fault Diagnosis Based on Intelligent Information Fusion. Huazhong University of Science and Technology, Wuhan (2005)
2. Han, S., Hao, R., Lee, J.: Inspection of insulators on high-voltage power transmission lines. IEEE Trans. Pow. Deliv. **24**(4), 2319–2327 (2009)
3. Li, L., Zhou, R.: Unmanned Aerial Vehicle Transmission Line Patrol Technology and its Application Research. Changsha University of Science and Technology, Changsha (2012)
4. Hosang, J., Benenson, R., Dollár, P.: What makes for effective detection proposals? IEEE Trans. Pattern Anal. Mach. Intell. **38**(4), 814 (2016)
5. Hosang, J., Benenson, R., Schiele, B.: How good are detection proposals, really? Comput. Sci. (2014)
6. Everingham, M.: The pascal visual object classes (VOC) challenge. Int. J. Comput. Vis. **88**(2), 303–338 (2010)
7. Deng, J., Dong, W., Socher, R., et al.: Imagenet: a large-scale hierarchical image database. In: 27th Computer Vision and Pattern Recognition. IEEE Press, Miami (2009)
8. Girshick, R., Donahue, J., Darrell, T., et al.: Rich feature hierarchies for accurate object detection and semantic segmentation. Comput. Sci. 580–587 (2014)
9. Van, S., Uijlings, R.: Segmentation as selective search for object recognition. In: IEEE Computer Society, pp. 1879–1886 (2011)
10. Krizhevsky, A., Sutskever, I.: Imagenet classification with deep convolutional neural networks. In: 19th International Conference on Neural Information Processing Systems, pp. 1097–1105. Springer Press, Doha (2012)
11. He, K., Zhang, X., Ren, S.: Spatial pyramid pooling in deep convolutional networks for visual recognition. IEEE Trans. Pattern Anal. Mach. **37**, 1904–1916 (2015)
12. Girshick, R.: Fast R-CNN. Comput. Sci. (2015)
13. Ren, S., He, K., Girshick, R.: Faster R-CNN: towards real-time object detection with region proposal networks. In: IEEE Trans. Pattern Anal. Mach. Intell. 1 (2016)
14. Zitnick, C.L., Dollár, P.: Edge boxes: locating object proposals from edges. In: Fleet, D., Pajdla, T., Schiele, B., Tuytelaars, T. (eds.) ECCV 2014. LNCS, vol. 8693, pp. 391–405. Springer, Cham (2014). https://doi.org/10.1007/978-3-319-10602-1_26
15. Cheng, M.M., Zhang, Z., Lin, W.Y.: BING: binarized normed gradients for objectness estimation at 300fps. In: 32th Computer Vision and Pattern Recognition, pp. 3286–3293. IEEE Press, Columbus (2014)
16. Zhao, Q., Liu, Z., Yin, B.: Cracking BING and beyond. In: 25th British Machine Vision Conference. Springer Press, Nottingham (2014)
17. Zhu, L., Chen, Y., Yuille, A.: Latent hierarchical structural learning for object detection. In: 28th Computer Vision and Pattern Recognition, pp. 1062–1069. IEEE Press, San Francisco (2010)

18. Mokhtarian, F.: Robust image corner detection through curvature scale space. IEEE Trans. Pattern Anal. Mach. Intell. **20**(12), 1376–1378 (2010)
19. Sun, J., Qiang, Q.: Contour corner detection based on curvature scale space. Opto-Electron. Eng. (2009)
20. Ai, W., Huang, X.: Outline of fast image recognition method in detail. Chinese patent: 200910100170 (2009)

Robust Tramway Detection in Challenging Urban Rail Transit Scenes

Cheng Wu, Yiming Wang[✉], and Changsheng Yan

School of Rail Transit, Soochow University,
8. Jixue Rd., Suzhou 215006, Jiangsu, China
ymwang@suda.edu.cn
http://jtxy.suda.edu.cn/

Abstract. With the rapid development of light rail transit, tramway detection based on video analysis is becoming the prerequisite and necessary task in driver assistance system. The system should be capable of automatically detecting the trackway using on-board camera in order to determine the train driving limit. However, due to the diversification of ground types, the diversity of weather conditions and the differences in illumination situations, this goal is very challenging. This paper presents a real-time tramway detection method that can effectively deal with various challenging scenarios in the real world of urban rail transit environment. It first uses an adaptive multi-level threshold to segment the ROI of the trolley track, where the local cumulative histogram model is used to estimate the threshold parameters. And then use the regional growth method to reduce the impact of environmental noise and predict the trend of tramway. We have experimentally proved that the method can correctly detect the tramway even in many undesirable situations and use less computational time to meet real-time requirements.

Keywords: Computer Vision · Track detection
Multilevel thresholding · Region growing
Intelligent Transportation System

1 Introduction

In recent years, public transport has been greatly advocated due to the desire of alleviating traffic congestion in metropolitan areas and the demand of reducing air emissions that induce climate change. Among various public transport modes, a tram is typically a light-rail public transport vehicle, which is faster than buses and much cheaper than rapid transit systems [9]. The term "*tram*" is called in Europe and also known as "streetcar" or "trolley-car" in North America. A tram vehicle often runs on tracks along city streets, sometimes on segregated rights-of-way (ROW) in public urban areas [16]. The kind of tram transport have a lot of benefits over bus and other rapid transit trains, such as lower construction cost, improving safety and reliability, better ridership and fewer carbon emissions etc.

Indeed, a tram transport system is technically different from bus systems and other high-capacity rapid rail transit systems. These differences include: (i) the

© Springer Nature Singapore Pte Ltd. 2017
J. Yang et al. (Eds.): CCCV 2017, Part I, CCIS 771, pp. 242–257, 2017.
https://doi.org/10.1007/978-981-10-7299-4_20

degree of separation of rail tracks, (ii) vehicle technology, and (iii) operating practices [5,7]. As one of the key features, the degree of separation of rail tracks shows the flexibility of rights-of-way. Most commonly, most of the trams travel in the middle of street strips or at roadsides. On the one hand, although tram routes are mixed with other traffic modes, tram vehicles still keep exclusive or semi-exclusive rights-of-way, which allow them much faster than the buses. On the other hand, the tram is different from a rapid transit line, which runs only on fully grade-separated rights-of-way, such as in tunnels or on elevated structures. A tram transport system often make use of *hybrid rights-of-way*. A usual design is to permit a tram line with a partial grade-separated rights-of-way and some grade crossings. For example, a tram running on the ground can over-pass or under-pass at heavy-traffic regions or overcrowding intersections and stop only at few street-level crossings. Such a design can reduce significantly construction costs [7,8].

However, the mode of hybrid rights-of-way is unavoidably bringing potential safety hazards to tram operating practices. Unlike other rapid rail transits, whose rail lines are not accessible and have no interaction with road users, the tram lines in certain areas, such as road crossings, are open and with potential dangers by littering and causing obstructions to the tramway line [15]. In addition, on a street-level crossing, a tram may collide with pedestrians, cyclists and drivers, as it cannot stop so quickly or avoid them. Most commonly, a tram traveling at just 40 kph needs around 50 m to stop. People may not fully realize the length of the braking distance of trams, and so it often ends in tragedy [12,14]. There are some other dangerous behaviors related to tram operations, such as crossing the tramway line from non-designated areas, entering a restricted area, interfering with the operation of the tramway or taking any action that would compromise the safety of the tram, and crossing the red light at the junction with the tramway and blocking the tramway without permission etc. [16].

Therefore, one of the key technical challenges for modern tram is to survey the track along a tram line with respect to obstacles. The track operated by a tram line is called *tramway*. Traditionally, a human tram driver uses visual perception to observe the tramway and triggers prompt actions, like whistling and braking, to avoid the occurrence of an accident. Recent studies and accident analysis show that only such the manual survey is unreliable and inefficient [13, 14,18]. It is mainly due to the following two factors. One is that the nature of a human tram driver's attention is hard to concentrate for a long time. Another is that human vision is limited under a narrow scale, especially in cases of bad weather or weak illumination. Hence, it is urgent to establish a fully-automated video-based surveillance system for modern tram.

Recently, driving assistance technologies in rapid rail transit have been paid so many attentions for improving rail safety. There are various methods and systems designed to help drivers for driving safety, but few publications are published specifically for the detection of tramway using on-board camera. The majority of such the papers been published used threshold segmentation or edge detection to extract the railways. Some methods used the iterative threshold

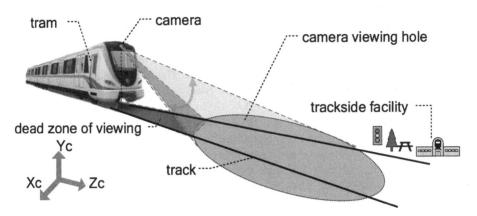

Fig. 1. A basic vision-based tramway detection model.

and *Otsu* method within the simple scenes, where two common edge detection operators: *Sobel* and *Canny* were used to extract the edge of track [3, 13]. These methods are difficult to extract the edges correctly in the complex background and are too parameter-sensitive to be estimated. Wohlfeil proposed a set of *LLPD* operators and used a *Hough* transform to merge several short linear segments into a continuous line [17]. The method based on *Hough* transform has an inhibitory effect on the noise and environmental disturbance, and is also applied in the situation when straight line breaks or part of pixels are lost [4, 17]. Espino use a sliding window approach to iteratively select local maxima in the gradient images to extract railway tracks for rubber-tyre trains [2]. Qi computed *HOG* features, constructed integral images and then extracted railway tracks by region-growing algorithm [10].

In fact, the issue of using on-board machine vision to assist train drivers to work and ensure safe operation of train, specially tram, is not easy to address. A basic vision-based railway detection system can be seen in Fig. 1. The core of the technical system in tram is tramway scene analysis. Tramway scene analysis consists of **tramway detection** and **obstacle detection**. Tramway detection is an essential and foregoing task, which includes the localization of tram rail tracks and the determination of traffic safety limits. Normally, the video frames acquisited by on-board camera often contain a large number of irrelevant entities, which would cause bad accuracy and low efficiency of obstacle detection. The region of interest (ROI) of obstacle detection is indeed around the tramways and within its surrounding limited area. So accurate tramway detection is extremely important. But a tram running in urban environment often has to be operated in different traffic environment scenarios and under varying weather and illumination conditions, as shown in Fig. 9. At the time, tramway detection becomes non trivial particularly with the presence of various ground types (e.g. Fig. 2(a), (b), (c) and (d)), the effects of weather conditions (e.g. Fig. 2(e), (f) and (g)) and the time of acquisition (e.g. Fig. 2(h), (i) and (j)) and the presence of obstacles occurrence (e.g. Fig. 2(k) and (l)). Its difficulty gets further accentuated in cases

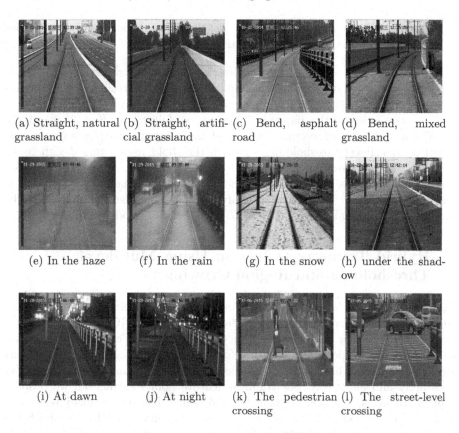

(a) Straight, natural grassland (b) Straight, artificial grassland (c) Bend, asphalt road (d) Bend, mixed grassland

(e) In the haze   (f) In the rain   (g) In the snow   (h) under the shadow

(i) At dawn   (j) At night   (k) The pedestrian crossing   (l) The street-level crossing

Fig. 2. Different traffic environment scenarios and varying weather and illumination conditions

of on-board camera movements, blur effects, and abrupt or unexpected actions of drivers, which are likely to distort the result of detection.

Aiming at these challenging situations, this paper proposed a real-time tramway detection method that deals with various challenging situations in real-world urban rail traffic scenarios. By efficiently using a on-board long-distance camera, our perception system figures out the tramway and its corresponding optimal safety limits. Our real-time tramway detection system is distinguished from related researches at the following points.

(i) Our approach can reliably deal with challenging urban environments, including various ground types, different weather and illumination conditions as well as varying time of acquisition.

(ii) An adaptive multilevel thresholding method is proposed for segmenting the ROIs of tramway, in which threshold parameters are estimated using local accumulated histogram.

(iii) The proposed method extracts the optimal tramway in front of the tram vehicle using region growing, instead of recognizing every pixel on every video frame, which is believe to be time-consuming and infeasible for autonomous driving assistance system.

The methods we presented here can work potentially as a cost-effective tram-driving assistance system. We use a simple setup with a surveillance camera plus adaptive algorithm since we don't want to rely on expensive equipment.

The remainder of the paper is organized as follows. In Sect. 2, we give a detailed introduction of recognition method, including track region segmentation based on multilevel thresholding, adaptive threshold parameters design based on local accumulation histogram, and feature points extraction based on pixel tracing. In Sect. 3, we evaluate our method using various challenging scenarios in real urban traffic environment and analyze some experimental results. Finally, in Sect. 4, we summarize the conclusions and give an outlook on future work.

2 Hybrid Tramway Detection Using Multilevel Thresholding and Region Growing

Tramway detection is mainly divided into two parts, *tramway ROI segmentation* and *tramway feature points extraction*. For tramway ROI segmentation, we present a method for ROI segmentation of tramway based on multilevel thresholding and a corresponding algorithm for adaptive threshold parameters using local accumulation histogram. It can extract the grey image of the tramway accurately and effectively. For tramway feature points extraction, we propose a pixel tracing and region growing method, which has strong anti-interference ability. The method then chooses the appropriate curve model to establish the tram equation, which can extract the tramway accurately and quickly.

2.1 Tramway ROI Segmentation Based on Multilevel Thresholding

The tramway in Suzhou are typically constructed with girder rails. Tram vehicle wheels ride on the rail surface and are held in place by a concave rail with the larger diameter, as shown in Fig. 3(a). The concave rail is called *flangeway*. In girder rails, the flangeway is part of the cast steel rail, as shown in Fig. 3(c) and (d), which is laid in the street concrete. The characteristics of the tramway we observe the images (e.g. Fig. 3(b)) are as follows:

(i) The inside of tramway is darker, the outside is lighter, and the difference of gray level is obvious;
(ii) The darker region is adjacent to the lighter region, and the boundary is clear and smooth.

Figure 3(e) is a flangeway segment that are sampled from the starting positions of left rail on the bottom of Fig. 3(b). Figure 3(f) is its corresponding gray histogram of Fig. 3(e). In these two figures, we can easily observe that the high-end in the range of grey values about [150–200] and low-end in about [0–50] have the obvious peaks.

(a) Tram wheels held in the flange-ways [14].

(b) Example of how the flangeway is installed along a tram line.

(c) Profile of flangeway cast in steel rail [14].

(d) Sample of actual flangeway.

(e) Sample of flingerway segment in real-world image

(f) Grey histogram of the sample segment.

Fig. 3. Flangeway and its gray histogram

According to the above characteristics, it is feasible to make use of the darker and the lighter gray values to segment the original image respectively and further extracts the ROIs of potential tramway using distance metric. The specific steps are as follows:

(i) **Image preprocessing**. The step includes gray scale processing and smoothing filtering. We define the original image after preprocessing as $I(x, y)$;

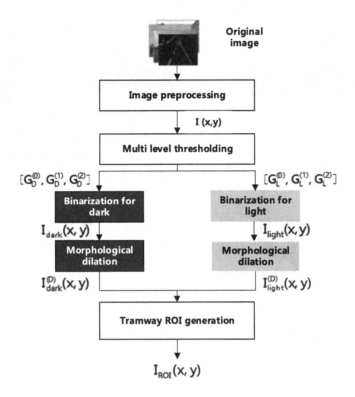

Fig. 4. Multilevel thresholding-based tramway ROI segmentation

(ii) **Multilevel thresholding.** We determine two sets of threshold values T_{dark} and T_{light} to deal with the darker and lighter regions, respectively. T_{dark} consists of the minimum threshold value T_{dark}^L and the maximum threshold value T_{dark}^H of the darker region, which can divide pixels into three groups:

$$
\begin{aligned}
G_D^{(0)} &= \{(x,y) \in I \mid 0 \le f(x,y) \le T_{dark}^L - 1\}; \\
G_D^{(1)} &= \{(x,y) \in I \mid T_{dark}^L \le f(x,y) \le T_{dark}^H - 1\}; \\
G_D^{(2)} &= \{(x,y) \in I \mid T_{dark}^H \le f(x,y) \le 255\}.
\end{aligned}
\tag{1}
$$

Similarly, we also have three groups of the lighter region using T_{light}^L and T_{light}^H:

$$
\begin{aligned}
G_L^{(0)} &= \{(x,y) \in I \mid 0 \le f(x,y) \le T_{light}^L - 1\}; \\
G_L^{(1)} &= \{(x,y) \in I \mid T_{light}^L \le f(x,y) \le T_{light}^H - 1\}; \\
G_L^{(2)} &= \{(x,y) \in I \mid T_{light}^H \le f(x,y) \le 255\}.
\end{aligned}
\tag{2}
$$

where $f(x,y)$ is the gray level of the point (x,y).

(iii) **Binarization using multilevel thresholding.** In the step, we obtain two binary images using T_{dark} and T_{light} for the darker region and the lighter region respectively, that is, $I_{dark}(x, y)$ and $I_{light}(x, y)$. They are constructed by Eqs. 3 and 4:

$$I_{dark}(x, y) = \begin{cases} 0, & (x, y) \in G_D^{(0)} \bigcup G_D^{(2)} \\ 255, & (x, y) \in G_D^{(1)} \end{cases} \tag{3}$$

$$I_{light}(x, y) = \begin{cases} 0, & (x, y) \in G_L^{(0)} \bigcup G_L^{(2)} \\ 255, & (x, y) \in G_L^{(1)} \end{cases} \tag{4}$$

(iv) **Morphological dilation.** The step is to connect the fracture ROIs of tramway and expand the width of potential tramway. $I_{dark}^{(D)}(x, y)$ and $I_{light}^{(D)}(x, y)$ are the morphological dilation of $I_{dark}(x, y)$ and $I_{light}(x, y)$.

(v) **Tramway ROI generation.** We generate the ROI of potential tramway through computing the intersection $I_{ROI}(x, y)$ of these two binary images after morphological dilation, $I_{dark}^{(D)}(x, y)$ and $I_{light}^{(D)}(x, y)$, as shown in Eq. 5.

$$I_{ROI}(x, y) = I_{dark}^{(D)}(x, y) \cap I_{light}^{(D)}(x, y) \tag{5}$$

A complete work flow about multilevel thresholding-based tramway ROI segmentation is described in Fig. 4.

2.2 Adaptive Threshold Parameters Estimation Using Local Accumulation Histogram

The difficulty of multilevel thresholding is to estimate the threshold parameters, that is, T_{light}^L, T_{light}^H, T_{dark}^L and T_{dark}^H. As is known to all, the tram operating environment is complex, traffic scenes are various and the intensity and angle of light always change. Hence, the range of gray levels is not the same, and it is hard to pre-assign a set of experienced thresholds to meet all the scenarios. In this section, we present an adaptive mechanism to calculate these thresholds using the statistic characteristic of local region. The specific steps are as follows:

(i) **The starting points of tramway positioning.** When the on-board camera is fixed in front of the locomotive of the tram, its viewing hole is also relatively fixed. The bottom pixels of the image acquisited by the camera are the nearest points to the tram. If we neglect the effect of dead zone (In fact, the length of dead zone (e.g. 30–50 m) can be neglected due to the fast speed of the tram (e.g. 40 mph)), the starting points of tramway always appear on several fixed pixels on the bottomline of the image. Hence, we can determine manually the starting points of tramway.

(ii) **Normalized histogram processing of local image.** Given the starting points of tramway, we can easily segment a tramway slice on the bottom of the image according to flangeway edges. Figure 5 gives an ideal illustration of tramway slice on pixel level. The height of local image is h, the widths of flangeway outside and inside are m and n respectively. $f_g(x, y)$ is the local gray image with intensity levels in the range $[0, L-1]$. The normalized histogram of $f_g(x, y)$ is a discrete function of intensity level r_k

$$p(r_k) = \frac{n_k}{\sum\limits_{l=0}^{L-1} n_l}, \tag{6}$$

where r_k is the kth intensity and n_k is the number of pixels of r_k, for $k = 0, 1, 2, \cdots, L-1$. $p(r_k)$ is an estimate of the probability of occurrence of intensity level r_k in the local image. Assume that the intensity levels of flangeway inside and outside are r_{fi} and r_{fo}, we have the probabilities of occurrence of r_{fi} and r_{fo}, that is,

$$p(r_{fi}) = \frac{n}{n+m}, \quad p(r_{fo}) = \frac{m}{n+m}. \tag{7}$$

(iii) **Threshold parameters estimation.** Threshold parameters can be estimated using the normalized histogram of local image. Figure 6 gives an illustration of normalized histogram. A typical normalized histogram of local image has two peaks on the intensity levels of flangeway inside and outside, that is r_{fi} and r_{fo}. Ideally, the probabilities at intensity levels of flangeway inside and outside are exactly $p(r_{fi})$ and $p(r_{fo})$. Actually, the value of intensity may fluctuate slightly due to the change of illumination and other environmental factors. We hence accumulate the probabilities of the neighboring intensities of r_{fi} and r_{fo} level by level till that Eqs. 8 and 9 are satisfied.

$$p(r_{fi}) + \sum_{i=1}^{i_T} (p(r_{fi} - i) + p(r_{fi} + i)) = \frac{n}{n+m}, \tag{8}$$

$$p(r_{fo}) + \sum_{j=1}^{j_T} (p(r_{fo} - j) + p(r_{fo} + j)) = \frac{m}{n+m}, \tag{9}$$

At the time, we obtain that

$$T_{dark}^L = r_{fi} - i_T, \quad T_{dark}^H = r_{fi} + i_T, \tag{10}$$

$$T_{light}^L = r_{fo} - j_T, \quad T_{light}^H = r_{fo} + j_T. \tag{11}$$

Compared with the traditional thresholding or edge detection methods [4,6,10,12,17] background interference and environmental noise can be greatly reduced when normalized histogram is locally accumulated in our approach. Particularly speaking, we inter multilevel thresholding based on the statistic characteristic of local accumulated histogram after normalization. It is the reason why the method can meet real-world requirements and deal with various challenging scenarios.

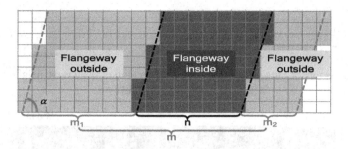

Fig. 5. Illustration of tramway slice on pixel level

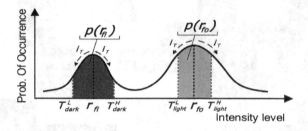

Fig. 6. Thresholds estimation on local normalized histogram.

2.3 Tramway Detection Using Region Growing

Given a binary image with potential tramway ROIs, the problem of tramway detection turns to link iteratively different potential tramway segments from some certain starting points. In each iteration, only the segments best to match the required tramway structure based on a set of given criteria would be kept. The *region growing* algorithm [1] is one of the commonly used methods for solving such the problem. The basic idea of region growing is to use various morphological gradient operators to extract the most approximate pixels in the neighborhood, and predict the position of a following segment using that of current segment [11].

In the section, we present our region growing method for tramway detection. The method starts from the binary image obtained by our tramway ROI segmentation. The basis of region growing is some growing seeds. In practice, the starting points in our local binary image are taken as growing seeds. At the beginning, these seeds grow from their exact pixel locations to adjacent pixels depending on a region detection criterion. Next, the detected adjacent pixels are examined using a tracing criterion. If they are similar with their parent seed, we classify them into the set of seeds. It is an iterative process until there are no changes in two successive stages. Finally, interferential lines constructed by sets of few seeds are removed according to prior knowledge.

The key of the method is to draw up a strong anti-interference detection and tracing criterion. The detection criterion we provide is to use some prior knowledge to search the region with potential seeds. The tracing criterion is to use the

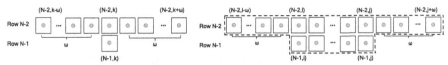

(a) Search region of single seed for next generation.

(b) Search region of transverse neighboring seeds for next generation.

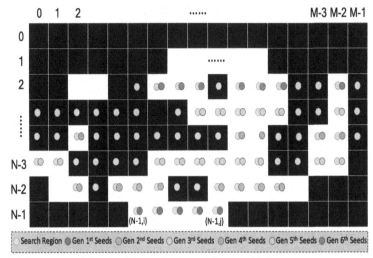

(c) Region growing on binary image with potential tramway segments.

Fig. 7. Illustration of region growing for tramway detection

location and connectivity of the found points to determine their membership to tramway.

The specific steps of region growing are as follows and Fig. 7 give an illustration in a sample binary image:

(i) **Growing seeds determining.** In a binary image of size $N \times M$ with tramway points, we search from the bottom row of the image, that is, the $N - 1th$ row. The starting points of tramway on the row is known, denoted by a range of pixel points $[(N - 1, i), (N - 1, j)]$. All the pixel points in the range are taken as the growing seeds. If there is no starting point in the $N - 1th$ row, the search moves to the $N - 2th$ row, and so on.

(ii) **Search region determining for growing seeds of next generation.** The growing seeds of next generation can be obtained from every starting points using our tracing criterion. Taking a pixel $(N - 1, k)$ as an example, its next generation can be found in the range of $[(N - 2, k - w), (N - 2, k + w)]$ of the $N - 2th$ row, where w is the parameter of pixel continuation, as shown in Fig. 7(a). Further, if growing seeds are in a transverse connected region $[(N - 1, i), (N - 1, j)]$ of the $N - 1th$

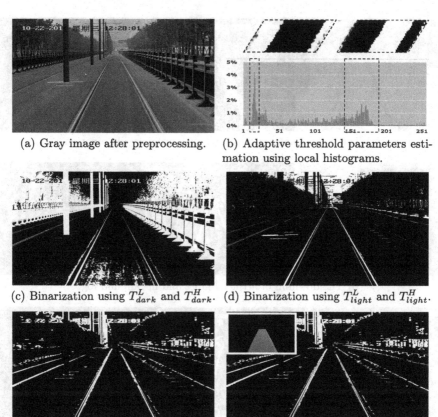

(a) Gray image after preprocessing.

(b) Adaptive threshold parameters estimation using local histograms.

(c) Binarization using T_{dark}^L and T_{dark}^H.

(d) Binarization using T_{light}^L and T_{light}^H.

(e) Tramway ROI generation after dilation.

(f) Region growing for tramway detection.

Fig. 8. Experiment validation

row, the next generation seeds in the $N - 2th$ row can be searched in the range of $[(N - 2, i - w), (N - 2, j + w)]$, as shown in Fig. 7(b).

(iii) **New generation of growing seeds.** Once the pixel points in the range $[(N - 2, i - w), (N - 2, j + w)]$ are the edge points, it can be considered as new growing seeds, and is also added to the set of trajectory points. Figure 7(c) shows the evolution of growing seeds. If there is no point in the $N - 2th$ row, the search along current trajectory line is waived. We repeat this step till all rows are traversed.

(iv) **Interferential lines removing.** After searching, several trajectory lines may be constructed. We need to remove interferential lines and pick up the optimal one as tramway. We consider the longest and most complete one of all trajectory lines as the tramway.

(iv) **Tramway equation establishing.** Considering the complexity of algorithm and real-time demand, least square piecewise polynomial fitting

(a) The bent of natural grassland with ordinary illumination.

(b) The downhill of artificial grassland in rain and fog.

(c) The uphill of asphalt road in the evening.

Fig. 9. Experimental results in different challenging scenarios

method is used to establish the equation of tramway. The extracted feature points are segmented in a sequence. The least square method is used to do quadratic fitting with $\mathcal{R}$ points as a piece. When the starting point or the ending point fails to meet the limits, the starting $\mathcal{R}$ points and the ending $\mathcal{R}$ points, respectively, use least square method to do quadratic fitting and extend until the limits. The specific values of $\mathcal{R}$ are as appropriate.

Our proposed region growing method can accurately predict the trend of tramway, neglecting the influence of traffic environment, such as the bent, the uphill and the downhill. It exhibits strong anti-interference, uses few calculation and meets the real-time demand.

3   Experiment Validation and Results

Our experiment platform was installed at *Tram Line* #1 in High-Tech District, Suzhou City, Jiangsu Province of China. The length of the tram line is 18.19 km, and the maximum operating speed of tram is 70 km/h. Our hardware used the camera *Hikvision DS-2CD6233F* and *Intel Core i7 2.5 GHz* Industrial computer with 4 GB memory. The frame size of the camera is 1080×1920 pixels and its frame rate is 10 fps. The scope of image collected by the camera is set in the range of [25 m, 250 m]. Our software system was developed in *Visual Studio 2010* platform with *EmguCV 2.9.0* vision library. The video datasets of all scenarios were collected in the field when the trams were actually running on the ground.

The original image, shown in Fig. 3(b), is taken as an example for experiment validation. Its gray image can be obtained by preprocessing, shown in Fig. 8(a). The preprocessing step consists of image gray-scale and Gaussian filtering. Figure 3(f) is the local image of tramway slice at the starting position of left rail. Its corresponding normalized histogram is obtained to execute our adaptive threshold parameter estimation. We first observe two peaks on the intensity levels of flangeway inside and outside from the normalized histogram, having $r_{fi} = 15$ and $r_{fo} = 165$. The adaptive method does automatically search and accumulate the neighbors of these two peaks level by level till the sum of histogram bars are equal to the given $p(r_{fi})$ and $p(r_{fo})$. The values of $p(r_{fi})$ and $p(r_{fo})$ can be induced using prior knowledge of flangeway structure. For *Suzhou Tram Line* #1, the flangeway outside has a typical width at the top of 37.5 mm plus 12.5 mm, and the flangeway inside has a width of 34 mm. We hence set $p(r_{fi}) = 0.405$ and $p(r_{fo}) = 0.595$. Using the adaptive algorithm, we get the thresholds: $T_{dark}^{L} = 9$, $T_{dark}^{H} = 21$, $T_{light}^{L} = 139$ and $T_{light}^{H} = 181$. Figure 8(b) shows the normalized histogram of local image of tramway slice, and the binarization results using T_{dark} and T_{light}.

We then conduct the binarization process using multilevel thresholding. Figures 8(c) and (d) are the binary images using T_{dark} and T_{light} for flangeway inside and outside respectively. In order to connect the fracture ROIs of tramway and expand the width of potential tramway, a morphological dilation operator is used. Finally, we generate the image with the ROIs of potential tramway through computing the intersection of these two dilated binary images, as shown in Fig. 8(e). Given a binary image with potential tramway ROIs, our region growing method is done for detecting the tramways. The initial growing seeds are picked up from the binary image. The parameter of pixel continuation is set to 7, that is $\omega = 7$. The red line in Fig. 8(f) is the induced tramway model through region growing and least square piecewise polynomial fitting.

To verify the effectiveness of the proposed method, we perform more experiments in which we apply our algorithm to different challenging scenarios. Figure 9 shows the experimental results in different scenes, including original image, binary image after multilevel thresholding segmentation and the region of interest between railways. The scenarios of Fig. 9(a), (b) and (c) are the bent of natural grassland with ordinary illumination, the downhill of artificial grassland in rain and fog, the uphill of asphalt road in the evening. The experimental results show that the proposed method can not only accurately detect the straight tramway, but also work for the curve, uphill and downhill. Further, it not only can be applied to the general scenario in ordinary illumination, but also has

Table 1. The comparison of computational time that tramway detection spends in different scenes (ms)

Scenario	Track Region Segmentation	Feature Points Extraction	Total Time
The straightaway of asphalt road in ordinary illumination	120.0 ± 6.4	25.9 ± 2.0	145.9 ± 4.1
The bent of natural grassland in ordinary illumination	124.4 ± 5.7	30.8 ± 2.2	155.2 ± 10.8
The downhill of artificial grassland in rain and fog	147.2 ± 5.9	56.8 ± 4.6	204.0 ± 8.3
The uphill of asphalt road in the evening	154.6 ± 12.2	44.3 ± 3.7	198.9 ± 6.8

good sensitivity and accuracy for varying weather, for example, at night, in rain and fog and on various ground types. We also investigate the performance on time consuming for practical application. Table 1 gives the comparison of computational time that our method spends in different scenarios. According to the table, the scenarios in the night, in rain and fog, and under other special scenes need more time to detect tramway. Thus it can be seen that accurate recognition in challenging scenarios is at the expense of a slight increase in time, which is in a controllable range. In the real-world system, we use the technique of dynamic thread pool, which can flexibly increase or lessen threads to satisfy real-time.

These results prove that the method can effectively detect tramway, accurately extract the ROI of tramway. The method shows good robustness, which can deal with different scenes and meets the demand of practical application.

4   Conclusion

Tramway detection is important but difficult due to various challenging situations in real-world urban rail traffic scenarios. In this paper, we propose a video-based tramway detection approach to address the issue. Our approach can efficiently segment the ROIs of tramway using a multilevel thresholding method with adaptive parameters estimation using local accumulated histogram. Instead of recognizing pixel by pixel on every video frame, which is believe to be time-consuming and infeasible for autonomous driving, our proposed method uses an improved region growing scheme to extract the optimal tramway. In addition, we use only a simple setup with a surveillance camera plus adaptive algorithm without rely on expensive equipment. Our method has been successfully tested in realistic urban environment, which proves that the accuracy and real-time performance of tram track detection in different challenging scenarios.

References

1. Adams, R., Bischof, L.: Seeded region growing. IEEE Trans. Pattern Anal. Mach. Intell. **16**(6), 641–647 (1994)

2. Espino, J.C., Stanciulescu, B.: Rail extraction technique using gradient information and a priori shape model. In: Proceedings of the IEEE Conference on Intelligent Transportation Systems, ITSC, pp. 1132–1136 (2012)
3. Jin, L.S., Tian, L., Wang, R.B., Guo, L., Chu, J.W.: An improved Otsu image segmentation algorithm for path mark detection under variable illumination. In: Proceedings of the IEEE Intelligent Vehicles Symposium 2005, pp. 840–844 (2005)
4. Kaleli, F., Akgul, Y.S.: Vision-based railroad track extraction using dynamic programming. In: Proceedings of the IEEE Conference on Intelligent Transportation Systems, ITSC, pp. 42–47 (2009)
5. Markova, O., Kovtun, H., Maliy, V.: Modelling train motion along arbitrary shaped track in transient regimes. Proc. Inst. Mech. Eng. Part F: J. Rail Rapid Transit **229**(1), 97–105 (2015)
6. Möckel, S., Scherer, F., Schuster, P.F.: Multi-sensor obstacle detection on railway tracks. In: Proceedings of the IEEE Intelligent Vehicles Symposium, pp. 42–46 (2003)
7. Mora, J.: A street named light rail. IEEE Spectr. **28**(2), 54–56 (1991)
8. Ning, J., Lin, J., Zhang, B.: Time-frequency processing of track irregularities in high-speed train. Mech. Syst. Signal Process. **66–67**, 339–348 (2016)
9. Novales, M., Orro, A., Bugarin, M.R.: Tram-train: new public transport system. Transp. Res. Rec. N **1793**(02), 80–90 (2002)
10. Qi, Z., Tian, Y., Shi, Y.: Efficient railway tracks detection and turnouts recognition method using HOG features. Neural Comput. Appl. **23**(1), 245–254 (2013)
11. Ben Romdhane, N., Hammami, M., Ben-Abdallah, H.: A comparative study of vision-based lane detection methods. In: Blanc-Talon, J., Kleihorst, R., Philips, W., Popescu, D., Scheunders, P. (eds.) ACIVS 2011. LNCS, vol. 6915, pp. 46–57. Springer, Heidelberg (2011). https://doi.org/10.1007/978-3-642-23687-7_5
12. Sinha, D., Feroz, F.: Obstacle detection on railway tracks using vibration sensors and signal filtering using bayesian analysis. IEEE Sens. J. **16**(3), 642–649 (2016)
13. Sun, R., Ochieng, W.Y., Feng, S.: An integrated solution for lane level irregular driving detection on highways. Transp. Res. Part C: Emerg. Technol. **56**, 61–79 (2015)
14. Teschke, K., Dennis, J., Reynolds, C.C.O., Winters, M., Harris, M.A.: Bicycling crashes on streetcar (tram) or train tracks: mixed methods to identify prevention measures. BMC Publ. Health **16**, 1–10 (2016)
15. Road Safety UAE: Dubai Tram (2016). http://www.roadsafetyuae.com/dubai-tram/
16. Wikipedia: Tram (2011). https://en.wikipedia.org/wiki/Tram
17. Wohlfeil, J.: Vision based rail track and switch recognition for self-localization of trains in a rail network. In: Proceedings of the IEEE Intelligent Vehicles Symposium (IV), pp. 1025–1030 (2011)
18. Yenikaya, S., Yenikaya, G., Düven, E.: Keeping the vehicle on the road: a survey on on-road lane detection systems. ACM Comput. Surv. **46**(1), 2:1–2:43 (2013)

Relevance and Coherence Based Image Caption

Tao Zhang, Wei Wang, Liang Wang, and Qinghua Hu[(✉)]

School of Computer Science and Technology, Tianjin University,
Tianjin 300350, China
{zhangtaorenyu,huqinghua}@tju.edu.cn, {wangwei,wangliang}@nlpr.ia.ac.cn

Abstract. The attention-based image caption framework has been widely explored in recent years. However, most techniques generate next word conditioned on previous words and current visual contents, while the relationship between the semantic and visual contents is not considered. In this paper, we present a novel framework which can explore the relevance and coherence at the same time. The relevance tries to explore the relationship between the semantic and visual contents in a semantic-visual embedding space, and the coherence is introduced to maximize the probability of generating the next word according to previous words and the current visual contents. The performance of our model is tested with three benchmark datasets: Flickr8k, Flickr30k and MS COCO. The experimental results show that the proposed approach can improve the performance of attention-based image caption method.

Keywords: Caption · Attention · Relevance · Coherence

1 Introduction

Image Caption is an integrated problem that integrates computer vision and natural language processing. The main task is given an image, the computer will generate a reasonable sentence to describe it. This task is easy for people but difficult for computers. Because it requires computers not only to know what objects and scene in the picture, but also to capture the relationships between objects and objects, also the relationships between objects and scene. Here, relationships include spatial location, attributes, actions and so on. Also, it needs a language model that is strong enough to generate reasonable sentences to describe a picture after fully understanding it. The generated sentences should be reasonable enough without any grammatical and logical errors. Image caption can be used in many scenes. For examples, it can be used to assist visually impaired people and report incident of surveillance. Moreover, it also can be applied to robotic vision and multi-media search tasks.

Figure 1 shows an example of attention-based image caption generation. Given an image, we first use a convolution neural network to obtain a set of feature maps (from low-level) which can preserve much visual information, then utilize attention model to generate a visual representation from those feature

J. Yang et al. (Eds.): CCCV 2017, Part I, CCIS 771, pp. 258–269, 2017.
https://doi.org/10.1007/978-981-10-7299-4_21

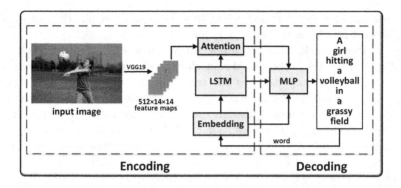

Fig. 1. An example of image caption.

maps and last we use Multi-Layer Perception to decode this visual representation to generate next word.

Recently, most existing image caption language models mainly maximize the probability of next word given the image and previous words. This describes how the image and previous words influences the next. The context relationships among the generated words can be guaranteed by this way. However, those models do not explore the relationships between the semantic and visual contents. As a result, the generated sentences from those existing approaches may be logically correct but the semantics (e.g. subjects, verbs or objects) in the sentences are wrong. For example, the sentence generated by existing approaches model for the image in Fig. 1 may be 'a man is hitting a volleyball', which is correct in logic but the subject 'man' ('woman' or 'girl' is correct) is not relevant to the image contents.

In [12], the context relationships mentioned above are defined as coherence, while the relationships between the semantic and visual contents are defined as relevance. In [12], the coherence and relevance were explored simultaneously for video description task. However, for image caption task, most existing methods only explored the coherence, while the relevance has not been explored. Specifically, as video description is similar to image description, so we first extend the work of [12] to image caption task which can explore the coherence and relevance simultaneously. Moreover, our method is also based on [21], which will be compared with, to validate the effectiveness of relevance.

2   Related Work

Recently, inspired by the successful use of sequence to sequence training with neural networks for machine translation, several methods based on deep neural networks have been proposed for image caption task. The first to use neural networks for image caption generation was Kiros et al. [8], who used a multi-modal log-bilinear model. Mao et al. [11] used a new approach to generate descriptions which is similar to [8], but Mao used a recurrent neural network as language model instead of a feed-forward one. Similarly, Vinyals et al. [19] and Donahue et al. [3] used LSTM as their language models.

All of above works encoder input image as a single feature vector. But Karpathy and Fei-Fei [6] proposed a method to create a joint embedding space to explore the similarity between the semantic and visual contents, and then generate description sentences. Fang et al. [4] proposed a model incorporating object detections. This model divided the caption generation into several parts: word detection through a CNN, caption candidates through a maximum entropy model, and sentence re-ranking through a deep multi-modal semantic model. Tran et al. [18] followed Fang's work [4]. They tried to address the challenges of describing images in the wild by adding an entity recognition model. The entity recognition model was used to identify celebrities and landmarks.

In [21], Xu et al. first added attention mechanisms to encoder-decoder image caption model conditions on the next word generation at each time step. Lu et al. [10] proposed an adaptive attention model via a visual sentinel. This model learnt to decide when and where to attend to the image for word generation at each time step.

Yao et al. [22] proposed a model named Long Short-Term Memory with Attributes (LSTM-A). LSTM-A presented attribute concept and added attributes into a based image caption framework [19]. More recently, as the exposure bias [14] and non-differentiable task metric issues can be addressed by Reinforcement Learning (RL) [17], in [14], Ranzato et al. used REINFORCE algorithm [20] to directly optimize non-differentiable and sequence-based test metrics to overcome both exposure bias and non-differentiable task metric issues. Rennie et al. [15] constructed a framework which used a new optimization approach called self-critical sequence training (SCST). Now this framework obtains state-of-the-art results on the MS COCO evaluation sever.

3   Model

Our goal is to generate good description sentences for images. In this section, we first describe the basic attention image caption model, and then we present a joint loss measuring the relevance and coherence simultaneously.

3.1   Word Embedding and Convolutional Networks

Our model takes a scaled image as input and generates a sentence $\mathbf{S}$ encoded as a sequence of words. We first encode each word as one-hot vector, thus the dimension of feature vector $\mathbf{x}_i$, i.e. V, is the vocabulary size.

$$\mathbf{S} = \{\mathbf{w_1}, \mathbf{w_2}, \cdots, \mathbf{w_L}\}, \mathbf{w_i} \in \mathbf{R^V}. \tag{1}$$

$$\mathbf{x}^t = \mathbf{E}\mathbf{w}_t, t \in \{1, 2, ..., L\}. \tag{2}$$

where, $\mathbf{E} \in \mathbf{R}^{m*V}$ is an embedding matrix. m is embedding dimensionality and L is the length of the sentence. Then we use a 2D Convolutional Neural Network

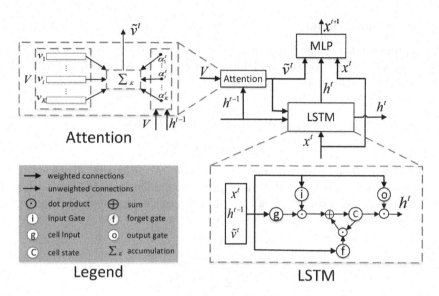

Fig. 2. An illustration of attention-based model.

to extract a set of features. The network produces K vectors, each of which is a D-dimensional representation corresponding to parts of the input image.

$$\mathbf{I} = \{\mathbf{v_1}, \mathbf{v_2}, \cdots, \mathbf{v_K}\}, \mathbf{v_i} \in \mathbf{R^D}. \tag{3}$$

3.2 Long Short Term Memory

We briefly introduce the standard Long Short-Term Memory (LSTM), a variant of RNN, which is effective and widely used in language generation model. LSTM incorporates a memory cell and non-linear gating units to effectively overcome the gradient vanishing and explosion problems. Our implementation of LSTM shown in Fig. 2. The formulas for LSTM forward pass are given below:

$$
\begin{aligned}
\mathbf{i}^t &= \sigma(\mathbf{W}_{ix}\mathbf{x}^t + \mathbf{U}_{ih}\mathbf{h}^{t-1} + \mathbf{V}_{iv}\tilde{\mathbf{v}}^t + \mathbf{b}_i) \\
\mathbf{f}^t &= \sigma(\mathbf{W}_{fx}\mathbf{x}^t + \mathbf{U}_{fh}\mathbf{h}^{t-1} + \mathbf{V}_{fv}\tilde{\mathbf{v}}^t + \mathbf{b}_f) \\
\mathbf{g}^t &= \phi(\mathbf{W}_{gx}\mathbf{x}^t + \mathbf{U}_{gh}\mathbf{h}^{t-1} + \mathbf{V}_{gv}\tilde{\mathbf{v}}^t + \mathbf{b}_g) \\
\mathbf{c}^t &= \mathbf{i}^t \odot \mathbf{g}^t + \mathbf{f}^t \odot \mathbf{c}^{t-1} \\
\mathbf{o}^t &= \sigma(\mathbf{W}_{ox}\mathbf{x}^t + \mathbf{U}_{oh}\mathbf{h}^{t-1} + \mathbf{V}_{ov}\tilde{\mathbf{v}}^t + \mathbf{b}_o) \\
\mathbf{h}^t &= \phi(\mathbf{c}^t) \odot \mathbf{o}^t
\end{aligned} \tag{4}
$$

where, $\mathbf{i}^t$, $\mathbf{f}^t$, $\mathbf{o}^t$, $\mathbf{g}^t$, $\mathbf{c}^t$ and $\mathbf{h}^t$ are input gate, forget gate, output gate, cell input, cell state and hidden state of LSTM respectively. σ is logistic sigmoid activation. ϕ is hyperbolic tangent activation. $\mathbf{W}_{**}$, $\mathbf{U}_{**}$, $\mathbf{V}_{**}$ and $\mathbf{b}_*$ are learned weight matrices and biases. The initializes of LSTM are given below:

$$\mathbf{c}^0 = f_{mlp}(\tfrac{1}{K}\textstyle\sum_i^K \mathbf{v}_i) \ and \ \mathbf{h}^0 = f_{mlp}(\tfrac{1}{K}\textstyle\sum_i^K \mathbf{v}_i). \tag{5}$$

3.3   Attention Model

The context vector $\tilde{\mathbf{v}}^t$ is a dynamic representation corresponding to parts of input image. Attention-based model learns a vector of weights $\alpha_i (i = 1, 2, ..., K)$ corresponding to the features extracted at different image locations from $\mathbf{v}_i$ and previous hidden state $\mathbf{h}^{t-1}$. α_i is a scalar between 0–1 and $\sum_{i=1}^{K} \alpha_i = 1$.

$$\mathbf{z}_i^t = \sigma((\mathbf{W}_{ah}\mathbf{h}^{t-1})\mathbf{1}^T + \mathbf{V}_{aI}\mathbf{I} + \mathbf{b}_a). \tag{6}$$

$$\alpha_i^t = soft\max(\mathbf{U}_{az}\mathbf{z}_i^t). \tag{7}$$

where, $\mathbf{1}$ is a vector with all elements set to 1. Additionally, attention-based model predicts a gating scalar β^t from previous hidden state $\mathbf{h}^{t-1}$ at each time step t. This gating variable allows the decoder to decide to whether put more emphasis on language model or the context at each time step.

$$\beta^t = \sigma(\mathbf{W}_\beta \mathbf{h}^{t-1}). \tag{8}$$

$$\tilde{\mathbf{v}}^t = \beta^t \sum_{i=1}^{K} \alpha_i^t \mathbf{v}_i. \tag{9}$$

In this work, we use MLP to compute the output word probability conditioned on the image (the context vector), the previously generated word, and the decoder hidden state $\mathbf{h}^t$. In formula 10, $\mathbf{W}_o \in \mathbf{R}^{V*n}$, n is LSTM hidden dimensionality, $\mathbf{W}_h \in \mathbf{R}^{m*n}$ and $\mathbf{V}_v \in \mathbf{R}^{m*D}$.

$$p(\mathbf{w}^t) = softmax(\mathbf{W}_o(\mathbf{W}_h\mathbf{h}^t + \mathbf{V}_v\tilde{\mathbf{v}}^t + \mathbf{x}^{t-1})). \tag{10}$$

3.4   Jointly Measuring Relevance and Coherence

We assume that a low dimensional embedding exists for the representations of image and sentence. To measure the relevance between the visual content and semantics, we compute their distance in the embedding space. Thus, we define the relevance loss as:

$$\mathbf{E}_r(\mathbf{v}, \mathbf{S}) = \sum_{t=1}^{L} \left\| \mathbf{V}_r\tilde{\mathbf{v}}^t - \mathbf{x}^t \right\|_2^2. \tag{11}$$

where $\mathbf{V}_v \in \mathbf{R}^{m*D}$, L is the length of sentence. Inspired by the recent success of probabilistic sequence models leveraged in machine translation, the coherence loss is defined as:

$$\mathbf{E}_c(\mathbf{v}, \mathbf{S}) = -\log \Pr(\mathbf{S}|\mathbf{v}) = -\sum_{t=1}^{L} \log \Pr(\mathbf{w}^t|\mathbf{v}, \mathbf{w}^1, \mathbf{w}^2, ..., \mathbf{w}^{t-1}). \tag{12}$$

By minimizing the coherence loss, the contextual relationship among the words in the sentence can be guaranteed, making the sentence coherent and smooth. In image caption task, both the relevance and coherence loss are estimated to complete the object function. The training of our model is performed by simultaneously minimizing the relevance loss and coherence loss. Therefore, the last object function is given below:

$$\mathbf{E}_l(\mathbf{v}, \mathbf{S}) = \mathbf{E}_c(\mathbf{v}, \mathbf{S}) + \lambda\mathbf{E}_r(\mathbf{v}, \mathbf{w}). \tag{13}$$

4 Experiments

In this section, we first describe our experimental method and then quantitatively analyse the experimental results.

Table 1. Performance on Flickr8k, Flickr30k and MS COCO datasets. BLEU-1,2,3,4/METEOR metrics compared to [21]. Higher is better in all columns. (-) indicates an unknown metric, bold font indicates the highest metric score.

Model	BLEU-1	BLEU-2	BLEU-3	BLEU-4	METEOR
(a) *Flickr8k dataset*					
CMU/MS research [2]	-	-	-	13.1	16.9
BRNN [6]	57.9	38.3	24.5	16.0	-
Google NIC [19]	63.0	41.0	27.0	-	-
Log bilinear [8]	65.6	42.4	27.7	17.7	17.31
Soft-attention [21]	67.0	44.8	29.9	19.5	18.9
Hard-attention [21]	67.0	45.7	31.4	21.3	20.3
Ours	**68.5**	**47.2**	**32.6**	**22.7**	**21.5**
(b) *Flickr30k dataset*					
CMU/MS research [2]	-	-	-	12.0	15.2
BRNN [6]	57.3	36.9	24.0	15.7	-
NIC [19]	66.3	42.3	27.7	18.3	-
Log bilinear [8]	60.0	38.0	25.4	17.1	16.88
ATT [23]	64.7	46.0	32.4	23.0	18.9
Soft-attention [21]	66.7	43.4	28.8	19.1	18.5
Hard-attention [21]	66.9	43.9	29.6	19.9	18.5
Ours	**68.7**	**46.9**	**33.0**	**23.2**	**20.4**
(c) *MS COCO dataset*					
CMU/MS research [2]	-	-	-	18.4	25.2
BRNN [6]	64.2	45.1	30.4	20.3	-
Google NIC [19]	66.6	46.1	32.9	24.6	-
Log bilinear [8]	70.8	48.9	34.4	24.3	20.03
ATT [23]	70.9	53.7	40.2	**30.4**	24.3
Soft-attention [21]	70.7	49.2	34.4	24.3	23.9
Hard-attention [21]	71.8	50.4	35.7	25.0	23.0
Ours	**75.5**	**55.3**	**40.8**	**30.4**	**25.5**

4.1 Experimental Preparation

Our experiments are validated on the widely used Flickr8k [5], Flickr30k [24] and MS COCO [9] datasets which have 8092, 31,783 and 123,287 images respectively.

For Flickr8k and Flickr30k datasets, each image is paired with 5 references. For Flickr8k dataset, We use the predefined splits containing 6,000 images for training, 1,000 images for validation and 1,000 images for test. For Flickr30k dataset, we use the publicly available splits[1] containing 29,000 images for training, 1,000 images for validation and 1,000 images for test.

For COCO dataset, some images have references in excess of 5, so we only keep 5 references for each image for consistency across our dataset. We use the same data split as in [6,21] containing 82,782 images for training, 5,000 images for validation and 5,000 images for test. However, a small part of images are not RGB format which are discard for feature extraction expediently. So the last number of images for training, validation and test are 113,079, 4989 and 4982 respectively.

We use only basic tokenization to Microsoft COCO which is same as the tokenization present in Flickr8k and Flickr30k. We don't convert any sentences to lowercase. And we keep all non-alphanumeric characters except double quotes. In our experiment, we use a fixed vocabulary size of 10000 and others are agreed to be marked as 'UNK'. For each sentence, we add a terminator '<**eos**>' to the end of a sentence.

We use the Oxford VGG19 [16] pretrained on ImageNet without finetuning for image feature extraction. Our model's hyper-parameters are all the same as [21]. Also, the hyper-parameters of models training on three datsets are the same. Our model is not end to end. We pre-extract the feature of images by VGG19 pretrained on ImageNet without finetuning. We only train the parameters of the encoder LSTM and decoder MLP. All learning parameters are initialized randomly and model uses ADAM [7] optimizer with an initial learning rate of 2×10^{-4}.

4.2 Experiments Analysis

Our model is based on [21] which we comparing with. By this way, we can verify that it can improve the performance of attention-based caption model significantly when considering the relationship of semantic and visual contents.

We report the results using the COCO captioning evaluation tool which reports the following metrics: BLEU [13] and Meteor [1]. Table 1 shows the results measured on Flickr8k, Flickr30k and MS COCO datasets. Compared with based-attention model [21] we can see that our model improves the performance of original attention based model significantly. When considering the relevance of visual and semantic contents, our model improves all metrics scores on Flickr8k, Flickr30k and MS COCO respectively. This means that the semantics (e.g. subjects, verbs or objects) in the sentence generated by our model are more precise. For Flickr8k dataset, our model improves BLEU-4 score from 21.3 to 22.7, METEOR score from 20.3 to 21.5 and increases by an average of 1.36 points for all metrics. BLEU-4 score improves from 19.9 to 23.2 and METEOR score from 18.5 to 20.4 on Flickr30k dataset. For COCO, BLEU-4 score increases

---

[1] http://cs.stanford.edu/people/karpathy/.

from 25.0 to 30.4, METEOR score from 23.0 to 25.5 and average score increases by 4.36 points for all metrics. On the other hand, we can see that the improvement of all score metrics is most obvious on the COCO dataset, followed by Flickr30k dataset. This proves that the richer the data and the wider the data distribution, the better the model works.

Fig. 3. Sentences 1–3 are ground-truth references. SAT: [21].

We also reproduce the method of [21], which we called SAT here, and generate descriptions for test-set images of MS COCO dataset (beam search size is fixed as 3). Then we compare the results with ours. Figure 3 shows some examples. We can see that the sentences, generating by our model, are more precise, reasonable and robust than SAT. In the first picture of Fig. 3, subject 'people' is right but 'women' is more precise and 'down a street' is more reasonable than 'on a beach'. In the third picture, verb 'grazing' is more suitable for 'walking across'. For the fourth picture, SAT generates a sentence with logical error but ours do not.

Through analysis, we also find that SAT and our model always produce the same sentence for similar pictures, both have the same trend but our model is relatively more flexible. We list some examples in Fig. 4. For 4982 test samples, the sentence of 'A man riding a snowboard down a snow covered slope' generated by SAT appears 36 times vs 10 times of ours, and 'A baseball player swinging a bat at a ball' generated by SAT appears 37 times vs 11 times of ours. From Fig. 4, we can see that the sentences generated by our model, which considers the relevance, are more diverse and precise than SAT.

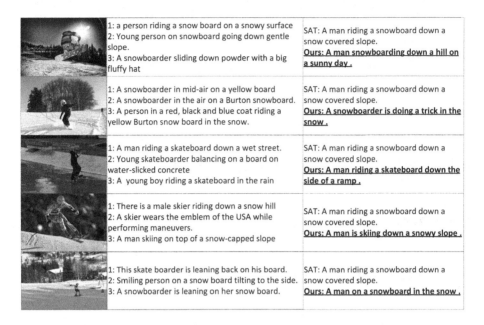

Fig. 4. Examples corresponding to the same sentence generated by SAT: [21].

In formula 13, we use tradeoff parameter λ for relevance. We fixed the value of beam size as 3. Then we illustrate the performance curves with different trade of parameter values in Fig. 5(a). To make all performance curves fall into a comparable scale, all BLEU-1,2,3,4 and METEOR scores are normalized as:

$$s'_\lambda = \frac{s_\lambda - \min_\lambda\{s_\lambda\}}{\max_\lambda\{s_\lambda\} - \min_\lambda\{s_\lambda\}}. \tag{14}$$

where s'_λ and s_λ denote normalized scores and original scores with a set of λ, respectively. When the value of λ exceeds 1.5, the corresponding value decreases rapidly. When $\lambda = 10$, the corresponding normalized score values decrease to 0.0 for all metrics scores. So in order to be easy to observe, We do not show corresponding values greater than 3 in the chart. From Fig. 5(a) we can see that the best performance is achieved when λ is about 0.8. When $\lambda = 0.8$, all metrics scores are more concentrated and larger than others relatively. This proves that it's reasonable to jointly learn the visual-semantic embedding space in the deep RNNs.

Then we explore the influences on all score metrics when changing the beam search size on the three benchmark datasets. In this experiment, we fix the value tradeoff parameter λ as 0.8. We illustrate the performance curves with different beam search sizes in Fig. 5(b)–(d). To make all performance curves fall into a comparable scale, all BLEU-1,2,3,4 and METEOR scores are normalized which is same as Eq. 14.

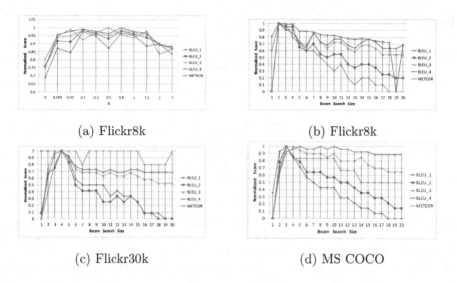

(a) Flickr8k

(b) Flickr8k

(c) Flickr30k

(d) MS COCO

Fig. 5. (a): The performances of different λ. (b)–(d): The performances of different beam search sizes.

From Fig. 5(b)–(d) we can see that different datasets and models have different best beam search sizes. For Flickr8k dataset, our model obtains the best performance when beam search size is 2. But 4 is best for Flickr30k dataset and 3 is best for COCO dataset. However, our experiment need to compare with [21] who fixed beam search size as 3 for all datasets and models in experiment. In order to be persuasive, we fix beam search size as 3 for all datasets and models.

5   Conclusion

In this paper, we present a new method, which can explore the learning of semantic-visual embedding and attention-based LSTM. In particular, the semantic-embedding space is incorporated with attention-based LSTM learning. We explore the local and global relationship between the semantic and visual contents. Experimental results show that this method can improve the performances of attention-based image caption model significantly on three benchmark datasets measured by the BLEU and METEOR score metrics. We also explore the influences of different hyper-parameters on model. We will continue to explore the relationship between the visual and semantic contents from different perspectives by different methods.

References

1. Satanjeev, B., Lavie, A.: Meteor: an automatic metric for MT evaluation with improved correlation with human judgments, pp. 228–231 (2005)
2. Chen, X., Lawrence Zitnick, C.: Mind's eye: a recurrent visual representation for image caption generation, pp. 2422–2431 (2015)
3. Donahue, J., Anne Hendricks, L., Guadarrama, S., Rohrbach, M., Venugopalan, S., Darrell, T., Saenko, K.: Long-term recurrent convolutional networks for visual recognition and description. In: CVPR, pp. 2625–2634 (2015)
4. Fang, H., Gupta, S., Iandola, F., Srivastava, R.K., Deng, L., Dollár, P., Gao, J., He, X., Mitchell, M., Platt, J.C., et al.: From captions to visual concepts and back. In: CVPR, pp. 1473–1482 (2015)
5. Hodosh, M., Young, P., Hockenmaier, J.: Framing image description as a ranking task: data, models and evaluation metrics. JAIR **47**, 853–899 (2013)
6. Karpathy, A., Fei-Fei, L.: Deep visual-semantic alignments for generating image descriptions. In: Proceedings of the IEEE Conference on Computer Vision and Pattern Recognition, pp. 3128–3137 (2015)
7. Kingma, D., Ba, J.: Adam: a method for stochastic optimization. arXiv preprint arXiv:1412.6980 (2014)
8. Kiros, R., Salakhutdinov, R., Zemel, R.: Multimodal neural language models. In: ICML, pp. 595–603 (2014)
9. Lin, T.-Y., Maire, M., Belongie, S., Hays, J., Perona, P., Ramanan, D., Dollár, P., Zitnick, C.L.: Microsoft COCO: common objects in context. In: Fleet, D., Pajdla, T., Schiele, B., Tuytelaars, T. (eds.) ECCV 2014. LNCS, vol. 8693, pp. 740–755. Springer, Cham (2014). https://doi.org/10.1007/978-3-319-10602-1_48
10. Lu, J., Xiong, C., Parikh, D., Socher, R.: Knowing when to look: adaptive attention via a visual sentinel for image captioning. arXiv preprint arXiv:1612.01887 (2016)
11. Mao, J., Xu, W., Yang, Y., Wang, J., Huang, Z., Yuille, A.: Deep captioning with multimodal recurrent neural networks (m-RNN). arXiv preprint arXiv:1412.6632 (2014)
12. Pan, Y., Mei, T., Yao, T., Li, H., Rui, Y.: Jointly modeling embedding and translation to bridge video and language. In: CVPR, pp. 4594–4602 (2016)
13. Papineni, K., Roukos, S., Ward, T., Zhu, W.J.: Bleu: a method for automatic evaluation of machine translation. In: ACL, pp. 311–318 (2002)
14. Ranzato, M.A., Chopra, S., Auli, M., Zaremba, W.: Sequence level training with recurrent neural networks. arXiv preprint arXiv:1511.06732 (2015)
15. Rennie, S.J., Marcheret, E., Mroueh, Y., Ross, J., Goel, V.: Self-critical sequence training for image captioning. arXiv preprint arXiv:1612.00563 (2016)
16. Simonyan, K., Zisserman, A.: Very deep convolutional networks for large-scale image recognition. Computer Science (2014)
17. Sutton, R.S., Barto, A.G.: Reinforcement Learning: An Introduction, vol. 1. MIT Press, Cambridge (1998)
18. Tran, K., He, X., Zhang, L., Sun, J., Carapcea, C., Thrasher, C., Buehler, C., Sienkiewicz, C.: Rich image captioning in the wild. In: CVPR, pp. 49–56 (2016)
19. Vinyals, O., Toshev, A., Bengio, S., Erhan, D.: Show and tell: a neural image caption generator. In: CVPR, pp. 3156–3164 (2015)
20. Williams, R.J.: Simple statistical gradient-following algorithms for connectionist reinforcement learning. Mach. Learn. **8**(3–4), 229–256 (1992)
21. Xu, K., Ba, J., Kiros, R., Cho, K., Courville, A., Salakhudinov, R., Zemel, R., Bengio, Y.: Show, attend and tell: neural image caption generation with visual attention. In: ICML, pp. 2048–2057 (2015)

22. Yao, T., Pan, Y., Li, Y., Qiu, Z., Mei, T.: Boosting image captioning with attributes. arXiv preprint arXiv:1611.01646 (2016)
23. You, Q., Jin, H., Wang, Z., Fang, C., Luo, J.: Image captioning with semantic attention. In: CVPR, pp. 4651–4659 (2016)
24. Young, P., Lai, A., Hodosh, M., Hockenmaier, J.: From image descriptions to visual denotations: new similarity metrics for semantic inference over event descriptions. ACL **2**, 67–78 (2014)

An Efficient Recoloring Method for Color Vision Deficiency Based on Color Confidence and Difference

Qing Xu[1], Xiangdong Zhang[1], Liang Zhang[2], Guangming Zhu[2], Juan Song[2], and Peiyi Shen[2(✉)]

[1] School of Telecommunications Engineering, Xidian University,
Xi'an 710071, Shaanxi, China
[2] School of Software Engineering, Xidian University, Xi'an 710071, Shaanxi, China
`pyshen@xidian.edu.cn`

Abstract. Human with color vision deficiency (CVD) cannot distinguish some colors under normal situation. And recoloring is a color adaptation procedure for human with CVD. This paper proposes an efficient recoloring method. **First**, key color confidence is introduced to judge and recolor unrecognizable colors sequentially instead of simultaneously to improve the execution speed. **Second**, color differences between one color and other colors with higher confidence are taken as the unrecognizable benchmark for this color without any pre-defined parameters. **Third**, these differences are also used in iterative recoloring procedure to maximize the contrast between different colors in the recolored image. Experimental results show that our proposed method outperforms state-of-the-art recoloring methods in terms of subjective, objective qualities and executive speed.

Keywords: Color vision deficiency · Recoloring · Confidence
Difference

1 Introduction

With the explosive growth of multimedia technology, perceiving color becomes more and more important for effective visual communication. Normal vision is based on the response to photons in L-, M- and S- cones located in the retina of the eyes. However, color vision deficiency (CVD) presents when there exists a deficiency in one of the cones, and it can be divided into protanopia, deuteranopia and tritanopia according to the deficient cone. In this paper, we focus on protanopia and deuteranopia since they occupy 98% of CVD, who cannot perceive red and green under normal situation [1]. We show an example of how human with CVD perceives colors in Fig. 1. The significant information disappears in Fig. 1(b) and (c), which are perceived by protanopia and deuteranopia, respectively. It has been demonstrated that CVD causes people considerable inconvenience, because the ability to distinguish colors is useful in many daily

© Springer Nature Singapore Pte Ltd. 2017
J. Yang et al. (Eds.): CCCV 2017, Part I, CCIS 771, pp. 270–281, 2017.
https://doi.org/10.1007/978-981-10-7299-4_22

activities such as recognizing road signs and traffic lights [2]. Therefore, it is important to enhance color perceptibility for the human with CVD.

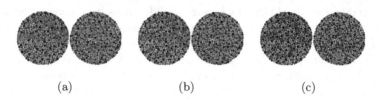

(a)   (b)   (c)

Fig. 1. An example of how human with CVD perceives colors. (a) The image that perceived by human with normal vision. (b) The image that perceived by human with protanopia. (c) The image that perceived by human with deuteranopia.

Various recoloring methods have been proposed to enhance color perceptibility for human with CVD. And these methods can be categorized into rule-based and optimization-based methods.

In the optimization-based methods [3–7], recoloring is performed by minimizing an objective function based on some metrics. Huang [3] and Huang [4] obtained optimal colors by minimizing the difference between original and recolored image in Lab color space. And Milić [5,6] found the optimal choice of discernible color bin for each color in the original image. These methods are easy to fall into a shallow local minimum of the objective function and cannot meet the real-time requirement because of exhaustive search for minimum.

The rule-based methods [8–12] need less time than optimization-based methods since they only require linear operation. There are two factors influencing the performance. One is to judge unrecognizable colors for human with CVD and the other is to recolor these unrecognizable colors using a mapping matrix. Doliotis [8] and Khurge [9] found unrecognizable colors by comparing the difference between original and simulated image with a pre-defined threshold. And Kim [10,11] found unrecognizable colors through a pre-defined red pixel mask for protanopia. These methods involve some pre-defined parameters, which has great impact on the performance. When recoloring unrecognizable colors, Khurge [9] proposed an experience-based matrix, but this matrix cannot guarantee all the unrecognizable colors are recolored accurately. Doliotis [8] and Kim [10] recolored the unrecognizable colors by multiplying a given matrix iteratively until the stopping iteration condition was satisfied. The stopping iteration condition is also a pre-defined parameter similar as unrecognizable colors judgment, which cannot guarantee high contrast between different colors in recolored images. And these methods recolored all the unrecognizable colors simultaneously, which need more iterations to make sure all the recolored colors are recognizable for human with CVD.

In this paper, an efficient rule-based recoloring method is proposed for human with CVD. **First**, key color confidence is introduced to judge and recolor one unrecognizable color at a time so that recoloring procedure can be done sequentially instead of simultaneously, which can speed up the execution time greatly.

Second, color differences between one color and other colors with higher confidence are taken as the unrecognizable benchmark for this color, which cannot involve any pre-defined parameters. **Third**, an optimal recolored color for this unrecognizable color is obtained when these differences have been maintained or enhanced after recoloring, which can guarantee high contrast between different colors in the recolored image.

The rest of this paper is organized as follows. Section 2 describes our proposed method in detail. Section 3 demonstrates the effectiveness of our proposed method in terms of subjective quality, objective quality and execution time. And Sect. 4 concludes this paper.

2   Proposed Algorithm

Figure 2 depicts a flowchart of our proposed recoloring method:

1. Key colors are extracted in RGB color space. RGB color space is a convenient color model for image processing because human visual system works in a similar way to it [13]. Since people cannot distinguish colors with small color difference under normal situation, key colors $C^0 = \{C_1^0, C_2^0, ..., C_N^0\}$ are extracted from an original image using the K-means [14], in which each key color C_i^0 represents a color of the cluster center. And then the key colors perceived by CVD $C^d = \{C_1^d, C_2^d, ..., C_N^d\}$ are obtained using a well-known color blind simulation algorithm proposed by Brettel [15,16];

2. The key color confidences are obtained according to the differences between C^0 and C^d, and all key color in C^0 are ordered according to their corresponding confidence, which is illustrated in Sect. 2.1 in detail;

3. From the highest to lowest confidence, judge whether one key color C_i^0 is unrecognizable for human with CVD by taking the differences between C_i^0 and other colors with higher confidence $C_j^0 (j < i)$ as judgment benchmark. If it is unrecognizable, go to step 4, otherwise go to step 5, which is illustrated in Sect. 2.2 in detail;

4. A unrecognizable color C_i^0 will be recolored to the recolored color O_i^0 by multiplying with a mapping matrix iteratively, until C_i^0 is not confused with those recolored colors with higher confidence $O_j^0 (j < i)$, which is also illustrated in Sect. 2.2 in detail;

5. Judge whether all key colors in C^0 have been recolored. If it is yes, then perform step 6, otherwise go back to step 3;

6. All pixels in original image are recolored through interpolation. We tend to maintain the perceptibility distance between a pixel and its corresponding key color after recoloring. Given a pixel P_j in original image and its corresponding key color C_i^0, the recolored color R_j for pixel P_j is obtained using Eq. (1).

$$R_j = O_i^0 + (P_j - C_i^0) \tag{1}$$

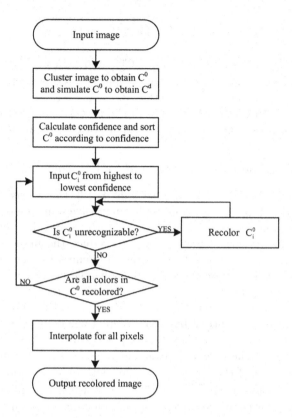

Fig. 2. The flowchart of our proposed algorithm.

2.1 Key Color Confidence Calculation

In the existing rule-based recoloring methods [8–12], all unrecognizable colors for human with CVD are judged and recolored simultaneously, which needs more iterations to make sure all unrecognizable colors are recolored accurately. To speed up execution time, we introduce key color confidence to judge and recolor one color at a time so that recoloring procedure can be done sequentially. And an optimal recolored color can be obtained as long as the recolored color is not confused with other recolored colors with higher confidence.

Key color confidence is the confidence whether one color is recognizable for human with CVD. A key color with higher confidence should be taken as the benchmark for recoloring other colors with lower confidence. To obtain the key color confidences, we first calculate the color differences $D = \{D_1, D_2, ..., D_N\}$ between C^0 and C^d. Then, one key color C_i^0 with smaller color difference D_i is assigned a higher confidence. Finally, all key colors in C^0 are ordered according to their corresponding confidence. We define color differences between C^0 and C^d using a well-known weighted Euclidean distance, as shown in Eq. (2), which approximates to the perceptibility distance of human [17].

$$D_i = \sqrt{3 * \Delta R_i^2 + 4 * \Delta G_i^2 + 2 * \Delta B_i^2} \tag{2}$$

where $\Delta R_i = C_{i,R}^0 - C_{i,R}^d$, $\Delta G_i = C_{i,G}^0 - C_{i,G}^d$ and $\Delta B_i = C_{i,B}^0 - C_{i,B}^d$ are the differences between three components of C^0 and C^d, respectively.

To speed up the execution time, we then judge and recolor one unrecognizable color for human with CVD at a time from the highest to lowest confidence, so that recoloring process can be done sequentially instead of simultaneously.

2.2   Unrecognizable Colors Judgment and Recoloring

In the existing rule-based recoloring methods [8–12], the unrecognizable colors judgement and the stopping iteration condition involve some pre-defined parameters. The performance is sensitive to these parameters and the contrast between different colors in recolored image is low. To reduce the influence of the pre-defined parameters, we propose to take the differences between one color and other colors with higher confidence as unrecognizable benchmark for this color. And we obtain an optimal recolored color for this unrecognizable color when these differences have been maintained or enhanced in the iterative recoloring, which can maximize the contrast between different colors in the recolored image.

The differences between one key color C_i^0 and other colors with higher confidence $C_j^0 (j < i)$ are the benchmark for judging whether C_i^0 is unrecognizable for human with CVD. If there exists one color with higher confidence $O_j^d(j < i)$ in O^d (recolored colors perceived by CVD), and when the difference between O_j^d and C_i^0 is smaller than the corresponding difference between C_j^0 and C_i^0, we regard C_i^0 as an unrecognizable color.

Then the unrecognizable color C_i^0 is recolored in LMS color space that specifies colors in terms of the relative excitations of L-, M- and S- cones. First, $[L_i^0, M_i^0, S_i^0]$ and $[L_i^d, M_i^d, S_i^d]$ are obtained by transforming C_i^0 and C_i^d from RGB to LMS color space. Then, the recolored color $[L_i', M_i', S_i']$ in LMS color space is obtained according to Eq. (3).

$$[L_i', M_i', S_i'] = [L_i^0, M_i^0, S_i^0] + \alpha * T * ([L_i^0, M_i^0, S_i^0] - [L_i^d, M_i^d, S_i^d]) \tag{3}$$

where T is a mapping matrix proposed in [16], as shown in Eq. (4). And α is a directional parameter that controls the direction of recoloring. If the recolored color O_i^0 has exceeded the range of RGB color space before obtaining an optimal recolored color, we move to the opposite direction of recoloring according to Eq. (5).

$$T = \begin{bmatrix} 0 & 2.02344 & -2.52581 \\ 0 & 1 & 0 \\ 0 & 0 & 1 \end{bmatrix} \tag{4}$$

$$\alpha = -0.75 * \alpha \tag{5}$$

Finally, the recolored color $[L_i', M_i', S_i']$ is transformed back to RGB color space to obtain the recolored key color O_i^0, and then O_i^0 is simulated to obtain the corresponding recolored color perceived by CVD O_i^d. If the difference between O_i^d and one of the colors with higher confidence $O_j^d (j < i)$ in O^d is lower than the corresponding difference between C_i^0 and C_j^0, we still regard O_i^0 as an unsatisfactory recolored color for C_i^0. Then we continue to perform recoloring until the differences between O_i^d and all recolored colors with high confidence perceived by CVD $O_j^d (j < i)$ are higher than the corresponding differences between C_j^0 and C_j^0.

3   Experimental Results and Analysis

Our proposed method is compared with algorithms including Doliotis [8], Huang [4] and Khurge [9] in terms of subjective quality, objective quality and execution speed. The test images consist of both natural images and Ishihara test images. We compile all the algorithms using Matlab 2013a on an Intel(R) Core(TM) i3-2310M CPU @ 2.10 GHz. In our proposed method, all images are clustered into 6 clusters in step 1 and the directional parameter is initialized to be 1 in step 4.

3.1   Subjective Quality Comparison

A recoloring algorithm can be considered to have a good performance if the resulting recolored images can preserve the overall characteristics of the original images and achieve a high contrast between different colors for human with CVD. First, all the algorithms are tested with two natural images and the experiment results are shown in Figs. 3, 4 and 5.

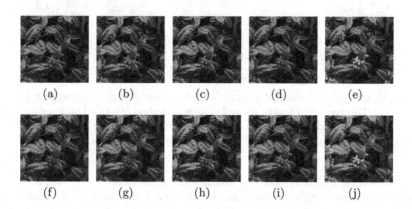

Fig. 3. (a) The original nature image. (b) The recolored image using our proposed method. (c) The recolored image using Doliotis's method [8]. (d) The recolored image using Huang's method [4]. (e) The recolored image using Khurge's method [9]. (f)–(j) The corresponding images perceived by protanopia. (Color figure online)

As shown in Fig. 3(a) and (f), the green leaves and red fruits are unrecognizable for human with protanopia. Doliotis [8] recolors unrecognizable colors iteratively until the stopping iteration condition is satisfied. As shown in Fig. 3(h), these two colors are still confused by human with protanopia, since the stopping iteration condition is not appropriate. Huang [4] proposes a optimization-based method. As shown in Fig. 3(i), the contrast between these two colors is still low for human with protanopia, since this method is likely to fall into the shallow local minimum. And the contrast between these two colors in Fig. 3(j) is also low for human with protanopia, since Khurge [9] recolors unrecognizable color according to an experience-based matrix.

As shown in Fig. 4(a) and (f), red is the only one unrecognizable color for human with protanopia. In Doliotis's [8] method, red have been recolored to light green, but it is confused with gray, as shown in Fig. 4(h). In Huang's [4] method, all colors in Fig. 4(a) are recolored regardless of whether they are recognizable for CVD. And the contrast between different colors becomes very low, as shown in Fig. 4(i). Similar to Doliotis, Khurge [9] recolors red to light pink, but it is confused with gray, as shown in Fig. 4(j).

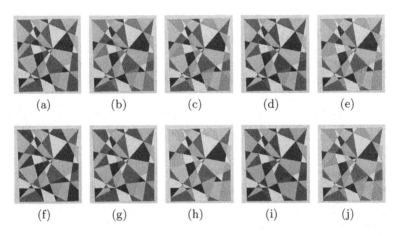

(a)   (b)   (c)   (d)   (e)

(f)   (g)   (h)   (i)   (j)

Fig. 4. (a) The original natural image. (b) The recolored image using our proposed method. (c) The recolored image using Doliotis's method [8]. (d) The recolored image using Huang's method [4]. (e) The recolored image using Khurge's method [9]. (f)–(j) The corresponding images perceived by protanopia. (Color figure online)

As shown in Fig. 5(a) and (f), red flower is confused with the green background for human with protanopia. In Doliotis's [8] method, red have been recolored to light green, but it is still easily confused with background, as shown in Fig. 5(h). In Huang's [4] method, red flower is recolored to dark green, but the contrast between flower and background is still lower than original image, as shown in Fig. 5(i). Similar to Doliotis, Khurge [9] recolors red flower and background to similar grey, so the contrast between flower and background is very low, as shown in Fig. 5(j).

Fig. 5. (a) The original Ishihara test image. (b) The recolored image using our proposed method. (c) The recolored image using Doliotis's method [8]. (d) The recolored image using Huang's method [4]. (e) The recolored image using Khurge's method [9]. (f)–(j) The corresponding images perceived by protanopia. (Color figure online)

By contrast, our proposed method only recolors unrecognizable colors sequentially and take the differences between one unrecognizable color and other colors with higher confidence as the stopping iteration benchmark, which can make sure the differences after recoloring is no lower than before. As shown in Figs. 3(g), 4(g) and 5(g), only unrecognizable colors have been recolored, and the contrast between different colors is higher for human with protanopia than other methods.

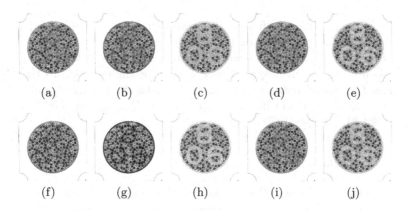

Fig. 6. (a) The original Ishihara test image. (b) The recolored image using our proposed method. (c) The recolored image using Doliotis's method [8]. (d) The recolored image using Huang's method [4]. (e) The recolored image using Khurge's method [9]. (f)–(j) The corresponding images perceived by protanopia.

Because the Ishihara test images are used in medical to diagnose whether a person is CVD, we also conduct a series of experiments using some Ishihara test images and the experimental results are shown in Figs. 6 and 7. Similar to the natural image, our proposed method obtains the highest contrast between different colors for human with CVD than other methods.

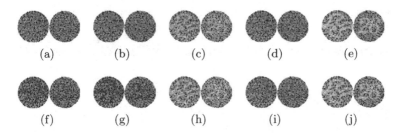

Fig. 7. (a) The original Ishihara test image. (b) The recolored image using our proposed method. (c) The recolored image using Doliotis's method [8]. (d) The recolored image using Huang's method [4]. (e) The recolored image using Khurge's method [9]. (f)–(j) The corresponding images perceived by protanopia.

3.2 Objective Quality Comparison

We now objectively assess the quality of the recolored images for human with CVD using $E_{contrast}$ which represents the average contrast between arbitrary pixel pairs. A higher score indicates better performance. The contrast calculation for one image is defined in Eq. (6).

$$E_{contrast} = \sum_i \sum_{j=i+1} D_{ij}/N \qquad (6)$$

where D_{ij} is the calculated difference between i-th and j-th pixels using (2) and N is the total number of the pixel pairs in the recolored image.

The $E_{contrast}$ of the various recolored images perceived by CVD are listed in Table 1. As it is shows, our proposed method achieves highest contrast than other algorithms. Compared with original image, Doliotis [8], Huang [4] and Khurge [9], our proposed algorithm increases the average $E_{contrast}$ by 6.56%, 21.80%, 9.4% and 20.49%, respectively.

Table 1. The $E_{contrast}$ obtained by different algorithms

Images	Original	Our	Doliotis [8]	Huang [4]	Khurge [9]
Figure 3	0.5967	0.6066	0.5750	0.5929	0.5915
Figure 4	0.9120	0.9550	0.8201	0.8422	0.8141
Figure 5	0.7020	0.7268	0.5968	0.6755	0.5761
Figure 6	0.7821	0.9144	0.6180	0.8246	0.6602
Figure 7	1.0631	1.1194	0.9384	1.0151	0.9453
Average	0.8112	0.8644	0.7097	0.7901	0.7174

3.3 Execution Speed Comparison

In this subsection, we compare the execution time of various algorithms. As shown in Table 2, Huang [4] requires more time to search optimal recolored colors for CVD than Doliotis [8], Khurge [9] and our proposed rule-based method. In addition, our proposed method further improves the execution time compared with Doliotis [8] because we judge and recolor unrecognizable colors sequentially instead of simultaneously. The execution time of our proposed method is approximately 5.4 times faster than Doliotis [8]. Khurge [9] recolors all pixels in the image by one multiplication operation, and our proposed method only recolors unrecognizable key colors. Therefore, if there exist a little unrecognizable colors in an image, our proposed method can execute faster than Khurge. For example, there exist only one unrecognizable color in the Fig. 4, so the execution time of our proposed method is less than Khurge.

Table 2. The execution speed of various algorithms (s)

Images	Size of images	Our proposed	Doliotis [8]	Huang [4]	Khurge [9]
Figure 3	360 * 348	0.9582	2.8995	8.9779	0.8683
Figure 4	381 * 430	0.7170	5.2996	14.6837	1.1291
Figure 5	231 * 157	0.6869	6.3350	10.6565	0.7986
Figure 6	380 * 372	0.9093	3.8395	24.5872	1.0470
Figure 7	228 * 111	0.5050	2.0064	6.8919	0.7738

4 Conclusion

In this paper, an efficient recoloring method is proposed for color vision deficiency. To better adapt to the recoloring task, we improve the rule-based recoloring methods from three aspects. **First**, key color confidence is introduced to judge and recolor unrecognizable colors sequentially, which can speed up the execution time greatly. **Second**, differences between one color and other colors with higher confidence are taken as the unrecognizable benchmark for this color, which cannot involve any pre-defined parameters. **Third**, these differences are also used in iterative recoloring to make sure the contrast between different colors after recoloring is no lower than before. Experimental results show that our proposed method can recolor unrecognizable colors for CVD effectively.

Acknowledgments. This work is partially supported by the China Postdoctoral Science Foundation (Grant No. 2016M592763), the Fundamental Research Funds for the Central Universities (Grant Nos. JB161001, JB161006), the Shaanxi Science and Technology Research Projects (Program D14013230107), and the National Natural Science Foundation of China (Grant Nos. 61401324, 61305109).

References

1. Sharpe, L.T., Stockman, A., Jagle, H., Nathans, J.: Color Vision: From Genes to Perception, Chapter Opsin Genes, Cone Photopigments, Color Vision, and Color Blindness. Cambridge University Press, Cambridge (1999). pp. 3–51
2. Kim, Y.K., Kim, K.W., Yang, X.: Real time traffic light recognition system for color vision deficiencies. In: 2007 International Conference on Mechatronics and Automation, ICMA 2007, pp. 76–81. IEEE (2007)
3. Huang, C.-R., Chiu, K.-C., Chen, C.-S.: Key color priority based image recoloring for dichromats. In: Qiu, G., Lam, K.M., Kiya, H., Xue, X.-Y., Kuo, C.-C.J., Lew, M.S. (eds.) PCM 2010. LNCS, vol. 6298, pp. 637–647. Springer, Heidelberg (2010). https://doi.org/10.1007/978-3-642-15696-0_59
4. Huang, J.B., Chen, C.S., Jen, T.C., Wang, S.J.: Image recolorization for the color-blind. In: IEEE International Conference on Acoustics, Speech and Signal Processing, pp. 1161–1164. IEEE (2009)
5. Milić, N., Belhadj, F., Dragoljub, N.: The customized daltonization method using discernible colour bins. In: Colour and Visual Computing Symposium (CVCS), pp. 1–6. IEEE (2015)
6. Milić, N., Hoffmann, M., Tómács, T., Novaković, D., Milosavljević, B.: A content-dependent naturalness-preserving daltonization method for dichromatic and anomalous trichromatic color vision deficiencies. J. Imaging Sci. Technol. **59**(1), 105041–1050410 (2015)
7. Shen, W., Mao, X., Hu, X., Wong, T.T.: Seamless visual sharing with color vision deficiencies. ACM Trans. Graph. **35**(4), 1–12 (2016)
8. Doliotis, P., Tsekouras, G., Anagnostopoulos, C.N., Athitsos, V.: Intelligent modification of colors in digitized paintings for enhancing the visual perception of color-blind viewers. In: Iliadis, Maglogiann, Tsoumakasis, Vlahavas, Bramer, (eds.) AIAI 2009. IFIPAICT, vol. 296, pp. 293–301. Springer, Boston (2009). https://doi.org/10.1007/978-1-4419-0221-4_35
9. Khurge, D.S., Peshwani, B.: Modifying image appearance to improve information content for color blind viewers. In: International Conference on Computing Communication Control and Automation, pp. 611–614. IEEE (2015)
10. Kim, H.J., Jeong, J.Y., Yoon, Y.J., Kim, Y.H.: Color modification for color-blind viewers using the dynamic color transformation. In: IEEE International Conference on Consumer Electronics, pp. 602–603. IEEE (2012)
11. Jeong, J.Y., Kim, H.J., Wang, T.S., Yoon, Y.J.: An efficient re-coloring method with information preserving for the color-blind. IEEE Trans. Consum. Electron. **57**(4), 1953–1960 (2011)
12. Orii, H., Kawano, H., Maeda, H., Kouda, T.: Color conversion algorithm for color blindness using self-organizing map. In: International Symposium on Soft Computing and Intelligent Systems, pp. 910–913. IEEE (2014)
13. Hunt, R.W.G.: The Reproduction of Colour. Wiley, Hoboken (2004)
14. Krishna, K., Murty, M.N.: Genetic k-means algorithm. IEEE Trans. Syst. Man Cybern. Part B Cybern. **29**(3), 433–439 (1999). A Publication of the IEEE Systems Man and Cybernetics Society
15. Brettel, H., Viénot, F., Mollon, J.D.: Computerized simulation of color appearance for dichromats. J. Opt. Soci. Am. A: Opt. Image Sci. Vis. **14**(10), 2647 (1997)

16. Mollon, J.D., Viénot, F., Brettel, H.: Digital video colourmaps for checking the legibility of displays by dichromats. Color: Res. Appl. **24**(4), 243–252 (1999)
17. Maurer, C.R.J., Qi, R., Raghavan, V.: A linear time algorithm for computing exact Euclidean distance transforms of binary images in arbitrary dimensions. IEEE Trans. Pattern Anal. Mach. Intell. **25**(2), 265–270 (2003)

A Saliency Based Human Detection Framework for Infrared Thermal Images

Xinbo Wang[1,2,3], Dahai Yu[1,2,3]([✉]), Jianfeng Han[1,2,3], and Guoshan Zhang[1,2,3]

[1] Tianjin University, Tianjin, China
wxb.tju@gmail.com, yudahai_dublin@hotmail.com
[2] Tianjin ShenQi Technology Co., Ltd., Tianjin, China
[3] Tianjin University of Commerce, Tianjin, China

Abstract. In this paper, a novel saliency framework for crowd detection in infrared thermal images is proposed. In order to obtain the optimal classifier from a large amount of data, the process of training consists of the following four steps: (a) a saliency contrast algorithm is employed to detect the regions of interest; (b) standard HOG features of the selected interest areas are extracted to represent the human object; (c) the extracted features, which are prepared for training, are optimized based on a visual attention map; (d) a support vector machine (SVM) algorithm is applied to compute the classifier. Finally, we can detect the human precisely after high-saliency areas of an image are input into the classifier. In order to evaluate our algorithm, we constructed an infrared thermal image database collected by a real-time inspection system. The experimental results demonstrated that our method can outperform the previous state-of-the art methods for human detection in infrared thermal images, and the visual attentional techniques can effectively represent prior knowledge for features optimization in a practicable system.

Keywords: Human detection · Saliency algorithm
Visual attention map · Infrared thermal images · Vision application

1 Introduction

With the development of imaging technology, infrared thermal (IRT) imaging systems have been widely used in many fields, such as local surveillance, fire detection and human inspection. In contrast to color imaging, IRT imaging is based on the temperature of objects, i.e. IRT imaging can actually represent the radiation range of infrared energy emitted, transmitted and reflected by any object. Therefore, surveillance integrated with IRT imaging can show a visual picture even in dark areas or blurred views, which makes it easier to detect the target, and this is why infrared target detection has received attention from many researchers in recent years. However, as a key technique, traditional target detection schemes have their limitations. For example, the threshold segmentation used in surveillance is sensitive to image content and methods based on background modelling cannot find stationary targets.

© Springer Nature Singapore Pte Ltd. 2017
J. Yang et al. (Eds.): CCCV 2017, Part I, CCIS 771, pp. 282–294, 2017.
https://doi.org/10.1007/978-981-10-7299-4_23

Enlightened by the fact that the object to be detected in an image is usually correlated with its salience, many researchers combine a salience model with object detection. Ren et al. [1] introduced a new saliency model based on a region-based solution and adopted it in encoding features for object recognition. Jung and Kim [2] proposed a novel unified spectral-domain approach to build a saliency map, and applied it in object segmentation. According to their experiments, object detection fused with a saliency model can achieve a higher precision. Therefore, we intend to utilize a saliency model in infrared object detection. In fact, there have been many works in this sphere so far. In view of the different visual characteristics between small targets and background clutter, Qi et al. [3] presented a new directional saliency-based method, which is competitive with the state-of-the-art methods, especially for images with various typical complex backgrounds. Another effective approach for small maritime object detection is described in [4] in 2011, it used local minimum patterns (LMP) to obtain the saliency map and segment the potential regions. Then a fast clustering algorithm was utilized for localizing objects from segmented regions. In the same year, Han et al. [5] improved the local contrast scheme for calculating the saliency map and then obtained the target region precisely by adopting a threshold operation along with a traversal mechanism.

From this short review, we notice that most of the focus in infrared object detection is on small targets. However, it is sometimes difficult to locate the targets accurately. The reason can be attributed to two causes: (1) small targets that occupy only a few pixels are short of distinct features and are easily contaminated by unknown noise; (2) the contrast between targets and background is low, thus making them difficult to distinguish. In this paper, we start from another assumption and propose a scheme for the detection of certain targets which is more applicable and operable than small target detection in surveillance. Here, the detection target is a human, and a saliency model integrated with visual attention[1] technique is embedded in the scheme.

It is well known that, saliency detection can be formulated into two categories: bottom-up and top-down. The principle of the bottom-up methods is based on the center-surround mechanism, which is in line with the performance of the human vision system. Due to this reason, the definition of the saliency model in a bottom-up method is varied and it is hard to compare the features between point and point, area and area in the image. In the model calculation, we extract the underlying characteristics of the image first, such as brightness, color, direction, etc. Then the contrast of these characteristics is calculated. Finally, the saliency map is obtained, which can show the degree of significance of each pixel in the image.

In 1998, Itti et al. built a representative saliency detection model that implements RGB color channels, intensity and orientation as a standard benchmark for comparison [6]. Then, Harel and Koch improved Ittis model by adopting a Markov chain, and named it graph-based visual saliency (GBVS) [7]. Later,

---

[1] The term visual attention refers to the human observer who concentrates to a specific area of the visual scene, and processing is performed in a serial fashion.

Hou and Zhang developed a spectral residual model whose computation time is competitive [8]. Achanta et al. used a frequency-based approach with low-level features in Lab color space to measure the saliency of each pixel [9]. In addition, Bi et al. (2010) combined visual attention guidance and a local descriptors representation for generic object detection [10]. In general, bottom-up methods are fast, simple and automatic, but they are not suitable for detecting homogeneous and quite large objects.

In contrast, the definition of saliency in a top-down model is usually related to the specific object and detection task. For example, the saliency map always depends on the shape, location and size of the objects to be detected. In other words, the identifiable high-level information is often treated as an indicator to measure the saliency of the image in a top-down model. In the state-of-the-art, the basis of top-down saliency models is first to learn the prior knowledge related to the goal or task, and then adopt the prior knowledge to guide the process of visual attention. Therefore, the top-down model contains two stages: learning and decision-making. In the learning stage, the main task is to extract the underlying characteristics of the image and learn the differences between the target and background in order to obtain a probabilistic statistical model of the training set. In the decision-making stage, the model extracts the underlying characteristics of the image in the testing set, and then utilizes the statistical model to guide the influence of the underlying characteristics for visual attention.

There have been three design principles for the top-down saliency model so far.

(1) The saliency model based on simple parameter estimation. Itti [11] mentioned a multi-scale feature integration method for specific target detection in his doctoral thesis. Frintrop et al. [12] developed a VOCUS system for extracting the saliency associated with a specific goal. In addition, Lee and Lee [13] established the relationship between specific knowledge and the goal, and then utilized the model to detect faces.

(2) The saliency model based on production models [14–20]. Its core is first to build a model according to the conditions of density and prior probability, and then get a posterior probability distribution. In 2006, Navalpakkam and Itti [14] defined the signal-to-noise ratio (SNR) as the ratio between degrees of significance of the target and background and then computed linear parameters for the fusion of the underlying features by maximizing the SNR so that the saliency of target objects is higher than that of the background. Analogously, Oliva et al. [15] and Torralba [16] defined the saliency as finding the probability of a series of characteristics in an image. Gao et al. [17] defined the question as a one-to-many classification, and in his view, the significant features are those features that can best separate targets from background faultlessly.

(3) The saliency model based on discriminant models. Peters and Itti [21] learned what they call the gist features of people in a particular task and then used these to predict the gaze area in the next frame. In addition, Judd et al. [22] computed the location of the focus for human eyes, and

Borji [23] introduced a new method that learns the weight coefficients of the feature fusion by classifier instead of by parameter estimation. Recently, Yang and Yang [24] also developed a top-down saliency model, which is based on a Conditional Random Field CRF with latent variables, to detect various objects.

Actually, top-down methods perform better in object detection compared with bottom-up models, but they are time-consuming and limited due to the lack of high-level information in some images. Therefore in this paper, we introduce a bottom-up model and top-down model simultaneously considering they have different characteristics for accurate human detection.

Firstly, we automatically segment the regions-of-interest (ROI) that may contain the target, through a bottom-up saliency algorithm. Then, we use histograms of oriented gradients (HOGs) as a feature extraction technique for the detection of humans. In order to select proper features and reduce redundant features without losing important local information, we introduce a novel method to optimize standard HOG features by using a visual attention map. The visual attention map represents the distribution effects of human fixation points that are collected by eye tracker equipment. In this regard the visual attention map plays a role as prior knowledge, so we can treat the feature optimization as a top-down method. Finally, Support Vector Machines (SVM) is applied for recognizing humans on the basis of the above results. The block diagram is depicted in Fig. 1.

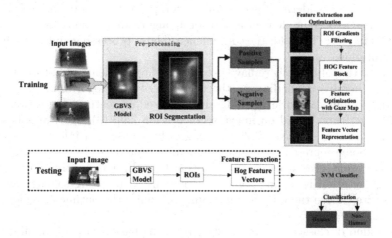

Fig. 1. Proposed framework.

The main contributions associated with this work are: (1) we practically analyse and develop a visual attention technique, which can be seen as a priori knowledge of the human subject, for feature optimization in a real-time object detection system; (2) we propose a new framework that combines a saliency model and a HOG based feature process method to detect a known object rather than unknown small objects in infrared thermal images.

The rest of this paper is organized as follows. In Sect. 2, the proposed detection model is described in detail. Section 3 demonstrates the effectiveness of our algorithm through a series of experiments. Finally, concluding remarks and discussions on the future work are drawn in Sect. 4.

2   The Detection Framework

The detection framework in this paper involves two main processes: training and testing. The training process includes three key steps: (1) pre-processing, (2) feature extraction and optimization, (3) classifier construction, while the testing includes (1) pre-processing, (2) feature extraction and (3) classification.

2.1   Pre-processing

Pre-processing, which appears in both image training and testing has the effect of a preliminary screening. It is inspired by the essence of IRT imaging. As we know, the intensity value of infrared images reflects the temperature of the objects in the scene, as shown in Fig. 2(a). It is very important to emphasize that the greater the intensity the higher the temperature. Therefore, the object that we detect usually is located in regions where the background has a strong contrast with the objects, i.e. regions where the temperature (intensity value) of a human is higher relative to the temperature (intensity value) of background in most cases. Considering the observers often pay attention to those significant regions, here we can implement a bottom-up algorithm to choose some potential regions of interest (ROI) for the convenience of later processing and the reduction of computation.

As a matter of fact, there have been many developments in bottom-up algorithms. Generally, depending on the principle of each algorithm, they can be divided into four categories: (1) algorithms based on feature integration [5,25–28]; (2) algorithms based on image segmentation [29]; (3) algorithms based on frequency domain analysis [7,30]; (4) algorithms based on information theory [6,31,32]. Some representative algorithms have been de-scribed in the Introduction so we wont give a detailed description here. In a word, the motivation of ROI segmentation is to find some local regions with strong saliency. For this purpose, we choose the GBVS algorithm, to acquire the outline of regions with high saliency.

As a kind of classic bottom-up algorithm, the basic process is similar to the Itti algorithm. The first step is to extract features at multiple spatial scales. For example, we can compute different level of Gaussian features based on luminance, R, G, B, Y, orientations, etc. Here, given the characteristics of the IRT image, we extract only two kinds of features viz. luminance and orientation. Then we consolidate the feature maps using Markov theory, and obtain a saliency map. Finally we normalize the unique saliency map. The result of Fig. 2(a) processed by GBVS is shown in Fig. 2(b).

Fig. 2. An example of an Infrared Thermal Image (a) and its saliency map of (b).

2.2 Feature Extraction and Optimization

Feature extraction is an essential module in object recognition. Many types of features can be extracted from an image. In this paper, we mainly adopt the Histogram of Oriented Gradient (HOG) descriptor to represent IRT images. HOG was proposed by Dalal et al. in 2005 [33], and its core is that the appearance and shape of local targets within an image can be represented by the direction density distribution of its gradients or edges. In detail, first the image is divided into small regions referred to as cells. Then for each pixel in each cell, we construct a histogram of gradients or edges. Finally, we combine these histograms to constitute a feature descriptor. In this paper, because the potential regions (ROIs) have been selected previously, in this stage we compute the HOG features only for the cells in each ROI.

Even though we have introduced ROI segmentation before feature extraction in order to reduce computation compared with the traditional algorithm, we still have to face the problem that the training is usually time-consuming. Thus we integrate a feature optimization into feature extraction to select proper features with important local information and reduce the computation workload for training further. The specific process is shown in Fig. 3.

It is worth to notice that the proposed algorithm adds a selection of the obtained descriptor blocks based on a gaze distribution map which can represent the objective visual attention of observers. We generate the gaze distribution map with the help of eye tracker equipment. The eye tracker can record the fixation points of observers when they watch the image. If the observer paid more attention to one area of the image, there will be more fixation points on this area. Based on this result, we represent the density of fixation points using different intensity values of RGB on the image to demonstrate the gaze distribution. The result of gaze distribution is referred to as a visual attention map. In detail, some examples of a gaze distribution map are shown in Figs. 4 and 5. The colored areas (red/yellow/green) in the map represent the attention distribution of single human subjects and multi human subjects respectively, and the stronger the intensity value, the more important the degree of visual attention. When we choose the important descriptor blocks, we first set a threshold, and then count the number of pixels located in the corresponding attention area of the

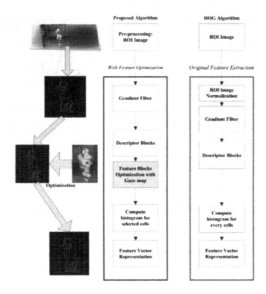

Fig. 3. The Process of feature extraction and optimization.

gaze distribution map for each cell. And only the blocks whose statistic is higher than a certain threshold are selected (the threshold is 40 for $[8 \times 8]$ cell in this paper). Finally, the selected blocks are weighted and normalized to get the HOG feature vectors for training.

2.3   SVM Classification

SVMs are a supervised learning mechanism, and its core is to find the optimal hyper-plane in a linear separable space which is mapped from high-dimensional inseparable space. Because SVMs are a very mature technique in object recognition, here we only describe the training data instead of elaborating its process at length. In this paper, the training data contains two sets. The positive set in which each image includes a human is selected manually from ROI images. And the negative set without a human is selected randomly from ROI images. In addition, the sizes of those sets are equal.

3   The Experiment

In order to demonstrate the effectiveness of our algorithm for human object detection in infrared thermal images, we collected 200 IRT images from website (60) and real-time inspection system (140) using a 640 by 380 resolution thermography camera. The number and size of human objects in the scene are varied. Specifically, we grouped all the images into two sets: training set and test set, with size 80 and 120 respectively. Further, 12 observers watched the image

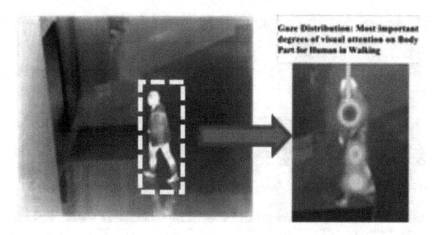

Fig. 4. A heat map of gaze distribution on single human object. (Color figure online)

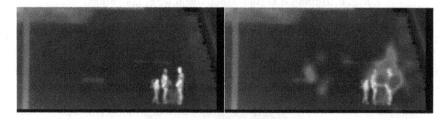

Fig. 5. A heat map of gaze distribution on multiple human objects. (Color figure online)

in a sequence of 80 training images one after another. Each image is shown on the screen about 3 s. During this processing, we collect the fixation results of each image using eye tracker equipment and its built-in software.

The experimental results for some example images are shown in Fig. 6. In the top group of images, there is only one hot object (person), while the rest con-tain many hot objects in different capture angles. Here we compare our method with four other algorithms, which are state-of-art saliency detection methods Itti and Koch [26], Yang and Yang [24], the traditional recognition methods [33], and an algorithm which is similar to ours apart from the absence of feature optimization. Obviously, the traditional method is not precise enough even for the single object detection. The results of our method and the other three methods are almost the same. Neverthe-less, it is worth to notice that our algorithm is superior to the other four algo-rithms whose results overlap or shows false positives. To sum up, our method performs best in the object detection.

To illustrate the validity of our algorithm at a deeper level, we computed the precision and error rate. In this regard, we manually selected the best region for each human for each testing image. Then we clustered all of the pixels into two groups: the correct pixels (the pixels that locate in the best region) and the error pixels (the pixels that locate in a region other than the best region), finally we

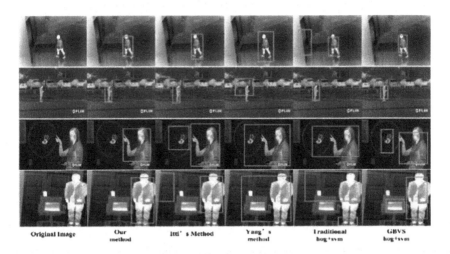

| Original Image | Our method | Itti's Method | Yang's method | Traditional hog+svm | GBVS hog+svm |

Fig. 6. Human detection comparison for single object.

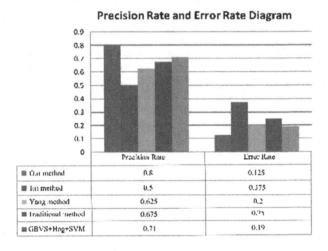

Fig. 7. The precision and error rate of different methods.

can cal-culate the precision rate and error rate respectively according to Eqs. (1) and (2). Figure 7 shows the statistical data. It is obvious that the precision rate of our algorithm is highest, and the order of other algorithms from high to low is the fourth algorithm (GBVS + HOG + SVM), the traditional method, Yangs method and Ittis method. Meanwhile, the order of the error rate is Ittis method, the tradi-tional method, Yangs method, the fourth kind of algorithm and our algorithm, which is the reverse of the order with the precision rate.

$$\text{precision rate} = \frac{\text{No. of correct pixels}}{\text{Total number of pixels in detection region}} . \qquad (1)$$

$$\text{error rate} = \frac{\text{No. of error pixels}}{\text{Total number of pixels in detection region}} . \tag{2}$$

In addition to this testing, we present more detection results using our algorithm under various conditions in IRT images (See Figs. 8 and 9). We can notice that all the people can be detected accurately. Finally, the capture sensor camera used was FLIR FC-S series. The eye tracker equipment used was Tobii X2 series. Our implementation tool is Matlab on a PC with Intel i5 CPU, 3G memory and Windows XP system. The average testing speed is 1.76 s for every 40 images.

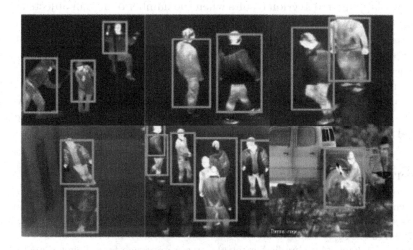

Fig. 8. Detection results of proposed method based on website collected images.

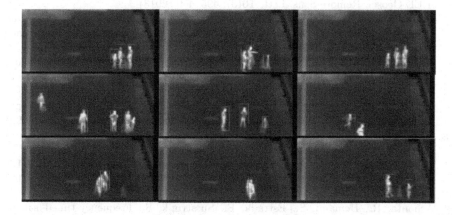

Fig. 9. Detection results of proposed method based on real-time images.

4   Conclusion

In this paper, we have proposed an efficient framework to solve the known object detection problem for infrared thermal images, which mainly relies on a saliency model and HOG feature optimization process. Evaluations on a number of real images captured from thermal cameras and comparisons with several state-of-the-art algorithms have demonstrated that the proposed method is effective and more accu-rate than other three other methods.

A few limitations of our method can be concluded as follows. First, there is some overlapping in detection results when the number of human objects is large. In our future work, we will examine how to adaptively classify overlapping objects to have better results. Second, the computational performance of the proposed algorithm is inadequate e for higher resolutions in real-time applications. Another task in the future is to improve the real-time performance using multi-resolution methods.

In order to extend the application fields of the proposed algorithm, we will put emphasis on extending the utilization of our algorithm in human body detection, animal detection, etc.

References

1. Ren, Z., Gao, S., Chia, L.T., Tsang, I.W.H.: Region-based saliency detection and its application in object recognition. IEEE Trans. Circ. Syst. Video Technol. **24**(5), 769–779 (2014)
2. Jung, C., Kim, C.: A unified spectral-domain approach for saliency detection and its application to automatic object segmentation. IEEE Trans. Image Process. **21**(3), 1272–1283 (2012)
3. Qi, S., Ma, J., Tao, C., Yang, C., Tian, J.: A robust directional saliency-based method for infrared small-target detection under various complex backgrounds. IEEE Geosci. Remote Sens. Lett. **10**(3), 495–499 (2013)
4. Qi, B., Wu, T., Dai, B., He, H.: Fast detection of small infrared objects in maritime scenes using local minimum patterns. In: 2011 18th IEEE International Conference on Image Processing (ICIP), pp. 3553–3556 (2011)
5. Han, J., Ma, Y., Zhou, B., Fan, F., Liang, K., Fang, Y.: A robust infrared small target detection algorithm based on human visual system. IEEE Geosci. Remote Sens. Lett. **11**(12), 2168–2172 (2014)
6. Itti, L., Koch, C., Niebur, E.: A model of saliency-based visual attention for rapid scene analysis. IEEE Trans. Pattern Anal. Mach. Intell. **20**(11), 1254–1259 (1998)
7. Harel, J., Koch, C., Perona, P.: Graph-based visual saliency. In: Advances in Neural Information Processing Systems, pp. 545–552 (2007)
8. Hou, X., Zhang, L.: Saliency detection: a spectral residual approach. In: 2007 IEEE Conference on Computer Vision and Pattern Recognition. CVPR 2007, pp. 1–8 (2007)
9. Achanta, R., Hemami, S., Estrada, F., Susstrunk, S.: Frequency-tuned salient region detection. In: 2009 IEEE Conference on Computer Vision and Pattern Recognition. CVPR 2009, pp. 1597–1604 (2009)

10. Bi, F., Bian, M., Liu, F., Gao, L.: A local descriptor based model with visual attention guidance for generic object detection. In: 2010 3rd International Congress on Image and Signal Processing (CISP), vol. 4, pp. 1599–1604 (2010)
11. Itti, L.: Models of bottom-up and top-down visual attention. Ph.D. thesis, California Institute of Technology (2000)
12. Frintrop, S., Backer, G., Rome, E.: Goal-directed search with a top-down modulated computational attention system. In: Kropatsch, W.G., Sablatnig, R., Hanbury, A. (eds.) DAGM 2005. LNCS, vol. 3663, pp. 117–124. Springer, Heidelberg (2005). https://doi.org/10.1007/11550518_15
13. Lee, Y.B., Lee, S.: Robust face detection based on knowledge-directed specification of bottom-up saliency. Etri J. **33**(4), 600–610 (2011)
14. Navalpakkam, V., Itti, L.: An integrated model of top-down and bottom-up attention for optimizing detection speed. In: 2006 IEEE Computer Society Conference on Computer Vision and Pattern Recognition, vol. 2, pp. 2049–2056 (2006)
15. Oliva, A., Torralba, A., Castelhano, M.S., Henderson, J.M.: Top-down control of visual attention in object detection. In: Proceedings of 2003 International Conference on Image Processing. ICIP 2003, vol. 1, pp. I-253 (2003)
16. Torralba, A.: Modeling global scene factors in attention. JOSA A **20**(7), 1407–1418 (2003)
17. Gao, D., Han, S., Vasconcelos, N.: Discriminant saliency, the detection of suspicious coincidences, and applications to visual recognition. IEEE Trans. Pattern Anal. Mach. Intell. **31**(6), 989–1005 (2009)
18. Torralba, A., Oliva, A., Castelhano, M.S., Henderson, J.M.: Contextual guidance of eye movements and attention in real-world scenes: the role of global features in object search. Psychol. Rev. **113**(4), 766 (2006)
19. Borji, A., Sihite, D.N., Itti, L.: Probabilistic learning of task-specific visual attention. In: 2012 IEEE Conference on Computer Vision and Pattern Recognition (CVPR), pp. 470–477 (2012)
20. Zhang, L., Tong, M.H., Marks, T.K., Shan, H., Cottrell, G.W.: Sun: a bayesian framework for saliency using natural statistics. J. Vis. **8**(7), 32 (2008)
21. Peters, R.J., Itti, L.: Beyond bottom-up: incorporating task-dependent influences into a computational model of spatial attention. In: 2007 IEEE Conference on Computer Vision and Pattern Recognition. CVPR 2007, pp. 1–8 (2007)
22. Judd, T., Ehinger, K., Durand, F., Torralba, A.: Learning to predict where humans look. In: 2009 IEEE 12th International Conference on Computer Vision, pp. 2106–2113 (2009)
23. Borji, A.: Boosting bottom-up and top-down visual features for saliency estimation. In: 2012 IEEE Conference on Computer Vision and Pattern Recognition (CVPR), pp. 438–445 (2012)
24. Yang, J., Yang, M.H.: Top-down visual saliency via joint CRF and dictionary learning. In: 2012 IEEE Conference on Computer Vision and Pattern Recognition (CVPR), pp. 2296–2303 (2012)
25. Le Meur, O., Le Callet, P., Barba, D., Thoreau, D.: A coherent computational approach to model bottom-up visual attention. IEEE Trans. Pattern Anal. Mach. Intell. **28**(5), 802–817 (2006)
26. Itti, L., Koch, C.: Computational modelling of visual attention. Nature Rev. Neurosci. **2**(3), 194 (2001)
27. Wai, W.K., Tsotsos, J.K.: Directing attention to onset and offset of image events for eye-head movement control. In: Proceedings of the 12th IAPR International Conference on Pattern Recognition, Conference A: Computer Vision and Image Processing, vol. 1, pp. 274–279 (1994)

28. Milanese, R., Bost, J.M., Pun, T.: A bottom-up attention system for active vision (1992)
29. Cheng, M.M., Mitra, N.J., Huang, X., Torr, P.H., Hu, S.M.: Global contrast based salient region detection. IEEE Trans. Pattern Anal. Mach. Intell. **37**(3), 569–582 (2015)
30. Guo, C., Zhang, L.: A novel multiresolution spatiotemporal saliency detection model and its applications in image and video compression. IEEE Trans. Image Process. **19**(1), 185–198 (2010)
31. Bruce, N., Tsotsos, J.: Saliency based on information maximization. In: Advances in Neural Information Processing Systems, pp. 155–162 (2006)
32. Hou, X., Zhang, L.: Dynamic visual attention: searching for coding length increments. In: Advances in Neural Information Processing Systems, pp. 681–688 (2009)
33. Dalal, N., Triggs, B.: Histograms of oriented gradients for human detection. In: 2005 IEEE Computer Society Conference on Computer Vision and Pattern Recognition. CVPR 2005, vol. 1, pp. 886–893 (2005)
34. Alpher, A.: Frobnication. J. Foo **12**(1), 234–778 (2002)

Integrating Color and Depth Cues for Static Hand Gesture Recognition

Jiaming Li[(✉)], Yulan Guo, Yanxin Ma, Min Lu, and Jun Zhang

College of Electronic Science and Engineering,
National University of Defense Technology, Changsha 410073, China
lijiaming15@nudt.edu.cn

Abstract. Recognizing static hand gesture in complex backgrounds is a challenging task. This paper presents a static hand gesture recognition system using both color and depth information. Firstly, the hand region is extracted from complex background based on depth segmentation and skin-color model. The Moore-Neighbor tracing algorithm is then used to obtain hand gesture contour. The k-curvature method is used to locate fingertips and determine the number of fingers, then the angle between fingers are generated as features. The appearance-based features are integrated to the decision tree model for hand gesture recognition. Experiments have been conducted on two gesture recognition datasets. Experimental results show that the proposed method achieves a high recognition accuracy and strong robustness.

Keywords: Hand gesture recognition · Skin-color model
Depth segmentation

1 Introduction

Hand gesture recognition is an active research problem in human-computer interaction, it provides a natural way for communication between humans and machines. Typical applications include sign language recognition, virtual reality and smart homes.

Existing hand gesture recognition methods can be divided into two categories: static and dynamic methods [20]. Although remarkable progress has been achieved in this area [7,11,20,25,27], several challenges are still faced by existing methods. A major challenge is the uncontrolled real-world environment. For static hand gesture recognition, the first problem is reliable hand segmentation. Some systems require users to wear an electronic glove to capture the key features of an hand. However, the device is usually costly and inconvenient [8]. Methods based on skin-color model [21,32,34] and hand shape model [12,13,31] have also been proposed. However, they are not robust in complex backgrounds and rely significantly on the models.

The first author (Jiaming Li) is a master student in National University of Defense Technology.

© Springer Nature Singapore Pte Ltd. 2017
J. Yang et al. (Eds.): CCCV 2017, Part I, CCIS 771, pp. 295–306, 2017.
https://doi.org/10.1007/978-981-10-7299-4_24

Recent development of depth cameras provides a new direction for static hand gesture recognition. In this work, we use a Kinect camera to capture color and depth images. We propose a static hand gesture recognition system based on depth segmentation and skin-color model (as shown in Fig. 1). We first use depth thresholding and skin-color model to segment hand from an RGB-D image, and then extract appearance-based features. Finally, the decision tree is used for hand gesture recognition. Experimental results show that the proposed system is highly accurate and efficient.

The rest of the paper is organized as follows. Related works are first reviewed in Sect. 2. The proposed method for hand segmentation, hand representation and hand gesture recognition is introduced in Sect. 3. Experimental results are discussed in Sect. 4. Finally, we conclude the paper in Sect. 5.

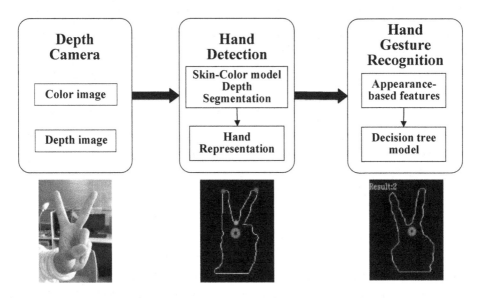

Fig. 1. The flowchart of the proposed hand gesture recognition method.

2 Related Works

A hand gesture recognition system includes data acquisition, hand localization, feature extraction, and gesture recognition [22]. Many vision-based hand gesture recognition algorithms have been proposed [7,10,20,27,29]. If depth information is available, the accuracy and robustness of hand gesture recognition can further be improved. Early hand segmentation methods mainly rely on features, such as chromatic distribution, which is sensitive to variations of skin colors. An appearance-based hand detection system is also proposed in [14], which is unreliable due to the complicated and unpredictable features. Therefore, depth information can be used to improve hand segmentation performance. For example, the point in the depth image closest to the camera is first detected, and its neighbors within a range of depth value can be used to segment the hand. The hand segmentation result is usually inaccurate since only a small portion of an image is occupied by the hand.

After hand segmentation, various hand features can be extracted from depth image. Li et al. [16] proposed a bag-of-3D-points feature for activity recognition from depth image sequences. Specifically, 3D points are sampled from the silhouettes of depth images. Ren et al. [24] represented the hand shape by time-series curve, with each finger corresponding to a segment of the curve. The time-series curve records the relative distance between each contour vertex to a center point and reserves the topological information as well. Besides, the Finger-Earth Mover's Distance (FEMD) was used in this method. However, this distance metric is sensitive to distortion. Wang et al. [30] proposed Superpixel Earth Mover's Distance (SP-EMD) which shows a promising way for hand gesture recognition. In this paper, the hand is located and segmented according to the depth and skeleton information from Kinect. Chen et al. [6] proposed image-to-class dynamic time warping (DTW) distance to distinguish the finger-let features of 3D hand contours. Liu et al. [18] explored the invariance from the hand contour and utilized Fourier descriptor, edge histogram, and boundary moment invariants for the feature extraction. Sorce et al. [26] proposed a neural network with backpropagation detecting the hand pose to recognize whether it is closed or not. An exponential weighted moving average noise reduction mechanism was used to suppress the noise effects of the neural network. Kuznetsova et al. [15] proposed an Ensemble of Shape Function (ESF) to represent the hand region. The ESF descriptor consists of concatenated histograms generated from randomly selected points in the point cloud. A Multi-Layered Random Forest (MLRF) was then trained to recognize the hand signs. Zhang et al. [33] defined a 3D facet as a 3D local support surface for each 3D point cloud and used the Histogram of 3D Facets (H3DF) to represent the 3D hand shape. H3DF can effectively represent the 3D shapes and structures of various hand gestures. The Support Vector Machines (SVM) with linear kernel was used for classification. SVM is the most popular classifier for the 3D static hand gesture recognition. Pugeault and Bowden [23] adopted linear SVM to predict the class label with the ASL data sets. For nonlinear kernels, Gallo et al. [9] trained the Radial Basis Function (RBF) classifiers to recognize four different hand postures in a 3D medical touchless interface. Bagdanov et al. [1] represented the hand by the concatenation of five Speeded Up Robust Feature (SURF) descriptors for a total of 640 dimensions. A SVM with a nonlinear RBF was trained for hand gesture recognition. This hand gesture recognition method is invariant to rotation, arm pose, background and distance from the sensor. The random decision forest has been experimentally compared with the SVM classification with various 3D hand features [19]. In general, the performance of RFs is dependent on the depth of the tree. The tradeoff of accuracy and speed always needs to be properly considered in its implementation.

3   Static Gesture Recognition

First, both depth and skin color are used for hand segmentation. Then, the appearance-based features are used to represent the hand. Finally, the decision tree model is trained for static hand gesture recognition.

3.1   Image Preprocessing

Preprocessing has to be performed before hand segmentation. Median filter is used to reduce salt and pepper noise. Compared to mean filter, median filter can an preserve more details around boundary structures. Besides, median filter is robust to noise.

3.2   Depth and Skin-Color Based Hand Detection

Both skin color and depth are used for hand detection. The hand is first segmented with depth threshold, histogram of the depth image is then used to distinguish the foreground from background. The depth data within a range are extracted and further combined with RGB images (using skin color segmentation) to remove unnecessary interference. The orthogonal color space $YCbCr$ is used in our paper, pixel values are considered as skin color if Cb is within the range of [77, 127] and Cr is within the range of [133, 173]. The depth range is set to [1.5 m, 1.7 m], pixels within the depth range are saved as $p_i(x_i, y_i)$. In this paper, the K-means clustering algorithm is used to classify all pixels into two clusters. Specifically, the initial K clustering centers $\{u_j\}$ are randomly selected, then the following calculation is repeated until the sample converges. For each pixel $x^{(i)}$, the class label $c^{(i)}$ is defined as

$$c^{(i)} = \mathrm{argmin}||x^{(i)} - u_j||^2 \tag{1}$$

where $x^{(i)}$ represents the point in sample space, the cluster center u_j is calculated as

$$u_j = \frac{\sum_{i=1}^m \mathbf{1}\left\{c^{(i)} = j\right\} x^{(i)}}{\sum_{i=1}^m \mathbf{1}\left\{c^{(i)} = j\right\}} \tag{2}$$

where is $\mathbf{1}\{c^{(i)} = j\}$=1 if $c^{(i)} = j$. Consequently, the pixels are divided into two classes: hand part and non-hand part. Figure 2 illustrates the hand segmented by our method.

3.3   Appearance Based Hand Representation

After hand segmentation, hand gesture contour is obtained using the Moore-Neighbor tracing algorithm [17]. Specifically, a pixel is compared to its 8 adjacent points, if all the adjacent points are internal points, the pixel is deleted. Otherwise, the pixel is considered as a hand contour point. This process is traversed over the entire image. Then, hand can be represented by the center of

Fig. 2. The depth map and the segmentation result.

the palm, the number of fingers and the angles between fingers. The center of the palm can be determined by calculating the maximum inscribed circle of the hand, which can effectively eliminate the impact caused by hand palm motion. The number of fingertips is obtained by the k-curvature algorithm [28]. A vector is first calculated from contour points, a candidate fingertip point is then determined as the point with the largest curvature, and then the following steps are used:

(1) A set is initialized by the points on the hand.
(2) Given a candidate fingertip p_i and its neighboring points p_{i-k} and p_{i+k}, two vectors are computed from these points to p_i, i.e., $\overrightarrow{p_i p_{i-k}}$ and $\overrightarrow{p_i p_{i+k}}$.
(3) θ is calculated using Eq. 3. If $\theta < 45°$, p_i can be determined as a fingertip. Otherwise, steps 2 and 3 are repeated with next unchecked point.

$$\theta = \frac{(p_i - p_{i-k})}{||p_i - p_{i-k}||} * \frac{(p_i - p_{i+k})}{||p_i - p_{i+k}||} \tag{3}$$

Consequently, the hand fingertips and the groove between two fingers can be obtained at the same time. The groove points are deleted using the cross product of the two vectors. This method can accurately find the position of fingertips for different gestures of different hands. The angle θ is relatively stable even if the hand position changes since the relative position of these three points is unchanged. Once fingertips are determined, angles between fingers are obtained by connecting the center of the palm and fingertips. Besides, the maximum angle between fingers is determined and the distances between the center of the palm and the fingertips are calculated.

3.4 Decision Tree Based Static Hand Gesture Recognition

Using the decision tree model, a complex multi-class classification problem can be transferred to a number of simple classification problems. A suitable tree structure should be carefully designed. That is, appropriate nodes and branches, the number of features and the appropriate decision rules should be determined.

The number of extracted fingers N and the maximum angle between fingers A are used as the decision tree node. The decision tree model used in our dataset is shown in Fig. 3.

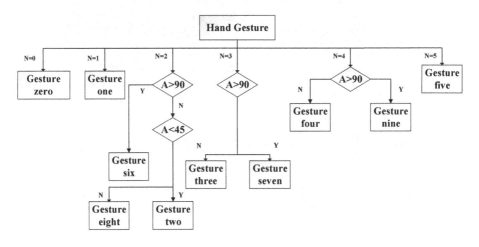

Fig. 3. The decision tree model of the proposed method.

Note that, gestures zero, one and five can be directly recognized at the root node by the number of fingers. When the number of fingers is same, a second-level node should be established to recognize gestures three, seven, four and nine using the maximum angle between fingers. Further, a third-level node should be established to distinguish gestures two, six and eight by dividing the maximum angle between fingers. An illustration of hand gesture recognition results is shown in Fig. 4.

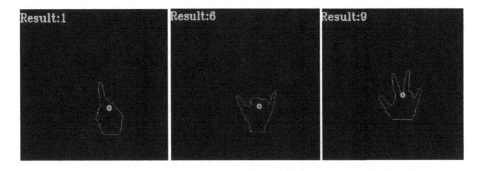

Fig. 4. An illustration of hand gesture recognition results.

4   Experiments

We evaluated the proposed gesture recognition approach on two datasets. The 10-Gesture dataset [24] contains both RGB and depth images of 10 gestures collected from 10 subjects. Each subject performs 10 different poses for each gesture. In total, the dataset has 1000 samples. To further evaluate our approach, we collected a digit dataset using a Kinect device. Our dataset contains 10 gestures of 5 subjects. For each subject, 10 different poses are performed for each gesture. Therefore, the dataset has 500 samples. Several gesture samples in our dataset are shown in Fig. 5. It should be noted that gesture 9 in the 10-Gesture dataset and our dataset is different, while the other gestures are the same. For fair comparison, we followed the experimental setup introduced in [24]. All experiments were conducted on a machine with an Intel Core i5-7200 CPU and 4 GB RAM. The proposed method was compared to existing methods in terms of mean accuracy and robustness.

Fig. 5. Gesture samples (0–9) captured by Kinect.

4.1   The 10-Gesture Dataset

The decision tree used for the 10-Gesture dataset is slightly different from that for our our dataset (Fig. 3). Specifically, the distances between the center of the palm and the fingertips rather than the angle are used to distinguish gestures 1 and 9 in the 10-Gesture dataset. As shown in Fig. 6, there are several false recognition results between gestures 1 and 9. That is because the number of fingers for gestures 1 and 9 is the same. For other gestures, a few errors are caused by individual difference for these gestures. For example, the angles between fingers for the same gesture may be different for different persons. The overall mean accuracy achieved by our method is 96.1%.

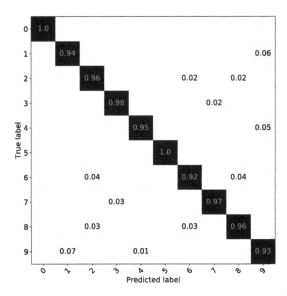

Fig. 6. The confusion matrix achieved by the proposed method on the 10-Gesture dataset.

We further compare our method to three existing methods on the 10-Gesture dataset, include Shape Context [5], Skeleton Matching [3] and FEMD [24]. Their mean accuracy results are shown in Table 1. It can be seen that the proposed method achieves the highest mean accuracy. The most confusing cases for FEMD are gestures 4 and 5, and gestures 1 and 9. That is because the fingers in those gestures are not accurately segmented due to image distortion. Moreover, different hand size can change the weight of each finger for FEMD, resulting in mismatching. For shape context based method, the most confusing cases are gestures 1 and 9, and gestures 8 and 9. That is because these gestures have similar contours. In some cases, two fingers may stick together due to distortion. That is why gesture 4 is falsely recognized as gesture 3. For skeleton matching based method, two different skeleton pruning methods can be used, including

Table 1. The mean accuracy results achieved by Shape Context, Skeleton Matching, FEMD, and our proposed method on the 10-gesture dataset.

Methods	Mean accuracy
Shape Context [5] (with bending cost)	90.9%
Shape Context [5] (without bending cost)	93.5%
Skeleton Matching [3] (DCE [4])	92.8%
Skeleton Matching [3] (DSE [2])	92.9%
FEMD [24]	94.1%
The proposed method	**96.1%**

Discrete Curve Evolution (DCE) [4] and Discrete Skeleton Evolution (DSE) [2]. DSE is more robust to small protrusions. Therefore, the recognition accuracy achieved by DSE is higher than DCE. The most confusing pairs are gestures 6 and 7, and gestures 2 and 7. That is because the pruned skeletons have similar global structures for these gestures. In summary, the proposed method achieves a higher recognition accuracy than existing methods.

4.2 The Kinect Dataset

The recognition results achieved on our Kinect dataset are shown in Fig. 7. For gestures 0, 1 and 5, the recognition rate is 100%. For other gestures, few false recognition can be found. The mean accuracy achieved by our method is 98.0%. The results show that the proposed method can achieve a high recognition accuracy.

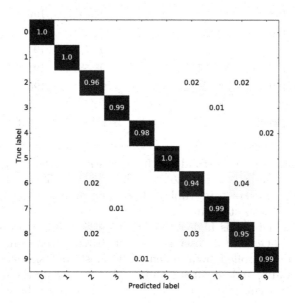

Fig. 7. The confusion matrix achieved by the proposed method on our Kinect dataset.

Hand gesture detection based on skin color can easily be affected by objects with the same color as skin. It is vulnerable to environment changes, such as illumination variation. Therefore, the combination of skin color and depth information can significantly improve the robustness of hand segmentation. To test the robustness of static gesture recognition in illumination changes and complex backgrounds, experiments were conducted under high illumination, low illumination and complex backgrounds, with each gesture being tested for 200 times. Experimental results are shown in Table 2. It can be seen that the recognition results under illumination changes are stable due to accurate hand segmentation. The accuracy drops slightly in complex background. That is mainly because hand segmentation is sensitive to background with skin color. In summary, our hand gesture recognition method is robust to illumination changes and complex background.

Table 2. The hand gesture recognition accuracy achieved in different circumstances.

Hand gesture	High illumination		Low illuminiation		Complex background	
	True positives	Mean accuracy	True positives	Mean accuracy	True positives	Mean accuracy
0	200	100%	200	100%	198	99%
1	200	100%	200	100%	196	98%
2	199	99.5%	198	99%	194	97%
3	197	98.5%	197	98.5%	195	97.5%
4	198	99%	195	97.5%	196	98%
5	196	98%	196	98%	194	97%
6	195	97.5%	198	99%	194	97%
7	197	98.5%	197	98.5%	195	97.5%
8	199	99.5%	198	99%	195	97.5%
9	198	99%	198	99%	196	98%
Mean	197.9	98.95%	197.7	98.85%	195.3	97.65%

We also compare our method to three existing methods on this dataset, including Shape Context [5], Skeleton Matching [3] and FEMD [24]. Their mean accuracy results are shown in Table 3. It can be observed that, the proposed method achieves the best mean accuracy. The mean accuracy of each method achieved on this dataset is higher than that on the 10-Gesture dataset. That is because the 10-Gesture dataset is more challenging than our dateset, as it is collected in uncontrolled environment. The most confusing cases for FEMD are gestures 4 and 5, and gestures 1 and 7. That is because the fingers in those gestures are not accurately segmented. For shape context based method, the most confusing cases are gestures 1 and 7, and gestures 8 and 9. The main reason

Table 3. The mean accuracy results achieved by Shape Context, Skeleton Matching, FEMD and our proposed method on our Kinect dataset.

Methods	Mean accuracy
Shape Context [5] (with bending cost)	94.7%
Shape Context [5] (without bending cost)	97.1%
Skeleton Matching [3] (DCE [4])	96.5%
Skeleton Matching [3] (DSE [2])	96.6%
FEMD [24]	96.8%
The proposed method	**98.0%**

is, these gestures have similar contours. For skeleton matching based method, the most confusing pairs are gestures 6 and 7, and gestures 2 and 7. That is because the pruned skeletons have similar global structures for these gestures. In summary, the proposed method achieves higher recognition accuracy than existing methods on our Kinect dataset.

5 Conclusions

In this paper, a hand gesture recognition method has been proposed using both depth and color information. The appearance features are extracted to represent a hand gesture, and the decision tree model is then used for hand gesture recognition. The effectiveness of the proposed method has been demonstrated on two datasets collected by Kinect. High mean accuracy and strong robustness has been achieved by our hand gesture recognition method. Compared to three state-of-the-art methods, the proposed method achieves the best hand gesture recognition performance.

Acknowledgement. This work was partially supported by the National Natural Science Foundation of China (Nos. 61602499 and 61471371), the National Postdoctoral Program for Innovative Talents (No. BX201600172), and China Postdoctoral Science Foundation.

References

1. Bagdanov, A.D., Del Bimbo, A., Seidenari, L., Usai, L.: Real-time hand status recognition from RGB-D imagery. In: ICPR, pp. 2456–2459. IEEE (2012)
2. Bai, X., Latecki, L.J.: Discrete skeleton evolution. In: Yuille, A.L., Zhu, S.-C., Cremers, D., Wang, Y. (eds.) EMMCVPR 2007. LNCS, vol. 4679, pp. 362–374. Springer, Heidelberg (2007). https://doi.org/10.1007/978-3-540-74198-5_28
3. Bai, X., Latecki, L.J.: Path similarity skeleton graph matching. IEEE TPAMI **30**(7), 1282–1292 (2008)
4. Bai, X., Latecki, L.J., Liu, W.Y.: Skeleton pruning by contour partitioning with discrete curve evolution. IEEE TPAMI **29**(3), 449–462 (2007)
5. Belongie, S., Malik, J., Puzicha, J.: Shape matching and object recognition using shape contexts. IEEE TPAMI **24**(4), 509–522 (2002)
6. Cheng, H., Dai, Z., Liu, Z.: Image-to-class dynamic time warping for 3D hand gesture recognition. In: IEEE ICME, pp. 1–6 (2013)
7. Cheng, H., Yang, L., Liu, Z.: Survey on 3D hand gesture recognition. IEEE TCSVT **26**(9), 1659–1673 (2016)
8. Dejmal, I., Zacksenhouse, M.: Coordinative structure of manipulative hand-movements facilitates their recognition. IEEE TBE **53**(12), 2455–2463 (2006)
9. Gallo, L.: Hand shape classification using depth data for unconstrained 3D interaction. J. Ambient Intell. Smart Environ. **6**(1), 93–105 (2014)
10. Han, J., Shao, L., Xu, D., Shotton, J.: Enhanced computer vision with microsoft kinect sensor: a review. IEEE Trans. Cybern. **43**(5), 1318–1334 (2013)
11. Hasan, H.S., Kareem, S.A.: Human computer interaction for vision based hand gesture recognition: a survey. AIR **43**(1), 1–54 (2015)

12. Hu, R.X., Jia, W., Zhang, D., Gui, J., Song, L.T.: Hand shape recognition based on coherent distance shape contexts. Pattern Recogn. **45**(9), 3348–3359 (2012)
13. Keskin, C., Kıraç, F., Kara, Y.E., Akarun, L.: Hand pose estimation and hand shape classification using multi-layered randomized decision forests. In: Fitzgibbon, A., Lazebnik, S., Perona, P., Sato, Y., Schmid, C. (eds.) ECCV 2012. LNCS, vol. 7577, pp. 852–863. Springer, Heidelberg (2012). https://doi.org/10.1007/978-3-642-33783-3_61
14. Kölsch, M., Turk, M.: Robust hand detection. In: IEEE FGR, pp. 614–619 (2004)
15. Kuznetsova, A., Leal-Taixé, L., Rosenhahn, B.: Real-time sign language recognition using a consumer depth camera. In: IEEE ICCV Workshops, pp. 83–90 (2013)
16. Li, W., Zhang, Z., Liu, Z.: Action recognition based on a bag of 3D points. In: IEEE CVPR Workshops, pp. 9–14. IEEE (2010)
17. Li, Y.: Hand gesture recognition using kinect. In: IEEE ICSESS, pp. 196–199 (2012)
18. Liu, Y., Yang, Y., Wang, L., Xu, J., Qi, H., Zhao, X., Zhou, P., Zhang, L., Wan, B., Ming, D.: Image processing and recognition of multiple static hand gestures for human-computer interaction. In: ICIG, pp. 465–470 (2013)
19. Minnenm, D., Zafrulla, Z.: Towards robust cross-user hand tracking and shape recognition. In: IEEE ICCV Workshops, pp. 1235–1241 (2011)
20. Mitra, S., Acharya, T.: Gesture recognition: a survey. IEEE TSMC **37**(3), 311–324 (2007)
21. Phung, S.L., Bouzerdoum, A., Chai, D.: Skin segmentation using color pixel classification: analysis and comparison. IEEE Trans. Pattern Anal. Mach. Intell. **27**(1), 148 (2005)
22. Plouffe, G., Cretu, A.M.: Static and dynamic hand gesture recognition in depth data using dynamic time warping. IEEE TIM **65**(2), 305–316 (2015)
23. Pugeault, N., Bowden, R.: Spelling it out: real-time asl fingerspelling recognition. In: IEEE ICCV Workshops, pp. 1114–1119 (2012)
24. Ren, Z., Yuan, J., Zhang, Z.: Robust hand gesture recognition based on finger-earth mover's distance with a commodity depth camera. In: ACMMM, pp. 1093–1096 (2011)
25. Shotton, J., Fitzgibbon, A., Cook, M., Sharp, T., Finocchio, M., Moore, R., Kipman, A., Blake, A.: Real-time human pose recognition in parts from single depth images. In: IEEE CVPR, pp. 1297–1304 (2011)
26. Sorce, S., Gentile, V., Gentile, A.: Real-time hand pose recognition based on a neural network using microsoft kinect. In: BWCCA, pp. 344–350 (2013)
27. Suarez, J., Murphy, R.R.: Hand gesture recognition with depth images: a review. In: RO-MAN, pp. 411–417 (2012)
28. Trigo, T.R., Pellegrino, S.R.M.: An analysis of features for hand-gesture classification. In: IWSSIP, pp. 412–415 (2010)
29. Wachs, J.P., Kölsch, M., Stern, H., Edan, Y.: Vision-based hand-gesture applications. Commun. ACM **54**(2), 60–71 (2011)
30. Wang, C., Liu, Z., Chan, S.C.: Superpixel-based hand gesture recognition with kinect depth camera. IEEE TMM **17**(1), 29–39 (2015)
31. Wu, Y., Lin, J., Huang, T.S.: Analyzing and capturing articulated hand motion in image sequences. IEEE TPAMI **27**(12), 1910–1922 (2005)
32. Yang, M.H., Ahuja, N., Tabb, M.: Extraction of 2D motion trajectories and its application to hand gesture recognition. IEEE TPAMI **24**(8), 1061–1074 (2002)
33. Zhang, C., Yang, X., Tian, Y.: Histogram of 3D facets: a characteristic descriptor for hand gesture recognition. In: FG, pp. 1–8 (2013)
34. Zhu, X., Wong, K.Y.K.: Single-frame hand gesture recognition using color and depth kernel descriptors. In: ICPR, pp. 2989–2992 (2013)

Accurate Joint Template Matching Based on Tree Propagating

Qian Kou[✉], Yang Yang, Shaoyi Du, Zhuo Chen, Weile Chen,
and Dexing Zhong

The School of Electronic and Information Engineering,
Xi'an Jiaotong University, Xi'an, China
kouqian@stu.xjtu.edu.cn, yyang@mail.xjtu.edu.cn

Abstract. Given a single template image, it is a big challenge to match all the target images accurately only by pairwise template matching. To handle this case, this paper introduces an accurate template matching method based on tree structure building to jointly match a set of target images. Our method aims to select well matched results for template updating and rescue badly matched images via the tree matching propagating. First, a novel similarity measure is given to evaluate the pairwise matching results. Then the joint matching is under an iterative framework which contains two main steps: (1) tree structure growing; and (2) matching propagating. When all the target images are included in the tree structure, the matching process is finished. Finally, experimental results demonstrate the improvement of the proposed method.

Keywords: Template matching · Joint · Tree propagating
Template updating

1 Introduction

Template matching is one of the important research issues in target tracking [1, 2], recognition [3], and 3D reconstruction. Generally speaking, a single template is given which contains a region of interest, then template matching method tries to find the object area in target image which may have different background and deformation.

In the past decade, many scholars focused on studying pairwise template matching methods [4–6]. These methods concentrated on how to use a given template to match one target image more accurately and fast. While with the development of the technology of image capturing, multi images matching is required in many tasks. A typical example is the non-rigid structure from motion, where we can't reconstruct a non-rigid shape only from two frames [7]. As template matching usually means matching a pair of images, our method is named as joint template matching because it is used to match a set of target images. While the given template is used to match a series of target images, some of them can't be well matched. Thus, for joint template matching, we distinguish

© Springer Nature Singapore Pte Ltd. 2017
J. Yang et al. (Eds.): CCCV 2017, Part I, CCIS 771, pp. 307–319, 2017.
https://doi.org/10.1007/978-981-10-7299-4_25

both well matched results and the bad ones firstly, and then we rescue the bad results by well mathed images.

Recent work has shown that leveraging multi-way information can dramatically improve matching results compared with pairwise matching. The early work of joint matching aims to select cycle consistent matches and identify incorrect matches from bad cycles [8], while this family of methods require sufficient number of correct matches which are difficult to find sometimes. Zhou et al. [9] proposed an algorithm to jointly bring poorly aligned images into pixel-wise correspondence by estimating a FlowWeb which represents the image set. The goal of this method is to establish global-consistence between all images, and this method built a correspondences graph between images. Another multi-image matching based on global optimization is proposed to jointly match a set of images, by simultaneously maximizing pairwise feature affinities and cycle consistency across multiple images [10]. This work aims to find global estimates from pairwise estimates such as rotation averaging and model fusion, and this method uses pairwise feature similarities as input. Besides, joint matching methods are also used to rescuing the bad alignment with their well aligned neighbors, Zhao et al. introduced a robust regularized re-fitting algorithm which exploits the appearance consistency between badly aligned images and its k well aligned nearest neighbors [11]. This method aims at face images with different facial features which have limits for matching other images in the wild. Differently, our work aims to match a set of images in the wild with a given template. And for rescuing bad matched results, we try to propose a matching method of tree building strategy.

In this paper, we propose a joint template matching method which aims to match a set of target images by a given template. Obviously, target images similar to the given template will be easy to match, but the remaining badly matched images can't be well matched by original template except a more similar template. By combining pairwise template matching with tree structure, and updating template at each level of tree simultaneously, the target images which are badly matched by the original template are rescued. In order to improve the accuracy of matched result, we proposed an optimal strategy to update template.

2   Overview of Our Joint Template Matching Method

Usually, the pairwise template matching method aims to match a target image by the given template. Many scholars have proposed different algorithms to get accurate matched results. While when given a template and a set of target images, pairwise method can't match all target images accurately. In order to solve this problem, the idea of joint matching methods are proposed. In this paper, we developed a pairwise template matching based on tree propagating to jointly register a set of target images, which aims to match all images more accurately.

In the following part, Fig. 1 intuitively illustrates overview of joint template matching based on tree propagating. Firstly we use the given template to match each target image by a pairwise method we proposed in our previous paper [6]. Next by comparing similarity between the original template and matched region in target images, we select the accurate results as the second level of tree which connect to root node directly. Then for rescuing bad results, the template need to be updated. The detail will be introduced in the following. Thus remaining target images are matched by new templates. Consequently, a tree is built across all target images by repeating matching badly matched images and updating templates. In the tree, root node is the original template, and each child node represent a target image, and similar images are connected as neighbors.

Simply speaking, our method includes two main stages: (1) tree structure growing; and (2) matching propagating. For tree growing, we match each target image with original template by pairwise template matching method, then based on defined evaluation, select good results to link to corresponding node of tree. For matching propagating, we get new templates based on the original template and current good matches, and then rescuing bad results by new templates. Finally, repeating two processing until all target images are added to tree structure.

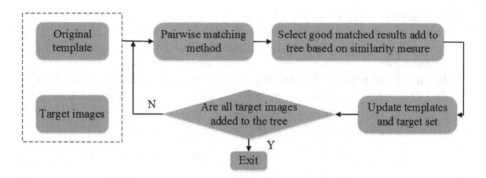

Fig. 1. Overview of the proposed joint template matching method based on tree structure.

3 Pairwise Template Matching

Given a template which usually is a region of interest in the source image, the goal of template matching is to find the template patch in a target image. We use the pairwise matching method based on BBS (Best-Buddies Pair) [6], this method is proposed to deal with target image with rotation, scale and translation transformations, it includes two main steps, firstly getting two feature point sets $P = \{p_i\}_{i=1}^{N}$ and $Q = \{q_j\}_{j=1}^{M}$ from template and target images base on BBS respectively, secondly computing transformation parameters including rotation, scale and translation by Scale Iterative Closest Point (SICP) algorithm [12]. Two main steps are repeated iteratively until the algorithm converges, see Eq. (1),

$$(s, \boldsymbol{R}, \boldsymbol{t}) = argmin \sum_{i=1}^{N} \| s\boldsymbol{R}\boldsymbol{p}_i + \boldsymbol{t} - \boldsymbol{q}_j \|^2 \tag{1}$$

The transformation parameters $\boldsymbol{R}$, s and $\boldsymbol{t}$ respectively represent rotation, scale and translation between two point sets, we can find the matched target region by these parameters. In this paper, for convenience we named this algorithm as SR-BBS.

4   Joint Template Matching Method

4.1   The Tree Structure Growing Based on Similarity Measure

In the tree, each target image is treated as a node, and well matched images will add to the tree. Therefore, how to select well matched results is the most important for the tree building. In this paper, to evaluate the matched results, we designed a similarity measure function by considering information of feature points and each pixels between given template and matched region of target image, it includes following baselines:

(1) The number of Best-Buddies Pair

In this paper we get two correspondence feature point sets P and Q from the template and target image based on Best-Buddies Similarity (BBS) method [5].

General definition of BBS: Suppose there are two point sets $P = \{\boldsymbol{p}_i\}_{i=1}^{N}$ and $Q = \{\boldsymbol{q}_j\}_{j=1}^{M}$, where $\boldsymbol{p}_i$, $\boldsymbol{q}_j \in \mathbb{R}^d$. If $\boldsymbol{p}_i$ is the nearest neighbor of $\boldsymbol{q}_j$ in set Q and vice versa, we call this pair of points $\{\boldsymbol{p}_j \in P, \boldsymbol{q}_j \in Q\}$ a BBP (Best-Buddies Pair). It defines as follow,

$$bb(\boldsymbol{p}_i, \boldsymbol{q}_j) = \begin{cases} 1 & NN(\boldsymbol{p}_i, Q) = \boldsymbol{q}_j \wedge NN(\boldsymbol{q}_j, P) = \boldsymbol{p}_i \\ 0 & otherwise \end{cases} \tag{2}$$

where $NN(\boldsymbol{p}_i, Q) = argmin\ d(\boldsymbol{p}_i, \boldsymbol{q})$, and $d(\boldsymbol{p}_i, \boldsymbol{q})$ is a distance measure between $\boldsymbol{p}_i$ and $\boldsymbol{q}$. The number of BBPs between P and Q is defined as Eq. (3),

$$N(P, Q) = \sum_{i=1}^{N} \sum_{j=1}^{M} bb(\boldsymbol{p}_i, \boldsymbol{q}_j, P, Q) \tag{3}$$

$N(P, Q)$ actually represents the total number of BBP between P and Q. When given a template, if we can get more BBPs between this template and the target image, it means the region we searched is more similar to the template. Otherwise, we may find the incorrect region or target image.

(2) Distribution consistent of feature points

By SR-BBS method, we get the transformation parameters $\boldsymbol{R}$, s and $\boldsymbol{t}$ respectively represent rotation, scale and translation between two point sets P and Q form template and target images respectively. Then we compute the distance of two point sets to evaluate the distribution consistent of feature points, defined as Eq. (4),

$$D_d = \|s\boldsymbol{R}P + \boldsymbol{t} - Q\|^2 \tag{4}$$

where D_d is the distance between P and Q, and two point sets have good distribution consistent when D_d is close to zero. If D_d is smaller, the more accurate matched result we get.

(3) Pixels difference

For measuring pixels difference, we compute square differences of RGB value defined as D_{rgb}, and Bhattacharyya distance based on histogram defined as D_{his} between template and the matched region in target images.

Thus based on these baselines, we designed similarity measure function as Eq. (5),

$$D = \frac{\|s\boldsymbol{R}P + \boldsymbol{t} - Q\|^2 \; D_{rgb}(T, I)}{\sum_{i=1}^{N} \sum_{j=1}^{M} bb(\boldsymbol{p}_i, \boldsymbol{q}_j, P, Q) \; D_{his}(T, I)} = \frac{D_d(P, Q) \; D_{rgb}(T, I)}{N(P, Q) \; D_{his}(T, I)} \tag{5}$$

where T represents template and I represents the matched region in target image, P is feature point set from template, Q is feature point set from target image, and D is the similarity measure between the template and the matched region in target image.

For demonstrating this combined similarity measure evaluation's improvement and effectiveness, we use four single baselines mentioned above as contrast, and the result show in the following. In Fig. 2, we use left template to match different frames of girl images by SR-BBS method, then we will select well matched images as the second level of tree. We compute the similarity between template and matched region in each target image (the region in red box), and if the value of similarity measure is under threshold we defined, we think this target image well matched and add it to tree.

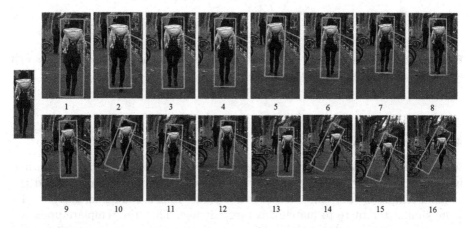

Fig. 2. Pairwise matching result by the same template (left). The sequence images 1–16 are different target images. Red boxes are matched results by pairwise method. (Color figure online)

In Fig. 3 we show five lines which represent D (combined method), D_d (Distribution consistent of feature points), $1/N$ (Derivative of Best-Buddies number), D_{rgb} (square differences of RGB value), $1/D_{his}$ (Derivative of histogram measure). For compared conveniently, we choose $1/N$ and $1/D_{his}$ to draw lines, then for all five criterions small value means well matched. In Fig. 3, although these five lines have similar distribution, for some frames $1/N$, D_{rgb} and $1/D_{his}$ can't give appropriate value. For example, it is obviously that the matched result of image 1 is better than image 10, 14, 15 and 16, but $1/D_{his}$ shows an abnormal value of image1. And another example, image 3, 4, 5, 6, 7 and 8 are well matched, D and D_d have small value for these images than other measure methods. Thus the measure methods D and D_d are better choice. Because D is the combination of these four single baselines and we also test other image sets and it proved that D is the most reliable evaluation, in this paper we use D as evaluation criterion to conduct tree building. Particularly, in the following parts of this paper similarity measure means D measure.

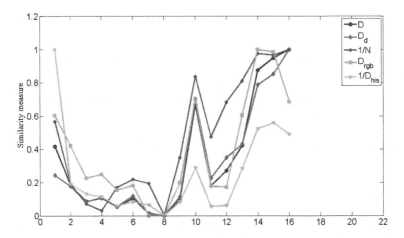

Fig. 3. The similarity measure by different baselines, the value close to zero means well matched result, the value close to one means badly matched result.

4.2 The Matching Propagating Based on Template Updating

Another key point of joint template matching based on tree is template updating, because when there are large deformations between the target images and the original template, they can't match well. Usually in this condition, there need a more similar template to match this target image. Thus the template updating method is necessary when building an accurate tree. For explaining our template updating method, we show the establishment process of a tree in Fig. 4.

In Fig. 4, at beginning, the original template (a) is used to match each target image (b)–(f) in target set. Then based on similarity measure, (b)–(c) are well matched and selected as second level of tree which link to root node (a) directly. Next, for rescuing (e)–(f), the template need be updated. The simple method is using the well matched region as new templates, So we obtain the red boxes in (b)–(d) as new templates directly for matching remaining target images, but we find that, when using this template updating method, more and more background region is matched with the level of tree increasing, see results of green boxes in 3rd level of tree in Fig. 4. The mismatch of background is resulted by the accumulated error which propagated by each level of tree, and the more level the tree has, the more background will be mismatched.

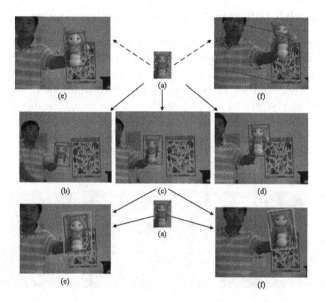

Fig. 4. The establishment process of three level tree based on template updating. (a) The original template. (b)–(f) are target images. (e) and (f) can't be well matched by original template, they are well matched by new template. In 3rd level of tree, green boxes are results with matched region as template, the red boxes are results with optimized template updating method. (Color figure online)

For improve the accuracy of matched result, we consider both matched region and original template as the new template. Firstly by using the matched region as new template, we can find an interest region in the target image roughly, which can regard as the initial values ($\boldsymbol{R}$, s and $\boldsymbol{t}$) of SR-BBS method. And then we use the original template to match the target images based on first step above, as the original template with less background pixels, we can remove mismatched region effectively, see red boxes of (e)–(f) in 3rd level of tree. And to distinguish, we named this strategy optimized template updating method.

5   Experiment Results

In this part, we test our method on datasets from an online object tracking benchmark [13]. The proposed tree structure is combined with SR-BBS algorithm and we name it joint template matching method (**JMM**). Specially, when tree structure only has two level, JMM is equal to using SR-BBS to match each target image. For convenience, **JMM-MR** represents JMM with template updated by matched region, **JMM-OT** represents JMM with optimized template updating. All the tests are finished by Matlab code.

5.1   Images with Different Scales and Viewing Angles

For demonstrating effectiveness of proposed method, we test 'doll' images with different scales and viewing angles. We get the original template by outlining a region of interest in a source image which is different from these target images, shown in Fig. 5(a). And all target images are from different frames of a video, some of them maybe are neighbors of the source image, some maybe are far away from the source image. This is to say, some target images are similar with the original template, and some are not. The experiment result of doll images is shown in Fig. 5.

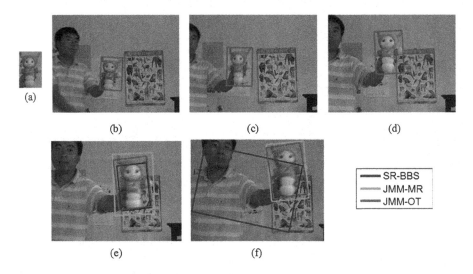

Fig. 5. The matching result of doll images. (a) The original template. (b)–(d) The target images in the 2nd level of tree, and at 2nd level, three methods get same results. (e)–(f) The target images in the 3rd level of tree. (Color figure online)

In Fig. 5(a) is the original template, (b)–(d) are well matched results by original template, and are selected as the 2nd level of tree. In (e)–(f), JMM-OT preforms well than other methods, this because target images (e)–(f) are badly matched by the original template, and rescued by new templates. Compared with JMM-MR, JMM-OT have more accurate matched results by optimized template updating method which can remove mismatched background region. And this also proves that JMM-OT can deal with accumulated and propagated error of tree structure.

We get the hierarchical tree based on similarity measure we defined, the well matched results are also selected by this evaluation. For explaining it further, we show the similarity measure results of 2nd level in Fig. 6(a), where all target images are matched by the original template. Figure 6(a) shows the similarity measures of target images (b)–(f) in sequence, and obviously target images (e) and (f) is over the threshold 0.1, which can't be selected as a child node to the 2nd level of tree. So target images (e) and (f) need to be rescued by new templates, and finally the hierarchical tree is built with each target images well matched.

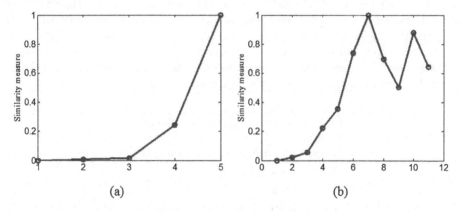

$$\text{(a)} \qquad\qquad\qquad\qquad \text{(b)}$$

Fig. 6. (a) The similarity measure of doll images in 2nd level of tree. (b) The similarity measure of David images in 2nd level of tree.

For comparing these three method quantitatively, we label these images and compute overlap rate to evaluate results. About computation of overlap rate, we explain it in the following, as shown in Fig. 7. In our experiment, we need compute a bidirectional rectangle overlap rate between two rotated and scale rectangles. Because when compute overlap rate between black rectangle and blue rectangle, you will find the black one will overlap the whole area of blue one almost, while blue one actually overlap part of black one. For avoiding this phenomenon, we compute bidirectional rectangle overlap rate between two rectangles. In Fig. 7 black rectangle is the ground truth we labeled, it obviously that the red rectangle has more overlapping area with black ones, so if overlap rate close to one value that means well matched result, and close to zero value means badly matched result.

The overlap rate of doll images is shown in Table 1, image (b)–(c) are well matched by given template and don't need to be rescued by new templates, that

Fig. 7. Explanation of overlap rate. Black rectangle is ground truth labeled. Blue box is result of SR-BBS. Green box is result of JMM-MR. Red box is result of JMM-OT. (Color figure online)

is to say, three methods get the same results at the second level. Strangely, it is seems image (e) can be well matched by given template, but in Fig. 5(e), it obviously that the matched result of JMM-OT (red box) is better than SR-BSS (blue box), and in Fig. 6(a) we also find that the similarity measure of (e) is over the threshold 0.1, which can't be selected as a child node to the 2nd level of tree. So target image (e) need to be rescued by new template. And this shows the effective of our similarity measure evaluation again. In Table 1, JMM-OT algorithm has higher overlap rate than others.

Table 1. The overlap rate of doll images.

Images	SR-BBS	JMM-MR	JMM-OT
(b)	**0.94**	\	\
(c)	**0.92**	\	\
(d)	**0.90**	\	\
(e)	0.80	0.78	**0.91**
(f)	0.24	0.88	**0.90**

[1] The bolder ones mean better.
[2] The \ means these target images are well matched by original template and selected as 2nd level of tree. For target images in 2nd level, three methods get same matched results.

5.2   Images with Different Non-rigid Transformations

In this part, we use 'David' images with different non-rigid transformations to verify the proposed method. The template and target images are obtained as Sect. 5.1.

In Fig. 8(a) is the original template. (b)–(d) is the 2nd level of tree which are well matched by original template. For other target images which badly matched with original template, we update template to rescue them. In image (e)–(l), the matched results of JMM-OT are better than other methods. We can find that, at 3rd and 4th level of tree, both three methods seem to have no obvious difference. However, as a result of tree's level increases, accumulated and propagated matching error also increases, three methods have obvious difference, and JMM-OT has good performance.

Fig. 8. The matching result of David images. (a) The original template. (b)–(d) The target images in the 2nd level of tree, and at 2nd level, three methods get same results. (e) The target image in the 3rd level of tree. (f) The target image in the 4th level of tree. (g) The target images in the 5th level of tree. (h)–(k) The target images in the 6th level. (l) The target image in the last level. (Color figure online)

In Table 2, we compute the overlap rate of matching results. Similarly, images (b)–(d) are well matched by given template and don't need to be rescued. At 2nd level, three methods have same result as second level. While some of remaining target images also have high overlap rate, but the matched result (blue box) in Fig. 8 don't satisfying, and their similarity measures are also over the given threshold shown in Fig. 6(b). So these images need to be rescued by new templates. In Table 2, we find that JMM-OT has higher overlap rate in most case.

Table 2. The overlap rate of David images.

Images	SR-BBS	JMM-MR	JMM-OT
(b)	**0.93**	\	\
(c)	**0.94**	\	\
(d)	**0.94**	\	\
(e)	0.88	0.89	**0.92**
(f)	0.83	0.80	**0.88**
(g)	0.68	0.75	**0.88**
(h)	0.59	0.68	**0.80**
(i)	0.75	0.67	**0.86**
(j)	**0.87**	0.65	0.86
(k)	0.66	0.69	**0.83**
(l)	0.74	0.67	**0.87**

[1] The bolder ones mean better.

6   Conclusion

This paper proposed a tree-based propagating joint template matching method to match a series of images with a given template. In our method, similarity measure evaluation is proposed for tree growing, and appropriate template updating method is proposed to match propagating and improve matching precision. In future work, we will try to use our method in video traching and other fields.

Acknowlegement. This work was in part supported by the National Natural Science Foundation of China (61627811, 61573274), Postdoctoral Science Foundation of China (2015M582661), Natural Science Foundation of Shaanxi, China (No. 2015JQ6257) and the Fundamental Research Funds for the Central Universities (xjj2017005, xjj2017067).

References

1. Huang, C., Wu, B., Nevatia, R.: Robust object tracking by hierarchical association of detection responses. In: Forsyth, D., Torr, P., Zisserman, A. (eds.) ECCV 2008. LNCS, vol. 5303, pp. 788–801. Springer, Heidelberg (2008). https://doi.org/10.1007/978-3-540-88688-4_58
2. Bae, J.S., Song, T.L.: Image tracking algorithm using template matching and PSNF-m. Int. J. Control Autom. Syst. **6**(3), 413–423 (2008)
3. Ahuja, K., Tuli, P.: Object recognition by template matching using correlations and phase angle method. Int. J. Adv. Res. Comput. Commun. Eng. **2**(3), 1368–1373 (2013)

4. Korman, S., Reichman, D., Tsur, G., Avidan, S.: Fast-match: fast affine template matching. In: Proceedings of the IEEE Conference on Computer Vision and Pattern Recognition, pp. 2331–2338 (2013)

5. Dekel, T., Oron, S., Rubinstein, M., Avidan, S., Freeman, W.T.: Best-buddies similarity for robust template matching. In: Proceedings of the IEEE Conference on Computer Vision and Pattern Recognition, pp. 2021–2029 (2015)

6. Luo, S., Yang, Y., Du, S., Kou, Q.: Robust image registration with rotation, scale and translation using Best-Buddies Pairs. In: International Joint Conference on Neural Networks (IJCNN), pp. 1728–1732. IEEE (2016)

7. Dai, Y., Li, H., He, M.: A simple prior-free method for non-rigid structure-from-motion factorization. Int. J. Comput. Vis. **107**(2), 101–122 (2014)

8. Nguyen, A., Ben Chen, M., Welnicka, K., Ye, Y., Guibas, L.: An optimization approach to improving collections of shape maps. Comput. Graph. Forum **30**(5), 1481–1491 (2011)

9. Zhou, T., Jae Lee, Y., Yu, S.X., Efros, A.A.: Flowweb: joint image set alignment by weaving consistent, pixel-wise correspondences. In: Proceedings of the IEEE Conference on Computer Vision and Pattern Recognition, pp. 1191–1200 (2015)

10. Zhou, X., Zhu, M., Daniilidis, K.: Multi-image matching via fast alternating minimization. In: Proceedings of the IEEE International Conference on Computer Vision, pp. 4032–4040 (2015)

11. Zhao, X., Chai, X., Shan, S.: Joint face alignment: rescue bad alignments with good ones by regularized re-fitting. In: Fitzgibbon, A., Lazebnik, S., Perona, P., Sato, Y., Schmid, C. (eds.) ECCV 2012. LNCS, vol. 7573, pp. 616–630. Springer, Heidelberg (2012). https://doi.org/10.1007/978-3-642-33709-3_44

12. Du, S., Zheng, N., Xiong, L., Ying, S., Xue, J.: Scaling iterative closest point algorithm for registration of m–D point sets. J. Vis. Commun. Image Represent. **21**(5), 442–452 (2010)

13. Wu, Y., Lim, J., Yang, M.: Online object tracking: a benchmark. In: Proceedings of the IEEE Conference on Computer Vision and Pattern Recognition, pp. 2411–2418 (2013)

Robust Visual Tracking Using Oriented Gradient Convolution Networks

Qi Xu[✉], Huabin Wang, Jian Zhou, and Liang Tao

Key laboratory of Intelligent Computing and Signal Processing of Ministry of Education, Anhui University, Hefei 230031, China
935821729@qq.com

Abstract. Convolutional networks have been successfully applied to visual tracking to extract some useful feature. However, deep networks are time-consuming to offline training and usually extract the feature from raw pixels. In this paper, we propose a two-layer convolutional network based on oriented gradient. The first layer is constructed by the convolution of the filter and an input image of oriented gradient, which is robust to the illumination variation and motion blur. Then, all of the feature maps of the simple layer are stacked to a complex feature map as the target representation. The complex feature map can encode the local structure feature which is robust to occlusion. The proposed approach is tested on nine challenging sequences in comparison with nine state-of-art trackers, and the result show that the proposed tracker achieves mean overlap rate of 0.75, which outperforms the secondary tracker by 26%.

1 Introduction

Visual Tracking is a major issue in computer vision with lots of applications, such as automatic supervision, human computer interaction and vehicle tracking. Given the true position of the first frame, the purpose of visual tracking is to find the position of the target in the successive frames. There are still some challenges because of occlusions, plane rotation, motion blur, illumination variation, etc.

The main influence factors of tracking include feature extractor, observation model, model update and motion model. Recently, lots of state-of-the-art algorithms are proposed, like LOT [1], Struck [2], SCM [3] and ASLA [4], and these methods focus on exploiting hand-crafted features. Although these algorithms perform well in the past, the hand-crafted features are not suitable for all generic objects.

Convolutional neural network (CNN) can exploit some useful features from raw data, so it has extensive applications, such as image classification [5] and object recognition [6] and segmentation [7]. However, traditional CNNs need too much time and samples to offline training for visual tracking. To solve these problems, Fan et al. [8] trained CNN by some auxiliary picture; Zhou et al. [9] proposed a tracking system using various CNNs; Li et al. [10,11] used multiple image cues to design a lightweight CNN; Zhang et al. [12] exploited a CNN with lightweight structure, which is fully feed-forward and achieves fast tracking even on a CPU.

© Springer Nature Singapore Pte Ltd. 2017
J. Yang et al. (Eds.): CCCV 2017, Part I, CCIS 771, pp. 320–330, 2017.
https://doi.org/10.1007/978-981-10-7299-4_26

All of above-mentioned CNN tracking methods pay particular attention to raw pixels for extracting features, and ignore some inherent features like oriented gradient. Gradient mainly exist in the brim of the target and some parts whose pixels change sharply. And oriented gradient is the most change rate of image gray value, which can reflect the salience of the target. The result of the convolution of a function and unit impulse is similar to copying the function in the position of the unit impulse, so only the value of the parts with similar oriented gradient is much larger when filters extracted from the oriented gradient convolve around the input image. Based on this observation, in this paper, we propose an oriented gradient convolutional network (COG) formulated within a particle filtering framework for object tracking. To build it, we first warp the image to a fixed size and get some patches by sliding windows. Each patch consists of oriented gradient about the object. Then, a overcomplete dictionary is studied from the image patches as a filter. Motivated by the work in [12], we propose tracking system using full feed-forward convolutional network, such that convolutional network is only simple two-layer. The first layer is simple layer, which is constructed by the convolution of the filter and image patches. And the complex layer consists of complex cell feature map that is a tensor, which stacks simple cell feature maps.

2   Particle Filter

Particle filter is an algorithm which could estimate the posterior distribution of state based on the Bayesian sequential importance sampling technique and Monte Carlo simulation. Because of estimating the state by some samples, it is applicable for non-linear system. And it has been widely used in visual tracking [3,4]. The process of particle filter includes two steps: predicting and updating. The predicting distribution $p(x_t|z_{1:t-1})$ which is given by all observations $z_{1:t-1} = \{z_1, z_2, \ldots, z_{t-1}\}$ up to time $t - 1$, can be computed as

$$p(x_t|z_{1:t-1}) = \int p(x_t|x_{t-1})p(x_{t-1}|z_{1:t-1})dx_{t-1} \qquad (1)$$

where x_t denotes the state variable which describe the affine motion parameters at time t. When the observation z_t is available at time t, the state vector can be updated

$$p(x_t|z_{1:t}) = \frac{p(z_t|x_t)p(x_t|z_{1:t-1})}{p(z_t|z_{1:t-1})} \qquad (2)$$

where $p(z_t|x_t)$ denotes the observation likelihood, and $p(z_k|z_{1:k-1})$ is a normalized constant which could be computed as

$$p(z_t|z_{1:t-1}) = \int p(z_t|x_t)p(x_t|z_{1:t-1})dx_t \qquad (3)$$

Considering that the integrals above-mentioned are difficult to be computed, the Monte Carlo simulation and importance sampling are utilized to approximate the $p(z_t|z_{1:t})$ by N samples which denote $\{x_t^i\}_{i=1}^N$. The weights of samples denoted w_t^i are updated as

$$w_t^i = w_{t-1}^i \frac{p(z_t|x_t^i)p(x_t^i|x_{t-1}^i)}{q(x_t|x_{1:t-1}, z_{1:t})} \tag{4}$$

where $q(x_t|x_{1:t-1}, z_{1:t})$ denotes importance distribution. Because of sample degenerating, the samples are resampled to duplicate the particles with high weight and abandon the low ones.

In this paper, the importance distribution is set to $q(x_t|x_{1:t-1}, z_{1:t}) = p(x_t|x_{t-1})$ and the weights become the observation likelihood $p(z_t|x_t)$. We also warp the input image to a fixed size with $m \times m$ pixels. Then, the state variable could be set to $x_t = (t_x, t_y, \alpha_1, \alpha_2, \alpha_3, \alpha_4)$, where $\{t_x, t_y\}$ are translation parameters and $\{\alpha_1, \alpha_2, \alpha_3, \alpha_4\}$ are the deformation parameters.

3   Oriented Gradient Convolutional Network

3.1   Preprocessing

After the input image is warped to a fixed size, the gradient of each pixel can be computed as

$$\boldsymbol{G}_x(x,y) = \boldsymbol{I}^{m \times m} \bigotimes \boldsymbol{I}_o, \boldsymbol{G}_y(x,y) = \boldsymbol{I}^{m \times m} \bigotimes \boldsymbol{I}_o^T \tag{5}$$

where $\boldsymbol{I}^{m \times m}$ is the wrapped image, which is preprocessed by gamma correction that is correspond to local brightness and contrast moralization, and $\boldsymbol{I}_0 = [-1, 0, 1]$. Then the oriented gradient can be formulated as

$$\alpha(x,y) = tan^{-1} \frac{\boldsymbol{G}_y(x,y)}{\boldsymbol{G}_x(x,y)} \tag{6}$$

Therefore, a given fixed input image is constructed by the oriented gradient. We sample $(m - n + 1) \times (m - n + 1)$ patches centered at each pixel location inside the $\hat{\boldsymbol{I}}^{m \times m}$ by sliding a window with $n \times n$. Finally, in each patch, we divide the oriented gradient into R sections, and get a feature vector of R dimensions denoted as $\boldsymbol{Y}_{K \times K}^R, K = (m-n+1)$ through counting the number of each section.

3.2   Design Filter

Inspired by the spare representation [13]: the most likely representation of the image can be achieved by over-complete sparse representation, which also can obtain higher resolution information than traditional non adaptive methods. So we use the overcomplete dictionary to represent the target region adaptively. Then, the filter can be formulated as

$$\boldsymbol{F}_{K \times W} = \boldsymbol{Y}_{K \times K}^R \boldsymbol{X}_{K \times W} \tag{7}$$

where $\boldsymbol{X}$ is a coefficient matrix, which should be as sparse as possible. The Eq. (7) can be rewritten as

$$\min_{\boldsymbol{F},\boldsymbol{X}} \{\| \boldsymbol{X}_i \|_o\} \qquad s.t. \| \boldsymbol{Y} - \boldsymbol{F}\boldsymbol{X} \|_F^2 \leq \varepsilon \tag{8}$$

To solve the above Equation, supposing the dictionary $\boldsymbol{F}$ is fixed, a method called Orthogonal Matching Pursuit is used to achieve sparse coding

$$\min_{\boldsymbol{x}_i} \{\| \boldsymbol{Y}_i - \boldsymbol{F}\boldsymbol{X}_i \|_2^2\} \qquad s.t. \| \boldsymbol{Y} - \boldsymbol{F}\boldsymbol{X} \|_F^2 \leq \varepsilon \tag{9}$$

where $\boldsymbol{X}$ is a sparse parameter. Then, $\boldsymbol{F}\boldsymbol{X}$ can be computed as

$$\boldsymbol{F}\boldsymbol{X} = \sum_{i=1}^{W} \boldsymbol{f}_i \boldsymbol{x}_i^T \tag{10}$$

where $\boldsymbol{f}_i$ is the i-th column in the $\boldsymbol{F}$ and $\boldsymbol{x}_i^T$ is the i-th row in the $\boldsymbol{X}$. Then we need update the dictionary through several iterations. When the k-th column is updated, the others are supposed to be fixed. Then, the Eq. (9) can be rewritten as

$$\begin{aligned} \min_{\boldsymbol{F}} \| \boldsymbol{Y} - \boldsymbol{F}\boldsymbol{X} \|_F^2 &= \min_{\boldsymbol{f}_k} \| \boldsymbol{Y} - \Sigma_{i=1}^{W} \boldsymbol{f}_i \boldsymbol{x}_i^T \|_F^2 \\ &= \min_{\boldsymbol{f}_k} \| (\boldsymbol{Y} - \Sigma_{i \neq k} \boldsymbol{f}_i \boldsymbol{x}_i^T) - \boldsymbol{f}_k \boldsymbol{x}_k^T \|_F^2 \\ &= \min_{\boldsymbol{f}_k} \| \boldsymbol{E}_k - \boldsymbol{f}_k \boldsymbol{x}_k^T \|_F^2 \end{aligned} \tag{11}$$

where $\boldsymbol{E}_k$ is a fixed error matrix. Then, two vectors should be found in $\boldsymbol{E}_k$ to update $\boldsymbol{f}_i$ and $\boldsymbol{x}_i^T$. In [14], author used the Singular Value Decomposition (SVD) method to solve this problem,

$$\boldsymbol{E}_k = \boldsymbol{U}\Lambda\boldsymbol{V}^T \tag{12}$$

where $\boldsymbol{U}$ and $\boldsymbol{V}$ are orthogonal basis, and Λ is a diagonal matrix. The vector of $\boldsymbol{U}$ which corresponds to the maximum value of Λ is accepted as $\boldsymbol{f}_i$. However, the sparsity of $\boldsymbol{X}$ may be corrupted if we choose the corresponding vector in $\boldsymbol{V}$ to update $\boldsymbol{x}_i^T$. The solution is to construct a new matrix $\Omega_{K \times L}$ which consists of the nonzero element in $\boldsymbol{x}_i^T$

$$\hat{\boldsymbol{E}}_k = \boldsymbol{E}_K \Omega_{K \times L}, \hat{\boldsymbol{x}}_k = \boldsymbol{x}_k^T \Omega_{K \times L} \tag{13}$$

Then, a sparse $\boldsymbol{x}_i^T$ can be obtained when using SVD to $\hat{\boldsymbol{E}}_k$.

3.3 Object Tracking

After getting the filter $\boldsymbol{F}$, the region of each particle is preprocessed, denoted as $\boldsymbol{P}$. Then, the simple layer can be defined as

$$S = \boldsymbol{F} \bigotimes \boldsymbol{P} \tag{14}$$

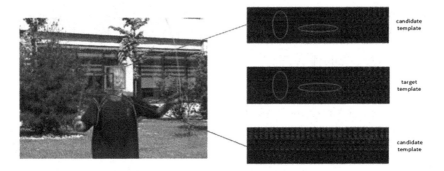

Fig. 1. The simple cell feature map can preserve the local structure of the target.

The simple layer S enhances the outline and structure of object region, which can be remained while illumination variation and motion blur So the position of the object can be discriminated according to the geometric layout information of the simple layer. To further strengthen the effect of representation, a 3D tensor is used to construct a complex feature map in [12],

$$C \in R^{K \times K \times N} \tag{15}$$

which can enhance the strength of local structural. So the complex cell feature map C is robust to occlusion and deemed as the candidate template. In this paper, the state transition distribution was model by a Gaussian distribution, and observation results are independent which could be computed as

$$p(z_t|x_t) = e^{-\|c_t - c_t^i\|_2^1} \tag{16}$$

where c_t and c_t^i are the target template and i-th candidate template respectively at frame t.

3.4   Model Update

Filter and target template c should be updated incrementally to accommodate appearance changes. The target template can be update as

$$c_t = (1 - \rho)c_{t-1} + \rho \hat{c}_{t-1} \tag{17}$$

where ρ is fixed parameter which is set to 0.05 in our experiments, c_t represent the target template at frame t, and $\hat{c}_{t-1}$ is the sparse representation of the target template at frame $t-1$, which can be easily achieved by a soft shrinkage function

$$\hat{c} = \text{sign}(vec(C)) \max(0, \text{abs}(vec(C)) - \text{median}(vec(C))) \tag{18}$$

The filter is updated as

$$F_{t+1} = \lambda F_o + \lambda F_t + (1 - 2\lambda)F_{t-1} \tag{19}$$

where F_o is the filter in first frame, λ is a fixed parameter which is set to 0.15 in our experiments, F_t is the filter at frame t, and F_{t-1} is the filter at frame $t-1$. To improve the speed, we update the filter every other 5 frames.

4   Experience and Discussion

4.1   Experimental Setup

Our tracker was implemented in matlab2016 on a PC with Intel Core i7-7700 CPU@ 3.5 GHz, and run at approximately 1 frame per second. To compare the robustness of our algorithm, we use the Visual Tracking Benchmark dataset [15] and its outstanding code library, including IVT [16], L1APG [17], ASLA [4], LOT [1], MTT [18] and SCM [3]. Furthermore, we also add the CNT algorithm, which could be downloaded on authors homepage. Nine challenging sequences from visual tracking benchmark are used for comparison, including blurcar1, boy, david3, fleetface, jumping, jogging-1, singer2, suv and trellis.

Several parameters are used in COG tracker. The variance of particle filter is set to $\lambda_t = 4 + (v_{t-1} + v_{t-2} + v_{t-3})/3$, where v_{t-1}, v_{t-2} and v_{t-3} are the targets movement speed at frame t, $t-1$ and $t-2$ respectively. Deformation parameters $\{\alpha_1, \alpha_2, \alpha_3, \alpha_4\}$ are set to $\{0.02, 0.005, 0.005, 0\}$. The size of the warped image is set to m = 32, and the size of sliding window is set to n = 6.

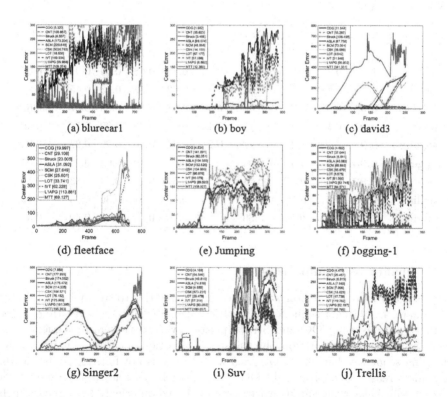

Fig. 2. Tracking results in terms of center error (in pixel). The COG is compared with 9 state-of-the-art algorithms on 9 challenging image sequences

Table 1. Average overlap rate of 10 algorithm. The best and second results are shown in bold-font and italic-font.

Sequences	COG	CNT	Struck	ASLA	SCM	CSK	LOT	IVT	L1APG	MTT
blurcar1	**0.81**	0.11	*0.79*	0.06	0.05	0.01	0.67	0.04	0.49	0.36
boy	**0.81**	0.40	*0.78*	0.38	0.39	0.68	0.55	0.27	0.74	0.51
david3	**0.73**	0.56	0.30	0.43	0.40	0.50	*0.67*	0.48	0.38	0.10
fleetface	**0.64**	0.58	*0.61*	0.57	0.60	0.59	0.57	0.46	0.46	0.51
jumping	**0.64**	0.08	*0.62*	0.23	0.12	0.05	0.58	0.12	0.15	0.10
jogging-1	**0.78**	*0.19*	0.17	0.18	0.18	0.18	0.09	0.18	0.17	0.18
singer2	**0.74**	0.04	0.04	0.04	0.17	0.04	*0.26*	0.04	0.04	0.04
suv	**0.83**	*0.80*	0.52	0.49	0.75	0.52	0.65	0.41	0.48	0.45
trellis	*0.78*	0.42	0.61	**0.80**	0.67	0.48	0.31	0.26	0.20	0.22
average	**0.75**	0.35	*0.49*	0.35	0.37	0.34	0.48	0.25	0.35	0.28

Table 2. Successful rate based on Fig. 3. The best and second results are shown in bold-font and italic-font.

Sequences	COG	CNT	Struck	ASLA	SCM	CSK	LOT	IVT	L1APG	MTT
blurcar1	**1**	0.05	*0.98*	0.05	0.03	0.01	0.78	0.04	0.61	0.49
boy	**1**	0.45	*1*	0.45	0.45	0.88	0.68	0.34	0.91	0.50
david3	**1**	0.68	0.33	0.51	0.48	0.62	*0.93*	0.63	0.45	0.11
fleetface	**0.91**	0.69	0.69	0.63	*0.77*	0.68	0.60	0.53	0.62	0.62
jumping	**0.91**	0.11	*0.80*	0.16	0.12	0.05	0.78	0.09	0.12	0.09
jogging-1	**0.98**	*0.22*	0.22	0.22	0.21	0.22	0.06	0.22	0.22	0.22
singer2	**0.95**	0.04	0.04	0.04	*0.17*	0.04	0.16	0.04	0.04	0.04
suv	**0.98**	0.70	0.57	0.58	*0.97*	0.57	0.78	0.44	0.53	0.53
trellis	**0.98**	0.24	0.78	*0.86*	0.85	0.59	0.31	0.30	0.15	0.19
average	**0.97**	0.35	*0.60*	0.39	0.45	0.40	0.56	0.29	0.40	0.31

4.2 Qualitative Comparisons

Motion Blur: Target region would become blurred due to the motion of target or camera. In blurcar1 sequences (Fig. 4(a)), camera moves so swift that target is blurred. Only the COG and Struck perform well in the entire sequence. As show in Fig. 4(e), The motion blur caused by drastic appearance change in the jumping sequence that only COG, Stuck and LOT is not failed after frame 16. The main reason that proposed COG successfully keeps track of the target with motion blur is that the simple feature map is based on oriented gradient, which is robust to motion blur (see. Fig. 1).

Plane Rotation: The target may be deformable while rotating in or out of the image plane. It also led to motion blur if its speed of rotation is too high. For the

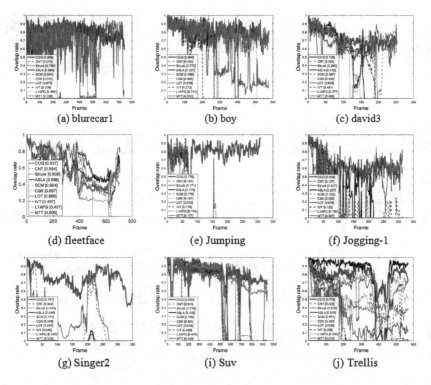

Fig. 3. Tracking results in terms of center error (in pixel). The COG is compared with 9 state-of-the-art algorithms on 9 challenging image sequences

boy sequence (Fig. 4(b)), the boy rotates both in and out of plane. Only COG and Stuck perform well in all sequence. In the fleetface sequence (Fig. 4(d)), there are both in-plane and out-of-plane rotations after frame 375. Except for COG and Stuck algorithms, the others all drift. The COG deals with plane rotation well because its online updated scheme is suit for appearance variation.

Illumination Variation: The reasons for illumination variation include different col-our and varying levels of light. In singer2 sequence (Fig. 4(g)), the colour of back-ground illumination change so drastic that only the COG algorithm can track the target stably in the entire sequence. In trellis sequences (Fig. 4(i)), only the COG and ASLA algorithms perform well while the strength of illumination is varied. The proposed COG algorithm well handles the situation with illumination variations as the feature is extracted from the normalized local filters with gamma correction and local brightness normalization.

Occlusion: The target may be partially or fully occluded by other object. In david3 sequence (Fig. 4(c)), only the COG and LOT algorithm perform well when the target is partially occluded by the tree (e.g., frame 84, frame 190). In jogging-1 sequences (Fig. 4(f)), the target is completely occluded by a lamppost (e.g., frame 74, frame 78). Only the COG is able to detect the person when the

(a) blurcar1

(b) boy

(c) david3

(d) fleetface

(e) jumping

(f) jogging-1

(g) singer2

(h) suv

(i) trellis

Fig. 4. Tracking results in terms of center error (in pixel). The COG is compared with 9 state-of-the-art algorithms on 9 challenging image sequences (Color figure online)

target appears again in the screen (e.g., frame 80). Our COG tracker achieves stable performance for occlusion due to employing the local features in complex feature maps.

4.3 Quantitative Comparisons

For quantitative comparison, the center position error plot and overlap rate plot are employed. The center position error is defined as

$$e = \sqrt{(x_t - x_s)^2 + (y_t - y_s)^2} \tag{20}$$

where x_t with y_t is the center position of the tracking result, and x_s with y_s is the ones of the true state. The overlap rate is defined as

$$S = \frac{Area(R_t \bigcap R_s)}{Area(R_t \bigcup R_s)} \tag{21}$$

where R_t represent the area of tracking bounding box, and R_s represent the area of true state. The tracking result is successful if the value of overlap rate is greater than 0.5 in current frame. And the successful rate is defined as the number of success frames divided by the number of all frames.

Figure 2 illustrates the center position error of each frame in terms of the results of 10 algorithms, while Fig. 3 shows the corresponding overlap rate plot. The results indicate that the center position errors of COG keep the value at relatively low level and the corresponding overlap rate keep higher level in all sequence. Table 1 shows the average overlap rate of each algorithm. Overall, the proposed COG achieves mean overlap rate of 0.75, which outperforms LOT by 26%. Meanwhile, in the success rate which shows in Table 2, its score is 0.97, outperforms significantly Struck by 37%.

5 Conclusion

In this paper, we have proposed and demonstrated an effective and robust tracking method based on the oriented gradient convolutional networks. The tracker is constructed by a simple two-layer convolutional network and formulated within a particle filtering framework. First, we warp the input images to a fixed size and extract a set of normalized patches of oriented gradient by sliding window, which can handle illumination variation. To obtain a spare representation of the target, we exploit the overcomplete dictionary from the normalized patches as filters. Then, the first layer is constructed from the convolution of filters and input images, which is robust to motion blur. Finally, we stack all the simple feature maps to construct the complex feature map as the representation of the target which can overcome drifts and occlusions. Furthermore, both the latest observations and the original filter are considered in our update scheme, which can deal with appearance change and the drift problem well. Compared with other nine state-of-the-art algorithms, the experimental results of quantitative and qualitative comparisons on 9 challenging image sequences show the robustness of proposed tracking algorithm.

References

1. Oron, S., Bar-Hillel, A., Levi, D., et al.: Locally orderless tracking. In: Computer Vision and Pattern Recognition, pp. 1940–1947. IEEE (2012)
2. Hare, S., Golodetz, S., Saffari, A., et al.: Struck: structured output tracking with kernels. IEEE Trans. Pattern Anal. Mach. Intell. **38**(10), 2096 (2016)
3. Yang, M.H., Lu, H., Zhong, W.: Robust object tracking via sparsity-based collaborative model. In: Computer Vision and Pattern Recognition, pp. 1838–1845. IEEE (2012)
4. Lu, H., Jia, X., Yang, M.H.: Visual tracking via adaptive structural local sparse appearance model. In: IEEE Conference on Computer Vision and Pattern Recognition. IEEE Computer Society, pp. 1822–1829 (2012)
5. Wu, J., Yu, Y., Huang, C., et al.: Deep multiple instance learning for image classification and auto-annotation. In: Computer Vision and Pattern Recognition, pp. 3460–3469. IEEE (2015)
6. Wang, A., Lu, J., Cai, J., et al.: Large-margin multi-modal deep learning for RGB-D object recognition. IEEE Trans. Multimed. **17**(11), 1887–1898 (2015)
7. Pan, H., Wang, B., Jiang, H.: Deep learning for object saliency detection and image segmentation. IEEE Trans. Neural Netw. Learn. Syst. **27**(6), 1135–1149 (2015)
8. Fan, J., Xu, W., Wu, Y., et al.: Human tracking using convolutional neural networks. IEEE Trans. Neural Netw. **21**(10), 1610–1623 (2010)
9. Zhou, X., Xie, L., Zhang, P., et al.: An ensemble of deep neural networks for object tracking. In: IEEE International Conference on Image Processing, pp. 843–847. IEEE (2015)
10. Li, H., Li, Y., Porikli, F.: Robust online visual tracking with a single convolutional neural network. In: Cremers, D., Reid, I., Saito, H., Yang, M.-H. (eds.) ACCV 2014 Part V. LNCS, vol. 9007, pp. 194–209. Springer, Cham (2015). https://doi.org/10.1007/978-3-319-16814-2_13
11. Li, H., Li, Y., Porikli, F.: Deep track: learning discriminative feature representations online for robust visual tracking. IEEE Trans. Image Process. **25**(4), 1834–1848 (2016). A Publication of the IEEE Signal Processing Society
12. Zhang, K., Liu, Q., Wu, Y., et al.: Robust tracking via convolutional networks without learning. IEEE Trans. Image Process. **25**(4), 1779–1792 (2015). A Publication of the IEEE Signal Processing Society
13. Lu, J., Zhang, Q., Xu, Z., et al.: Image super-resolution by dictionary concatenation and sparse representation with approximate L_0 norm minimization. Comput. Electr. Eng. **38**(5), 1336–1345 (2012)
14. Elad, M., Aharon, M.: Image denoising via sparse and redundant representations over learned dictionaries. IEEE Trans. Image Process. **15**(12), 3736–3745 (2006)
15. Wu, Y., Lim, J., Yang, M.H.: Online object tracking: a benchmark. In: Computer Vision and Pattern Recognition, pp. 2411–2418. IEEE (2013)
16. Ross, D.A., Lim, J., Lin, R.S., et al.: Incremental learning for robust visual tracking. Int. J. Comput. Vis. **77**(1), 125–141 (2008)
17. Ji, H.: Real time robust L1 tracker using accelerated proximal gradient approach. In: IEEE Conference on Computer Vision and Pattern Recognition, pp. 1830–1837. IEEE Computer Society (2012)
18. Zhang, T., Ghanem, B., Liu, S., et al.: Robust visual tracking via multi-task sparse learning. In: Computer Vision and Pattern Recognition, pp. 2042–2049. IEEE (2012)

Novel Hybrid Method for Human Posture Recognition Based on Kinect V2

Bo Li[1], Baoxing Bai[2(✉)], Cheng Han[1], Han Long[1], and Lin Zhao[1]

[1] School of Computer Science and Technology,
Changchun University of Science and Technology, Changchun, Jilin Province, China
bolivar.bo.li@hotmail.com, hchwork@sina.com, 1543112670@qq.com,
liteccc@gmail.com
[2] College of Optical and Electronical Information,
Changchun University of Science and Technology, Changchun, Jilin Province, China
19546615@qq.com

Abstract. In recent years, human posture recognition based on Kinect gradually has been paid more attention. However, the current researches and methods have drawbacks, such as low recognition accuracy and less recognizable postures. This paper proposed a novel method. The method utilized image processing technique, BP neural network technique, skeleton data and depth data captured by Kinect v2 to recognize postures. We distinguished four types of postures (sitting cross-legged, kneeling or sitting, standing, and other postures) by using the natural ratios of human body parts, and judged the kneeling and sitting postures by calculating the 3D spatial relation of the feature points. Finally, we applied BP neural network to recognize the lying and bending postures. The experimental results indicated that the robustness and timeliness of our method was strong, the recognition accuracy was 98.98%.

Keywords: Human posture recognition · Kinect v2 · Depth data
Skeleton data · Image processing · BP neural network

1 Introduction

Human posture recognition plays a very important role in the field of human-computer interaction, as the low-cost somatosensory device Kinect was provided by Microsoft, many researches and experiments based on Kinect have been conducted by several institutions or scholars, aim to find a better way for human posture recognition or human-computer Interaction. These researches achieved good effect.

Some human posture recognition methods used depth image. The authors of [13,14] predicted 3D positions of body joints from a single depth image. The paper [15] combined several image processing techniques by using the depth images from a Kinect to recognize the five distinct human postures. Literature [12] estimated human postures of the upper limbs and upper limbs motion by using depth images.

© Springer Nature Singapore Pte Ltd. 2017
J. Yang et al. (Eds.): CCCV 2017, Part I, CCIS 771, pp. 331–342, 2017.
https://doi.org/10.1007/978-981-10-7299-4_27

Some human posture recognition methods used skeleton data. In the paper [5], four types of features (lying, sitting, standing, bending) were extracted from skeleton data, used to recognize the body posture. The paper [10] proposed a human posture detection and recognition algorithm based on geometric features and 3D skeleton data. The postures are classified by Support Vector Machines classifier.

Some human posture recognition methods used depth image and skeleton data. Literature [11] utilized human anatomy marks and human skeleton model, estimated human posture by using depth image and the Geodesic Distance. The authors of [16] obtained 3D body features from 3D coordinate information using depth image, identified 3D human posture human with the models of skeleton joint.

Kinect v2 improved a lot contrast to Kinect v1, but the skeleton data from Kinect v2 is not entirely accurate, for instance, the skeleton data may be incorrect when some skeleton joints overlapped the others, as shown in Fig. 1(a) and (b). In order to solve this problem, some researchers proposed several methods. The authors of [17] proposed a repair method, but this method only solve the problem of the occlusion of single joint; The literatures [7,12] recognized posture only by using depth data, and the paper [15] did not use Kinect SDK for human posture recognition. However, the methods have limitations, such as less recognizable postures and low recognition accuracy. A novel hybrid method was proposed in this paper. Depth images and skeleton data were obtained with the method by using Kinect SDK 2.0. Combined with other methods and anthropometry, six postures (standing, sitting cross-legged, sitting, kneeling, bending and lying) could be recognized.

This paper is organized as follows: Sect. 2 introduces our method. Section 3 describes and discusses the experimental content, steps and results. Finally, the conclusion of our method is described in Sect. 4.

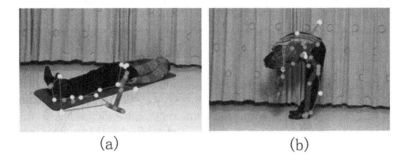

(a)   (b)

Fig. 1. Skeleton data captured by Kinect v2 and its SDK

2   Our Method

2.1   Generation of the Body Height and the Body Center of Gravity

Kinect v2 can recognize and directly output the depth data of humanoid area, without the need to distinguish the foreground and background [2]. We extracted the humanoid area by using Kinect SDK 2.0, and calculated the body center of gravity (x_c, y_c), x_c is the average value of all of the x-axis values of the null-black pixels, and y_c is the average value of all of the y-axis values of the null-black pixels.

We obtained human contour by Candy Operator, as shown in Fig. 2, the red rectangle in (a) is the center of gravity of the human body area, the red rectangle in (b) is the center of gravity of the human contour, and the height of the body contour is the difference value (on the y-axis) between the highest point and the lowest point of the body contour. When the center of gravity is not in the human body area, draw two lines (the horizontal line and the vertical line) at the origin point (the center of gravity), as shown in Fig. 2(c), there is a line has two points of intersection with the human contour, the new center of gravity is the blue point (the center position of the two points of intersection).

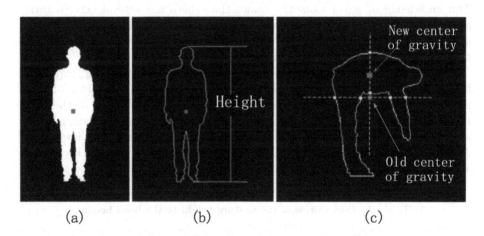

Fig. 2. The center of gravity in human body area and the human contour (Color figure online)

2.2   Head Localization and Calculation of the Head Height

Our method firstly distinguished two types of the postures (1. The head is on the top of the body; 2. The head is not on the top of the body) by the head position. Because the skeleton data may be incorrect, so we positioned the head by using the depth image and the skeleton data simultaneously. Note that the resolutions of the depth image and the skeleton image were both set to 512 * 424, because our method required accurate pixel values. The general judging process according to (1) whether the head point is in the human body area, and (2) whether the slope of connection-line (connect the head coordinate and the neck coordinate) is less than −1 (or greater than 1).

When the head is not on the top of the body, the posture recognition is processed by BP neural network (see Sect. 2.5), otherwise, the postures will be recognized according to the natural ratios of human body parts.

In order to obtain the head information and the neck information, we measured the related data of 100 (50 males, 50 females) healthy person (18 to 60 years old) when they facing the Kinect at different directions. We calculated the equation by statistical analysis, as follows:

$$H_h = 1.8726 \times L_{hn} . \tag{1}$$

where H_h is the head height (data is from the depth image), L_{hn} is the height difference of the head point and the neck point (data is from the skeleton image).

2.3 Estimation of Human Posture When the Head Is on the Top of the Body

The knowledge of anthropometry shows that there are several certain ratios among all of the human body parts. Chinese National Standard GB10000-88 [1] and the related research literature [18] proved that the ratios among the human body parts changed only a little in the past 20 years.

We calculated the ratio of the posture height to the head height (data is from [1]) in every percentile, as shown in Table 1, the average values of these ratios are shown in the P column. It proved that the differences of the three ratios are relatively large. However, it is difficult to judge the kneeling height and the sitting height. The way of distinguishing the postures as follows: (1) If $P \geqslant 6.5$, the posture is standing; (2) If $6.0 > P \geqslant 5.0$, the posture is kneeling or sitting; (3) If $4.5 > P \geqslant 3.3$, the posture is sitting cross-legged.

Table 1. Table ratios of the posture height to the head height

		1%	5%	10%	50%	90%	95%	99%	P
H/H	Male	7.7538	7.6844	7.6381	7.5247	7.4009	7.3651	7.2851	7.5516
	Female	7.5078	7.4200	7.4039	7.2685	7.1930	7.1509	7.1004	7.3222
S/H	Male	6.0704	6.0243	5.9952	5.9238	5.8481	5.8340	5.7912	5.9538
	Female	5.8031	5.7550	5.7586	5.7269	5.6579	5.6293	5.5941	5.7265
Sc/H	Male	4.2010	4.1650	4.1429	4.0717	3.9959	3.9751	3.9317	4.0845
	Female	4.0881	4.0450	4.0345	3.9583	3.9079	3.8836	3.8494	3.9817

P.S.: **H/H** = the height/the head height; **S/H** = the sitting height/the head height; **Sc/H** = the sitting cross-legged height/head height.

2.4 Distinguishing the Postures of Kneeling or Sitting by Specific Points

We obtained two coordinates, $P_h(x_h, y_h, z_h)$ is corresponding to the central point D_h of the head, and $P_e(x_e, y_e, z_e)$ is corresponding to the bottom point D_e of the human area. Our method distinguished the kneeling or sitting posture by the 3D spatial relation of the two coordinates, and the head radius was set as:

$$R_h = \frac{1}{2} \times \frac{1}{3} \times (H_h + W_h + L_h) . \tag{2}$$

where H_h is the head height, W_h is the head width, L_h is the head length.

In the 2D image, D_h is the center of the head in the human body area; In the 3D space, P_h is the center point of the head corresponds to D_h (the coordinate of P_h is (x_h, y_h, z_h)), and P_{hs} is the point of the head surface corresponds to D_h (the coordinate of P_{hs} is (x_{hs}, y_{hs}, z_{hs})), P_h can be calculated by the following equation:

$$P_h(x_h, y_h, z_h) = (x_{hs}, y_{hs}, z_{hs} - R_h) . \tag{3}$$

As shown in Fig. 3(a), P_e is the point corresponds to the bottom point of the human body area in the 2D image. $P_{h'}$ is the projected point of P_h in X-Z plane, and the coordinate of $P_{h'}$ is (x_h, y_e, z_h). The distance between P_e and $P_{h'}$ is calculated by Eq. (4).

$$L_{eh'} = \sqrt{(x_e - x_h)^2 + (z_e - z_h)^2} . \tag{4}$$

P_e can be confirmed as the foot when $L_{eh'}$ exceeds a threshold (one head height H_h). However, at this time, the posture maybe the kneeling or sitting posture, as shown in Fig. 3(b) and (c). We proposed two ways to judge the two postures as follows:

(I) Draw a circle, the circle center is at the right above P_e, the radius is H_h, the distance between P_e and the circle center is also H_h. If there are no human body parts in the circle, then the posture is kneeling, otherwise, the posture is sitting.

(II) Draw a circle, the circle center is at the right above $P_{h'}$, the radius is R_h, and the distance between $P_{h'}$ and the circle center is H_h. If there are no human body parts in the circle, then the posture is sitting, otherwise, the posture is kneeling.

2.5 Posture Recognition by Using BP Neural Network

BP Neural Network
BP neural network is based on error backpropagation (BP) algorithm for its training, and this neural network is a kind of feed-forward multilayer network. BP neural network comprises one input layer, one or more hidden layer and one output layer. Numerous hidden layers require considerable training time.

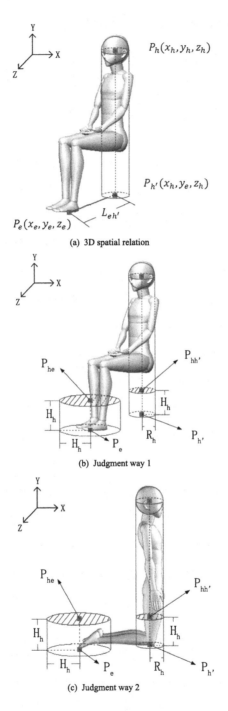

(a) 3D spatial relation

(b) Judgment way 1

(c) Judgment way 2

Fig. 3. Judge the sitting posture or the kneeling posture

According to Kolmogorov theory, a three-layer BP network (input, hidden and output layers) can approximate any continuous function [3,6] in the condition of reasonable structure and appropriate weights. Therefore, we selected the three-layer BP neural network.

Extraction of Feature Vectors

We chose and transformed (from the Cartesian coordinate system to the Polar coordinate system) the feature vectors by using the methods of Literature [15]. As shown in Fig. 4(a). In order to distinguish the feature vectors better, we need to specify the order of the feature vectors. Therefore, we set a disk with four regions, as shown in Fig. 4(b).

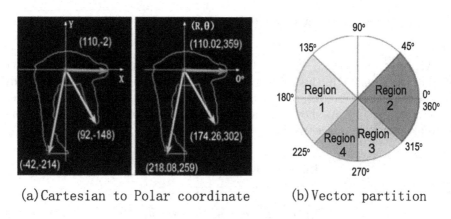

(a) Cartesian to Polar coordinate   (b) Vector partition

Fig. 4. Extraction of feature vectors

Because the feature vectors of the lying or bending posture must be in Region 1, 2, 3 and 4, so the order of the feature vectors can be arranged as follows:

(1) In Region 1, if there is no feature vectors, we set $L_1 = 0$ and $T_1 = 0$; If a vector angle is closer to $180°$ than the others, then this feature vector is called V_1, and we set $L_1 = R_1$ and $T_1 = \theta_1$.

(2) In Region 2, if there is no feature vectors, we set $L_2 = 0$ and $T_2 = 0$; If a vector angle is closer to $360°(0°)$ than the others, then this feature vector is called V_2, and we set $L_2 = R_2$ and $T_2 = \theta_2$.

(3) In Region 3, if there is no feature vectors, we set $L_3 = 0$ and $T_3 = 0$; If a vector angle is closer to $315°$ than the others, then this feature vector is called V_3, and we set $L_3 = R_3$ and $T_3 = \theta_3$.

(4) In Region 4, if there is no feature vectors, we set $L_4 = 0$ and $T_4 = 0$; If a vector angle is closer to $225°$ than the others, then this feature vector is called V_4, and we set $L_4 = R_4$ and $T_4 = \theta_4$. The specifying process of the order of the feature vectors is finished.

Figure 5 shows the ordered feature vectors.

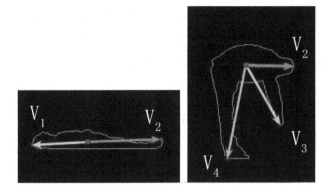

Fig. 5. The ordered feature vectors

We set $\overline{R_i} = L_i/R_{max}$, $\overline{\theta_i} = T_i/360$, where R_{max} denotes the maximum value of all of the R value. In this case, no matter what the height of the human body area is, the feature vectors are ratio values. Therefore, our method can be applied to different persons. As shown in Table 2, the eight feature vectors are the input neurons for BP neural network.

Table 2. The final feature vectors

N_1	N_2	N_3	N_4	N_5	N_6	N_7	N_8
$\overline{R_1}$	$\overline{\theta_1}$	$\overline{R_2}$	$\overline{\theta_2}$	$\overline{R_3}$	$\overline{\theta_3}$	$\overline{R_4}$	$\overline{\theta_4}$

The Training of the BP Neural Network

We conducted many experiments based on several empirical formulas in [3,4,6], and confirmed that the number of hidden layer neurons is 5. As shown in Fig. 6 (a), the eight types of the postural samples are the training data.

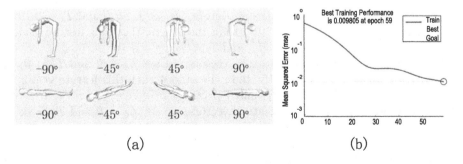

(a)   (b)

Fig. 6. The training of BP neural network

In this paper, the learning step size is set to 0.01, the accuracy of training target is set to 0.001; as shown in Fig. 6(b), the best training performance is 0.009805 at epoch 59.

3 Experimental Results and Discussion

In this paper, the hardware environment as follows: Lenovo E51 laptop computer with $Intel \textregistered Core^{TM}$ i7-6500 - CPU@2.60 Hz and 16G memory, Kinect 2.0 for Windows; System environment: the 64 - bit Windows 10 enterprise edition; Software environment: Visual Studio.net 2012, Kinect SDK 2.0.

We recruited four healthy adults (2 male and 2 female) who have different heights and weights. Among them, one male and one female are not only the participants of all the experiments, but also the models for the training samples of BP neural network. The distance of the ground and the Kinect v2 is 80 cm, and the distances of the Kinect v2 and the participants are 2.0 m–4.2 m.

Two participants (1 male and 1 female) posed two postures (bending and lying) in four directions (±45° and ±90°), and each participant posed 5 times in every direction. The eight types (as shown in Fig. 6(a)) of input samples are obtained and input to the BP neural network for training, and then the training result set is saved.

We classified all the participants into three sets, Set1 is the set of the models who posed postures for training samples, and Set2 is the set of the other participants who did not pose postures for training samples, Set3 is the set of all the participants. We recognized six postures (standing, sitting cross-legged, kneeling, sitting, lying, and bending) of each participant in two environments (indoor and outdoor). Each participant oriented to the Kinect v2 in different direction, and then posed six postures several times. The directions and real scenes of postures are shown as Fig. 7. The experimental statistic values are shown in Table 3, where Average denotes the average accuracy of six postures in one experiment; Accuracy i ($i = 1, 2, 3$) denotes the accuracy of one posture in the $i - th$ set.

Table 3. Accuracy of human posture recognition

	Standing	Sitting CL	Kneeling	Sitting	Lying	Bending	Average
Set1	120/120	100/100	78/80	109/112	96/96	80/80	583/588
Accuracy1	100%	100%	97.50%	97.32%	100%	100%	99.15%
Set2	120/120	100/100	77/80	110/112	94/96	80/80	581/588
Accuracy2	100%	100%	96.25%	98.21%	97.91%	100%	98.81%
Set3	240/240	200/200	155/160	219/224	190/192	160/160	1164/1176
Accuracy3	100%	100%	96.88%	97.77%	98.96%	100%	98.98%

Our program comprises of five parts, Part 1-Part 4 of our program corresponds to the contents of the Sects. 2.1–2.4; Part 5 corresponds to the process of BP neutral network (the recognition of bending and lying). Table 4 shows the running time of the program. The average runtime is very short (within 30 ms), and meets the demands of real-time.

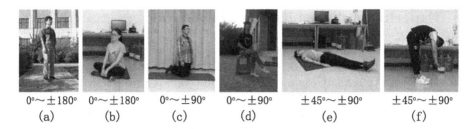

0°~±180°	0°~±180°	0°~±90°	0°~±90°	±45°~±90°	±45°~±90°
(a)	(b)	(c)	(d)	(e)	(f)

Fig. 7. Experimental postures. (a) Standing. (b) Sitting cross-legged. (c) Kneeling. (d) Sitting. (e) Lying. (f) Bending.

Table 4. Average running time of each stage of each posture (unit: ms)

	Standing	Sitting CL	Kneeling	Sitting	Lying	Bending
Part 1	14.642	13.213	13.146	13.545	13.893	15.714
Part 2	0.048	0.049	0.055	0.055	0.021	0.043
Part 3	0.574	0.576	0.582	0.55	0	0
Part 4	0	0	0.054	0.053	0	0
Part 5	0	0	0	0	4.614	5.523
Total	15.264	13.838	13.837	14.203	18.528	21.28

Literature [15] utilized the ratio of the upper body width to the lower body width to distinguish kneeling posture, and the accuracy is 99.69%, but the person subject must pose kneeling posture in some fixed direction (±45°, ±90°). The authors of [9] recognized human postures by using four methods, the best accuracy is 100% when they used BP neutral network, but the method only can recognize three postures (standing, lying and sitting), because the method of [9] could not extract effective feature vectors of the other postures. The methods of [10] and [8] were respectively proposed to recognize 10 postures. The three postures (standing, sitting and bending) in our paper are similar to the three postures (standing, sitting on chair, and bending forward) in [10] and the three postures (standing up, sitting, and picking up) in [8]. Literature [10] utilized a polynomial kernel SVM classifier and the angular features of the postures, the accuracy is 95.87%. The method of [8] was based on HMM (hidden Markov model) and 3D joints from depth images.

We compared these methods with our method, as shown in Table 5.

Table 5. Accuracies of several methods and our method

	Standing	Sitting CL	Kneeling	Sitting	Lying	Bending	Average
This paper	100%	100%	96.88%	97.77%	98.96%	100%	98.98%
Paper [15]	99.38%	-	99.69%	97.25%	99.53%	99.22%	
Paper [9]	100%	-	-	100%	100%	-	
Paper [10]	95.87%	-	-	95.87%	-	95.87%	
Paper [8]	93.50%	-	-	91.50%	-	97.50%	
Total	15.264	13.838	13.837	14.203	18.528	21.28	

However, our method has limitations. For example, when the person orients to the Kinect and poses bending and lying postures in some specific directions, such as 0°, maybe lead to misjudgment about the two postures, because the feature vector is unobvious and cannot be extracted precisely.

4   Conclusions

This paper proposed a novel method which innovatively used the related knowledge of anthropometry and the BP neural network, this method can recognize six postures (standing, sitting cross-legged, kneeling, sitting, lying, and bending) when the person orients to the Kinect in different direction. The average recognition accuracy is 98.98%. We preliminarily judged four postures which the head is on the top of the body by using the ratio of the body parts; the feature vectors of the other two postures (lying and bending) were extracted, and the two postures were recognized by BP neural network. However, our method can only recognized the lying and bending postures in a small number of directions (from ±45° to ±90°), because a few of body joints are obscured when we used only one Kinect, lead to extract feature vectors difficultly. In the further work, we will use two or more Kinect, thus when the person orients to any Kinect in any the directions, more human postures can be recognized.

Acknowledgments. This project is supported by the National Natural Science Foundation of China (Grant No. 61602058).

References

1. China_national_institute_of_standardization: Human Dimensions of Chinese Adults (GB10000-88). Standards press of China (1988)
2. Hua, L., Chao, Z., Wei, Q., Cheng, H., Hong-Yu, Z., Ting-Ting, L.: An automatic matting algorithms of human figure based on kinect depth image. J. Changchun Univ. Sci. Technol. **39**, 81–84 (2016)
3. Hua-Yu, S., Zhao-Xia, W., Cheng-Yao, G., Juan, Q., Fu-Bin, Y., Wei, X.: Determining the number of BP neural network hidden layer units. J. Tianjin Univ. Technol. **24**, 13–15 (2008)

4. Jadid, M.N., Fairbairn, D.R.: The application of neural network techniques to structural analysis by implementing an adaptive finite-element mesh generation. AI EDAM : Artif. Intell. Eng. Design Anal. Manuf. **8**(3), 177–191 (1994)

5. Le, T.L., Nguyen, M.Q., Nguyen, T.T.M.: Human posture recognition using human skeleton provided by kinect. In: International Conference on Computing, Management and Telecommunications, pp. 340–345 (2013)

6. Li-Qun, H.: Artificial Neural Network Tutorial. Beijing University of Posts and Telecommunications Press (2006)

7. Lu, X., Chia-Chih, C., Aggarwal, J.K.: Human detection using depth information by kinect. In: Computer Vision and Pattern Recognition Workshops, pp. 15–22 (2011)

8. Lu, X., Chia-Chih, C., Aggarwal, J.K.: View invariant human action recognition using histograms of 3D joints. In: Computer Vision and Pattern Recognition Workshops, pp. 20–27 (2012)

9. Patsadu, O., Nukoolkit, C., Watanapa, B.: Human gesture recognition using kinect camera. In: International Joint Conference on Computer Science and Software Engineering, pp. 28–32 (2012)

10. Pisharady, P.K., Saerbeck, M.: Kinect based body posture detection and recognition system. In: Proceedings of SPIE - The International Society for Optical Engineering, vol. 8768, no. 1, pp. 87687F–87687F-5 (2013)

11. Schwarz, L.A., Mkhitaryan, A., Mateus, D., Navab, N.: Human skeleton tracking from depth data using geodesic distances and optical flow. Image Vis. Comput. **30**(3), 217–226 (2012)

12. Shih-Chung, H., Jun-Yang, H., Wei-Chia, K., Chung-Lin, H.: Human body motion parameters capturing using kinect. Mach. Vis. App. **26**(7–8), 919–932 (2015)

13. Shotton, J., Fitzgibbon, A., Cook, M., Sharp, T., Finocchio, M., Moore, R., Kipman, A., Blake, A.: Real-time human pose recognition in parts from single depth images. In: IEEE Conference on Computer Vision and Pattern Recognition, pp. 1297–1304 (2011)

14. Shotton, J., Kipman, A., Finocchio, M., Finocchio, M., Blake, A., Moore, R., Moore, R.: Real-time human pose recognition in parts from single depth images. Commun. ACM **56**(1), 116–124 (2013)

15. Wen-June, W., Jun-Wei, C., Shih-Fu, H., Rong Jyue, W.: Human posture recognition based on images captured by the kinect sensor. Int. J. Adv. Robot. Syst. **13**(2), 1 (2016)

16. Xiao, Z., Meng-Yin, F., Yi, Y., Ning-Yi, L.: 3D human postures recognition using kinect. In: International Conference on Intelligent Human-Machine Systems and Cybernetics, pp. 344–347 (2012)

17. Xin-Di, L., Yun-Long, W., Yan, H., Guo-Qiang, Z.: Research on the algorithm of human single joint point repair based on kinect (a chinese article). Tech. Autom. App. (a Chinese journal) **35**(4), 96–98 (2016)

18. Yan, Y., Jie, Y., Chun-Sheng, M., Jin-Huan, Z.: Analysis on difference in anthropomety dimensions between East and West human bodies. Stand. Sci. **7**, 10–14 (2015)

Visibility Estimation Using a Single Image

Qin Li and Bin Xie[✉]

School of Information Science and Engineering, Central South University,
Changsha 410083, China
xiebin@csu.edu.cn

Abstract. In this paper, we propose a novel method for visibility estimation using only a single image as input. An assumption is proposed:
the extinction coefficient of light is approximately a constant in clear
atmosphere. Using this assumption with the theory of atmospheric radiation, the extinction coefficient in clear atmosphere can be estimated.
Based on the dark channel prior, ratio of two extinction coefficients
in current and clear atmosphere is calculated. By multiplying the clear
extinction coefficient and the ratio, we can estimate the extinction coefficient of the input image and then obtain the visibility value. Compared
with other methods that require the explicit extraction of the scene, our
method needs no constraint and performs well in various types of scenes,
which might open a new trend of visibility measurement in meteorological research. Moreover, the actual distance information can also be
estimated as a by-product of this method.

Keywords: Clear atmosphere · Extinction coefficient
Dark channel prior · Visibility

1 Introduction

Visibility is an indicator of atmosphere transparency, which is closely related
to our daily lives in many ways, like road safety applications, meteorological
observation, airport safety and monitoring pollution in urban areas.

Little work has been done on visibility estimation based on outdoor cameras. There are two classes of methods. The first one is to estimate the distance
to the most distant object on the road surface having a contrast above 5%.
Pomerleau [19] measures the contrast attenuation of road markings at various
distances in front of a moving vehicle for visibility estimation, which requires
detecting road markings. Hautière et al. [7,9] estimate two kinds of visibility
distances: meteorological visibility and mobilized visibility. The meteorological
visibility is the greatest distance at which a black object of a suitable dimension
can be seen in the sky on the horizon. The method [11] consists of dynamically
implementing Koschmieder's law [16] and estimate the visibility based on local
contrasts above 5%. The mobilized visibility represents the distance to the fur-
thest pixel that still displays a contrast greater than 5%, which is very close to

© Springer Nature Singapore Pte Ltd. 2017
J. Yang et al. (Eds.): CCCV 2017, Part I, CCIS 771, pp. 343–355, 2017.
https://doi.org/10.1007/978-981-10-7299-4_28

the meteorological visibility. They combine computations of a depth map and local contrasts above 5% to estimate the mobilized visibility [10]. This class of methods depends on an accurate geometric calibration of the camera. A second one correlates the contrast in the scene with the visual range estimated by reference additional sensors [8]. The geometric calibration is not necessary. The contrast can be simply calculated by a Sobel filter or a high-pass filter. Once the contrast is calculated, a linear regression is then performed to estimate the mapping function [14,21]. Such methods can be seen as data-driven approaches. Hautière et al. propose a probabilistic model-driven approach, which performs a simplified depth distribution model of the scene and applies a non-linear fitting to the data [3]. This class of methods needs a learning phase to estimate the mapping function, which requires collecting a large number of ground truth data. All of existing methods based on outdoor cameras rely on human assistance, like geometric calibration of the camera, road markings extraction or ground truth data collection for building mapping functions. Moreover, strict constraints for scene environments are essential for almost all methods, some of which even require the presence of just road and sky in the scene.

This paper proposes a novel method to estimate visibility just using a single image as input. According to the haze imaging model [12], a hi-quality haze-free image can be recovered, which makes it possible to obtain the ratio of two extinction coefficients respectively in current atmosphere and clear atmosphere by a single image. With the aim to calculate current atmosphere extinction coefficient, our work is inspired by the success of proposing an important prior-the clear atmosphere extinction coefficient is approximately a constant. It can be calculated by the theory of atmospheric radiation. By multiplying the clear atmosphere extinction coefficient and the ratio, the extinction coefficient of the input image is obtained and then the visibility is finally computed. Our method succeeds in estimating visibility in various application scenes by just a single image. The feasible results show high accuracy and robustness of this method, which might open a new trend of visibility measurement in meteorological research.

The remainder of this paper is organized as follows. Section 2 introduces our approach in details. Section 3 provides experimental results in various scenes and comparisons with existing approaches. A discussion and summary concludes this paper in Sect. 4.

2   Our Approach

The visibility calculation formula is proposed as follows [11]:

$$V_{met} = \frac{1}{\beta} ln \frac{1}{\varepsilon} \tag{1}$$

where ε is contract threshold and equal to 5%. β is extinction coefficient which indicates the scene radiance attenuated by the turbid atmosphere from a scene point to the sensor. The procedure of our method is depicted in Fig. 1. Based on

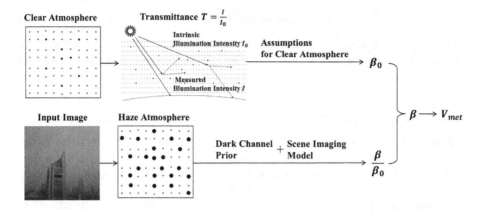

Fig. 1. Procedure of our method

a key observation of the clear atmosphere, we find that distributions of particles in clear atmosphere remain almost unchanged. Based on this, a conclusion is proposed that the clear atmosphere extinction coefficient β_0 is usually a constant. By introducing transmittance T, β_0 can be calculated combined with assumptions on clear atmosphere, which is developed by the theory of atmospheric radiation. Meanwhile, we advance the scene imaging model with dark channel prior to estimate β/β_0 of the input image. By multiplying β_0 and β/β_0 , the extinction coefficient β is obtained and then the visibility is successfully calculated by Eq. (1). Moreover, the scene distance information of the input image also can be calculated as a by-product.

2.1 β_0 Calculation

The atmosphere is mainly composed of air molecules and aerosol particles. Air particles consist of nitrogen, oxygen, noble gas, carbon dioxide, water and other impurities. Aerosol is a suspension system of liquid or solid particles in the air. Only large-scale aerosol particles (e.g., dust, carbon, smoke particles) are sensitive to terrain, wind direction and other geographical factors [5], which affect visibility, climate and our health a lot. In the clear atmosphere, air molecules accounts for about 99% of the total atmosphere composition and the remaining 1% are small-scale aerosols. Experiments prove that the distribution of air particles remains unchanged within the height of 100 km [18]. Therefore, a reasonable assumption is proposed in this paper: the extinction coefficient in clear atmosphere is approximately a constant. We introduce transmittance T to calculate β_0. Transmittance T indicates the degree of solar radiation weakened in vertical direction [15]. The transmittance in clear atmosphere is defined by

$$T_0(\lambda, L) = e^{-\int_0^L \beta_0(\lambda, L)dl} \tag{2}$$

where $\lambda = 0.55\,\mu\text{m}$, L is the path length of solar radiation. The clear extinction coefficient β_0 consists of two parts: clear air molecular extinction coefficient β_{R_0} and clear aerosol extinction coefficient β_{α_0}

$$\beta_0 = \beta_{R_0} + \beta_{\alpha_0} \tag{3}$$

where β_{R_0} is a known constant [18]: 1.159×10^{-5}(at $\lambda = 0.55\,\mu$m, and the unit is m^{-1}).With the aim to calculate β_{α_0} , the clear aerosol transmittance is described by

$$T_{\alpha_0}(\lambda, L) = e^{\int_0^L \beta_{\alpha_0}(\lambda, L)dl} \tag{4}$$

We introduce two assumptions on clear atmosphere to simplify the Eq. (4)

– Assume that the effective height of the atmosphere H is 10^4m. 10^4m is the average height of troposphere including 80% of atmospheric particles and almost all the water vapor particles. In the height range higher than m, density coefficient of particle quantity is irrelevant to meteorological optical range [4].
– Assume the clear aerosol extinction coefficient is invariable. In the clear atmosphere, aerosol particle distributions are unaffected by airflow activities like convection and turbulence [1].

Based on above two assumptions, Eq. (4) is simplified as

$$T_{\alpha_0} = e^{-\beta_{\alpha_0} L} \tag{5}$$

By Eq. (5)

$$\beta_{\alpha_0} = -\frac{\log(T_{\alpha_0})}{L} \tag{6}$$

where $L = m_0 H$, $H = 10^4 m$. m_0 is relative optical mass [20], which is the ratio of path length of solar radiation and the effective height of the atmosphere(see Fig. 2). m_0 is calculated by

$$m_0 = \frac{1}{\cos\theta_s + 0.15 * (93.885 - \theta_s \frac{180}{\pi})^{-1.253}} \tag{7}$$

The solar zenith angle θ_s indicates the position relationship of the sun and the earth, which is determined by the observation information: date, time and location [15].

$$t = t_s + 0.170 \sin(\frac{4\pi(J - 80)}{373}) - 0.129 \sin(\frac{2\pi(J - 8)}{355}) + \frac{12(SM - L)}{\pi} \tag{8}$$

$$\delta = 0.4093 \sin(\frac{2\pi(J - 81)}{368}) \tag{9}$$

$$\theta_s = \frac{\pi}{2} - \arcsin(\sin l \sin\delta - \cos l \cos\delta \cos\frac{\pi t}{12}) \tag{10}$$

where t_s is observation time in decimal hours, J is Julian date (the day of the year as an integer in the range 1–365), L is site longitude in radians, SM is standard meridian for the time zone in radians.

With the aim to calculate β_{α_0}, we introduce the clear aerosol transmittance formula proposed by Ångström [2]

$$T_{\alpha_0} = e^{-\beta_A m_0 \lambda^{-\alpha_w}} \tag{11}$$

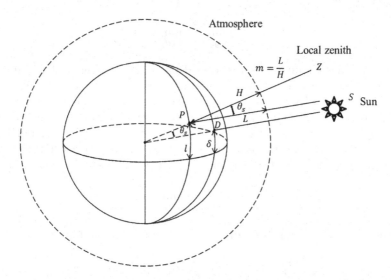

Fig. 2. Diagram of relative optical mass. P is the site point, D is the under-solar point, l is site latitude in radians, δ is solar declination, θ_s is solar zenith angle.

where β_A is Ångströms turbidity coefficient indicating the level of atmospheric turbidity. The value range of β_A is $0.01 \leq \beta_A \leq 0.2$ [2]. α_w is wavelength exponent reflecting the distribution characteristics of aerosol particle spectrum. The value range of α_w is $0.1 \leq \beta_A \leq 4$ [6]. When the proportion of small particles increases, the value of also grows. In pure atmosphere, air molecules accounts for about 99% of atmosphere constituents, which has almost reached a limit condition of atmosphere. So we take the minimum value of β_A and the maximum value of α_w to Eq. (11). Combining with Eq. (11), Eq. (6) becomes

$$\beta_{\alpha_0} = \frac{\beta_A m_0 \lambda^{-\alpha_w}}{L} \tag{12}$$

where $L = m_0 H$

$$\beta_{\alpha_0} = \frac{\beta_A \lambda^{-\alpha_w}}{H} \tag{13}$$

where m_0 is significantly omitted, which proves that β_{α_0} is independent on observation information (date, time and location) and always a constant. Based on Eq. (3), we finally calculate $\beta_0 : 2.2518 \times 10^{-5} m^{-1}$.

2.2 β/β_0 Calculation

In this paper, we combine the dark channel prior [12] and the improved scene imaging model to deduce β/β_0. The dark channel prior is a kind of statistics of haze-free outdoor images–in clear atmosphere, most local patches contain some pixels which have very low intensities in at least one color channel. For an image I, the dark channel is defined by

$$I_{\Omega(x)}^{dark} = \min_{c\in\{r,g,b\}} \left(\min_{y\in\Omega(x)} I^c(y) \right) \tag{14}$$

where c is a color channel. $\Omega(x)$ is a local patch centered at x. The patch size is 15×15 in this paper. Based on the statistics of 1000 outdoor images, we find that I^{dark} increase with the aggravation of the atmosphere turbidity. The patch with the minimum I^{dark} at other levels of atmosphere turbidity corresponds to the minimum one in clear atmosphere, which satisfies the dark channel prior most. Therefore, we take the min operation on Eq. (10) to select the minimal dark channel patch $\Omega_{min}(x)$ in the input image I (see Fig. 3[1st column]):

$$I_{\hat{\Omega}_{min}(x)}^{dark} = \min_{x\in I}(\min_{c\in\{r,g,b\}}(\min_{y\in\hat{\Omega}(x)}(I^{dark}(y)))) \tag{15}$$

We refine $\Omega_{min}(x)$ using guided filter [13](see Fig. 3[2nd column])

$$I_{\hat{\Omega}_{min}(x)}^{dark} = guidedfilter(I_{\hat{\Omega}_{min}(x)}^{dark}) \tag{16}$$

Figure 3 [2nd column] shows that patches in $\Omega_{min}(x)$ may be not at the same distance, and $I_{\hat{\Omega}_{min}(x)}^{dark}$ are probably unequal in a patch due to the block effects, which will influence the following calculation for β/β_0. Therefore, we need to improve $\Omega_{min}(x)$. The specific process is as follows:

- Select the minimal value V_{min} of $I_{\hat{\Omega}_{min}(x)}^{dark}$.
- Retrieve all connected regions of $\Omega_{min}(x)$ and find the $\Omega_{min}(x_{min})$ containing V_{min}. Note that the number of regions in $\Omega_{min}(x_{min})$ is always more than one.
- Traverse the $\Omega_{min}(x_{min})$ and calculate $\left| I_{\hat{\Omega}_{min}(x_{min})}^{dark}(y) - V_{min} \right|$ where $y = 1, 2$ $\cdots n$, $n \in \Omega_{min}(x_{min})$. If $\left| I_{\hat{\Omega}_{min}(x_{min})}^{dark}(y) - V_{min} \right| \geq T$, y is supposed as a false pixel and omitted. We fix the T to 0.2 in this paper.
- Count the number of each remaining regions in $\Omega_{min}(x_{min})$ and eliminate the noise region whose quantity of pixels is less than N_{num}. We fix the N_{num} to 5% of the total number of $\Omega_{min}(x)$ in this paper.

The resulting regions are almost at the same distance and satisfy the dark channel prior most, which are supposed to be the *ROI* ((see Fig. 3[3rd column])). Next we will calculate β/β_0 of this *ROI*.

For an outdoor image, a model [17] is widely used to describe the imaging process (see Eq. (13)). Note that the atmosphere is homogenous

$$I = \rho I_\infty e^{-\beta d} + I_\infty(1 - e^{-\beta d}) \tag{17}$$

where I is scene image, ρ is scene albedo. I_∞ is air light, which is invariant with β and determined only by the illumination intensity. ρI_∞ is scene radiance, I_∞ is medium transmission. Based on Eq. (13), the scene imaging in clear atmosphere is easily derived by

$$J = \rho I_\infty e^{-\beta_0 d} + I_\infty(1 - e^{-\beta_0 d}) \tag{18}$$

Fig. 3. Procedure of β/β_0 Calculation. [1st column] result of $\Omega_{\min}(x)$ and $I_{\Omega_{\min}(x)}^{d\hat{a}rk}$. [2nd column] result of $I_{\Omega_{\min}(x)}^{dark}$. [3rd column] result of ROI. [4th column] the above result is $e^{-(\beta-\beta_0)d}(y)$ and the below result is $e^{-\beta_0 d}(y), y \in ROI$. [5th column] result of β/β_0.

where J is intrinsic scene image. Combine Eq. (13) with Eq. (14), the advanced scene imaging model is obtained:

$$J = Je^{-(\beta-\beta_0)d} + I_\infty(1 - e^{-(\beta-\beta_0)d}) \tag{19}$$

By Eq. (15), we find that when β decreases to β_0, scene image I is recovered to J. Compared with previous model-driven methods for haze removal, this restoring process retains the fundamental cue for human to perceive depth. So we use Eq. (15) to recover J and calculate β/β_0. According to the dark channel prior, the dark channel of scene radiance tends to be 0.

$$\min_{c\in\{r,g,b\}}\left(\min_{y\in ROI}(\rho I_\infty^c(y))\right) = 0 \tag{20}$$

Note that the clear atmosphere is composed of air molecules and a few small-scale aerosol particles. So the dark channel of the intrinsic scene intensity is close to 0.

$$\min_{c\in\{r,g,b\}}\left(\min_{y\in ROI}(J_\infty^c(y))\right) = \mu \tag{21}$$

where μ should be very close to 0, so we fix it to 0.001 in this paper. $e^{-\beta_0 d}$ and $e^{-(\beta-\beta_0 d)}$ are obtained by

$$e^{-\beta_0 d}(y) = guidedfilter(1 - \min_{c\in\{r,g,b\}}(\min_{y\in ROI}(\frac{J^c(y)}{I_\infty^c}))) \tag{22}$$

$$e^{(\beta-\beta_0)d}(y) = guidedfilter((1 - \min_{c\in\{r,g,b\}}(\min_{y\in ROI}(\frac{I^c(y)}{I_\infty^c})))/(1 - \frac{\mu}{I_\infty^c})) \tag{23}$$

where I_∞ can be estimated by [12] and J is recovered by:

$$J(y) = \frac{I(y) - I_\infty}{\max(e^{-(\beta-\beta_0)d}(y), t_0)} + I_\infty, y \in ROI \tag{24}$$

where t_0 is a lower bound 0.1 in case $e^{-(\beta-\beta_0)d}$ decrease down to 0. Figure 3[4th column] shows the results of $e^{(\beta-\beta_0)d}$ and $e^{-\beta_0 d}$. Then $\beta/\beta_0(y)$ is computed by

$$\frac{\beta}{\beta_0}(y) = \frac{\ln(e^{-(\beta-\beta_0)d})}{\ln(e^{-\beta_0 d})}(y) + 1 \tag{25}$$

We take the average value of $\beta/\beta_0(y)$ in ROI (see Fig. 3[5th column]). Based on Eq. (1), visibility V_{met} is finally obtained by

$$V_{met} = \frac{1}{\beta_0 \times \frac{\beta}{\beta_0}} \ln \frac{1}{\varepsilon} \tag{26}$$

3 Experimental Results

In our experiment, there are three kinds of scenes: city scene, road scene and nature scene. We use the visibility V_{met}^{ob} provided by Yahoo!Weather as a criterion of the experimental results. We define the error rate E by

$$E = \frac{\left|V_{met} - V_{met}^{ob}\right|}{V_{met}^{ob}} \tag{27}$$

If error rate is less than 20%, the result is acceptable.

Figure 4 shows the experimental images in city scenes at different levels of atmosphere turbidity. As can be seen, the variance of atmosphere turbidity in city scene is mainly caused by air pollution. Visibility is a good indicator of city's air quality. Figures 5 and 6 show the experimental images in road scenes and nature scenes at different levels of atmosphere turbidity. Tables 1, 2 and 3 shows the visibility results of Figs. 4, 5 and 6. Error rates in different kinds of scenes are less than 20%, which demonstrates the high accuracy of our method. Moreover, our method preforms well without human assistance. [More results can be found at http://airl.csu.edu.cn/airquality.html]

Table 1. Visibility results of Fig. 4

Instance	F4-1	F4-2	F4-3	F4-4
V_{met}[m]	19645.0	8782.0	2895.5	985.0
V_{met}^{ob}[m]	24000	8000	3000	1000.0
E[%]	18.1	9.8	3.5	1.5

Table 2. Visibility results of Fig. 5

Instance	F5-1	F5-2	F5-3	F5-4
V_{met}[m]	20972.6	7436.2	319.6	31.5
V_{met}^{ob}[m]	24000.0	8000.0	300.0	30.0
E[%]	12.6	7.1	6.5	5.0

Table 3. Visibility results of Fig. 6

Instance	F6-1	F6-2	F6-3	F6-4
$V_{met}[\text{m}]$	19267.7	16205.3	9529.5	5840.8
$V_{met}^{ob}[\text{m}]$	24000.0	15000.0	9000.0	6000.0
$E[\%]$	19.72	8.04	5.88	2.65

Fig. 4. Experimental images in different city scenes

Fig. 5. Experimental images in the road scene

Fig. 6. Experimental images in nature scenes

This paper also adopts Hautière N's two popular methods [9,11] as comparison objects (see Fig. 7). Method [9] based on in-vehicle cameras describes two specific onboard techniques to estimate the meteorological visibility and the mobilized visibility. We only consider the meteorological visibility in this paper. Method [11] consists of dynamically implementing Koschmieder's law and estimate the visibility based on local contrasts above 5%. V_{met}^{H} represents the visibility estimated by Hautière N's methods. In Table 4, error rates of our

Fig. 7. Comparisons with Hautière N's work [9, 11]

method are all less than 20%. Our results are comparable with Hautière N's in the road scene. Furthermore, our method is also applicable to other types of scenes, which is better than Hautière N's methods. Our approach even works for distance estimation. Having obtained β , we can calculate the distance from the scene point to the sensor according to Eq. (28)

$$d = \frac{\log(e^{-(\beta - \beta_0)d})}{\beta_0 - \beta} \tag{28}$$

Table 4. Experimental results compared with Hautière N's methods

Instance	V_{met}[m]	V_{met}^{ob}[m]	V_{met}^{H}[m]	E[%]
F7-1	4078.8	5000.0	4753.7	18.42
F7-2	2291.0	2000.0	1903.7	14.55
F7-3	351.9	400.0	436.0	12.03
F7-4	42.5	40.0	43.0	6.25

In order to prove the above algorithm, we collect a specific scene(see Fig. 8). The building marked by a red circle is the China Central Television in Beijing, which is about 200 m far from the camera. The building marked by a green circle

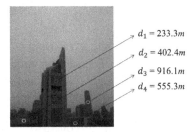

$d_1 = 233.3m$

$d_2 = 402.4m$

$d_3 = 916.1m$

$d_4 = 555.3m$

Fig. 8. Distance results in the same scene at different scene points

is the China World Trade Center Tower 3, which is about 1000 m far from the sensor. Figure 8 shows the distance results at four different scene points. Note that distances for reference at the four scene points are as follows: $d_1 \approx 200$ m, $d_2 \approx 400$ m, $d_3 \approx 1000$ m and $d_4 \approx 500$ m. Results show that calculated distances are close to the reference data, which proves the rationality of this method. Unlike other camera calibration methods, our method merely depends on an outdoor image, which is simple and cost-effective. This method might have an important propelling effect on field measurement in outdoor scenes.

4   Conclusion

In this paper, we propose a novel method for visibility estimation using a single image. A key assumption is developed - the extinction coefficient is approximately a constant in clear atmosphere which is composed of air molecules and a few small-scale aerosol particles. Using this assumption with the dark channel prior, the extinction coefficient of the input image can be calculated and the visibility is finally obtained. Without human assistance, the feasible results prove high accuracy and robustness, which might open a new trend in meteorological research on visibility measurement.

Compared with existing visibility measurements, our method has the following advantages:

- This method proposes an assumption for calculating the clear extinction coefficient and realizes the separation of extinction coefficient β and distance d in depth map βd, which makes it true that the visibility estimation no longer relies on the actual distance information.
- This method depends on a single image and successfully accomplishes fully automatic visibility estimation without human assistance, which performs well in various types of scene.
- This method makes it possible to measure scene distance using a single image, which might open a new trend in meteorological research about scene distance measurement.

Our method proposes a framework to estimate the visibility, which is based on an algorithm of single image haze removal. With the development of haze removal technology, other new methods may be proposed. Therefore, our future work is to explore the possibility of other methods to calculate β/β_0.

Acknowledgments. This research was supported by the National Natural Science Foundation of China(No. 61602520), the Fundamental Research Funds for the Central Universities of Central South University(No. 2017zzts500)

References

1. Anderson, G.P., Berk, A., Acharya, P.K., et al.: MODTRAN4: radiative transfer modeling for remote sensing. Proceedings of SPIE - The International Society for Optical Engineering **3866**(1), 1029–1032 (1999)

2. Ångström, A.: The parameters of atmospheric turbidity. Tellus **16**(1), 64–75 (1964)

3. Babari, R., Hautière, N., Dumont, E., Papelard, J.P.: Computer vision for the remote sensing of atmospheric visibility. In: 2011 IEEE International Conference on Computer Vision Workshops (ICCV Workshops), pp. 219–226. IEEE (2011)

4. Berk, A., Anderson, G.P., et al.: MODTRAN 5: a reformulated atmospheric band model with auxiliary species and practical multiple scattering options: update. In: Defense and Security, pp. 662–667. International Society for Optics and Photonics (2005)

5. Bohren, C.F., Huffman, D.R.: Absorption and Scattering of Light by Small Particles. Wiley, Hoboken (2008)

6. Calinoiu, D.G., Stefu, N., Paulescu, M., et al.: Evaluation of errors made in solar irradiance estimation due to averaging the Angstrom turbidity coefficient. Atmos. Res. **150**(1), 69–78 (2014)

7. Gallen, R., Hautière, N., Cord, A., et al.: Supporting drivers in keeping safe speed in adverse weather conditions by mitigating the risk level. IEEE Trans. Intell. Transp. Syst. **14**(4), 1558–1571 (2013)

8. Hallowell, R., Matthews, M., Pisano, P.: An automated visibility detection algorithm utilizing camera imagery. In: 23rd Conference on Interactive Information and Processing Systems for Meteorology, Oceanography, and Hydrology (IIPS) (2007)

9. Hautière, N., Aubert, D., Dumont, E., et al.: Experimental validation of dedicated methods to in-vehicle estimation of atmospheric visibility distance. IEEE Trans. Instrum. Meas. **57**(10), 2218–2225 (2008)

10. Hautière, N., Labayrade, R., Aubert, D.: Estimation of the visibility distance by stereovision: a generic approach. IEICE Trans. Inf. Syst. **89**(7), 2084–2091 (2006)

11. Hautière, N., Tarel, J.P., Lavenant, J., Aubert, D.: Automatic fog detection and estimation of visibility distance through use of an onboard camera. Mach. Vis. App. **17**(1), 8–20 (2006)

12. He, K., Sun, J., Tang, X.: Single image haze removal using dark channel prior. IEEE Trans. Pattern Anal. Mach. Intell. **33**(12), 2341–2353 (2011)

13. He, K., Sun, J., Tang, X.: Guided image filtering. IEEE Trans. Pattern Anal. Mach. Intell. **35**(6), 1397–1409 (2013)

14. Liaw, J.-J., Lian, S.-B., Huang, Y.-F., Chen, R.-C.: Atmospheric visibility monitoring using digital image analysis techniques. In: Jiang, X., Petkov, N. (eds.) CAIP 2009. LNCS, vol. 5702, pp. 1204–1211. Springer, Heidelberg (2009). https://doi.org/10.1007/978-3-642-03767-2_146

15. Liou, K.N.: An Introduction to Atmospheric Radiation, vol. 84. Academic press, Cambridge (2002)

16. Middleton, W.E.K.: Vision through the atmosphere. In: Bartels, J. (ed.) Geophysik II / Geophysics II. HPEP, vol. 10/48, pp. 254–287. Springer, Heidelberg (1957). https://doi.org/10.1007/978-3-642-45881-1_3

17. Narasimhan, S.G., Nayar, S.K.: Vision and the atmosphere. Int. J. Comput. Vis. **48**(3), 233–254 (2002)

18. Patterson, M.S., Chance, B., Wilson, B.C.: Time resolved reflectance and transmittance for the noninvasive measurement of tissue optical properties. Appl. opt. **28**(12), 2331–2336 (1989)

19. Pomerleau, D.: Visibility estimation from a moving vehicle using the Ralph vision system. In: IEEE Conference on Intelligent Transportation System, ITSC 1997, pp. 906–911. IEEE (1997)
20. Preetham, A.J., Shirley, P., Smits B.: A practical analytic model for daylight. In: Proceedings of the 26th Annual Conference on Computer Graphics and Interactive Techniques, pp. 91–100. ACM Press/Addison-Wesley Publishing Co. (1999)
21. Xie, L., Chiu, A., Newsam, S.: Estimating atmospheric visibility using general-purpose cameras. In: Bebis, G., et al. (eds.) ISVC 2008. LNCS, vol. 5359, pp. 356–367. Springer, Heidelberg (2008). https://doi.org/10.1007/978-3-540-89646-3_35

A Text-Line Segmentation Method for Historical Tibetan Documents Based on Baseline Detection

Yanxing Li[1,2], Longlong Ma[3], Lijuan Duan[1,4(✉)], and Jian Wu[1,3]

[1] Faculty of Information Technology, Beijing University of Technology, Beijing, China
liyanxing15@outlook.com, ljduan@bjut.edu.cn, wujian@iscas.ac.cn
[2] Beijing Key Laboratory of Trusted Computing, Beijing, China
[3] Chinese Information Processing Laboratory, Institute of Software,
Chinese Academy of Sciences, Beijing, China
[4] Beijing Key Laboratory on Integration and Analysis of Large-Scale Stream Data,
Beijing, China
longlong@iscas.ac.cn

Abstract. Text-line segmentation is an important task in the historical Tibetan document recognition. Historical Tibetan document images usually contain touching or overlapping characters between consecutive text-lines, making text-line segmentation a difficult task. In this paper, we present a text-line segmentation method based on baseline detection. The initial positions for the baseline of each line are obtained by template matching, pruning algorithms and closing operation. The baseline is estimated using dynamic tracing within pixel points of each line and the context information between pixel points. The overlapping or touching areas are cut by finding the minimum width stroke. Finally, text-lines are extracted based on the estimated baseline and the cut position of touching area. The proposed algorithm has been evaluated on the dataset of historical Tibetan document images. Experimental result shows the effectiveness of the proposed method.

Keywords: Historical Tibetan document · Text-line segmentation
Baseline detection

1 Introduction

As original documents that contain important information about person, place, or event, historical documents are usually stored in libraries or museums. Unlike others, most historical Tibetan documents are kept in temples. Besides, most historical Tibetan documents exist in the form of scriptures rather than books. As a result, those documents are extremely hard for accessing. At present, there is a growing trend towards the digitization to make these significant documents easier to be preserved and accessed. Considering that the number of historical Tibetan documents and the limitation of technology and human resources, manual transcription is not a reasonable solution. To recognize the historical Tibetan document automatically is a proper way to solve the problem.

© Springer Nature Singapore Pte Ltd. 2017
J. Yang et al. (Eds.): CCCV 2017, Part I, CCIS 771, pp. 356–367, 2017.
https://doi.org/10.1007/978-981-10-7299-4_29

There are four steps in the document recognition. Firstly, the document images are preprocessed by noise removing, correction of image orientation, normalization and binarization. Secondly, the layout analysis module segments the preprocessed image into text regions, pictures and tables. Text areas are obtained by removing borders and margins that may interfere with later operations. Thirdly, the text areas are further segmented into text-lines by text-line segmentation methods. Finally, text-lines are sent into a character segmentation and recognition system, which converts digitized documents into text files, usually in ASCII format. After digitalization, the documents are not only convenient for accessing, but also are protected properly. During the process of document recognition, in particular, text-line segmentation is a significant stage to ensure better performance. What's more, text-line segmentation is also an important process for some document analysis tasks. However, historical Tibetan document images usually contain touching or overlapping characters between consecutive text-lines, making text-line segmentation a difficult task.

The main objective of our work is to extract text-lines in historical Tibetan documents. Unlike many documents written by other languages, the layout structure of historical Tibetan documents is more complex than printed document. Text-line extraction is more challenging on Tibetan historical document because of the curved lines and the touching components in the documents.

The rest of this paper is organized as follows. Section 2 gives a brief introduction about the basic methods in text-line extraction and some discussions about the touching components in Tibetan text-lines. Section 3 gives the main methodology. Section 4 shows the experimental results. The conclusion is presented in Sect. 5.

2 Related Works

Few researches have been done for text-line extraction of historical Tibetan documents. Considering that Tibetan is a spelling language, we refer some works which are used for text-line extraction of historical Latin script. Generally, text-line segmentation can be classified into five categories [1] projection-based method, smearing method, grouping method, Hough-based method, and repulsive attractive network (RA-network) method.

Projection-based method [2] is most commonly used for the printed document. This method regards the pixel in the foreground as 1 while other is 0. The projection value is computed by summing the values in horizontal axis of each line. Nevertheless, this method is very effective on printed or slightly sticky document. Considering the overlapping and touching components usually occurs in historical Tibetan documents and the text-line is curved, we cannot employ this method directly on text-line segmentation of historical Tibetan documents. However, according to the writing habit of humankind, the direction of the text-lines in the former columns is approximately parallel. Because of this, the method can be used to find the number of text-lines.

Smearing method [1] fills the white space between black pixels if the distance of consecutive white pixels is within a predefined threshold. This method is used to text-line segmentation of historical Tibetan documents, and the performance is far from satisfaction. The main reason is that some Tibetan vertical strokes can be so long that smearing the image horizontal could connect the stroke with vowels, which belong to the next text-line. As a consequence, it could produce much more touching components. Grouping method [1] regards connected components, blocks or other features as units, aggregating these units to form alignments. Taking into account lots of touching components in Tibetan document, to define some proper rules for obtaining the units and join the units together is very difficult. Thus, this method cannot be employed. Hough-based method [3] is proper to detect text-lines because text-lines are usually parallel in certain areas. RA-network [4] method constructs the base-line from the top of the image to bottom one by one. The extracted baseline acts as repulsive forces while pixels of the image act as attractive forces.

In historical Tibetan documents, touching or overlapping occurs more often than other languages. Touching and overlapping components between consecutive text-lines is the main challenges for the text-line extraction. According to the writing habits and the language features, Tibetan touching strokes can be generally grouped into three categories: SC1 (see Fig. 2) touching with the baseline of the next text-line, Fig. 1; FC (see Fig. 2) touching with the baseline of next text-line, Fig. 1(c) and the FC (see Fig. 2) touching with vowels in the next text-line Fig. 1(b).

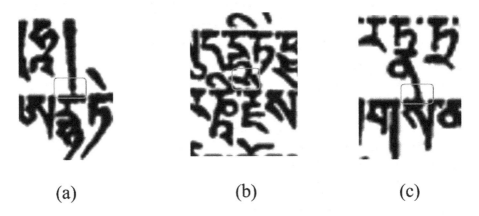

(a)   (b)   (c)

Fig. 1. Tibetan touching examples (a) SC1 touching with the baseline of the next text-line. (b) FC touching with vowels in the next text-line (c) FC touching with the baseline of the next text-line.

The detail of Tibetan scripts is shown in Fig. 2. Just like other spelling languages, Tibetan script consists of 30 consonants and 5 vowels, which are combined to produce Tibetan syllables according to the spelling rule. Each syllable has a base consonant (BC). According to the relative position to base consonant, other consonants are called prefix consonant (PC), foot consonant (FC), the first suffix consonant (SC1) and the second suffix consonant (SC2) [5] (see Fig. 2). In Fig. 2, the baseline is a fictitious line which follows and joins the consonants at the higher part of Tibetan character bodies. Most consonants lie under the baseline while only few of upper vowel letters lie above the baseline.

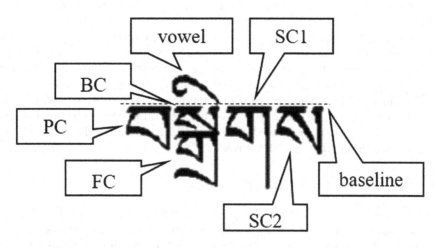

Fig. 2. The Tibetan syllable, which consist of base consonant (BC), vowel, prefix consonant (PC), foot consonant (FC), the first suffix consonant (SC1) and the second suffix consonant (SC2) [5].

3   Methodology

Here is the architecture we extract text-lines from Tibetan historical documents shown in Fig. 3. The input image is composed of the uniformed text areas of the Tibetan historical documents and is obtained by the layout segmentation method [6]. In this method, text areas are extracted based on connected component analysis and corner point detection. Firstly, document images are equally divided into $ESize * ESize$ grids. Based on the classification information of CCs and corner points, the grids are filtered by the predefined rules. By analysing vertical and horizontal projections of remaining grids, the approximate text area is located and further corrected. Based on the results of layout analysis, text-lines are extracted by our proposed text-line segmentation method.

Fig. 3. The text-line segmentation process

Our method includes three stages:

1. Baseline initial position detection: Partial information is extracted from the left N pixel columns of input images. Then, the upper vowels and FC are removed based on this extracted information. Finally, the initial position is detected using the projection method. At the same time, the number of text-lines is obtained.
2. Estimating the baseline: Based on the above detected initial position, the baseline is estimated by dynamic tracing the pixel points from left to right direction. During dynamic tracing process, the context information between pixels is used to solve the baseline estimation of the curve text-lines.
3. Touching area detection and segmentation: For two consecutive baselines, the image between two baselines is divided into several patches. The patch-based touching area is detected based on the prior information and are further cut by finding the minimum width stroke.

3.1   Baseline Initial Position Detection

The objective of the baseline initial position detection is to find the estimated baseline initial position information and the text-line number of the document. The image for detecting initial position is the left N pixel columns of input image. Denote the image as *image A* (see Fig. 4(a)). Assuming the text-lines in *image A* are approximately parallel. Due to the touching strokes between consecutive text-lines (see Fig. 1), it will be useful to remove the vowels and FCs before finding the position. Figure 4 shows the steps of estimating the initial position.

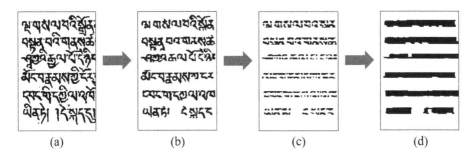

Fig. 4. The step for detecting the initial position of the input image. (a) The image for detecting the initial position. (*image A*) (b) The image that extracted by template matching. (*image B*) (c) The image that has been pruned salient strokes. (*image C*) (d) The image that has been performed closing operation. (*image D*)

The baseline initial position detection method includes three steps. The first step is to establish the detection template database and generate text blocks by template matching. Firstly, select randomly several input images and split them into patches by a sliding window. Secondly, the patches which contain the baseline of the syllable at the top are picked up manually as the detection templates (see Fig. 5). The detection template database with 80 patches is constructed. Thirdly, the text blocks are generating by the template matching method. The features of these patches are extracted using Principal Component Analysis (PCA [7]). *image A* with the same size of sliding window are obtained. Calculate the similarity between the patches in the sliding window and the templates using the computed PCA model. The similarity is calculated as follows:

$$similarity(x, y) = \frac{x \cdot y}{\|x\| \|y\|} \tag{1}$$

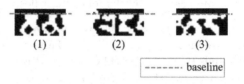

(1)   (2)   (3)

- - - - - - - baseline

Fig. 5. Samples of the detection templates

Keep the patch if the similarity greater than a predefined threshold. After that, we can obtain a new image which has been removed FCs and vowels, denote it as *image B* (see Fig. 4(b)).

Secondly, pruning salient strokes. In *image B*, FC and most vowels in Tibetan syllables have been removed, although few vowels are remained incorrectly (see Fig. 6). Refer the block covering method [8], the *image B* is cut into text blocks based on horizontal projection profile. In each text block, recalculating the value of the projection profile by:

$$len(x) = \begin{cases} x & x > \frac{\max(block)}{2} \\ 0 & other \end{cases} \tag{2}$$

Max (block) is the max profile value in each text block. Set the pixels of the row to 0 whose profile value equals to 0 in *image B*, denote it as *image C* (see Fig. 4(c)).

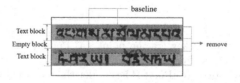

Fig. 6. Pruning salient strokes in text blocks of *image B*

Finally, we obtain *image D* (see Fig. 4(d)) by closing operation on *image C*. The initial position of the baseline is the upper boundary of each text block in *image D*. The number of text-lines equals to that of the text-blocks.

3.2   Estimating the Baseline

In this stage, we need to estimate the baseline. Because the baseline of document image is not strictly horizontal, the baselines are established from left to right by dynamic tracing within the pixel points of each line. We assume the value of foreground pixel as 1 and background pixel as 0. Based on the initial position of the baseline, in each line from left to right, we pick up the five key points every N pixels based on the previous key tracing point, which are P, U1, U2, D1 and D2 for picking up the key tracing point. U1 and U2 are the point that lie 1 and 2 pixels above P. D1 and D2 are the point that lie 1 and 2 pixels below P. Regarding the baseline initial position as the starting key tracing point. The next key tracing points are selected by the following rules:

- If the values of P, U1, U2, D1 and D2 are the same, P is considered as a back-ground point. (see Fig. 7(a)). Select P as the next key tracing point.
- If the value of P equals 0 and D1 or D2 equals 1, P is considered as a point that lies above the baseline. D1 is selected as the next key tracing point (see Fig. 7(b)).
- If the value of P equals 1 and U1 or U2 equals 1, P is considered as a point that lies under the baseline, U1 is selected as the next key tracing point (see Fig. 7(c)).
- Else, P is selected as the next key tracing point.

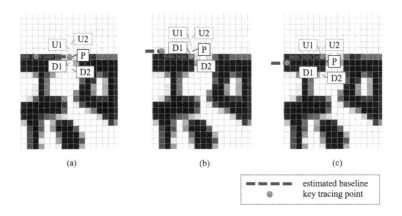

Fig. 7. Examples of selecting key tracing point. (a) Point P is the background point. Select P as the next key tracing point. (b) Point P lies above the baseline. Select D1 as the next key tracing point. (c) Point P lies under the baseline. Select U1 as the next key tracing point.

The baselines are estimated by joining the tracing key point together horizontally. Then, the input image is divided into several strips base on these baselines.

3.3 Touching Area Detection and Segmentation

The document image has been split into some horizontal strips based on the estimated baselines. However, the vowels in the syllable cannot be properly segment. In this stage, we present a method to locate the touching area and cut it by finding the minimum width stroke. Firstly, the strips are divided into a number of patches with a fixed step. In accordance with the information of connected components, these patches are classified into two classes by finding the bounding box (BB) of the connected component (CC). If there exists a BB has the same height with the patch, the patch is labelled as a touching patch (Fig. 8(b)), else label the patch as a normal patch (Fig. 8(a)). The segmentation for the normal patch can be easily found: if the CCs touch with the baseline of the next text-line, the CCs belong to the next text-line.

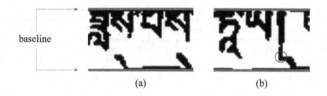

Fig. 8. Patch types: (a) normal patch (b) touching patch

For the touching patches, by observing the composition of Tibetan syllables, we can conclude that the touching strokes generally exist in the bottom 1/2 part of them and the stroke width of the touching area is short. With the method proposed by Epshtein et al. [9], we transform the patch by calculating the width of the stroke horizontally and vertically. The cut position of the patch is computed based on the projection profile of the transformed patch to Y-axis with the following criteria:

- If there is one minimum, select the corresponding row in the patch as the position for segment.
- If there are a few minima, select the corresponding row which is the nearest to the middle of bottom half of the patch as the position for segment.

After located the position for cutting, we need to verify if the position could segment the patch properly. If the value of cutting position in projection profile is greater than a predefined threshold, it shows that touching area is complex. A further research is needed to solve this problem. In this paper, these patches are cut with a simple method: if the text strokes lie under the cutting position, they belong to the next text-line.

4 Experiments and Results

The experimental dataset are provided by Qinghai Nationalities University. Limited by a variety of factors, we only acquired 120 high quality images for experiment. The method presented in this paper is implemented in python. The PCA model established base on the project of scikit-learn [10]. We use the library of opencv-python for image input and output and pre-processing, and scikit-image [11] for processing the connected components and bonding box.

Fig. 9. The original Tibetan document image

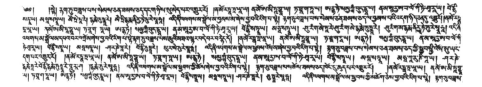

Fig. 10. The extracted image for text-line segmentation

Fig. 11. The input image is divided into strips based on the baselines

Figure 9 is an original historical Tibetan document image. The method for layout analysis and text area extracting is proposed by Zhang et al. [6]. Figure 10 is the extracted text image that has been normalized to a width of 1300px with a scaled height and threshold for further processing. We extract 150 pixels columns from the input image for detecting baseline initial position. The width of the sliding window is 30px and the height is 15px. Two document samples are selected to establish the PCA model. They are split into patches with the same size of the sliding window. In the similarity calculation, 10-d is suitable. Figure 11 shows the text-line strips. The main part of a syllable has been segmented in each strip while the vowels are not properly labelled. It needs for further processing before extracting. Figure 12 gives the result, different with Fig. 11, the vowels are labelled correctly. To evaluate the performance of the presented method, we use the same evaluation method as that used in ICDAR2013 Handwritten Segmentation Contest [12]. It is based on counting the number of matches between

Fig. 12. Patch types: (a) normal patch (b) touching patch

the entities detected by the algorithm and the entities in the ground truth. The evaluation method uses a MatchScore (3) table to detect the matches whose values are computed by the intersection of the on pixel set of the result and the ground truth.

$$MatchScore(i, j) = \frac{T(G_j \cap R_i \cap I)}{T((G_j \cup R_i) \cap I)} \tag{3}$$

Let I be the set of all images points, G_j the set of all points inside the ground truth region, R_i the set of all points inside the i result region and $T(s)$ is a function that counts the points of set s. A region pair is considered as a one-to-one match only if the matching score is equal to or above a threshold. Let N be count of ground-truth elements, M be the count of result elements, and $o2o$ be the number of one-to-one matches, the detection rate (DR) and the recognition accuracy (RA) are defined as follows:

$$DR = \frac{o2o}{N}, RA = \frac{o2o}{M} \tag{4}$$

A performance metric FM can be extracted if we combine the values of detection rate (DR) and recognition accuracy (RA):

$$FM = \frac{2\,DR\,RA}{DR + RA} \tag{5}$$

Table 2 shows the performance of our method. Comparing with performance of project-based method in Table 1, our method has a considerable improvement in accuracy. As the stage of processing touching area is not very exactly, the

Table 1. The performance with project-based method

T(s)	M	N	o2o	DR	FM
0.90	769	770	552	67.79%	67.83%
0.91	769	770	446	57.92%	57.96%
0.92	769	770	360	46.75%	46.78%
0.93	769	770	269	34.94%	34.96%
0.94	769	770	197	25.58%	25.60%
0.95	769	770	135	17.53%	17.54%

Table 2. The performance with our method

T(s)	M	N	o2o	DR	FM
0.90	770	770	750	97.40%	97.40%
0.91	770	770	740	96.10%	96.10%
0.92	770	770	723	93.90%	93.90%
0.93	770	770	691	89.74%	89.74%
0.94	770	770	641	83.25%	83.25%
0.95	770	770	553	71.82%	71.82%

presented method does not perform so well when $T(s)$ increases. We will improve the accuracy in a further research.

5   Conclusion

This paper presents a text line segmentation method for historical Tibetan documents. The initial baseline position is obtained by template matching, pruning the salient strokes and closing operation. To solve the text-line segmentation of the curve text-lines, the baseline is estimated by joining together a number of key tracing points which are located by the context information. At the end, the touching area is detected and cut by analysing the patches between the two baselines. The patches are classified into touching patches and normal patches according to the decision if they have CCs that have the same height with the patches. The touching patches are cut by finding the minimum stroke width. However, at the stage of processing the touching area, the method presented by this paper is not suitable for complex areas (Fig. 1(b)). In the future research, we will find a better way to locate and remove the vowels. In addition, we need to adapt this method on the Tibetan document with non-horizontal direction of text-line.

Acknowledgments. This work was supported by the Science and Technology Project of Qinghai Province (no. 2016-ZJ-Y04) and the Basic Research Project of Qinghai Province (no. 2016-ZJ-740). The authors would like to thank Qilong Sun, the Department of Computer Science, Qinghai Nationalities University for providing the experimental dataset of historical Tibetan document images.

References

1. Likforman-Sulem, L., Zahour, A., Taconet, B.: Text line segmentation of historical documents: a survey. Int. J. Doc. Anal. Recogn. **9**(2), 123–138 (2007)
2. Manmatha, R., Rothfeder, J.: A scale space approach for automatically segmenting words from historical handwritten documents. IEEE Trans. Pattern Anal. Mach. Intell. **27**(8), 1212–1225 (2005)

3. Louloudis, G., et al.: Text line detection in handwritten documents. Pattern Recogn. **41**(12), 3758–3772 (2008)
4. Oztop, E., et al.: Repulsive attractive network for baseline extraction on document images. Sig. Process. **75**(1), 1–10 (1999)
5. Huang, H., Da, F.: General structure based collation of Tibetan syllables. J. Inf. Comput. **6**(5), 1693–1703 (2010)
6. Zhang, X., Duan, L., Ma, L.-L.: Text extraction for historical Tibetan document images based on connected component analysis and corner point detection. In: Yang, J., et al. (eds.) CCCV 2017, Part I. CCIS, vol. 771, pp. 545–555. Springer, Heidelberg (2017)
7. Wold, S., Esbensen, K., Geladi, P.: Principal component analysis. Chemometr. Intell. Lab. Syst. **2**(1–3), 37–52 (1987)
8. Zahour, A., et al.: Overlapping and multi-touching text-line segmentation by block covering analysis. Pattern Anal. Appl. **12**(4), 335–351 (2009)
9. Epshtein, B., Ofek, E., Wexler, Y.: Stroke width transform. In: Computer Vision and Pattern Recognition (2010)
10. Pedregosa, F., et al.: Scikit-learn: machine learning in Python. J. Mach. Learn. Res. **12**(8), 2825–2830 (2011)
11. van der Walt, S., et al.: Scikit-image: image processing in Python. PeerJ **2**, e453 (2014)
12. Stamatopoulos, N., et al.: ICDAR 2013 handwriting segmentation contest. In: 2013 12th International Conference Document Analysis and Recognition (ICDAR), pp. 1402–1406. IEEE press (2013)

Low-Light Image Enhancement Based on Constrained Norm Estimation

Tan Zhao[1], Hui Ding[1,2,3]([✉]), Yuanyuan Shang[1,2,3], and Xiuzhuang Zhou[1,2,4]

[1] College of Information Engineering Capital Normal University,
Beijing 100048, China
dhui@cnu.edu.cn
[2] Engineering Research Center of High Reliable Embedded System,
Beijing 100048, China
[3] Key Laboratory of Electronic System Reliability Technology, Beijing 100048, China
[4] Collaborative Innovation Center for Mathematics and Information of Beijing,
Beijing 100048, China

Abstract. Low-light images often suffer from poor quality and low visibility. Improving the quality of low-light image is becoming a highly desired subject in both computational photography and computer vision applications. This paper proposes an effective method to constrain the illumination map t by estimating the norm and constructing the constraint coefficients, which called LieCNE. More specifically, we estimate the initial illumination map by finding the maximum value of R, G and B channels and optimize it by norm estimation. We propose a function t^γ to contain the exponential power γ in order to optimize the enhancement effect under different illumination conditions. Finally, a new evaluation criterion is also proposed. We use the similarity with the true value to determine the enhanced effect. Experimental results show that LieCNE exhibits better performance under a variety of lighting conditions in enhancement results and image spillover prevention.

Keywords: Low-light image enhancement · Illumination optimization
Norm estimate · Quadtree algorithm

1 Introduction

The visibility of the image as an important criterion for evaluating the quality of the images reflects the detail, clarity and contrast of the image. Some natural scenes such as sunny days, rainy days or night scenes always lack light or external light conditions. The quality and illumination of images taken from these scenes will be reduced. Improving the quality of images has critical positions in the field of image fusion [1], feature extraction [2] and machine recognition [3].

This work was supported by National Natural Science Foundation of China (Nos. 61303104, 61373090, 61203238), Beijing Natural Science Foundation of China (4132014), and also supported by Youth Innovative Research Team of Capital Normal University.

© Springer Nature Singapore Pte Ltd. 2017
J. Yang et al. (Eds.): CCCV 2017, Part I, CCIS 771, pp. 368–379, 2017.
https://doi.org/10.1007/978-981-10-7299-4_30

For low-light image, the key to improving image quality is improving contrast and resolution. A number of methods have been developed to address visibility enhancement for low-light scenes from a single image. The simplest way is increasing the brightness of images directly. But the brighter area will overflow. Image gray scale dynamic range can be balanced with histograms, but the details will be lost [4–7]. The multi-scale retinex algorithm with color restoration will reduce the image contrast and keep the edge detail unsatisfactory [8,9].

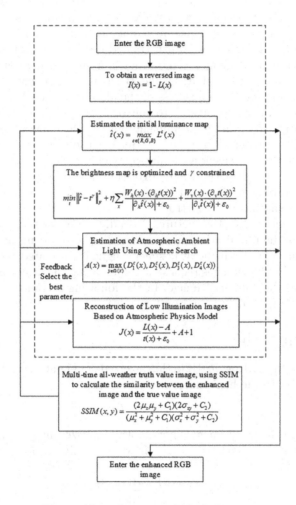

Fig. 1. Algorithm general block diagram.

Building the geometric relationship between the original image and reconstructed image can be constructed recovery function. Fu et al. establishes a color estimation model for transforming the image from night to day [10]. In [11], the observed image can be decomposed into the product of the desired scene and the illumination image. Dong et al. applies an enhancement method based on dark channel prior algorithm [12,13].

Based on the dark channel a prior algorithm, Zhang et al. gives the weight to each illumination component on the dark channel to optimize the illumination map [14]. In [15], the illumination map is estimated by using the illumination component in the HSI space, but it will bring color distortion. Recent years, scholars applied convolution neural networks into the dehaze algorithm, estimating illumination map by training samples [16]. Although the treatment results are improved, it also bringing new problems with high time costs.

This paper proposes a low-light image enhancement algorithm based on the atmospheric physical model with a constrained norm estimation so as to adapt to different illumination images. The initial illumination map is estimated by finding the maximum value of R, G and B channels and refinement by norm estimation. Considering the robustness of the algorithm, we introduce the parameter γ into the power exponent t^γ to constrain the brightness map t. We use quadtree algorithm to estimate the ambient light A, which has a faster processing effect than the dark channel proposed by He. We also propose a new evaluation criteria to verify the effectiveness of our approach.

To this end, we present the whole algorithm flow in Fig. 1, including image reversal, image inversion, illumination map estimation and constraint, ambient light estimation and image reconstruction.

First of all, we reverse the picture. Secondly we extracte the maximum in RGB channel as the illumitation map and use the norm estimation to optimize the propagation map. Then the atmospheric light is estimated by the quadtree algorithm. Finally, the image is reconstructed by using the atmospheric physical model to obtain the enhanced image. In the following sections, we will discuss in detail the algorithm details and performance.

2   Our Approach

With statistics supporting in [12], the inverted low-light image on the RGB channel is similar to the haze image:

$$I(x) = 1 - L(x) \tag{1}$$

where I indicates inverted image which looks like haze image, L represents the input low-light image. The model as follows is widely used in haze removing [13,20]:

$$R(x) = \frac{I(x) - A(1 - t(x))}{t(x)} \tag{2}$$

where R represents the reconstructed haze image, I is the haze image, A is the global atmospheric light in haze image, but it is also called ambient light in low-light image. t represents the medium transmission. Reconstructing the inverted low illumination image as a haze image, we substitute Eq. (1) into Eq. (2):

$$J(x) = \frac{L(x) - A}{t(x)} + A + 1 \tag{3}$$

where $J = 1 - R$ is the enhanced image, L and represents the input low-light image. A is the ambient light and t indicates illumination map.

For preventing oversaturation, Eq. (4) is introduced as:

$$J(x) = \frac{L(x) - A}{t(x) + \varepsilon_0} + A + 1 \tag{4}$$

where ε_0 approaches zero infinity. Figure 2 provides an example, we can see input image and the corresponding inverted image as well as enhanced image.

$$\begin{array}{ccc} \text{(a)} & \text{(b)} & \text{(c)} \end{array}$$

Fig. 2. The enhanced results by our approach. (a) Input low-light image. (b) Inverted image. (c) Image after enhance by our approach.

2.1   Initial Illumination Map Estimation

The medium transmission map t* describes the light portion that attenuate with the increase of depth of field. t* is defined as:

$$t^*(x) = e^{-\beta d(x)} \tag{5}$$

where β is the scattering coefficient of the atmosphere and $d(x)$ is scene depth. In low-light image, the illumination map does not attenuate with the increase of depth of field. Therefore, the method of defogging to estimating the illumination map may not be applicable. This paper estimates the initial illumination map by finding the maximum value of R, G and B channels [11]:

$$\hat{t}(x) = \max_{c \in \{R,G,B\}} L^c(x) \tag{6}$$

Initial illumination map $\hat{t}$ contains the luminance information of the image but does not contain the texture information of the image.

2.2   Illumination Map Optimization by Norm Estimation

Initial estimation can promote the global illumination. But the enhancement processing is required to preserve more structures. In order to solve this problem, we use the following refinement method for propagation maps [19]:

$$\min_{t} \left\|\hat{t} - t\right\|_F^2 + \eta \|W \cdot |(\partial_h t)_x + (\partial_v t)_x|\|_1 \tag{7}$$

where t is the optimization illumination map, $\|\|_F$ is the Frobenious-norm and $\|\|_1$ represents the l_1-norm respectively. The first term takes into account the degree of fusion and smoothness between the initial map $\hat{t}$ and the refined map t while the second one squares up the structure and texture information. η (the experience value is 0.2) designates a coefficient that balances the weights between two terms. x indexes 2D pixels. ∂_x and ∂_y are the partial derivatives in two directions. Further, W is the weight matrix, we defined it as:

$$W = \sum_{y \in \Omega(x)} \frac{G_\sigma(x,y)}{\left|\sum_{y \in \Omega(x)} G_\sigma(x,y) * (\partial_h \hat{t})_y\right| + \varepsilon_0} \tag{8}$$

where $G_\sigma(x,y)$ is a Gaussion filter with standard deviation σ and $*$ is the convolution operator. W can be estimated by the known $\hat{t}$ rather than through the iteration of t, thus reducing the calculation time. We transform the nonlinear part in Eq. (7) into a linear equation for calculation:

$$\min_{t} \left\|\hat{t} - t\right\|_F^2 + \eta \sum_x \frac{W_h(x) \cdot (\partial_h t(x))^2}{\left|\partial_h \hat{t}(x)\right| + \varepsilon_0} + \frac{W_v(x) \cdot (\partial_v t(x))^2}{\left|\partial_v \hat{t}(x)\right| + \varepsilon_0} \tag{9}$$

After simplifying the equation and resolving the quadratic term, we get the optimization illumination map t. The different illumination maps are shown in the top row of Fig. 3 and the bottom row is reconstructed images with different illumination maps. We can see the advance of our method. The reconstructed image has more details and more suitable brightness.

2.3   Constraint Illumination Map by Exponential Power

We have discussed the method of estimating the propagation pattern with the norm in Sect. 2.2. Furthermore, the concentration of fog images have the different in thick-fog and light-fog. Similarly, the illumination intensity of the low-light images are not stable.

We propose a new method called LieCNE to accommodate a wider range of illumination conditions. The estimation of propagation map t needs to be constrained. We propose the parameter γ to make the constraint of t by:

$$T = t^\gamma \tag{10}$$

γ is defined as:

$$\gamma = min(\sqrt{\frac{K}{\Delta}}, \mu) \tag{11}$$

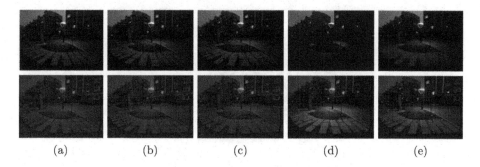

(a)   (b)   (c)   (d)   (e)

Fig. 3. Comparison of different illumination maps and corresponding enhanced results. (a) Initial. (b) Xu [18]. (c) RTA [19]. (d) LIME [11]. (e) With norm estimation.

where K is a threshold (empirical value of 80), when the image illumination is low, K may take a smaller value, and when the image illumination is high, K may take a larger value. $\mu \in (0,1]$ is a given constant (0.8 by experience). Δ is given by the following equation:

$$\Delta = \left| \min_c I^c(x) - \min_c A^c \right| \tag{12}$$

where A is the ambient light and we will discuss in Sect. 2.4. We obtain a more adaptable illumination map T, Eq. (4) can be written as:

$$J(x) = \frac{L(x) - A}{T(x) + \varepsilon_0} + A + 1 \tag{13}$$

2.4 Quadtree Algorithm for Estimating Ambient Light

In this paper, the quadtree algorithm is used to estimate the value of ambient light. The quadtree algorithm and the algorithm [13] have almost the same effect on the atmospheric light estimation, but the quadtree algorithm has a faster processing speed. Quadtree algorithm to estimate the process of atmospheric light are shown in Fig. 4.

The channel minimum map of image is firstly computed, which aims to avoid the local image in a channel while there is a maximum value of the atmospheric light caused by the error estimates.

$$D(x) = \min_{c \in \{r,g,b\}} I^c(x) \tag{14}$$

Secondly, the image is divided into four blocks. We calculate the average of each block image and take the mean of the largest block to four equal parts thirdly. The iterations are repeated until the resulting image block size is less than a given threshold. The maximum value of the pixels in the resulting image block is taken as the value of the ambient light.

$$A(x) = \max_{y \in \Omega(x)} \left(D_1^c(x), D_2^c(x), D_3^c(x), D_4^c(x) \right) \tag{15}$$

Fig. 4. Schematic figure of quad-tree algorithm.

3   Experimental Results

3.1   Algorithm Performance Comparison

The proposed method was tested on several 816×612 pixel images with various degrees of darkness and noise levels. In order to further verify the enhancement effect of the proposed algorithm, we compare the algorithm with LIME [11], Dong et al. [12], Hu et al. [15] and Ren et al. [17] in Fig. 5. Moreover, all the codes are carried out on the MATLAB2014a platform to ensure the algorithm fairness and time results based on Windows 10 OS with 8 G RAM and 3.2 GHz CPU.

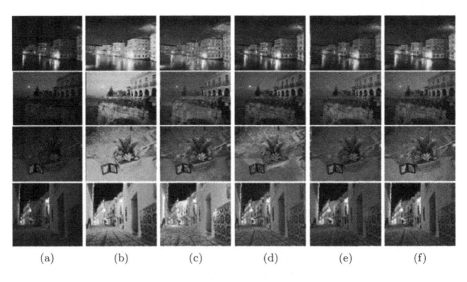

(a)   (b)   (c)   (d)   (e)   (f)

Fig. 5. Comparison of different method. (a) input. (b) Hu [15]. (c) Dong [12]. (d) LIME [11]. (e) Ren [17]. (f) LieCNE.

[12,15] reduce the fidelity of the image. Our method recovers the structures without sacrificing the fidelity of the colors. In order to further prove the performance of our algorithm, we compare the proposed algorithm with the state-of-the-art using the Mean Squared Error (MSE), Peak Signal-to-Noise Ratio (PSNR), Entropy, Absolute Mean Brightness Error (AMBE) and Lightness order Error (LOE) metrics. (image size is 816×612).

Table 1. The average results of MSE, PSNP, E, AMBE, LOE, times on the reconstructed.

Metric	Hu et al.	Dong et al.	LIME	Ren et al.	LieCNE
MSE	0.067	0.049	0.026	0.014	**0.016**
PSNP	59.882	61.220	63.910	66.568	**65.988**
E	7.169	6.005	6.778	6.972	**6.650**
AMBE	0.186	0.154	0.134	0.091	**0.090**
LOE (10^{-3})	573.49	606.36	582.53	516.89	**527.34**
Time (s)	2.79	10.5	1.78	5.71	**1.64**

In Table 1, LieCNE works well on several objective indicators. Although the entropy of LieCNE and Ren's entropy has a difference of 0.32, the remaining indicators are better than other algorithms. It is worth mentioning that, although the Rens method and LieCNE are close to most of the indicators, our method just using 1.64 s, runs 3.48 times faster than the Rens.

3.2 A New Evaluation Criteria

When dealing with low-light images, it is necessary not only to enhance the brightness of the image as much as possible, but also to avoid the image pixel value of the overflow phenomenon, especially with a strong light source of low-light images. The enhancement effect needs to consider from two aspects: one is the enhanced intensity that is the degree of improvement of the brightness contrast, and the other is the statistics of the spillover intensity near the bright spot or the local strong light source. So the principle of enhancement is trying to avoid too much overflow while enhancing the lower part of the brightness.

Figure 6(a) shows RGB weight contrast between different algorithms which reflects the overall enhanced strength. In Fig. 6(b), the degree of spillover of the image after reconstruction by several algorithms can be seen more intuitively. Gray level of 256 is considered as overflow and the more pixels gather in this gray-scale the more overflow appear in the image. LieCNE has the better overall enhancement effect and also has advantages in suppressing spillover result. Due to the lack of true values, we collected images of different lighting scenes in the same scene as samples. We fixed the camera on a tripod to ensure the reliability of the experimental data and take the images in the same scene in 14:00, 16:00,

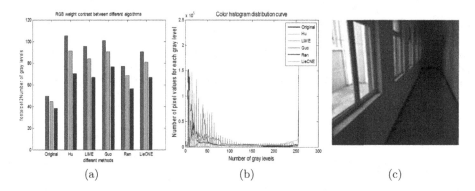

Fig. 6. Comparison of intensity enhancement among different algorithms. (a) RGB weight contrast between different algorithms. (b) Color histogram distribution curve. (c) The image under normal light (filmed at 14:00) (Color figure online).

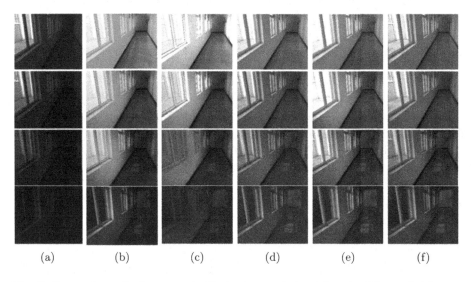

Fig. 7. Comparison of enhancement effects of several algorithms in different brightness of the same scene. (a) Input. (b) Hu [15]. (c) Dong [12]. (d) LIME [11]. (e) Ren [17]. (f) LieCNE

17:00, 17:30, 18:00 respectively. We take at 14:00 as the true value of the data comparison, as shown in Fig. 6(c).

Figure 7 shows the enhancement effects of several algorithms for the same scene at different light intensities. Row (1)–(4) are the original images filmed at 16:00, 17:00, 17:30 and 18:00 and their correspond enhanced images respectively. Furthermore, we propose a new evaluation method, which contrasts the enhancement effect of the algorithm under different illumination intensities. The image enhancement effect was evaluated by computing the structural similarity (SSIM) between the reconstructed low image and the normal illumination.

$$SSIM(x, y) = \frac{(2\mu_x\mu_y + C_1)(2\sigma_{xy} + C_2)}{(\mu_x^2 + \mu_y^2 + C_1)(\sigma_x^2 + \sigma_y^2 + C_2)} \tag{16}$$

Table 2. The $SSIM$ with normal image under several illuminations.

$SSIM$	Original	Hu et al.	Dong et al.	LIME	Ren et al.	LieCNE
14:00	0.8298	0.5942	0.6242	0.6716	**0.7206**	0.7058
16:00	0.6672	0.6779	0.6875	0.7131	0.6987	**0.7422**
17:00	0.3182	0.5937	0.6458	0.7278	0.7549	**0.7579**
18:00	0.2375	0.3625	0.4216	0.4792	0.5182	**0.5542**

In Table 2, the $SSIM$ enhanced by LieCNE under different lighting conditions is closer than other methods to the normal light of the image. Figure 8 shows the line graph.

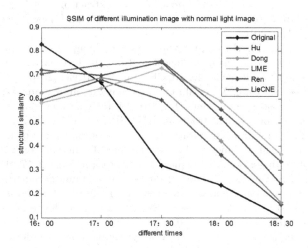

Fig. 8. The comparison of $SSIM$ between enhanced image and original image under different brightness.

Figure 8 reflects the comparison results more directly. First, as the decreasing of image brightness, $SSIM$ with normal light is also reduced. Second, LieCNE exhibits better $SSIM$ to the original image in each illumination. Finally, it is found that the enhancement of image photographed at 17:00 has a higher $SSIM$ with the original image. As described in Sect. 2.2, it is necessary to vary the estimates of the illumination map appropriately for different light intensities.

4   Discussions and Conclusions

This paper has presented an effective approach to enhance low-light images for boosting the visual quality. Inspired by the existing atmospheric haze imaging model and low illumination image enhancement algorithms, the illumination map can be estimated from the image color channel and optimized using the norm, ambient light is determined by the quadtree algorithm. In addition, this paper proposes a method of constraining the illumination map for different illumination images as well as a new evaluation method, that is, determining the enhancement effect of the algorithm by comparing the quality of the image under different light intensity. The experimental results reveal that LieCNE achieves higher efficiency and better enhancing effects a variety of light intensity. But some problems still exist. Low-light images have a lot of noise and it will be amplified during enhancement processing. Filtering noise while ensuring the image enhancement effect is not affected is the next issue of concern.

References

1. Liu, Y., Chen, X., Ward, R.K., et al.: Image fusion with convolutional sparse representation. Sig. Process. Lett. **23**(12), 1882–1886 (2016)
2. Kang, J., Lu, S., Gong, W., et al.: Image feature extraction based on spectral graph information. In: 2016 IEEE International Conference on Image Processing (ICIP), pp. 46–50. IEEE (2016)
3. Wei, M., Ma, X., Li, H., et al.: Application of image recognition algorithm in satellite interference transmitter location. In: 2016 International Conference on Communication Problem-Solving (ICCP), pp. 1–2. IEEE (2016)
4. Yao, Z., Zhou, Q., Yang, X., et al.: Quadrants histogram equalization with a clipping limit for image enhancement. In: 2016 8th International Conference on Wireless Communications Signal Processing (WCSP), pp. 1–5. IEEE (2016)
5. Nithyananda, C.R., Ramachandra, A.C.: Review on histogram equalization based image enhancement techniques. In: International Conference on Electrical, Electronics, and Optimization Techniques (ICEEOT), pp. 2512–2517. IEEE (2016)
6. Yao, Z., Lai, Z., Wang, C.: Brightness preserving and non-parametric modified bi-histogram equalization for image enhancement. In: 2016 12th International Conference on Natural Computation, Fuzzy Systems and Knowledge Discovery (ICNC-FSKD), pp. 1872–1876. IEEE (2016)
7. Youshi, Y., Qiang, S., Lei, S., et al.: A adaptive dual-platform deep space infrared image enhancement algorithm based on linear gray scale transformation. In: 2014 33rd Chinese Control Conference (CCC), pp. 7434–7438. IEEE (2014)
8. Okuno, T., Nishitani, T.: Efficient multi-scale retinex algorithm using multi-rate image processing. In: 2009 16th IEEE International Conference on Image Processing (ICIP), pp. 3145–3148. IEEE (2009)
9. Fu, X., Zhuang, P., Huang, Y., et al.: A retinex-based enhancing approach for single underwater image. In: 2014 IEEE International Conference on Image Processing (ICIP), pp. 4572–4576. IEEE (2014)
10. Fu, H., Ma, H., Wu, S.: Night removal by color estimation and sparse representation. In: 2012 21st International Conference on Pattern Recognition (ICPR), pp. 3656–3659. IEEE (2012)

11. Guo, X.: LIME: a method for low-light image enhancement. J. Nerv. Ment. Dis. **167**(7), 402–409 (1979)
12. Dong, X., Wang, G., Pang, Y., et al.: Fast efficient algorithm for enhancement of low lighting video. In: 2011 IEEE International Conference on Multimedia and Expo (ICME), pp. 1–6. IEEE (2011)
13. He, K., Sun, J., Tang, X.: Single image haze removal using dark channel prior. Trans. Pattern Anal. Mach. Intell. **33**(12), 2341–2353 (2011)
14. Zhang, X., Shen, P., Luo, L., et al.: Enhancement and noise reduction of very low light level images. In: 2012 21st International Conference on Pattern Recognition (ICPR), pp. 2034–2037. IEEE (2012)
15. Hu, Y., Shang, Y., Fu, X., et al.: A low illumination video enhancement algorithm based on the atmospheric physical model. In: 2015 8th International Congress on Image and Signal Processing (CISP), pp. 119–124. IEEE (2015)
16. Cai, B., Xu, X., Jia, K., et al.: Dehazenet: an end-to-end system for single image haze removal. IEEE Trans. Image Process. **25**(11), 5187–5198 (2016)
17. Ren, W., Liu, S., Zhang, H., Pan, J., Cao, X., Yang, M.-H.: Single image dehazing via multi-scale convolutional neural networks. In: Leibe, B., Matas, J., Sebe, N., Welling, M. (eds.) ECCV 2016. LNCS, vol. 9906, pp. 154–169. Springer, Cham (2016). https://doi.org/10.1007/978-3-319-46475-6_10
18. Xu, L., Lu, C., Xu, Y., et al.: Image smoothing via L0 gradient minimization. ACM Trans. Graph. (TOG) **30**(6), 174 (2011). ACM
19. Xu, L., Yan, Q., Xia, Y., et al.: Structure extraction from texture via relative total variation. ACM Trans. Graph. (TOG) **31**(6), 139 (2012)
20. Tan, R.T.: Visibility in bad weather from a single image. In: 2008 IEEE Conference on Computer Vision and Pattern Recognition, CVPR 2008, pp. 1–8. IEEE (2008)

FPGA Architecture for Real-Time Ultra-High Definition Glasses-Free 3D System

Ran Liu[1,2(✉)], Mingming Liu[1], Dehao Li[1], Yanzhen Zhang[2], and Yangting Zheng[1]

[1] College of Computer Science, Chongqing University, Chongqing 400044, China
Ran.liu_cqu@qq.com
[2] College of Communication Engineering, Chongqing University, Chongqing 400044, China

Abstract. This paper presents an FPGA architecture for real-time ultra-high definition (UHD) glasses-free 3D system by solving high bandwidth requirement and high complexity of the system problems. Video + Depth (V + D) video format and the depth-image-based rendering (DIBR) technique are supported by the system to reduce the requirement of bandwidth. In addition, an asymmetric shift-sensor camera setup is introduced to reduce the hardware cost as well as the complexity of the system. We also simplify the microlens array weight equations so as to reduce the complexity of subpixel rearrangement coefficients calculation for glasses-free 3D image creation. Experiments results show that the proposed architecture can support the resolution of 4K for real-time UHD glasses-free 3D display.

Keywords: Ultra high definition · Glasses-free 3D display
Depth-image-based rendering · Multiview rendering · FPGA

1 Introduction

In recent years, Glasses-free 3D technology has gradually entered the public view. However, the lack of glasses-free 3D content and the shortage of transmission bandwidth for this content hinder the popularity of the glasses-free 3D system. To overcome these obstacles, the V + D video format begins to be applied to the system. Each frame in this video format consists of a 2D color image with a depth map, where the depth map is a gray image with the value varying from 0 to 255, which is usually obtained by depth estimation algorithm or depth camera. In addition, only one channel of video and its depth map are needed to transmit for this format, hence it requires lower bandwidth in comparison with the traditional 3D video. Once each frame has been received and decoded, the color and depth image are recovered and rescaled, and the virtual multiviews required for glasses-free 3D display can be generated from color and depth images by depth-image-based rendering (DIBR) technique [1,2]. Finally, each display view is synthesized by image fusion method, and then the synthesized view is

J. Yang et al. (Eds.): CCCV 2017, Part I, CCIS 771, pp. 380–392, 2017.
https://doi.org/10.1007/978-981-10-7299-4_31

displayed in a set of subpixels, each of which is viewable from a specific angle through a fine lenticular overlay placed over the LCD display [3].

However, this video format has some defects when it is used in a glasses-free 3D system. First, big holes may appear in the generated virtual view (destination image) after DIBR; hence high-performance hole-filling algorithm should be applied to fix these holes. Second, implementation of DIBR and hole-filling algorithm by software is time-consuming and cannot meet the requirement of real-time processing of video stream. In virtue of the power of field programmable gate array (FPGA) in real-time video processing [4], using FPGA to implement real-time glasses-free 3D system has become the consensus of current research. Until now, only a few studies focus on FPGA implementation of the real-time glasses-free 3D system. Most of them only discussed the traditional 3D video format, and the resolution of the 3D system is lower than ultra-high definition resolution, which cannot meet the requirement of the newest ultra-high definition display.

In this paper, we propose an FPGA architecture for real-time ultra-high definition glasses-free 3D system. We present the asymmetric shift-sensor camera setup to simplify the hardware implementation of multiview rendering module of the system. In addition, as this camera setup is applied, no DDR is needed for our FPGA architecture, hence the traditional scheme of using DDR for pipelining is avoided [5]. We also proposed a view fusion method which can rearrange the subpixels from the generated views to form a single glasses-free 3D image. The proposed architecture is implemented by Verilog HDL on the Xilinx Virtex-7 platform and simulated with ModelSim 10.2.

2   Asymmetric Shift-Sensor Camera Setup

As mentioned above, a glasses-free 3D system needs multiple views of the same scene to form a 3D image. In our system, these views are generated by DIBR technique using specified parameters in multiview rendering module. Note that the V + D video used in this paper has no camera calibration parameters, so it is necessary to modify the 3D image warping equations to adapt to the case without camera calibration parameters [6]. In this paper, a scheme of "fusing" 8 views into a glasses-free 3D image is adopted. Positions of these 8 viewpoints/virtual cameras determine the effect of glasses-free 3D display. As shown in Fig. 1, I_r is a reference image, U_r is a point on the reference image, $U_r = [u_r, v_r, 1]^T \in I_r$. The view of the same scene captured at different viewpoints is labeled as $I_n(1 \leq n \leq 7)$; point $U_i = [u_i, v_i, 1]^T \in I_n$. Point $U = [X_w, Y_w, Z_w, 1]^T$ (expressed in normalized homogeneous coordinates in the world coordinate system) is the intersection of u_r and u_i when they are mapped to the scene using the projection matrices of I_r and I_n. B represents the length of the baseline between two adjacent virtual cameras (units: millimeter), Plane z_c ($z_c > 0$) is the zero-parallax setting (*ZPS*) plane. By setting the plane z_c, we can generate a virtual with positive or negative parallax. Note that I_r is taken as one of the 8 views and placed on the rightmost. We call this camera setup asymmetric shift-sensor camera setup. Plane z_c is determined by (1).

$$z_c = \frac{f \cdot s_x \cdot B_n}{2h_n} \qquad 1 \leq n \leq 7 \tag{1}$$

where S_x is the number of pixels per unit length in the x axis, f is the focal length of the camera, B_n represents the length between the camera corresponding to I_r and I_n, $B_n = n \times B$. h_n denotes the horizontal sensor shift (measured in pixels) of I_n when setting the ZPS plane.

Note that the seven virtual views correspond to the same ZPS plane (plane z_c), and their 3D image warping equations can be written as follows [2].

$$\begin{cases} u_i - u_r = (-1)^\alpha \left[2h_n - \frac{f \cdot s_x \cdot B_n}{z_w} \right] \\ v_i = v_r \end{cases} \qquad 1 \leq n \leq 7. \tag{2}$$

where is defined as follows: if I_n is located on the right side of I_r, $\alpha = 0$; otherwise, $\alpha = 1$. As for our camera setup, all virtual views are located on the left side of I_r, hence $\alpha = 1$. Parameter z_w is the depth value of point u_r; it is determined by:

$$z_w = \frac{1}{\frac{D(u_r, v_r)}{255} \times \left(\frac{1}{z_{min}} - \frac{1}{z_{max}} \right) + \frac{1}{z_{max}}} \tag{3}$$

where $D(u_r, v_r)$ is the depth value of the point (u_r, v_r) in depth map, $D(ur, vr) \in [0, 225]$. z_{min} is the near clipping plane, z_{max} is the far clipping plane.

In the case of no camera calibration parameters, in order to simplify the calculation, the near shear plane z_{min} is usually set to 1, and the far shear plane z_{max} is set to infinity [6]. Therefore, Eq. (3) can be simplified as follows.

$$z_w = \frac{255}{D(u_r, v_r)} \tag{4}$$

Putting Eq. (1) into Eq. (2) yields the following equation:

$$\begin{cases} u_i - u_r = -n \cdot f \cdot s_x \cdot B \left(\frac{1}{z_c} - \frac{1}{z_w} \right) \\ v_i = v_r \end{cases} \qquad 1 \leq n \leq 7. \tag{5}$$

Usually, the depth map stores depth information as 8-bit gray values with the gray level 0 specifying z_{max} and the gray level 255 defining z_{min} [7]. According to (4), (5) can be simplified to the following (6).

$$\begin{cases} u_i - u_r = -n \cdot p \left(D_{zps} - D(u_r, v_r) \right) \\ v_i = v_r \end{cases} \qquad 1 \leq n \leq 7. \tag{6}$$

where $p = f \cdot s_x \cdot B/255$ is the scaling factor, $D_{zps} = 255/z_c$, $D(u_r, v_r) = 255/z_w$. D_{zps} are depth value of ZPS plane, $D_{zps} \in [0, 255]$. According to [7,8], the maximum horizontal sensor parallax $d(u_i - u_r)$ which is comfortable for viewing is approximately 5% of the width of an image. Hence we have:

$$|d| \leq W \times 5\% \tag{7}$$

where W is the width of I_r, and both the units of d and W are pixels. Eq. (7) is applied to (6) to get (8).

$$0 \leq p \cdot n \cdot |D_{zps} - D(u_r, v_r)| \leq W/20 \qquad 1 \leq n \leq 7. \tag{8}$$

$$0 \leq |D_{zps} - D(u_r, v_r)| \leq 255 \tag{9}$$

Therefore, the range of p value can be deduced from (8) and (9): $0 \leq p \leq W/35700$.

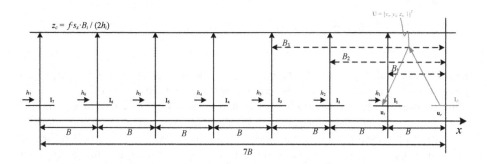

Fig. 1. Asymmetric shift-sensor camera setup

To facilitate hardware implementation, we introduce the integer arguments p', $p' = p \times 2^{15}$. Thus, Eq. (6) becomes as follows.

$$\begin{cases} u_i - u_r \approx -n \cdot \frac{p'}{2^{15}} \left(D_{zps} - D(u_r, v_r)\right) \\ v_i = v_r \end{cases} \qquad 1 \leq n \leq 7. \tag{10}$$

Note the resolution of the ultra-high definition image is 3840×2160, hence $W = 3840$, the maximum possible value of p is $\lfloor 3840 \times 32768/35700 \rfloor = 3524$. When p is the maximum, the maximum absolute value of parallax d is $\lceil 7 \times 3524 \times 255/32768 \rceil = 192$, hence $d \in [-192, 192]$, which satisfies the requirement of Eq. (7).

3   Hardware Architecture

Our hardware architecture is based on Xilinx Virtex-7 FPGA development platform (XC7V585T). The overall system architecture is illustrated in Fig. 2. The system consists of 4 modules: video input, multiview rendering, view fusion and video output. Each module can be divided into several submodules. In multiview rendering module, the current row should be initialized before their parallax values are calculated. So the clock frequency of the system is chosen to reach more than twice the pixel clock frequency of 4K video. In order to reduce the

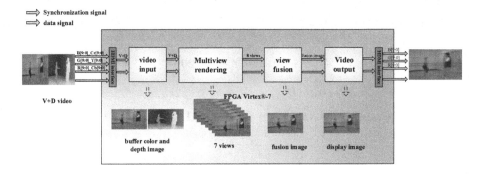

Fig. 2. Hardware architecture of the system

transmission bandwidth of the video, V + D video format begins is applied to the system. Each frame in V + D video is split, stretched and buffered in video input module. Video input module output a color image and its corresponding depth map whose resolutions are both 3840×2160. Multiview rendering module is composed of seven DIBR submodules that are responsible for generating seven different virtual views. View fusion module receives the reference image and the generated virtual views and synthesizes a glasses-free 3D image [9]. Video output module buffers the glasses-free 3D images and output them to the HDMI interface for display.

3.1 Video Input Module

The horizontal resolution of each frame in V + D format video is converted from 3840 to 7680 by video input module using column copy mode. Then, the reference image of each frame and its corresponding depth map is stored into different Block RAMs. In this way, the resolution of the reference image and depth map is expanded to 3840×2160. The architecture of video input module is shown in Fig. 3, which can be divided into system boot, stretching, splitting and buffer module.

Note that stretching module extends the input image data to twice the width of the original image using column replication, and then stores the data correctly into buff_ram. Splitting module contains three Block RAMs: buff_ram, col_ram and dep_ram, where buff_ram's data input port and data output port bit width is 60 bits and 480 bits respectively. The width of data selected here should ensure that the output of data is expected to be completed when the data valid signal is negative. In addition, buff_ram's input clock is the pixel clock, while its output clock is internal working clock of the system, so that the switch of clock domain is realized. Col_ram and dep_ram's input data bit width is 480 bits and its output data bit width is 30 bits and 8 bits respectively, which main function is to achieve the splitting of reference image and depth map. The data buffer module consists of four Block RAMs. By using three Block RAMs, the color data are delayed two lines by ping-pong operation and then entered into the subsequent DIBR module.

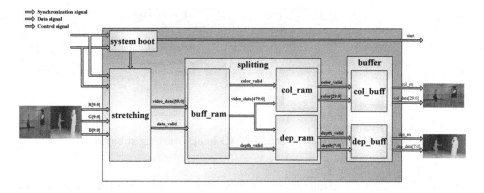

Fig. 3. Hardware architecture of video input module

3.2 Multiview Rendering Module

The multiview rendering module consists of 7 DIBR submodules, which are in charge of generating seven virtual views with different viewing angles. Therefore, this section will focus on DIBR working principle and its mainly design, as shown in Fig. 4. The DIBR module implements two algorithms: 3D image warping and hole filling algorithm mentioned in [10]. An asymmetric shift-sensor camera setup is achieved in 3D image warping. The reason for data valid signal extension is to complete the output of the last two rows of the disparity map after buffering. Small holes of virtual depth map are filtered by median filter in filter module. After hole filling, depth map without hole is used to generate novel view in pixel copy module. 3D image warping and hole-filling submodules are described in detail later.

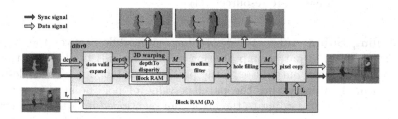

Fig. 4. Hardware architecture of DIBR

3D Image Warping Submodule. As shown in Fig. 5, the 3D image warping submodule receives depth value D outputted by data valid signal extension submodule, calculates the parallax value M of each pixel in the reference image according to D, and outputs M to disparity map filtering submodule. As three rows of parallax value are needed for median filtering, the 3D image warping submodule will buffer these data in the Block RAM (see Fig. 5), and output three parallax values in each clock to the median filter mask in disparity map filtering submodule. Three parallax values are read from the three rows respectively.

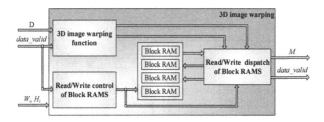

Fig. 5. Hardware architecture of 3D image warping submodule

Note that the destination image generated by 3D image warping may contain holes. In this paper, the parallax of a hole pixel is set to -256 (the parallax cannot achieve at -256 in order to avoid visual fatigue). In 3D image warping submodule, Block RAM is used to store parallax values which is initialized to -256 before processing a depth map. Hence, the subsequent submodules will be able to distinguish which pixels are holes.

In 3D image warping submodule, the Look-Up-Table (LUT) *depth2disparity* is used to realize the depth-to-disparity conversion and it is implemented by *ROM*. The size of the LUT is 256×9 bits, and the values stored in it are calculated in advance outside the system. These values are loaded by parameter setting submodule during the Vertical Blanking Interval (VBI). By this way, the function of (10) can be realized by the LUT *depth2disparity*. In addition, the depth adjustment can be achieved by changing the values in the *depth2disparity*. Another important aspect of the submodule is that the input sequence of the depth map is from right to left, top to bottom. In this way, occlusion-compatible results can be obtained with this sequence, therefore the folds can be eliminated without any extra hardware resources [11].

Hole-Filling Submodule. The hole-filling submodule receives the parallax value RX_M from disparity map filtering submodule, fills the holes in the disparity map, and then outputs it to the pixel-copy submodule. The hole-filling is implemented by modifying the parallax value of hole point (i.e. -256) in the disparity map. According to the algorithm mentioned in [10], the hardware architecture of hole-filling submodule is implemented in Fig. 6. It is shown that hole-filling submodule consists of four function modules: detect_hop, mark_holes, fill_holes and dilate_big_holes module.

The responsibility of detect_hop module is distinguished from foreground and background. Then, the background parallax value of mutation is output. The mark_holes module marks the different categories of holes in the data stream for the subsequent module to fill holes. At last, different methods are used to fill small and big holes in the fill_holes and dilate_big_holes modules, respectively. The small holes are filled with adjacent pixels. After dilation, the big holes are filled with the background parallax value which comes from mutation of detect_hop outputting.

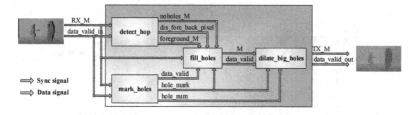

Fig. 6. Hardware architecture of hole-filling submodule

3.3   View Fusion Module

The view fusion module is responsible for subpixels rearrangement of the eight views. Then, subpixels in accordance with arrangement rules of the cylindrical lens glasses-free 3D displays are arranged to generate a stereoscopic image and sent to the HDMI interface [12]. In different glasses-free 3D displays, the subpixel arrangement requirements are different, even if the size of the screen is same. The subpixel arrangement of grating can be divided into integer arrangement and floating-point arrangement. In particular, the integer arrangement can be seen as a special case of floating-point arrangement. The model used in this paper support 4K floating-point cylindrical lens gratings. In addition, lenticular screens are typically slanted at a small angle to reduce a vertical banding effect known as the "picket fence" (an artifact of the microlenses magnifying the underlying LCD mask). This tilt angle is usually divided into two cases, counterclockwise and clockwise. This paper uses an anticlockwise tilt angle. Here we use the floating-point arrangement and counterclockwise tilt angle as an example to introduce the realization of view fusion module. Figure 7 illustrates a slanted lenticular layout with RGB stripe subpixels. According to the physical characteristics of 4K glasses-free displays, the RGB pixel values of the composite image are calculated using the formula (11), and the microlens array weights are calculated using (12) and (13). Note that (13) is further optimized on the basis of previous method in [9].

$$R_{i,j} = \sum_{n=1}^{N} F_{i,j}^n \times r_{i,j}^n \qquad G_{i,j} = \sum_{n=1}^{N} F_{i,j}^n \times g_{i,j}^n \qquad B_{i,j} = \sum_{n=1}^{N} F_{i,j}^n \times b_{i,j}^n \quad (11)$$

$$F_{i,j}^n = \begin{cases} 1 - (k_{i,j} - \lfloor k_{i,j} \rfloor) & if \; n = \lfloor k_{i,j} \rfloor \\ k_{i,j} - \lfloor k_{i,j} \rfloor & if \; n = \lfloor k_{i,j} \rfloor + 1 \\ 0 & otherwise \end{cases} \quad (12)$$

$$K_{i,j} = \frac{(i - i_{off} - 3j \times tan\alpha)\%pitch}{pitch} \times N \quad (13)$$

where i and j denote the horizontal and vertical coordinates, i_{off} is the horizontal offset, $i_{off} = 0$, $i \in [0, 3840 \times 3)$ and $j \in [0, 2160)$, because of RGB three channel. N denote the number of views ($N = 8$), $pitch$ denote the ratio of pixels per inch and per inch, which is a constant ($pitch = 8.01$). Each pixel coefficient

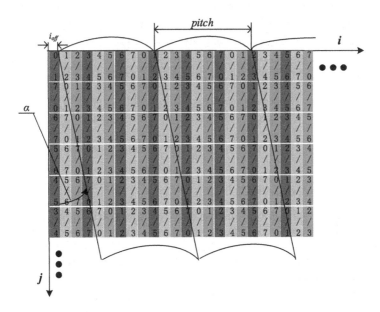

Fig. 7. Slanted lenticular layout with RGB stripe subpixels: two different numbers in one rectangle denote the two participating subpixels from which viewpoint

of the composite image is calculated directly by (13), which greatly consumes hardware resources and increases the complexity of the hardware implementation. Therefore, we make an equivalent transformation to get (14), which has been verified by computer.

$$K_{i,j} = \frac{(i - i_{off} - 3j \times tan\alpha) \times N}{pitch} \% N \tag{14}$$

By observing (14), it can be seen that once the RGB subpixel coefficients of the first pixel are determined, the subsequent subpixel coefficients can be obtained by adding or subtracting the constant from the previous subpixel coefficients. The subpixel coefficients calculation equation in the horizontal and vertical direction can be obtained respectively by (14).

$$K_{i+3,j} = (K_{i,j} + 3\rho)\% N \qquad \rho = N/pitch \tag{15}$$

$$K_{i,j+1} = (K_{i,j} - 3\rho tan\alpha)\% N \qquad \rho = N/pitch \tag{16}$$

According to the above deduction, the coefficient K can be obtained by simple addition or subtraction and one modular operation, which greatly saves the resource consumption and reduces the technical difficulty of the implementation. In the design of this module, the first subpixel coefficients of each image are fixed constants, which can be stored in registers, such as $K_{0,0} = 0$, $K_{1,0} = 0.9988$, $K_{2,0} = 1.9975$, $tan\alpha = 0.2344$, $\rho = 0.9988$. In addition, we adopted IEEE 754 to governs binary floating-point arithmetic, which can greatly guarantee the accuracy of floating-point calculation. The view fusion module consist of three submodules: buffer module, coefficients calculate module and subpixel rearrangement module, and among which the most difficult is coefficients calculate module. Figure 8 demonstrates the coefficients calculation module at the circuit-level.

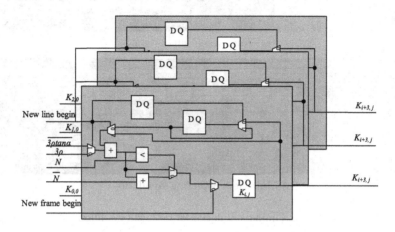

Fig. 8. Subpixel coefficients calculation circuit

4   System Implementation

We used Verilog HDL to implement the design module of system. A proof-of-concept implementation was synthesized and simulated for a Xilinx XC7VX330T FPGA using Xilinx Vivado 2014 toolset and ModelSim 10.2. According to the HDMI 1.4, the 4K ultra-high definition video has a transmission rate of 297 MHz on the HDMI interface. Therefore, in view of the performance of the hardware platform and the processing efficiency of whole system, 666 MHz clock frequency was adopted in the system.

The timing sequence of video input module is shown in Fig. 9(a). We can see that the module implemented the splitting, stretching, and buffering of input image correctly. The reference image and depth map are output by 30 bits and 8 bits respectively.

Timing simulation of multiview rendering module is shown in Fig. 9(b). From this figure, we can see that the module has instantiated 7 DIBR submodules and produced expected timing.

Until now, few studies focus on 8 viewpoints real-time UHD glasses-free 3D system. Hence it is difficult for us to find an appropriate design for comparison. In this paper, we compared the symmetrical shift-sensor camera setup, which was proposed in our previous study [13], with asymmetric shift-sensor camera setup for multiview rendering module. Table 1 summarizes the detailed device utilization results. Compared with a previous design, our innovative architecture saves 1799 slice registers and 1526 slice LUTs for 4K@30 Hz video.

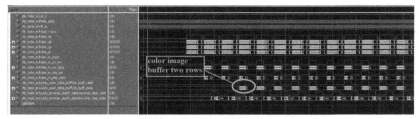

(a) Video input simulation

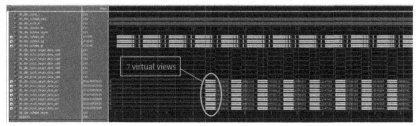

(b) Multiview rendering simulation

Fig. 9. Timing simulation of main submodules;

Table 1. FPGA utilization summary*

Resource utilization	Symmetrical shift-sensor		Asymmetrical shift-sensor		Saved
	Used	Utilization	Used	Utilization	
Slice registers	13424	3%	11625	2%	1799
Slice LUTs	12690	6%	11164	5%	1526
Fully used LUT-FF pairs	8441	47%	7461	42%	980
Bonded IOBs	283	47%	252	42%	31
Block RAM/FIFO	121	16%	113	15%	8
BUFG/BUFCTRLs	3	9%	3	9%	0

*Does not include the recourse consumption of the image fusion module.

Figure 10 shows the original view and the seven virtual views generated by the multiview rendering module. The viewing angles (viewpoints) of those views are slightly different from each other, hence the contents are slightly different. The difference is marked with a red vertical line in the same position of each view in Fig. 10. Go through the red vertical lines from Fig. 10(a) to (h) we can see that the foreground object (man) shifted gradually towards the right side, which indicates that the multiview rendering module can generate correct virtual views. Figure 10(i) shows the fused glasses-free 3D image, which has a good stereoscopic effect when displayed on a 4 K glasses-free 3D display (Fig. 10(j)).

As for view fusion, a real-time system that supports view fusion functionality was presented in [9]. The system in [9] supports 1080p@60 Hz and 4k@30 Hz for 4/6/8/16-viewpoint lenticular-based 3DTV displays. Comparing the design in [9], our design implements a lightweight view fusion module that only supports

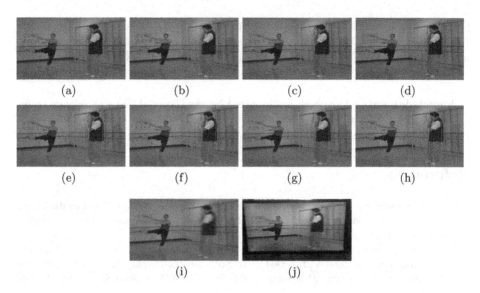

Fig. 10. Eight different views and the glasses-free 3D image fused from them. (a) Reference image I_r; From (b) to (h): I_1–I_7; (i) Glasses-free 3D image; (j) Stereo effect when (i) was displayed in a 4K glasses-free 3D display.

8-viewpoint fusion for 4k display, hence hardware cost is reduced significantly. In addition, the hardware cost is further reduced since some floating-point multiply operations are avoided in the sub-pixel rearrangement coefficients calculation in comparison with the method in [9].

5   Conclusion

This paper presented an FPGA architecture for real-time ultra-high definition glasses-free 3D system. The main contributions of this study are:

- a hardware architecture that supports the V + D format. As the V + D video format is used, the system can generate multiviews that required for glasses-free 3D display using DIBR technique. As a result, only one channel of video and its depth map are needed to transmit for this format, hence it requires lower bandwidth in comparison with the traditional 3D video.
- an asymmetric shift-sensor model. With this camera setup, the multiview rendering module in the system can generate 7 virtual views line by line with a few of Block RAMs. Hence the traditional scheme of using DDR for pipelining is avoided. As a result, the hardware cost, as well as the complexity of the system, is greatly reduced. In addition, as the system with the proposed camera setup can generate glasses-free 3D images, it helps to solve the problem of 3D content shortage.

Our proof-of-concept FPGA implementation supports 4K@30 Hz for real-time UHD glasses-free 3D display. A further extension to support 16-viewpoint (or more) autostereoscopic view synthesis can be easily achieved by duplicating the proposed design.

Acknowledgments. This work is supported by the National Natural Science Foundation of China (No. 61201347) and the Chongqing foundation & advanced research project (cstc2016jcyjA0103).

References

1. Liu, R., Tan, W., Wu, Y., Tan, Y., Li, B., Xie, H., Tai, G., Xu, X.: Deinterlacing of depth-image-based three-dimensional video for a depth-image-based rendering system. J. Electron. Imaging **22**(3) (2013)
2. Fehn. C.: Depth-image-based rendering (DIBR), compression and transmission for a new approach on 3D-TV. In: 2004 SPIE Stereoscopic Displays and Virtual Reality Systems XI, 19 January–21 January 2004, San Jose, CA, United States, pp. 93–104 (2004)
3. Neil, A.: Autostereoscopic 3D displays. Computer **8**, 32–36 (2005)
4. Jin, L., Liang, C., Ying, L., Xie, Y.: Design of spaceborne SAR imaging processing and fast verification based on FPGA. In: IET International Radar Conference 2013, pp. 1–5 (2013)
5. Li, Z., Ye, X.S., Zhang, H., Lu, L., Lu, C., Cheng, L.C.: Real-time 3D video system based on FPGA. In: International Conference on Consumer Electronics, Communications and Networks, pp. 469–472 (2014)
6. Lin, T.-C., Huang, H.-C., Huang, Y.-M.: Preserving depth resolution of synthesized images using parallax-map-based DIBR for 3D-TV. IEEE Trans. Consum. Electron. **56**(Compendex), 720–727 (2010)
7. Liu, R., Xie, H., Tai, G., Tan, Y., Guo, R., Luo, W., Xu, X., Liu, J.: Depth adjustment for depth-image-based rendering in 3D TV system. J. Inf. Comput. Sci. **8**(16), 4233–4240 (2011)
8. Zhang, L., Tam, W.J.: Stereoscopic image generation based on depth images for 3D TV. IEEE Trans. Broadcast. **51**(2), 191–199 (2005)
9. Ren, P., Zhang, X., Bi, H., Sun, H., Zheng, N.: Towards an efficient multiview display processing architecture for 3D TV. IEEE Trans. Circuits Syst. II Express Briefs **PP**(99), 1 (2016)
10. Liu, R., Deng, Z., Yi, L., Huang, Z., Cao, D., Xu, M., Jia, R.: Hole-filling based on disparity map and inpainting for depth-image-based rendering. Int. J. Hybrid Inf. Technol. **9**(5), 145–164 (2016)
11. Liu, R., Xie, H., Tai, G., Tan, Y.: A DIBR view judgment-based fold-elimination approach. J. Tongji Univ. **41**(1), 142–147 (2013)
12. Zhao, W.X., Wang, Q.H., Li, D.H., Tao, Y.H., Wang, F.N.: Sub-pixel arrangement for multi-view autostereoscopic display based on step parallax barrier. Sichuan Daxue Xuebao **41**(6), 216–218+236 (2011)
13. Tan, W.: Research on key technologies of glasses-free 3D display. Chongqing University (2014)

Quantum Video Steganography Protocol Based on MCQI Quantum Video

Zhiguo Qu$^{(\boxtimes)}$, Siyi Chen, and Sai Ji

Jiangsu Collaborative Innovation Center of Atmospheric Environment and
Equipment Technology, Nanjing University of Information Science and Technology,
Nanjing 210044, China
qzghhh@126.com, ycyzcsy@163.com, jisai@nusit.edu.cn

Abstract. A secure quantum video steganography protocol with large
payload based on the video strip encoding method called as MCQI
(Multi-Channel Quantum Images) is proposed in this paper. The new
protocol exploits to embed quantum video as secret information for
covert communication. As a result, its capacity are greatly expanded
compared with the previous quantum steganography achievements. The
new protocol achieves good security and imperceptibility by virtue of
the randomization of embedding positions and efficient use of redundant
frames. Furthermore, the receiver enables to extract secret information
from stego video without retaining the original carrier video, and restore
the original quantum video as a follow. The simulation and experiment
results prove that the algorithm not only has good imperceptibility, high
security, but also has large payload.

Keywords: Quantum information · Quantum image watermark
Continued-fraction algorithm · Invisibility · Capacity

1 Introduction

As one of important research branches of information hiding, quantum steganography is to hide secret information in the form of classical bits or quantum qubits
into carriers in the public channel for covet communications. At present, quantum multi-media mainly includes quantum image, quantum video and quantum
audio, etc. Due to quantum image has been widely used in quantum network, its
representation methods and the related researches on quantum image processing are extensively exploited in recent decades [5–8]. Based on Moire pattern,
Ref. [1] circularly and progressively carried out the recursive operation of embedding and extracting processes in term of layered quantum circuits. Combining
with LSB (Least Significant Bit), Ref. [2,3] embedded secret information into
frequency coefficients of quantum Fourier transforming in frequency domain for
covert communication.

In recent years, with the rapid development of quantum secure communication technology, quantum representation of digital video and the related processing techniques have made great achievements for high level information security

© Springer Nature Singapore Pte Ltd. 2017
J. Yang et al. (Eds.): CCCV 2017, Part I, CCIS 771, pp. 393–403, 2017.
https://doi.org/10.1007/978-981-10-7299-4_32

consequently. For example, in 2015, Fei et al. [4], proposed a method for quantum video encryption and decryption. So far, there is no achievement to consider to embed secret quantum video as secret information for covert communication, at present.

In this paper, we propose a novel quantum steganography protocol which achieves to embed a special digital video as secret information into quantum video carrier. As a result it greatly expands its capacity. In addition, the new protocol effectively improves its security by randomly embedding secret video into quantum video carrier.

2   The Steps of the New Protocol

The flow chart of the protocol is shown as Fig. 1, which consists of four parts including the processes of preprocessing, embedding, extraction and recovery. The detailed step diagram of the protocol is shown as Fig. 2. The two parts of preprocessing and video embedding are implemented by the sender, and the two parts of video extraction and recovery are implemented by the receiver. With the aid of Fig. 4, the new protocol can be explained in detail as follows.

Here the carrier video is presented as:

$$|S_C(m,n)>= \frac{1}{2^{m/2}} \sum_{s=0}^{2^m} |F_s(n)> \otimes |s> \tag{1}$$

$$|F_s(n)>= \frac{1}{2^{n+1}} \sum_{i=0}^{2^{2n}-1} |C_{RGB\alpha}^{s,i}> \otimes |i> \tag{2}$$

$$\begin{aligned}|C_{RGB\alpha}^{s,i}>= &\cos\theta_R^{s,i}|000> + \cos\theta_G^{s,i}|001> + \cos\theta_B^{s,i}|010> + \cos\theta_\alpha^{s,i}|011> \\ &+ \sin\theta_R^{s,i}|100> + \sin\theta_G^{s,i}|101> + \sin\theta_B^{s,i}|110> + \sin\theta_\alpha^{s,i}|111>\end{aligned} \tag{3}$$

And the secret video is denoted as:

$$|S_M(p,q)>= \frac{1}{2^{p/2}} \sum_{s=0}^{2^p} |I(q)> \otimes |s> \tag{4}$$

$$|I(q)>= \frac{1}{2^{q+1}} \sum_{i=0}^{2^{2q}-1} |C_{RGB\alpha}^{s,i}> \otimes |i> \tag{5}$$

$$\begin{aligned}|C_{RGB\alpha}^{s,i}>= &\cos\varphi_R^{s,i}|000> + \cos\varphi_G^{s,i}|001> + \cos\varphi_B^{s,i}|010> + \cos\varphi_\alpha^{s,i}|011> \\ &+ \sin\varphi_R^{s,i}|100> + \sin\varphi_G^{s,i}|101> + \sin\varphi_B^{s,i}|110> + \sin\varphi_\alpha^{s,i}|111>\end{aligned} \tag{6}$$

Here, $|S_C(m,n)>$, $|S_M(p,q)>$ are the quantum carrier video and the quantum secret video respectively. $|S_C(m,n)>$ is composed of a series of 2^m quantum frames, and $|S_M(p,q)>$ is composed of a series of 2^p quantum frames. $|s>$ is the position of each frame in the video. $|C_{RGB\alpha}^{s,i}>$ and $|i>$ are used to encoded the color and position information of quantum image. $|F_s(n)>$ is a $2^n \times 2^n$ quantum image of the s-th frame in carrier video. $|I_s(q)>$ is a $2^q \times 2^q$ quantum image of the s-th frame in secret video. $\theta_X^{s,i}$ and $\varphi_R^{s,i}$ represent the phase value of X color channel of the i-th pixel of s-th frame, where $X \in \{R, G, B, \alpha\}$.

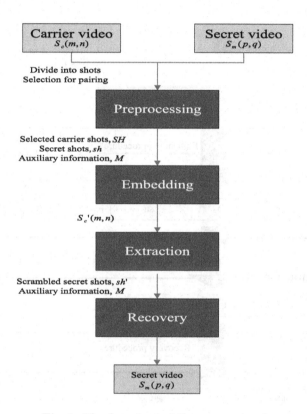

Fig. 1. The flow chart of the new protocol

2.1 Preprocessing

The preprocessing process includes three steps which are presented in details as follows.

The first step is shot division and encoding. The video can be divided into several shots according to the key-frames. The position of key-frames in the carrier video as the key L are predetermined by the sender and the receiver. The position of key-frames in the secret video are determined only by the sender. In term of the key-frames, the sender divides video into different shots. After dividing carrier video into shots, the sender separates the key-frames from other frames to compose of a group except for $|F_t>$, as a separate entity for the next

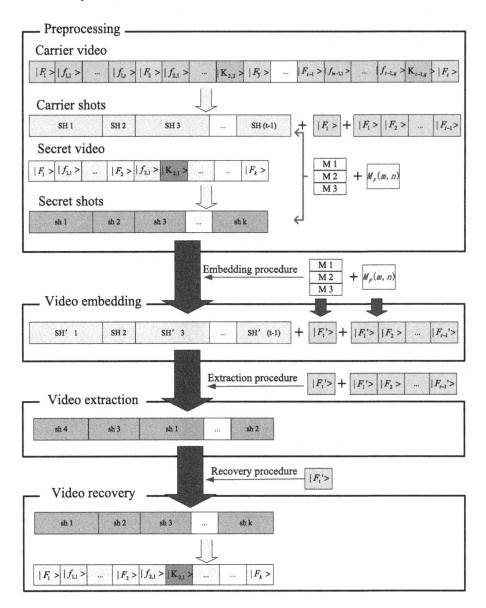

Fig. 2. The detailed step diagram of the new protocol

step. The illustration of the preprocessing process is as shown in Fig. 3. Here, the key-frames, makeup frames and viewing frames are $|F_m>$, $|K_{m,q}>$ and $|f_{m,q}>$ respectively in m-shots.

The second step is shot matching. The sender firstly chooses one carrier shot randomly to compare with the first secret shot. If the length of the carrier shot chose is longer than that of secret shots, this carrier shot is selected as an carrier

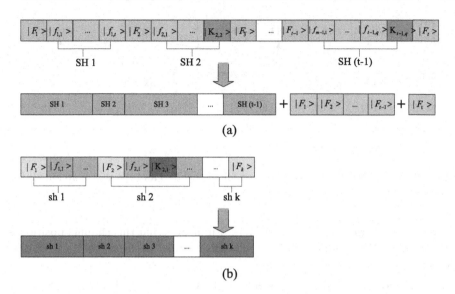

Fig. 3. (a) Dividing carrier video into shots and decoding; (b) dividing secret video into shots and decoding.

shot to embed the first secret shot. Otherwise, another shot will be chose from the rest of carrier shots until finding out an appropriate carrier shot. And so on, all the matching carrier shots are selected for each of secret shot.

The third step is auxiliary information management. There are two kinds of auxiliary information in the procotol. The first one is shot matching information. It consists of three different parts, $M1$, $M2$ and $M3$. Here, $M1$ is the sequence information of the selected carrier shots. $M2$ is the sequence information of the secret shots which are mapped to the selected carrier shots. $M3$ is the number of frames contained by each secret shot. The relation between them is shown as Fig. 4. The other one is embedded position information $M_F(m, n)$. The size of carrier frame is $2^n \times 2^n$ and the size of secret frame is $2^q \times 2^q$. Here, $n > q$. The sender selects $2^q \times 2^q$ pixels on RGB channels respectively in the key-frames of the selected carrier shots for the next step. The position information selected is denoted as $M_F(m, n)$.

2.2 Embedding

The embedding process includes two parts: auxiliary information embedding and secret video embedding. The concrete process are presented in details as follows.

Auxiliary information embedding. The sender firstly converts the shot matching information $M1$, $M2$ and $M3$ into binary sequences, and then embeds them into least significant bits on RGB channels respectively for each pixel in $|F_t>$. Next, the sender embeds the embedded position information $M_F(m, n)$ by modifying least significant bits on RGB channels respectively in corresponding

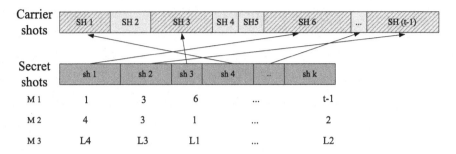

Fig. 4. Secret shots, carrier shots mapped (shaded shots) and the shot matching information $M1$, $M2$, $M3$

key-frames of carrier shots mapped. Here, 1 indicates that this channel position is embedding location and 0 indicates that this channel position is not embedding location.

Secret video embedding. The sender embeds secret video based on shot units. In term of the position information $M_F(m,n)$, each frame of secret shots is embedded into selected carrier shots in sequence.

Table 1. Division of the angle value θ_X^i and the corresponding grayscales G_X^i

θ_X^i	G_X^i
$[0, 1.1°)$	0,1
$[1.1°, 1.8°)$	2,3
$[1.8°, 2.5°)$	4,5
$\dots$	$\dots$
$[88.6°, 89.3°)$	252,253
$[89.3°, 90°]$	254,255

The division of the interval of angle value θ_X^i and the corresponding grayscales G_X^i is presented in Table 1. The θ_X^i is the angle value of X channel in the i-th pixel, $\theta_X^i \in [0°, 90°]$ and G_X^i denotes the grayscale value of X channel in the i-th pixel l in 8 bits, $G_X^i \in [0, 255]$. Here, $X \in \{R, G, B, \alpha\}$.

Referring to the Table 1 to define the interval $[a, b)$ satisfying $\theta_X^{s,i} \in [a, b)$. Here, a is the left endpoint of the interval and b is the right endpoint. $\theta_X^{s,i}$ is a value in the interval $[a, b)$, and ε is a spillover angle.

$$\theta_X^{s,i} = a + \frac{\varepsilon}{90} \times 0.7 \quad \theta_X^{s,i} \in [a, b), \varepsilon \in [0, 90°] \tag{7}$$

After embedding the color information of the secret video $\varphi_X^{s,i}$ into the pixels of carrier video $\theta_X^{s,i}$, the color information of the i-th pixel in the s-th stego frame, $\theta_X^{s,i'}$ is shown as the Eq. (14). Here, $X \in R, G, B, \alpha$.

$$\theta_X^{s,i'} = a + \frac{\varphi_x^{s,i}}{90} \times 0.7 \quad \theta_X^{s,i'} \in [a,b] \tag{8}$$

The shifting angle can be calculated as:

$$\theta_{Xr}^{s,i} = \theta_X^{s,i'} - \theta_X^{s,i} \tag{9}$$

It adopts COI [4] operation to complete the above steps:

$$COI_X = U_X \otimes I^{\otimes 2n}, \quad X \in \{R, G, B\} \tag{10}$$

$$U_X = C^2 R_y(2\theta_{Xr}^{s,i}) \tag{11}$$

The COI operation is realized by using $C^2 R_y(2\theta)$ gate, where $\theta_{Xr}^{s,i}$ is the shifting angle. The concrete introduction of the quantum circuits of U_X are $C^2 R_y(2\theta)$ gate are given in [4].

After embedding secret shots, the sender transmits the embedded carrier video $S_C'(m, n)$ to the receiver. Here, the carrier video is presented as:

$$|S_C'(m,n) >= \frac{1}{2^{m/2}} \sum_{s=0}^{2^m} |F_s'(n) > \otimes |s > \tag{12}$$

$$|F_s(n) >= \frac{1}{2^{n+1}} \sum_{i=0}^{2^{2n}-1} |C_{RGB\alpha}^{s,i}{}' > \otimes |i > \tag{13}$$

2.3   Extraction

The extraction process of secret video includes two parts: auxiliary information extraction and secret shot extraction. The concrete steps are presented as follows.

Auxiliary information extraction. After receiving the embedded carrier video $S_C'(m,n)$, the receiver divides the embedded carrier video into shots according to the key L. And then the receiver extract the auxiliary information $M1$, $M2$ and $M3$ from the least significant bit of RGB channels respectively of $|F_t' >$. In term of $M1$, the receiver finds out the carrier shots which embedded the secret information. Finally, the receiver extract the embedded position information $M_F(m, n)$ from the least significant bit of RGB channels of corresponding key-frames.

Secret shot extraction. Firstly, the receiver find out the selected carrier shots which embedded the secret information based on the auxiliary information $M1$. And then, the receiver confirm the number of frames contained by each secret shot based on the auxiliary information $M3$. Next, the receiver confirm the embedded position in each stego frame based on the auxiliary information $M_F(m, n)$. Finally, the receiver extracts secret shot frames from carrier shots mapped.

In specific, the receiver firstly confirms the interval $[a, b)$ satisfying $\theta_X^{s,i'} \in [a, b)$ by referring to the Table 1. Then he calculates the secret pixel $\varphi_X^{s,i}$ according to $\theta_X^{s,i'}$ and a:

$$\varphi_X^{s,i} = \frac{(\theta_X^{s,i'} - a)}{0.7} \times 90 \quad \theta_X^{s,i'} \in [a, b) \tag{14}$$

After recursively implementing the above steps for every carrier shot mapped, the receiver can obtain all the secret shots consequently.

2.4 Recovery

Because the secret shots extracted from carrier shots are disorder, in term of the auxiliary information $M2$, the receiver can regain the correct sequence of secret shots to restore the complete secret video.

3 Simulations and Analysis

There are three main performance parameters for quantum steganography, including capacity, security and imperceptibility. In this subsection, these performance parameters of the new algorithm are analyzed in details.

The simulation are carried out on a desktop computer with Intel(R), Pentium(R) G860 3.00 GHz CPU, 2.00 GB RAM, and 32bit operating system equipped with the MATLAB R2012a environment. A dataset of ten videos are used for the simulation as shown in Fig. 5. The five videos in Fig. 5(a) are used as carrier videos, while the other five videos in Fig. 5(b) are used as secret videos in the simulation. The details of each video are given in Table 2.

Fig. 5. (a) The videos used as carrier video in the simulation; (b) the videos used as secret video in the simulation

Table 2. The videos used in the simulation

Video ID	Name	Information bits	Resolution	Numbers of frames	Numbers of key-frames
Vc1	Climber	50.2 MB	512 × 256	280	15
Vc2	Figure	36.0 MB	352 × 288	260	15
Vc3	MGM	65.4 MB	352 × 288	310	15
Vc4	Earth	72.0 MB	436 × 344	320	20
Vc5	Box	96.4 MB	560 × 420	420	25
Vs1	Trtris	13.4 MB	256 × 256	150	12
Vs2	Clock	4.2 MB	176 × 144	90	10
Vs3	Building	7.6 MB	176 × 144	220	15
Vs4	Cycling	17.4 MB	325 × 245	160	8
Vs5	Man	27.8 MB	325 × 245	200	15

3.1   Capacity

Capacity usually is defined as the maximum bits of secret information embedded. In this paper, for convenience, we redefine it as the ratio between the maximum bits of secret information embedded with that of original carrier video. It is shown as Eq. (15).

$$\text{Capacity} = \frac{Maximum\ bits\ of\ \text{secret information embedded}}{\text{Information}\ bits\ \text{of}\ original\ \text{carrier video}} \times 100\% \qquad (15)$$

In the new protocol, after embedding five secret videos into five carrier videos respectively, their capacity are obtained as presented in Table 3.

Table 3. The capacity of each video for embedding

Video ID	Vs1	Vs2	Vs3	Vs4	Vs5	Average
Vc1	26.8%	8.3%	15.2%	34.7%	43.4%	25.68%
Vc2	37.2%	11.5%	21.6%	48.2%	60.2%	38.74%
Vc3	31.2%	9.7%	17.7%	40.4%	50.5%	29.9%
Vc4	13.9%	4.3%	7.9%	18.0%	22.5%	13.32%
Average	25.92%	8.02%	14.8%	33.56%	41.96%	24.852%

From Table 3, we can see that the average capacity is 24.852%. In term of Table 2, it can be concluded that embedding capacity is determined by the embedding method and carrier video's inherent characteristics. Although there are some capacity ratios of existing protocols based on quantum image are greater than that of the new protocol, the information bit of secret video embedded in the new protocol is greatly larger than that of them, due to that one video usually contains hundreds of images.

3.2 Imperceptibility

Imperceptibility is one of most significant performance factors for quantum steganography. It's usually introduced to evaluate effects on carriers caused by embedding secret information. In this paper, for convenience, we use SSIM (Structural Similarity Index Measurement) to evaluate imperceptibility.

Given two frames, x and y. The SSIM of them can be calculated as follow.

$$SSIM(x,y) = \frac{(2\mu_x\mu_y + c_1)(2\sigma_{xy} + c_2)}{(\mu_x^2 + \mu_y^2 + c_1)(\sigma_x^2 + \sigma_y^2 + c_2)} \tag{16}$$

Here, μ_x is the average of x and μ_y is the average of y. σ_x^2 is the variance of x, σ_y^2 is the variance of y and σ_{xy} is the standard deviation between x and y. $c_1 = (k_1 l)^2$ and $c_2 = (k_2 l)^2$ are constant to maintain stable. l is the dynamic range of pixel values. In this case, it is 255. $k_1 = 0.01$ and $k_2 = 0.03$. The range of SSIM is 0 to 1. If two images are the same, the SSIM value between them is equal to 1.

After embedding five secret videos into five carrier videos respectively, the SSIMs are calculated as shown in Fig. 6.

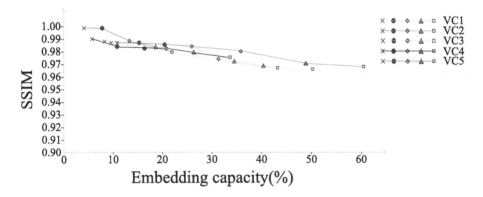

Fig. 6. SSIM of the encrypted videos

It's found that the distribution of data in Fig. 6 is centralized among 0.98 1.00, and the lowest value is also more than 0.96. These results clearly indicate that it is almost impossible to tell whether carrier video was encrypted or not. And then, the imperceptibility of the new protocol is proved.

3.3 Security

The new protocol achieves good security by virtue of unique features of quantum video. The concrete partition of shot is based on key-frames. In the new protocol, the key denotes the position of key-frames in carrier video, which are predetermined by the sender and the receiver. Even if the eavesdropper obtain the embedded video, he cant divide it into correct shots to extract secret video. Besides that, the randomization of embedding shots effiectively enhances its security. Furthermore, each encrypted pixel in carrier frames contanins partial channel information of secret pixel. Even the eavesdropper gains partial encrypted pixels, he still disables to obtain secret pixel information.

Compared with classical steganography, quantum video is expressed as a normalized quantum superposition state. When eavedropping, the quantum state will be collopsed and the sender will be known. So we can ensure the high security of the new protocol.

3.4   Conclusion

By taking full advantages of unique features of video frames, the new protocol realizes to embed a secret video into quantum carrier video randomly. It capacity are greatly expanded compared with the previous quantum steganography achievements. Meanwhile, the new algorithm also achieves good security and imperceptibility by virtue of the randomization of embedding positions and efficient use of redundant frames.

Taking into account a basic fact that quantum video has been widely used in quantum networks, we will extend the proposed quantum video steganography in the frequency and consider introducing coding theory for better security and larger capacity.

Acknowledgments. This work was supported by the National Natural Science Foundation of China (Nos. 61373131, 61303039, 61232016, 61501247), PAPD and CICAEET funds.

References

1. Jiang, N., Wang, L.: A novel protocol for quantum image steganography based on Moire patten. Int. J. Theor. Phys. **54**(3), 1021–1032 (2015)
2. Wang, S., Sang, J.Z., Song, X.H.: LSQb information hiding algorithm for quantum image. Measurement **73**, 352–359 (2015)
3. Jiang, N., Zhao, N., Wang, L.: LSB based quantum image steganography algorithm. Int. J. Theor. Phys. **55**(1), 107–123 (2016)
4. Yan, F., Iliyasu, A., Venegas-Andraca, S.E., Yang, H.: Video encryption and decryption on quantum computers. Int. J. Theor. Phys. **54**(8), 2893–2904 (2015)
5. Sun, B., Iliyasu, A., Yan, F., Dong, F., Hirota, K.: An RGB multi-channel representation for images on quantum computers. J. Adv. Comput. Intell. Intell. Inf. **17**(3), 404–417 (2013)
6. Zhang, Y., Lu, K., Gao, Y., Wang, M.: NEQR: a novel enhanced quantum representation of digital images. Quantum Inf. Process. **12**(8), 2833–2860 (2013)
7. Le, P.Q.: Flexible representation and processing transformations for quantum images, and their application. 09D35210(2012)
8. Le, P., Dong, P., Hirota, K.: Flexible representation of quantum images and its computational complexity analysis, pp. 146–149. ISIS, Busan Korea (2009)

Video Question Answering Using a Forget Memory Network

Yuanyuan Ge, Youjiang Xu, and Yahong Han[✉]

School of Computer Science and Technology,
Tianjin University, Tianjin 300350, China
{geyuanyuan,yjxu,yahong}@tju.edu.cn

Abstract. Visual question answering combines the fields of computer vision and natural language processing. It has received much attention in recent years. Image question answering (Image QA) targets to automatically answer questions about visual content of an image. Different from Image QA, video question answering (Video QA) needs to explore a sequence of images to answer the question. It is difficult to focus on the local region features which are related to the question from a sequence of images. In this paper, we propose a forget memory network (FMN) for Video QA to solve this problem. When the forget memory network embeds the video frame features, it can select the local region features that are related to the question and forget the irrelevant features to the question. Then we use the embedded video and question features to predict the answer from multiple-choice answers. Our proposed approaches achieve good performance on the MovieQA [21] and TACoS [28] dataset.

Keywords: Image QA · Video QA · Forget memory network

1 Introduction

Question answering (QA) is a hot topic in artificial intelligence. It focuses on building a system that can automatically answer questions posed by humans in a natural language. Textual question answering (Textual QA) [3,11,19,22] has made great progress. Kumar et al. [11] proposed the dynamic memory network (DMN) to solve textual question answering. The dynamic memory network has four modules. The input module is used to encode raw text inputs from the task into distributed vector representations. And the question module is used to encode the question of the task into a distributed vector representation. The episodic memory module can search the inputs and retrieve relevant facts according to the attention mechanism. The answer module can generate an answer from the final memory vector of the memory module. Sukhbaatar et al. [19] introduced a novel recurrent neural network (RNN) architecture where the recurrence reads from a possibly large external memory multiple times before outputting a symbol to solve textual question answering. The model can be considered as a continuous form of the memory network [22]. And it can also

© Springer Nature Singapore Pte Ltd. 2017
J. Yang et al. (Eds.): CCCV 2017, Part I, CCIS 771, pp. 404–415, 2017.
https://doi.org/10.1007/978-981-10-7299-4_33

be considered as a version of RNNsearch [1] with multiple computational steps per output symbol. Due to the good performance of deep learning applications in image classification [5,10,18,20], object detection [15,16], image captioning [4,8,24] and video captioning [12,26], researchers begin to utilize deep learning techniques to solve the image question answering (Image QA) task. Image QA combines the fields of computer vision and natural language processing. It can deeply understand the content of an image and answer questions according to the visual content of an image. Many approaches have been proposed to sovle the task of Image QA [7,14,23,25,29]. Jiang et al. [7] proposed a Compositional Memory for an end-to-end training framework. They advance the study of the VQA by taking the advantage of the recent progresses in image captioning [4,8], natural language processing [11], and computer vision. Noh et al. [14] proposed a dynamic parameter prediction network (DPPnet) to solve Image QA. The dynamic parameter prediction network is composed of the classification network and the parameter prediction network. The parameter prediction network predicts candidate weights through a GRU model. The classification network has a dynamic parameter layer whose weights are mapped by a hashing trick from the candidate weights. Xiong et al. [23] proposed improved dynamic memory networks (DMN+) and introduced a new input module for images. The input module splits an image into small local regions and considers each region equivalent to a sentence. There are some improvements to the memory and input modules. Yang et al. [25] proposed the stacked attention networks (SANs) that learn to answer natural language questions from images. The stacked attention networks allow multi-step reasoning for image QA. And It can be viewed as an extension of the attention mechanism that has been successfully applied in image captioning [24] and machine translation [1].

Image QA task only needs to explore one image's visual content to answer the question. But due to the complexities of the video data, Video QA usually needs to explore a sequence of images to answer the question. The research of Video QA is still at its beginning. Only a few methods have been proposed for Video QA [21,27,28]. Tapaswi et al. [21] collected the MovieQA dataset which includes diverse of sources of information. And they provided several intelligent baselines and extended existing QA techniques to solve movie question answering. Zeng et al. [27] proposed a scalable approach to learn video question answering. They first collected a large-scale video QA dataset with 18100 videos and 175076 candidate QA pairs by using a state-of-the-art question generation method [6]. Then the Video QA model takes a video and the generated questions as input and outputs the corresponding answers. Zhu et al. [28] also collected a large scale dataset for Video QA tasks. And they proposed an encoder-decoder framework to learn temporal structures of videos and use a dual-channel ranking loss to predict the answer from multiple-choice answers. When these proposed video QA approaches embed the video features, they utilize all video frame features extracted by CNNs and get the video features by mean-pooling all video frame features or inputting them to the GRU model. These methods take all video frame features into consideration. But not all features are related to the

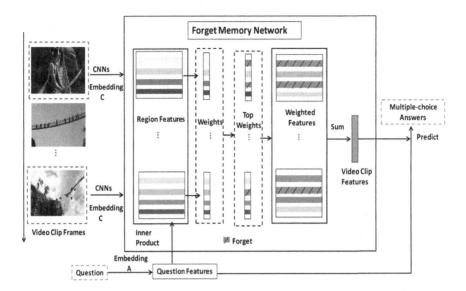

Fig. 1. Overview of our forget memory network (FMN) for video question answering. Firstly we use CNNs and embedding matrix C to extract region features for each video clip frame. Then the FMN model can select useful region features and forget irrelevant region features according to the question features, and get video clip features for a video clip. Finally we use video clip features and question features to predict the correct answer from multiple-choice answers.

question. The question usually has relation with some local region features of video frames. When they utilize the whole features of video frames to embed the video features, the irrelevant region features may weaken the useful region features for the question. So how to find the useful region features and remove the irrelevant features for the question needs to be solved. Therefore, we propose a Video QA framework based on forget memory network to solve this problem.

Figure 1 depicts our framework of the FMN for Video QA. Firstly, we utilize convolutional neural networks and embedding matrix C to get region features for each video clip frame. And the question is embedded via embedding matrix A. Then we fuse question features with each region features by taking the inner product and obtain a weight for each region features of all video clip frames. In order to find out the most relative region features for the question, we set the top k-th weight of all weights as a threshold. The weights below the threshold are set to zero, which are used to forget irrelevant region features for the question. Then we get the weighted region features by using these weights multiply the corresponding region features. We sum all weighted features to get the final video clip features to represent the video clip. Finally we fuse the video clip features and question features to predict the correct answer from multiple-choice answers.

This paper mainly makes two contributions: Firstly, we propose a forget memory network (FMN) to solve Video QA. It can select the useful features for the question and forget the irrelevant information. Secondly, our proposed approach achieves good performance on the MovieQA [21] and TACoS [28] dataset.

2   The Proposed Approach

In this section, we describe our proposed approach in detail. Firstly, we introduce how to get the representation of questions. Secondly, we review the memory network. Then we detail how to use the forget memory network (FMN) to get the video clip features for a video clip. And we also introduce an extension of the FMN. Finally, we introduce two different ways to predict the correct answer from multiple-choice answers.

2.1   Feature Representation of Questions

Question feature representation plays an important role in video question answering. In this paper, we use the following method to represent the question features. For each word of the question, we use d_1-dimensional word2vec [13] vectors to represent the word. Suppose that a question has n words, we can use the following Eq. (1) to represent it:

$$q_w = \{w_1, w_2, \ldots, w_n\}, w_i \in R^{d_1} \tag{1}$$

Then we average the word2vec features of all words and use an embedding matrix $A \in R^{d_2 \times d_1}$ to get the final question feature representation q.

$$\bar{q_w} = \frac{1}{n} \sum_{i=1}^{n} w_i, w_i \in R^{d_1} \tag{2}$$

$$q = A\bar{q_w}, q \in R^{d_2} \tag{3}$$

2.2   Review of Memory Network

As our forget memory network is based on memory network [22], we first introduce how to use the memory network to solve question answering task. The memory network was used to solve text question answering at the beginning. And the MemN2N proposed by [19] improved the memory network. We review the MemN2N model in the single layer case. The MemN2N model takes an sentence set $x_1, x_2, \ldots, x_i$ and a question as inputs. The entire set of $\{x_i\}$ are encoded as two different sentence representation sets $\{m_i\}$ and $\{c_i\}$ using two different word embeddings. And the question is encoded as a vector y using a word embedding matrix. The encoded features have the same dimension. Then we compute a probability for each inputs by taking the inner product by a *Softmax* between the sentence representation m_i and the question representation y.

$$p_i = Softmax(y^T m_i) \tag{4}$$

where $Softmax(z_i) = e^{z_i} / \sum_j e^{z_j}$. Then we multiply the probability p_i with another sentence representation c_i and sum them together to get the final representation vector o of the sentence set $\{x_i\}$.

$$o = \sum_i p_i c_i \tag{5}$$

The final representation vector o and the embedding question y is passed through a final weight W and a $Softmax$ to predict the correct label.

$$\hat{a} = Softmax(W(o + y)) \tag{6}$$

The MemN2N was not designed for multi-choice answering. Tapaswi et al. [21] proposed two improvements to make the network suitable for the movie question answering. The first approach uses an additional embedding layer map the multi-choice answer a_j to a vector g_j. Then we use the following equation to get an score vector for answers g. And the answer which has the highest score $\hat{a}$ in a is the correct answer.

$$a = Softmax((o + y)^T g) \tag{7}$$

The second improvement is using one word embedding matrix to replace the different word embedding matrixes which are used to encode the sentence set and question.

2.3   Forget Memory Network

In this part, we describe how to use the forget memory network (FMN) to obtain the video clip features. Firstly, we extract video clip frames from the original video clip with the rate of 1 frame per second and select T frames to represent the whole video clip. Secondly, we resize the video frame and use a convolutional neural network to extract features for the video frame. Suppose that the output feature size of each video clip frame is $H \times W \times D$. H and W are the height and width of the output feature map. And D is the number of feature maps. The original frame can be recognized as being splited into $H \times W$ grids. Each grid is a local region. And we use D dimensional vector to represent each region feature $r_{ij} \in R^D, i \in \{1, 2, \ldots, T\}, j \in \{1, 2, \ldots, H \times W\}$. Therefore, we can get $H \times W$ region features for each frame and $H \times W \times T$ region features for a video clip. Then we use embedding matrix $C \in R^{d_2 \times D}$ to get the embedded feature u_{ij} for each local region of a video clip frame.

$$u_{ij} = Cr_{ij}, i \in \{1, 2, \ldots, T\}, j \in \{1, 2, \ldots, H \times W\}, u_{ij} \in R^{d_2} \tag{8}$$

Then we use the question feature q inner product each embedded region feature u_{ij} and obtain a weight for each region of the video clip.

$$w_{ij} = u_{ij}^T q, i \in \{1, 2, \ldots, T\}, j \in \{1, 2, \ldots, H \times W\} \tag{9}$$

We know that not all video clip frame features are useful to the question and not all region features in one frame are relevant to the question. The question may only have relation with several region features in some video frames. So some irrelevant local region features need to be forgot. We select the top k-th weight k_w from all weights as the threshold using the Eq. (7).

$$G = \{w_{ij} | i \in \{1, 2, \ldots, T\}, j \in \{1, 2, \ldots, H \times W\}\} \tag{10}$$

$$k_w = max_k (G, k), k \in \{1, 2, \ldots, H \times W \times T\} \tag{11}$$

Then the weights below the threshold k_w are set to zero, so that we can forget some region features which are not related to the question and remain the useful region features to the question. Finally we use the processed weight w_{ij} multiply the corresponding features to get the weighted features. We obtain the frame feature representation f_i by sum the weighted features of one video frame. And we sum all frame representation of a video clip to get the final video clip feature representation.

$$w_{ij} = \begin{cases} w_{ij}, & w_{ij} \geq k_w; \\ 0, & \text{else.} \end{cases} \tag{12}$$

$$f_i = \sum_{j=1}^{W \times H} w_{ij} \times u_{ij}, i \in \{1, 2, \ldots, T\} \tag{13}$$

$$v = \sum_{i=1}^{T} f_i \tag{14}$$

2.4   Extended Forget Memory Network

The forget memory network (FMN) is a basic structure to extract video clip features. In order to exploit the temporal information of video clip frames and find better representation of the video clip, we put forward the extended FMN, which utilize a GRU [2] model to deal with the video clip frame features. The states in a GRU can be updated using the following equations:

$$z_t = \sigma(W_z x_t + U_z h_{t-1}) \tag{15}$$

$$r_t = \sigma(W_r x_t + U_r h_{t-1}) \tag{16}$$

$$\widetilde{h_t} = tanh(W_h x_t + U_h(r_t \odot h_{t-1})) \tag{17}$$

$$h_t = (1 - z_t) \odot h_{t-1} + z_t \odot \widetilde{h_t} \tag{18}$$

where x_t is the input. z_t and r_t respectively denote the update gate and reset gate. σ is the sigmoid function and $\odot$ is the element-wise multiplication of two vectors. We use $f_i, i \in \{1, 2, \ldots, T\}$ to represent the video clip frame features. T is the number of frames sampled from the video clip. We input a sequence of video clip frame features $f_1, f_2, \ldots, f_T$ into the GRU model. The output of the last time step h_t can be used to represent the video clip features v.

2.5   Predicting the Answer from Multiple Answers

In this section, we introduce two different approaches to predict the answer from multiple-choice answers. The first approach to predict the answer is as follows. Suppose that each question has N candidate answers, only one of which is correct. We use the method similar to obtain the question feature representation to get each answer feature representation $a_i \in R^{d_2}, i \in \{1, 2, \ldots, N\}$. Then we fuse the question and video feature representation $q \in R^{d_2}, v \in R^{d_2}$ by sum them together. Then each answer can get a score $s_i, i \in \{1, 2, \ldots, N\}$ via a inner product between each answer and fused video question features. Finally we use the $Softmax$ function to get a probability for each answer. The correct answer has the highest probability.

$$s_i = (v + q)^T a_i, i \in \{1, 2, \ldots, N\} \tag{19}$$

$$S = [s_1, s_2, \ldots, s_N] \tag{20}$$

$$\hat{p_a} = Softmax(S) \tag{21}$$

The second approach to predict the answer can be simplified to a classification task. We collect all answers of the training set and create a candidate answer dictionary. The length of the candidate answer dictionary is the number of final classes the models need to learn to predict. We use a fully connected layer followed by dropout and $softmax$ function to do classification. Then we find the probability of the N answers of a question and select the most confident one as the correct answer.

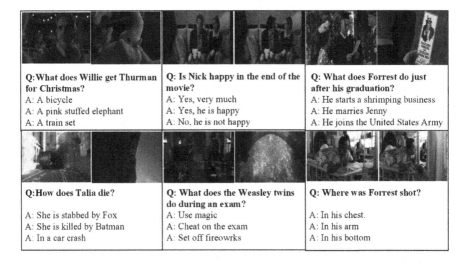

Fig. 2. Examples of multiple-choice QA from the MovieQA dataset. Each question has 5 multiple-choice answers. We only show the correct answer (blue) and two wrong answers for each question. (Color figure online)

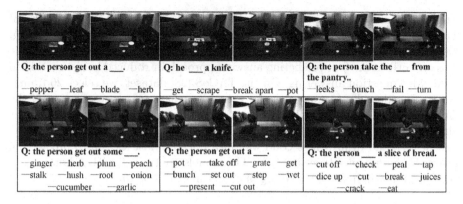

Fig. 3. Examples of multiple-choice QA from the TACoS dataset. The first row shows easy level QAs, each of which has 4 multiple-choice answers. The second row shows hard level QAs, each of which has 10 multiple-choice answers. The blue one is correct, and the others are wrong. (Color figure online)

3   Experiments

3.1   Dataset

We evaluate our forget memory network (FMN) on the MovieQA [21] and TACoS [28] datasets. The MovieQA [21] dataset contains multiple sources of information such as video clips, plots, subtitles, scripts and DVS [17]. It consists of 14,944 questions about 408 movies. For each movie, there are a set of questions. Each question has 5 multiple-choice answers, only one of which is correct. But not all movies have video clips. There are only 140 movies (6,462 QAs) that have video clips. In this paper, we only use video clips to solve movie question answering. Figure 2 showes some examples of the multiple-choice QA from the MovieQA dataset.

The TACoS [28] dataset consists of 127 long videos with total 18227 video clips in the cooking scenario. It contains easy and hard level QAs. There are 3 tasks for each level questions, describing the past, present, and future. And there are 3 splits for each task. For easy level QAs, each question has 4 multiple-choice answers. And For hard level QAs, there are 10 multiple-choice answers for each question. Two level QAs both have only one correct answer and the others are wrong. Figure 3 shows some examples from the TACoS dataset. The first row is easy level QAs and the second row is hard level QAs.

3.2   Experimental Setup

Firstly, we extract video frames from video clips with the rate of 1 frame per second. And the frames are resized to 224 × 224. We use GoogLeNet to extract the last concat layer inception_5b/output features for each video frame. The output feature size is 7 × 7 × 1024. On the MovieQA [21] dataset, we select 32 frames from all movie clips to represent a movie. For the word representation of the MovieQA dataset, we utilize the word2vec model supported by [21] to obtain 300 dimensional word embeddings. In this dataset, we only focus on the movie clips and utilize the first predicting answer approach introduced in last section to solve video question answering. On the TACoS [28] dataset, we select 26 frames from each video clip to represent the video. For the word representation of the TACoS dataset, we represent the word features with a 512 dimensional vector which is initialized randomly. In this dataset, we only focus on describing the present task and utilize the second way introduced in last section to predict the answer from multiple-choice answers. The models are trained with Adam update rule [9], mini-batch size 32, and a global learning rate of 10^{-3}.

3.3   Evaluation Metrics

Our forget memory network (FMN) solves video question answering which has multiple-choice answers. In our experiment, we use accuracy to describe our experimental results. The accuracy is defined as the ratio between the number of correct QAs and all QAs. The formula is as follows:

$$acc = \frac{\{\sharp \text{ correct QAs}\}}{\{\sharp \text{ all QAs}\}} \tag{22}$$

3.4   Experimental Results

Firstly, we compare our experimental results with SSCB all clips [21] and MemN2N all clips [21] performance on the MovieQA dataset. On the MovieQA dataset, we only utilize the movie clips to sovle the MovieQA task.

Table 1. Comparison with different methods on the MovieQA dataset.

Model	Accuracy
SSCB all clips [21]	21.6%
MemN2N all clips [21]	23.1%
FMN	35.4%
Extended FMN	36.6%

From Table 1, we can see that our model performance outperforms the methods proposed in [21]. With only video features to MovieQA task, we achieve 35.4% accuracy on the validation set using the FMN model. Compared to the SSCB all clips and MemN2N all clips methods, it increases 13.8% and 12.3%. And With the extended FMN model, we can achieve 36.6% accuracy, with an improvement of 15% and 13.5% compared to the SSCB all clips and MemN2N all clips method.

Table 2. Comparison with different methods on the TACoS dataset.

Models	Easy				Hard			
	Split 1	Split 2	Split 3	Mean	Split 1	Split 2	Split 3	Mean
CCA [28]	67.1%	64.9%	63.2%	65.1%	-	-	-	-
ConvNets [28]	77.7%	78.3%	72.9%	76.3%	-	-	-	65.5%
GRUmodel [28]	79.1%	81.9%	78.1%	79.7%	66.9%	66.2%	68.2%	67.1%
FMN	83.8%	83.2%	81.1%	82.7%	77.2%	78.2%	76.7%	77.3%

Secondly, we show our experimental results on the TACoS dataset, and compare our model's performance with CCA [28], ConvNets [28], and GRUmodel [28] methods.

From Table 2, we can observe that our FMN model achieves good performance on the easy level QAs and hard level QAs. On the easy level QAs, our FMN model outperforms CCA model with a large margin. Compared to the ConvNets and GRUmodel, our FMN model has 6.4% and 3% improvements. On the hard level QAs, our FMN models has 11.8% and 10.2% improvements compared to the ConvNets and GRUmodel.

Finally, we compare the FMN and extended FMN models performance on the TACoS dataset. The extended FMN can exploit the temporal information of the video and obtain a better video feature representation. From Table 3, we can see that the extended FMN model achieves 83.9% accuracy outperforming the FMN model's performance by 1.2% on the easy level QAs. On the hard level QAs, the extended FMN model improves 0.7% than the FMN model.

Table 3. Comparison between the FMN and extended FMN models on the TACoS dataset.

Models	Easy				Hard			
	Split 1	Split 2	Split 3	Mean	Split 1	Split 2	Split 3	Mean
FMN	83.8%	83.2%	81.1%	82.7%	77.2%	78.2%	76.7%	77.3%
Extended FMN	85.4%	84.6%	81.9%	83.9%	78.1%	78.6%	77.4%	78.0%

4   Conclusion

In this paper, we propose a forget memory network to solve video question answering. The forget memory network can forget some irrelevant video frame features to the question, and select the most useful video frame features to the question. We also extend the forget memory network by taking a GRU model to deal with the video clip frame features. These approaches achieve good performance on the MovieQA and TACoS datasets.

Acknowledgments. This work was supported by the NSFC (under Grant U1509206, 61472276).

References

1. Bahdanau, D., Cho, K., Bengio, Y.: Neural machine translation by jointly learning to align and translate. In: International Conference on Learning Representations (ICLR) (2015)
2. Cho, K., Van Merriënboer, B., Gulcehre, C., Bahdanau, D., Bougares, F., Schwenk, H., Bengio, Y.: Learning phrase representations using RNN encoder-decoder for statistical machine translation. arXiv preprint arXiv:1406.1078 (2014)
3. Das, R., Zaheer, M., Reddy, S., McCallum, A.: Question answering on knowledge bases and text using universal schema and memory networks. arXiv preprint arXiv:1704.08384 (2017)
4. Donahue, J., Anne Hendricks, L., Guadarrama, S., Rohrbach, M., Venugopalan, S., Saenko, K., Darrell, T.: Long-term recurrent convolutional networks for visual recognition and description. In: Proceedings of the IEEE Conference on Computer Vision and Pattern Recognition, pp. 2625–2634 (2015)
5. He, K., Ren, X.Z.S., Sun, J.: Deep residual learning for image recognition. In: Proceedings of the IEEE Conference on Computer Vision and Pattern Recognition, pp. 770–778 (2016)
6. Heilman, M., Smith, N.A.: Good question! statistical ranking for question generation. In: Human Language Technologies: The 2010 Annual Conference of the North American Chapter of the Association for Computational Linguistics, pp. 609–617. Association for Computational Linguistics (2010)
7. Jiang, A., Wang, F., Porikli, F., Li, Y.: Compositional memory for visual question answering. arXiv preprint arXiv:1511.05676 (2015)
8. Karpathy, A., Fei-Fei, L.: Deep visual-semantic alignments for generating image descriptions. In: Proceedings of the IEEE Conference on Computer Vision and Pattern Recognition, pp. 3128–3137 (2015)
9. Kingma, D., Ba, J.: Adam: a method for stochastic optimization. arXiv preprint arXiv:1412.6980 (2014)
10. Krizhevsky, A., Sutskever, I., Hinton, G.E.: Imagenet classification with deep convolutional neural networks. In: Advances in Neural Information Processing Systems, pp. 1097–1105 (2012)
11. Kumar, A., Irsoy, O., Ondruska, P., Iyyer, M., Bradbury, J., Gulrajani, I., Zhong, V., Paulus, R., Socher, R.: Ask me anything: dynamic memory networks for natural language processing. In: International Conference on Machine Learning, pp. 1378–1387 (2016)
12. Li, G., Ma, S., Han, Y.: Summarization-based video caption via deep neural networks. In: Proceedings of the 23rd ACM International Conference on Multimedia, pp. 1191–1194. ACM (2015)
13. Mikolov, T., Sutskever, I., Chen, K., Corrado, G.S., Dean, J.: Distributed representations of words and phrases and their compositionality. In: Advances in Neural Information Processing Systems, pp. 3111–3119 (2013)
14. Noh, H., Hongsuck Seo, P., Han, B.: Image question answering using convolutional neural network with dynamic parameter prediction. In: Proceedings of the IEEE Conference on Computer Vision and Pattern Recognition, pp. 30–38 (2016)
15. Redmon, J., Farhadi, A.: YOLO9000: better, faster, stronger. arXiv preprint arXiv:1612.08242 (2016)

16. Ren, S., He, K., Ross, G., Sun, J.: Faster R-CNN: towards real-time object detection with region proposal networks. In: Advances in Neural Information Processing Systems, pp. 91–99 (2015)

17. Rohrbach, A., Rohrbach, M., Tandon, N., Schiele, B.: A dataset for movie description. In: Proceedings of the IEEE Conference on Computer Vision And Pattern Recognition, pp. 3202–3212 (2015)

18. Simonyan, K., Zisserman, A.: Very deep convolutional networks for large-scale image recognition. arXiv preprint arXiv:1409.1556 (2014)

19. Sukhbaatar, S., Arthur, S., Weston, J., Fergus, R.: End-to-end memory networks. In: Advances in Neural Information Processing Systems, pp. 2440–2448 (2015)

20. Szegedy, C., Liu, W., Jia, Y., Sermanet, P., Reed, S., Anguelov, D., Erhan, D., Vanhoucke, V., Rabinovich, A.: Going deeper with convolutions. In: Proceedings of the IEEE Conference on Computer Vision and Pattern Recognition, pp. 1–9 (2015)

21. Tapaswi, M., Zhu, Y., Stiefelhagen, R., Torralba, A., Urtasun, R., Fidler, S.: MovieQA: understanding stories in movies through question-answering. In: Proceedings of the IEEE Conference on Computer Vision and Pattern Recognition, pp. 4631–4640 (2016)

22. Weston, J., Chopra, S., Bordes, A.: Memory networks. arXiv preprint arXiv:1410.3916 (2014)

23. Xiong, C., Merity, S., Socher, R.: Dynamic memory networks for visual and textual question answering. In: International Conference on Machine Learning, pp. 2397–2406 (2016)

24. Xu, K., Ba, J., Kiros, R., Cho, K., Courville, A., Salakhudinov, R., Zemel, R., Bengio, Y.: Show, attend and tell: neural image caption generation with visual attention. In: International Conference on Machine Learning, pp. 2048–2057 (2015)

25. Yang, Z., He, X., Gao, J., Li, D., Smola, A.: Stacked attention networks for image question answering. In: Proceedings of the IEEE Conference on Computer Vision and Pattern Recognition, pp. 21–29 (2016)

26. Yang, Z., Han, Y., Wang, Z.: Catching the temporal regions-of-interest for video captioning. In: Proceedings of the ACM International Conference on Multimedia (ACM MM) (2017)

27. Zeng, K.H., Chen, T.H., Chuang, C.Y., Liao, Y.H., Niebles, J.C., Sun, M.: Leveraging video descriptions to learn video question answering. In: AAAI, pp. 4334–4340 (2017)

28. Zhu, L., Xu, Z., Yang, Y., Hauptmann, A.G.: Uncovering temporal context for video question and answering. arXiv preprint arXiv:1511.04670 (2015)

29. Zhu, Y., Groth, O., Bernstein, M., Li, F.F.: Visual7W: grounded question answering in images. In: Proceedings of the IEEE Conference on Computer Vision and Pattern Recognition, pp. 4995–5004 (2016)

Three Dimensional Stress Wave Imaging Method of Wood Internal Defects Based on TKriging

Xiaochen Du, Hailin Feng$^{(\boxtimes)}$, Mingyue Hu, Yiming Fang, and Jiajie Li

School of Information Engineering, Zhejiang Agricultural and Forestry University,
Hangzhou, China
sealinfeng@gmail.com

Abstract. In order to detect the size, shape and degree of decay inside wood, a three dimensional stress wave imaging method based on TKriging is proposed. The method uses sensors to obtain the stress wave velocity data sets by hanging around the timber randomly, and reconstructs the image of internal defect with those data sets. TKriging optimized structural relationship between interpolation point and reference point in space firstly. The searching radius is used to select the reference points accordingly. Top-k query method is introduced to find the k value with relevant points. The values of the estimated points are calculated and three dimensional image of the internal defect inside wood is reconstructed. The results show the effectiveness of the method and the accuracy rate of sample with one hole is higher than the Kriging method.

Keywords: TKriging · Three dimensional stress wave imaging · Top-k

1 Introduction

Stress wave tomography technique has been widely used in non-destructive testing of wood. Ross et al. [8] are the earliest researchers who used stress wave detection technology on red oak decayed area for testing. The image observed by the stress wave imaging software can be obtained the rotten woods interior location, size and extent of decay [3]. Sun and Wang [10] combined the diagnostic method of stress waves with the two dimensional X-ray/CT image to determine the internal decay region of logs efficiently and accurately. Qi [6] used the X-ray pattern to filter and sharpen which made the detailed wood internal defects more obvious.

These studies are seldom focusing on the stress wave imaging algorithm but application of existing commercial imaging software for two dimensional imaging research. Feng [2] presented an image reconstruction algorithm which used the speed around the points to estimate the value of unknown grid points. The test results demonstrated the feasibility of this method. Du [1] used spatial interpolation for trees and timber-based approach to plot the internal defect analysis and the timber internal defect image. However, there are few researches in three dimensional stress wave imaging algorithm of defect detecting in wood.

© Springer Nature Singapore Pte Ltd. 2017
J. Yang et al. (Eds.): CCCV 2017, Part I, CCIS 771, pp. 416–427, 2017.
https://doi.org/10.1007/978-981-10-7299-4_34

The basic idea of Kriging is to predict the unknown samples by introduced the weight of each sample and computed a weighted average, base on the differences of the correlation and the spatial location between the each samples [7]. Kriging algorithm designed simulation study was first used in mine exploration and soil mapping. The method makes use of geostatistical methods. Asumptions a smooth mathematical function can not be simulated because of space attribute is irregular changing continuously, but can be read by random surface representation [11].

The algorithm existed the following problems when applied to the timber stress wave three dimensional imaging of internal defects: first, increase the complexity of the algorithm because of the predicted results are related to all known points in the neighborhood; second, neglect the spatial structure of interpolation points because of the results of the prediction points are related to the distance between all known points in the neighborhood. Therefore, we propose a TKriging (Top-k Kriging) algorithm which can be extended to estimate the relationship between the neighborhoods of three dimensional space. We increase the Top-k query within the search radius to find the point with the greatest impact on the estimated k known points, which calculated the value of the estimated point and stress wave three dimensional imaging. We use four samples to do the experimental comparison of TKriging algorithm and Kriging algorithm for three dimensional stress wave imaging.

2 Proposed Method

Regionalized variable, the variable spatial distribution, can reflect the specific spatial distribution of various special properties. For example, those exist in space, meteorological factors, soil information, eco-hydrology, geological environment, maritime climate, and so on. They not only have a specific spatial attributes but also have a dual nature of the variable region. Then we can assume a random field $Z(x)$ to represent the numerical of regionalized variables observed before, after the observation, it is a value of a function that determines the spatial points [5]. There are two major characteristics in regionalized variables:

(1) $Z(x)$, a random uncertain function, represents the amount of change in the region that main feature is random, localized and abnormal function.
(2) To some extent, the self-correlation should be: when x is a variable, the attribute at point $Z(x)$ and spatial distance deviation of the point $x + h$ property at $Z(x + h)$ have some relationship, and its relationship has correlation with the characteristics of variable and the distance between two points.

In this paper, we propose a stress wave imaging algorithm using TKriging. We optimize the spatial structural about estimating points and known points in their neighborhoods, and increase the Top-k query technology to improve the calculation accuracy about estimating points.

2.1 TKriging Interpolation Algorithm

It is supposed that samples $p_i'(x_i, y_i, z_i)$ $(i = 1,2, \ldots, n)$ is any point at the region of study, which satisfies the second-order stationary assumptions or intrinsic hypothesis. Thereinto, $Z(q_i')$ is the point that have been measured as $q_i'(x_i, y_i, z_i)(i = 1,2, \ldots, m)$, and λ_i is the weighted coefficient, for an arbitrary predicted value $Z*(p_i')$ of the estimating point are as follows:

$$Z * (p_i') = \sum_{i=1}^{m} \lambda_i Z(q_i') \tag{1}$$

In the equations above, λ_i, the weighted coefficient of the known point, is not singly decided by the distance, which is calculated based on variogram at minimum variance unbiased condition. $Z*(p_1')$ is good or bad by how to choose the weight coefficient λ_i.

In order to ensure the $Z*(p_1')$ is calculated based on the unbiased estimate of $Z(q_1')$, which means the deviation of the mathematical expectation is zero; if the sum, square of the difference between the estimated value $Z*(p_1')$ and the actual value $Z(q_1')$, is the minimum, the sum should satisfies the conditions:

$$E[Z * (p_1') - Z(p_1')] = 0 \tag{2}$$

In order to obtain the minimum variance of the estimated under the above constraints, λ_i needs to be constructed by the function of the Lagrange multiplier method, and is the Lagrange multiplier:

$$F = E \left\{ [Z * (p_1') - Z(p_1')]^2 - 2\mu(\sum_{i=1}^{m} \lambda_i - 1) \right\} \tag{3}$$

If there is a set of values λ_i to meet the requirements, the equations can be:

$$\begin{cases} \sum_{i=1}^{m} \lambda_i = 1 \\ \sum_{i,j=1}^{m} \lambda_i Cov(q_i', q_j') - \mu = Cov(p_1', q_j') \end{cases} \tag{4}$$

The $m + 1$ equations group is ordinary Kriging equations. According to the above formula, the matrix equations can be obtained:

$$[K][\lambda] = [M] \tag{5}$$

The K, M, are respectively expressed as:

$$[K] = \begin{bmatrix} C_{11} & C_{12} & \ldots & C_{1n} & 1 \\ C_{21} & C_{22} & \ldots & C_{2n} & 1 \\ \ldots & \ldots & \ldots & \ldots & \ldots \\ C_{n1} & C_{n2} & \ldots & C_{nn} & 1 \\ 1 & 1 & 1 & 1 & 0 \end{bmatrix} \quad [\lambda] = \begin{bmatrix} \lambda_1 \\ \lambda_2 \\ \ldots \\ \lambda_n \\ -\mu \end{bmatrix} \quad [M] = \begin{bmatrix} C_{01} \\ C_{02} \\ \ldots \\ C_{0n} \\ 1 \end{bmatrix} \tag{6}$$

The result of estimated variance is minimum by ordinary Kriging method, which is expressed as follows:

$$\sigma^2 = Var[Z * (p_1') - Z(p_1')]^2 = Cov(p_1', p_1') - \sum_{i=1}^{m} \lambda_i Cov(p_1', q_i') + \mu \quad (7)$$

Wherein variogram function $\gamma(q_i', q_j')$ as follows:

$$\gamma(q_i', q_j') = \gamma(q_i' - q_j') = \frac{1}{2} E[Z(q_i') - Z(q_j')]^2 \quad (8)$$

Wherein, the variation function $\gamma(q_i', q_j')$ substitutes for the covariance $Cov(q_i', q_j')$, and brings into the equation to obtain the equation as follows:

$$\begin{cases} \sum_{i=1}^{m} \lambda_i = 1 \\ \sum_{i,j=1}^{m} \lambda_i \gamma(q_i', q_j') - \mu = \gamma(p_1', q_j') \end{cases} \quad (9)$$

And the corresponding ordinary Kriging variance can be expressed as follows:

$$\sigma 2 = \sum_{i=1}^{m} \lambda_i \gamma(p_1', q_j') + \mu - \gamma(p_1', p_1') \quad (10)$$

2.2 Top-k Query-based Method

Top-k discover technology is used mainly to search the most relevant k results from a large amounts of data [4,12], Kriging algorithm uses the Top-k query technology to find out the most influential k known points in the neighborhood range of estimated points and their properties. Top-k query is defined as follows: given M tuples as a collection T, each tuple has its own attribute. Then store the collection T as a column files collection, each column file is a binary array, wherein representing the object identifier and object property values at the property. The storage way of each column file is a monotonous decreasing sequence of each tuples property value. That is defined as the scoring function F. The formula is as follows [9]

$$F(a) = \sum_{i=1}^{m} (\lambda_i \times a.u_i) \quad (11)$$

λ_i is a weight which is defined on the attribute values, u_i by the scoring function. Usually, F is a monotonic function. The k subset of each tuples in the set T can be queried by Top-k query technology. Subsets each element in a group, by reading the m column which is a collection of files in descending order of columns S, read the sequence in order. When tuples appear, get the corresponding attribute value in another column file by random reading mode, and then calculate their score value. If the value of k is the largest one ever seen, put the tuple and its relevant information to the optimal queue. For each column sequence, set its current reading position, and set threshold. When the value of meta component k in priority queue is not less than the minimum, the query is finished. Equation (1) follows the deformation below

$$Z_{p_i} = \sum_{i=1}^{\delta} \lambda_{pi} Z_{qi} \tag{12}$$

Z_{qi} is the property values about known points. λ_{pi} is the weight defined on the property value. Z_{pi} is the final attribute value. So F_{pi} is the scoring function as follows:

$$F(P_i) = \sum_{i=1}^{\delta} \lambda_{pi} Z_{qi} \tag{13}$$

2.3 The Steps of TKriging Imaging Algorithm

Finally, the steps of three dimensional stress wave imaging using TKriging method are as follows:

Step 1: Enter the velocity data of stress wave collected from the sensor.

Step 2: Enter the sensor coordinate $X(z,y,z)$. Calculate out the line velocity matrix Vm^*n by the time matrix Tm^*n of collected.

Step 3: Set estimating point X to 2000. Calculate the known set of points in the neighborhood M and the variation function.

Step 4: Match the variation function and select the spherical model.

Step 5: Ordinary Kriging equation calculation, count the Weight of interpolation point.

Step 6: Take the weight counted into the Eq. (1).

Step 7: When the search is complete, compare score.

Step 8: According to the property value $F(p_i)$ in formula (13), the estimating points could be assigned different color and make the three dimensional visualization.

3   Experimental Results

This study uses the team self-developed portable timber tomographic imaging apparatus. When the timber is detected, the sensors are needed to install randomly at different heights to measure wood. Each time we collect data by knocking 12 sensors on the same sample twice, so the data are collected about 24 groups. The collection of experimental data is shown in Fig. 1(a).

3.1   Materials and Methods

We select four samples as subjects. The types of sample are pecan trees, paulownia trees and phoenix trees. As it shown in Fig. 1, that b and f are pecan tree
samples, a is paulownia tree sample, c is the sample of firmiana tree. Samples a
and d are the artificial dig holes to simulate natural decay situation. Samples b
and c are the natural decay defect. We used a tape to measure the circumference
of each sample. Sample a's circumference is 100 cm, sample b's circumference
is 116 cm, sample c's circumference is 139 cm, and sample d's circumference is
93 cm. The maximum height difference between the sensor and the surrounding
wood in the measured samples is 20 cm, 30 cm, 15 cm, 20 cm.

KT-R heavy hammer Wood Moisture Meter, Production of Italy Klortner
company, is used to measure the moisture content in the experiment. Through
testing, the water content of sample a is 36.5%, the water content of sample
b is 23.5%, the water content of sample c is 33.5%, and the water content of
sample d is 20.3%. Volume of measured samples: the sample volume equals to
the measured height multiplied by the sample cross-sectional area. Wherein the
measured volume of sample a is $15923.57 \, cm^3$, the measured volume of sample
b is $32140.13 \, cm^3$, the measured volume of sample c is $23074.44 \, cm^3$, and the
measured volume of sample d is $13772.29 \, cm^3$. The measured defect volume of
sample a is $4019.2 \, cm^3$, the measured defect volume of sample b is $2601.12 \, cm^3$,
the measured defect volume of sample c is $2418.39 \, cm^3$, and the measured defect
volume of sample d is $1538.6 \, cm^3$.

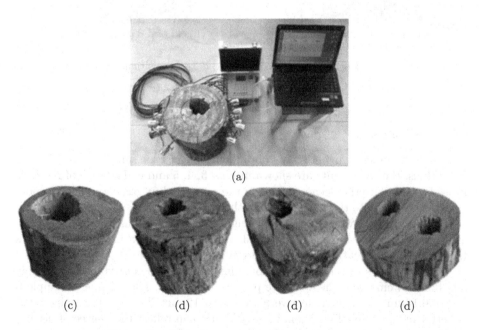

(a)

(c)   (d)   (d)   (d)

Fig. 1. Imaging equipment and four test samples

The time matrix of stress wave propagation in wood is obtained by self-made portable timber tomographic imaging instrument. 12 sensors of the instrument are used to get enough input data for the proposed method. The arrangement of sensors fixation and data acquisition is shown in Fig. 2, the 12 sensors are fixed at different heights and rounded the test sample, six sensors are fixed on one side and other six sensors are fixed on another side. There are two data acquisitions, and the positions of the sensors in second data acquisition are different from that in first data acquisition. After merging the two data, 24 groups of stress wave propagation time matrix as the original experimental data can be can collected. We archive the input data and opened it in text mode and organize the data to get the stress wave propagation time matrix. Each value in the matrix is the feedback value of the stress wave propagation time which received by a certain pair of sensors. The main role of propagation time matrix is to calculate the velocity matrix of stress wave by the time matrix and the corresponding distance matrix.

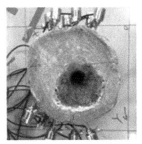

(a) First data acquisition   (b) Second data acquisition

Fig. 2. Sensors fixation and data acquisition

3.2   Comparison

Compare the imaging experiments of different defect type samples from the two algorithms, the test results are shown in Figs. 3, 4, 5 and 6. The area of red dots is used to simulate the decadent area in the samples, but green and yellow spots are used to simulate the health area in the samples.

Figure 3 is the two dimensional imaging results of regular defect samples. The test result shows that TKriging algorithm is better in imaging. It reflects the approximate location and the cases of defects.

Figure 4 shows the three dimensional imaging results. The volume of sample a is $15923.57\,cm^3$, and the defective part is $4019.2\,cm^3$. The volume of sample b is $32140.13\,cm^3$, and the defective part is $2601.12\,cm^3$. The TKriging algorithm is better in the three dimensional imaging that can reflect the degree of decay.

Figure 5 is the test results for the sample c. By the test results, we can see that TKriging algorithm is better in imaging. It reflects the approximate location and

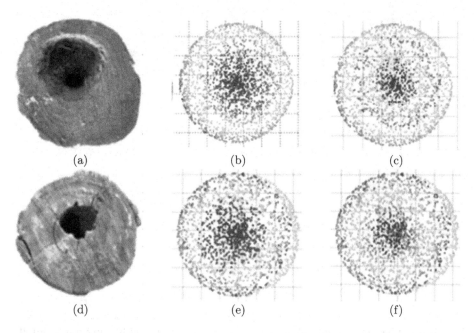

Fig. 3. The test results of 2D regular defect samples. (a) Sample a, (b) Kriging result, (c) TKriging result, (d) sample b, (e) Kriging result, (f) TKriging result. (Color figure online)

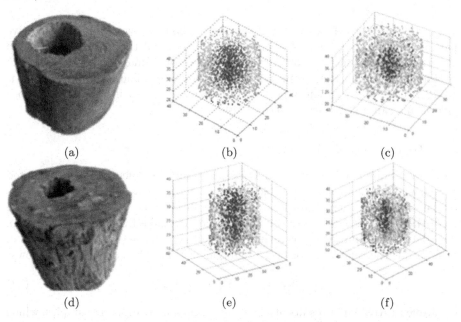

Fig. 4. 3D regular defect samples test results. (a) Sample a, (b) Kriging result, (c) TKriging result, (d) sample b, (e) Kriging result, (f) TKriging result. (Color figure online)

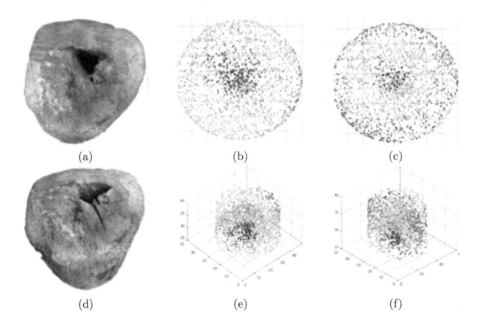

(a)   (b)   (c)

(d)   (e)   (f)

Fig. 5. Irregular defect sample test results. (a) Sample c, (b) 2D Kriging result, (c) 2D TKriging result, (d) sample c, (e) 3D Kriging result, (f) 3D TKriging result. (Color figure online)

the cases of defects. The volume of sample c is $23074.44\,\mathrm{cm}^3$, and the defective part is $2418.39\,\mathrm{cm}^3$. The experimental results show that the TKriging algorithm is better than Kriging algorithm in three dimensional imaging.

Figure 6 is the test results for the sample d. The sample volume is $13772.29\,\mathrm{cm}^3$, and the defect portion is $1538.6\,\mathrm{cm}^3$. In such sample tests, both algorithms can not detect the second hole.

3.3   Imaging Analysis

The defect volume of the actual measurement and the defect volume calculated by two algorithms are show in Table 1, which can be used to verify the correctness of the test results. The data in column 3 is the volume calculated by Kriging algorithm. The data in column 4 is the volume calculated by TKriging algorithm. The data in column 5 is the relative error of Kriging algorithm. The data in column 6 is the relative error of TKriging algorithm. The formula is as follows:

$$\Delta t_i = |V_i - V|/V i = 1, 2 \tag{14}$$

Table 1 reveals the results of defect detection among different samples, which include measured defect volume, calculated defect volume and relative error. Based upon the data of the table, the relative error range of Kriging algorithm is between 8.70% and 195.08%, the relative error range of TKriging algorithm is

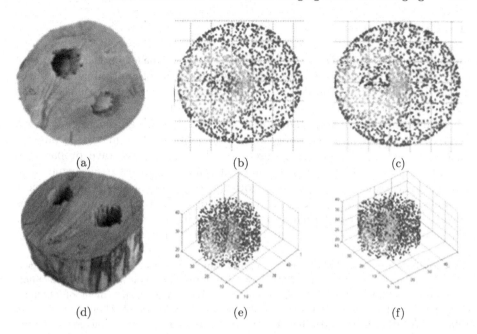

Fig. 6. Test results of the sample with two holes. (a) Sample d, (b) 2D Kriging result, (c) 2D TKriging result, (d) sample d, (e) 3D Kriging result, (f) 3D TKriging result. (Color figure online)

Table 1. Sample defect detection results

Sample	Actual measured defective volume (cm^3)	Kriging calculated defective volume (cm^3)	TKriging calculated defective volume (cm^3)	Kriging relative error rate	TKriging relative error rate
a	4019.2	8025.48	4681.53	99.68%	16.48%
b	2601.12	7675.32	5110.28	195.08%	96.46%
c	2418.39	4707.19	2584.34	94.64%	6.86%
d	1538.6	1404.77	1349.68	8.70%	12.28%

between 6.86% and 96.46% and the relative error of TKriging for sample a–c is all less than that of Kriging. It has been proved that TKriging algorithm has a higher imaging precision. Relative error of Sample d is 12.28%, the reason may be the two artificial holes in the sample. As indicated in Fig. 6, the accuracies of the two algorithms are not high in this case, and only one hole in the sample is calculated by both methods. Therefore, lower relative error can not reflect that Kriging performs better in this case. According to the results of the experiment, the average relative error rate of Kriging algorithm is 99.53%, while the average relative error rate of TKriging algorithm is 33.02%.

4   Conclusion

In this paper, an imaging method is proposed which is based on the TKriging of the three dimensional stress wave in the internal defects of wood. In addition, the three dimensional imaging of wood defects is studied. Compared with the results of Kriging algorithm, the algorithm not only can better reflect the internal decay of wood in qualitative and quantitative, but also has a high imaging accuracy. To verify the algorithms feasibility of the three dimensional imaging of internal defects in the wood, sensors are used to collect the stress wave propagation velocity around wood, through the different samples of defect experiments. The average relative error rate of the algorithm, detection of single empty defect samples, is only 33.02%, while the accuracy of the sample test for the porous hole is not high. It can basically reflect the position of the sample defect, but the second hole can not be detected, we still need further study.

Acknowledgements. This work is jointly supported by National Natural Science Foundation of China (61272313), Zhejiang province key science and technology projects, China (2015C03008), Science and Technology Department of Zhejiang Province, China (2014C31044), Natural Science Foundation of Zhejiang Province, China (LY15F020034) and Zhejiang Agricultural and Forestry University, China (2013200036).

References

1. Du, X.C., Li, S.Z., Li, G.H., Feng, H.L., Chen, S.Y.: Stress wave tomography of wood internal defects using ellipse-based spatial interpolation and velocity compensation. Bioresources **10**(3), 3948–3962 (2015)
2. Feng, H.L., Li, G.H., Fu, S., Wang, X.P.: Tomographic image reconstruction using an interpolation method for tree decay detection. Bioresources **9**(2), 3248–3263 (2014)
3. Gilbert, E.A., Smiley, E.T.: Picus sonic tomography for the quantification of decay in white oak (quercus alba) and hickory (carya spp.). J. Arboric. **30**(5), 277–281 (2004)
4. Le, T.M.N., Cao, J.L., He, Z.: Top-k best probability queries and semantics ranking properties on probabilistic databases. Data Knowl. Eng. **88**(6), 248–266 (2013)
5. Liu, S., Anh, V., Mcgree, J., Kozan, E., Wolff, R.C.: A new approach to spatial data interpolation using higher-order statistics. Stoch. Environ. Res. Risk Assess. **29**(6), 1679–1690 (2015)
6. Qi, D.W.: A study on image processing system of non-destructive log detection. Sci. Silvae Sinicae **37**(6), 92–96 (2001)
7. Quirante, N., Javaloyes, J., Caballero, J.: Rigorous design of distillation columns using surrogate models based on Kriging interpolation. AIChE J. **61**(7), 2169–2187 (2015)
8. Ross, R.J., Ward, J.C., Tenwolde, A.: Stress wave nondestructive evaluation of wetwood. For. Prod. J. **44**(7), 79–83 (1994)
9. Shaikh, S.A., Kitagawa, H.: Top-k outlier detection from uncertain data. Int. J. Autom. Comput. **11**(2), 128–142 (2014)

10. Sun, T.Y., Wang, L.H.: Non-destructive testing of log internal decay based on two-dimensional CT images of stress wave and X-ray testing. For. Eng. **27**(6), 26–29 (2011)
11. Webster, R., Burgess, T.M.: Optimal interpolation and isarithmic mapping of soil properties III changing drift and universal Kriging. Eur. J. Soil Sci. **31**(3), 505–524 (1980)
12. Zheng, J., Jia, W.J., Wang, G.J.: DDT-based Top-k queries protocol in wireless sensor networks. Chin. J. Sens. Actuators **23**(9), 1340–1346 (2010)

A Novel Color Image Watermarking Algorithm Based on QWT and DCT

Shaocheng Han[1(✉)], Jinfeng Yang[2], Rui Wang[3], and Guimin Jia[2]

[1] Basic Experimental Center, Civil Aviation University of China,
Tianjin 300300, China
schan@cauc.edu.cn, schan_cauc2012@163.com
[2] College of Electronic Information and Automation,
Civil Aviation University of China, Tianjin 300300, China
{jfyang,gmjia}@cauc.edu.cn
[3] College of science, Civil Aviation University of China, Tianjin 300300, China
r-wang@cauc.edu.cn

Abstract. A novel color image watermarking algorithm is proposed based on quaternion wavelet transform (QWT) and discrete cosine transform (DCT) for copyright protection. The luminance channel Y of the host color image in YCbCr space is decomposed by QWT to obtain four approximation subimages. A binary watermark is embedded into the mid-frequency DCT coefficients of two randomly selected subimages. The experimental results show that the proposed watermarking scheme has good robustness against common image attacks such as adding noise, filtering, scaling, JPEG compression, cropping, image adjusting, small angle rotation and so on.

Keywords: Image watermarking · QWT · DCT · Chaotic map
YCbCr space

1 Introduction

With the rapid development of information technology, the reproduction, distribution and tampering of digital products have become common nowadays [1]. How to protect the copyright of these digital products effectively has been a hot topic in recent years. As a copyright protection technology, digital watermarking draws a lot of attentions. Because it can embed authentication information into the host image, audio and video data and also ensure the integrity and security of these data [2].

Image watermarking algorithms can be classified into spatial domain algorithms and transform domain algorithms. For the spatial domain algorithms, the watermark is usually embedded into the host image by modifying its pixels directly [3,4]. Although the spatial domain algorithms have lower computational complexity, they are fragile against some image attacks. For transform domain

© Springer Nature Singapore Pte Ltd. 2017
J. Yang et al. (Eds.): CCCV 2017, Part I, CCIS 771, pp. 428–438, 2017.
https://doi.org/10.1007/978-981-10-7299-4_35

watermarking algorithms, the host image is firstly decomposed by some transforms, and then the watermark is embedded by modifying the coefficients after the transforms. The transform domain algorithms usually have better robustness than the spatial domain algorithms, so a lot of researchers pay more attentions to the transform domain algorithms. In recent years, the popular transform domain watermarking algorithms include Discrete Cosine Transform (DCT) [5], Discrete Fourier Transform (DFT) [6], Discrete Wavelet Transform (DWT) [7], Discrete Shearlet Transform (DST) [8] and so on.

A watermarking scheme using subsampling based on DCT was proposed by Chu et al. [9]. Many subsampling-based watermarking algorithms have been developed. In [10], a binary watermark image was permuted using Baker chaotic map. Then it was embedded into the DCT coefficients of two subimages which were selected according to a random coordinates sequence generated by a secret key. As a new multi-scale analysis tool for image representation, QWT was proposed based on Hilbert transform [11]. QWT has already been successfully applied in the field of image processing such as image denoising [12], image fusion [13], object recognition [14], and digital watermarking [15].

In this paper, a novel image watermarking scheme is proposed based on QWT and DCT. A binary watermark is scrambled by Arnold map and iterated sine chaotic system. The processes of watermark embedding and extracting are implemented on the luminance component of a host color image in YCbCr space. Instead of using subsampling, we obtain four approximation subimages of the luminance component using QWT. The signs and amplitudes of mid-frequency DCT coefficients are relatively stable to some attacks, so these coefficients are used to embed the encrypted watermark. The experimental results demonstrate the proposed scheme is more robust and effective to protect the copyright of color images.

2 Quaternion Wavelet Transform

QWT overcomes the common drawbacks of wavelet transforms by shift-invariant feature. Besides of the shift-invariant magnitude, QWT has three phases. The first two phases represent local image shifts, and the third one denotes the texture information [12]. Let $f(x, y)$ be 2-D image signal, the quaternion analytic signal is defined as follows:

$$f^q(x, y) = f(x, y) + i f_{H_x}(x, y) + f_{H_y}(x, y) + k f_{H_x H_y}(x, y), \qquad (1)$$

where $f_{H_x}(x, y)$ and $f_{H_y}(x, y)$ are partial Hilbert Transform respectively, $f_{H_x H_y}(x, y)$ is the total Hilbert Transform.

Suppose $\phi(x, y)$ is the scaling function, $\psi^D(x, y)$, $\psi^V(x, y)$ and $\psi^H(x, y)$ are respectively mother wavelets. If $\phi(x, y)$ and $\psi(x, y)$ are separable, the 2D QWT of $f(x, y)$ according to the quaternion analytic is defined as follows:

$$\begin{cases} \phi(x,y) = \phi_h(x)\phi_h(y) + i\phi_g(x)\phi_h(y) + j\phi_h(x)\phi_g(y) + k\phi_g(x)\phi_g(y) \\ \psi^D(x,y) = \psi_h(x)\psi_h(y) + i\psi_g(x)\psi_h(y) + j\psi_h(x)\psi_g(y) + k\psi_g(x)\psi_g(y) \\ \psi^V(x,y) = \phi_h(x)\psi_h(y) + i\phi_g(x)\psi_h(y) + j\phi_h(x)\psi_g(y) + k\phi_g(x)\psi_g(y) \\ \psi^H(x,y) = \psi_h(x)\phi_h(y) + i\psi_g(x)\phi_h(y) + j\psi_h(x)\phi_g(y) + k\psi_g(x)\phi_g(y) \end{cases} . \quad (2)$$

In other words, the coefficients of the image decomposed by QWT constitute a matrix F,

$$F = \begin{bmatrix} LL_{\phi_h(x)\phi_h(y)} & LH_{\phi_h(x)\psi_h(y)} & HL_{\psi_h(x)\phi_h(y)} & HH_{\psi_h(x)\psi_h(y)} \\ LL_{\phi_g(x)\phi_h(y)} & LH_{\phi_g(x)\psi_h(y)} & HL_{\psi_g(x)\phi_h(y)} & HH_{\psi_g(x)\psi_h(y)} \\ LL_{\phi_h(x)\phi_g(y)} & LH_{\phi_h(x)\psi_g(y)} & HL_{\psi_h(x)\phi_g(y)} & HH_{\psi_h(x)\psi_g(y)} \\ LL_{\phi_g(x)\phi_g(y)} & LH_{\phi_g(x)\psi_g(y)} & HL_{\psi_g(x)\phi_g(y)} & HH_{\psi_g(x)\psi_g(y)} \end{bmatrix} . \quad (3)$$

Each column of matrix F is corresponding to a quaternion. The first column is corresponding to the approximation subimage, and the other three columns are respectively corresponding to horizontal, vertical and diagonal subimages [14].

3   Watermarking Scheme Based on DCT and QWT

In this section, the novel watermarking embedding and extracting schemes are detailed based on DCT and QWT.

3.1   Watermark Embedding

The steps of embedding watermark into the original host color image are described as follows:

Step 1. The original watermark W is permuted by Arnold map controlled by the scrambling times K_1 to obtain the scrambled watermark W_1.

Step 2. A random sequence $X_1 = \{x_n \,|\, n = 1, 2, \dots, L\}$ is generated by the iterated sine chaotic map which is defined as

$$x_{n+1} = \sin\left(\varepsilon \arcsin \sqrt{|x_n - 1|}\right), \quad (4)$$

and then X_1 is converted into a binary sequence G_1 by comparing with the threshold T.

Step 3. W_1 is converted into a vector $\widetilde{W}$ by zigzag scanning. Then we do XOR operation on $\widetilde{W}$ with G_1 to obtain the vector $\overline{W}$. $\overline{W}$ is divided into $\overline{W}_1$ and $\overline{W}_2$ as the final watermark information. The above process is shown as

$$\begin{cases} \overline{W} = \widetilde{W} \oplus G_1 \\ \begin{cases} \overline{W}_1 = \overline{W}(1 : L/2) \\ \overline{W}_2 = \overline{W}(L/2 + 1 : L) \end{cases} \end{cases} . \quad (5)$$

Step 4. The original host image I is converted from RGB space into YCbCr space. The luminance component Y is decomposed by 2-level QWT transform to obtain one magnitude low-frequency subimage and three phase low-frequency subimages, denoted by D_1, D_2, D_3 and D_4.

Step 5. D_1, D_2, D_3 and D_4 are decomposed by 2D-DCT transform, and the transformed coefficients are zigzag scanned into four vectors respectively, denoted by V_1, V_2, V_3 and V_4.

Step 6. Two random sequences X_2 and X_3 are generated following the step2, where the length of X_2 and X_3 are both $L/2$. For every x_n of X_2 or X_3, we set $g_n = \text{ceil}(4 * x_n)$ to obtain G_2 and G_3. According to G_2 and G_3, we construct a sequence of random number pairs $Z_1 = \{(i,j) \mid i \neq j \text{ and } i,j \in \{1,2,3,4\}\}$, where the length of Z_1 is $L/2$.

Step 7. According to Z_1, a complementary sequence with random number pairs $Z_2 = \{(x,y)\}$ is determined. For instance, if $(i,j) = (1,4)$ in Z_1, then (x,y) must be $(2,3)$ or $(3,2)$ in Z_2. Z_1 and Z_2 are used to select which two vectors of V_1, V_2, V_3 and V_4 to embed the watermark $\overline{W}_1$ and $\overline{W}_2$ respectively.

Step 8. Because the process of embedding $\overline{W}_1$ into V_i and V_j is same to that of embedding $\overline{W}_2$ into V_x and V_y, we take embedding $\overline{W}_1$ into V_i and V_j as an example to describe the process of watermark embedding:

Firstly, we select two vectors V_i and V_j based on Z_1 so that $V_i(k) \geqslant V_j(k)$ if $\overline{W}_1(k) = 1$ and $V_i(k) < V_j(k)$ if $\overline{W}_1(k) = 0$:

$$Z_1'(k) = \begin{cases} Z_1(k)\{i \leftrightarrow j\} & \text{if } \overline{W}_1(k) = 1 \ \& \ V_i(k) < V_j(k) \\ Z_1(k)\{i \leftrightarrow j\} & \text{if } \overline{W}_1(k) = 0 \ \& \ V_i(k) \geq V_j(k) \end{cases}, \quad (6)$$

where $Z_1(k)\{i \leftrightarrow j\}$ means swapping i and j for the k-th number pairs in Z_1.

Secondly, we re-select the k-th pair coefficients based on Z_1', and calculate the following equations

$$d_1 = |V_i(a_1 + k)| + |V_j(a_1 + k)|, \quad d_2 = |V_i(a_1 + k) - V_j(a_1 + k)|. \quad (7)$$

Finally, the watermark embedding algorithm is given as

$$\begin{cases} \begin{cases} V_i'(a_1 + k) = V_i(a_1 + k) + \alpha \left(2\overline{W}_1(k) - 1\right) d_1 \\ V_j'(a_1 + k) = V_j(a_1 + k) - \alpha \left(2\overline{W}_1(k) - 1\right) d_1 \end{cases} & \text{if } d_2/d_1 \leq \beta \\ V_i'(k) = V_i(k), \ V_j'(k) = V_j(k) & \text{else} \end{cases}, \quad (8)$$

where a_1 means the first position of the selected coefficients, α and β are two strength parameters.

Step 9. Inverse zigzag scanning and inverse DCT are applied to V_1', V_2', V_3' and V_4' to obtain four watermarked low-frequency sub-images. After the inverse QWT, the watermarked luminance component Y' is obtained. Then we compose Y' with the other two components to obtain the watermarked image I_1.

3.2   Watermark Extraction

The detailed steps of watermark extracting are described as follows:

Step 1. Suppose the watermarked image which undergone some attacks is I_2. I_2 is converted from RGB space to YCbCr space. Then we obtain the luminance component of watermarked image, denoted by Y_1.

Step 2. Y_1' is decomposed by 2-levels QWT transform to obtain one magnitude low-frequency subimage and three phase low-frequency subimages, denoted by D_1^*, D_2^*, D_3^* and D_4^*.

Step 3. D_1^*, D_2^*, D_3^* and D_4^* are decomposed by 2D-DCT transform, and the transformed coefficients are zigzag scanned into four vectors respectively, denoted by U_1, U_2, U_3 and U_4.

Step 4. Two vectors U_i and U_j are selected according to Z_1 or Z_2. The extracted watermark $\hat{W}_1$ or $\hat{W}_2$ can be obtained as follows:

$$\hat{W}_l(k) = \begin{cases} 1 & \text{if } U_i(a_1 + k) \geqslant U_j(a_1 + k) \\ 0 & \text{else} \end{cases}, \tag{9}$$

where $l = 1, 2$.

Step 5. $\hat{W}_1$ is combined with $\hat{W}_2$. Then the final extracted watermarked W' is obtained

$$W' = \text{Arnold}^{-1}\left(\text{zigzag}^{-1}\left(\hat{W} \oplus G_1\right)\right). \tag{10}$$

4   Experimental Results

The experimental results of the proposed watermarking scheme are presented in this section. Three 512×512 RGB color images in Fig. 1(a)–(c) are used as test images. A 64×64 binary logo image in Fig. 1(d) is used as the original watermark. In the experiments, we set the initial values x_0 and factors ε to 0.6345, 0.5872, 0.4536 and 2.356, 2.123, 2.678 for X_1, X_2, X_3 respectively. The threshold T is set to 0.5. The parameter a_1 in Eqs. (7) and (8) is set to 200. The parameters α and β in Eq. (8) are set to 0.4 and 0.2.

Fig. 1. (a) Host Tiffany image; (b) Host Sailboat image; (c) Host House image; (d) Original watermark.

4.1   Imperceptibility Test

The imperceptibility means that the human visual quality of original host image should not be affected much even after watermark embedding [5]. The peak value signal to noise ratio (PSNR) is used as objective criteria to evaluate the quality of the watermarked image. The PSNR for color image can be calculated as follows:

$$\text{PSNR} = \frac{1}{3} \sum_{c} 10 \log_{10} \frac{N^2 P^2}{\sum_{x=1}^{N} \sum_{y=1}^{N} \left(I(x,y) - I'(x,y)\right)^2}, \tag{11}$$

where I and I' are the host image and watermarked image respectively, N denotes the size of I and I', P is their peak value, and $c \in \{R, G, B\}$ denotes each component of color image.

The proposed scheme shows high imperceptibility, as shown in Fig. 2, where high PSNR values are achieved for the host test images, such as Tiffany image (55.4122 dB), Sailboat image (48.4961 dB), House image (49.7479 dB).

Fig. 2. (a) Watermarked Tiffany image; (b) Watermarked Sailboat image; (c) Watermarked House image.

4.2   Robustness Test

Robustness is an important requirement for any watermarking system. In this paper, the normalized cross-correlation (NC) is employed to evaluate the robustness of the algorithms. The NC is defined as

$$\text{NC}(W, W') = \frac{\sum_{i=1}^{M} \sum_{j=1}^{M} W(i,j)W'(i,j)}{\sqrt{\sum_{i=1}^{M} \sum_{j=1}^{M} W^2(i,j)} \sqrt{\sum_{i=1}^{M} \sum_{j=1}^{M} W'^2(i,j)}}, \tag{12}$$

where M represents the size of the watermark, W and W' indicate the original watermark and the extracted watermark. The NC value ranges from 0 to 1. If the NC value is near to 1, it means that the extracted watermark is strong correlate to the original watermark. But if it is near to 0, it means that the extracted watermark is almost uncorrelated. If the similarity between the original watermark and the extracted watermark is higher, the NC value is larger, which indicates that the algorithm is more robust. Generally, the NC value is considered acceptable if it is not less than 0.75.

The proposed scheme has strong robustness. This can be proved by extracting the watermark image and calculating NC between the original watermark and the extracted watermark after image attacks. In Fig. 3, we show some watermarked Tiffany images which have undergone the attacks of (a) Gaussian noise (0.03), (b) Speckle noise (0.1), (c) Pepper & Salt noise (0.1), (d) Median filter (7 × 7), (e) Scaling (512–64–512), (f) JPEG compression $Q = 5$, (g) Cropping (1/32 upper left corner), (h) Rows deleting (101–130), (i) Right translation by two columns, (j) Image adjusting ([0.3 0.7] → [0 1]), (k) Brightening (−80), and (l) Turn right rotation (1°). Compared with the original Tiffany image, the watermarked Tiffany images in Fig. 3 have changed a lot after these attacks. The corresponding extracted watermarks from the attacked images in Fig. 3 are shown in Fig. 4. It is shown that the proposed scheme has strong robustness, since the extracted watermarks in Fig. 4 are basically clear.

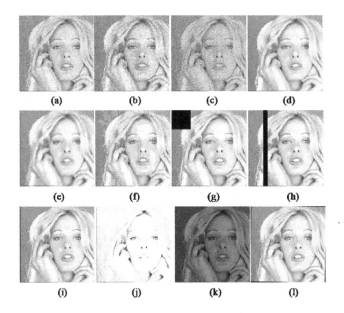

Fig. 3. Watermarked Tiffany images under different attacks

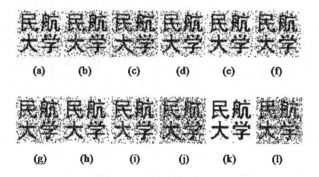

(a)   (b)   (c)   (d)   (e)   (f)

(g)   (h)   (i)   (j)   (k)   (l)

Fig. 4. Extracted watermarks from watermarked Tiffany images under different attacks

4.3   Comparative Analysis and Discussion

The binary watermark is embedded into the luminance component in YCbCr space and YUV space respectively in [16,17]. The host images are 512×512 gray-scale images in [10]. In our experiments, the binary watermark is embedded into the luminance components of host color images in YCbCr space by Lu's method in [10]. In Table 1, Figs. 5 and 6, the proposed method shows the strongest robustness, especially in adding noise, filtering, scaling, JPEG compression and rotation attacks for different color images.

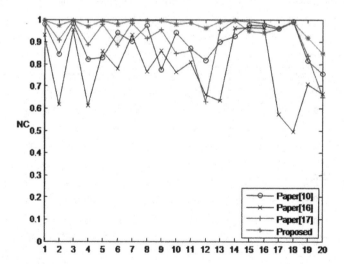

Fig. 5. Robustness comparison in different schemes for Sailboat image

Table 1. NC values comparison in different schemes for Tiffany image

No.	Attacks	Paper [10]	Paper [16]	Paper [17]	Proposed
1	Gaussian noise (0.001)	0.9418	0.7244	0.8828	0.9908
2	Gaussian noise (0.03)	0.7745	0.6231	0.8071	0.9487
3	Pepper & Salt noise (0.02)	0.9442	0.7473	0.8902	0.9908
4	Pepper & Salt noise (0.1)	0.7381	0.6171	0.7972	0.9345
5	Median filter (3 × 3)	0.8273	0.7452	0.8652	0.9859
6	Median filter (5 × 5)	0.9063	0.7108	0.8101	0.9558
7	Wiener filter (3 × 3)	0.8852	0.7675	0.8723	0.9959
8	Wiener filter (5 × 5)	0.9551	0.7275	0.8344	0.9927
9	Average filter (3 × 3)	0.7406	0.7481	0.8441	0.9895
10	Average filter (5 × 5)	0.8967	0.7118	0.7894	0.9721
11	Scaling (512–128–512)	0.8261	0.7371	0.8029	0.9743
12	Scaling (512–64–512)	0.7713	0.6448	0.5978	0.9548
13	JPEG compression $Q = 20$	0.8738	0.5658	0.8386	0.9840
14	JPEG compression $Q = 50$	0.9001	0.7412	0.8798	0.9911
15	Cropping (1/16 upper left corner)	0.9792	0.7308	0.8829	0.9105
16	Columns deleting (101–130)	0.9587	0.7439	0.8814	0.9474
17	Image adjusting ([0.3 0.7]→[0 1])	0.8961	0.5537	0.9399	0.8637
18	Brightening (+50)	0.9416	0.5956	0.9278	0.9233
19	Rotation (Right 0.5°)	0.8132	0.6574	0.7589	0.9265
20	Rotation (Right 1°)	0.7533	0.6218	0.6011	0.8746

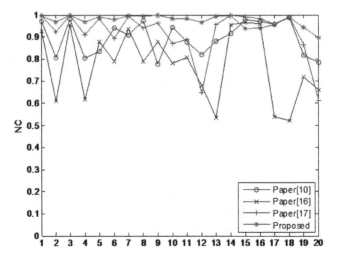

Fig. 6. Robustness comparison in different schemes for House image

5   Conclusion

In this paper, a novel color image watermarking algorithm has been proposed based on QWT and DCT. The Arnold map and iterated sine chaotic system are used to encrypt the watermark to increase the security of this algorithm. The luminance component of host color image in YCbCr space is decomposed by QWT firstly. Then the encrypted watermark has been embedded into the mid-frequency DCT coefficients of two random approximation subimages. The experimental results have shown that the proposed method attains more excellent imperceptibility and has stronger robustness against some attacks than the watermarking schemes in [10,16,17]. In future, we will extend the proposed watermarking scheme to the video watermarking.

Acknowledgments. This work was partially supported by the National Natural Science Foundation of China (Grant Nos. 61379102, U1433105, U1433120 and 61502498).

References

1. Hu, H.T., Hsu, L.Y.: A mixed modulation scheme for blind image watermarking. AEU - Int. J. Electron. Commun. **70**(2), 172–178 (2015)
2. Leung, H.Y., Cheng, L.M.: Robust watermarking scheme using wave atoms. EURASIP J. Adv. Sig. Process. **2011**(1), 1–9 (2011)
3. Karybali, I.G., Berberidis, K.: Efficient spatial image watermarking via new perceptual masking and blind detection schemes. IEEE Trans. Inf. Forensics Secur. **1**(2), 256–274 (2006)
4. Nasir, I., Weng, Y., Jiang, J., Ipson, S.: Multiple spatial watermarking technique in color images. SIViP **4**(2), 145–154 (2010)
5. Das, C., Panigrahi, S., Sharma, V.K., Mahapatra, K.K.: A novel blind robust image watermarking in DCT domain using inter-block coefficient correlation. AEU - Int. J. Electron. Commun. **68**(3), 244–253 (2014)
6. Urvoy, M., Goudia, D., Autrusseau, F.: Perceptual DFT, watermarking with improved detection and robustness to geometrical distortions. IEEE Trans. Inf. Forensics Secur. **9**(7), 1108–1119 (2014)
7. Keyvanpour, M.R., Bayat, F.M.: Blind image watermarking method based on chaotic key and dynamic coefficient quantization in the DWT domain. Math. Comput. Model. **58**(1–2), 56–67 (2013)
8. Mardanpour, M., Chahooki, M.A.Z.: Robust transparent image watermarking with Shearlet transform and bidiagonal singular value decomposition. AEU - Int. J. Electron. Commun. **70**(6), 790–798 (2016)
9. Chu, W.C.: DCT-based image watermarking using subsampling. IEEE Trans. Multimedia **5**(1), 34–38 (2003)
10. Lu, W., Lu, H., Chung, F.: Robust digital image watermarking based on subsampling. Appl. Math. Comput. **181**(2), 886–893 (2006)
11. Bayro-Corrochano, E.: Multi-resolution image analysis using the quaternion wavelet transform. Numer. Algorithms **39**(1), 35–55 (2005)
12. Yin, M., Liu, W., Shui, J., Wu, J.: Quaternion wavelet analysis and application in image denoising. Math. Probl. Eng. **2012**(1), 587–612 (2012)

13. Pang, H., Zhu, M., Guo, L.: Multifocus color image fusion using quaternion wavelet transform. In: International Congress on Image and Signal Processing, pp. 543–546. IEEE (2012)
14. Priyadharshini, R.A., Arivazhagan, S.: A quaternionic wavelet transform-based approach for object recognition. Def. Sci. J. **64**(4), 350–357 (2014)
15. Lei, B., Ni, D., Chen, S., Wang, T., Zhou, F.: Optimal image watermarking scheme based on chaotic map and quaternion wavelet transform. Nonlinear Dyn. **78**(4), 2897–2907 (2014)
16. Su, Q., Niu, Y., Wang, Q., Sheng, G.: A blind color image watermarking based on DC component in the spatial domain. Optik - Int. J. Light Electron Opt. **124**(23), 6255–6260 (2013)
17. Liu, N., Li, H., Dai, H., Chen, D.: Robust blind image watermarking based on chaotic mixtures. Nonlinear Dyn. **80**(3), 1329–1355 (2015)

A Novel Camera Calibration Method for Binocular Vision Based on Improved RBF Neural Network

Weike Liu[1], Ju Huo[2], Xing Zhou[2], and Ming Yang[1(✉)]

[1] Control and Simulation Center, School of Astronautics,
Harbin Institute of Technology, Harbin 150001, China
myang_csc@163.com
[2] School of Electrical Engineering, Harbin Institute of Technology,
Harbin 150001, China

Abstract. Considering the problems that camera imaging model is complex and operation is complicated, a binocular camera calibration method of RBF neural network based on k-means and gradient method is proposed in this paper. The data center selection method based on the law of clustering error function can obtain hidden nodes and data centers of RBF network accurately. Dynamic learning of data centers, spread constants and weight values based on gradient method can contribute to improving the precision. Experimental results show that the proposed method has high precision and can be well applied in machine vision.

Keywords: RBF neural network · Binocular vision
Camera calibration · K-means · Gradient method

1 Introduction

Machine vision is becoming more and more widely applied in many fields, such as robot vision navigation and positioning, 3-D reconstruction, etc., while camera calibration is the most important part of machine vision. Traditional camera calibration methods are generally based on precision machined 2-D or 3-D targets. Direct linear transformation (DLT) calibration method [1] proposes the relationship between the two-dimensional plane image and the actual object coordinates in the three-dimensional space. This method is based on the ideal pinhole model. The parameters involved are relatively few and the calculation process is simple and fast, but the nonlinear distortion caused by the imaging process is not taken into account, so the precision is poor. Tsai proposed a two-step method based on radial constraint [2], combining the traditional linear method and nonlinear optimization method, but the highprecision three-dimensional calibration block is difficult to process and maintain in practice. In the literature [3], Kruppa equation is established based on the epipolar geometry and curvilinear geometry theory which is used in camera self-calibration technology and the calibration

© Springer Nature Singapore Pte Ltd. 2017
J. Yang et al. (Eds.): CCCV 2017, Part I, CCIS 771, pp. 439–448, 2017.
https://doi.org/10.1007/978-981-10-7299-4_36

speed is fast. It is suitable for onsite calibration, but the precision is low and the stability is poor. Zhang proposed a flexible calibration method [4] according to the plane target images shot by camera from different locations to solve the camera parameter model. This method is widely used, but it also needs to establish a complex mathematical model.

Considering the problems that mathematical model constructed is complex, operation is complicated and precision is poor in traditional camera calibration methods, this paper proposes a camera calibration method based on improved RBF neural network, which has high precision and strong real-time, and the method can be applied to more complex environments.

2   Camera Imaging Model

The projection of three-dimension spatial point D on camera imaging plane is usually obtained by ideal pinhole model, which can be written as

$$
s \begin{bmatrix} u \\ v \\ 1 \end{bmatrix} = \begin{bmatrix} f_u & 0 & u_0 & 0 \\ 0 & f_v & v_0 & 0 \\ 0 & 0 & 1 & 0 \end{bmatrix} \begin{bmatrix} R & T \\ 0^T & 1 \end{bmatrix} \begin{bmatrix} X \\ Y \\ Z \\ 1 \end{bmatrix} = A \begin{bmatrix} R & T \\ 0^T & 1 \end{bmatrix} \begin{bmatrix} X \\ Y \\ Z \\ 1 \end{bmatrix} \tag{1}
$$

where $X = (X_w^T, 1)^T$ is the homogeneous coordinate of the three-dimensional spatial point D in the world coordinate system. $(u, v, 1)$ is the homogeneous coordinate of the projection point on the image plane. R and T are the rotation matrix and the translation vector that the world coordinate system transforms into the camera coordinate system. (u_0, v_0) is the optical center of the camera. f_u and f_v are the scale factors in u and v axis directions. With the impact of the camera lens optical distortion, as shown in Fig. 1, the actual imaging point is not the intersection of the image plane and the joint line between three-dimensional point and the optical center, but with a certain offset $(\Delta u, \Delta v)$. The distortion of the camera mainly includes radial distortion and tangential distortion.

Binocular stereo vision imaging principle is shown in Fig. 2. The binocular camera imaging equation is expressed as

$$
\left\{ s_l \begin{bmatrix} u_l \\ v_l \\ 1 \end{bmatrix} = A_l \begin{bmatrix} R_l & T_l \end{bmatrix} \begin{bmatrix} X_w \\ Y_w \\ Z_w \\ 1 \end{bmatrix} \right., s_r \begin{bmatrix} u_r \\ v_r \\ 1 \end{bmatrix} = A_r \begin{bmatrix} R_r & T_r \end{bmatrix} \begin{bmatrix} X_w \\ Y_w \\ Z_w \\ 1 \end{bmatrix} \tag{2}
$$

where s is the scale factor. Subscript l and r represent left camera and right camera, respectively. For binocular camera calibration, in addition to calibrating the internal parameters of the cameras, the rotation matrix and the translation matrix between two cameras are also needed.

Above the mentioned, the traditional calibration method based on the camera imaging mathematical model involves a large number of unknown parameters, which takes many factors into account and constructs complexly. While the calibration method based on neural network calibration is able to overcome these problems [5, 6].

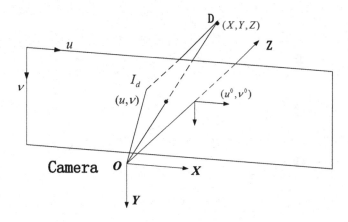

Fig. 1. Camera imaging model of monocular vision

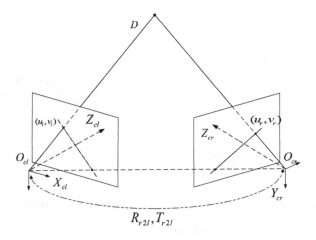

Fig. 2. Camera imaging model of binocular vision

3   RBF Neural Network Based on K-means and Gradient Method

3.1   The Structure of RBF Neural Network

RBF neural network is generally three-tier structure. As shown in Fig. 3, the RBF neural network structure is $n - h - m$, that means, the network has n inputs, h hidden nodes and m outputs. $x = (x_1, x_2, \cdots, x_n)^T \in R^n$ is the network input vector. $\omega \in R^{h \times m}$ is the output weight matrix. $y = [y_1, y_2, \cdots, y_m]^T$ is the network out-put vector. $\phi(\bullet)$ is the activation function of the i-th hidden node. In this paper, the Euclidean distance function is used as the basis function represented as $\|\bullet\|$ and the Gauss function is used as the activation function, which is as follow

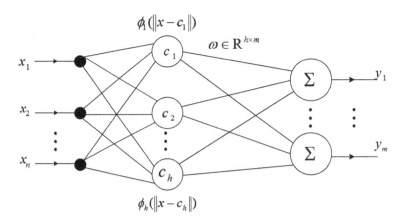

Fig. 3. The structure of RBF neural network

$$\phi(u) = e^{-\frac{u^2}{\delta^2}} \tag{3}$$

where the spread constant of RBF $\delta > 0$. The smaller δ is, the smaller the width of the RBF is and more selective it becomes. In Fig. 3, the j-th output of the RBF neural network can be given as

$$y_j = \sum_{i=1}^{h} \omega_{ij}\phi_i(\|x - c_i\|), \quad 1 \le j \le m \tag{4}$$

3.2   The Learning Algorithm of RBF Neural Network

Clustering method is the most classic RBF neural network learning algorithm [7,8]. The idea is to use the unsupervised learning method to obtain the data centers of the hidden nodes in RBF neural network, and to determine the spread constants according to the distance between the data centers. Then we can use the supervised learning method to train the output weight values of the hidden nodes.

In this paper, we propose a novel method of clustering number selection based on k-means clustering error law to obtain the number of hidden nodes and data centers. In the training, we construct the objective function based on multi-output error and calculate the gradient to update the data centers, the spread constants and the weight values dynamically.

The basis of the k-means algorithm is error sum of squares [9]. If N_i represents the sample size of the i-th class Γ_i, m_i is the mean value of the samples in Γ_i, as follow $m_i = \frac{1}{N_i} \sum_{y \in \Gamma_i} y$. The error sum of squares for each class Γ_i is calculated and the error sum of squares for all classes are added, as follow $J_e = \sum_{i=1}^{c} \sum_{y \in \Gamma_i} \|y - m_i\|^2$. J_e is called the clustering criteria based on error sum of

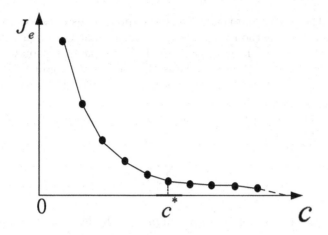

Fig. 4. The law of clustering error

squares. The basic premise of the k-means algorithm is that the clustering number is given in advance, which is not satisfied in unsupervised learning problems such as camera calibration.

For the initial partitioning after a given clustering number, we can get good results by dividing according to the natural order of the samples (the order of the feature points on the calibration board).

Obviously, the clustering error function decreases monotonically with the clustering number increasing. When the clustering number is equal to the size of all samples, we can obtain $J_e(c) = 0$. Thus, each sample itself becomes a class. If there are c^* much clustered classes in the data, will decrease rap-idly when increases from 1 to c^*. But the rate of decreasing will be significantly slower when outnumbers c^*, because the original intensive samples will be separated [10]. Figure 4 shows the curve that represents the change of $J_e(c)$ following c. Based on this law, in order to obtain a better clustering number, we can start from 1 until we select a preferable number. Formula (5) is given below as a criterion of judgment

$$|J_e(c^*) - J_e(c^* - 1)| \leq \alpha \bullet J_e(c^* - 1), \quad c^* = 2, \cdots, c \tag{5}$$

where α is the convergence factor, the range is $\alpha = 0.2 - 0.3$. When the formula (5) is satisfied, the constructed neural network can achieve the fitting accuracy, and there will be not too many nodes in the hidden layer that lead to overfitting and affect generalization ability.

By using k-means clustering algorithm which selects the number of clusters automatically, the hidden nodes and data centers of RBF neural networks can be determined. The spread constants of RBF neural network can be determined by $\delta = \frac{d}{\sqrt{2h}}$, where d is the maximum distance of all classes and h is the clustering number of RBF neural network.

When hidden nodes, data centers and the spread constants are initially determined, the gradient method is adopted to train the output weight values of hidden nodes [11]. The gradient method used in this paper not only updates the output weight values, but also updates the data centers and the spread constants dynamically on the basis of the k-means method to ensure the global optimization of the parameters.

We construct an objective function based on the errors of multiple output nodes. The objective function is defined as

$$\varepsilon = \frac{1}{2} \sum_{n=1}^{N} \|e_n\|_2^2 \tag{6}$$

where N is the number of training samples of the learning process. e_n is the error signal corresponding to the J output nodes and $J \times 1$ is the vector of n dimension.

The j-th component of the error vector (that is the error of the j-th output node) is established as

$$_je_n = {_jd_n} - {_j}\left[\sum_{i=1}^{I} \omega_{ij}\phi_i(\|X_n - c_i\|) \right] \tag{7}$$

where I is the number of hidden nodes. d represents the expected output. The left subscript represents the component of the vector.

We aim at finding the parameters ω_{ij}, c_i and δ_i that minimize ε. For the Gauss activation function used in this paper, the gradient of the objective function to each parameter is given as

$$\begin{cases} \frac{\partial \varepsilon(m)}{\partial \omega_{ij}(m)} = -\sum_{n=1}^{N} {_je_n}(m)\phi_i(\|X_n - c_i(m)\|) \\ \frac{\partial \varepsilon(m)}{\partial c_i(m)} = -\sum_{j=1}^{J}\sum_{n=1}^{N} \frac{2\omega_{ij}(m)}{\delta_i^2(m)} {_je_n}(m)\phi_i(\|X_n - c_i(m)\|)\|X_n - c_i(m)\| \\ \frac{\partial \varepsilon(m)}{\partial \delta_i(m)} = -\sum_{j=1}^{J}\sum_{n=1}^{N} \frac{2\omega_{ij}(m)}{\delta_i^3(m)} {_je_n}(m)\phi_i(\|X_n - c_i(m)\|)\|X_n - c_i(m)\|^2 \end{cases} \tag{8}$$

The update law is given as

$$\begin{cases} \omega_{ij}(m+1) = \omega_{ij}(m) + \eta_1 \frac{\partial \varepsilon(m)}{\partial \omega_{ij}(m)} \\ c_i(m+1) = c_i(m) + \eta_2 \frac{\partial \varepsilon(m)}{\partial c_i(m)} \\ \delta_i(m+1) = \delta_i(m) + \eta_3 \frac{\partial \varepsilon(m)}{\partial \delta_i(m)} \end{cases} \tag{9}$$

where m represents the number of iterations. η_1, η_2 and η_3 are different convergence factors.

In the camera calibration experiment based on improved RBF neutral network, we let the image coordinates of the feature points, (u_l, v_l) and

(u_r, v_r), as inputs and the spatial coordinates in the world coordinate system, (X_W, Y_W, Z_W), as outputs. A RBF neural network with 4 input nodes and 3 output nodes is constructed. The hidden nodes and data centers are determined automatically and the data centers, spread constants and weight values are updated dynamic until the parameters satisfy the precision requirement.

4   Experiment and Analysis

The calibration experiment utilizes a binocular vision system as shown in Fig. 5. The camera in measuring experiment is from EoSens® 3CL series of Mikrotron company, whose model number is MC3010 and resolution is 1280 1024 pixel. The pixel size is 0.008 mm/pixel and the lens model number is AF Zoom-Nikkor 24–85 mm/1:2.8–4D.

Fig. 5. Binocular vision system

A flat plate with round feature points is used as a calibration board and the points on the calibration board are 10×9, which arrange in a certain order. We place the calibration board vertically on the linear motion platform and establish the X-axis, Y-axis of the world coordinate system on the plane of the calibration board, with Z-axis perpendicular to the plane of the calibration board. The calibration board moves along Z-axis 30 mm, 60 mm, 90 mm and 120 mm. We use two cameras to capture simultaneously 5 pairs of images at 0, 30 mm, 60 mm, 90 mm and 120 mm. When Z is 0, the images acquired by the left camera and right camera are shown in Fig. 6.

Zernike moments algorithm [12] is used to extract feature points, and 370 sets of data are obtained. These sets of data were divided into two groups, of

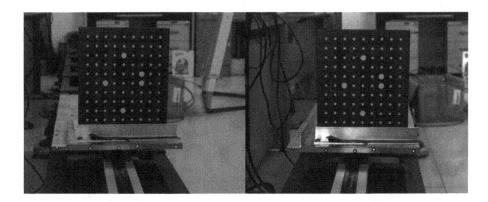

Fig. 6. The images acquired by left camera and right camera

which 280 sets of data were used as training data for the network, and the other 80 groups were used for testing.

RBF network with 4-10-3 structure is trained according to 370 sets of data. The formula for calculating the error of each spatial point is given as

$$e = \sqrt{\left(\tilde{X}_W - X_W\right)^2 + \left(\tilde{Y}_W - Y_W\right)^2 + \left(\tilde{Z}_W - Z_W\right)^2} \qquad (10)$$

where $(\tilde{X}_W, \tilde{Y}_W, \tilde{Z}_W)$ expresses the reconstructed coordinates of the spatial points according to the trained RBF neural network. The results of the reconstruction of some points are shown in Table 1.

Table 1. The test result of some points (Unit: mm)

Spatial points			Reconstructed coordinates			Error
X_W	Y_W	Z_W	$\tilde{X}_W$	$\tilde{Y}_W$	$\tilde{Z}_W$	
125	0	0	124.9691	0.0112	0.1913	0.1940
25	50	30	25.0415	50.0821	30.1745	0.2151
200	50	30	200.0829	50.0475	30.0988	0.1375
125	100	60	124.9302	100.0949	60.1935	0.2265
50	150	60	49.9193	148.0450	60.1221	0.1531
200	175	90	200.0505	175.1134	90.1213	0.1735
75	175	90	75.0387	175.0554	90.0869	0.1101
100	50	120	100.0984	50.0955	120.1154	0.1792

The average of the errors of all 80 testing points is 0.1719 mm. It can be seen from Table 1 that the precision of reconstruction in direction X and Y direction is higher than that in direction Z, because there is an error in the installation of the calibration board on motion platform and the precision of the motion platform is less than that of the calibration board.

The precision of our method is verified by measuring the length of the segment between the two feature points as shown in Fig. 7.

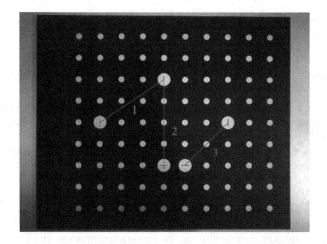

Fig. 7. Three segments that need to be measured

We place the calibration board in any position within the range of the field of view, and take the average of the actual length of the segment between five measurements. Zhang method [4] is used as a comparison. The results are shown in Table 2. It can be seen that the precision of our new method is better than that of Zhang method.

Table 2. Comparision of two methods (Unit: mm)

No.	Actual length	Measurement result		Error	
		Our method	Zhang	Our method	Zhang
1	90.1388	90.3712	90.5723	0.2324	0.4335
2	100	100.2918	100.3652	0.2918	0.3652
3	70.7107	70.8991	71.1634	0.1884	0.4527

5   Conclusion

The camera calibration method based on improved RBF neural network do not need to consider the impact of the lens distortion and environmental factors. Our method can reduce the error caused by the imperfect mathematical model of the traditional calibration method and contribute to improving measurement precision. The reconstruction error of spatial points is 0.1719 mm in this paper. Measuring experiment shows that the precision of our new method is better than that of Zhang method. The data center selection method and dynamic learning of data centers, spread constants and weight values based on the gradient method further can improve the precision.

Acknowledgements. This work was supported by the National Natural Science Foundation of China [grant number 61473100].

References

1. Abdelaziz, Y.I.: Direct linear transformation from comparator coordinates in close-range photogrammetry. In: ASP Symposium on Close-Range Photogrammetry in Illinois (1971)
2. Tsai, R.Y.: A versatile camera calibration technique for high-accuracy 3D ma-chine vision metrology using off-the-shelf TV cameras and lenses. In: Radiometry. Jones and Bartlett Publishers Inc. (1992)
3. Faugeras, O.D., Luong, Q.-T., Maybank, S.J.: Camera self-calibration: theory and experiments. In: Sandini, G. (ed.) ECCV 1992. LNCS, vol. 588, pp. 321–334. Springer, Heidelberg (1992). https://doi.org/10.1007/3-540-55426-2_37
4. Zhang, Z.: A flexible new technique for camera calibration. IEEE Trans. Pattern Anal. Mach. Intell. **22**(11), 1330–1334 (2000)
5. Qi, Z., Wang, Z., Huang, Z.: Research on system calibration of structured-light measurement based on neural network. Acta Photonica Sinica **45**(5), 512002 (2016)
6. Tian, X.C., Zhong-Ke, L.I.: Binocular vision 3D measurement system calibrated by dual plane method. Electron. Opt. Control (2015)
7. Zhang, Z.Z., Qiao, J.F.: Design RBF neural network architecture based on online subtractive clustering. Control Decis. **27**(7), 997–1002 (2012)
8. Guan, S., Gao, J.W., Zhang, B., Liu, X., Leng, Z.W.: Traffic flow prediction based on k-means clustering algorithm and RBF neural network. J. Qingdao Univ. (2014)
9. Wu, S.: Survey on k-means algorithm. N. Technol. Libr. Inf. Serv. **29**(3), 433–439 (2011)
10. Ripley, B.D.: Pattern Recognition and Neural Networks. Cambridge University Press, Cambridge (2009)
11. Wei, H., Li, Q., Song, W.: Gradient learning dynamics of radial function networks. Control Theory Appl. **24**(3), 355–356 (2017)
12. Cui, J., Tan, J.: Algorithm for edge subpixel location based on Zernike moment. Opt. Tech. **31**(5), 775–779 (2005)

Collaborative Representation Based Neighborhood Preserving Projection for Dimensionality Reduction

Miao Li[1], Lei Wang[1](✉), Hongbing Ji[1], Shuangyue Chen[1], and Danping Li[2]

[1] School of Electronic Engineering, Xidian University, Xi'an 710071, China
`leiwang@mail.xidian.edu.cn`
[2] School of Telecommunications Engineering, Xidian University, Xi'an 710071, China

Abstract. Collaborative graph-based discriminant analysis (CGDA) has been recently proposed for dimensionality reduction and classification. It uses available samples to construct sample collaboration via L2 norm minimization-based representation, thus showing great computational efficiency. However, CGDA only constructs the intra-class graph, so it only takes into account local geometry and ignores the separability for samples in different classes. In this paper, we propose a novel method termed as collaborative representation based neighborhood preserving projection (CRNPP) for dimensionality reduction. By incorporating the intra-class and inter-class discriminant information into the graph construction of collaborative representation coefficients, CRNPP not only maintains the same level of time cost as CGDA, but also preserves both global and local geometry of the data simultaneously. In this way, the collaborative relationship of the data from the same class is strengthened while the collaborative relationship of the data from different classes is inhibited in the projection subspace. Experiments on benchmark face databases validate the effectiveness and efficiency of the proposed method.

Keywords: Collaborative representation · Face recognition
Neighborhood preserving projection · Dimensionality reduction
Discriminant analysis

1 Introduction

Sparse representations of signals have received a great deal of attentions in recent years [1]. Many classification and dimensionality reduction approaches have been proposed to explore the sparsity of the data. For classification, Wright et al. [2] presented a sparse representation based classification (SRC) scheme. Mi and Liu

This work was partially supported by National Natural Science Foundation of China (Nos. 61203137, 61401328), Natural Science Foundation of Shaanxi Province (Nos. 2014JQ8306, 2015JM6279), and the Fundamental Research Funds for the Central Universities (No. K5051301007).

© Springer Nature Singapore Pte Ltd. 2017
J. Yang et al. (Eds.): CCCV 2017, Part I, CCIS 771, pp. 449–460, 2017.
https://doi.org/10.1007/978-981-10-7299-4_37

[3] suggest performing SRC on the K selected classes. The L1-norm used in SRC has shown certain robustness to noise in the data. For dimensionality reduction, Qiao et al. [4] proposed a sparsity preserving projections (SPP) and Ly et al. [5] proposed a sparse graph-based discriminant analysis (SGDA). Both of them aim to preserve the sparse construction relationship of the data. Besides sparse representation, sparse subspace learning methods, which seek sparse projection directions, also enjoy the benefits of sparsity. The famous sparse PCA (SPCA) [6] and sparse linear discriminant analysis (SLDA) [7] both impose a sparseness constraint on projection vectors. Generally, the above methods have achieved outstanding performance in various tasks. However, solving L1-norm minimization makes the sparsity based classification schemes very expensive, especially for large-scale datasets.

While SRC emphasizes the role of sparsity on representation coefficients, it has been shown in [8,9] that collaborative representation mechanism is more important to the success of SRC and a collaborative representation based classification (CRC) is then proposed. Following it, Yang et al. [10] proposed a collaborative representation based projection (CRP) which aims at preserving the CR based reconstruction relationship of the data. But like CRC, CRP is also an unsupervised method. To utilize the label information, collaborative graph-based discriminant analysis (CGDA) [11] was proposed to use all training samples from the same class to reconstruct each training sample by CR. It then seeks a discriminant subspace by minimizing the intra-class scatter matrix. The motivation of all these methods is to emphasize "collaborative" instead of "competitive" nature of relationships. At the same time, these methods enjoy computational efficiency because a closed-form solution is available when estimating the representation coefficients, due to the used L2-norm based optimization.

Nevertheless, CGDA only constructs the intra-class graph which takes into account local geometry and ignores global geometry. How to construct supervised graph and embed local and global structures into subspace learning is still a challenging problem in the field of dimensionality reduction (DR). The classic graph-construction DR methods in literature include linear discriminant analysis (LDA) [12], locality preserving projection (LPP) [13,14], neighborhood preserving projection (NPE) [15], neighborhood preserving discriminant embedding (NPDE) [16], and Marginal Fisher Analysis (MFA) [17]. For these methods, it is commonly believed that, if two data points lay closely in the original space, their intrinsic geometry distribution should be preserved in the new subspace.

Building on the success of CR and graph-construction methods, we propose a novel DR method called collaborative representation based neighborhood preserving projection (CRNPP). It incorporates the intra-class and inter-class discriminant information into the graph construction of collaborative representation coefficients. To be specific, the main merits of CRNPP lie in three folds:

(1) When deriving the representation coefficients of input samples, we strengthen the collaborative relationship of data from the same class, as well as inhibit the collaborative relationship of data from different classes. This step also avoids the difficulty of choosing parameters in graph construction;

(2) By constructing the intra-class and inter-class discriminant graphs of reconstruction coefficients, the global and local geometry is preserved so that
samples in the same class are as compact as possible while samples in different classes are as separable as possible in the projection subspace.

(3) A closed-form solution can be obtained and CRNPP does not involve the
iterative calculation as in sparse presentation based methods. Experiments
on ORL, UMIST and BANCA face databases demonstrate the effectiveness
and efficiency of the proposed approach.

The rest of paper is organized as follows. In Sect. 2, we briefly review collaborative representation and CGDA. In Sect. 3, we describe our proposed method
CRNPP in detail and in Sect. 4, the experimental results and discussions are
provided. Finally, conclusions are given in Sect. 5.

2   Collaborative Graph-Based Discriminant Analysis

2.1   Collaborative Representation

Suppose $X = \{x_1, x_2, \ldots, x_n\}$ is a training set of n samples. The mechanism
of CR [8,9] is to represent every sample x by a linear combination of the rest
samples via L2-norm regularized least squares. The collaborative representation
coefficient α for x with L2-norm optimization is to be found:

$$\operatorname*{argmin}_{\alpha} \|\alpha\|_2$$
$$s.t. \quad x = X\alpha. \tag{1}$$

The constrained optimization problem in Eq. (1) can be formulates as:

$$\operatorname*{argmin}_{\alpha} \|x - X\alpha\|_2^2 + \lambda\|\alpha\|_2^2, \tag{2}$$

where λ is an adjusting parameter which controls the relative importance of
L2-regularization term $\|\alpha\|_2^2$ and error term $\|x - X\alpha\|_2^2$.

The role of L2-regularization term $\|\alpha\|_2^2$ is two-folds. First, it makes the
least square solution stable, particularly when X is under-determined; second, it
introduces a certain amount of sparsity to α, although this sparsity is generally
weaker than that by L1-norm regularization. The solution to Eq. (2) can be
obtained as:

$$\alpha = (X^T X + \lambda I)^{-1} X^T x, \tag{3}$$

where I is an identity matrix.

2.2   Collaborative Graph-Based Discriminant Analysis

We introduce CGDA [11] briefly. Suppose that we have a training set $X = \{X_1, X_2, \ldots, X_c\}$ of n samples. There are $N_i(i = 1, 2, \ldots, C)$ samples in the ith

class. In CGDA, the collaborative representation vector is calculated by solving a L2-norm optimization problem,

$$\underset{\alpha_j^i}{\operatorname{argmin}} \|\alpha_j^i\|_2$$
$$s.t. \quad x_j^i = X_i \alpha_j^i. \tag{4}$$

In (4), x_j^i denotes the jth samples in the ith class and X_i denotes the samples from the ith class except x_j^i. Note that Eq. (4) can be further written as:

$$\underset{\alpha_j^i}{\operatorname{argmin}} \|x_j^i - X_i \alpha_j^i\|_2^2 + \lambda \|\alpha_j^i\|_2^2, \tag{5}$$

where λ is an adjusting parameter. The solution of Eq. (5) is

$$\alpha_j^i = (X_i^T X_i + \lambda I)^{-1} X_i^T x_j^i, \tag{6}$$

We define $W_i = [\alpha_1^i, \alpha_2^i, \ldots, \alpha_{N_i}^i]$ as the graph weighted matrix of size $N_i \times N_i$ whose column α_j^i is the collaborative representation vector corresponding to x_j^i. Note that the diagonal elements in W_i are set to be zero. Thus, the intra-class weight matrix W_s can be expressed as:

$$W_s = \begin{bmatrix} W_1 & 0 & \cdots & 0 \\ 0 & W_2 & \cdots & 0 \\ \vdots & \vdots & \ddots & \vdots \\ 0 & 0 & \cdots & W_c \end{bmatrix}. \tag{7}$$

According to the graph-embedding based dimensionality reduction framework, the aim is to find a $m \times d$ projection matrix P (with $d \ll m$) which results in a low-dimensional subspace $Y = P^T X$. The objective of CGDA is to maintain the collaborative representation in the low-dimensional space, which can be formulated as:

$$P^* = \underset{P^T X L_p X^T P = I}{\operatorname{argmin}} \sum_{i \neq j} \|P^T x_i - P^T \sum_j W_{ij} x_j\|^2$$
$$= \underset{P^T X L_p X^T P = I}{\operatorname{argmin}} tr(P^T X L X^T P), \tag{8}$$

where $L = (I - W_s)^T (I - W_s)$ and $L_p = I$. The optimal projection matrix P can be obtained as:

$$P^* = \underset{P}{\operatorname{argmin}} \frac{|P^T X L X^T P|}{|P^T X L_p X^T P|}, \tag{9}$$

which can be solved as a generalized eigenvalue decomposition problem,

$$X L X^T p = \Lambda X L_p X^T p. \tag{10}$$

For a $m \times d$ projection matrix P, it is constructed by the d eigenvectors corresponding to the d smallest nonzero eigenvalues.

3 Collaborative Representation Based Neighborhood Preserving Projection

The aforementioned CGDA only constructs the intra-class graph and ignores the separability for samples in different classes. To cope with this, we propose to incorporate the intra-class and inter-class discriminant information into the graph construction of collaborative representation coefficients simultaneously. We formally state the proposed CRNPP method in detail here.

3.1 Calculating Coefficients and Weight Matrix

We construct local weight matrix like in CGDA. In the collaborative representation, for intra-class graph construction, every sample can be represented as a linear combination of remaining samples from the same class as in Eq. (4). Then we can get the intra-class weight matrix as in Eq. (7).

On the other hand, every sample can be represented as a linear combination of samples from different classes,

$$\operatorname*{argmin}_{\beta_j^i} \|\beta_j^i\|_2 \tag{11}$$
$$s.t. \quad x_j^i = X^i \beta_j^i,$$

where x_j^i denotes the jth samples in the ith class and X^i denotes the samples except the ones from the ith class. Note that Eq. (11) can be further written as:

$$\operatorname*{argmin}_{\beta_j^i} \|x_j^i - X^i \beta_j^i\|_2^2 + \lambda \|\beta_j^i\|_2^2, \tag{12}$$

where λ is an adjusting parameter. The solution of Eq. (12) is:

$$\beta_j^i = ((X^i)^T X^i + \lambda I)^{-1} (X^i)^T x_j^i, \tag{13}$$

We can get the optimal inter-class collaborative representation vectors as $\beta_j^i = (b_{j,1}^i, b_{j,2}^i, \dots, b_{j,i-1}^i, 0, b_{j,i+1}^i, \dots, b_{j,c}^i)^T$. Similar to the intra-class weight vector, the inter-class weight vector for the ith class is in the form of $B_k^i = [(b_{1,k}^i)^T, (b_{2,k}^i)^T, \dots, (b_{N_i,k}^i)^T]$, which denotes the weight vector of kth class to reconstruct the ith class. The element of inter-class weight matrix $W_b^{(ij)}$ can be expressed as:

$$W_b = \begin{cases} B_i^j, & if \quad i \neq j \\ 0, & if \quad i = j \end{cases}. \tag{14}$$

Finally, we can get the inter-class weight matrix as:

$$W_b = \begin{bmatrix} 0 & B_1^2 & B_1^3 & \dots & B_1^c \\ B_2^1 & 0 & B_2^3 & \dots & B_2^c \\ \vdots & \vdots & \vdots & \ddots & \vdots \\ B_c^1 & B_c^2 & B_c^3 & \dots & 0 \end{bmatrix}. \tag{15}$$

3.2   Computing the Neighborhood Preserving Projection

To get the linear projection in the low-dimensional subspace, we aim to minimize the local compactness, which can be defined as:

$$\underset{P}{\operatorname{argmin}} J_1(P) = \|P^T x_i - \sum_j W_s^{(ij)} P^T x_j\|^2 \tag{16}$$
$$= tr(P^T X M_s X^T P),$$

where $M_s = (I - W_s)^T (I - W_s)$ is the Laplacian matrix of the graph associated with the intra-class weight matrix.

At the same time, in the projection subspace, we also want to maximize the distance of samples between different classes. The objective function can be defined as:

$$\underset{P}{\operatorname{argmax}} J_2(P) = \|P^T x_i - \sum_j W_b^{(ij)} P^T x_j\|^2 \tag{17}$$
$$= tr(P^T X M_p X^T P),$$

where $M_p = (I - W_b)^T (I - W_b)$ is Laplacian matrix of the graph associated with the inter-class weight matrix.

Combining Eqs. (16) and (17) together, we can get the following optimization problem,

$$\underset{P}{\operatorname{argmax}} J(P) = J_2(P) - \gamma J_1(P)$$
$$= tr(P^T X M_p X^T P) - \gamma tr(P^T X M_s X^T P) \tag{18}$$
$$= tr(P^T X M X^T P),$$

where $M = M_p - \gamma M_s$, and γ is a parameter to balance the inter-class and intra-class information of the data.

Generally, the dimension of the sample is much larger than number of training samples. Thus, we replace the quotient criterion in previous DR methods [12–17] with the difference criterion, which can overcome the small sample size problem. This criterion is similar to the generalized version of conventional MMC criterion [18]. In this way, we transform the data from original space to some suitable place, which aims at maximizing the margin of different classes in the reduced space.

Equation (18) can be solved as a standard eigenvalue decomposition problem,

$$X M X^T p = \Lambda p, \tag{19}$$

where Λ is the eigenvalue and p is the corresponding eigenvector. Suppose that the first d largest eigenvalues are $\Lambda_1, \Lambda_2, \ldots, \Lambda_d$ and their corresponding eigenvectors are $p_1, p_2, \ldots, p_d$, then we can get a $m \times d$ projection matrix $P = [p_1, p_2, \ldots, p_d]$.

The proposed CRNPP algorithm is summarized as follows:

CRNPP Algorithm

Input: Training set $X = \{x_i\}_{i=1}^n$ for C classes, and parameters λ, γ.

Output: Optimal projection matrix W.

Step1: Use PCA to project the original data into a lower dimensional subspace. The projection matrix is denoted by W_{PCA}.

Step2: Construct the weight matrix W_s and W_b based on collaborative representation (7) and (15), respectively.

Step3: Construct the intra-class neighborhood preserving graph and interclass neighborhood preserving graph according to Eq. (16) and (17), respectively. Then construct the final neighborhood preserving graph by Eq. (18).

Step4: Obtain the d eigenvectors corresponding to first d largest nonzero eigenvalues according to Eq. (19). Denote the obtained projection matrix as $P = [p_1, p_2, \ldots, p_d]$.

Step5: Obtain the whole projection matrix $W = W_{PCA}P$.

In summary, the key idea for the proposed CRNPP is to incorporate the reconstruction coefficients of collaborative representation into subsequent graph construction. Then, the optimal projection is obtained by minimizing the local compactness and maximizing the global discriminant information simultaneously. After the projection, the Nearest Neighbors (NN) algorithm can be used to identify the class label of each testing sample.

4 Experimental Verification

We evaluate the proposed CRNPP algorithm on three well-known face databases (ORL, UMIST and BANCA) and compare its performance with CRC [8], NPE [15], SPP [4], MFA [17] and CGDA [11]. In the experiments, we keep 99% energy in the pre-processing step as in [13] and use the NN classifier for final classification due to its simplicity. The average recognition accuracy is reported over 10 random splits.

Dataset preparation. The ORL face database contains 400 images of 40 individuals under various facial expressions and lighting conditions. In our experiments, each image is manually cropped and resized to 32×32 pixels. The UMIST database consists of 20 people with totally 564 images. Each image is downsampled to 56×46. The BANCA database took 208 people at different times, status, quality, light and expressions of the standard face images. We randomly selected 52 people which are of different ages and different genders. This subset contains 520 pictures with each picture being down-sampled to 56×46.

Performance Comparison. First, we compare the classification performance of the proposed CRNPP method with the related methods. For all ORL, UMIST and BANCA datasets, we randomly choose 4, 5 and 6 images per person for training, respectively, and the corresponding rest images are for test. Tables 1, 2 and 3 show the top classification accuracy corresponding to optimal dimension of all methods on ORL, UMIST and BANCA datasets, respectively. From the results, it can be seen that with increasing number of training samples, the accuracy of all methods increases. CRNPP consistently outperforms the other methods.

Table 1. The top recognition accuracy (%) of the six approaches and the corresponding optimal dimension on the ORL database

	CRC	NPE	SPP	MFA	CGDA	CRNPP
4train	79.00(80)	88.17(66)	85.71(68)	91.21(50)	89.25(44)	**93.92(52)**
5train	81.30(100)	91.90(64)	89.15(76)	94.10(62)	92.45(62)	**96.45(50)**
6train	84.50(100)	94.50(76)	91.00(58)	96.06(84)	96.00(68)	**97.56(80)**

Table 2. The top recognition accuracy (%) of the six approaches and the corresponding optimal dimension on the UMIST database

	CRC	NPE	SPP	MFA	CGDA	CRNPP
4train	73.01(60)	77.89(36)	74.09(60)	84.90(42)	78.39(60)	**87.54(20)**
5train	80.59(70)	86.36(42)	80.80(70)	90.78(56)	86.55(64)	**92.20(18)**
6train	82.92(80)	90.32(62)	85.16(84)	92.79(38)	89.46(88)	**94.19(72)**

Table 3. The top recognition accuracy (%) of the six approaches and the corresponding optimal dimension on the BANCA database

	CRC	NPE	SPP	MFA	CGDA	CRNPP
4train	46.03(60)	66.63(52)	54.04(52)	73.01(56)	67.37(58)	**77.08(60)**
5train	45.92(60)	72.54(42)	57.65(54)	77.62(58)	73.73(56)	**82.08(60)**
6train	47.74(80)	78.13(56)	60.72(58)	80.67(56)	79.47(60)	**85.94(60)**

Further, we plot the average accuracy of the DR methods versus the reduced dimensions in Fig. 1, where the case of 4 training samples are considered. CRC, which is not a DR method, is also presented as a baseline. We have the following observations. First, SPP is inferior to NPE and CGDA, due to the fact that SPP is an unsupervised method. Second, both NPE and CGDA do not explicitly encode the discriminant information of data points which have different labels, so MFA, which considers both global and local information, shows a little performance improvements. Finally, with further preserving the collaborative relationship of the data, the proposed CRNPP performs the best.

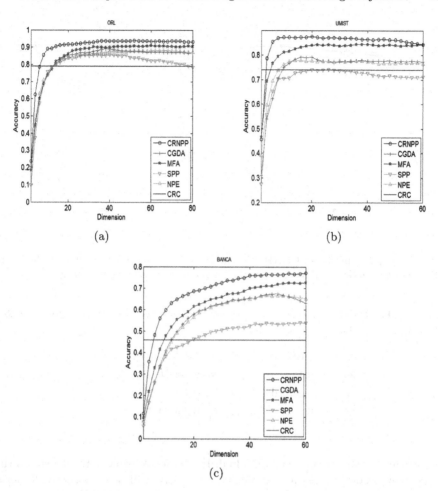

Fig. 1. The recognition accuracy versus the number of reduced dimensions on three databases, when 4 training samples are considered. (a) ORL database; (b) UMIST database; (c) BANCA database.

To further investigate the performance of our method against the used adjusting parameters, we also discuss the recognition accuracies of CRNPP with the variations of λ and γ, respectively. Figure 2(a) shows the recognition accuracies of CRNPP on the UMIST dataset when λ is tuned from $\{10^{-5}, 10^{-4}, \ldots, 10^{0}\}$, while γ is fixed as 15. Figure 2(b) gives the recognition accuracies of CRNPP when γ is turned from $\{10, 11, \ldots, 20\}$, while λ is set to 0.1. From Fig. 2, we can see that the performance of CRNPP is not very sensitive to the chosen parameters and it is more stable on γ than on λ.

Finally, we compare the average computational time of related methods in the training stage. Table 4 gives the average time cost (in seconds) of all the methods on the ORL, UMIST and BANCA datasets, respectively, where 4 training

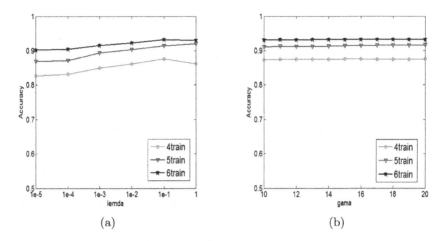

Fig. 2. The performance of CRNPP on the UMIST database with different choices of adjusting parameters. (a) average accuracies versus λ; (b) average accuracies versus γ.

Table 4. The average time cost (in seconds) in training stage on the ORL, UMIST and BANCA databases

	CRC	NPE	SPP	MFA	CGDA	CRNPP
ORL	0.193	0.576	6.496	0.409	0.749	0.825
UMIST	0.243	0.417	1.901	0.450	0.521	0.479
BANCA	0.261	0.343	12.482	0.399	1.451	1.412

samples are selected from per individual. It can be seen that CRNPP shows the same level of computational complexity as CGDA, which is another collaborative representation based DR method. It should be noted that CRNPP consumes much less time than SPP because CRNPP provides a closed-form solution while SPP is an iterative method. This experiment shows the advantage of L2-norm based representation against L1-norm based one, in terms of computational cost.

5　Conclusion

In this paper, we have presented a novel dimensionality reduction (DR) method for image classification, named as collaborative representation based neighborhood preserving projection (CRNPP). On one hand, it incorporates the intra-class and inter-class discriminant information into the graph construction, so the global and local geometry of the data can be preserved in the projection subspace. On the other hand, it can take advantage of the collaborative representation to mining the collaborative relationship of the data, thus the proposed method avoids the difficulty of choosing the number of nearest neighbors in weight matrix construction as in traditional DR methods. Moreover, by using

L2-norm minimization-based optimization, our method can provide a closed-form solution, therefore it is as fast as CRC alike methods. Comparisons to most used DR techniques show the superiority of the proposed algorithm in terms of both recognition accuracy and computional cost.

Kernel methods can map an input feature space into a high dimensional feature space [19–21], where the problem of weak linear separability could be solved. Kernel trick has been widely used without explicitly evaluating a non-linear mapping function. It is interesting to further extend our CRNPP into a kernel version, which is able to extract nonlinear discriminant features in kernel-included spaces. We left it as our future work.

References

1. Huang, K., Aviyente, S.: Sparse representation for signal classification. In: Advances in Neural Information Processing Systems, pp. 609–616 (2007)
2. Wright, J., Yang, A.Y., Ganesh, A., Sastry, S.S., Ma, Y.: Robust face recognition via sparse representation. IEEE Trans. Pattern Anal. Mach. Intell. **31**(2), 210–227 (2009)
3. Mi, J.X., Liu, J.X.: Face recognition using sparse representation-based classification on k-nearest subspace. PLoS ONE **8**(3), e59430 (2013)
4. Qiao, L., Chen, S., Tan, X.: Sparsity preserving projections with applications to face recognition. Pattern Recognit. **43**(1), 331–341 (2010)
5. Ly, N.H., Du, Q., Fowler, J.E.: Sparse graph-based discriminant analysis for hyperspectral imagery. IEEE Trans. Geosci. Remote Sens. **52**(7), 3872–3884 (2014)
6. Zou, H., Hastie, T., Tibshirani, R.: Sparse principal component analysis. J. Comput. Graph. Stat. **15**(2), 265–286 (2006)
7. Clemmensen, L., Hastie, T., Witten, D., Ersboll, B.: Sparse discriminant analysis. Technometrics **53**(4), 406–413 (2011)
8. Zhang, L., Yang, M., Feng, X.: Sparse representation or collaborative representation: which helps face recognition? In: 2011 IEEE International Conference on Computer Vision (ICCV), pp. 471–478. IEEE (2011)
9. Zhang, L., Yang, M., Feng, X., Ma, Y., Zhang, D.: Collaborative representation based classification for face recognition. Technical report, arXiv 1204.2358 (2012)
10. Yang, W., Wang, Z., Sun, C.: A collaborative representation based projections method for feature extraction. Pattern Recognit. **48**(1), 20–27 (2015)
11. Ly, N.H., Du, Q., Fowler, J.E.: Collaborative graph-based discriminant analysis for hyperspectral imagery. IEEE. Sel. Top. Appl. Earth Obs. Remote Sens. **7**(6), 2688–2696 (2014)
12. Belhumeur, P.N., Hespanha, J.P., Kriegman, D.J.: Eigenfaces vs. fisherfaces: recognition using class specific linear projection. IEEE Trans. Pattern Anal. Mach. Intell. **19**(7), 711–720 (1997)
13. He, X., Yan, S., Hu, Y., Niyogi, P., Zhang, H.J.: Face recognition using laplacianfaces. IEEE Trans. Pattern Anal. Mach. Intell. **27**(3), 328–340 (2005)
14. He, X., Niyogi, P.: Locality preserving projections. In: Advances in Neural Information Processing Systems, pp. 153–160 (2004)
15. He, X., Cai, D., Yan, S., Zhang, H.J.: Neighborhood preserving embedding. In: 2005 Tenth IEEE International Conference on Computer Vision (ICCV), vol. 2, pp. 1208–1213. IEEE (2005)

16. Han, P.Y., Jin, A.T.B., Abas, F.S.: Neighbourhood preserving discriminant embedding in face recognition. J. Vis. Commun. Image Represent. **20**(8), 532–542 (2009)
17. Yan, S., Xu, D., Zhang, B., Zhang, H.J., Yang, Q., Lin, S.: Graph embedding and extensions: a general framework for dimensionality reduction. IEEE Trans. Pattern Anal. Mach. Intell. **29**(1), 40–51 (2007)
18. Li, H., Jiang, T., Zhang, K.: Efficient and robust feature extraction by maximum margin criterion. In: Advances in Neural Information Processing Systems, pp. 97–104 (2004)
19. Ding, M., Tian, Z., Xu, H.: Adaptive kernel principal component analysis. Sig. Process. **90**(5), 1542–1553 (2010)
20. Zeng, W.J., Li, X.L., Zhang, X.D., Cheng, E.: Kernel-based nonlinear discriminant analysis using minimum squared errors criterion for multiclass and undersampled problems. Sig. Process. **90**(8), 2333–2343 (2010)
21. Yu, X., Wang, X., Liu, B.: Supervised kernel neighborhood preserving projections for radar target recognition. Sig. Process. **88**(9), 2335–2339 (2008)

Iterative Template Matching with Rotation Invariant Best-Buddies Pairs

Zhuo Chen, Yang Yang[⊠], Weile Chen, Qian Kou, and Dexing Zhong

The School of Electronic and Information Engineering, Xi'an Jiaotong University,
Xi'an, China
yyang@mail.xjtu.edu.cn

Abstract. In this paper, we propose a new method for template matching method with rotation invariance. Our template matching can not only find the location of the object, but also annotate its rotation angle. The key idea is to firstly rectify the local rotation patches according to their intensity centroids, and then to find the corresponding patch-features between template and target images under an iterative matching framework. We adopt the coarse-to-fine search ways, so the patch size should be updated accordingly, which is time-consuming. To tackle this problem, we use the integral image to update the intensity centroid to accelerate the computing speed. The corresponding feature matching is based on the Best-Buddies Pairs (BBPs), which is robust to the non-rigid transform of local range and outliers. Experimental results demonstrate the effectiveness and robustness of the proposed algorithm.

Keywords: Template matching · Rotation invariance
Intensity centroid · Integral image

1 Introduction

Template matching is an essential work in the computer vision field, which has many applications in the object tracking, 3D reconstruction and image registration. Corresponding techniques has rapidly developed in recent years, which can be generally divided into two types: the similarity measure algorithm and local feature matching algorithm.

The former type relies mostly on the global search matching. Some good similarity measure functions such as the Sum of Squared Differences (SSD) [1], Sum of Absolute Differences (SAD) [2] and Histogram has been used to judge the difference be-tween the template and target images. Mattoccia et al. [3] proposed to use the bounded partial correlation (BPC) with the similarity measure, which reduced the computation cost by comparing the correlation between template and target images. Furthermore, to solve the matching problem when rotation occurs, some methods [4,5] combined the circular projection transformation (CPT) or radial projections [6,7] to the conventional methods. However, when the region of interest has high-level outliers like object's non-rigid transformation, background noise and partial occlusion. These global similarity methods

© Springer Nature Singapore Pte Ltd. 2017
J. Yang et al. (Eds.): CCCV 2017, Part I, CCIS 771, pp. 461–471, 2017.
https://doi.org/10.1007/978-981-10-7299-4_38

can not automatically distinguish the outliers. Recently, a new similarity measure called Best-Buddies Similarity (BBS) [8] was proposed. The key idea is to count the number of a subset of nearest neighbor patch-pairs – the Best-Buddies Pairs (BBPs) between two point sets. BBPs are robust to outliers. But, since the BBS is under the slide-window searching, it cannot annotate the angle of the object of interested.

The other type of algorithm is to extract local invariant features between two images. The Scale-Invariant Feature Transform (SIFT) [9] is a popular method in this type. The SIFT is robust to the light change in the scene, various object deformation and other complex conditions. SURF [10] is another famous feature matching method. It is 10 times faster than SIFT with little decline of the performance. Feature line based method [11] is also a local feature instead feature points in SIFT or SURF. The line structure provides more information to consists more accurate descriptors and achieve better matching results. However, these methods are dependent much on the assumption that enough edges or corner points in the object could be distinguished from the background. This limits the application in the wild.

This paper concentrates on how to achieve rotation invariance during the matching process, as well as to be robust to the outliers, partial occlusion, local rigid or non-rigid deformation of the object in the wild. Our purpose can not only find the location of the object, but also annotate its rotation angle. We proposed to matching rotation invariant BBPs under an iterative matching framework, which combines the advantages of the above two types of methods. The key idea is to firstly rectify the local rotation patches according to their intensity centroids, and use the integral image to accelerate the computing speed. And then we find the corresponding BBP patch-features and update the rotation and translation parameter to update the location. Compared with the similar work proposed by Luo [12], our method has a great superiority on the template matching with large rotation angles. Experimental results demonstrate the proposed method is effectiveness and robustness.

2   Proposed Method

In template matching, rotation of the object in target images differs in various scenes. How to reduce the error brought by the rotation is our first concern in template matching. Intensity centroid is chosen in our method to overcome the problem and achieve matching by improved Best-Buddies Similarity algorithm.

Figure 1 shows the process of our proposed method. The algorithm mainly includes following steps: 1. divide template image and target image into k * k patches and utilize intensity centroid to realize rotation invariance; 2. find corresponding patch-feature in processed template and target image; 3. calculate the matching parameters of position and locate the target object in this iteration.

In each iteration, we need to calculate matching parameter by acquired corresponding patch-features in template and target image to locate the target object and move the template image for next iteration. The corresponding patch-features are two point sets $P = \{p_i\}_{i=1}^{N}$ and $Q = \{q_j\}_{j=1}^{M}$, where P is the template

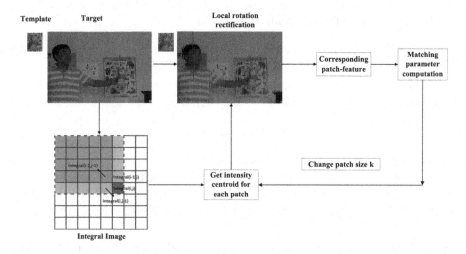

Fig. 1. Matching process.

set and Q is the target set, and pi, qj represents the coordinate. The matching parameters are rotation matrix R and translation vector t are computed between P and Q by minimizing the following distance function:

$$(R, t) = argmin \sum_{i=1}^{M} \| R * p_i + t - q_i \|. \tag{1}$$

The template is set at the upper left corner in target image as its first iteration position. In each iteration, the position of the template changes according to the parameter. Meanwhile the patch size k also changes for matching from coarse to precise. However, different k means different intensity centroid for each patch in two images which will cost more time in computing. We choose integral image to help improve computing speed. The following subsections will discuss the details of our method.

2.1   Local Rotation Rectification

To obtain rotation invariance for our method, we use an effective measure of orientation-the intensity centroid.

Intensity centroid. The intensity centroid defines that a patch's intensity is offset from its geometrical center, thus the vector from the geometrical center to the intensity centroid can be used to get the orientation. Rosin [13] defines the moments of a patch as:

$$m_{pq} = \sum_{x,y} x^p y^q I(x, y). \tag{2}$$

where x, y is the coordinate of each pixel and I(x, y) is the grey value.

Then we can find the centroid:

$$C = (\frac{m_{10}}{m_{00}}, \frac{m_{01}}{m_{00}}). \tag{3}$$

We can construct a vector from a patch's geometrical center, O, to the intensity centroid, $\overrightarrow{OC}$. The orientation of the patch is:

$$\theta = atan2(m_{01}, m_{10}). \tag{4}$$

where atan2 is the quadrant-aware version of arc tan.

In template matching, template image and target image are divided into k * k distinct patches. For each path, we need to find the centroid to get an orientation. We choose integral image to complete it simply and quickly.

Integral image. The integral image is also called summed area table (SAT). It is a fast and efficient data structure and algorithm for calculating the sum of a rectangular sub-region of a grid. The value of each point (x, y) in the integral image, I, is the sum of all the gray-scale values, i, in the upper left corner of the corresponding position in the image:

$$I(x, y) = \sum_{x' \leq x, y' \leq y} i(x', y'). \tag{5}$$

Integral image has a good property that it can be calculated by traversing the whole image only once. Equation can be rewritten as:

$$I(x, y) = i(x, y) + I(x - 1, y) + I(x, y - 1) - I(x - 1, y - 1). \tag{6}$$

In our experiment, we construct three integral images for each image. The first one is the ordinary integral image corresponding to m_{00} in Eq. (2). The second one is calculated after multiplying gray-scale values of each position at (x, y) by x-coordinate. It corresponds to m_10. Analogously, the last one corresponds to m_{01} obtained by multiplying the y-coordinate.

2.2   Corresponding Patch-Features

Finding correlation points is an essential step in template matching. Feature based approach such as scale-invariant feature transform (SIFT) extract feature points by local information like edges and corners from images, but it will fail to find enough correlation points in images with less edges or corners.

Our method chooses a new proposed similarity measure called Best-Buddies Similarity (BBS) to get correlation points. BBS is robust to outliers from the background and complex deformations. These advantages enable it to obtain plenty of feature points. The key of BBS is to compute corresponding features between two sets of points. The corresponding feature is relied on the Best-buddies Pairs (BBPs) extracted from the two sets of points.

BBPs. There are two sets of points $P = \{p_i\}_{i=1}^N$ and $Q = \{q_j\}_{j=1}^M$ from two images. Each point has two kind of information. RGB shows its pixel appearance and (x, y) shows its pixel location. A pair of points from P and Q respectively can be regarded as BBP when they satisfy the following equation:

$$bb(p_i, q_j, P, Q) = \begin{cases} 1 & NN(p_i, Q) = q_j \wedge NN(P, q_j) = p_i \\ 0 & otherwise \end{cases} \tag{7}$$

The equation means that the nearest neighbor of p_i in set Q is q_j, and vice versa. $NN(p_i,Q)$ and $NN(P,q_j)$ is calculated by:

$$NN(p_i, Q) = argmin_{q \in Q} d(p_i, q). \tag{8}$$

$$NN(P, q_j) = argmin_{p \in P} d(p, q_j). \tag{9}$$

$$d(p_i, q_j) = (x_i - x_j)^2 + (y_i - y_j)^2 + \lambda * [(r_i - r_j)^2 + (g_i - g_j)^2 + (b_i - b_j)^2]. \tag{10}$$

r, g, b is the RGB value of the point. λ is a weight number and we choose $\lambda = 2$ in our experiments.

2.3 Implementation Detail

Our method aims to achieve template matching with rotation invariance. The algorithm procedure is concluded as follows:

1. Given template image, T and target image, I, calculate the integral images as mentioned in Sect. 2.1;
2. Break the image regions into k * k distinct patches and update the pixel RGB value in each patch by rotating a degree which is got from intensity centroid;
3. Calculate $d(p_i, q_j)$ to obtain BBPs sets P and Q in I, T;
4. Get matching parameters R, t to obtain the computing position for next iteration;
5. Iterate steps 2 to 4 until the variation of parameters R and t reach the threshold or iteration ends.

The innovation of our algorithm is that we use intensity centroid combined with integral image to achieve rotation invariance for each patch and change the patch size k as iteration time increases to accelerate the computing process.

Specifically, each k * k patch in the image contains k * k pixels. Using the three integral images, we can simply get m_{01}, m_{10} and m_{00} to compute the intensity centroid for a patch when we know the coordinates of the lower right corner coordinates of the patch in the original image. Then the rotation angle for each patch is got from the vector $\overrightarrow{OC}$ as mentioned before and all pixels in the patch rotate to get new RGB value. When computing $d(p_i, q_j)$ the patch is regarded as a 'point'– the center coordinates of the patch denote the location and all pixels' new RGB value in the patch denote appearance. For different k,

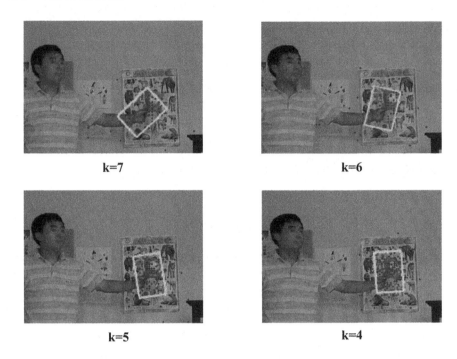

Fig. 2. Results under different patch size k from 7 to 4. (Color figure online)

we can just look up in the integral image to calculate the intensity centroid for the patch, which is much less time-consuming.

In practice, we change the patch size k in the iteration. The k is getting smaller with the iteration time increases. According to Eq. (10), in the first iteration, the BBPs of target image I mostly locate in upper left corner in I. With iteration time increases, they will gradually converge to the correct matching region. Initially, the size k is set larger. This will accelerate the convergence speed to make the BBPs in I approach an approximate region. Then the k is getting smaller $(k > 3)$ for accurate matching. Smaller k will help find more BBPs and better matching result with more time consumed. Figure 2 shows the different results under different patch size k.

In Fig. 2 blue dots are the BBPs in the target image and red rectangular is the matching result. When size k is too big, the matching speed is faster but less BBPs will be got. While k is small, the speed will decrease but more BBPs are got. We combine the two properties of different size to adjust size k during the iteration process, so we can get enough BBPs to help matching and spend less time.

3   Experimental Results

Experiments are conducted on a computer with 3.10 GHz CPU and 8G memories. The images are chosen randomly. To verify the better performance of our algorithm, testing images in our experiments are taken from real-world scenarios and some are from the BBS work [9]. To demonstrate effectiveness of the matching method, we compare the results with original BBS algorithm and an improved algorithm (B+S) by Luo [12], which can also duel with rotation in template matching.

We set up two groups of experiments. First one uses images which are rotated by 0°, 90°, 180°, 270° respectively. The second one uses images which contain rotated template in the target image. As shown in Fig. 4, our results outperform other two algorithms. For each template with rotation of 90° or 270°, the BBS fails in most cases because the sliding window search can only find locations of the objects. Since the BBS doesn't have rotation invariant features, it loses the object in Fig. 4(4). The B+S method also doesn't use rotation invariant features. So it cannot get accurate results when large rotation happens.

We adopt the overlap ratio to quantitatively compare the results, which is the percentage of the overlapping region with the ground truth. The overlap results are shown in Fig. 3. We can see that our method get the highest overlap under every large rotation condition.

Template ID	Overlap											
	0°			90°			180°			270°		
	Ours	B+S	BBS	Ours	B+S	BBS	Ours	B+S	BBS	Ours	B+S	BBS
(1)	0.95	0.94	0.98	0.89	0.74	0	0.93	0.81	0.06	0.87	0.88	0
(2)	0.93	0.92	0.89	0.89	0.73	X	0.88	0.91	0.54	0.90	0.93	X
(3)	0.95	0.95	0.95	0.93	0.93	0.71	0.92	0.83	0.90	0.96	0.87	0.68
(4)	0.96	0.94	0	0.96	0.53	0	0.93	0.82	0	0.95	0.47	0.41
Mean	0.95	0.94	0.71	0.92	0.73	0.24	0.92	0.84	0.38	0.92	0.79	0.36

Fig. 3. Overlap of experiment 1

To further testify the rotation invariance of our method, we repeat the experiment after rotating the images by every 30°. Figure 5 shows the detailed results.

Under every rotation condition, overlap-rate represents the area ratio of the matching result covering the ground truth.

$$OverlapRate = \frac{area(Result \cap GT)}{area(GT)}. \tag{11}$$

The rotation accuracy is the results' rotating angle compared to the current image rotation angle:

$$Rotation accuracy = \frac{\theta_{Result}}{\theta_{GT}}. \tag{12}$$

where the θ_{Result} is the rotation angle which is got from matching result and θ_{GT} is the ground truth angle we annotated manually.

Template   **Matching Result**

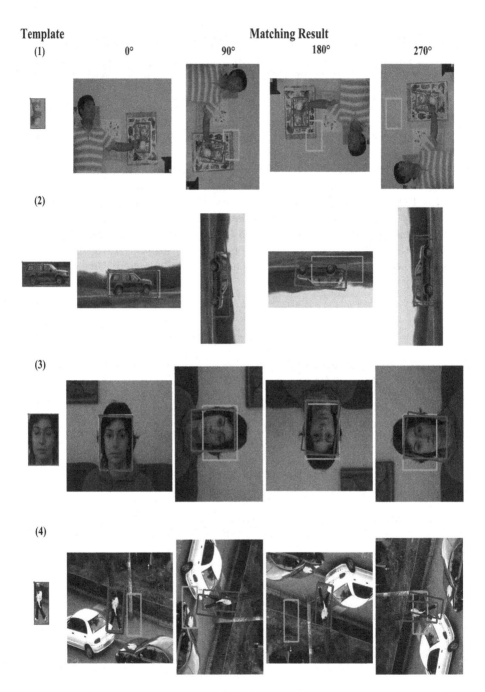

Fig. 4. Template matching results of different rotated images in experiment 1. The green rectangular is BBS's [9] result; blue one is B+S [12] result; red one is ours. (Color figure online)

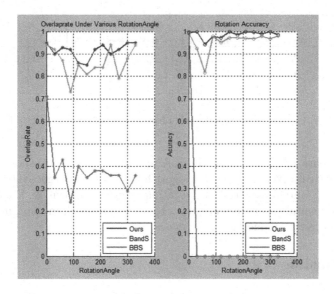

Fig. 5. The overlap-rate of compared three methods is shown on the left and the rotation accuracy is on the right.

The two parameters are between thresholds from 0 to 1. Bigger value means better result. As shown in Fig. 5, the proposed scheme has the highest overlap rate and rotation accuracy which demonstrate its good performance and rotation invariance under different angles.

In the experiment 2, we compute the rotation error (RE) as Eq. (13) by comparing the matching angle with the ground truth. The smaller RE is, the more accurate results are.

$$RE = \frac{\theta_{Result} - \theta_{GT}}{\theta_{GT}}. \tag{13}$$

Fig. 6. Template matching results of different rotated template in target image in experiment 2. The green rectangular is BBS's [9] result; blue one is B+S [12] result; red one is ours. (Color figure online)

Template ID	Overlap			Rotation angle				Rotation error		
	Ours	B+S	BBS	Ours	B+S	BBS	GT	Ours	B+S	BBS
(1)	0.91	0.89	0	45°	36°	0°	47°	-4.3‰	-23.4‰	-1
(2)	0.86	0.93	0.73	-20°	-20°	0°	-22°	-9.1‰	-9.1‰	-1
(3)	0.97	0.62	0.70	-40	-5°	0°	-44°	-9.1‰	-88.6‰	-1

Fig. 7. Result of experiment 2

The compare RE results and the corresponding images are shown in Figs. 5 and 6, respectively. Among the three algorithms, our method gets the best results in the overlap and most accurate rotation angle as shown in Fig. 7. All the above experimental results demonstrate that the proposed method has the best rotation-invariance among the three methods.

4 Conclusion

This paper proposed a template matching method with robustness and rotation in-variance. The method employed the intensity centroid to rectify the local rotation patches and used integral image to reduce the computing time. Then BBPs is used to get corresponding points and the rigid transform parameters are updated iteratively. Our method is demonstrated to have better performance under large rotation angles compared with other methods in the experimental results. Future work will introduce the scaling invariance and more robust features in our methods.

Acknowledgement. This work was in part supported by the National Natural Science Foundation of China (61627811, 61573274), Postdoctoral Science Foundation of China (2015M582661), Natural Science Foundation of Shaanxi, China (No. 2015JQ6257) and the Fundamental Research Funds for the Central Universities (xjj2017005, xjj2017067)

References

1. Stefano, L.D., Mattoccia, S.: Fast template matching using bounded partial correlation. Mach. Vis. Appl. **13**(4), 213–221 (2003)
2. Li, W., Salari, E.: Successive elimination algorithm for motion estimation. IEEE Trans. Image Process. **4**, 105–107 (1995)
3. Mattoccia, S., Tombari, F., Stefano, L.D.: Fast full-search equivalent template matching by enhanced bounded correlation. IEEE Trans. Image Process. **17**(4), 528–538 (2008)
4. Lee, W.C., Chen, C.H.: A fast template matching method with rotation invariance by combining the circular projection transform process and bounded partial correlation. IEEE Signal Process. Lett. **19**(11), 737–740 (2012)
5. Lin, Y.H., Chen, C.H., Wei, C.C.: New method for sub-pixel image matching with rotation invariance by combining the PT method and the RPT process. Opt. Eng. **45**(6), 067202(1)–067202(9) (2006)

6. Choi, M.S., Kim, W.Y.: A novel two stage template matching method for rotation and illumination invariance. Pattern Recogn. **35**, 119–129 (2002)
7. Kim, Y.H.: Rotation-discriminating template matching based on Fourier coefficients of radial projections with robustness to scaling and partial occlusion. Pattern Recogn. **43**(3), 859–872 (2010)
8. Dekel, T., Oron, S., Rubinstein, M., Avidan, S., Freeman, W.T.: Best-buddies similarity for robust template matching. In: Computer Vision and Pattern Recognition, pp. 2021–2029 (2015)
9. Lowe, D.: Distinctive image features from scale-invariant key points. Int. J. Comput. Vis. **60**(2), 91–110 (2004)
10. Bay, H., Tuytelaars, T., Van Gool, L.: SURF: speeded up robust features. In: Leonardis, A., Bischof, H., Pinz, A. (eds.) ECCV 2006. LNCS, vol. 3951, pp. 404–417. Springer, Heidelberg (2006). https://doi.org/10.1007/11744023_32
11. Zhang, Y., Qu, H.S.: Rotation invariant feature lines transform for image matching. J. Electron. Imaging **23**(5), 053002 (2014)
12. Luo, S., Yang, Y., Du, S.Y., Kou, Q.: Robust image registration with rotation, scale, translation using best-buddies pairs. In: International Joint Conference on Neural-Networks, pp. 1728–1732 (2016)
13. Rosin, P.L.: Measuring corner properties. Comput. Vis. Image Underst. **73**(2), 291–307 (1999)

Image Forgery Detection Based on Semantic Image Understanding

Kui Ye[1,2], Jing Dong[1,3(✉)], Wei Wang[1,2,4], Jindong Xu[1], and Tieniu Tan[1]

[1] Center for Research on Intelligent Perception and Computing,
Institute of Automation, Chinese Academy of Sciences, Beijing 100190, China
kui.ye@cripac.ia.ac.cn, {jdong,wwang,jindong.xu,tnt}@nlpr.ia.ac.cn
[2] State Key Laboratory of Cryptology, Chinese Academy of Sciences,
P.O. Box 5159, Beijing 100878, China
[3] State Key Laboratory of Information Security,
Institute of Information Engineering, Chinese Academy of Sciences,
Beijing 100093, China
[4] Shenzhen Key Laboratory of Media Security, Shenzhen University, Shenzhen
518060, China

Abstract. Image forensics has been focusing on low-level visual features, paying little attention to high-level semantic information of the image. In this work, we propose the framework for image forgery detection based on high-level semantics with three components of image understanding module, the normal rule bank (NR) holding semantic rules that comply with our common sense, and the abnormal rule bank (AR) holding semantic rules that don't. Ke et al. [1] also proposed a similar framework, but ours has following advantages. Firstly, image understanding module is integrated by a dense image caption model, with no need for human intervention and more hierarchical features. secondly, our proposed framework can generate thousands of semantic rules automatically for NR. Thirdly, besides NR, we also propose to construct AR. In this way, not only can we frame image forgery detection as anomaly detection with NR, but also as recognition problem with AR. The experimental results demonstrate our framework is effective and performs better.

Keywords: Image forensics · Image understanding module · NR · AR
Deep learning

1 Introduction

Manipulating and editing digital images are becoming increasingly easy with unimpeded access to advanced image processing softwares, such as Photoshop and Gimp. The ubiquity of forged images erodes our trust in photography and furthermore affects national security, commerce, the media and so on, which subsequently gives rise to the emergence of the field of image forensics that aims to help restore some trust in photography.

© Springer Nature Singapore Pte Ltd. 2017
J. Yang et al. (Eds.): CCCV 2017, Part I, CCIS 771, pp. 472–481, 2017.
https://doi.org/10.1007/978-981-10-7299-4_39

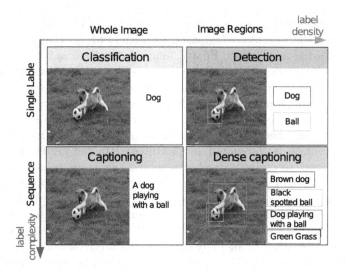

Fig. 1. We implement the image understanding module with a dense image caption model (bottom right).

Traditionally the problem of detecting image forgery is addressed in several aspects. On the pixel level, analyses detect pixel correlations that are introduced by cloning [2], re-sizing [3], or non-linear filtering [4]. Analyses w.r.t the camera sensor can detect inconsistencies in chromatic aberrations [5], color filter array interpolation [6], or sensor noise [7]. On the scene level, analysis can detect inconsistencies in reflections [8], lighting [9], or shadows [10]. The above detection approaches are mainly based on general low-level visual features such as texture, light, color, shape, etc. Analysing the high-level semantic content of the image can also have a crucial role in image forgery detection [1]. Because every authentic picture is a organic whole, involving a certain degree of significance, and forgery images forged from different pictures express a fictional content, it's feasible to detect image forgery by analysing the semantic content w.r.t the image such as clothing, movements, season climate, physical environment and so on.

Although image semantics have been widely used for object detection, object tracking, image retrieval and so on, there is little research on applying high-level semantic information to the field of image forensics. Inspired by Johnson et al. [11], we propose a framework of image forgery detection based on high-level semantics, which consists of image understanding module, NR, and AR. Firstly, semantic instances and their relationship are extracted from the real and forgery image dataset, mapped to captions or words, and stored as rules of NR and AR respectively, both to be used in the next step. Secondly, a test image is processed by the image understanding module. Last, image forgery detection is completed based on the result of image understanding module, NR, and AR.

However, we are not the first to do image forgery detection based on semantics. Ke et al. [1] also proposed a similar framework consisting of three components including image recognition, semantic logic reasoning engine and generation of semantic rules. Compared with Ke et al. [1], our proposed framework has following advantages. Firstly, image recognition in Ke et al. proposed framework needs to segment foreground and background manually and extract hand-crafted features from them before recognition, whereas image understanding module in our proposed framework is achieved with a dense image caption model that integrates object detection, image captioning and soft spatial attention into one single neural network using deep learning, in which the need for human intervention is eliminated and features extracted are more hierarchical and sophisticated. secondly, instead of generating semantic rules manually when constructing common sense knowledge base in Ke et al. proposed framework, our proposed framework is capable of generating thousands of semantic rules automatically for NR. Thirdly, besides NR, similar to the common sense base in Ke et al. [1], we also propose to construct AR. As such, not only can we frame image forgery detection as anomaly detection with NR, but also as recognition problem with AR. Fourthly, our proposed framework is pretty straightforward and simple due to the integration of image understanding module.

Our paper is arranged as follows. Section 2 introduces related work. Section 3 introduces our proposed framework and how it works. Section 4 talks about experiment results, and demonstrate the effectiveness of our framework. In Sect. 5, conclusions are drawn.

2   Related Work

Ke et al. [1] put forward a semantic based framework consisting of three components including image recognition, semantic logic reasoning engine and generation of semantic rules. They complete image forgery detection by first building a common sense knowledge base, then extracting semantic information from the testing image, and finally reaching a conclusion based on the semantic logic reasoning result. However, the framework sadly needs frequent human intervention e.g. in the process of segmentation and of constructing the knowledge base.

Li et al. [15] propose a keypoint-based forgery detection method, which first segments the test image into semantically independent patches. And for each patch, instead of extracting visual or texture features, it only extracts locations of keypoints to estimate the affine transform matrix in order to find the suspicious copy-move regions. Pun et al. [15] push a step forward by proposing a scheme that integrates both block-based and keypoint-based forgery detection methods, but the features used are largely hand-crafted.

Recently, deep convolutional [12] and recurrent networks [13] for images have yielded promising results in image understanding and caption. Johnson et al. [11] introduce the dense captioning model, which takes an image as input and outputs dozens of captions, and which is able to recognize the semantic instances along with their relationships in an image and map such instances and relationships into captions or words. Standing on the shoulder of giants, we adopt the dense captioning model that integrates object detection, image captioning and soft spatial attention into one single neural network for image understanding.

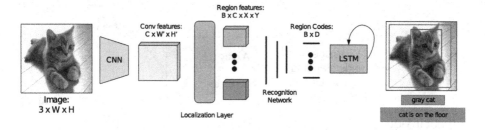

Fig. 2. The structure of the image understanding module.

3   Framework

The framework of image forgery detection based on semantic image understanding consists of three components: image understanding module, NR, and AR. Firstly, while training the image understanding module i.e. the dense image caption model, NR is meanwhile constructed from real image dataset, and AR from forgery image dataset. Secondly, a test image is processed by the image understanding module to map its semantic contents into captions or words. Last, image forgery detection is completed based on the result of image understanding, NR and AR.

3.1   Image Understanding Module

The image understanding module is achieved with a dense image caption model, with the goal to jointly localize regions of interest and then describe each with natural language. Standing on the shoulder of giants, we adopt the dense captioning model proposed by Johnson et al. [11] that integrates object detection, image captioning and soft spatial attention into a single neural network. As shown in Fig. 2, the model takes an image as input and outputs dozens of captions. Firstly, the appearance of the image is encoded by a convolutional network at a set of uniformly sampled image locations. Secondly, a fully convolutional localization layer is used to identify spatial regions of interest and smoothly extracts a fixed-sized representation from each region, the output of which is later processed by a fully-connected neural network before fed to RNN language model implemented with LSTM units. Last, the model produces dozens of captions, each related to a corresponding region of interest. During training the ground truth consists of positive boxes (regions of interest) and descriptions, and binary logistic losses, L1 loss, and cross-entropy loss are used for training. For more detail, please refer to Johnson et al. [11].

As shown in Fig. 1, instead of using image caption model e.g. [14] that takes an image as input and output just one caption, the **dense** image caption model in Johnson et al. [11] is capable of mapping an image to dozens of captions, which can not only produce rich snippet descriptions, but also align the captions and corresponding bounding boxes in the image, leading to better understanding of the image than image caption model.

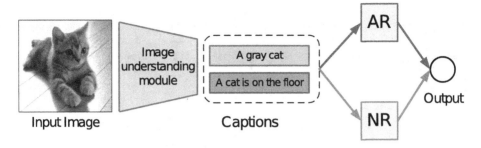

Fig. 3. Forgery detection pipeline. The standardization process is left out from the diagram.

3.2 NR and AR

When trained, the image understanding module i.e. the dense image caption model takes positive boxes (regions of interest in an image) as input and descriptions as supervision. The dataset used to construct NR is Visual Genome (VG) region caption dataset [15] that contains over 90k images and over 4 million snippets of text, each corresponding to a region of an image. The dataset is extremely large, but all it contains are genuine images and normal captions that comply with our common sense; so normal captions are utilized as the semantic rules of NR. Unlike manually adding rules to common sense knowledge base as in Ke et al. [1], the construction of NR is completed automatically.

Next, AR is constructed when further train or fine-tune the dense image caption model with a self-designed dataset that includes forgery images with captions as well, and data in which are organized in the form like what's in the VG dataset. The self-designed dataset contains 113 photoshopped images with bounding box coordinates and their corresponding captions, the latter of which contrast with our common sense and are utilized as semantic rules of AR. As later experiments would demonstrate, AR also contributes to the detection of image forgeries.

3.3 Forgery Detection

Forgery detection. After the image understanding module i.e. dense image caption model is trained and fine-tuned, and thus both NR and AR constructed, the process of the forgery detection is ready to begin. As Fig. 3 shows, for a test image, it is first processed by the dense image caption model to produce dozens of captions, of which the top k (e.g. k = 3) captions with highest confidence after removing duplicated captions are selected and then go through the standardization process, a process that strips all the adjectives and articles in a caption, and restores all the remaining words to their base form using senna tool[1] and natural language tool kit (NLTK). For example, after standardization, "a dog

---

[1] Available at http://ml.nec-labs.com/senna/.

is on the green grass" becomes "dog be on grass", and "a person riding a bear" becomes "person ride bear". The reason for standardization is that it facilitates the later search process while reserving the vital semantic information. The k captions after standardization later search for matches in NR and AR. Finally, we frame image forgery detection as anomaly detection with NR, and as recognition problem with AR. More specifically, considering the rules hold by NR and AR, if **at least one** of k standardized captions don't match with semantic rules in NR or if **at least one** of k standardized captions match with semantic rules in AR, the image is considered a forgery.

4   Experiments

4.1   Implementation

Semantic rules before added to NR when the dense image caption model is being trained, also go through standardization process. Captions longer than 10 words in VG dataset before standardization are ignored as are in Johnson et al. [11], and so are semantic rules (after standardization) shorter than three words, because they barely contain useful information. Of over four million snippets of captions in the dataset, about 300k rules are extracted and standardized to constitute NR, which is much larger than that of the common sense knowledge base manually built in Ke et al. [1]. When NR is small, the time spent on a test image mainly comes from the image understanding module i.e. the dense image caption model. When NR grows large, the time complexity is approximately $O(n)$, where n is the number of rules in NR.

Table 1. Experiments on self-designed dataset. "True Positive" stands for correctly identifying a forgery, "False Alarm" misclassifying a forgery as a authentic image, "densecap" dense image caption model, "ft" fine-tuning, and "−k" the value of k.

Method	True positive	False alarm
densecap-5	0.66	0
densecap+tf-5	0.76	0

In order to build AR, we fine-tune the model with a extra self-designed dataset that includes forgery images with captions, and data in which are organized in the form like what's in the VG dataset. The self-designed dataset contains 113 photoshopped images with information of bounding box coordinates and their corresponding captions, and we only draw one box in the image, and correspondingly label one caption with the length no longer than 10 words(since Johnson et al. [11] discard all annotations with more than 10 words), and replicate them five times before used for fine-tuning. Example captions from our self-designed dataset include "a dog is in the air", "a person riding a bear",

Table 2. Experiments on public dataset.

Method	Detection accuracy
Ke et al. [1]	70.51%
densecap-5	65.8%
densecap+tf-1	47.0%
densecap+tf-5	77.4%
densecap+tf-10	79.6%

etc. When the dense image caption model is fine-tuned, AR is also constructed, containing 113 semantic rules.

The rule for an image to be considered as a forgery as mentioned in Sect. 3.3 is that **at least one** of k standardized captions for the image don't match with semantic rules in NR or **at least one** of k standardized captions match with semantic rules in AR.

4.2 Experiments on Self-designed Dataset

Contributions of AR. We examine our framework on a test set, which contains 50 forged images and 50 genuine images. The results are shown in Table 1. It's worth noting that AR is constructed when the dense image caption model is being fine-tuned. When only NR is only constructed before fine-tuning, 66% of forged images are detected correctly, leaving the remaining 34% undetected. And none of the genuine images is detected as a forgery, indicating no false alarm. After AR is constructed i.e. after fine-tuning, true positive rate reaches 76%, improving by 10%. And there is no false alarm as well. Therefore it testifies the previous claim that AR contributes to the image forgery detection.

Importance of fine-tuning. To dig a little deeper, we show in Fig. 4 how the dense image caption model performs on several forged images from the test set. The dense image caption model before fine-tuned is already powerful enough to recognise the semantic instances in an image and their relative relationships. For example, as shown at the top of the Fig. 4, the model is able to recognise the semantic instances like elephant, sky, and grass, and capable of mapping their relative relationships into captions, albeit with a ignorable defect that the model mistakes one elephant as many, which is probably due to the multiple legs bounded by the orange box.

The dense image caption model before fine-tuned is, on one hand, able to enhance the recognition of rarely seen semantic instances like "desert" which the model before fine-tuned struggles to recognize. On the other hand, it also adapts to the self-designed dataset and produces specific captions for forgery detection more easily. For the image at the bottom of Fig. 4, the model before fine-tuned just recognizes the desert as dirt, and fails to understand its sitting action, whereas the model after fine-tuned recognizes the desert and action quite well.

Fig. 4. At the top, displayed in each image are the top ten boxes in color with highest confidence, each of which corresponds to a caption. For brevity, only the key caption are attached to the bottom of each image. At the bottom, only top five boxes are drawn in the image and the key caption attached to the bottom of the image, both for clarity (Color figure online)

4.3 Experiments on Public Dataset

For better comparison, we also test 500 images obtained from Berkeley image dataset [16] and Columbia Image Splicing Detection Evaluation Dataset[2] as did in Ke et al. [1]. As Table 2 shows, without fine-tuning, our framework performs a little worse than that in Ke et al. [1] (k = 5). But it's worth noting that the detection accuracy Ke et al. [1] achieved is on the basis of the correct identification of objects in the image, whereas our framework doesn't need the basis, and directly takes an image as input. After fine-tuning and at the same time AR is constructed, performance varies as value k increases. The following analyses are based on the rule for an image to be considered as a forgery as mentioned in Sect. 4.1. When k = 1 i.e. only the standardized caption with top confidence is considered, the low detection accuracy is because the framework

---

[2] Credits for the use of the Columbia Image Splicing Detection Evaluation Dataset are given to the DVMM Laboratory of Columbia University, CalPhotos Digital Library and the photographers listed in http://www.ee.columbia.edu/ln/dvmm/downloads/ AuthSplicedDataSet/photographers.htm.

tends to first extract semantic contents complying with our common sense, thus only NR is at work. When k equals to 10, the detection accuracy of our framework reaches 79.6%, the best detection accuracy. But there are also many false alarms. Because the more captions considered, the more likely there are output captions not existed in NR(anomaly), and the more likely there are output captions existed in AR. When k = 5, our proposed framework reaches a appropriate balance of utilizing both NR and AR, and in our experiments achieves good detection accuracy without many false alarms.

5   Conclusion

In this work, we propose the framework of image forgery detection based on high-level semantics which consists of three components including image understanding module, NR, and AR. A test image is first processed by the image understanding module to produce captions, which after standardization search for matches with semantic rules in NR and AR. And the search result constitutes the basis for reaching a conclusion about whether it is a forgery or not. We push the work in Ke et al. [1] forward by a moderate step. Firstly, we eliminate all the human interventions in the process of constructing NR much like the common sense knowledge base in Ke et al. [1], and in the process of image forgery detection. In addition, besides NR that considers semantic rules complying with our common sense, we also propose to build AR that takes into account semantic rules contrasting with our common sense, which, as experiments turns out, also helps to boost detection accuracy. It's a very important research approach to detecting image forgery based on semantics, since it can circumvent forgeries based on low-level features. However, the prototype framework also has limitations, and is still at the research stage that needs improving in the following aspects: (a) developing more powerful dense image caption mode for the image understanding module; (b) automating the construction of self-designed dataset; (c) developing more efficient matching strategies.

Acknowledgement. This work is supported by NSFC (Nos. U1536120, U1636201, 61502496), the National Key Research and Development Program of China (No. 2016YFB1001003) and China Postdoctoral Science Foundation funded project (No. 2016M601168).

References

1. Ke, Y., Min, W., Qin, F., Shang, J.: Image forgery detection based on semantics. Int. J. Hybrid Inf. Technol. **7**(1) (2014)
2. Fridrich, A.J., Soukal, B.D., Luk, A.J.: Detection of copy-move forgery in digital images. In: Proceedings of Digital Forensic Research Workshop (2003)
3. Popescu, A.C., Farid, H.: Statistical tools for digital forensics. In: Fridrich, J. (ed.) IH 2004. LNCS, vol. 3200, pp. 128–147. Springer, Heidelberg (2004). https://doi.org/10.1007/978-3-540-30114-1_10

4. Lin, Z., Wang, R., Tang, X., Shum, H.-V.: Detecting doctored images using camera response normality and consistency. In: Proceedings of Computer Vision and Pattern Recognition (2005)
5. Johnson, M.K., Farid, H.: Exposing digital forgeries through chromatic aberration. In: Proceedings of ACM Multimedia and Security, Workshop, pp. 48–55 (2006)
6. Popescu, A.C., Farid, H.: Exposing digital forgeries in color filter array interpolated images. IEEE Trans. Sig. Process. **53**(10), 3948–3959 (2005)
7. Luk, J., Fridrich, J., Goljan, M.: Detecting digital image forgeries using sensor pattern noise. In: Proceedings of SPIE Electronic Imaging Security Steganography Watermarking of Multimedia Contents VIII, vol. 6072, pp. 0Y1–0Y11 (2006)
8. O'Brien, J., Farid, H.: Exposing photo manipulation with inconsistent reflections. ACM Trans. Graph. **31**(1), 1–11 (2012)
9. Kee, E., Farid, H.: Exposing digital forgeries from 3-D lighting environments. In: 2010 IEEE International Workshop on Information Forensics and Security, pp. 1–6. IEEE (2010)
10. Kee, E., O'Brien, J.F., Farid, H.: Exposing photo manipulation with inconsistent shadows. ACM Trans. Graph. (ToG) **32**(3), 28 (2013)
11. Johnson, J., Karpathy, A., Fei-Fei, L.: Densecap: fully convolutional localization networks for dense captioning, arXiv preprint arXiv:1511.07571 (2015)
12. Krizhevsky, A., Sutskever, I., Hinton, G.E.: Imagenet classification with deep convolutional neural networks. In: Advances in Neural Information Processing Systems, pp. 1097–1105 (2012)
13. Lipton, Z.C., Berkowitz, J., Elkan, C.: A critical review of recurrent neural networks for sequence learning, arXiv preprint arXiv:1506.00019 (2015)
14. Donahue, J., Anne Hendricks, L., Guadarrama, S., Rohrbach, M., Venugopalan, S., Saenko, K., Darrell, T.: Long-term recurrent convolutional networks for visual recognition and description. In: Proceedings of the IEEE Conference on Computer Vision and Pattern Recognition, pp. 2625–2634 (2015)
15. Krishna, R., Zhu, Y., Groth, O., Johnson, J., Hata, K., Kravitz, J., Chen, S., Kalantidis, Y., Li, L., Shamma, D.A., et al.: Visual genome: connecting language and vision using crowdsourced dense image annotations, arXiv preprint arXiv:1602.07332 (2016)
16. Martin, D., Fowlkes, C., Tal, D., Malik, J.: A database of human segmented natural images and its application to evaluating segmentation algorithms and measuring ecological statistics. In: Proceedings of 8th International Conference on Computer Vision, vol. 2, pp. 416–423, July 2001

Relative Distance Metric Leaning Based on Clustering Centralization and Projection Vectors Learning for Person Re-identification

Tongguang Ni[1], Zongyuan Ding[1], Fuhua Chen[2], and Hongyuan Wang[1(✉)]

[1] School of Information Science and Engineering,
Changzhou University, Changzhou, Jiangsu, China
hywang@cczu.edu.cn
[2] Department of Nature Science and Mathematics,
West Liberty University, West Liberty, WV, USA

Abstract. Existing projection-based person re-identification methods usually suffer from long time training, high dimension of projection matrix, and low matching rate. In addition, the intra-class instances may be much less than the inter-class instances when a training dataset is built. To solve these problems, a novel relative distance metric leaning based on clustering centralization and projection vectors learning (RDML-CCPVL) is proposed. When constructing training dataset, the images of a same target person are clustering centralized with FCM. The training datasets are built by these clusters in order to alleviate the imbalanced data problem of the training datasets. In addition, during learning projection matrix, the resulted projection vectors can be approximately orthogonal by using an iteration strategy and a conjugate gradient projection vector learning method to update training datasets. The advantage of this method is its quadratic convergence, which can promote the convergence. Experimental results show that the proposed algorithm has higher efficiency. The matching rate can be significantly improved, and the time of training is much shorter than most of existing algorithms of person re-identification.

Keywords: Person re-identification · Distance centralization
Metric learning · Projection vectors · Conjugate gradient

1 Introduction

Recently, surveillance systems have become ubiquitous in public places like airports, railway stations, college campuses, and office buildings [1]. There are a large number of cameras in the surveillance systems and they provide huge amounts of video data. The analysis of the computer vision obtained in a surveillance system often requires the ability to track people across multiple cameras. Therefore, person re-identification (Re-ID) has attracted more and more interests [2–4]. Re-ID is defined as a process of establishing correspondence between

© Springer Nature Singapore Pte Ltd. 2017
J. Yang et al. (Eds.): CCCV 2017, Part I, CCIS 771, pp. 482–493, 2017.
https://doi.org/10.1007/978-981-10-7299-4_40

images of a person taken from different cameras. In the past five years, a large number of models have been proposed for Re-ID systems [5,6], which can be categorized generally into two types: (1) designing discriminative, descriptive and robust visual descriptors to characterize a person's appearance [7,8]; (2) learning suitable distance metrics that maximize the chance of a correct correspondence [9–11]. In this paper, we focus on the second type of person re-identification, that is, given a set of features extracted from each person image, we seek to quantify and differentiate these features by learning the optimal distance measure that is most likely to give correct matches.

Many metric learning algorithms have been proposed to act with the distances or similarity functions of person re-identification features. For example, Pedagadi et al. applied the Local Fisher Discriminate Analysis (LFDA) [12] to solve person re-identification problems. The authors in [13] introduced the KISSME method from equivalence constraints based on a statistical inference perspective. Dikmen et al. proposed a metric learning framework to obtain a robust Mahalanobis metric for Large Margin Nearest Neighbour classification with Rejection (LMNN) [14]. [15] presented an information-theoretic approach to learn a Mahalanobis distance function. Zheng et al. proposed the Relative Distance Comparison (RDC) [16] approach to maximize the likelihood of a pair of true matches which having a relatively smaller distance than that of a wrongly matched pair in a soft discriminate manner. Chen [17] formulated an asymmetric distance model for learning camera-specific projections to transform the unmatched features of each view to a common space. Chen [18] presented a relevance metric learning method with listwise constraints (RMLLCs) by adopting listwise similarities using predefined similarity lists.

However, the existing Re-ID methods discussed above all have two shortcomings.

(1) As we all known, for the person re-identification datasets, there are usually much more inter-class person image pairs than intra-class ones. Imbalanced datasets would drastically affect the modelling of machine learning methods.
(2) The training set for learning the model consists of images of matched people across different camera views. In order to capture the large intra and inter variations, this represents a large scale learning problem that challenges existing machine learning algorithms.

To address the problems of the existing Re-ID methods, a novel relative distance metric leaning based on clustering centralization and projection vectors learning for person re-identification (RDML-CCPVL) is proposed. First, there are usually much more inter-class instances than intra-class instances when traditional metric learning methods collect training dataset. Constructing counterexamples of each instance needs to compute the distances of all the other instances, so the training time is greatly increased. Using FCM [19], the number of counterexamples of each instance is decreased by divided into clusters and the important structural information is retained, so the overfitting problem caused by class imbalance is relieved.

Second, traditional matrix projection learning methods usually have greater storage and computing complexity. In this work, we decomposed the projection matrices into low rank ones by eigenvalue decomposition for projection matrices. According to our iterative optimization method, updating the distance vectors of instance features only need to learn a new projection vector using the updated training dataset each time and stop when achieving a good enough accuracy. The conjugate gradient method is used to learn the projection vector, which only needs to compute the initial gradient one time. For the quadratic function, the conjugate gradient method can converge to the target precision soon due to quadratic termination. Our method can effectively reduce the computational complexity and storage. In addition, our algorithm can approximately ensure to keep the orthogonal characteristics of the vectors after eigenvalue decomposition.

2 Relative Distance Metric Leaning Based on Clustering Centralization and Projection Vectors Learning

2.1 Learning Function Based on Clustering Centralization

The person re-identification problem can be casted into a distance comparison problem [16]. Suppose we have a set of m training pedestrian images $\mathbf{D}^k = [\mathbf{X}, \mathbf{Y}]_1^m$ with a feature dataset $\mathbf{X} = \{x_i, i = 1, ..., m\}, x_i \in \mathbf{R}^d$, where d is the dimension of the features and $\mathbf{Y} = \{y_i, i = 1, ..., m\}$ is the input label dataset. For an instance x_a of person A, we want to find another instance x_b of pedestrian A captured elsewhere in space and time by learning the distance $dis(x_a, x_b)$ of these two instances. As we all known, the distances of intra-class instances should be smaller in general compared with the distances of inter-class instances, so that $dis(x_a, x_b) < dis(x_a, x_c)$, where x_c is an instance of any other pedestrian except A. Here we construct a pairwise dataset $\mathbb{S} = \{\mathbb{S}_t = (\boldsymbol{d}_t^{pos}, \boldsymbol{d}_t^{neg})\}_{t=1}^m$ to describe the distances of the instances, where $\boldsymbol{d}_t^{pos}$ is the distance of x_t with intra-class instances and $\boldsymbol{d}_t^{neg}$ is the distance of x_t with inter-class instances.

However, as we all known, the number of inter-class instances is much larger than the number of intra-class instances. So the pairwise dataset $\mathbb{S} = \{\mathbb{S}_t = (\boldsymbol{d}_t^{pos}, \boldsymbol{d}_t^{neg})\}_{t=1}^m$ will be a class imbalanced dataset. Training with such a dataset will lead to over fit-ting of the inter-class instances and under fitting of the intra-class instances, which will decrease the performance of the learning algorithm. Here we choose FCM [19] to alleviate the class imbalanced data problem. The principle of our clustering centralization is shown in Fig. 1. From Fig. 1, the traditional training datasets constructing methods [16] (shown with green lines) include much more inter-class instances, while our method (shown with red lines) with clustering centralization built training datasets by these clusters, which effectively alleviate the class imbalanced problem of the existing Re-ID algorithms. The other advantage of our training clustering centralization method is the number of clusters can adjust for different Re-ID datasets to get the optimal performance, which will be discussed in Sect. 4.3 lately. After clustering centralization, we have $\mathbb{S} = \{\mathbb{S}_t = (\boldsymbol{d}_t^{pos}, \ddot{\boldsymbol{d}}_t^{neg})\}_{t=1}^m$, $\ddot{\boldsymbol{d}}_t^{neg}$ is the distance of x_t

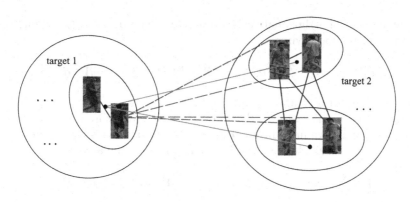

Fig. 1. The principle of the clustering centralization (Color figure online)

with the centers of clusters of its counterexamples. In order to maximize the inter-class distance and minimize the intra-class distance at the same time, we can formulate it into a minimization problem as following:

$$dis(\boldsymbol{d}_t^{pos}, \ddot{\boldsymbol{d}}_t^{neg}) = g(\boldsymbol{d}_t^{pos}) - g(\ddot{\boldsymbol{d}}_t^{neg}), \tag{1}$$

where $g()$ is a distance function. The function in Eq. (1) is unbounded, so it cannot guarantee convergence during iteration. Here, we transform it to a continuous sigmoid function as following:

$$dis(\boldsymbol{d}_t^{pos}, \ddot{\boldsymbol{d}}_t^{neg}) = (1 + \exp(g(\boldsymbol{d}_t^{pos}) - g(\ddot{\boldsymbol{d}}_t^{neg}))^{-1}, \tag{2}$$

Considering the convenience of computing, Eq. (1) is then transformed to a logistic form as following:

$$f = -\log(\prod_{\mathbb{S}_t} dis(\boldsymbol{d}_t^{pos}, \ddot{\boldsymbol{d}}_t^{neg})) = \sum_{\mathbb{S}_t} \log(1 + \exp(g(\boldsymbol{d}_t^{pos}) - g(\ddot{\boldsymbol{d}}_t^{neg}))), \tag{3}$$

We can see minimizing Eq. (1) is equivalent to maximizing Eq. (2), maximizing Eq. (2) is equivalent to minimizing Eq. (3).

Because the Mahalanobis matrix of the Mahalanobis distance function has good projective property and learning property, here we choose the Mahalanobis distance function as the distance function g:

$$g(\boldsymbol{d}_t^{pos}) = (\boldsymbol{d}_t^{pos})^T \mathbf{M}(\boldsymbol{d}_t^{pos}), \tag{4}$$

where $\mathbf{M}$ is a semi-definite matrix. Our goal becomes to learn $\mathbf{M}$ in Eq. (4) by minimizing the functional defined in Eq. (3). By performing eigenvalue decomposition on $\mathbf{M}$, we can find $\mathbf{M} = \mathbf{P}\mathbf{P}^T$, $\mathbf{P}$ is a matrix of column orthogonal vectors. The number of orthogonal bases may be smaller than the rank of matrix $\mathbf{M}$, therefore, $\mathbf{P} \in \mathbf{R}^{n*d'}$ can be regard as a dimension reduction matrix, where

d' is the number of orthogonal basis after dimension reduction. Equation (4) can be transformed as following:

$$g(d_t^{pos}) = (d_t^{pos})^T M(d_t^{pos}) = (d_t^{pos})^T PP^T(d_t^{pos}) = \left\| P^T d_t^{pos} \right\|^2, \qquad (5)$$

In addition, for a small dataset, the function shown in Eq. (3) may be over-fitting. In order to alleviate the risk of over fitting and ensure the sparsity of the projection matrix, we introduce $r\|P\|^2$ as a regularization term, where r is the regularization factor. The distance function can be formulated as:

$$f = \sum_{\mathbb{S}_t} \log(1 + \exp(\left\| P^T d_t^{pos} \right\|^2 - \left\| P^T \ddot{d}_t^{neg} \right\|^2)) + r\|P\|_2^2, \qquad (6)$$

2.2 An Iterative Optimization Algorithm for Projection Vector Learning

In this paper, we choose a similar iterative optimization method of [16] to learn an optimal P.

Step 1. Assume that after $l - 1$ iterations a set of orthogonal vectors $p_1, p_2, ..., p_{l-1}$ have been learned, to the next vector p, let

$$d_t^{s,l} = d_t^{s,l-1} - \frac{p_{l-1} p_{l-1}^T}{\left\| p_{l-1} + u \right\|^2} d_t^{s,l-1}, \qquad (7)$$

where $s \in \{pos, neg\}$, $t \in 1, ..., |\mathbb{S}|$.

Step 2. After obtain $d_t^{s,l}$ from Eq. (7), let $\mathbb{S}^l = \{\mathbb{S}_t^l = (d_t^{pos,l}, \ddot{d}_t^{neg,l})\}$. Then, we use conjugate gradient method to learn projection vectors of the Eq. (8).

$$f(p_l^k) = \sum_{d_t \in \mathbb{S}^l} \log(1 + \exp(\left\| (p_l^k)^T d_t^{pos} \right\|^2 - \left\| (p_l^k)^T \ddot{d}_t^{neg} \right\|^2)) + r\|p_l^k\|_2^2, \qquad (8)$$

The optimal projection vector after kth iteration is defined as following:

$$p_l^{k+1} = p_l^k + \alpha_k q_k, \qquad (9)$$

where α_k is computed by $f(p_l^k + \alpha q_k)$ using one-dimensional accurate search, q_k is the search direction of projection vector after kth iteration, and the conjugate direction is computed by PRP equation as following.

$$q_k = -g_k + \beta_{k-1} q_{k-1}, \beta_{k-1} = \begin{cases} \frac{g_k^T(g_k - g_{k-1})}{g_{k-1}^T g_{k-1}}, & k > 1 \\ 0, & k = 1 \end{cases}, \qquad (10)$$

where g_l is:

$$\frac{\partial f}{\partial p_l} = \sum_{d_t \in \mathbb{S}^l} \frac{2 \exp(\left\| (p_l^k)^T d_t^{pos} \right\|^2 - \left\| (p_l^k)^T \ddot{d}_t^{neg} \right\|^2)}{1 + \exp(\left\| (p_l^k)^T d_t^{pos} \right\|^2 - \left\| (p_l^k)^T \ddot{d}_t^{neg} \right\|^2)}$$
$$\times (d_t^{pos} d_t^{pos^T} - \ddot{d}_t^{neg} \ddot{d}_t^{neg^T}) p_l^k + 2r p_l^k, \qquad (11)$$

If $|f(\boldsymbol{p}_l^k) - f(\boldsymbol{p}_l^{k-1})| < \varepsilon_g$, the iteration is terminated.
The initial value of $\boldsymbol{p}_l$ is formulated as following:

$$\boldsymbol{p}_l^0 = \frac{1}{\|\mathbb{S}_{pos}^l\|} \sum_{\mathbb{S}_{pos}^l} \boldsymbol{d}_i^{pos} - \frac{1}{\|\mathbb{S}_{neg}^l\|} \sum_{\mathbb{S}_{neg}^l} \ddot{\boldsymbol{d}}_i^{neg}, \tag{12}$$

Different from the iterative optimization algorithm of [16], a smaller perturbation term u is added into Eq. (7), which make each projection spaces preserving a relation with each other and are more suitable to the real-world learning
problem.

According to Eqs. (9) and (11), we can see $\boldsymbol{p}_l \in span\{\boldsymbol{d}_i^{s,l}\}$, where $span\{\boldsymbol{d}_i^{s,l}\}$
is a range space of $\{\boldsymbol{d}_i^{pos,l}\} \cup \{\boldsymbol{d}_i^{neg,l}\}$, $s \in \{pos, neg\}$, $i \in 1, ..., |\mathbb{S}|$. According to
Eq. (7), we know that $\boldsymbol{p}_j^T \boldsymbol{d}_i^{s,j+1} \approx 0$ where $j = 1, ..., l-1$ and $\boldsymbol{p}_l \in span\{\boldsymbol{d}_i^{s,l}\}$,
where $span\{\boldsymbol{d}_i^{s,l}\} \subseteq span\{\boldsymbol{d}_i^{s,l-1}\} \subseteq ... \subseteq span\{\boldsymbol{d}_i^{s,0}\}$, so $\boldsymbol{p}_l$ and $\boldsymbol{p}_j, j = 1, ..., l-1$
is approximately orthogonal.

3   Learning Algorithm for RDML-CCPVL

Based on the above iteration, the learning algorithm of the proposed RDML-
CCPVL is presented in Algorithm 1.

Algorithm 1. learning algorithm for RDML-CCPVL

Input: $\mathbf{X} = \{(x_i, y_i)\}_{i=1}^m$, r , u , ε_o , ε_g
Output: $\mathbf{P} = [\boldsymbol{p}_1, \boldsymbol{p}_2, ..., \boldsymbol{p}_l]$
  1: build $\mathbb{S} = \{\mathbb{S}_t = (\boldsymbol{d}_t^{pos}, \boldsymbol{d}_t^{neg})\}_{t=1}^m$ with $\mathbf{X}$ using the method of feature-clustered
  2: $\mathbf{P} = \boldsymbol{p}_0 = 0$, $l = 0$
  3: **while** not convergence **do**
  4:     update $\mathbb{S}^l$ with Eq. (7)
  5:     compute $\boldsymbol{p}_l^0$ using Eq. (12)
  6:     $k = 0$
  7:     **while** not convergence **do**
  8:         compute $\boldsymbol{p}_l^k$ by Eq. (9)
  9:     **end while**
 10:     $\boldsymbol{p}_l = \boldsymbol{p}_l^k$, $\mathbf{P} = [\mathbf{P}, \boldsymbol{p}_l]$
 11: **end while**

---

4   Experimental and Analysis

4.1   Datasets and Experimental Setting

Four popular datasets are selected for our experiments: VIPeR [20], CUHK01
[21], 3DPeS [22], CAVIAR4REID [23], TownCenter [24] and Market-1501 [25].
Datasets have different problems of angle, lighting, occlusion and low resolution,

Fig. 2. Illustration of different dataset, which from left to right are VIPeR, CUHK01, 3DPeS, CAVIAR4REID, Town Centre and Market-1501

etc. For each image, we use six type of features descriptor, such as RGB, YCbCr, HSV, Lab, YIQ and Gabor [26]. Then we use PCA to compress them into 2688-dimensional feature vectors.

We compare the proposed RDML-CCPVL with seven existing person re-identification works: ITML [15], LMNN [14], KISSME [13], PRDC [16], LFDA [12], CVDCA [17], RMLLC [18]. The performance of all the methods is evaluated in terms of cumulative matching characteristic (CMC), which is a standard measurement for Re-ID. The CMC curve represents the probability of finding the correct match over the top r in the gallery image ranking, with r varying from 1 to 30. In order to evaluate the efficiency of our algorithm, in this paper, the comparison of the training time between our algorithm with other existing algorithms is also shown. Since the number c of the clustering centralization is a important parameter for our algorithm, Normalized Discounted Cumulative Gain (NDCG) [27] is choose to evaluate the performance of RDML-CCPVL with varying number of clusters c. We run all the benchmarking algorithms with MATLAB 7 on a 1.90 GHz machine with 8G RAM.

4.2   Comparisons of Performance on Different Re-ID Datasets

Comparisons of the CMC and training time on different Re-ID datasets are shown in Tables 1, 2, 3, 4, 5 and 6 with optimal value of c, respectively. Since clustering centralization is used to solve the imbalanced data problem of the training datasets, it can be seen that our algorithm is considerably superior to other algorithms. The CMC of RDML-CCPVL is at least 3.2%–4.5% higher than other algorithms at rank 30. The advantage of clustering centralization is the reduction of the training datasets, we can see the training time of RDML-CCPVL is much less than other algorithms except method of KISSME, because this algorithm is a one-pass metric learning method.

Table 1. Comparisons of performance on dataset VIPeR with $c = 2(\%)$

Methods	Rank 1	Rank 10	Rank 20	Rank 30	Training time (s)
RDML-CCPVL	**13.33**	**60.00**	**83.33**	**91.67**	**77.69**
CVDCA	16.68	54.11	67.73	87.5	187.56
RMLLC	15.22	47.37	59.52	85.33	502.63
LFDA	11.45	40.98	56.03	82.77	624.37
PRDC	10.36	37.62	54.14	78.35	465.26
KISSME	7.21	30.68	44.46	75.24	52.14
ITML	9.50	36.74	54.66	80.00	985.67
LMNN	8.46	34.58	50.25	74.98	872.53

Table 2. Comparisons of performance on dataset CUHK01 with $c = 2(\%)$

Methods	Rank 1	Rank 10	Rank 20	Rank 30	Training time (s)
RDML-CCPVL	**18.52**	**57.14**	**79.89**	**86.67**	**98.73**
CVDCA	17.77	53.26	68.72	84.46	225.23
RMLLC	16.02	54.85	73.33	83.87	324.75
LFDA	12.47	51.15	66.33	81.72	486.8
PRDC	14.58	52.60	68.50	82.67	502.43
KISSME	10.65	48.78	62.45	79.63	86.48
ITML	9.24	45.05	61.36	75.52	1095.4
LMNN	10.46	47.76	58.45	72.00	924.06

Table 3. Comparisons of performance on dataset 3DPeS with $c = 10(\%)$

Methods	Rank 1	Rank 10	Rank 20	Rank 30	Training time (s)
RDML-CCPVL	**24.58**	**54.19**	**67.6**	**82.12**	**203.86**
CVDCA	22.54	50.44	64.71	77.64	1072.54
RMLLC	21.04	51.25	63.45	75.73	1521.82
LFDA	12.24	48.75	59.07	71.98	1795.15
PRDC	17.33	50.04	61.72	73.66	1302.25
KISSME	7.12	29.02	46.36	58.79	153.44
ITML	10.65	31.44	54.47	61.73	4686.34
LMNN	8.42	26.7	50.09	57.93	2410.95

4.3   Experiments with Vary Number of Clusters

In this section, we report the change tendency of the performance of RDML-CCPVL by running them on the different datasets with varying number of clusters c. There are two pictures of each person in dataset VIPeR and four

Table 4. Comparisons of performance on dataset CAVIAR4REID with $c = 15(\%)$

Methods	Rank 1	Rank 10	Rank 20	Rank 30	Training time (s)
RDML-CCPVL	**13.19**	**56.88**	**73.44**	**89.69**	**207.83**
CVDCA	12.36	54.3	70.05	84.81	1104.71
RMLLC	10.78	53.65	71.26	88.52	1589.13
LFDA	8.67	44.29	65.15	80.38	1645.65
PRDC	9.54	54.12	70.97	82.46	1147.87
KISSME	5.75	36.12	58.98	75.18	141.13
ITML	4.15	29.71	60.08	77.17	4963.28
LMNN	2.18	26.56	62.17	59.6	2553.46

Table 5. Comparisons of performance on dataset Town Centre with $c = 2(\%)$

Methods	Rank 1	Rank 10	Rank 20	Rank 30	Training time (s)
RDML-CCPVL	**62.22**	**86.36**	**92.11**	**95.33**	**174.41**
CVDCA	57.09	83.45	90.74	92.1	1076.05
RMLLC	55.97	80.84	88.86	93.04	1516.84
LFDA	46.89	77.82	80.92	89.15	1748.15
PRDC	52.46	81.15	86.81	90.43	1489.94
KISSME	38.47	68.17	75.52	84.63	122.19
ITML	25.58	46.25	59.76	71.04	5364.69
LMNN	22.92	39.24	46.82	64.8	2986.14

Table 6. Comparisons of performance on dataset Market-1501 with $c = 2(\%)$

Methods	Rank 1	Rank 10	Rank 20	Rank 30	Training time (s)
RDML-CCPVL	**26.95**	**78.49**	**89.94**	**94.86**	**1525.62**
CVDCA	24.02	75.4	84.82	90.54	2588.65
RMLLC	23.93	72.86	81.15	85.9	3134.43
LFDA	18.59	62.04	75.17	80.06	4912.77
PRDC	20.22	68.4	79.27	84.64	5974.45
KISSME	15.52	59.87	70.93	75.48	303.87
ITML	12.07	52.43	68.55	70.32	7059.74
LMNN	11.25	45.89	62.24	69.58	8208.03

pictures in dataset CUHK01, so we choose two clusters of each person of these two datasets. From the Fig. 3 we can see that the proposed RDML-CCPVL is considerably sensitive to c for all the datasets except Town Centre. The performance on 3DPeS, CAVIAR4REID and Market-1501 of will become better as

c increases because more clusters mean more information of the datasets. But when c becomes too large, the curves in Fig. 2 start to go down because we find that c has an optimal range of value for different datasets and too many clusters also can not generate useful distribution information. For Town Centre, which include only images of Video sequences, so how many clusters of intra-class instances of the same person has little difference. When $c = 2$, the proposed algorithm achieves the optimal performance on Town Centre.

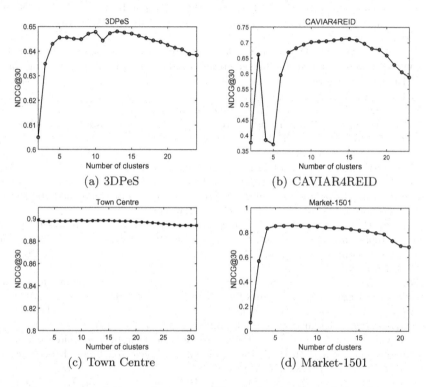

Fig. 3. Comparison of the performance with varying number of the clusters c. (a) 3DPeS; (b) CAVIAR4REID; (c) Town Centre; (d) Market-1501.

5   Conclusion

In this work, we proposed a relative distance metric leaning algorithm based on clustering centralization and projection vectors learning for person re-identification problem. We have shown that cluster centralization can improve the performance and efficiency in person re-identification and reduce both the computational complexity and the storage space. In addition, the conjugate gradient method is used in the projection vector learning. The proposed approach shows a significant improvement over other existing algorithms.

Acknowledgements. This work was supported in part by the National Natural Science Foundation of China under Grants No. 61502058 and 61572085.

References

1. Agrawal, P., Narayanan, P.: Person re-identification in videos. IEEE Trans. Circuits Syst. Video Technol. **21**(3), 299–310 (2011)
2. Cai, X., Wang, C., Xiao, B., Chen, X., Zhou, J.: Deep nonlinear metric learning with independent subspace analysis for face verification. In: Proceedings of the 20th ACM International Conference on Multimedia, pp. 749–752. ACM Press, New York (2012)
3. Cui, Z., Li, W., Xu, D., Shan, S., Chen, X.: Fusing robust face region descriptors via multiple metric learning for face recognition in the wild. In: Proceedings of the IEEE Conference on Computer Vision and Pattern Recognition, pp. 3554–3561. IEEE Press, New York (2013)
4. Ji, S., Xu, W., Yang, M., Yu, K.: 3D convolutional neural networks for human action recognition. IEEE Trans. Pattern Anal. Mach. Intell. **35**(1), 221–231 (2013)
5. Cao, Z., Yin, Q., Tang, X., Sun, J.: Face recognition with learning-based descriptor. In: Proceedings of the IEEE Conference on Computer Vision and Pattern Recognition, pp. 2707–2714. IEEE Press, New York (2010)
6. Hu, J., Lu, J., Tan, Y.P.: Discriminative deep metric learning for face verification in the wild. In: Proceedings of the IEEE Conference on Computer Vision and Pattern Recognition, pp. 1875–1882. IEEE Press, New York (2014)
7. Liao, S., Hu, Y., Zhu, X., Li, S.Z.: Person re-identification by local maximal occurrence representation and metric learning. In: Proceedings of the IEEE Conference on Computer Vision and Pattern Recognition, pp. 2197–2206. IEEE Press, New York (2015)
8. An, L., Kafai, M., Yang, S., Bhanu, B.: Person re-identification with reference descriptor. IEEE Trans. Circuits Syst. Video Technol. **26**(4), 776–787 (2016)
9. An, L., Yang, S., Bhanu, B.: Person re-identification by robust canonical correlation analysis. IEEE Signal Process. Lett. **22**(8), 1103–1107 (2015)
10. Loy, C.C., Liu, C., Gong, S.: Person re-identification by manifold ranking. In: Proceedings of the 20th IEEE International Conference on Image Processing, pp. 3567–3571. IEEE Press, New York (2013)
11. Roth, P.M., Hirzer, M., Köstinger, M., Beleznai, C., Bischof, H.: Mahalanobis distance learning for person re-identification. In: Gong, S., Cristani, M., Yan, S., Loy, C.C. (eds.) Person Re-Identification. ACVPR, pp. 247–267. Springer, London (2014). https://doi.org/10.1007/978-1-4471-6296-4_12
12. Pedagadi, S., Orwell, J., Velastin, S., Boghossian, B.: Local Fisher discriminant analysis for pedestrian re-identification. In: Proceedings of the IEEE Conference on Computer Vision and Pattern Recognition, pp. 3318–3325. IEEE Press, New York (2013)
13. Koestinger, M., Hirzer, M., Wohlhart, P., Roth, P.M., Bischof, H.: Large scale metric learning from equivalence constraints. In: Proceedings of the 2012 IEEE Conference on Computer Vision and Pattern Recognition, pp. 2288–2295. IEEE Press, New York (2012)
14. Dikmen, M., Akbas, E., Huang, T.S., Ahuja, N.: Pedestrian recognition with a learned metric. In: Kimmel, R., Klette, R., Sugimoto, A. (eds.) ACCV 2010. LNCS, vol. 6495, pp. 501–512. Springer, Heidelberg (2011). https://doi.org/10.1007/978-3-642-19282-1_40

15. Davis, J.V., Kulis, B., Jain, P., Sra, S., Dhillon, I.S.: Information theoretic metric learning. In: Proceedings of the 24th International Conference on Machine Learning, pp. 209–216. ACM Press, New York (2007)
16. Zheng, W.S., Gong, S., Xiang, T.: Reidentification by relative distance comparison. IEEE Trans. Pattern Anal. Mach. Intell. **35**(3), 653–668 (2013)
17. Chen, Y.C., Zheng, W.S., Lai, J.H., Yuen, P.: An asymmetric distance model for cross-view feature mapping in person re-identification. IEEE Trans. Circuits Syst. Video Technol. 1 (2016)
18. Chen, J.X., Zhang, Z.X., Wang, Y.L.: Relevance metric learning for person reidentification by exploiting listwise similarities. IEEE Trans. Image Process. **24**(12), 4741–4755 (2015)
19. Bezdek, J.C.: Pattern recognition with fuzzy objective function algorithms. Adv. Appl. Pattern Recognit. **22**(1171), 203–239 (1981)
20. Gray, D., Tao, H.: Viewpoint invariant pedestrian recognition with an ensemble of localized features. In: Forsyth, D., Torr, P., Zisserman, A. (eds.) ECCV 2008. LNCS, vol. 5302, pp. 262–275. Springer, Heidelberg (2008). https://doi.org/10.1007/978-3-540-88682-2_21
21. Li, W., Zhao, R., Wang, X.: Human reidentification with transferred metric learning. In: Lee, K.M., Matsushita, Y., Rehg, J.M., Hu, Z. (eds.) ACCV 2012. LNCS, vol. 7724, pp. 31–44. Springer, Heidelberg (2013). https://doi.org/10.1007/978-3-642-37331-2_3
22. Baltieri, D., Vezzani, R., Cucchiara, R.: 3DPeS: 3D people dataset for surveillance and forensics. In: Joint ACM Workshop on Human Gesture and Behavior Understanding, pp. 59–64. ACM Press, New York (2011)
23. Cheng, D.S., Cristani, M., Stoppa, M., Bazzani, L., Murino, V.: Custom pictorial structures for re-identification. In: Proceedings of the British Machine Vision Conference, pp. 1–11. Piscataway Press, NJ (2011)
24. Benfold, B., Reid, I.: Stable multi-target tracking in real-time surveillance video. In: Proceedings of the IEEE Conference on Computer Vision and Pattern Recognition, pp. 3457–3464. IEEE Press, New York (2011)
25. Zheng, L., Shen, L., Tian, L., Wang, S., Wang, J., Tian, Q.: Scalable person re-identification: a benchmark. In: Proceedings of the 14th IEEE International Conference on Computer Vision, pp. 1116–1124. IEEE Press, New York (2015)
26. Chen, Y., Zheng, W., Lai, J.: Mirror representation for modeling view-specific transform in person re-identification. In: Proceedings of the 24th International Joint Conference on Artificial Intelligence, pp. 3402–3408. AAAI Press, Palo Alto (2015)
27. Wang, Y., Wang, L., Li, Y., He, D., Chen, W., Liu, T.Y.: A theoretical analysis of NDCG type ranking measures. J. Mach. Learn. Res. **30**, 25–54 (2013)

Multi-context Deep Convolutional Features and Exemplar-SVMs for Scene Parsing

Xiaofei Cui[1], Hanbing Qu[1,2,3]($\boxtimes$), Songtao Wang[2,3], Liang Dong[2,3], and Ziliang Qi[1]

[1] School of Computer Science and Engineering, Hebei University of Technology, Tianjin 300401, China
dlaicxf@126.com, quhanbing@gmail.com, 18233183254@163.com
[2] Beijing Institute of New Technology Applications, Beijing 100094, China
wangsongtao1983@163.com, dongliang@bionta.cn
[3] Beijing Academy of Science and Technology, Beijing 100089, China

Abstract. Scene parsing is a challenging task in computer vision field. The work of scene parsing is labeling every pixel in an image with its semantic category to which it belongs. In this paper, we solve this problem by proposing an approach that combines the multi-context deep convolutional features with exemplar-SVMs for scene parsing. A convolutional neural network is employed to learn the multi-context deep features which include image global features and local features. In contrast to hand-crafted feature extraction approaches, the convolutional neural network learns features automatically and the features can better describe images on the task. In order to obtain a high class recognition accuracy, our system consists of the exemplar-SVMs which is training a linear SVM classifier for every exemplar in the training set for classification. Finally, multiple cues are integrated into a Markov Random Field framework to infer the parsing result. We apply our system to two challenging datasets, SIFT Flow dataset and the dataset which is collected by ourselves. The experimental results demonstrate that our method can achieve good performance.

Keywords: Scene parsing · Convolutional neural network Exemplar-SVMs · Markov Random Field

1 Introduction

In the computer vision field, scene parsing is one of the core problems. This paper researches the problem of scene parsing. Scene parsing aims at assigning a label to each pixel in images with the semantic category it belongs to. This work is very challenging, and it involves all kinds of problems, e.g., detecting object categories problems, segmentation problems, and multi-label recognition problems of the image, and so on. In the last decade, scene parsing had been comprehensive studied by several researchers.

© Springer Nature Singapore Pte Ltd. 2017
J. Yang et al. (Eds.): CCCV 2017, Part I, CCIS 771, pp. 494–505, 2017.
https://doi.org/10.1007/978-981-10-7299-4_41

For lots of problems in computer vision field including scene parsing, it is necessary to extract features for each image. Features are the expressions of images for easy calculation and classification easily. Hence extracting good features is the key of computer vision tasks. For image representation, the conventional approaches use hand-crafted features by designing carefully, e.g., HOG [3], SIFT [14], GIST [16], etc. Since the introduction by LeCun [11], deep CNN has been applied to a wide range of computer vision tasks such as hand-written digit classification and face detection. Recently, the latest generation of CNNs have substantially outperformed traditional approaches in many recognition tasks [2,5,10,17,18]. Deep learning is advantageous for large image regions with complex variations, because its deep architectures can automatically learn the hierarchies of features representation.

The issue of scene parsing is improving the recognition rate of the object classes. Some of the previous methods [4,12,21] had low recognition rate for small objects which occupied only a few pixels. To increase the accuracy rate of thing samples, Tighe and Lazebnik [22] proposed using per-exemplar detector which is training a linear SVM classifier for every exemplar in the training set. In addition, Yang et al. [23] expanded the retrieval set using rare class exemplars to achieve more balanced superpixel classification results and combined both global and local semantic context information to boost the performance of their system. Our approach is inspired by the work of [22], so we use exemplar-SVMs [15] which are identical to Tighe and Lazebnik [22] for improving the accuracy rate of rare classes.

In this work, we present a scene parsing approach which combines the multi-context deep convolutional features with exemplar-SVMs. This method can improve the overall accuracy and recognize the rare classes better. Our system has the following steps. We first retrieves the most similar images to query images, as shown in Fig. 1, in an annotated dataset using the global features which are obtained from the convolutional neural network of the GoogLeNet [20]. Then, we segment the images into superpixels and compute the likelihood scores of superpixels of the query images using local features, which incorporate the exemplar Support Vector Machine for classification. Finally, our system integrates multiple cues into a Markov Random Field framework to infer the parsing result.

The rest of this paper is organized as follows. In Sect. 2, we review some relevant works for scene parsing. Section 3 presents our approach which combines the multi-context deep convolutional features with exemplar-SVMs for scene parsing. Section 4 provides the experimental results and compares the performance with other approaches, followed by a conclusion in Sect. 5.

2   Related Work

Most recently, as the development of artificial intelligence, many researchers use the deep learning approach which is one of the machine learning algorithms to solve a lot of issues in computer vision field, including object recognition,

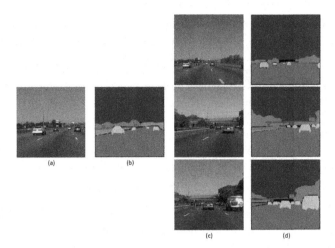

Fig. 1. For a query image (a), our system retrieves the most similar images (c) (three images are shown here) from training set using the global features which are obtained from the convolutional neural network. The ground-truth annotation of (a) and (c) are shown in (b) and (d), respectively.

object detection, simultaneous image segmentation and labeling. The studies have shown that convolution neural network has achieved good results in these works. In recent years, a lot of methods for scene parsing had been proposed. The goal of this work is to assign each pixel of the image a semantic label. Grangier et al. [8] presented a method that did not use any graphical model for scene parsing. The deep convolutional network was used for this paper to model the complex scene label structures, depending on a supervised greedy learning approach. Farabet et al. [5] learned the different levels features applying a multi-scale convolutional neural network handling on original pixels. Their system depended on those features representation to think about context for each decision. Another innovation point of this paper was that they employed the optimal cover to obtain the most consistent region from the tree, taking the place of graph cuts and other inference approaches. Pinheiro and Collobert [17] applied the recurrent convolutional neural networks to scene labeling. The advantage of this paper was that it was no need segmentation and task-specific features compared to other approaches. It was trained in an end-to-end way over raw pixels, and the inference cost was low. In order to improve the overall performance, Bu et al. [2] combined the convolutional neural network architecture with a graphical model. The neural network can learn the hybrid features which could provide the hierarchical information and the spatial relationship information.

From what has been discussed above, the deep learning method was applied to the scene parsing work had achieved impressive results. But training network model of deep learning need a lot of data. The following will introduce some other methods that they were the traditional and did not apply deep learning. For instance, Liu et al. [12] proposed a nonparametric approach for scene parsing using a dense scene alignment to label transfer. The SIFT flow algorithm was used to match the structures within two images. This system did not require any training, but inference via the SIFT flow was very complex and computationally expensive. With the consideration of these, Tighe and Lazebnik [21] presented a scalable nonparametric image parsing with superpixels. This method also required no training and it was easy to implement. It established a new benchmark for scene parsing. The system in Zhang et al. [24] proposed an ANN bilateral matching scheme to match the test images and retrieved the training images. Eigen and Fergus [4] used per-descriptor weights to minimize the classification error and to improve performance on rare classes, the paper used a context-driven adaptation of the training set for each query image. Different from other paper, this system did not use SVMs for classification. Tighe and Lazebnik [22] incorporated region-level features with per-exemplar sliding window detectors outputs for image parsing to achieve broad coverage across hundreds of object categories. Yang et al. [23] used thing class exemplars extending the retrieval set, to obtain more balanced superpixel classifications and combined the global with local semantic context information for refining the image retrieval and superpixel matching. The experimental results compared with previous methods have improved. To boost the label likelihood estimates at superpixels George [6] integrated the likelihood scores forming different probabilistic classifiers and in the parsing process incorporated the semantic context through global label costs. The newest paper, a large and open vocabulary was presented by Zhao et al. [25] for scene parsing, which aimed at parsing images in the wild. Image pixels and word concepts connecting with semantic relations were merged to the parsing framework. To parse a 2D image and recover the 3D scene structures, Liu et al. [13] presented an attribute grammar, which contained a set of production rules, each of which described a series of spatial relation among planar surfaces in 3D scenes.

3   Approach

In this section, we propose an overview of our scene parsing system which combines the convolutional neural network with exemplar-SVMs. Our approach builds on the framework of [22]. Firstly, the GoogLeNet [20] is used to extract the global image features for computing the similar images of training set to the query image. Then we divide the image into superpixels. The features of superpixels are extracted to obtain the probability of class labels for each query image. Finally, we incorporate the superpixel parsing results with exemplar-SVMs parsing results and integrate them to the Markov Random Field framework for obtaining the final parsing. The common architecture of the system which our presented is shown in Fig. 2.

3.1   Feature Extraction with CNN

The deep Convolutional Neural Network (CNN) [11], which showed its powerfulness in extracting high-level feature representations [7] recently, can solve

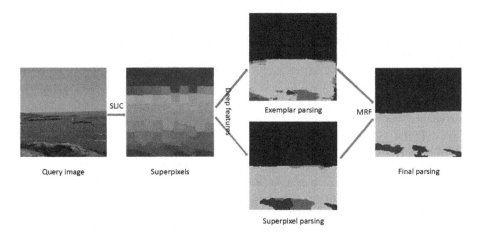

Fig. 2. The general framework of our system. Given a query image, we extract super-pixels using the SLIC method. Then, we use the multi-context deep convolutional features including global and local deep features to describe superpixels and compute the superpixel likelihood ratio scores to obtain the superpixel parsing. We use exemplar detectors to obtain the exemplar parsing. Finally, in order to obtain the final parsing, we combine the exemplar parsing results with superpixel parsing results into the MRF framework.

previous some problems well. Deep CNN aims to mimic the functions of neo-cortex in human brain as a hierarchy of filters and nonlinear operations. The major goal of deep learning is learning the feature representations. The good network structure types should be selected to study features. In this paper, we choose the convolutional neural network of GoogLeNet [20]. The multi-context deep convolutional features which are introduced in this paper refer to the global features and the local features.

Global Features. For each query image I_q, we extract the 1024-dimensional feature vector of GoogLeNet [20] the pool5 layer output for computing similar images which include similar classes and spatial layouts of the training dataset. Then we rank all the training images in increasing order according to the Euclidean distance between training images and the query image. It is similar to the previous works. We select the top K similar images to constitute the retrieval set. The number of images in retrieval set is less than that in training dataset. So we process images in retrieval set instead of those in training set for reducing the computational cost. The process of extracting image features using the convolutional neural network of GoogLeNet is presented in Fig. 3.

Local Features. According to the computer vision applications, superpixels are becoming increasingly popular for use. To reduce the problem space and complexity, we also use superpixels as computing element, which can provide

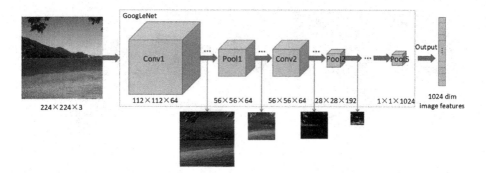

Fig. 3. The process of extracting image features using the convolutional neural network of GoogLeNet. For a query image, the input size of the network is $224 \times 224 \times 3$. Though convolution and pooling operation, the visualization results of some layers are shown in the figure. We extract the pool5 layer output of GoogLeNet as our deep global features which are 1024-dimensional.

better spatial support for aggregating image features that may pertain to a single object than pixels. The image superpixels are extracted by the SLIC [1] method which is easy to implement. For each superpixel, we use the convolutional network of GoogLeNet to extract features for describing it. For each query image, we pad the image and select a superpixel-centered context window. The input of the local-context network is the superpixel-centered context window which is upsample to $224 \times 224 \times 3$. The local feature shares the same deep structure with the global feature. Other deep structures can also be used to extract the local image features. The process is shown in Fig. 4. Those local features produce a log-likelihood ratio score $L(r_i, c)$ for class label c at superpixel r_i:

$$L\left(r_i, c\right) = \log\frac{P\left(r_i|c\right)}{P\left(r_i|\overline{c}\right)} = \sum_k \log\frac{P\left(f_i^k|c\right)}{P\left(f_i^k|\overline{c}\right)} \tag{1}$$

where f_i^k is the feature; $P\left(f_i^k|c\right)$ (resp. $P\left(f_i^k|\overline{c}\right)$) is the likelihood of feature type k for region r_i given class c (resp. all classes but c) and here $k = 1$.

3.2 Exemplar-SVMs

In order to improve the rare class recognition rate, we incorporate the per-exemplar framework of [15], which is similar to the previous work [22]. In our dataset, the per-exemplar detector is trained for each labeled class instance. As the same with the detector training procedure of [15], all objects which include "thing" classes and "stuff" classes are trained. Each exemplar-SVMs creates the detection windows in a sliding fashion, and an exemplar co-occurence-based mechanism is applied for suppressing the redundant responses.

The leave-one-out method is employed to train SVM classifier on the training set. Each feature dimension is normalized by its standard deviation and the

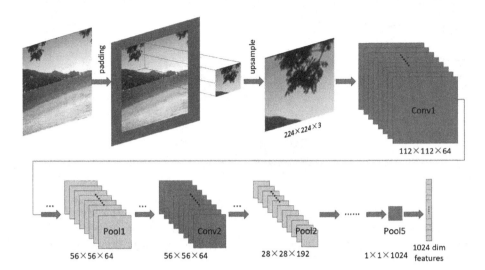

Fig. 4. Process of extracting local features using GoogLeNet. For a query image, we pad regions exceeding boundaries with mean pixel value. Then a superpixel-centered large context window is selected, and warped to $224 \times 224 \times 3$ as input. We extract the pool5 layer output of GoogLeNet as our deep local features.

regularization constant is computed using fivefold cross validation. The RBF kernel is applied for SVM training. At each pixel, the output results of SVMs produce Y responses. The $D_{SVM}(c_i, r_i)$ is defined to denote the response of the SVM for class c_i at superpixel r_i.

3.3 Inference

To assign labels to all of superpixels r_i in the query image I_q, the Markov Random Field model is incorporated into our system. The scene parsing problem is formulated as minimization of a standard MRF energy function, which combines the unary term with the smoothness term to obtain the final parsing result. We define the MRF energy function of the superpixel class labels $\mathbf{c}$ as:

$$\mathbf{E}(\mathbf{c}) = \sum_{r_i \in I_q} D(r_i, \mathbf{c}) + \sum_{r_i \in I_q} \max\left[0, M - D_{SVM}(c_i, r_i)\right] + \lambda \sum_{(r_i, r_j) \in N} \psi\left(c_{r_i}, c_{r_j}\right)$$

(2)

where $D(r_i, \mathbf{c})$ is the data term; once we obtain the likelihood $L(r_i, c)$, $D(r_i, \mathbf{c}) = -\omega_i log L(r_i, c)$; M is the highest expected value of SVM response; N is the set of adjacent pixels; The smoothing constant is λ. The smoothness term ψ is similar with which defined in [21,23] according to the probabilities of label c_i and c_j co-occurring in the training images:

$$\psi\left(c_{r_i}, c_{r_j}\right) = -\log\left[\left(\mathbf{P}\left(c_{r_i}|c_{r_j}\right) + \mathbf{P}\left(c_{r_j}|c_{r_i}\right)\right)/2\right] \delta\left[c_{r_i} \neq c_{r_j}\right]$$

(3)

where $\mathbf{P}\left(c_{r_i}|c_{r_j}\right)$ is the conditional probability of class label c_{r_i} of the pixel r_i, and its adjacent pixel r_j has class label c_{r_j}. We use the α-expansion algorithm [9] to optimize the MRF energy function.

4   Experiments

4.1   Dataset

The system results are reported on two datasets: the SIFT Flow dataset [12] containing 2688 images which were divided into 2488 training images, 200 test images, and 33 classes, and the per-pixel frequency counts of object categories in test images are presented in the left of Fig. 5. Another is video images dataset collected from the actual video surveillance scene, which included 500 images, among 450 training images, 50 test images, and 20 classes, and the per-pixel frequency counts of object categories are presented in the right of Fig. 5. As can be seen from the graph, some classes do not exist in the test set.

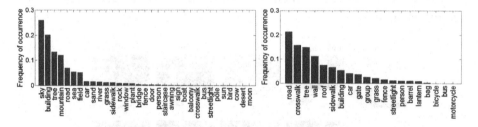

Fig. 5. The per-pixel frequency counts of the object categories on the SIFT Flow dataset (left) and video images dataset of we gathering in the actual video surveillance scene (right).

4.2   The Experimental Results

There are two measurement function for evaluating our scene parsing system: the per-pixel classification accuracy rate (the percentage of pixels that were correctly labeled of test images) and per-class classification rate (the average of all classes recognition rates). For SIFT Flow dataset, the size of the retrieval set is set as 200, namely $K = 200$. For the video images which are collected by ourselves, the size of the retrieval set is set as 50. The smoothing constant λ is 1. The highest expected value of SVM response M is approximately 10. The per-class recognition rates of SIFT Flow dataset and our collected video images dataset are presented in the left and right of Fig. 6, respectively. On the SIFT Flow dataset, the top five categories of recognition rate are sun, sky, building, sea and mountain.

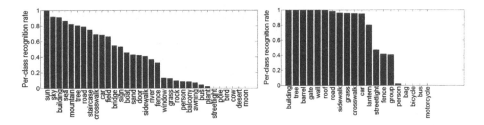

Fig. 6. Left: the per-class recognition rate on the SIFT Flow dataset. Right: the per-class recognition rate on the video images dataset of we gathering in the actual video surveillance scene.

Table 1 shows that the results of our system in contrast to the state-of-the art methods about per-pixel and per-class recognition rate on the SIFT Flow dataset. Although the recognition rate of presenting method is not the highest, the per-pixel accuracy increases by around 4% and the per-class accuracy increase about 17% over the method of [21]. The per-pixel accuracy of our approach is less than George [6] about 6% and the per-class recognition rate is less approximately 3%.

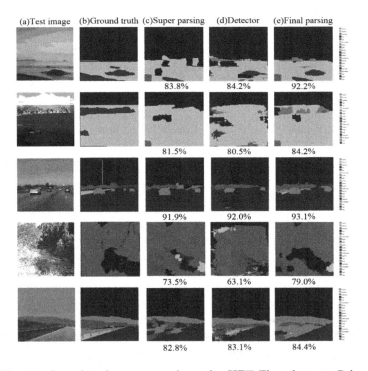

Fig. 7. The visual results of our approach on the SIFT Flow dataset. Column (a) is the test images; the ground truth of (a) is shown in (b); (c) is the superpixel parsing results; the exemplar parsing results are displayed in (d) and (e) is the final parsing result, which combines (c) with (d).

Table 1. A comparison of our system to the state-of-the-art methods about per-pixel and per-class classification accuracy rate (%) on the SIFT Flow dataset.

Approach	Per-pixel	Per-class
Liu et al. [12]	74.75	N/A
Tighe and Lazebnick [21]	76.90	30.10
Farabet et al. [5]	78.50	29.50
Tighe and Lazebnick [22]	78.60	39.20
Yang et al. [23]	79.80	48.70
Bu et al. [2]	80.40	35.80
Shuai et al. [19]	81.20	45.50
George [6]	81.70	50.10
Our approach	81.16	47.91

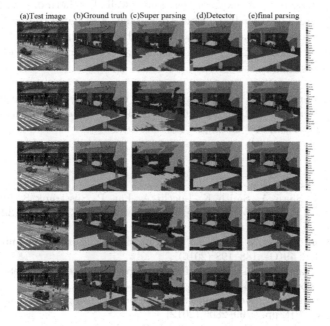

Fig. 8. The visual results of our approach on the video images. Column (a) is the test images; the ground truth of (a) is shown in (b); (c) is the superpixel parsing results; the exemplar parsing results are displayed in (d) and (e) is the final parsing result, which combines (c) with (d).

The part of visual samples results of our method on the SIFT Flow dataset and the video images are shown on Figs. 7 and 8, respectively. In Fig. 7, (a) expresses the test images; The ground truth of (a) is shown in (b); (c) is super-pixel parsing results and the recognition rates are displayed under the images. The exemplar parsing results are shown in (d). The final parsing results are displayed in (e). The Fig. 8 is similar to Fig. 7, but does not have the accuracy of images. As can be seen from Fig. 8, the parsing results of video images are impressive.

The computing environment of our system is MATLAB 2012b and on a four-core 3.4 GHz CPU with Intel Core i7-6700 and 8G of RAM. Training time and parsing time are influenced by feature dimension. The higher the dimension and amount of features are, the longer the training time costs.

5   Conclusions

This paper has proposed a method for scene parsing to improve the quality of retrieval sets and the class recognition rate of combining the deep convolutional features with exemplar-SVMs. Our approach takes advantage of a retrieval set of similar objects and scenes for reducing computational complexity. The deep features we extracted can express the image well. Furthermore, multiple cues are integrated into a MRF framework to segment and parse the test image. On the SIFT Flow dataset, the experimental results show that the method achieve promising results. This system is used to actual scene images and obtained the ideal parsing results.

Acknowledgments. This research was supported by Beijing Municipal Special Financial Project (PXM2016_178215_000013), Beijing Urban System Engineering Research Center's Independent Project (2017C010).

References

1. Achanta, R., Shaji, A., Smith, K., Lucchi, A., Fua, P., Süsstrunk, S.: Slic superpixels compared to state-of-the-art superpixel methods. IEEE TPAMI **34**(11), 2274–2282 (2012)
2. Bu, S., Han, P., Liu, Z., Han, J.: Scene parsing using inference embedded deep networks. PR **59**(C), 188–198 (2016)
3. Dalal, N., Triggs, B.: Histograms of oriented gradients for human detection. IEEE CVPR **1**, 886–893 (2005)
4. Eigen, D., Fergus, R.: Nonparametric image parsing using adaptive neighbor sets. In: IEEE CVPR, pp. 2799–2806 (2012)
5. Farabet, C., Couprie, C., Najman, L., LeCun, Y.: Scene parsing with multiscale feature learning, purity trees, and optimal covers. arXiv preprint arXiv:1202.2160 (2012)
6. George, M.: Image parsing with a wide range of classes and scene-level context. In: IEEE CVPR, pp. 3622–3630 (2015)
7. Girshick, R., Donahue, J., Darrell, T., Malik, J.: Rich feature hierarchies for accurate object detection and semantic segmentation. In: IEEE CVPR, pp. 580–587 (2014)
8. Grangier, D., Bottou, L., Collobert, R.: Deep convolutional networks for scene parsing. In: ICML 2009 Deep Learning Workshop, vol. 3 (2009)
9. Kolmogorov, V., Zabin, R.: What energy functions can be minimized via graph cuts? IEEE TPAMI **26**(2), 147–159 (2004)

10. Krizhevsky, A., Sutskever, I., Hinton, G.E.: Imagenet classification with deep convolutional neural networks. In: NIPS, pp. 1097–1105 (2012)

11. LeCun, Y., Boser, B., Denker, J.S., Henderson, D., Howard, R.E., Hubbard, W., Jackel, L.D.: Backpropagation applied to handwritten zip code recognition. Neural Comput. **1**(4), 541–551 (1989)

12. Liu, C., Yuen, J., Torralba, A.: Nonparametric scene parsing: label transfer via dense scene alignment. In: IEEE CVPR, pp. 1972–1979 (2009)

13. Liu, X., Zhao, Y., Zhu, S.C.: Single-view 3D scene reconstruction and parsing by attribute grammar. IEEE TPAMI **PP**(99), 1 (2017)

14. Lowe, D.G.: Distinctive image features from scale-invariant keypoints. IJCV **60**(2), 91–110 (2004)

15. Malisiewicz, T., Gupta, A., Efros, A.A.: Ensemble of exemplar-SVMs for object detection and beyond. In: IEEE ICCV, pp. 89–96 (2011)

16. Oliva, A., Torralba, A.: Modeling the shape of the scene: a holistic representation of the spatial envelope. IJCV **42**(3), 145–175 (2001)

17. Pinheiro, P., Collobert, R.: Recurrent convolutional neural networks for scene labeling. In: ICML, pp. 82–90 (2014)

18. Sermanet, P., Eigen, D., Zhang, X., Mathieu, M., Fergus, R., LeCun, Y.: Overfeat: integrated recognition, localization and detection using convolutional networks. arXiv preprint arXiv:1312.6229 (2013)

19. Shuai, B., Zuo, Z., Wang, G., Wang, B.: Scene parsing with integration of parametric and non-parametric models. IEEE Trans. Image Process. **25**(5), 2379–2391 (2016)

20. Szegedy, C., Liu, W., Jia, Y., Sermanet, P., Reed, S., Anguelov, D., Erhan, D., Vanhoucke, V., Rabinovich, A.: Going deeper with convolutions. In: IEEE CVPR, pp. 1–9 (2015)

21. Tighe, J., Lazebnik, S.: SuperParsing: scalable nonparametric image parsing with superpixels. In: Daniilidis, K., Maragos, P., Paragios, N. (eds.) ECCV 2010. LNCS, vol. 6315, pp. 352–365. Springer, Heidelberg (2010). https://doi.org/10.1007/978-3-642-15555-0_26

22. Tighe, J., Lazebnik, S.: Finding things: image parsing with regions and per-exemplar detectors. In: IEEE CVPR, pp. 3001–3008 (2013)

23. Yang, J., Price, B., Cohen, S., Yang, M.H.: Context driven scene parsing with attention to rare classes. In: IEEE CVPR, pp. 3294–3301 (2014)

24. Zhang, H., Fang, T., Chen, X., Zhao, Q., Quan, L.: Partial similarity based nonparametric scene parsing in certain environment. In: IEEE CVPR, pp. 2241–2248 (2011)

25. Zhao, H., Puig, X., Zhou, B., Fidler, S., Torralba, A.: Open vocabulary scene parsing. arXiv preprint arXiv:1703.08769 (2017)

How Depth Estimation in Light Fields Can Benefit from Angular Super-Resolution?

Mandan Zhao[1], Gaochang Wu[2(✉)], Yebin Liu[3], and Xiangyang Hao[1]

[1] Information Engineering University, Zhengzhou, China
mandanzhao@163.com, xiangyanghao2004@163.com
[2] Northeastern University, Shenyang, China
ahwgc2009@163.com
[3] Tsinghua University, Beijing, China
liuyebin@mail.tsinghua.edu.cn

Abstract. With the development of consumer light field cameras, the light field imaging has become an extensively used method for capturing the 3D appearance of a scene. The depth estimation often require a dense sampled light field in the angular domain. However, there is an inherent trade-off between the angular and spatial resolution of the light field. Recently, some studies for novel view synthesis or angular super-resolution from a sparse set of have been introduced. Rather than the conventional approaches that optimize the depth maps, these approaches focus on maximizing the quality of synthetic views. In this paper, we investigate how the depth estimation can benefit from these angular super-resolution methods. Specifically, we compare the qualities of the estimated depth using the original sparse sampled light fields and the reconstructed dense sampled light fields. Experiment results evaluate the enhanced depth maps using different view synthesis approaches.

Keywords: Light field · Angular super-resolution · View synthesis
Depth estimation

1 Introduction

Light field imaging [4,6] has emerged as a technology allowing to capture richer information from our world. Rather than a limited collection of 2D image, the light field camera is able to collect not only the accumulated intensity at each pixel but light rays from different directions. Recently, with the introduction of commercial and industrial light field cameras such as Lytro [1] and RayTrix [2], light field imaging has been one of a most extensively used method to capture 3D information of a scene.

However, due to restricted sensor resolution, light field cameras suffer from a trade-off between spatial and angular resolution. To mitigate this problem,

---

M. Zhao is student author.

J. Yang et al. (Eds.): CCCV 2017, Part I, CCIS 771, pp. 506–517, 2017.
https://doi.org/10.1007/978-981-10-7299-4_42

researchers have focused on novel view synthesis or angular super-resolution using a small set of views [7,8,10,15,16] with high spatial resolution. Typical view synthesis or angular super-resolution approaches first estimate the depth information, and then warp the existing images to the novel view based on the depth [12,16]. However, the depth-based view synthesis approaches rely heavily on the estimated depth, which is sensitive to noise, textureless and occluded regions. In recent years, some studies based on convolutional neural network (CNN) aiming at maximizing the quality of the synthetic views have been presented [5,14].

In this paper, we investigate how the depth estimation can benefit from these angular super-resolution methods. Specifically, we compare the qualities of the estimated depth using the original sparse sampled light fields and the reconstructed dense sampled light fields. Experiment results evaluate the enhanced depth maps using different view synthesis approaches.

2   Depth Estimation Using Angular Super-Resolved Light Fields

In this section, we describe the idea that uses super-resolved light fields in angular domain for depth estimation. We first investigate several angular super-resolution and view synthesis approaches. Then several depth estimation approaches are introduced using the super-resolved light field.

2.1   Angular Super-Resolution for Light Fields

Two angular super-resolution (view synthesis) approaches are investigated in the paper, which were proposed by Kalantari et al. [5] and Wu et al. [14]. Kalantari et al. [5] proposed a learning-based approach to synthesize novel views using a sparse set of input views. Specifically, they break down the process of view synthesis into disparity and color estimation and used two sequential CNNs to model them. In the disparity CNN (see Fig. 1), all the input views are first warped (backwarped) to the novel view with disparity range of $[-21, 21]$ and level of 100. Then the mean and standard deviation of all the warped input images are computed at each disparity level to form a feature vector of 200 channels.

In the color CNN, the feature vector is consisted of warped images, the estimated disparity and the position of the novel view, where the disparity is applied to occlusion boundaries detection and information collection from the adjacent regions, and the position of the novel view is used to assign the warped images with appropriate weights. The networks contain 4 convolutional layers with kernel sizes decreased from 7×7 to 1×1, where each layer is followed by a rectified linear unit (ReLU); the networks were trained simultaneously by minimizing the error between synthetic and ground truth views.

Unlike Kalantari et al. [5] that super-resolves light fields directly using images, Wu et al. [14] super-resolve light fields using EPIs. They indicated that

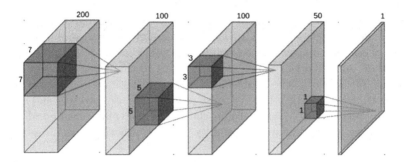

Fig. 1. The disparity CNN consists of four convolutional layers with decreasing kernel sizes. All the layers are followed by a rectified linear unit (ReLU). The color CNN has a similar architecture with different number of input and output channels.

the sparse sampled light field super-resolution involves information asymmetry between the spatial and angular dimensions, in which the high frequencies in angular dimensions are damaged by under-sampling. Therefore, they model the light field super-resolution as a learning-based angular high frequencies restoration on EPI.

Specifically, they first balance the information between the spatial and angular dimension by extracting the spatial low frequencies information. This is implemented by convolving the EPI with a Gaussian kernel. It should be noted that the kernel is defined in 1D space because only the low frequencies information in the spatial dimension are needed to be extracted. The EPI is then up-sampled to the desired resolution using bicubic interpolation in the angular dimension. Then a residual CNN is employed, which they called "detail restoration network" (see Fig. 2), to restore the high frequencies in the angular dimension. Different with the CNN proposed by Kalantari et al. [5], the detail restoration network is trained specifically to restore the high frequency portion in the angular dimension, rather than the entire information. Finally, a non-blind deblur is applied to recover the high frequencies depressed by EPI blur. Compared with the approach by Kalantari et al., Wu et al.'s approach has more flexible super-resolution factor; moreover, because of the depth-free framework, their approach achieves higher performance especially in occluded and transparent regions and non-Lambertian surfaces.

The architecture of the detail restoration network of Wu *et al.* is outlined in Fig. 2. Consider an EPI that is convolved with the blur kernel and up-sampled to the desired angular resolution, denoted as $\mathbf{E}'_L$ for short, the desired output EPI $f(\mathbf{E}'_L)$ is then the sum of the input $\mathbf{E}'_L$ and the predicted residual $\mathcal{R}(\mathbf{E}'_L)$:

$$f(\mathbf{E}'_L) = \mathbf{E}'_L + \mathcal{R}(\mathbf{E}'_L). \tag{1}$$

The network for the residual prediction comprises three convolution layers. The first layer contains 64 filters of size $1 \times 9 \times 9$, where each filter operates on 9×9 spatial region across 64 channels (feature maps) and used for feature extraction. The second layer contains 32 filters of size $64 \times 5 \times 5$ used for non-linear mapping.

The last layer contains 1 filter of size $32 \times 5 \times 5$ used for detail reconstruction. Both the first and the second layers are followed by a rectified linear unit (ReLU). Due to the limited angular information of the light field used as the training dataset, we pad the data with zeros before every convolution operations to maintain the input and output at the same size.

The desired residuals are $\mathbf{r} = \mathbf{e}' - \mathbf{e}'_L$, where $\mathbf{e}'_L$ are the blurred and interpolated low angular resolution sub-EPIs. The goal is to minimize the mean squared error $\frac{1}{2}\|\mathbf{e}' - f(\mathbf{e}'_L)\|^2$. However, due to the residual network we use, the loss function is now formulated as follows:

$$L = \frac{1}{n}\sum_{i=1}^{n}\|\mathbf{r}^{(i)} - \mathcal{R}(\mathbf{e}'^{(i)}_L)\|^2, \tag{2}$$

where n is the number of training sub-EPIs. The output of the network $\mathcal{R}(\mathbf{e}'_L)$ represents the restored detail, which must be added back to the input sub-EPI $\mathbf{e}'_L$ to obtain the final high angular resolution sub-EPI $f(\mathbf{e}'_L)$.

They apply this residual learning method for the following reasons. First, the undersampling in the angular domain damages the high frequency portion (detail) of the EPIs; thus, only that detail needs to be restored. Second, extracting this detail prevents the network from having to consider the low frequency part, which would be a waste of time and result in less accuracy.

Compared with the approach by Kalantari et al., Wu et al.'s approach has more flexible super-resolution factor; moreover, because of the depth-free framework, their approach achieves higher performance especially in occluded and transparent regions and non-Lambertian surfaces.

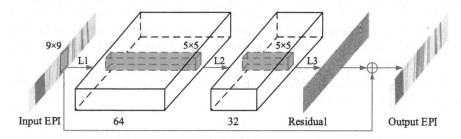

Fig. 2. Detail restoration network proposed by Wu et al. [14] is composed of three layers. The first and the second layers are followed by a ReLU. The final output of the network is the sum of the predicted residual (detail) and the input.

2.2 Depth Estimation for Light Fields

The Approach for Depth Measurement. In this subsection, we investigate several depth estimation approaches for light field data.

Tao et al. [9] proposed a depth estimation approach that combines depth cues from both defocus and correspondence using EPIs extracted from a light field. Since the slope of a line in an EPI is equivalent to a depth of a point in the scene [12], the EPIs are sheared to several possible depth values for computing defocus and correspondence cues responses. For a shear value α, a contrast-based measurement $\bar{L}_\alpha$ is performed at each pixel by averaging the intensity values in the angular dimension of the EPI. Then the defocus response D_α is measured by weighting the contrast-based measurements in a window in spatial dimension of the EPI. For the correspondence cue of a shear value α, the variance of each pixel in spatial dimension σ_α is computed, then the correspondence response C_α is the average of the variance values in a patch. After the computations of defocus and correspondence cue, a MRF global optimization is performed to obtain the final depth map.

Wang et al. [11] developed a depth estimation approach that treats occlusion explicitly. Their key insight is that the edge separating occluder and correct depth in the angular patch correspond to the same edge of occluder in the spatial domain. With this indication, the edges in the spatial image can be used to predict the edge orientations in the angular domain. First, the edges on the central pinhole image are detected using Canny operation. Based on the work by Tao et al. [9] and the occlusion theory described above, the initial local depth estimation is performed on the two regions in the angular patch of the sheared light field. In addition, a color consistency constraint is applied to prevent obtaining a reversed patch which will lead to incorrect depth estimation. Finally, the initial depth is refined with global regularization.

The Approach for 3D Measurement. In this subsection, we're going to focus on the 3D measurement of the light field imaging refocus: (a) We propose a novel digital refocusing algorithm to realize the refocus of the light field imaging. (b) We choose and refine the clarity-evaluation-function and interpolation method to establish the system of the evaluation and the refinement about light field imaging. (c) We take the corrected arithmetic mean filtering to refine the results of the 3D measurement.

(a) Refocus of the light field: Light fields transfer one plane to the other, the position coordinate can be defined as

$$[u', v', s', t'] = \begin{bmatrix} 1 & 0 & d & 0 \\ 0 & 1 & 0 & d \\ 0 & 0 & 1 & 0 \\ 0 & 0 & 0 & 1 \end{bmatrix} \cdot [u, v, s, t]^T, \tag{3}$$

where, $T_c = \begin{bmatrix} 1 & 0 & d & 0 \\ 0 & 1 & 0 & d \\ 0 & 0 & 1 & 0 \\ 0 & 0 & 0 & 1 \end{bmatrix}$, $[u, v]$ represents the plane position of the camera array, $[s, t]$ represents the direction information.

First, we transform the light field information to the standard description of the light field. pix is the size of the pixel, and f_m is the focal length of the camera. We can get:

$$\begin{bmatrix} s \\ t \end{bmatrix} = \frac{pix}{f_m} \begin{bmatrix} p_x \\ p_y \end{bmatrix}, \tag{4}$$

$$\begin{bmatrix} u \\ v \end{bmatrix} = pix \begin{bmatrix} grid_x \\ grid_y \end{bmatrix}, \tag{5}$$

where, $[grid_x, grid_y]^T$ is the center of the camera pixel coordinate.

(b) Principle of focus ranging: Light field camera still follow the principle of optical imaging. On the contrary, if the object deviating from the ideal image plane, beyonding the scope of the depth of field, which can cause the focal and imaging blur.

$$\frac{1}{f} = -\frac{1}{u} + \frac{1}{v}, \tag{6}$$

where, f represents the focal length of the light field camera. u represents the object distance. v represents the image distance. Here we use virtual image for distance measurement, so the value of the object distance is negative. Since light field camera focal length is known, and therefore is obtained by heavy focus on clear imaging like distance, which can calculate the distance. The above is the basic principle of focusing range.

(c) Corrected arithmetic mean filtering: Due to the light distribution or upheaval gaussian noise image gradient, so we choose the adjusted arithmetic average filtering processing.

$$f(x, y) = \frac{1}{\sqrt{m^2 + n^2}} \sum_{(s,t) \in s_{xy}} g(s, t), \tag{7}$$

where, s_{xy} represents rectangular window in the coordinate (x, y), which size is $m \times n$. $g(x, y)$ represents the jamming image.

3   Experimental Result

In this section, the proposed idea is evaluated on synthetic scenes and real-world scenes. We evaluate the quality of super-resolved light fields by measuring the PSNR values of synthetic views against ground truth images. And the quality of estimated depth maps using super-resolved light fields is compared with those using low angular resolution light fields. For synthetic scenes, ground truth depth maps are further applied for numerical evaluations.

3.1   Synthetic Scenes

The synthetic light fields in HCI datasets [13] are used for the evaluations. The input light fields have 3×3 views, where each view has a resolution of 768×768 (same as the original dataset); and the output angular resolution is 9×9 for comparison with the ground truth images.

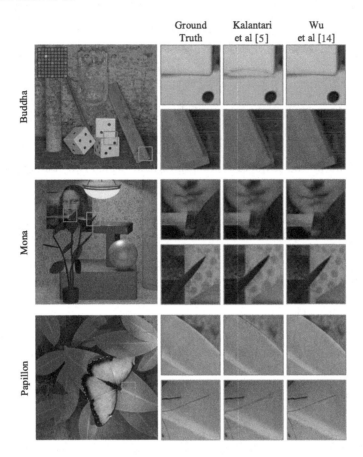

Fig. 3. Comparison of synthetic views produced by Kalantari et al.'s approach [5] and Wu et al.'s approach [14] on synthetic scenes. The ground truth views are shown on the left column, and the right columns show the close-up versions of the ground truth views, views produced by Kalantari et al.'s approach [5] and by Wu et al.'s approach [14], respectively.

Table 1. Quantitative results of reconstructed light fields on the synthetic scenes of the HCI datasets. The angular resolutions of input light fields are set to 3×3, and the output angular resolutions are 9×9.

	Buddha	Mona	Papillon
Kalantari [5]	34.05	32.53	28.26
Wu [14]	**43.20**	**44.37**	**48.55**

Table 1 shows a quantitative evaluation of the super-resolution approaches on synthetic scenes. The approach by Wu et al. [14] produces light fields of higher quality than those yielded by Kalantari et al. [5], because the CNNs in the latter approach are specifically trained on real-world scenes. Figure 3 shows the synthetic images in a certain viewpoint. The approach by Wu et al. [14] has better performance especially in the occluded regions, e.g., the board in the *Buddha* case, the leaves in the *Mona* case and the antenna in the *Papillon* case.

The numerical results of depth maps using the approaches by Tao et al. [9] and Wang et al. [11] are tabulated in Table 2. And Fig. 4 demonstrates the depth maps estimated by Wang et al.'s approach [11] on the *Buddha* using input low angular resolution (3 × 3) light field, ground truth high resolution (9 × 9) light field and super-resolved light fields (9 × 9) by Kalantari et al. [5] and Wu et al. [14], respectively. The depth estimation using super-resolved light fields show prominent improvement when compared with the results using input low resolution light fields. In addition, due to the better quality of synthetic views

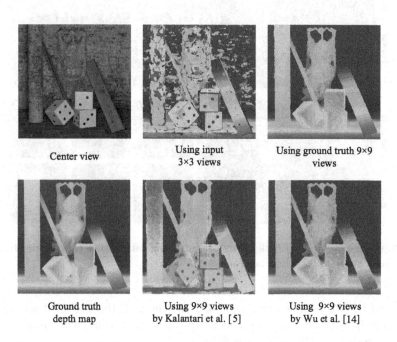

Fig. 4. Comparison of depth maps estimated by Wang et al.'s approach [11] using light fields of different angular resolutions on the *Buddha*.

Table 2. RMSE values of the estimated depth using the approaches by Tao et al. [9]/Wang et al. [11] on synthetic scenes of HCI datasets.

	Buddha	*Mona*	*Papillon*
Input views	0.2642/0.2926	0.2115/0.2541	0.1871/0.1533
Kalantari [5]	0.1721/0.1576	0.0876/0.0829	0.1665/0.1430
Wu [14]	0.0550/0.0401	0.0678/0.0517	0.0610/0.0532
GT Light Fields	0.0543/0.0393	0.0652/0.0529	0.0583/0.0455

Fig. 5. Comparison of synthetic views produced by Kalantari et al.'s approach [5] and Wu et al.'s approach [14] on real-world scenes.

produced by Wu et al.'s approach [14], especially in the occluded regions, the estimated depth maps are more accurate than those using super-resolved light fields by Kalantari et al.'s approach [5].

3.2   Real-World Scenes

The Stanford Lytro Light Field Achieve [3] is used for the evaluation on real-world scenes. The dataset is divided into several categories including occlusions, and refractive and reflective surfaces, which are challenge cases to test the robustness of the approaches. We use 3×3 views to reconstruct 7×7 light fields.

Table 3 lists the numerical results of the super-resolution approaches on the real-world scenes. The approach by Wu et al. [14] shows better performance in terms of PSNR. Figure 5 shows some representative cases that contains complex occlusions or non-Lambertian surfaces. The networks proposed by Kalantari

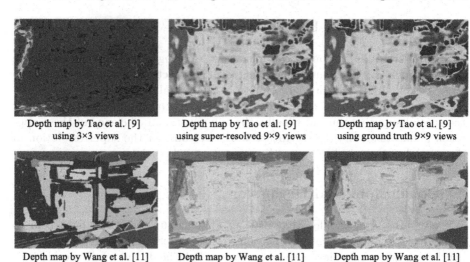

Fig. 6. Comparison of depth maps estimated by Tao et al. [9] and Wang et al. [11] using light fields of different angular resolutions on the *Reflective surfaces* 29.

et al. [5] were specifically trained for Lambertian regions, thus tend to fail in the reflective surfaces, such as the pot and pan in the *Reflective* 29 case. In addition, due to the depth estimation-based framework, the synthetic views have ghosting and tearing artifacts in the occlusion boundaries, such as the branches in the *Occlusion* 16 case and the boundary between air-condition and grass in the *Flowers & plants* 7 case.

Figure 6 shows the depth maps estimated by Tao et al.'s approach [9] and Wang et al.'s approach [11] using input low angular resolution (3 × 3) light field, super-resolved light fields (7 × 7) by Wu et al. [14] and ground truth high resolution (7 × 7) light field, respectively. The quality of estimated depth maps are significantly improved using super-resolved light fields.

Table 3. Quantitative results of reconstructed light fields on the real-world scenes.

	Kalantari [5]	Wu [14]
Occlusions 2	28.90	**38.12**
Occlusions 16	32.24	**38.86**
Flowers & plants 7	26.70	**38.70**
Flowers & plants 12	34.97	**42.27**
Reflective surfaces 17	28.84	**42.28**
Reflective surfaces 29	37.70	**46.10**

4   Discussion and Future Work

There are several limitations of this idea. First, existing algorithms cannot handle the large parallax. They may occur the tearing artifacts when the parallax is too large. Second, light field super-resolution in the spatial domain is not considered. We believe that we can get the better results in depth estimation when jointing the spatial super-resolution. Finally, the computational efficiency of these approach is unsatisfactory. In the future work, we consider the super-resolution in the spatial domain and focus on the computational efficiency.

5   Conclusions and Discussion

We have presented an idea that uses an angular super-resolved light field to improve the quality of depth estimation. A straightforward way is to estimate a depth map using input low angular resolution light field, and render novel views using DIBR techniques. However, this approach always leads to error accumulation when recompute depth map using synthetic views. We therefore investigate approaches that directly minimize the quality of super-resolved light fields rather than depth maps. We evaluate this idea on synthetic scenes as well as real-world scenes, which contains non-Lambertian and reflective surfaces. The experimental results demonstrate that the quality of depth map is significantly improved using angular super-resolved light field.

References

1. Lytro. https://www.lytro.com/
2. RayTrix: 3D light field camera technology. http://www.raytrix.de/
3. Stanford Lytro Light Field Archive. http://lightfields.stanford.edu/
4. Ihrke, I., Restrepo, J.F., Mignard-Debise, L.: Principles of light field imaging: briefly revisiting 25 years of research. IEEE Signal Process. Mag. **33**(5), 59–69 (2016). https://doi.org/10.1109/MSP.2016.2582220
5. Kalantari, N.K., Wang, T.C., Ramamoorthi, R.: Learning-based view synthesis for light field cameras. ACM Trans. Graph. (TOG) **35**(6), 193–202 (2016)
6. Levoy, M., Hanrahan, P.: Light field rendering. In: Siggraph, pp. 31–42 (1996). https://doi.org/10.1145/237170.237199
7. Pujades, S., Devernay, F., Goldluecke, B.: Bayesian view synthesis and image-based rendering principles. In: CVPR, pp. 3906–3913 (2014)
8. Shi, L., Hassanieh, H., Davis, A., Katabi, D., Durand, F.: Light field reconstruction using sparsity in the continuous fourier domain. ACM TOG **34**(1), 12 (2014)
9. Tao, M.W., Hadap, S., Malik, J., Ramamoorthi, R.: Depth from combining defocus and correspondence using light-field cameras. In: ICCV, pp. 673–680 (2013)
10. Vagharshakyan, S., Bregovic, R., Gotchev, A.: Image based rendering technique via sparse representation in shearlet domain. In: ICIP, pp. 1379–1383. IEEE (2015)
11. Wang, T.C., Efros, A.A., Ramamoorthi, R.: Occlusion-aware depth estimation using light-field cameras. In: ICCV, pp. 3487–3495 (2015)
12. Wanner, S., Goldluecke, B.: Variational light field analysis for disparity estimation and super-resolution. IEEE TPAMI **36**(3), 606–619 (2014)

13. Wanner, S., Meister, S., Goldlücke, B.: Datasets and benchmarks for densely sampled 4D light fields. In: Vision, Modeling & Visualization, pp. 225–226 (2013)
14. Wu, G., Zhao, M., Wang, L., Chai, T., Dai, Q., Liu, Y.: Light field reconstruction using deep convolutional network on EPI. In: CVPR (2017)
15. Yoon, Y., Jeon, H.G., Yoo, D., Lee, J.Y., So Kweon, I.: Learning a deep convolutional network for light-field image super-resolution. In: CVPRW, pp. 24–32 (2015)
16. Zhang, Z., Liu, Y., Dai, Q.: Light field from micro-baseline image pair. In: CVPR, pp. 3800–3809 (2015)

Influence Evaluation for Image Tampering Using Saliency Mechanism

Kui Ye[1,2], Xiaobo Sun[1], Jindong Xu[2], Jing Dong[2,3(✉)], and Tieniu Tan[2]

[1] School of Automation, Harbin University of Science and Technology,
Harbin 150080, China
`kui.ye@cripac.ia.ac.cn`
[2] Center for Research on Intelligent Perception and Computing,
Chinese Academy of Sciences, Beijing 100190, China
`jdong@nlpr.ia.ac.cn`
[3] State Key Laboratory of Information Security,
Institute of Information Engineering, Chinese Academy of Sciences,
Beijing 100093, China

Abstract. Due to the immediacy and the easy way to understand the image content, the applications of digital image have brought great opportunity to the development of social networks. However, there exist some serious problems. Some visual content has been maliciously tampered to achieve illegal purpose, while some modifications are benign, just for fun, for enhancing artistic value, or effectiveness of news dissemination. So beyond the tampering detection, how to evaluate the influence of image tampering is on schedule. In this paper, with the help of forensic tools, we study the problem of automatically assessing the influence of image tampering by examining whether the modification affects the dominant visual content, and utilize saliency mechanism to assess how harmful the tampering is. The experimental results demonstrate the effectiveness of our method.

Keywords: Influence evaluation · Image tampering · Saliency
Image forensics

1 Introduction

Nowadays, images play a vital role on online social networks (OSN), since an image says more than a thousand words, and is not restricted by the barrier of language. However, the easy access to image editing software tools makes it extremely simple to alter the content of images, or to create new ones, so that material may be manipulated to serve personal ambition or propaganda, e.g. reusing past images from unrelated contexts, or explicitly tampering with media content. As a result, seeing is no longer believing [1].

Image forensics, which focuses on physical content, is an effective technique to detect whether images are modified. But many image operations are benign,

© Springer Nature Singapore Pte Ltd. 2017
J. Yang et al. (Eds.): CCCV 2017, Part I, CCIS 771, pp. 518–527, 2017.
https://doi.org/10.1007/978-981-10-7299-4_43

performed for fun or for artistic value, such as visible watermarking in Fig. 1(a), and photo retouching and repair in Fig. 1(b). Most images on the social networks usually have been processed to enhance posting effect or resize, such as image enhancement, median filtering and JPEG compressing, all of which are regarded as doctored images in the field of image forensics, even though the essential content of these images are not changed. It is different from malicious tampering in rumor images. Therefore, how to differentiate innocent operation and malicious tampering is our main focus, that is to say, we want to evaluate the influence of image tampering.

Fig. 1. Two examples of benign modification. (left) An image with visible watermarking. (right) An old photo and its repair version.

So what features are best for evaluating the importance of tampering content? The first is whether the virtual content is changed, i.e., whether the core and main content are changed. The saliency features are usually seen as important content of an image [2,3]. So in this paper, based on image forensics techniques, we use saliency mechanism to evaluate the influence of image tampering. Meanwhile, we set some evaluation indexes and corresponding thresholds. If the index of altered content in the saliency map exceeds preset threshold, it is seen as malicious operation, otherwise, it is seen as innocent or uncertain.

In the rest of the paper, after introducing related works about information credibility and image forensics, and our motivation, we describe the proposed method about influence evaluation of image tampering content, and finally we present experimental results and discussion.

2 Related Work

Some microblogs and websites have the function of identifying false information, such as [4–6], and most are human-based. However, for today's OSN big data, human-based method is not effective. Recently there have been many efforts to study information credibility in microblogs, ranging from automated algorithms to model and predict credibility of users [7] and messages [8–10]. However, with the exception of [11], little research has focused on image credibility. Also, there is yet no work on content-based image credibility. Image credibility verification on OSN is a fairly challenging problem, and it is based on image forensic technology. In the field of image forensics, there are many excellent algorithms to detect image operations, such as seam carving, copy-move [12–14], near-duplicate [15], median filtering [16], JPEG compression [17], and image composition [18]. However, most of these methods do not evaluate the degree of importance of tampered parts.

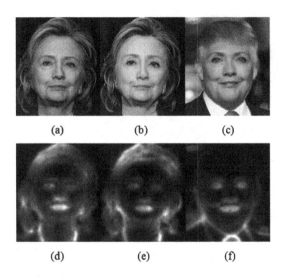

Fig. 2. Images and their saliency maps. (a) Original image; (b) retouching image; (c) malicious tampered image; (d)–(f) are saliency maps of (a)–(c).

In Fig. 2, (a) and (c) are downloaded from Internet [19]. Obviously, the (c), which has been retweeted excessively [20], is the malicious doctored version of (a) and (b) is the benign retouching version of (a) and (d)–(f) are saliency maps of (a)–(c) by CA attention algorithm [3], respectively. We can see that innocent retouching alters the image saliency at a tiny scale, while malicious tampering alters the image saliency enormously. Therefore, we will evaluate the influence of image tampering via how much operations affect saliency content of an image.

3   Proposed Method

As shown in the previous parts, we will use saliency mechanism to evaluate the influence of image tampering. Firstly, we give assumptions of our model, then we present 3 indexes to evaluate the importance of modified content, and at last the evaluation pipeline is proposed.

3.1 Assumptions of the Model

The importance evaluation of image tampering is based on following assumptions.

(1) The bigger area of modified content, the stronger influence of the content;
(2) The more salient pixels of modified content, the stronger influence of the content.

3.2 Evaluation Indexes

Based on the assumptions in Sect. 3.1, 3 indexes are proposed in this part. Supposing we have an image of size $M \times N$ which has been modified, $p(i, j)$ is the probability of corresponding pixel value in position (i, j) and $v_s(i, j)$ is the saliency value of the pixel in position (i, j).

(1) **Ratio of modified area R_a**

$$R_a = \frac{C'}{C}$$

where C' is modified area, and C is original image area.

(2) **Ratio of saliency value per unit area E_s**

$$E_s = (\frac{v'_s}{v_s})/R_a$$

where

$$v_s = \sum_{i,j}^{M,N} v_s(i, j)$$

$$v'_s = \sum_{i,j}^{M,N} v'_s(i, j)$$

$v'_s(i, j)$ is the saliency value of modified region in position (i, j).

(3) **Ratio of saliency information amount per unit area E_{cs}**

$$E_{cs} = (\frac{I'_{cs}}{I_{cs}})/R_a$$

where

$$I_{cs} = \sum_{i,j}^{M,N} I_c(i, j)v_s(i, j) \sum_{i,j}^{M,N} -log_2 p(i, j)v_s(i, j)$$

$$I'_{cs} = \sum_{i,j}^{M,N} I'_c(i, j)v'_s(i, j) \sum_{i,j}^{M,N} -log_2 p'(i, j)v'_s(i, j)$$

$p'(i, j)$ is the probability of corresponding pixel value of modified region in position (i, j).

The tampering influence of an image is defined as a numeric value, and a threshold is used to determine whether it is benign or malicious. For index R_a, the bigger R_a, the stronger influence of the content. For indexes R_s and R_{cs}, we set confidence value according to confidence level 95%. When the index is in interval $[0, 0.95)$, the image is benign; when in $[0.95, 1.05]$, uncertain; and when in $(1.05, \beta]$, malicious, where $\beta > 1.05$.

3.3 Evaluation Pipeline

Given an image, the evaluation pipeline is as follows:

(1) Detecting modified regions by corresponding forensics algorithm;
(2) Generating saliency map via a series of attention models;
(3) Computing the indexes presented in Sect. 3.2;
(4) Classification and analyzing the influence of image tampering.

4 Experiments

In this part, we use forgery detection dataset [14] to evaluate the performance of our method, and utilize 5 classical attention models (Itti [2], CA [3], GBVS [21], SIM [22], SUN [23]) to generate saliency maps respectively. First, we ask 6 forensic experts to assess the tampering images in the dataset and average the scores. Then we use coincidence degree between human-based assessments and evaluation indexes to test the effectiveness of the proposed method.

4.1 Impact of Different Modified Area

In the experimental dataset, the R_a lies in the interval $[1.3\%, 13.71\%]$. Sorting R_a in ascending order, the human-based assessment score is shown in Fig. 2. According to Sect. 3.1 assumption, the bigger area of modified content, the higher influence of the tampering. However, as Fig. 3 shows, the line chart does not have a explicit tendency or direction. That is to say, in this dataset, there is no direct relationships between the influence of different tampering images and R_a.

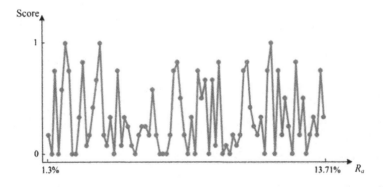

Fig. 3. Line chart of human assessment sore under sorted R_a.

4.2 Coincidence Degree

The coincidence degree between human-based assessments and evaluation indexes using different attention models is shown in Table 1.

Table 1. Coincidence degree under different attention models.

Indexes	Itti	CA	GBVS	SIM	SUN
E_s	71.25%	80.00%	65.00%	72.50%	68.75%
E_{cs}	83.75%	87.50%	81.50%	81.25%	86.25%

As shown in Table 1, in general, E_s and E_{cs} all have fine coincidence degree with human-based tampering assessments using these 5 classical attention models, and the index E_{cs} has higher coincidence degree than E_s, and CA attention model is better than the other 4 attention models. Then we utilize these 5 attention algorithms on Fig. 2(a), (b) and (c), and get their saliency maps shown in Fig. 4. We can see that CA saliency map of the retouching image is most similar to that of the original image, and also could differentiate malicious tampering one.

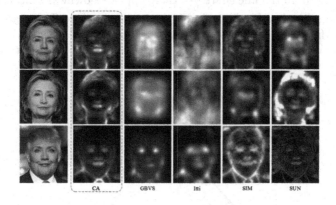

Fig. 4. Images and their saliency maps.

4.3 Different-Level Modification Within One Image

To further demonstrate the effectiveness of our proposed method, we alter an image with different modifications as shown in Fig. 5. There are 9 images with different modification from Fig. 5. Top left image (E_s and $E_{cs} > 1$) is gotten from dataset [14], other 8 images are modified with different number of ships by copy-move operation.

Fig. 5. Modified images with different number of ships.

Then CA algorithm are utilized to detect saliency features, and corresponding line chart of indexes R_a, $\frac{v_s'}{v_s}$ and $\frac{I_{cs}'}{I_{cs}}$ are shown in Fig. 6.

As Fig. 6 shows, when E_s and $E_{cs} > 1$, the bigger R_a, $\frac{v_s'}{v_s}$ and $\frac{I_{cs}'}{I_{cs}}$, the higher level modification and the greater influence. This suggests that we can assign an influence score via indexes to each image such that greater influence corresponds to a higher score. In the line chart $\frac{I_{cs}'}{I_{cs}}$, there appears downward inflection points. The corresponding images shown in Fig. 6 are not easy to differentiate by human eyes at the first glimpse. That is to say, the modification is uncertain about innocent or malicious. Of course, this is just a specific case not covering all situations.

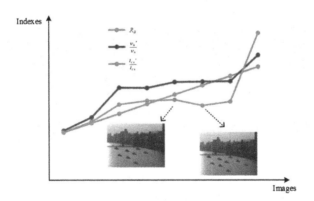

Fig. 6. Line chart of indexes R_a, $\frac{v_s'}{v_s}$ and $\frac{I_{cs}'}{I_{cs}}$.

4.4   Classic Doctored Image Testing

In this part, we select two famous doctored images whose semantic content has been heavily altered, and get their saliency maps by CA algorithm, as shown in Fig. 7. We can see that the tampering part (the dashed ellipses) is conspicuous in their saliency maps. That is to say, the saliency content of these two forgery images are both heavily tampered, and it testifies that saliency mechanism could evaluate the influence of image tampering.

Fig. 7. Two famous tampered images and their saliency maps.

5   Conclusion and Future Work

Information with maliciously tampered images on OSN will encourage the propagation of rumors. Giving early alerts of fake news can prevent further spreading of malicious content on social media. Although image tampering analysis cannot solve all problems of rumor dissemination, it could address the issue of information with doctored images as early as possible.

In this work, we attempt to evaluate the importance of tampering content via saliency mechanism and achieve some results. The performance of the method is affected by human-based assessment with a prior knowledge, image content description, subjective factor and so forth. Also, whether attention models could catch the "saliency content" affects the experimental results. In the future work, we will study image tampering influence in the real situation, generating saliency map based on the context inspiring state-of-the-art caption algorithms and text analysis methods to connect image and text.

Acknowledgement. This work is supported by China Postdoctoral Science Foundation funded project (No. 2016M601168), Science and technology research project of Heilongjiang Education Department (No. 12521092), NSFC (No. U1536120, U1636201, 61502496) and the National Key Research, Development Program of China (No. 2016YFB1001003).

References

1. Farid, H.: Digital doctoring: how to tell the real from the fake. Significance **3**(4), 162–166 (2006)
2. Itti, L., Koch, C., Niebur, E.: A model of saliency-based visual attention for rapid scene analysis. IEEE Trans. Pattern Anal. Mach. Intell. **20**(11), 1254–1259 (1998)
3. Goferman, S., Zelnik-Manor, L., Tal, A.: Context-aware saliency detection. IEEE Trans. Pattern Anal. Mach. Intell. **34**(10), 1915–1926 (2012)
4. http://py.qianlong.com. Accessed 9 May 2017
5. http://service.account.weibo.com. Accessed 9 May 2017
6. www.snopes.com. Accessed 9 May 2017
7. Kumar, S., Morstatter, F., Zafarani, R., Liu, H.: Whom should I follow?: identifying relevant users during crises. In: Proceedings of the 24th ACM Conference on Hypertext and Social Media, pp. 139–147. ACM (2013)
8. Alrubaian, M., Alqurishi, M., Hassan, M., Alamri, A.: A credibility analysis system for assessing information on twitter. IEEE Trans. Dependable Secure Comput. (2016). https://doi.org/10.1109/TDSC.2016.2602338
9. Jin, Z., Cao, J., Zhang, Y., Luo, J.: News verification by exploiting conflicting social view-points in microblogs. In: Thirtieth AAAI Conference on Artificial Intelligence, pp. 2972–2978. AAAI (2016)
10. Sikdar, S., Kang, B., Odonovan, J., Höllerer, T., Adalı, S.: Understanding information credi-bility on Twitter. In: International Conference on Social Computing, pp. 19–24. IEEE (2013)
11. Gupta, A., Lamba, H., Kumaraguru, P., Joshi, A.: Faking Sandy: characterizing and identifying fake images on Twitter during Hurricane Sandy. In: International Conference on World Wide Web, pp. 729–736 (2013)
12. Cozzolino, D., Poggi, G., Verdoliva, L.: Efficient dense-field copy-move forgery detection. IEEE Trans. Inf. Forensics Secur. **10**(11), 2284–2297 (2015)
13. Ardizzone, E., Bruno, A., Mazzola, G.: Copy-Move forgery detection by matching triangles of keypoints. IEEE Trans. Inf. Forensics Secur. **10**(10), 2084–2094 (2015)
14. Cozzolino, D., Poggi, G., Verdoliva, L.: Copy-move forgery detection based on PatchMatch. In: International Conference on Image Processing, pp. 5312–5316. IEEE (2014)
15. Oikawa, M.A., Dias, Z., Rocha, A.R., Goldenstein, S.: Manifold learning and spectral clus-tering for image phylogeny forests. IEEE Trans. Inf. Forensics Secur. **11**(1), 5–18 (2016)
16. Fan, W., Wang, K., Cayre, F., Xiong, Z.: Median filtered image quality enhancement and anti-forensics via variational deconvolution. IEEE Trans. Inf. Forensics Secur. **10**(5), 1076–1091 (2015)
17. Wang, W., Dong, J., Tan, T.N.: Exploring DCT coefficient quantization effects for local tampering detection. IEEE Trans. Inf. Forensics Secur. **9**(10), 1653–1666 (2014)
18. Peng, B., Wang, W., Dong, J., Tan, T.N.: Optimized 3D lighting environment estimation for image forgery detection. IEEE Trans. Inf. Forensics Secur. **12**(2), 479–494 (2017)
19. http://dailyhive.com/toronto/places-in-toronto-to-watch-u-s-election-2016. Accessed 9 May 2017
20. https://twitter.com/CNNPolitics/status/726144622083358720. Accessed 9 May 2017

21. Schölkopf, B., Platt, J., Hofmann, T.: Graph-based visual saliency. In: Advances in Neural Information Processing Systems (NIPS), pp. 545–552 (2006)
22. Murray, N., Vanrell, M., Otazu, X., Parraga, C.A.: Saliency estimation using a non-parametric low-level vision model. In: IEEE Conference on Computer Vision and Pattern Recognition (CVPR), pp. 433–440 (2011)
23. Zhang, L., Tong, M., Marks, T., Shan, H., Cottrell, G.: SUN: a Bayesian framework for sali-ency using natural statistics. J. Vis. **8**(7), 1–20 (2008)

Correlation Filters Tracker Based on Two-Level Filtering Edge Feature

Dengzhuo Zhang[1], Donglan Cai[1], Yun Gao[1,2(✉)], Hao Zhou[1], and Tianwen Li[3]

[1] School of Information Science and Engineering,
Yunnan University, Kunming, China
gausegao@163.com
[2] Kunming Institute of Physical, Kunming, China
[3] Faculty of Science, Kunming University of Science and Technology,
Kunming, China

Abstract. In recent years, correlation filter frame has attracted more attention in visual object tracking, providing a real-time way. However, with the increase of the computing complexity of feature extractor, the trackers lost the real-time advantage of correlation filters. Moreover, correlation filters model drift can result in tracking failure. In order to solve these problems, a novel and simple correlation filters tracker based on two-level filtering edge feature (ECFT) was proposed. ECFT extracted a low-complexity edge feature based on two-level filtering for object representation. For reducing model drift as much as possible, an object model is updated adaptively by the maximum response value of correlation filters. The comparative experiments of 7 trackers on 20 challenging sequences showed that the ECFT tracker can perform better than other trackers in terms of AUC and Precision.

Keywords: Object tracking · Correlation filters · Two-level filtering
Edge feature · Model updating

1 Introduction

Visual object tracking has been a popular problem in computer vision, with proposed various tracking algorithms [1–5]. With the development and application of machine learning and deep learning, visual object tracking based on the discriminant modeling method has been widely used, and the visual tracking performance has improved significantly. The most common idea was to convert the tracking problem into a binary classification problem, and the use of online learning technology to dynamically update the model, such as TLD [3], MIL [9], CT [10], CNT [7] and many advanced algorithms.

At the same time, the tracking algorithm based on correlation filters has also been very popular. Bolme et al. proposed the minimum output squared error sum algorithm named MOSSE [5]. The algorithm used the discriminant model of the minimum output squared error sum as the discriminant model of the target

© Springer Nature Singapore Pte Ltd. 2017
J. Yang et al. (Eds.): CCCV 2017, Part I, CCIS 771, pp. 528–538, 2017.
https://doi.org/10.1007/978-981-10-7299-4_44

and the background, and used the discrete Fourier transform to the frequency domain to calculate, which significantly reduced the computing complexity. Then CSK [2] adopted a circulant matrix to obtain a large number of training samples. Then, in the literature [12], Kiani Galoogahi et al. proposed in the correlation filters using multi-channel features, the experimental results are great. Various features were also of great concern in the correlation filters, such as gray, HOG, CNN [6] and so on. For the HOG feature, it has many advantages. First of all, because the HOG is on the image of the local grid unit operation, so it the image geometric and optical can keep good invariance, the deformation of the two deformation will only appear on the larger space. As long as pedestrians are generally able to maintain upright posture, they allow pedestrians to have some minor body movements. Therefore, HOG feature are particularly suitable for human detection in the image. In KCF [1], HOG feature improved CSK based on Gaussian kernel and HOG feature, which improved the performance of the CSK algorithm. Li and Zhu proposed the adaptive Kernel Correlation Filters Tracker with feature integration in the reference [13]. For the gray feature, It only contains brightness information and no color information, so the tracker handles it very quickly. There are correlation filters trackers algorithms [14–20] have a good effect. The correlation filters trackers based on CNN feature [6] can further improve the tracking performance. However, there are some disadvantages in these feature. the HOG feature is very sensitive to noise because of the nature of the HOG gradient. The gray feature is too simple to deal with the video sequence in the complex environment. with the increase of the computing complexity of feature extractor, the trackers lost the real-time advantage of correlation filters. Wang et al. proposed 5 stages of the object tracking process in reference [8], and proposed the feature extraction was the most important stage. So in the correlation filters tracking frame, a simple and effective feature needed to be researched. Moreover, model drift of the correlation filters can result in tracking failure.

In this paper, a novel and simple correlation filter tracker based on two-level filtering edge feature (ECFT) was proposed. In Sect. 2, correlation filters tracking frame was introduced. In Sect. 3, ECFT was detailed, which extracted a low-complexity edge feature based on two-level filtering for object representation and updated an object model adaptively by the maximum response value of correlation filters. Experiments in Sect. 4 showed the performance of ECFT by 7 trackers on 20 challenging sequences in terms of AUC and Precision. And Sect. 5 concluded this paper.

2   Correlation Filters Tracking Frame

In this section, we introduce the basic principle of the correlation filters and the circular matrix in the correlation filters. The advantages and disadvantages of the correlation filters are also analyzed.

2.1   The Basic Principle of Correlation Filters

The first step of correlation filters frame is to learn a convolution filters h_i as the Formula (1).

$$h_i = g_i + f_i \tag{1}$$

where g_i abides by a two-dimensional Gaussian distribution about the center position (x_i, y_i) as the Formula (2). And the f_i denotes input image.

$$g_i(x, y) = e^{\frac{(x - x_i)^2 + (y - y_i)^2}{\delta^2}} \tag{2}$$

For the second step of the correlation filters frame, the filter h_i and the image f_i can get a correlation response map g_i by the convolution operation, and the convolution operation is define as the Formula (3).

$$g_i' = f_i \otimes h_i \tag{3}$$

Finally, in the correlation response map g_i', the maximum response value is the current position of a tracked object.

2.2   Circulant Matrices in the Correlation Filters

Under the correlation filters frame, the cyclic matrix is used for sampling to accelerate.

Let $x = [x_1, x_2, ..., x_n]^T$ be an n-dimensional column vector, and a $n \times n$ cyclic matrix for dense sampling based on can be constructed as X^c by the Formula (4).

$$\begin{bmatrix} (P^0 x)^T \\ \vdots \\ (P^{n-1} x)^T \end{bmatrix} \tag{4}$$

where the matrix P is the permutation matrix, P_i and means that the row vectors in the P matrix are cyclized i times. The matrix P is expressed as the Formula 5.

$$\begin{bmatrix} 0 & 0 & 0 & \cdots & 1 \\ 1 & 0 & 0 & \cdots & 0 \\ 0 & 1 & 0 & \cdots & 0 \\ & \vdots & & \ddots & \\ 0 & 0 & \cdots & 1 & 0 \end{bmatrix} \tag{5}$$

So there is $Px = [x_n, x_1, x_2, ..., x_{n-1}]^T$, and the vector x is a base samples vector in cyclic matrix. $C(x)$ means that vector x performs cyclic shift operations as the Formula (6).

$$X^c = C(x) = \begin{bmatrix} x_1 & x_2 & \cdots x_n \\ x_n & x_1 & \cdots x_{n-1} \\ \vdots & & \ddots \\ x_2 & x_3 & \cdots x_1 \end{bmatrix} \tag{6}$$

All circulant matrices can be diagonalization in discrete Fourier space, and the diagonalization formula is the Formula (7).

$$X^c = F diag(\hat{x}) F^H \tag{7}$$

where $\hat{x}$ is a Discrete Fourier Transform (DFT) for x, and $F^H F = 1$. This method avoids the complex process of matrix inversion and decreases the computing complexity.

We summarize the advantages of the correlation filters to several following aspects. First, in the Fourier domain as an element product, effectively calculating the spatial correlation, the correlation filters tracker can achieve a fast tracking. Second, the correlation filters considers the information around an object, thus having more discrimination than the appearance model only on an object. Third, the correlation filters can be described as one ridge regression problem, where the input feature of a cyclic shift is returned to a soft label. Moreover, the tracker on correlation filters does not need to distribute a positive label or a negative label for a spatial correlation sample. For the disadvantages of the correlation filters, The correlation filters are very sensitive to image noise, which can lead to inaccurate tracking in complex environment. In this paper, based on the correlation filters tracking frame, we research a simple and effective feature extracting method and a model updating method.

3   Proposed Method

In this section, the proposed correlation filters tracker based on two-level filtering edge feature is given, abbreviated as ECFT, which is improved based on KCF [1]. In ECFT, a candidate object can be represented by the edge feature vector based on two-level filtering. In the model updating stage, an object model is updating adaptively.

3.1   Edge Feature Based on Two-Level Filtering

The edge feature of a candidate object is extracted based on two-level filtering. In the first-level filtering, for reducing the noise disturbing of a frame image effectively, the median filtering on an input frame is performed based on non-linear smoothing technique. The median filtering value $g(x, y)$ at (x, y) can be calculated by the Formula (8).

$$g(x, y) = med\{f(x - k, y - l), (k, l \in W)\} \tag{8}$$

where $f(x, y)$ is the pixel value at (x, y), and W is a two-dimensional template. In the second-level filtering, for extracting the edge characteristic, the median filtering image is convoluted by a Laplace operator as a filter. The Laplace operator is as the Formula (9), and it can be calculated by the Formula (10) in the practical operation.

$$\triangle^2 g = \partial^2 g(x, y)/\partial x^2 + \partial^2 g(x, y)/\partial y^2 \tag{9}$$

$$\Delta^2 g = [g(x+1,y) + g(x-1,y) + g(x,y+1) + g(x,y-1)] - 4g(x,y) \quad (10)$$

The Laplace filtering value at can be calculated by the Formula (11).

$$h(x,y) = g(x,y) + c(\Delta^2 g) \quad (11)$$

where $c(\bullet) = 1$ when a mask centre is positive and $c(\bullet) = -1$ when a mask centre is negative.

3.2   Model Updating Adaptively

In the common model updating, an object model always is updated by the current tracking result for every frame. However, when the current tracking result is not credible, updating model can result in model drift. For reducing model drift as much as possible, an object model is updated based on the Formula 12.

$$C_t = \begin{cases} (1-\rho)C_{t-1} + \rho\hat{C}_t & , a_n - th \le r_t \le a_n + th \\ C_{t-1} & , otherwise \end{cases} \quad (12)$$

where C_t is the object model at frame t, C_{t-1} is the object model at frame $t-1$, $\hat{C}_t$ is the object feature of the current tracking result at frame t, and ρ is the updating rate. Only when the maximum response value r_t for correlation filter at frame t is satisfied with the adaptive condition $a_n - th \le r_t \le a_n + th$, C_t can be updated, otherwise C_t maintains the object model C_{t-1} at frame $t-1$, where a_n is the average of the maximum response values for n previous frames as the Formula (13), and th is a threshold for model updating.

$$a_n = \sum_{i=t-1}^{t-n} r_i/n \quad (13)$$

4   Experiments

For evaluating the overall performance of ECFT, the ECFT tracker was compared with 6 state-of-the-art trackers on 20 sequences. The 20 sequences were publicly available from OTB100 datasets [11]. For fair comparisons, the initial tracking positions and the ground truth positions of 20 sequences were publicly available. The 6 compared trackers were KCF [1], CSK [2], CXT [4], TLD [3], CNT [7], and MOSSE [5]. ECFT was implemented in MATLAB, running on the hardware platform with Intel Quad-Core i5-4590 3.3 GHz CPU and 4 GB memory. The median filter was a smooth space filter. The updating rate was set as 0.09. 10 previous response values were used for calculating a_n. The threshold was set as 0.2. The running speed of ECFT was approximately 150 frames per second. The comparisons were performed from both quantitative evaluation and visual evaluation. The experimental results showed that ECFT performed favorably against other 6 trackers on 20 challenging sequences in terms of accuracy and robustness.

4.1   Quantitative Evaluation

We performed the quantitative evaluations with 2 evaluation metrics: AUC and Precision [11]. An object in a frame was successfully tracked if the score is not less than $t = area(ROI_T \cap ROI_G)/area(ROI_T \cup ROI_G)$, where ROI_T and ROI_G were respectively the tracking bounding box and the ground truth bounding box. AUC was the area under curve of a success plot. Precision (20 px) was the precision when center location errors were less than 20 pixels. The Fig. 1 showed the success plots of OPE and the precision plots of OPE for 7 trackers on 20 sequences. From Fig. 1, AUC and Precision of ECFT were respectively 0.712 and 0.930, which were better than those of other 6 tracker.

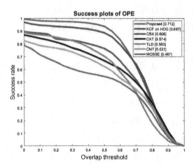
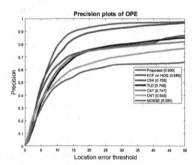

Fig. 1. Success plots and precision plots

Tables 1 and 2 respectively showed the evaluation results in terms of AUC and Precision for 7 trackers on 20 sequences. From Table 1, ECFT had the best AUC value for 10 sequences, and either KCF or CNT had the best AUC value only for 3 sequences. Moreover, ECFT had the best overall AUC on 20 sequences. From Table 2, ECFT had the best overall precision on 20 sequences, and KCF ranked the second.

We note that KCF has the highest performance in KCF, CSK, MOSSE, and MOSSE has the lowest performance, in part due to the complexity of features. However, the proposed algorithm (ECFT) does significantly improve performance.

The speed of the tracker is also critical in the performance comparison. The running speed of ECFT was approximately 150 frames per second, It's about the same speed as KCF. The running speed of TLD was approximately 20 frames per second, and the running speed of CNT was only approximately 1.5 frames.

4.2   Visual Evaluation

We have done a lot of experiments and listed the positioning results of the seven trackers in different videos. And the screenshots for some sample tracking results on several sequences was shown in Fig. 2. This involved in the results of multiple

Table 1. AUC. **Bold** fonts indicated the best performances

Algorithm	ECFT	KCF	CSK	CXT	TLD	CNT	MOSSE
BlurCar4	**0.818**	0.798	0.826	0.752	0.707	0.365	0.813
Board	0.608	**0.653**	0.445	0.298	0.093	0.612	0.119
Boy	**0.763**	0.762	0.646	0.539	0.653	0.485	0.303
Car2	0.680	0.678	0.685	0.855	0.714	**0.882**	0.609
Coke	**0.601**	0.547	0.565	0.423	0.399	0.386	0.425
Crowds	**0.797**	0.778	0.741	0.091	0.759	0.624	0.695
Dancer2	0.777	0.762	0.775	0.714	0.660	0.769	**0.778**
David2	0.625	0.813	0.807	**0.866**	0.684	0.843	0.478
Deer	**0.737**	0.612	0.733	0.690	0.590	0.030	0.689
Dudek	0.700	0.720	0.708	0.719	0.640	**0.770**	0.518
FaceOcc2	0.763	0.739	**0.767**	0.731	0.611	0.575	0.391
Fish	0.753	**0.824**	0.222	0.770	0.796	0.365	0.132
Football	**0.577**	0.547	0.545	0.541	0.489	0.484	0.568
Jumping	**0.745**	0.283	0.051	0.522	0.655	0.068	0.112
Man	**0.888**	0.813	0.861	0.837	0.735	0.832	0.853
Mhyang	0.785	0.783	0.782	**0.834**	0.627	0.733	0.423
MountainBike	0.694	0.700	0.702	0.223	0.198	**0.719**	0.700
Singer2	0.671	**0.721**	0.046	0.068	0.223	0.042	0.039
Skater2	**0.601**	0.563	0.588	0.407	0.362	0.448	0.122
Sylvester	0.652	0.636	0.625	0.594	**0.666**	0.589	0.574
Overall AUC	**0.712**	0.687	0.606	0.574	0.563	0.531	0.467

environments. In the experimental results, our tracker performed well especially when other trackers were drifting. For example, in Board sequence (e.g., #0036, #0069, #0119, #0276 and #0445), Fish sequence (e.g., #0027, #0044, #0084, #0136, #0161 and #0470), Jumping sequence (e.g., #0099, #0158, #0290, #0274, #0307 and #0313) and so on, our tracker can be still able to track continuously due to updating object model adaptively. In Sylvester sequence, before the No. 716 frame, all trackers can track the object well. Then, the model drifted from the tracked object in Sylvester (e.g., #1148, #1259, #1259, #1276, #1317 and #1345), and KCF, CXT, CNT and MOSSE lost the object. In Singer2 sequence (e.g., #0085, #150, #0246, #0349 and #0366), Skater2 sequence (e.g., #0247, #0337, #0406, #0415 and #0435), Board sequence (e.g.,#0036, #0069, #0119, #0276, #0445), and so on, Not only that, but in video Coke and Boy sequence (e.g., #0134, #0165, #0222, #0254 and #261) and (e.g., #0402, #0417, #551, #0557 and #593) have this characteristic. only ECFT can well overcome the model drift and keep a continuous tracking. In the Coke sequence, Before 134 frames, each tracker was able to track accuracy, and after about 165 frames, there

Table 2. Precision. **Bold** fonts indicated the best performances

Algorithm	ECFT	KCF	CSK	CXT	TLD	CNT	MOSSE
BlurCar4	**0.997**	**0.997**	1.000	0.803	**0.997**	0.105	0.995
Board	**0.686**	0.656	0.089	0.011	0.113	0.655	0.017
Boy	**1.000**	**1.000**	0.844	**1.000**	0.937	0.666	0.367
Car2	**1.000**	**1.000**	1.000	1.000	0.964	**1.000**	0.816
Coke	0.869	0.838	**0.873**	0.684	0.653	0.543	0.385
Crowds	**1.000**	**1.000**	1.000	1.000	0.127	**1.000**	1.000
Dancer2	**1.000**	**1.000**	1.000	0.853	**1.000**	**1.000**	1.000
David2	**1.000**	**1.000**	1.000	1.000	1.000	**1.000**	0.998
Deer	**1.000**	0.817	1.000	0.732	**1.000**	0.028	0.930
Dudek	0.779	0.877	0.807	0.597	0.822	**0.878**	0.575
FaceOcc2	**1.000**	0.972	1.000	0.856	**1.000**	0.631	0.482
Fish	**1.000**	**1.000**	0.042	1.000	1.000	0.118	0.055
Football	0.801	0.796	0.798	**0.804**	0.796	0.798	0.793
Jumping	**1.000**	0.339	0.051	1.000	0.997	0.102	0.179
Man	**1.000**	**1.000**	1.000	1.000	0.985	**1.000**	1.000
Mhyang	**1.000**	**1.000**	1.000	0.978	**1.000**	**1.000**	0.430
MountainBike	**1.000**	**1.000**	1.000	0.259	0.281	**1.000**	1.000
Singer2	0.866	**0.945**	0.036	0.071	0.063	0.036	0.036
Skater2	0.657	**0.694**	0.655	0.379	0.345	0.478	0.011
Sylvester	**0.949**	0.843	0.910	0.949	0.852	0.856	0.830
Overall Precision	**0.930**	0.889	0.755	0.749	0.747	0.645	0.595

was a gradual drift away from the other tracking devices we tracked. Moreover, The MOSSE, CNT and the CSK trackers started to drift early, but our algorithms have been tracking very high accuracy.

In the reference [14], Ma et al. proposed an online random fern classifier to re-detect objects in case of tracking failure,and it can realize long-term correlation tracking. This is an undeniable fact. In our method, we proposed the adaptive model updating technology, which also can realize long-term tracking. The feature of the proposed method can be well adapted to the correlation filters. For example, in Jumping sequence, the KCF tracker with a HOG feature fails after 158 frames. For the CNT tracker with deep feature, it not only processes very slowly, but the tracking success rate in Jumping sequence is very low. However, as shown in Fig. 2, our tracker can keep tracking of high accuracy all the time.

Fig. 2. Screenshots for some sample tracking results

5   Conclusions

In this paper, we proposed a correlation filters tracker based on two-level filtering edge feature (ECFT), improved based on KCF in the frame of correlation filters. For feature extractor, a candidate object can be represented by the edge feature vector based on two-level filtering. For model updating, an object model was updated adaptively when a current tracking result was credible. The comparative experiments of 7 trackers on 20 challenging sequences showed that the proposed ECFT tracker can perform well in terms of AUC and Precision.

Acknowledgement. The work is supported by Application Foundation Project of Yunnan Province of China (Grant No.: 2016FB103), National Natural Science Foundation of China (Grant Nos.: 61262067 and 11663007) and Personnel Training Project of Yunnan Province of China (Grant No.: KKSY201407074).

References

1. Henriques, J.F., Caseiro, R., Martins, P., et al.: High-speed tracking with kernelized correlation filters. IEEE Trans. Pattern Anal. Mach. Intell. **37**(3), 583–596 (2015)
2. Dinh, T.B., Vo, N., Medioni, G.: Context tracker: exploring supporters and distracters in unconstrained environments. In: IEEE Conference on Computer Vision and Pattern Recognition (CVPR), Colorado Springs, pp. 1177–1184 (2011)
3. Kalal, Z., Matas, J., Mikolajczyk, K.: PN learning: bootstrapping binary classifiers by structural constraints. In: IEEE Conference on Computer Vision and Pattern Recognition (CVPR), San Francisco, pp. 49–56 (2010)
4. Henriques, J.F., Caseiro, R., Martins, P., Batista, J.: Exploiting the circulant structure of tracking-by-detection with kernels. In: Fitzgibbon, A., Lazebnik, S., Perona, P., Sato, Y., Schmid, C. (eds.) ECCV 2012. LNCS, vol. 7575, pp. 702–715. Springer, Heidelberg (2012). https://doi.org/10.1007/978-3-642-33765-9_50
5. Bolme, D.S., Beveridge, J.R., Draper, B.A., et al.: Visual object tracking using adaptive correlation filters. In: IEEE Conference on Computer Vision and Pattern Recognition (CVPR), San Francisco, pp. 2544–2550 (2010)
6. Kristan, M., Matas, J., Leonardis, A., et al.: The visual object tracking VOT2015 challenge results. In: IEEE International Conference on Computer Vision Workshop, pp. 564–586 (2016)
7. Zhang, K., Liu, Q., Wu, Y., Yang, M.H.: Robust visual tracking via convolutional networks without training. IEEE Image Process. **25**(4), 1779–1792 (2016)
8. Wang, N., Shi, J., Yeung, D.Y., et al.: Understanding and diagnosing visual tracking systems. In: IEEE Conference on Computer Vision (ICCV), Santiago, pp. 3101–3109 (2015)
9. Babenko, B., Yang, M.H., Belongie, S.: Visual tracking with online multiple instance learning. In: IEEE Conference on Computer Vision and Pattern Recognition (CVPR), USA, pp. 983–990 (2009)
10. Zhang, K., Zhang, L., Yang, M.-H.: Real-time compressive tracking. In: Fitzgibbon, A., Lazebnik, S., Perona, P., Sato, Y., Schmid, C. (eds.) ECCV 2012. LNCS, vol. 7574, pp. 864–877. Springer, Heidelberg (2012). https://doi.org/10.1007/978-3-642-33712-3_62
11. Wu, Y., Lim, J., Yang, M.H.: Object tracking benchmark. IEEE Trans. Pattern Anal. Mach. Intell. **37**(9), 1834–1848 (2015)
12. Kiani Galoogahi, H., Sim, T., Lucey, S.: Multi-channel correlation filters. In: IEEE Conference on Computer Vision (ICCV), Sydney, pp. 3072–3079 (2013). (CVPR), Portland, pp. 2291–2298 (2013)
13. Li, Y., Zhu, J.: A scale adaptive kernel correlation filter tracker with feature integration. In: Computer Vision (ECCV), Zurich, pp. 254–265 (2014)
14. Ma, C., Yang, X., Zhang, C., Yang, M.H.: Long-term correlation tracking. In: IEEE Conference on Computer Vision and Pattern Recognition (CVPR), Boston, pp. 5388–5396 (2015)

15. Naresh Boddeti, V., Kanade, T., Vijaya Kumar, B.V.K.: Correlation filters for object alignment. In: IEEE Conference on Computer Vision and Pattern Recognition (CVPR), Portland, pp. 2291–2298 (2013)
16. Vedaldi, A., Gulshan, V., Varma, M., Zisserman, A.: Multiple kernels for object detection. In: IEEE Conference on Computer Vision (ICCV), Kyoto, pp. 606–613 (2009)
17. Liu, T., Wang, G., Yang, Q.: Real-time part-based visual tracking via adaptive correlation filters. In: IEEE Conference on Computer Vision and Pattern Recognition (CVPR), Boston, pp. 4902–4912 (2015)
18. Danelljan, M., Hager, G., Khan, S.F., Felsberg, M.: Learning spatially regularized correlation filters for visual tracking. In: IEEE Conference on Computer Vision and Pattern Recognition (CVPR), Boston, pp. 4310–4318 (2015)
19. Ma, C., Huang, J.B., Yang, X., Yang, M.H.: Hierarchical convolutional features for visual tracking. In: IEEE Conference on Computer Vision and Pattern Recognition (CVPR), Boston, pp. 3074–3082 (2015)
20. Zhang, K., Zhang, L., Liu, Q., Zhang, D., Yang, M.-H.: Fast visual tracking via dense spatio-temporal context learning. In: Fleet, D., Pajdla, T., Schiele, B., Tuytelaars, T. (eds.) ECCV 2014. LNCS, vol. 8693, pp. 127–141. Springer, Cham (2014). https://doi.org/10.1007/978-3-319-10602-1_9

An Accelerated Superpixel Generation Algorithm Based on 4-Labeled-Neighbors

Hongwei Feng, Fang Xiao, Qirong Bu$^{(\boxtimes)}$, Feihong Liu,
Lei Cui, and Jun Feng$^{(\boxtimes)}$

School of Information and Technology, Northwest University, Xi'an 710127, China
fenghw@foxmail.com, {boqirong,fengjun}@nwu.edu.cn,
{xiaofang,child,chrislei}@stumail.nwu.edu.cn

Abstract. Superpixels are perceptually meaningful atomic regions that could effectively improve efficiency of subsequent image processing tasks. Simple linear iterative clustering (SLIC) has been widely used for superpixel calculation due to outstanding performance. In this paper, we propose an accelerated SLIC superpixel generation algorithm using 4-labeled neighbor pixels called 4L-SLIC. The main idea of 4L-SLIC is that the labels are assigned to a portion of the pixels while the others that associated with certain cluster are restrained by adjacent four labeled pixels. In this way, the average number of distance calculated times of pixels are effectively reduced and the similarity between adjacent pixels ensures a better segmentation effect. The experimental results confirm that 4L-SLIC achieved a speed up of 25%–30% without declining accuracy sharply compared to SLIC. In contrast to the method published on CVIU 2016, 4L-SLIC has an acceptable increase in the cost of time, in the mean time, there is a significant ascension to the accuracy of the segmentation.

Keywords: Superpixels · SLIC · Acceleration · 4-labeled neighbors

1 Introduction

Superpixel algorithms [1] can effectively extract the features of perceptually meaningful regions in images. Simple linear iterative clustering (SLIC) [2] is the most popular superpixel framework. Follow the pipeline of SLIC, there are many state-of-the-art superpixel algorithms which are widely used in various computer vision applications, such as image segmentation [1], object recognition [3], motion segmentation [4], 3D reconstruction [5] and so on.

However, the superpixel algorithms can't be applied to some tasks since the time consumption is very high. In order to satisfy the requirement of real-time processing, Choi and Oh proposed an accelerative strategy by introducing 2-labeled neighbors (2L-SLIC) verification which can reduce the processing time as less as a half [6]. But this method suffers from the accuracy since all four neighbors are similar to their central pixel with equal probability, and the usage of only 2-labeled neighbors may lose half part of information.

© Springer Nature Singapore Pte Ltd. 2017
J. Yang et al. (Eds.): CCCV 2017, Part I, CCIS 771, pp. 539–550, 2017.
https://doi.org/10.1007/978-981-10-7299-4_45

In this paper, we propose an accelerated version of superpixel algorithm in SLIC framework based on 4-labeled neighbors which is called 4L-SLIC. Using the similarity in neighbors, 4L-SLIC only searches the half pixels directly, and neglects the remaining half which labels are estimated by their neighbors. In this way, 4L-SLIC reduces the maximum search times near halfly while keeps the accuracy of general SLIC. Our algorithm is verified on the public Berkeley segmentation dataset and benchmark. The experiments shows that the speed of 4L-SLIC is increased almost 25%–30% than that of SLIC and slightly slower than 2L-SLIC. In the mean time, the proposed algorithm has a competitive accuracy with SLIC and is more accurate than 2L-SLIC.

2   A New SLIC Algorithm Based on 4-Labeled Neighbors

In this section, we firstly review the basic SLIC algorithms and analysis of computational redundancy in the algorithm. Afterwards, we propose a novel superpixel algorithm called 4L-SLIC based on 4-labeled neighbors to more efficiently generate superpixels. Through this algorithm, half of the pixels in the image are calculated by a drastically decrease. In the final, we analysis for how much the 4L-SLIC reduces the number of cluster searches has made compared to SLIC.

2.1   The Analysis of Computational Redundancy in SLIC

In general, SLIC can be roughly divided into four steps: initialization, cluster assignment, update and postprocessing. Given an input image T of size $R \times C = N$ and the only default parameter of SLIC is K which represent the number of superpixels.

In the initialization step, K initial cluster centers $f_{C_k} = [l_k, a_k, b_k, x_k, y_k]^T$ are sampled on a regular grid spaced S regions, $S = \sqrt{N/K}$ represent the distance between the adjacent initial clusters. f_{C_k} is a feature vector consisting of both three colors in the CIELAB color space and position. Similarly, a pixel i could be expressed as $f_i = [l_i, a_i, b_i, x_i, y_i]^T$.

In the clustering step, each pixel i is associated with the nearest cluster center whose search region overlaps its location [7]. The size of the search region was set to $2S \times 2S$ around the cluster center like Fig. 1(a). Achanta et al. use $d_5(i, C_k)$ denotes the similarity between the pixel i and cluster C_k in the $labxy$ color-image plane space [2], which is defined as :

$$d_5(i, C_k) = \sqrt{d_3(i, C_k)^2 + \lambda^2 \cdot d_2(i, C_k)^2} \tag{1}$$

$$d_3(i, C_k) = \sqrt{(l_k - l_i)^2 + (a_k - a_i)^2 + (b_k - b_i)^2} \tag{2}$$

$$d_2(i, C_k) = \sqrt{(x_k - x_i)^2 + (y_k - y_i)^2} \tag{3}$$

Which $d_3(i, C_k)$ and $d_2(i, C_k)$ respectively indicate normalized Euclidean distances in color and spatial space. And $\lambda = m/S$, to combine the two distances into a single measure, m represents the normalization factor to control the compactness.

When all the pixels are labeled, the update step is executed. Each f_{C_k} is updated in accordance with their average of pixels in the cluster. The iteration is continued until the residual error E between the cluster center of this iteration and the previous cluster center is less than the threshold T_e.

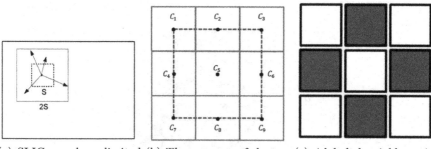

(a) SLIC searchs a limited region. (b) The coverage of clustering center in SLIC. (c) 4-labeled neighbor pixels.

Fig. 1. Different search regions between SLIC and 4L-SLIC. (Color figure online)

Although SLIC through limiting the size of search region to significantly reduce the number of distance calculations, we found that there still has computational redundancy [8]. For example, In Fig. 1(b), each blue region represents the initial superpixel block, while the red region represents the search range of the $2S \times 2S$ of C_5. For a pixel i belonging to C_5, it is likely to be covered by the search range of many clusters. However, the most extreme case is when i and C_5 are coincide, i will be covered by the search of nine clusters C_1 to C_9. In this case, i requires nine distance calculations. Thus, in SLIC, the number of calculations per pixel is $1 \sim 9$ times. However, the interpixel correlation is not considered in the SLIC algorithm.

Generally, there is a considerable spatial correlation between adjacent pixels, so-called interpixel correlation, which is one of the important characteristic of images that is often overlooked. The interpixel correlation can be expressed as if a pixel i is associated with a certain cluster, then its neighboring pixels have a highly trend of belonging to this cluster. For example, in Fig. 1(c), the block indicate a pixel. If the four red pixels are already labeled, and the middle white pixel are likely to be the same as one of the four red pixels according to the interpixel correlation. If this is done, the number of calculations per pixel will be reduced to a maximum of four times. So our algorithm could increase the speed of generating superpixels.

2.2 SLIC Acceleration with 4-Labeled Neighbor Pixels

In this subsection, we propose a more efficient labeling strategy in the clustering step based on interpixel correlation. In our algorithm, half of the pixels greatly reduced the number of distance calculations, resulting in faster generation of superpixel.

Before describing the proposed algorithm in detail, we first declare a few notations. According to a certain rule, divide all the pixels in the image into two parts as α and β respectively. Let θ_i, ω_i, $L(i)$ and $P(C_k)$ denote a set of cluster centers whose search domains contain i, the set of 8-connected neighbor pixels around i, a label associated with i, and the set of all cluster centers, respectively.

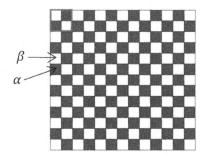

Fig. 2. α:red blocks, β:white blocks, and α:$\beta = 2$:2 for each four blocks e.g. the blue region. (Color figure online)

The process and accelerated theory are described below:

(1) At first, as demonstrated in Fig. 2, where a block indicate a pixel, from both left to right or top to bottom, we subsampled every second pixel from the first pixel as β, and the remaining pixels as α.

(2) Then, compared to the SLIC algorithm which labeled all pixel in the image, our algorithm only label α to their closest clusters by calculating distance of $d_5(i, C_k)$, but at this time β has no label yet. Therefore, except for the first row and column, the last row and column, the others pixels in β have four neighbor pixels that are already labeled. It is noteworthy that the number of

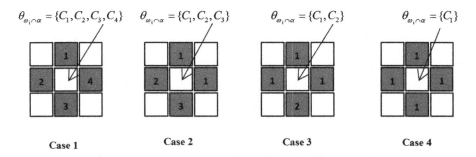

Fig. 3. Case 1: the four-connected pixel labels of i are not all the same. Case 2: three different labels around i. Case 3: only two different labels around i. Case 4: four labels are the same.

label $|L(\omega_i \cap \alpha)|$ is usually less than $|\theta_i|$ in which $i \in \beta$. Because the number of searches for θ_i is usually $1 \sim 9$ times and $|L(\omega_i \cap \alpha)|$ is 4 at most.

(3) Subsequently, the label is assigned to β using 4-labeled neighbor pixels as show in Fig. 3. It is guaranteed that, for each pixel $i \in \beta$, some of its four-connected neighboring pixels have been associated with a cluster. The figures in the red block indicates the label of this pixel. Then, for the in-between white pixel i, the label of its four-connected neighboring pixels $\theta_{\omega_i \cap \alpha}$ can have four different cases. In the Case 1, the four-connected pixels of i have four different labels. The distance metric is only need to calculate four times between i and cluster centers $\{C_1, C_2, C_3, C_4\}$ in which i is assigned to the nearest cluster center. It can be found that the number of distance calculations for each pixel is changed from 9 times in SLIC algorithm to 4 times at most. Likewise, Case 2 and Case 3 only need to calculate distance three times and twice. This has achieved drastically computation reduction. However, Case 4 always occurs inside a superpixel and all the pixels have the identical label in α. In this case, i is associated with the certain cluster without any distance calculation. It is notable that this situation usually occurs and can further significantly reduce execution time.

(4) The first cluster has been completed when all the pixels belonging to β are labeled. Executing the same update and iteration as SLIC until the convergence conditions are satisfied T_e.

2.3   Complexity Analysis of 4L-SLIC

In the previous section, we point out that our algorithm reduces the number of pixels calculation. In this subsection, we analyze how much the 4L-SLIC reduces the number of cluster searches compared to SLIC.

We use mathematical expectation to represent the number of distance calculations per pixel in each algorithm. In SLIC, the cluster search is performed the same number of times as $|\theta_i|$, even $|\theta_i| = 1$ is available. However, in the 2L-SLIC and 4L-SLIC, only one quarter and one half of the pixels are calculated by the same number times as SLIC while the other pixels only need to calculate $0 \sim 4$ times. The expectation of SLIC, 2L-SLIC and 4L-SLIC is simply expressed as Eqs. (4), (5) and (6) respectively :

$$E_{SLIC} = E[|\theta_i|] = \sum_{c=1}^{9} p(|\theta_i| = c) \cdot c \tag{4}$$

$$E_{2L-SLIC} = \frac{|\alpha|}{|N|} E_{SLIC} + \frac{|\beta|}{|N|} E[|\theta_{\omega_i \cap \alpha}|] \tag{5}$$

$$= \frac{1}{4} E[|\theta_i|] + \frac{3}{4} E[|\theta_{\omega_i \cap \alpha}|]$$

$$= \frac{1}{4} \sum_{c=1}^{9} p(|\theta_i| = c) \cdot c + \frac{3}{4} \sum_{c=2}^{4} p(|\theta_{\omega_i \cap \alpha}| = c) \cdot c$$

$$E_{4L-SLIC} = \frac{|\alpha|}{|N|}E_{SLIC} + \frac{|\beta|}{|N|}E[|\theta_{\omega_i \cap \alpha}|] \tag{6}$$

$$= \frac{1}{2}E[|\theta_i|] + \frac{1}{2}E[|\theta_{\omega_i \cap \alpha}|]$$

$$= \frac{1}{2}\sum_{c=1}^{9} p(|\theta_i| = c) \cdot c + \frac{1}{2}\sum_{c=2}^{4} p(|\theta_{\omega_i \cap \alpha}| = c) \cdot c$$

Based on these three formulas, we can compute the average number of calculations for each pixel in a different algorithm. The complexity of three algorithms is linear in the number of pixels, which can be expressed as $O(NIE_{SLIC})$, $O(NIE_{2L-SLIC})$, $O(NIE_{4L-SLIC})$ respectively, where I is the number of iterations required for convergence [9]. From the complexity we can seen, the smaller expectation, the more efficient of algorithm.

The proposed 4L-SLIC algorithm is summarized in Algorithm 1.

3   Experimental Results

We implemented 4L-SLIC in Matlab and tested it on a PC with an Intel I5-4590 CPU (3.60 GHz) and 8 GB RAM. We tested with the Berkeley database which containing three-hundred 321×481 images data with ground truth segmentation. Due to the literature [2] has demonstrated that SLIC already overcomed other all conventional algorithms. So in this paper, we mainly compared 4L-SLIC with SLIC and 2L-SLIC algorithm.

3.1   The Number of Pixels in Clustering Calculation

In order to demonstrate how much computing redundancy that 4L-SLIC has reduced, statistics of the calculations number for each pixel in SLIC and 4L-SLIC were summarized at Tables 1 and 2 respectively. Each row showed the ratio of between the number of pixels counted to c and the total number of pixels with K block. The last column E showed the expectation calculated by Eqs. (4) and (6), which representing the average number of calculations for each pixel in the image.

According to the statistics in Table 1, when the traditional cluster search is used, an average of 81.57% pixels needed to be calculated at $4 \sim 6$ times. More clusters should be inspected for some pixels within complex patterned regions, especially for pixels located along the image boundary. Finally, the average number of calculations per pixel in the SLIC is $E = 4.37$.

The statistical information for the number of pixels are calculated in 4L-SLIC was shown in Table 2. It was notable that about 92.92% of the pixels only need to be calculated by 0 or 2 times which drastically reduces the execution times. Therefore, in the 4L-SLIC, the cluster inspection for each pixel $i \in \beta$ is only performed $E = 0.965$.

Algorithm 1. 4L-SLIC

Input: image T,initial cluster interval S,regularization factor m

/*Initialization*/
1. Initialize cluster centers $f_{C_k} = [l_k, a_k, b_k, x_k, y_k]^T$ by sampling pixels at regular grid steps S.
2. Move cluster centers to the lowest gradient position within their 3∗3 neighborhood.
3. Set label $L(i) = -1$ for each pixel i.

/*Cluster Assignment*/
for $i \in \alpha$, **do**
    4. $L(i) \leftarrow C_{k*} \leftarrow \underset{C_k \in \theta_i}{\operatorname{argmin}} d_5(i, C_k)$
end for
for $i \in \beta$, **do**
    if $|\theta_{\omega_i \cap \alpha}|$ is greater than 1, **then**
        for $unique(\theta_{\omega_i \cap \alpha})$, **do**
            $5 - 1.$ $L(i) \leftarrow C_{k*} \leftarrow \underset{C_k \in \theta_{\omega_i \cap \alpha}}{\operatorname{argmin}} d_5(i, C_k)$
    end for
    else
        $5 - 2.$ $L(i) \leftarrow L(j)$, where $j \in \omega_i \cap \alpha$
    end if
end for

/*Update*/
6. Update f_{C_k} with $mean(P(C_k)), k = 1, \cdots, K$
7. Compute residual error E.
8. Repeat Steps 4, 5 until $E \leq T_e$. // T_e is a threshold for iteration

/*Postprocessing*/
9. Merge small clusters into one of the adjacent clusters. termination.

Output: superpixel results $P(C_k), k = 1, \cdots, K$

Table 1. Numbers of computation for different superpixel scales in SLIC.

c	1	2	3	4	5	6	7	8	9	E
$K = 100$	0.018	0.130	0.177	0.405	0.166	0.099	0.005	0	1.5E−05	3.888
$K = 300$	0.006	0.073	0.113	0.464	0.184	0.147	0.011	0.002	0	4.242
$K = 500$	0.003	0.054	0.099	0.465	0.183	0.181	0.012	0.003	0	4.377
$K = 800$	0.002	0.040	0.070	0.459	0.198	0.203	0.022	0.006	0	4.538
$K = 1000$	0.002	0.042	0.075	0.520	0.172	0.175	0.011	0.003	4.5E−06	4.402
$K = 1500$	0.001	0.027	0.045	0.418	0.213	0.242	0.039	0.015	0	4.772
Avg	0.005	0.061	0.097	0.455	0.186	0.175	0.017	0.005	3.3E−06	4.370

Table 2. Numbers of computation for different superpixel scales in 4L-SLIC.

c	1	2	3	4	E
$K = 100$	0.699	0.268	0.032	0.001	0.636
$K = 300$	0.602	0.344	0.052	0.002	0.852
$K = 500$	0.557	0.377	0.063	0.003	0.955
$K = 800$	0.513	0.407	0.077	0.003	1.057
$K = 1000$	0.502	0.415	0.081	0.004	1.083
$K = 1500$	0.453	0.442	0.100	0.005	1.204
Avg	0.554	0.376	0.067	0.003	0.965

According to Eqs. (4), (5) and (6), we can conclude the number of operations in these three algorithms as follows.

$$E_{SLIC} = 4.37$$

$$E_{2L-SLIC} = \frac{1}{4} \times 4.37 + \frac{3}{4} \times 0.965 = 1.8163$$

$$E_{4L-SLIC} = \frac{1}{2} \times 4.37 + \frac{1}{2} \times 0.965 = 2.6675$$

We can see that $E_{2L-SLIC}$ and $E_{4L-SLIC}$ have a great improvement compared to E_{SLIC}. Although $E_{4L-SLIC}$ was more than $E_{2L-SLIC}$, the two algorithms was still on the same order of magnitude in computation time. Moreover, 4L-SLIC segmentation effect was much better than 2L-SLIC which will be confirmed in the following experimental results.

We demonstrated the validity of the algorithm from a visual view point in Fig. 4. We used four different colors (Yellow: 1, Green: 2, Red: 3, Blue: 4) to depict the number of pixels are calculated that belong to β on the image after ten iterations running the 4L-SLIC with setting $K = 100$ and $m = 10$. From Fig. 4, we can clearly observe that almost all pixels inside superpixels are represented in Yellow, whereas it was hard to find Blue ones. Green and Red pixels usually appeared along superpixel boundaries and on junctions with three branches, respectively.

3.2   Computational Efficiency of 4L-SLIC

Figure 5 showed the 4L-SLIC and 2L-SLIC contrast to SLIC in terms of segmentation speed improvement. In order to verify the pure algorithm performance, all algorithms runed without any parallel hardware and compiler optimization.

Since three algorithms were all based on iterative clustering algorithm, the execution time was related to the number of iterations, and both quantities changed with similar tendency. As shown in Fig. 5, although the 2L-SLIC was

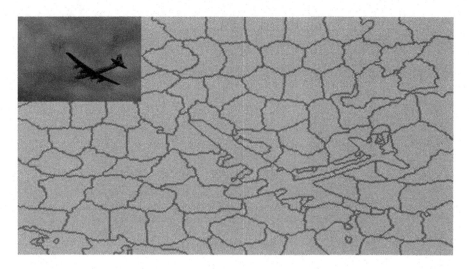

Fig. 4. Visual effect of the number of operations in 4L-SLIC. (Color figure online)

faster than the 4L-SLIC, these two algorithms were still in the same order of magnitude and the acceleration ratio was both about 30% compared to SLIC.

We can found further that with the increase in the number of superpixels K, the execution time of 4L-SLIC was getting closer to 2L-SLIC. The reason was that the more the number of superpixels, the more boundaries will be separated. That was to say there were more pixels need to be calculated many times. Therefore, the calculating time of the two algorithms was getting closer.

3.3   Segmentation Performance of 4L-SLIC

Superpixels were commonly used as a preprocessing step in image segmentation. A good superpixel algorithm should improve the performance of segmentation. So boundary adherence was also an important factor for evaluating the performance of superpixel segmentation algorithms.

Boundary Recall measured the fraction that ground truth boundaries correctly recovered by the superpixel boundaries. A high BR indicated that very few true boundaries are missed. Figure 6 showed the boundary recall curve. When $K < 1000$, 4L-SLIC was slightly lower than SLIC, whereas when $K > 1000$, 4L-SLIC has a better scores. However, compared with these two algorithms, the segmentation effect of 2L-SLIC was not as good as ours.

As shown in Figs. 7 and 8 where visual comparison of superpixels generated by all three algorithms. Figure 7 showed the result of segmentation on a simple image and Fig. 8 on a complex image. It was hard to find any difference result between SLIC and 4L-SLIC. But the boundary adherence of the regions that were circled in red is relatively poor at Fig. 7(b).

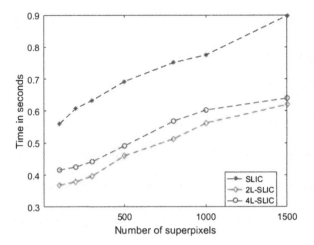

Fig. 5. Comparison of three algorithmic execution times.

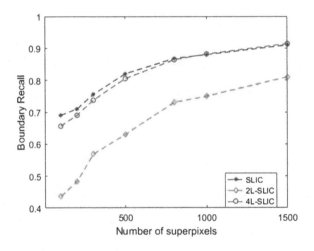

Fig. 6. Comparison of boundary recall of three algorithm.

(a) SLIC   (b) 2L-SLIC   (c) 4L-SLIC

Fig. 7. Segmentation effect in simple image.

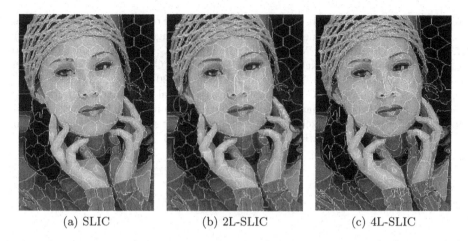

(a) SLIC   (b) 2L-SLIC   (c) 4L-SLIC

Fig. 8. Segmentation effect in complex image.

4 Conclusion

We proposed an accelerated version of superpixel algorithm which is named 4L-SLIC. 4L-SLIC reduces half search times without reducing any accuracy. We evaluated the experiments on the Berkeley public dataset. The time consumption results show that the speed of 4L-SLIC ups of 25%–30% than SLIC and 3%–4% to 2L-SLIC. As for the accuracy results, they show that our algorithm has a competitive performance with SLIC and is more accurate than 2L-SLIC. Our future work will focus on the accelerative vision of 4L-SLIC by parallel implementation.

Acknowledgments. This work was supported by National Natural Science Foundation of China (No. 61602380).

References

1. Malik, J.: Learning a classification model for segmentation. In: Proceedings of International Conference on Computer Vision, vol. 1, pp. 10–17 (2003)
2. Achanta, R., Shaji, A., Smith, K., et al.: SLIC superpixels compared to state-of-the-art superpixel methods. IEEE Trans. Pattern Anal. Mach. Intell. **34**(11), 2274 (2012)
3. Li, Z., Wu, X.M., Chang, S.F.: Segmentation using superpixels: a bipartite graph partitioning approach. In: IEEE Conference on Computer Vision and Pattern Recognition, pp. 789–796. IEEE Computer Society (2012)
4. Ayvaci, A., Soatto, S.: Motion segmentation with occlusions on the superpixel graph. In: IEEE International Conference on Computer Vision Workshops, pp. 727–734. IEEE (2009)
5. Mičušík, B., Košecká, J.: Multi-view superpixel stereo in urban environments. Int. J. Comput. Vis. **89**(1), 106–119 (2010)

6. Choi, K.S., Oh, K.W.: Subsampling-based acceleration of simple linear iterative clustering for superpixel segmentation. Comput. Vis. Image Underst. **146**, 1–8 (2016)
7. Liu, Y.J., Yu, C.C., Yu, M.J., et al.: Manifold SLIC: a fast method to compute content-sensitive superpixels. In: IEEE Conference on Computer Vision and Pattern Recognition, pp. 651–659. IEEE Computer Society (2016)
8. Lv, J.: An Improved SLIC superpixels using reciprocal nearest neighbor clustering. Int. J. Signal Process. Image Process. Pattern Recognit. 8 (2015)
9. Li, Z., Chen, J.: Superpixel segmentation using linear spectral clustering. In: Computer Vision and Pattern Recognition, pp. 1356–1363. IEEE (2015)

A Method of General Acceleration SRDCF Calculation via Reintroduction of Circulant Structure

Xiaoxiang Hu[✉] and Yujiu Yang

Shenzhen Key Laboratory of Broad-band Network and Multimedia,
Graduate School at Shenzhen, Tsinghua University, Beijing, China
huxx15@mails.tsinghua.edu.cn, yang.yujiu@sz.tsinghua.edu.cn

Abstract. Discriminatively learned correlation filters (DCF) have been widely used in online visual tracking filed due to its simplicity and efficiency. These methods utilize a periodic assumption of the training samples to construct a circulant data matrix, which is also introduces unwanted boundary effects. Spatially Regularized Correlation Filters (SRDCF) solved this issue by introducing penalization on correlation filter coefficients. However, which breaks the circulant structure used in DCF. We propose Faster SRDCF (FSRDCF) via reintroduction of circulant structure. The circulant structure of training samples in the spatial domain is fully used, more importantly, we exploit the circulant structure of regularization function in the Fourier domain, which allows the problem to be solved more directly and efficiently. Our approach demonstrates superior performance over other non-spatial-regularization trackers on the OTB2013 and OTB2015.

Keywords: Object tracking · Correlation filter · Circulant structures

1 Introduction

Visual tracking is one of the core problems in the field of computer vision with a variety of applications. Generic visual tracking is to estimate the trajectory of a target in an image sequence, given only its initial state. It is difficult to design a fast and robust tracker from a very limited set of training samples due to various critical issues in visual tracking, such as occlusion, fast motion and deformation.

Recently, Discriminative Correlation Filter (DCF) [1] has been widely used in visual tracking because of its simplicity and efficiency, and there are many improvements [2,5–7,12] about DCF to address the above mentioned problems. These methods learn a correlation filer from a set of training samples to encode the targets appearance. Nearly all correlation filter based trackers utilize the circulant structure of training samples proposed in work [10]. The circulant structure allows the correlation filter training and target detection computation efficiently. However, this structure also introduces unwanted boundary effects that

© Springer Nature Singapore Pte Ltd. 2017
J. Yang et al. (Eds.): CCCV 2017, Part I, CCIS 771, pp. 551–562, 2017.
https://doi.org/10.1007/978-981-10-7299-4_46

leads to an inaccurate appearance mode. To address the boundary effects problem, Danelljan et al. propose Spatially Regularized Correlation Filters (SRDCF) [6]. The SRDCF introduces penalization to force the correlation filters to concentrate on center of the training patches. This penalization allows the tracker to be trained on a larger area without the effect of background, so the SRDCF can handle some challenging cases such as fast target motion. However, the penalization makes the correlation filters complex to solve, which is unacceptable for an online visual tracking situation. In this work, we revisit the SRDCF, in our formulation, we make full use of the circulant structure to simplify the problem.

In this paper, we propose Faster Spatially Regularized Discriminative Correlation Filters (FSRDCF) for tracking. The circulant structure of training data matrix in the spatial domain is utilized, besides, we exploit the circulant structure of regularization matrix in the Fourier domain. Our approach more computation efficient without any significant degradation in performance and more suitable for online tracking problems. To validate our approach, we preform comprehensive experiments on the most popular benchmark datasets: OTB-2013 [18] with 50 sequences and OTB-2015 [19] with 100 sequences. Our approach obtains a more than twice faster running speed and faster startup time than the baseline tracker SRDCF and achieves state-of-the-art performance over other non-spatial-regularization trackers.

2   Spatially Regularized DCF

After Bolme et al. [1] first introduced the MOSSE filter, lots of notable improvements [5–7,12,13] are proposed from different aspects to strengthen the correlation filter based trackers. New features have been widely used, such as HOG [4], Color-Name [17] and deep features [13,15]; feature integration is also used [11]. To address occlusion, part-based trackers [12] are widely adapted. However, the periodic assumption also produced unwanted boundary effects. Galoogahi et al. [8] investigate the boundary effect issue, their method removes the boundary effects by using a masking matrix to allow the size of training patches larger than correlation filters. They use Alternative Direction Method of Multipliers (ADMM) to solve their problem and have to make transitions between spatial and Fourier domain in every ADMM iteration, which increasing the trackers computational complexity. To get rid of those transitions, Danelljan et al. [6] propose the spatially Regularized Correlation filters (SRDCF), they introduce a spatial weight function to penalize the magnitude of the correlation filter coefficients, and use Gauss-Seidel method to solve the filters, in this way, both the boundary effects and the transitions in [8] are avoided. The work [3] also use a spatial regularization like the SRDCF and derived a simplified inverse method to get a closedform solution. However, these methods still have a relatively high computational complexity. In our proposal, we apply a spatial regularization like [3,6], by exploiting circulant structure of train samples in the spatial domain and regularization matrix in the Fourier domain, our formulation has no problem transition, correlation filters needed are solved directly.

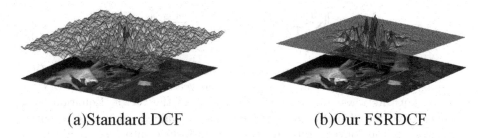

(a)Standard DCF   (b)Our FSRDCF

Fig. 1. Visualization of the filter coefficients trained by the standard DCF (a) and our FSRDCF (b). The top layer is the learned filter corresponding to the bottom layer training patch. Target is outlined by the green rectangle. As we can see, our filter puts more attention on the object than the standard DCF.

Standard SRDCF Training and Detection. The way to handle the boundary effects in the SRDCFis most popular in literature. So we give some details of the convolution filters training and Detection after introduced the regularization. In this section, we use the term convolution, because the SRDCF are modeled with convolution instead of correlation, we will give some key differences used in our proposal in Sects. 3 and 4. Convolution filters are learned from a set of training samples $\{(x_k, y_k)\}_{k=1}^t$. Every training sample $x_k \in \mathbb{R}^{d \times M \times N}$ consists of a d-channel feature map with spatial size of $M \times N$ extracted from a training image patch. We use x_k^l to represent the lth feature layer of x_k. y_k is the optimal convolution output corresponding to training sample x_k. The Spatially Regularized Correlation Filters (SRDCF) is obtained from the convex problem,

$$\min_f \sum_{k=1}^t \alpha_k \|S_f(x_k) - y_k\|^2 + \sum_{l=1}^d \|w \odot f^l\|^2. \tag{1}$$

where the $\alpha_k \geq 0$ is the weight of every training sample x_k, spatial regularization is introduced by w, which is a Gaussian shaped function with smaller values in center area and bigger values in marginal area, $\odot$ denotes the element-wise multiplication. $S_f(x_k)$ is the convolution function,

$$S_f(x_k) = \sum_{l=1}^d x_k^l * f^l. \tag{2}$$

where $*$ denotes the circular convolution. With the use of Parseval's theorem and convolution property, the Eq. 1 is transformed into Fourier domain,

$$\min_{\hat{f}} \sum_{k=1}^t \alpha_k \left\| \sum_{l=1}^d \hat{x}_k^l \odot \hat{f}^l - \hat{y}_k \right\|^2 + \sum_{l=1}^d \left\| \frac{\hat{w}}{MN} * \hat{f}^l \right\|^2. \tag{3}$$

where the hat denotes Discrete Fourier Transformed (DFT) of a variable. For convenience, all variables in Eq. 3 are vectorized, convolution is transformed into matrix multiplication,

$$\min_{\hat{f}} \sum_{k=1}^{t} \alpha_k \left\| \sum_{l=1}^{d} \mathcal{D}(\hat{x}_k^l)\hat{f}^l - \hat{y}_k \right\|^2 + \sum_{l=1}^{d} \left\| \frac{\mathcal{C}(\hat{w})}{MN}\hat{f}^l \right\|^2. \qquad (4)$$

Here, bold letters are the corresponding variables' vectorization form, $\mathcal{D}(\mathbf{v})$ is a diagonal matrix with the elements of the vector in its diagonal. $\mathcal{C}(\hat{\mathbf{w}})$ is a matrix with its rows consist of all of the shift of the vector. Equation 4 is a complex convex problem, because the DFT of a real-valued function is Hermitian symmetric, so the convex problem (4) can transformed into a re-al-valued one by a unitary matrix $B \in \mathbb{R}^{MN \times MN}$,

$$\min_{\tilde{f}} \sum_{k=1}^{t} \alpha_k \left\| \sum_{l=1}^{d} D_k^l \tilde{f}^l - \tilde{y}_k^l \right\|^2 + \sum_{l=1}^{d} \left\| C\tilde{f}^l \right\|^2. \qquad (5)$$

Here, $D_k^l = B\mathcal{D}(\hat{\mathbf{x}}_k^l)B^{\mathrm{H}}$, $\tilde{\mathbf{f}}^l = B\hat{\mathbf{f}}^l$, $\tilde{\mathbf{y}}_k = B\hat{\mathbf{y}}_k$ and $C = \frac{1}{MN}BC(\hat{\mathbf{w}})B^{\mathrm{H}}$, where the ^{H} denotes the conjugate transpose of a matrix. Then concatenate all layers of training data and convolution filters, in other words, $\tilde{\mathbf{f}} = ((\tilde{\mathbf{f}}^1)^T, \cdots, (\tilde{\mathbf{f}}^d)^T)^T$ and $D_k = (D_k^1, \cdots, D_k^d)$, Eq. 5 is simplified as,

$$\min_{\tilde{f}} \sum_{k=1}^{t} \alpha_k \left\| D_k \tilde{\mathbf{f}} - \tilde{\mathbf{y}}_k \right\|^2 + \left\| W \tilde{\mathbf{f}} \right\|^2. \qquad (6)$$

where $W \in \mathbb{R}^{dMN \times dMN}$ is a block diagonal matrix with its diagonal blocks being equal to C. Letting the derivative of Eq. 6 with respected to $\tilde{\mathbf{f}}$ be zero,

$$\left(\sum_{k=1}^{t} \alpha_k D_k^{\mathrm{H}} D_k + W^{\mathrm{H}}W \right)\tilde{\mathbf{f}} = \sum_{k=1}^{t} \alpha_k D_k^{\mathrm{H}} \tilde{\mathbf{y}}_k. \qquad (7)$$

Due to the sparsity of D_k and W, problem (7) can be efficiently solved by Gauss-Seidel method with the computational complexity of $\mathcal{O}((d + K^2)dMNN_{GS})$, where the K is the number of non-zero entries in $\hat{w}$, the N_{GS} is the number of Gauss-Seidel iterations.

Excluding the feature extraction, the total computational complexity of SRDCF tracker is $\mathcal{O}(dSMN\log(MN) + SMNN_{NG} + (d+K^2)dMNN_{GS})$. Here, S denotes the number of scales in the scaling pool, N_{NG} is the number of Newton iterations in sub-grid detection. It's worth noting that the result takes none of the transformations into consideration, especially from Eqs. 4 to 5, which including high dimensional matrix multiplication. In reality, those transformations are time consuming. In our approach, all of them will be by-passed. We'll directly get the correlation filters in Eq. 8.

3   Faster SRDCF

We revisit spatially regularization correlation filters for tracking from Ridge Regression viewpoint. In our proposal, problem is solved more directly by exploiting both circulant structure in training data and regularization function.

Our proposal is to find a function $g(z) = \boldsymbol{f}^{\mathrm{T}} z$ to minimizes the squared error over all training samples x_k (d-channel feature map) and their regression target y_k,

$$\min_{\boldsymbol{f}} \sum_{k=1}^{t} \alpha_k \left\| \sum_{l=1}^{d} X_k^l \boldsymbol{f}^l - \boldsymbol{y}_k^l \right\|^2 + \sum_{l=1}^{d} \left\| \mathcal{D}(\boldsymbol{w}) \boldsymbol{f}^l \right\|^2 \qquad (8)$$

For simplicity, we let $\boldsymbol{y}_k^l = \boldsymbol{y}_k$. In general Ridge Regression problem, each row of X_k^l is a vectorized training sample, here, rows of X_k^l consist of all circular shift of $\boldsymbol{x}_k^l$, $X_k^l \boldsymbol{f}^l$ is the correlation between $\boldsymbol{x}_k^l$ and $\boldsymbol{f}^l$, it's worth noting that $X_k^l \boldsymbol{f}^l \neq vec(x_k^l * \boldsymbol{f}^l)$, where $vec(v) = \boldsymbol{v}$. Now we can directly take derivative of Eq. 1 with respected to $\boldsymbol{f}$ and let the derivative be zero, then we get,

$$\sum_{l=1}^{d} \left(X_k^{l\,\mathrm{H}} X_k^l + \mathcal{D}(\boldsymbol{w})^{\mathrm{H}} \mathcal{D}(\boldsymbol{w}) \right) \boldsymbol{f}^l = \sum_{l=1}^{d} X_k^{l\,\mathrm{H}} \boldsymbol{y}_k^l \qquad (9)$$

Because all variables in Eq. 8 are real-valued, so we use $(*)^{\mathrm{T}}$ and $(*)^{\mathrm{H}}$ equivalently. Due to the circulant structure of X_k^l, we have,

$$X_k^l = \ddot{\mathrm{F}} \mathcal{D}(\hat{\boldsymbol{x}}_k^l) \ddot{\mathrm{F}}^{\mathrm{H}} \qquad (10)$$

where $\ddot{\mathrm{F}} = \mathrm{F} \otimes \mathrm{F}$ is two-dimensional DFT matrix for vectorized two-dimensional signals, F is known as DFT matrix, $\otimes$ denotes the Kronecker product. Both $\ddot{\mathrm{F}}$ and F are constant matrix and unitary. We apply Eqs. 10 to 9,

$$\sum_{l=1}^{d} \mathrm{F} \left(\mathcal{D}(\hat{\boldsymbol{x}}_k^{l\,*} \odot \hat{\boldsymbol{x}}_k^l) + \mathrm{F}^{\mathrm{H}} \mathcal{D}(\boldsymbol{w})^{\mathrm{H}} \mathcal{D}(\boldsymbol{w}) \mathrm{F} \right) \mathrm{F}^{\mathrm{H}} = \sum_{k=1}^{t} \sum_{l=1}^{d} \mathrm{F} \mathcal{D}(\hat{\boldsymbol{x}}_k^{l\,*}) \ddot{\mathrm{F}}^{\mathrm{H}} \boldsymbol{y}_k^l \qquad (11)$$

We can see that because of the introduction of regularization w, a simple closed-form solution can't be obtained from Eq. 11 like the way in work [10]. So far, we only use the circulant structure of training data matrix X_k^l in the spatial domain. From now on, we will further exploit the circlulant structure of the regularization matrix $\mathcal{D}(\boldsymbol{w})$.

From Eq. 10, we can know that a spatially circulant matrix can be diagonalized by the matrix of $\ddot{\mathrm{F}}$ in Fourier domain, however, from another point of view, we can also get,

$$\ddot{\mathrm{F}}^{\mathrm{H}} X_k^l \ddot{\mathrm{F}} = \mathcal{D}(\hat{\boldsymbol{x}}_k^l) \qquad (12)$$

The first row of X_k^l is equal to $\mathcal{F}^{-1}(\hat{\boldsymbol{x}}_k^l)$, all other rows are the circular shifts of $\mathcal{F}^{-1}(\hat{\boldsymbol{x}}_k^l)$. So if we have a diagonal matrix, then we can transform it to a circulant matrix by $\ddot{\mathrm{F}}$ In Eq. 11, $\mathcal{D}(\boldsymbol{w})$ is a real-valued diagonal matrix, so we have

$$\mathrm{F}^{\mathrm{H}} \mathcal{D}(\boldsymbol{w})^{\mathrm{H}} \mathcal{D}(\boldsymbol{w}) \mathrm{F} = \mathrm{F} \mathcal{D}(\boldsymbol{w})^{\mathrm{H}} \mathcal{D}(\boldsymbol{w}) \mathrm{F}^{\mathrm{H}} = \mathrm{R}^{\mathrm{H}} \mathrm{R} \qquad (13)$$

where R is circulant matrix constructed from $\mathcal{F}(\boldsymbol{w})$, where $\mathcal{F}$ denotes DFT. For a real-valued function, unitary DFT and IDFT have the same results, here, we treat $\mathcal{D}(\boldsymbol{w})$ as a spatial domain signal, so we use DFT instead of IDFT. If we choose a real-valued even regularization function w, therefore, $\mathcal{F}^{-1}(\boldsymbol{w})$ is a real-valued vector, then we will get a real-valued regularization matrix. Applying Eqs. 13 to 11, we have,

$$\sum_{k=1}^{t}\sum_{l=1}^{d}\mathrm{F}\left(\mathcal{D}(\hat{\boldsymbol{x}}_k^{l\,*}\odot\hat{\boldsymbol{x}}_k^l)+\mathrm{R}^{\mathrm{H}}\mathrm{R}\right)\ddot{\mathrm{F}}^{\mathrm{H}}\boldsymbol{f}^l=\sum_{k=1}^{t}\sum_{l=1}^{d}\mathrm{F}\mathcal{D}(\hat{\boldsymbol{x}}_k^{l\,*})\ddot{\mathrm{F}}^{\mathrm{H}}\boldsymbol{y}_k^l \qquad (14)$$

Here, we call R the potential circulant structure of regularization function w in the Fourier domain. By using the unitary property, we can further get,

$$\sum_{k=1}^{t}\sum_{l=1}^{d}\left(\mathcal{D}(\hat{\boldsymbol{x}}_k^{l\,*}\odot\hat{\boldsymbol{x}}_k^l)+\mathrm{R}^{\mathrm{H}}\mathrm{R}\right)(\ddot{\mathrm{F}}\ddot{\mathrm{F}})^{\mathrm{H}}\hat{\boldsymbol{f}}^l=\sum_{k=1}^{t}\sum_{l=1}^{d}\mathcal{D}(\hat{\boldsymbol{x}}_k^{l\,*})(\ddot{\mathrm{F}}\ddot{\mathrm{F}})^{\mathrm{H}}\hat{\boldsymbol{y}}_k^l \qquad (15)$$

where $\ddot{\mathrm{F}}\ddot{\mathrm{F}}=(\mathrm{F}\otimes\mathrm{F})(\mathrm{F}\otimes\mathrm{F})=(\mathrm{FF})\otimes(\mathrm{FF})$ is a permutation matrix. To simplify Eq. 15, we define $\hat{\boldsymbol{f}}=\left((\hat{\boldsymbol{f}}^l)^{\mathrm{T}},\cdots,(\hat{\boldsymbol{f}}^d)^{\mathrm{T}}\right)^{\mathrm{T}}$, $\hat{\boldsymbol{x}}_k=\left((\hat{\boldsymbol{x}}_k^l)^{\mathrm{T}},\cdots,(\hat{\boldsymbol{x}}_k^d)^{\mathrm{T}}\right)^{\mathrm{T}}$, then the Eq. 15 can be equivalently expressed as,

$$\sum_{k=1}^{t}\left(\mathcal{D}(\hat{\boldsymbol{x}}_k^*\odot\hat{\boldsymbol{x}}_k)+\mathrm{B}^{\mathrm{H}}\mathrm{B}\right)\mathcal{P}(\hat{\boldsymbol{f}})=\sum_{k=1}^{t}\hat{\boldsymbol{x}}_k^*\odot\mathcal{P}(\hat{\boldsymbol{y}}_k^l) \qquad (16)$$

where B is $dMN\times dMN$ block diagonal matrix with each diagonal block being equal to R, $\mathcal{P}(\cdot)$ is a permutation function according to $\ddot{\mathrm{F}}\ddot{\mathrm{F}}$. In Sect. 4, we will find that we just need to find the solution of $\mathcal{P}(\hat{\boldsymbol{f}})=\hat{\boldsymbol{f}}_p$ instead of $\hat{\boldsymbol{f}}$, so what we really used equation is,

$$\sum_{k=1}^{t}\left(\mathcal{D}(\hat{\boldsymbol{x}}_k^*\odot\hat{\boldsymbol{x}}_k)+\mathrm{B}^{\mathrm{H}}\mathrm{B}\right)\hat{\boldsymbol{f}}_p=\sum_{k=1}^{t}\hat{\boldsymbol{x}}_k^*\odot\mathcal{P}(\hat{\boldsymbol{y}}) \qquad (17)$$

Because we use the same regression targets for all frames, we define $\hat{\boldsymbol{y}}=\left((\hat{\boldsymbol{y}}^1)^{\mathrm{T}},\cdots,(\hat{\boldsymbol{y}}^d)^{\mathrm{T}}\right)^{\mathrm{T}}$. Equation 17 defines a $dMN\times dMN$ linear system of equations, its coefficients matrix is real-valued, so we can solve it directly. We can see that for each frame what we need to do is element-wise multiplication. The regularization part $\mathrm{B}^{\mathrm{H}}\mathrm{B}$ and regression targets part $\mathcal{P}(\hat{\boldsymbol{y}})$ is constant for all frames.

In Eq. 17, if we choose $\mathrm{B}^{\mathrm{H}}\mathrm{B}=\lambda I$, which is equivalent to choose a regularization matrix w with all elements being equal to $\sqrt{\lambda}$, we get a standard DCF. In this paper, w is a real-valued even Gaussian shaped function, profiting from its smooth property, we get a sparse coefficients matrix for Eq. 17, which makes significant difference for optimization. The Gauss-Seidel method is used to solve Eq. 17. Figure 1 shows the visualization of the filter solved from our proposal.

4   Our Tracking Framework

In this section, we describe our tracking framework according to the Faster Spatially Regularized Discriminative Correlation Filters proposed in Sect. 3.

4.1   Training

At the first frame, to give a better initial point for Gauss-Seidel methods, we obtain a precise solution of Eq. 17 by,

$$\hat{\mathbf{f}}_{p1} = \left(\mathcal{D}(\hat{\mathbf{x}}_1^* \odot \hat{\mathbf{x}}_1) + \mathbf{B}^H\mathbf{B}\right)^{-1}\left(\hat{\mathbf{x}}_1^* \odot \mathcal{P}(\hat{\mathbf{y}})\right) \tag{18}$$

In the subsequent frames, the starting point in current frame (time t) is the optimization results in last frame (time $t-1$). For simplicity, we define A_t and b_t as,

$$A_t = \sum_{k=1}^{t}\left(\mathcal{D}(\hat{\mathbf{x}}_k^* \odot \hat{\mathbf{x}}_k) + \mathbf{B}^H\mathbf{B}\right) \tag{19}$$

$$\mathbf{b}_t = \sum_{k=1}^{t}\hat{\mathbf{x}}_k^* \odot \mathcal{P}(\hat{\mathbf{y}}) \tag{20}$$

Equation 17 can be rewrite as $A_t\hat{\mathbf{f}}_p = \mathbf{b}_t$, we split A_t into data part D_t and regularization part $\mathbf{B}^H\mathbf{B}$, split $\mathbf{b}_t$ into $\mathbf{d}_t$ and $\mathcal{P}(\hat{\mathbf{y}})$, then we update our model by,

$$A_t = (1-\gamma)D_{t-1} + \gamma\mathcal{D}(\hat{\mathbf{x}}_t^* \odot \hat{\mathbf{x}}_t) + \mathbf{B}^H\mathbf{B} \tag{21}$$

$$\mathbf{b}_t = \left((1-\gamma)\mathbf{d}_{t-1} + \gamma\hat{\mathbf{x}}_t^*\right) \odot \mathcal{P}(\hat{\mathbf{y}}) \tag{22}$$

where $D_1 = \mathcal{D}(\hat{\mathbf{x}}_1^* \odot \hat{\mathbf{x}}_1)$, $\mathbf{d}_1 = \hat{\mathbf{x}}_1^*$. In Eqs. 21 and 22, $\mathbf{B}^H\mathbf{B}$, $\mathcal{P}(\hat{\mathbf{y}})$ are constant during model updating. We just need to precompute once for a sequence. To get new correlation filters $\hat{\mathbf{f}}_p$, a fixed N_{GS} numbers of Gauss-Seidel iterations are conducted after model updating Eqs. 21 and 22.

4.2   Detection

At the detection stage, according Eq. 8, the location of the target is estimated by finding the peak correlation between correlation filters $\hat{\mathbf{f}}_p$ and new feature maps,

$$\max \sum_{l=1}^{d} X_t^l \mathbf{f}^l = \max \sum_{l=1}^{d} x_t^l \star f^l \tag{23}$$

where $\star$ denotes the correlation operator. For the sake of computational efficiency, we use convolution instead of correlation,

$$\max \sum_{l=1}^{d} x_t^l \star f^l = \max \sum_{l=1}^{d} x_t^l * (-f^l) \tag{24}$$

where $-$ denotes 180° rotation operator. Both x_t and f are real-valued, so their DFT is Hermitian symmetric. Computing Eq. 24 in Fourier domain, finally we get,

$$\max \mathcal{F}^{-1}\left(\sum_{l=1}^{l}\hat{x}_t^l \odot \hat{\mathbf{f}}_p^l\right) \tag{25}$$

So in Eq. 16, we directly solve $\mathcal{P}(\hat{\mathbf{f}}) = \hat{\mathbf{f}}_p$ instead of $\hat{\mathbf{f}}$. At detection stage, we also use the scaling pool technique in paper [16] and Fast Sub-grid Detection in work [6].

5   Experiments

On benchmark datasets: OTB2013, OTB2015. To make a fair comparison with the baseline tracker SRDCF [6], we use most of the parameters used in SRDCF throughout all our experiments. Because we need to get real-valued DFT coefficients of the regularization function w and use Matlab as our implementation tool, so we reconstructed w from function $w(m, n) = \mu + \eta(m/\sigma_1)^2 + \eta(n/\sigma_2)^2$, where $[\sigma_1, \sigma_2] = \beta[P, Q]$, $P \times Q$ is the size of target. $m = [-(M/2) : (M-2)/2)]$, $n = [-(N/2) : (N-2)/2)]$. In our experiments, β is set to 0.8. Our source code is fully available at the githuab website[1].

5.1   Baseline Comparison

We do comprehensive comparison between our approach and the baseline tracker SRDCF [6]. Accuracy, robustness and speed are taken into consideration. In this section, all experiments are performed on a standard desktop computer with Intel Core i5-6400 processor. For the baseline tracker, we use the Matlab implementation provided by authors.

Our comparison follows the protocol proposed in [18]. One-Pass Evaluation (OPE), Temporal Robustness Evaluation (TRE) and Spatial Robustness Evaluation (SRE) are performed. TRE runs trackers on 20 different length subsequences segmented from the original sequences, SRE run trackers with 12 different initializations constructed from shifted or scaled ground truth bounding box. After running the trackers, we report the overall results using the area under the curve (AUC) based on success plot and mean overlap precision (OP). Besides, attribute-based evaluation results are also reported. The OP is calculated as the percentage of frames where the intersection-over-union overlap with the ground truth exceeds a threshold of 0.5. Attributes are including scale variation (SV), occlusion (OCC), deformation (DEF), fast motion (FM), in-plane-rotation (IPR), out-plane-rotation (OPR), background cluster (BC) and low resolution (LR), illumination variation (IV), out of view (OV), motion blur (MB).

In Table 1, our tracker runs at 11.4 fps, 11.1 fps on OTB-2013 and OTB-2015 datasets respectively, which are twice faster than the SRDCF on both datasets. At the same time, our approach start up much faster which is very important when tracker needs to be initialized frequently. Table 2 shows OPE results on OTB-2013 and OTB-2015, for clarity, we reported 9 attribute-based evaluation results. For overall performance, our approach outperforms the baseline tracker by 1.1%, 2.7% in AUC and OP respectively on OTB-2013 dataset and achieves

---

[1] https://github.com/KnockKnock13/FSRDCF.git.

Table 1. The comparison of speed and start-up time on OTB-2013 and OTB-2015. Our approach runs more than twice fast as the baseline tracker SRDCF and a faster startup time.

	OTB-2013		OTB-2015	
	Speed (fps)	Start-up (s)	Speed (fps)	Start-up (s)
SRDCF	5.7	1.27	5.3	1.36
FSRDCF	11.4	0.51	11.1	0.50

Table 2. Robustness evaluation comparison on OTB-2013 and OTB-2015 datasets. Both trackers achieve equivalent results in TRE on OTB-2015 dataset.

		OTB-2013									Overall	
		SV	OCC	DEF	OV	IPR	OPR	BC	LR	IV	AUC	OP
OTB13	SRDCF	59.5	62.7	63.5	55.5	57.3	60.4	58.7	42.6	57.6	63.0	78.9
	FSRDCF	61.8	63.6	65.4	54.9	59.6	63.0	61.9	42.1	58.6	64.4	81.6
OTB15	SRDCF	56.9	55.7	54.7	46.1	54.6	55.1	58.4	48.1	60.9	59.9	73.1
	FSRDCF	56.4	56.1	54.3	45.6	56.4	56.3	59.0	46.8	60.8	59.5	73.4

equivalent performance on OTB-2015 dataset. For attribute-based evaluation, our method wins in most attribute sub-datasets on OTB- 2013 dataset. Both trackers have no significant difference on OTB-2015 dataset. Table 3 shows the robustness evaluation results. Except for our tracker is a little bit more sensitive to the background due to bigger derivation parameters, we think two trackers have the same robustness performance.

5.2 OTB-2013 and OTB-2015 Datasets

Finally, we perform a comprehensive comparison with 9 recent state-of-art trackers: DLSSVM [16], SCT4 [2], MEEM [20], KCF [10], DSST [5], SAMF [11], LCT [14], Struck [9] and the baseline tracker SRDCF [6].

State-of-the-Art Comparison. We show the results of comparison with state-of-the-art trackers on OTB-2013 and OTB-2015 datasets over 100 videos in Table 4, only the results for the top 8 trackers are reported in consideration of space. The results are presented in OP and ranking according to performance on OTB-2015 dataset. In Table 4, we also give the running speed of trackers. The best results on both datasets are obtained by our tracker with mean OP of 81.6%, and 73.4%, outperforming the best non-spatial regularization trackers by 8.4% and 6.3% respectively. From the perspective of running speed, our approach runs at 11.1 frames per second, which is more than twice faster than the tracker ranking the second. Our tracker gets a better balance between accuracy and efficiency.

Table 3. Comparison with baseline tracker on OTB-2013 and OTB-2015. The results in the table are based on success plot and all reported in percent.

	OTB-2013		OTB-2015	
	SRE	TRE	SRE	TRE
SRDCF	0.569	0.647	0.542	0.613
FSRDCF	0.557	0.650	0.522	0.607

Table 4. State-of-the-art trackers comparison on OTB-2013 and OTB-2015 datasets in OP (in percent). The best two results are shown in red and blue respectively. Our approach achieves the best results on both datasets and have a balanced perform on accuracy and speed.

	DSST	SCT4	DLSSVM	MEEM	LCT	SAMF	SRDCF	FSRDCF
OTB-2013	66.7	73.9	72.5	70.6	73.8	73.2	78.9	81.6
OTB-2015	61.3	62.0	62.4	62.7	62.9	67.1	73.1	73.4
Speed (fps)	29.4	32.2	9.5	8.2	20.3	16.8	5.3	11.1

Figure 3 shows the success plots on OTB-2013 and OTB-2015 datasets. The trackers are ranked according to the area under the curve (AUC) and displayed in the legend. Our tracker ranks the first on OTB-2013 with a AUC of 64.4%, outperforming the best non-spatial regularization tracker by 5.1%, and ranks the second with a AUC of 59.5%, outperforming the best non-spatial-regularization tracker by 4.3% on OTB-2015.

Robustness Comparison. Like in Sect. 5.1, we perform SRE and TRE to compare the robustness of our tracker to the state-of-the-art trackers. Figure 2 shows success plots for SRE and TRE on OTB-2015 dataset with 100 videos. Our approach outperforms the best non-spatialregularization tracker 1% and 2% in SRE and TRE respectively.

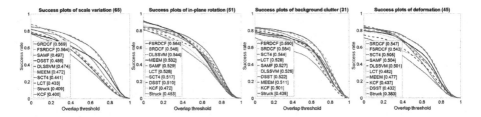

Fig. 2. Attribute-based evaluations of our approach on OTB-2015 dataset. Number in bracket of each plot title is the videos in corresponding sub-dataset. Our tracker demonstrates superior performance compared to other non-spatial regularization trackers.

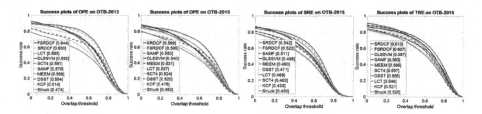

Fig. 3. Success plots showing a comparison with state-of-the-art trackers on OTB-2013 and OTB-2015 datasets. Our FSRDCF ranks the first on OTB-2013 and the second on OTB-2015. Robustness to initialization comparison on the OTB-2015 dataset. Success plots for both SRE and TRE are shown, our tracker achieves state-of-the-art performance.

Attribute Based Comparison. We performed attribute-based evaluations of our approach on OTB-2015 and compare to other state-of-the-art trackers. Our approach wins on 10 attribute sub-datasets compared to other non-spatial-regularization trackers, Fig. 2 shows the success plots of 4 different attributes on OTB-2015 dataset. Due to the using of spatial regularization, the spatially regularized trackers can learn more discriminative filters and detect targets from a larger area than standard DCF, so our tracker have big advantages in situations such as occlusion, background cluster and fast motion over other non-spatial-regularization trackers.

6   Conclusion

We introduce a new formulation of Spatially Regularized Discriminative Correlation Filters (FSRDCF) to efficiently learn a spatially regularized correlation filer. The use of circulant structure of data matrix in the spatial domain and circulant structure of regularization function in the Fourier domain significantly simplify the problem construction and solving. In our approach, both problem construction and solving are in the spatial domain. Our approach validated on the OTB-2013 and OTB-2015 datasets, and obtains a more than twice faster running speed and much faster start-up time than the baseline tracker SRDCF without any performance degradation. At the same time, our approach demonstrates superior performance compared to other non-spatial-regularization trackers.

Acknowledgments. This work was supported in part by the ShenZhen Research Fund for the development of strategic emerging industries (Nos. JCYJ20160301151844537 and CXZZ20150504163316885). In addition, we thank the anonymous reviewers for their careful read and valuable comments.

References

1. Bolme, D.S., Beveridge, J.R., Draper, B.A., Lui, Y.M.: Visual object tracking using adaptive correlation filters. In: IEEE CVPR, pp. 2544–2550 (2010)

2. Choi, J., Chang, H.J., Jeong, J., Demiris, Y., Choi, J.Y.: Visual tracking using attention-modulated disintegration and integration. In: IEEE CVPR, pp. 4321–4330 (2016)

3. Cui, Z., Xiao, S., Feng, J., Yan, S.: Recurrently target-attending tracking. In: IEEE CVPR, pp. 1449–1458 (2016)

4. Dalal, N., Triggs, B.: Histograms of oriented gradients for human detection. In: IEEE CVPR, pp. 886–893 (2005)

5. Danelljan, M., Häger, G., Khan, F.S., Felsberg, M.: Accurate scale estimation for robust visual tracking. In: IEEE BMVC (2014)

6. Danelljan, M., Häger, G., Khan, F.S., Felsberg, M.: Learning spatially regularized correlation filters for visual tracking. In: IEEE ICCV, pp. 4310–4318 (2015)

7. Danelljan, M., Häger, G., Khan, F.S., Felsberg, M.: Adaptive decontamination of the training set: a unified formulation for discriminative visual tracking. In: IEEE CVPR, pp. 1430–1438 (2016)

8. Galoogahi, H.K., Sim, T., Lucey, S.: Correlation filters with limited boundaries. In: IEEE CVPR, pp. 4630–4638 (2015)

9. Hare, S., Saffari, A., Torr, P.H.S.: Struck: structured output tracking with kernels. In: IEEE ICCV, pp. 263–270 (2011)

10. Henriques, J.F., Caseiro, R., Martins, P., Batista, J.: High-speed tracking with kernelized correlation filters. IEEE Trans. PAMI **37**(3), 583–596 (2015)

11. Li, Y., Zhu, J.: A scale adaptive kernel correlation filter tracker with feature integration. In: IEEE ECCV, pp. 254–265 (2014)

12. Liu, T., Wang, G., Yang, Q.: Real-time part-based visual tracking via adaptive correlation filters. In: IEEE CVPR, pp. 4902–4912 (2015)

13. Ma, C., Huang, J., Yang, X., Yang, M.: Hierarchical convolutional features for visual tracking. In: IEEE ICCV, pp. 3074–3082 (2015)

14. Ma, C., Yang, X., Zhang, C., Yang, M.: Long-term correlation tracking. In: IEEE CVPR, pp. 5388–5396 (2015)

15. Nam, H., Han, B.: Learning multi-domain convolutional neural networks for visual tracking. In: IEEE CVPR, pp. 4293–4302 (2016)

16. Ning, J., Yang, J., Jiang, S., Zhang, L., Yang, M.: Object tracking via dual linear structured SVM and explicit feature map. In: IEEE CVPR, pp. 4266–4274 (2016)

17. van de Weijer, J., Schmid, C., Verbeek, J.J., Larlus, D.: Learning color names for real-world applications. IEEE TIP **18**(7), 1512–1523 (2009). https://doi.org/10.1109/TIP.2009.2019809

18. Wu, Y., Lim, J., Yang, M.: Online object tracking: a benchmark. In: IEEE CVPR, pp. 2411–2418 (2013)

19. Wu, Y., Lim, J., Yang, M.: Object tracking benchmark. IEEE Trans. PAMI **37**(9), 1834–1848 (2015). https://doi.org/10.1109/TPAMI.2014.2388226

20. Zhang, J., Ma, S., Sclaroff, S.: MEEM: robust tracking via multiple experts using entropy minimization. In: IEEE ECCV, pp. 188–203 (2014)

Using Deep Relational Features
to Verify Kinship

Jingyun Liang, Jinlin Guo$^{(\boxtimes)}$, Songyang Lao, and Jue Li

School of Information System and Management,
National University of Defense Technology, Changsha 410072, China
gjlin99@nudt.edu.cn

Abstract. Kinship verification from facial images is a very challenging research topic. Differing from most of previous methods focusing on calculating a similarity metric, in this work, we utilize convolutional neural network and autoencoder to learn deep relational features for verifying kinship from facial images. Specifically, we firstly train a convolutional neural network to extract representative facial features, which derive from the last fully-connected layer in network. Then, facial features from two person are set as two ends of an autoencoder respectively, and relational features are extracted from the middle layer of the trained autoencoder. Finally, SVM classifiers are adopted to verify kinship (e.g., Father-Son). Experimental results on two public datasets show the effectiveness of the approach proposed in this work.

Keywords: Kinship verification · Convolutional neural network
Autoencoder · Relational feature

1 Introduction

Compared with other human identification features, face is much more direct and humanly recognizable. People can learn about considerable information from human faces, such as gender, age and expression. In computer vision, face analysis and recognition is also one of the hotspots, attractive but challenging. And great progress has been made in recent years, including face recognition [1], expression recognition [2], age estimation [5], etc.

According to knowledge of genetics, individual genes are inherited from parents. Therefore, offsprings tend to resemble parents with respect to their appearance, especially face appearance, as shown in Fig. 1. Kinship verification aims to analyzing the consanguinity from facial images automatically. In practice, it can be used widely, such as helping to build family tree, search lost children and estimate appearance.

However, kinship verification is very challenging and only a few attempts, e.g. [4,10,15,16,18,20], have been made in recent years. Besides extracting features and recognizing faces for merely one person in traditional face recognition tasks, kinship verification is supposed to compare two images and calculate the

© Springer Nature Singapore Pte Ltd. 2017
J. Yang et al. (Eds.): CCCV 2017, Part I, CCIS 771, pp. 563–573, 2017.
https://doi.org/10.1007/978-981-10-7299-4_47

Defined Father-Son kinship

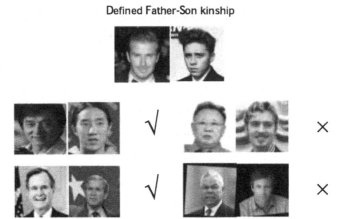

Fig. 1. Examples of Father-Son kinship. Top images give an example of Father-Son kinship. Bottom-left images are positive samples with Father-Son kinship, and bottom-right images are negative samples without kinship.

similarity of them comprehensively. The key problem lies in the judgment of a relationship, more specifically, a defined kinship. Due to the impact of action, illumination, expression and backgrounds, extracting stable and useful features become extremely hard. When it comes to a pair of images, gender and age can be influential factors as well. Moreover, qualifying the similarity of two person under uncontrolled conditions is also very difficult. Representing a special kinship requires a superior feature. A specific kinship and its corresponding image pairs is a typical one-to-many relationship, and for image pairs who share an identical kinship, the relational images are not completely consistent with each other.

Recently, deep learning has become increasingly popular in artificial intelligence and achieved notably significant success in computer vision. In well-known ImageNet Large Scale Visual Recognition Competition [12], methods based on deep learning report the best performance. Two typical models are convolutional neural network and autoencoder. In this work, we put forward combining these two models to learn relational features for kinship verification. Firstly, we build and train a convolutional neural network using YouTube Faces Dataset [14]. Then, it is used to extract facial features from KinFaceW-I and KinFaceW-II datasets [8], which contain distinctive characteristics and maintain useful information from a facial image. After processing all paired images, image pairs are transformed to facial feature pairs. A designed autoencoder is then called on top of that. Relational features can be learned during the training of the autoencoder. Such relational features are distinctive for indicating a defined kinship. Finally, supported vector machine (SVM) classifiers are utilized to verify if two persons exist kinship relation from their facial image pair. Contributions of this work are as follows: (1) we propose a method for extracting deep relational feature, and

(2) we put forward combining convolutional neural network and autoencoder for kinship verification.

The remainder of the paper is organized as follows: In Sect. 2, we give a brief review of related work in kinship verification. Section 3 describes the main method proposed in this work. Section 4 presents experiments used to evaluate the approach, and furthermore, comparison with several existed models are also provided. Finally, we summarize our work in Sect. 5.

2   Related Work

2.1   Kinship Verification

Kinship verification from facial images is a very challenging research topic. It was first proposed by Fang *et al.* [3], and most existing datasets at that time were of small sizes and lacking standard evaluation protocols. In 2014, Lu *et al.* developed the first standardized benchmark in the first kinship verification competition [8]. Performance of different methods can be compared from then on.

Many papers deal with the challenge by features. The main idea is to extract discriminative features and use classifiers in succession. In [3], Fang *et al.* combines low level features, including color, facial parts, facial distance, gradient histograms, to a feature vector and applies K-Nearest-Neighbors and Support Vector Machine methods to train the classifier. Zhou *et al.* use Viola-Jones face detector [20] while Guo *et al.* use salient features like DAISY descriptor [4].

Other researchers try to measure similarities of face samples by subspace or metric learning. In an early attempt, Xia *et al.* develop a transfer subspace learning based algorithm and reduce the significant differences between children and the elderly [16]. Lu *et al.* [8] propose a neighborhood repulsed metric learning method to exploit more discriminative information. Later, in [17], Yan *et al.* highlight kin relations by using correlation similarity metric rather than Euclidian similarity metric. Zhou *et al.* improve the method by salable similarity learning [22], and make it better using ensemble similarity learning [21]. Still, there are shortcomings in the framework. Using low-level features and shallow learning model, it is not powerful or robust enough to reveal the intrinsic similarity between parents and children. There is very few attempts to verify kinship using deep learning, such as [19]. Thus, we try to solve the problem by making use of deep learning's strong ability of feature learning.

2.2   Deep Learning

The conception of deep learning was first proposed by [6] in 2006, which kicks off the wide usage in artificial intelligence and pattern recognition. In 2012, [7] builds a convolution neural network called AlexNet in LSVRC [12]. AlexNet outperforms all other models dramatically and leads to explosive development of deep learning.

Convolutional neural network (CNN) takes the lead in computer vision in recent years owing to its effectiveness. In terms of facial images, [13] learns a set of high-level feature taken from the last hidden layer neuron activation of their network, with verification accuracy even better than human. Features learned are learned automatically from data rather than human-designed, which allows CNN to extract features effectively and intrinsically. Due to its incredible learning ability and robustness, we use CNN as a feature extractor in our model. Only [19] has attempt to study kinship verification with CNN.

Autoencoder is an unsupervised learning algorithm and often used to learn an approximation to the identity function [11]. Based on the encoder-decoder architecture, the model can reconstruct the input and extract hierarchical features from hidden layers. In view of kinship verification, which focus on relational feature learning, we creatively use two different images as the input and output of the autoencoder respectively. Hopefully, autoencoder will transform one image to another image compulsively. The parameters in autoencoder will then reflect the relationship of two images.

3   Method

In this section, we give a detailed description of our approach by dividing the model into three parts. The first part is extracting facial features (FF) from raw images to represent individuals through convolutional neural network. The second part is learning deep relational feature (DRF) from facial feature pairs $((FF^{(1)}, FF^{(2)}))$ by autoencoder, which is the core and emphasis of our method. Finally, a SVM classifier is trained to verify kinship. Figure 2 gives the flowchart of learning deep relational features.

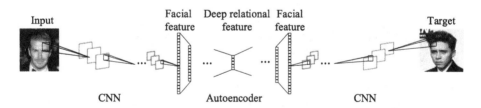

Fig. 2. The flowchart for learning relational features using our model. The input and output are two different images with or without kinship. Facial features are extracted using our designed convolutional neural network respectively. Relational feature can then be drawn from the autoencoder.

3.1   Facial Feature Extraction

Given facial images of two person, in order to eliminate noises and accelerate the process of relational feature learning, image features FF needs to be extracted to represent original images. Here, convolutional neural network (CNN) is adopted to learn image features.

The core of CNN is the operation of convolution and pooling [7]. The output $y[m,n]$ of two-dimensional convolution is formulated as

$$y[m,n] = x[m,n] * h[m,n]$$
$$= \sum_{j=-\infty}^{+\infty} \sum_{i=-\infty}^{+\infty} x[m,n] \cdot h[m-i,n-j] \qquad (1)$$

where $x[m,n]$ is the input signal and $h[m,n]$ is the convolution kernel. Such treatment perfectly suits the structure of image processing. And CNN is a kind of special deep neural network, with connections between neurons are not fully connected and some wights are shared. Sparse connectivity and shared weight can reduce computation complexity and get a certain space invariance.

Pooling computes the statistic value of a particular feature over a region of the image. Max pooling, for example, can be computed as follows:

$$y[m,n] = \max x[i,j] \qquad (2)$$

where $x[i,j]$ comes from a particular region. This aggregation operation can decrease computation complexity and avoid over-fitting. Desirable translation invariant can also be achieved. In the practical construction of CNN, we often use convolution layer and pooling layer alternately. At the end, a fully-connected layer is added generally. With the deepening of level calculation, initial data is transformed to more abstract features. High-level features can then represent the input and be regarded as facial feature. Our CNN is constructed in Fig. 3.

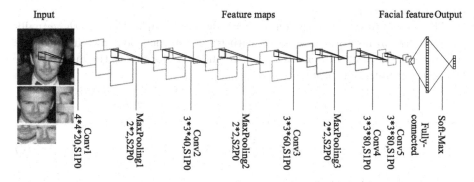

Fig. 3. The structure of the designed convolutional neural network. The setting of convolution and pooling in each layer is displayed at the bottom. 'Conv1 4 * 4 * 20, S1P0' means that the layer is a convolution layer generated after convolving the previous layer via 20 kernels of size 4 * 4, stride 1 and padding 0. 'MaxPooling1 2 * 2, S2P0' demonstrates a max pooling layer with a neighboring region size of 2 * 2, stride 2 and padding 0. Details of intermediate feature maps are not shown for the sake of clarity.

Due to the small amount of data in KFW-I and KFW-II [8], we use YouTube Faces Dataset (YTF) [14] to train the network. To achieve better results, we clip the images to six patches and train six CNNs. The original images firstly cropped to one basic image (64 * 64), one central clipping (40 * 40) and four corner clippings (20 * 20). To suit the input size, patches should then be scaled. Each kind of input can get an individual network and extract features separately. For each input, after six convolution layers and three pooling layers, the fully-connected layer outputs a 160-dimensional vector and sends it to a soft-max layer. In detail, we adopt ReLU function as activation function and use back propagation algorithm to train the network with iteration set to 400. Images are preprocessed with a simple rescaling to the range $[0, 1]$ and a standardization of zero-mean and unit-variance. The weights are initialized with a Gaussian distribution with zero mean and a standard deviation of 0.01. Learning rate ranges from 0.1 to 0.0001 as the training goes on. Other parameters including bias, weight decay, batch size are set as 0.1, 0.005 and 100 respectively.

In training, we use a soft-max layer at last to achieve supervised feature learning. However, in feature extraction, the last layer is removed and activation value of the fully-connected layer is used as the extracted feature. When extracting, images of kinship follow the same procession in training. The features extracted from each CNNs are joined to a 960-dimensional feature FF.

3.2   Relational Feature Learning and Classification

After extracting facial features from different people, we need to find the relationship between them. The main idea of autoencoder is to reconstruct target value using input value. During the process of reconstruction, intermediate activation value are used to represent the compact relationship between the first layer and the last layer.

The structure of our designed autoencoder is shown in Fig. 4. It is actually a 5-layer neural network. Compared with typical autoencoder owning only one training samples, our autoencoder has two ends. We formulate the model as the following optimization problem:

$$
min \quad J(W, b) = \frac{1}{2} \left\| h_{W,b}(FF^{(1)}) - FF^{(2)} \right\|_2
$$
$$
+ \frac{\lambda}{2} \sum_{l=1}^{L-1} \sum_{i=1}^{s_l} W_i^{(l)} \tag{3}
$$

where $b^{(l)}$ denotes bias, $W_i^{(l)}$ denotes weight and λ denotes regularization parameter. $h_{W,b}(FF^{(1)})$ means the hypothesis of $FF^{(1)}$. The first part of $J(W, b)$ is quadratic sum of error, and the second part is a regularization term.

Here, back propagation can stilled be used to train the network. We first calculate the activation $a^{(l)}$ of the network in the forward pass. After that, an error term $\delta_i^{(l)}$ for node i in layer l can be calculated in the backward pass. Then, we compute desired partial derivatives and update weight $W_i^{(l)}$ as well as bias $b^{(l)}$. Detailed steps are described in Algorithm 1.

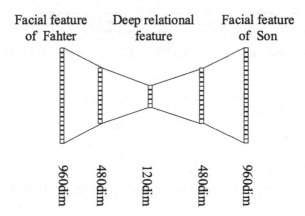

Fig. 4. The structure of the autoencoder. Taking Father-Son kinship for example, two ends of the network are 960-dimensional facial features of son and father respectively. There are three hidden layers with less units used to extract features. A 120-dimensional relational feature can then be draw from the third layer.

As an unsupervised learning algorithm, autoencoder is simple and efficient. After iterations, cost function reduces and the algorithm converges. We choose the activation value of the third layer as the deep relational feature, which can reflect the kinship of two person. The max iteration is 2000, learning rate is 0.01, and regularization is 0.0001.

When relational features are extracted from facial pairs that are with or without kinship, we get a correspondence between relational feature and kinship. To classify them, we employ a RBF-kernel support vector machine (SVM) to finish the task. An SVM model separate samples with a high-dimensional hyperplane, which ensures that different categories are divided by a clear gap that is as wide as possible. Thus, samples can be divided into two classes with or without kinship.

4   Experiments and Results

Experiments are divided into two steps generally following the descriptions in Sect. 3. The first step is training CNN using YouTube Faces Dataset [14] and extracting facial features for facial pairs from KinFaceW-I (KFW-I) and KinFaceW-II (KFW-II) [8]. The second part concentrates on extracting deep relational features using our autoencoder and classifying them using SVM.

4.1   Datasets and Settings

Three public datasets are used to train and evaluate our model. YouTube Faces Dataset contains clips from 3,425 videos of 1,595 different people under unconstrained environment. Each frame might differ in background, scene, expression,

---

Algorithm 1. Back propagation of our designed autoencoder.

Input:
    Facial feature pairs of two person, $(FF^{(1)}, FF^{(2)})$; Weight, $W^{(l)}$; Bias, $b^{(l)}$;

Output:
    Deep relational feature, DRF;

1: Initialize $W^{(l)}$ and $b^{(l)}$ with a Normal distribution;
2: **repeat**
3:    Perform a feed forward pass, compute activation $a^{(l)}$,
        $a^{(l+1)} = f(z^{(l)})$, $z^{(l)} = W^{(l)}a^{(l)} + b^{(l)}$, $f(z) = \frac{1}{1+e^{-z}}$;
4:    Compute error term $\delta^{(L)}$ of the last layer L,
        $\delta^{(L)} = -(FF^{(2)} - a^{(L)}) \cdot f'(z^{(L)})$;
5:    Perform a feed forward pass from penultimate layer to the first layer, compute
    error term $\delta^{(l)}$,
        $\delta^{(l)} = (W^{(l)T}\delta^{(l+1)} \cdot f'(z^{(L)})$;
6:    Compute partial derivatives,
        $\nabla_{W^{(l)}} J(W, b; FF^{(1)}, FF^{(2)}) = \delta^{(l+1)}(a^{(l)})^T$,
        $\nabla_{b^{(l)}} J(W, b; FF^{(1)}, FF^{(2)}) = \delta^{(l+1)}$;
7:    Update parameters,
        $W^{(l)} = W^{(l)} - \alpha(\nabla_{W^{(l)}} J(W, b; FF^{(1)}, FF^{(2)}) + \lambda W^{(l)})$,
        $b^{(l)} = b^{(l)} - \alpha\nabla_{b^{(l)}} J(W, b; FF^{(1)}, FF^{(2)})$;
8: **until** Reach max iteration;
9: **return** $DRF = f(z^{(3)})$;

---

pose, lighting, perspective, etc. The diversity brings great difficulty as well as strong robustness. Preprocessing includes rotation correction, scaling, and clipping. After that, we obtain six samples for each original face images and train six CNNs for different kind of samples. We randomly choose 31,900 images as training set and 7,975 images as testing set.

KFW-I and KFW-II consist of four representative types of kinships: Father-Son (F-S), Father-Daughter (F-D), Mother-Son (M-S) and Mother-Daughter (M-D). KFW-I, acquired from different photos, has 156, 134, 116, and 127 pairs of images for the four relations. KFW-II, acquired from the same photo, contains 250 pairs for each relation. Using the convolutional neural network designed in Fig. 3, we firstly train it on YouTube Faces Dataset. Same as preprocessing in YouTube Faces Dataset, we obtain six samples from each image. We use the pretrained model as fixed feature extractor to output facial features. Then, using pairs of facial features, we train them to extract corresponding relational features from autoencoders. Finally, a SVM is trained to classify using relation features and their labels.

4.2   Results and Analysis

For the training on YouTube Faces Dataset, the final test accuracy is 94.1%. It means that the convolutional neural network has been trained well and can extract useful facial features. It is the foundation for next steps. By contrast, the accuracy of our method is compared with 7 popular methods. HumanA and HumanB are classification results from two volunteers [9]. CML, MNRML, DMML are abbreviations for Concatenated Metric Learning [18], Multifeature Neighborhood Repulsed Metric Learning [8] and Discriminative Multi-Metric Learning [18] respectively. BasicCNN is the convolutional neural networks proposed in [19], which uses an end-to-end method.

Table 1 shows comparison between different methods in KFW-I. Our method performs better than human and other traditional models in all four kinships, which proves the effectiveness of deep relational feature. Compared with Basic-CNN, ours performs better in F-S kinship and slightly worse in other three kinships. Table 2 reveals different methods' performance in KFW-II. Our method achieves high accuracy than other methods in F-S and F-D kinships. And only lower than BasicCNN in M-S and M-D kinships. What's more, there is an improvement from KFW-I to KFW-II, which indicates that larger data size can boost the model's performance.

Table 1. Accuracy (%) of kinship verification using different methods in KFW-I.

Method	F-S	F-D	M-S	M-D
HumanA [9]	58.0	61.0	70.0	66.0
HumanB [9]	65.0	67.0	77.0	75.0
CML [18]	65.5	69.5	72.0	64.5
MNRML [8]	66.5	72.5	72.0	66.2
DMML [18]	69.5	74.5	75.5	69.5
BasicCNN [19]	70.8	75.7	79.4	73.4
Ours	**71.2**	**74.3**	**77.2**	**73.3**

Table 2. Accuracy (%) of kinship verification using different methods in KFW-II.

Method	F-S	F-D	M-S	M-D
HumanA [9]	61.0	61.0	73.0	69.0
HumanB [9]	68.0	70.0	80.0	78.0
CML [18]	73.0	73.5	76.5	76.0
MNRML [8]	74.3	76.9	77.6	77.4
DMML [18]	76.5	78.5	79.5	78.5
BasicCNN [19]	79.6	84.9	88.5	88.3
Ours	**80.3**	**85.2**	**83.3**	**82.6**

The results have demonstrated the effectiveness of our model. The relational feature is more powerful than traditional methods and comparable with Basic-CNN. However, BasicCNN tends to be overfitting during training. Our method utilizes a totally different architecture with BasicCNN and overcomes its short-coming. In addition, unlike traditional methods, our model do not need to design specific features or add extra intervention in experiments.

5   Conclusion

Kinship verification from facial images is a very challenging research topic. In this work, we propose to use deep learning for verifying kinship from facial images. Specifically, convolutional neural networks are firstly used to extract facial features and then deep relational features are learned by autoencoder. Finally, a trained SVM classifier is employed to verify kinship with learned deep relational features. Experiment results on public datasets demonstrate that our approach outperforms other low-level models. In future, we are going to optimize our model in kinship verification and expand it to other fields. Collecting larger-size datasets will be also our research focus.

References

1. Ashalatha, M.E., Holi, M.S., Mirajkar, P.R.: Face recognition using local features by LPP approach. In: Proceedings of International Circuits, Communication, Control and Computing (I4C) Conference, pp. 382–386, November 2014
2. Eckert, M., Gil, A., Zapatero, D., Meneses, J., Ortega, J.F.M.: Fast facial expression recognition for emotion awareness disposal. In: 2016 IEEE 6th International Conference on Consumer Electronics - Berlin (ICCE-Berlin) (2016)
3. Fang, R., Tang, K.D., Snavely, N., Chen, T.: Towards computational models of kinship verification. In: Proceedings of IEEE International Conference on Image Processing, pp. 1577–1580, September 2010
4. Guo, G., Wang, X.: Kinship measurement on salient facial features. IEEE Trans. Instrum. Meas. **61**(8), 2322–2325 (2012)
5. Hackman, L., Black, S.M.: Age estimation in the living. In: Encyclopedia of Forensic & Legal Medicine, pp. 34–40 (2016)
6. Hinton, G.E., Salakhutdinov, R.R.: Reducing the dimensionality of data with neural networks. Science **313**(5786), 504–507 (2006)
7. Krizhevsky, A., Sutskever, I., Hinton, G.E.: ImageNet classification with deep convolutional neural networks. In: Advances in Neural Information Processing Systems, vol. 25, no. 2 (2012)
8. Lu, J., Zhou, X., Tan, Y.P., Shang, Y., Zhou, J.: Neighborhood repulsed metric learning for kinship verification. IEEE Trans. Pattern Anal. Mach. Intell. **36**(2), 331–345 (2014)
9. Lu, J., Hu, J., Zhou, X., Shang, Y., Tan, Y.P., Wang, G.: Neighborhood repulsed metric learning for kinship verification. In: IEEE Conference on Computer Vision and Pattern Recognition, pp. 2594–2601 (2012)
10. Lu, J., Hu, J., Zhou, X., Zhou, J., Castrillon-Santana, M., Lorenzo-Navarro, J., Kou, L., Shang, Y., Bottino, A., Vieira, T.F.: Kinship verification in the wild: the first kinship verification competition. In: IJCB 2014–2014 IEEE/IAPR International Joint Conference on Biometrics (2014)

11. Ranzato, M., Huang, F.J., Boureau, Y., Lecun, Y.: Unsupervised learning of invariant feature hierarchies with applications to object recognition. In: IEEE Conference on Computer Vision and Pattern Recognition, CVPR 2007, pp. 1–8 (2007)
12. Russakovsky, O., Deng, J., Su, H., Krause, J., Satheesh, S., Ma, S., Huang, Z., Karpathy, A., Khosla, A., Bernstein, M.: Imagenet large scale visual recognition challenge. Int. J. Comput. Vis. **115**(3), 211–252 (2015)
13. Sun, Y., Wang, X., Tang, X.: Deep learning face representation from predicting 10,000 classes, pp. 1891–1898 (2014)
14. Wolf, L., Hassner, T., Maoz, I.: Face recognition in unconstrained videos with matched background similarity. In: Proceedings of IEEE Conference on Computer Vision and Pattern Recognition (CVPR), pp. 529–534, June 2011
15. Xia, S., Shao, M., Fu, Y.: Kinship verification through transfer learning. In: IJCAI International Joint Conference on Artificial Intelligence, pp. 2539–2544 (2011)
16. Xia, S., Shao, M., Luo, J., Fu, Y.: Understanding kin relationships in a photo. IEEE Trans. Multimedia **14**(4 PART1), 1046–1056 (2012). https://doi.org/10.1109/TMM.2012.2187436
17. Yan, H.: Kinship verification using neighborhood repulsed correlation metric learning. Image Vis. Comput. (2016). https://doi.org/10.1016/j.imavis.2016.08.009
18. Yan, H., Lu, J., Deng, W., Zhou, X.: Discriminative multimetric learning for kinship verification. IEEE Trans. Inf. Forensics Secur. **9**(7), 1169–1178 (2014)
19. Zhang, K., Huang, Y., Song, C., Wu, H., Wang, L.: Kinship verification with deep convolutional neural networks. BMVC **2015**, 1–12 (2015)
20. Zhou, X., Hu, J., Lu, J., Shang, Y., Guan, Y.: Kinship verification from facial images under uncontrolled conditions. In: International Conference on Multimedia 2011, Scottsdale, AZ, USA, 28 November - 01 December, pp. 953–956 (2011)
21. Zhou, X., Shang, Y., Yan, H., Guo, G.: Ensemble similarity learning for kinship verification from facial images in the wild. Inf. Fusion **32**, 40–48 (2016). https://doi.org/10.1016/j.inffus.2015.08.006
22. Zhou, X., Yan, H., Shang, Y.: Kinship verification from facial images by scalable similarity fusion. Neurocomputing **197**, 136–142 (2016). https://doi.org/10.1016/j.neucom.2016.02.039

Deep Neural Network

Hyperspectral Image Classification Using Spectral-Spatial LSTMs

Feng Zhou, Renlong Hang, Qingshan Liu$^{(\boxtimes)}$, and Xiaotong Yuan

Jiangsu Key Laboratory of Big Data Analysis Technology,
School of Information and Control, Nanjing University of Information
Science and Technology, Nanjing 210044, China
`qsliu@nuist.edu.cn`

Abstract. In this paper, we propose a hyperspectral image (HSI) classification method using spectral-spatial long short term memory (LSTM) networks. Specifically, for each pixel, we feed its spectral values in different channels into Spectral LSTM one by one to learn the spectral feature. Meanwhile, we firstly use principle component analysis (PCA) to extract the first principle component from a HSI, and then select local image patches centered at each pixel from it. After that, we feed the row vectors of each image patch into Spatial LSTM one by one to learn the spatial feature for the center pixel. In the classification stage, the spectral and spatial features of each pixel are fed into softmax classifiers respectively to derive two different results, and a decision fusion strategy is further used to obtain a joint spectral-spatial results. Experiments are conducted on two widely used HSIs, and the results show that our method can achieve higher performance than other state-of-the-art methods.

Keywords: Deep learning · Long short term memory · Decision fusion
Hyperspectral image classification

1 Introduction

With the development of hyperspectral sensors, it is convenient to acquire images with high spectral and spatial resolutions simultaneously. Hyperspectral data is becoming a valuable tool to monitor the Earth's surface. The classification of hyperspectral image (HSI) has become one of the most important tasks for many applications including both commercial and military domains.

Many methods have been proposed to deal with HSI classification. Traditional methods, such as k-nearest-neighbors and logistic regression, often use the high-dimensional spectral information as features, thus suffering from the issue of "curse of dimensionality". To address this issue, dimensionality reduction methods are widely used. These methods include principal component analysis

---

F. Zhou—Currently working toward the Master degree in the School of Information and Control, Nanjing University of Information Science and Technology.

J. Yang et al. (Eds.): CCCV 2017, Part I, CCIS 771, pp. 577–588, 2017.
https://doi.org/10.1007/978-981-10-7299-4_48

(PCA) [14,21] and linear discriminant analysis (LDA) [1,8]. In [20], a promising method called support vector machine (SVM) was successfully applied to HSI classification. It exhibits low sensitivity to the data with high dimensionality and small sample size. In most cases, SVM-based classifiers can obtain superior performance as compared to other methods. However, SVM is still a shallow architecture. As discussed in [7], these shallow architectures have shown effectiveness in solving many simple or well-constrained problems, but their limited representational power is insufficient in complex cases.

In the past few years, with the advances of the computing power of computers and the availability of large-scale datasets, deep learning techniques [19] have gained great success in a variety of machine learning tasks. Among these techniques, CNN [6,18] has been recognized as a state-of-the-art feature extraction method for various computer vision tasks [16,22] owing to its local connections and weight sharing properties. Besides, recurrent neural network (RNN) and its variants have been widely used in sequential data modeling such as speech recognition [9,10] and machine translation [5,23].

Recently, deep learning has been introduced into the remote sensing community especially for HSI classification [26]. For example, in [3], a stacked autoencoder model was proposed to extract high-level features in an unsupervised manner. Inspired from it, Tao et al. proposed an improved autoencoder model by adding a regularization term into the energy function [24]. In [4], deep belief network (DBN) was applied to extract features and classification results were obtained by logistic regression classifier. For these models, inputs are high-dimensional vectors. Therefore, to learn the spatial feature from HSIs, an alternative method is flattening a local image patch into a vector and then feeding it into them. However, this method may destroy the two-dimensional structure of images, leading to the loss of spatial information. To address this issue, a two dimensional CNN model was proposed in [27]. Due to the use of the first principal component of HSIs as input, two dimensional CNN may lose the spectral information. To simultaneously learn the spectral and spatial features, three-dimensional CNN considers the local cube as inputs [2].

Since hyperspectral data are densely sampled from the entire spectrum, they are expected to have dependencies between different spectral bands. First, it is easy to observe that for any material, the adjacent spectral bands tend to have very similar values, which implies that adjacent spectral bands are highly dependent on each other. In addition, some materials also demonstrate long-term dependency between non-adjacent spectral bands [25]. In this paper, we regard each hyperspectral pixel as a data sequence and use long short term memory (LSTM) [13] to model the dependency in the spectral domain. Similar to spectral channels, pixels of the image also depend on each other in the spatial domain. Thus, we can also use LSTM to extract spatial features. The extracted spectral and spatial features for each pixel are then fed into softmax classifiers. The classification results can be combined to derive a joint spectral-spatial result.

The rest of this paper is structured as follows. In the following section, we will present the proposed method in detail. The experiments are reported in Sect. 3, followed by the conclusion in Sect. 4.

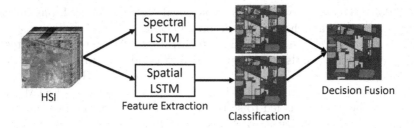

Fig. 1. Flowchart of the proposed SSLSTMs.

2 Methodology

The flowchart of the proposed spectral-spatial LSTMs (SSLSTMs) is shown in Fig. 1. From this figure, we can observe that SSLSTMs consist of two important components: Spectral LSTM (SeLSTM) and Spatial LSTM (SaLSTM). For each pixel in a given HSI, we feed its spectral values into the SeLSTM to learn the spectral feature and then derive a classification result. Similarly, for the local patch of each pixel, we feed it into a SaLSTM to extract the spatial feature and then obtain a classification result. To fuse the spectral-spatial results, we finally combine these two classification results in a weighted sum manner. In the following subsections, we will introduce these processes in detail.

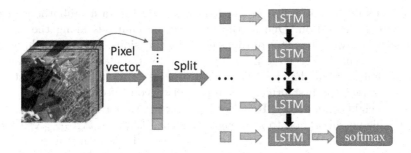

Fig. 2. Flowchart of SeLSTM.

2.1 Spectral LSTM

Hundreds of spectral bands in HSIs provide different spectral characteristics of the object in the same location. Due to the complex situation of lighting, rotations of the sensor, different atmospheric scattering conditions and so on, spectra have complex variations. Therefore, we need to extract robust and invariant features for classification. It is believed that deep architectures can potentially lead to progressively more abstract features at higher layers, and more abstract features are generally invariant to most local changes of the input. In this paper, we consider the spectral values in different channels as an input sequence and use LSTM discussed above to extract spectral features for HSI classification. Figure 2 shows the flowchart of the proposed classification scheme with spectral features. First, we choose the pixel vector $x_i \in \mathbf{R}^{1 \times K}$ where K indicates the number of spectral bands from a given HSI. Second, we transform the vector to a K-length sequence $\{x_i^1, \cdots, x_i^k, \cdots, x_i^K\}$ where $x_i^k \in \mathbf{R}^{1 \times 1}$ indicates the pixel value of k-th spectral band. Then, the sequence is fed into LSTM one by one and the last output is fed to softmax classifier. We set the loss function to cross entropy and optimize it by Adam algorithm [15]. Finally, we can obtain the probability value $P_{spe}(y = j \mid x_i), j \in \{1, 2, \cdots, C\}$ where C indicates the number of classes.

Fig. 3. Flowchart of SaLSTM.

2.2 Spatial LSTM

To extract the spatial feature of a specific pixel, we take a neighborhood region of it into consideration. Due to the hundreds of channels along the spectral dimension, it always has tens of thousands of dimensions. A large neighborhood region will result in too large input dimension for the classifier, containing too large amount of redundancy [3]. Motivated by the works in [3,27], we firstly use PCA to extract the first principle component. Second, for a given pixel x_i, we choose a neighborhood $\mathbf{X}_i \in \mathbf{R}^{S \times S}$ centered at it. After that, we transform the rows in this neighborhood to a S-length sequence $\{X_i^1, \cdots, X_i^l, \cdots, X_i^S\}$ where X_i^l indicates the l-th row of $\mathbf{X}_i$. Finally, we feed the sequence into LSTM to extract the spatial feature of x_i. Similar to spectral features-based classification, we use the last output of LSTM as an input to the softmax layer and achieve the probability value $P_{spa}(y = j \mid x_i), j \in \{1, 2, \cdots, C\}$. The configurations of loss function and optimization algorithm in SaLSTM is the same as those of SeLSTM. The overall flowchart of the proposed spatial features-based classification method is demonstrated in Fig. 3.

2.3 Joint Spectral-Spatial Classification

The above two subsections introduce the classification methods based on spectral and spatial features respectively. With the development of imaging spectroscopy technologies, current sensors can acquire HSIs with very high spatial resolutions. Therefore, the pixels in a small spatial neighborhood belong to the same class with a high probability. For a large homogeneous region, the pixels may have different spectral responses. If we only use the spectral features, the pixels will be classified into different subregions. On the contrary, for multiple neighboring regions, if we only use the spatial information, these regions will be classified as the same one. Thus, for accurate classifications, it is essential to take into account the spatial and spectral information simultaneously [12]. Based on the posterior probabilities $P_{spe}(y = j \mid x_i)$ and $P_{spa}(y = j \mid x_i)$, an intuitive method to combine the spectral and spatial feature is to fuse these two results in a weighted sum manner, which can be formulated as $P(y = j \mid x_i) = w_{spe} P_{spe}(y = j \mid x_i) + w_{spa} P_{spa}(y = j \mid x_i)$, where w_{spe} and w_{spa} are fusion weights that satisfy $w_{spe} + w_{spa} = 1$. For simplicity, we use uniform weights in our implementation, i.e., $w_{spe} = w_{spa} = \frac{1}{2}$.

3   Experimental Results

3.1   Datasets

We test the proposed method on two famous HSI datasets, which are widely used to evaluate classification algorithms.

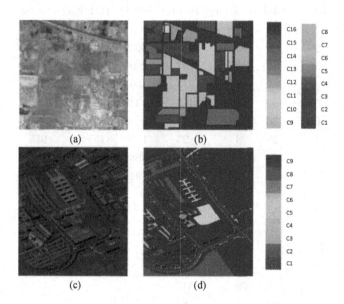

Fig. 4. False-color composite images and ground-truth maps of (a)–(b) IP, (c)–(d) PUS.

Indian Pines Scene (IP): The first dataset was acquired by the AVIRIS sensor over the Indian Pine test site in northwestern Indiana, USA, on June 12, 1992 and it contains 224 spectral bands. We utilize 200 bands after removing four bands containing zero values and 20 noisy bands affected by water absorption. The spatial size of the image is 145 × 145 pixels, and the spatial resolution is 20 m. The false-colour composite image and the ground-truth map are shown in Fig. 4(a)–(b). The available number of samples is 10249 ranging from 20 to 2455 in each class, which is reported in Table 1.

Pavia University Scene (PUS): The second dataset was acquired by the ROSIS sensor during a flight campaign over Pavia, northern Italy, on July 8, 2002. The original image was recorded with 115 spectral channels ranging from 0.43 μm to 0.86 μm. After removing noisy bands, 103 bands are used. The image size is 610 × 340 pixels with a spatial resolution of 1.3 m. A three band false-colour composite image and the ground-truth map are shown in Fig. 4(c)–(d). In the ground-truth map, there are nine classes of land covers with more than 1000 labeled pixels for each class shown in Table 1.

Table 1. Number of pixels for training/testing and the total number of pixels for each class in IP and PUS ground truth maps.

IP				PUS		
Class	Total	Training	Testing	Total	Training	Testing
1	46	5	41	6631	548	6083
2	1428	143	1285	18649	540	18109
3	830	83	747	2099	392	1707
4	237	24	213	3064	524	2540
5	483	48	435	1345	265	1080
6	730	73	657	5029	532	4497
7	28	3	25	1330	375	955
8	478	48	430	3682	514	3168
9	20	2	18	947	231	716
10	972	97	875			
11	2455	246	2209			
12	593	59	534			
13	205	21	184			
14	1265	127	1138			
15	386	39	347			
16	93	9	84			

3.2 Experimental Setup

To demonstrate the effectiveness of the proposed LSTM-based classification method, we quantitatively and qualitatively evaluate the performance of

SeLSTM, SaLSTM and SSLSTMs. Besides, we compare them with several state-of-the-art methods, including PCA, LDA, NWFE [17], RLDE [28], MDA [11] and CNN [2]. We also directly use the original pixels as a benchmark. For LDA, the within-class scatter matrix $\mathbf{S}_W$ is replaced by $\mathbf{S}_W + \varepsilon \mathbf{I}$, where $\varepsilon = 10^{-3}$, to alleviate the singular problem. The optimal reduced dimensions for PCA, LDA, NWFE and RLDE are chosen from $[2, 30]$. For MDA, the optimal window size is selected from a given set $\{3, 5, 7, 9, 11\}$. For CNN, the number of layers and the size of filters are the same as the network in [2]. For LSTM, we only use one hidden layer, and the number of optimal hidden nodes are selected from a given set $\{16, 32, 64, 128, 256\}$.

For IP dateset, we randomly select 10% pixels from each class as the training set, and the remaining pixels as the testing set. For PUS dataset, we randomly choose 3921 pixels as the training set and the rest of pixels as the testing set [11]. The detailed numbers of training and testing samples are listed in Table 1. In order to reduce the effects of random selection, all the algorithms are repeated five times and the average results are reported. The classification performance is evaluated by overall accuracy (OA), average accuracy (AA), per-class accuracy, and Kappa coefficient κ. OA defines the ratio between the number of correctly classified pixels to the total number of pixels in the testing set, AA refers to the average of accuracies in all classes, and κ is the percentage of agreement corrected by the number of agreements that would be expected purely by chance.

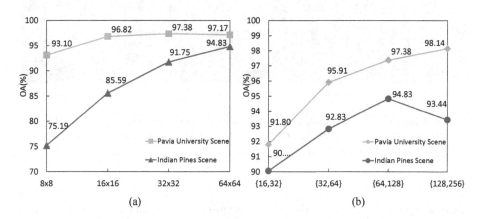

Fig. 5. (a) OAs of the SSLSTMs with different size of neighborhood regions. (b) OAs of SeLSTM and SaLSTM with different numbers of hidden nodes.

3.3 Parameter Selection

There are two important parameters in the proposed classification framework, including the size of neighborhood regions and the number of hidden nodes. Firstly, we fix the number of hidden nodes and select the optimal region size from a given set $\{8 \times 8, 16 \times 16, 32 \times 32, 64 \times 64\}$. Figure 5(a) demonstrates OAs of the SSLSTMs method on two datasets. From this figure, we can observe that as the region size increases, OA will firstly increase and then decrease on PUS dataset. Therefore, the optimal size is chosen as 32 × 32. For IP dataset, OA will increase as the size increases. However, larger sizes significantly increase the computation time. Thus, we set the optimal size as 64 × 64 for IP dataset.

Secondly, we fix the region size and search for the optimal number of hidden nodes for SeLSTM and SaLSTM from four different combinations {16, 32}, {32, 64}, {64, 128} and {128, 256}. As shown in Fig. 5(b), when the number of hidden nodes for SeLSTM and SaLSTM are set to 64 and 128 respectively, the SSLSTMs method achieves the highest OA on IP dataset. Similarly, we can see that SSLSTMs obtains the highest OA on PUS dataset when the number of hidden nodes for SeLSTM and SaLSTM are set to 128 and 256 respectively.

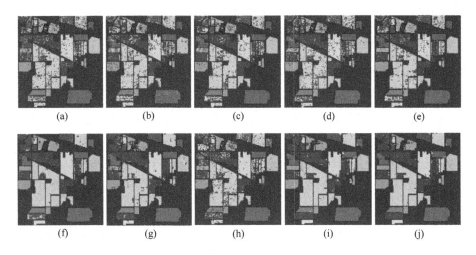

Fig. 6. Classification maps on the IP dataset. (a) Original. (b) PCA. (c) LDA. (d) NWFE. (e) RLDE. (f) MDA. (g) CNN. (h) SeLSTM. (i) SaLSTM. (j) SSLSTMs.

3.4   Performance Comparison

Table 2 reports the quantitative results acquired by ten methods on IP dataset. From these results, we can observe that PCA achieves the lowest OA among ten methods, mainly because PCA directly extracts spectral features for classification without considering spatial features. Although LDA and NWFE are still spectral-based methods, they achieve better results than PCA due to the use of label information in training samples. Besides, MDA achieves better performance than the other LDA-related methods which consider spectral information only, because it can extract spatial and spectral features simultaneously. This indicates the importance of spatial features for HSI classification. So, as spatial based methods, CNN and SaLSTM perform better than other spectral-based methods. However, they only use the first principal component of all spectral

Table 2. OA, AA, per-class accuracy (%), and κ performed by ten methods on IP dataset using 10% pixels from each class as the training set.

Label	Original	PCA	LDA	NWFE	RLDE	MDA	CNN	SeLSTM	SaLSTM	SSLSTMs
OA	77.44	72.58	76.67	78.47	80.97	92.31	90.14	72.22	91.72	95.00
AA	74.94	70.19	72.88	76.08	80.94	89.54	85.66	61.72	83.51	91.69
κ	74.32	68.58	73.27	75.34	78.25	91.21	88.73	68.24	90.56	94.29
C1	56.96	59.57	63.04	62.17	64.78	73.17	71.22	25.85	85.85	88.78
C2	79.75	68.75	72.04	76.27	78.39	93.48	90.10	66.60	89.56	93.76
C3	66.60	53.95	57.54	59.64	68.10	84.02	91.03	54.83	91.43	92.42
C4	59.24	55.19	46.58	59.83	70.80	83.57	85.73	43.94	90.61	86.38
C5	90.31	83.85	91.76	88.49	92.17	96.69	83.36	83.45	88.60	89.79
C6	95.78	91.23	94.41	96.19	94.90	99.15	91.99	87.76	90.81	97.41
C7	80.00	82.86	72.14	82.14	85.71	93.60	85.60	23.20	51.20	84.80
C8	97.41	93.97	98.74	99.04	99.12	99.91	97.35	95.40	99.02	99.91
C9	35.00	34.00	26.00	44.00	73.00	63.33	54.45	30.00	38.89	74.44
C10	66.32	64.18	60.91	69.18	69.73	82.15	75.38	71.29	88.64	95.95
C11	70.77	74.96	76.45	77.78	79.38	92.76	94.36	75.08	94.62	96.93
C12	64.42	41.72	67.45	64.05	72.28	91.35	78.73	54.49	86.10	89.18
C13	95.41	93.46	96.00	97.56	97.56	99.13	95.98	91.85	90.11	98.48
C14	92.66	89.45	93.79	93.49	92.36	98.22	96.80	90.37	98.10	98.08
C15	60.88	47.77	65.54	58.50	67.10	87.84	96.54	30.49	88.59	92.85
C16	87.53	88.17	83.66	89.03	89.68	94.29	81.90	62.86	64.05	87.86

bands, leading to the loss of spectral information. Therefore, the performance obtained by CNN or SaLSTM is inferior to that by MDA. Nevertheless, if we combine the spectral information and spatial information together, SSLSTMs can significantly improve the performance as compared to SeLSTM and SaL-STM. Additionally, as a kind of neural network, SSLSTMs is able to capture the non-linear distribution of hyperspectral data, while the linear method MDA may fail. Therefore, SSLSTMs obtains better results than MDA. Figure 6 demonstrates classification maps achieved by different methods on the IP dataset. It can be observed that SSLSTMs obtains a more homogeneous map than other methods.

Similar conclusions can be observed from the PUS dataset in Table 3 and Fig. 7. Again, MDA, CNN, and LSTM-based methods achieve better performance than other methods. Specifically, OA, AA and κ obtained by CNN are almost the same as MDA, and SSLSTMs obtains better performance than CNN and MDA. It is worth noting that the improvement of OA, AA and κ from MDA or CNN to SSLSTMs is not remarkable as those on IP dataset, because CNN and MDA have already obtained a high performance and a further improvement is very difficult.

Table 3. OA, AA, per-class accuracy (%), and κ performed by ten methods on PUS dataset using 3921 pixels as the training set.

Label	Original	PCA	LDA	NWFE	RLDE	MDA	CNN	SeLSTM	SaLSTM	SSLSTMs
OA	89.12	88.63	84.08	88.73	88.82	96.95	96.55	93.20	94.98	98.48
AA	90.50	90.18	87.23	90.38	90.45	96.86	97.19	93.13	94.86	98.51
κ	85.81	85.18	79.59	85.31	85.43	95.93	95.30	90.43	92.84	97.56
C1	87.25	87.07	82.91	86.86	87.20	96.69	96.72	91.33	92.20	96.83
C2	89.10	88.38	80.68	88.50	88.40	97.76	96.31	94.58	95.86	98.74
C3	81.99	81.96	69.21	82.20	81.69	90.69	97.15	83.93	92.42	96.57
C4	95.65	95.14	95.99	95.27	95.79	98.44	96.16	97.78	91.59	98.43
C5	99.76	99.76	99.90	99.81	99.87	100.00	99.81	99.46	98.70	99.94
C6	88.78	88.06	89.53	88.16	88.67	96.26	94.87	91.73	96.91	99.43
C7	85.92	85.32	81.11	86.57	86.06	97.95	97.44	90.76	98.74	99.31
C8	86.14	86.06	85.81	86.13	86.42	93.98	98.23	88.78	94.79	97.98
C9	99.92	99.92	99.92	99.89	99.94	100.00	98.04	99.83	92.54	99.39

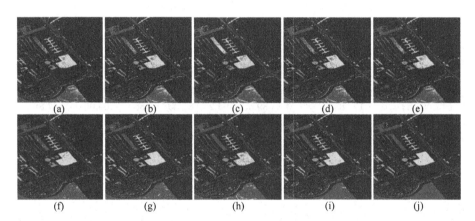

Fig. 7. Classification maps on the PUS dataset. (a) Original. (b) PCA. (c) LDA. (d) NWFE. (e) RLDE. (f) MDA. (g) CNN. (h) SeLSTM. (i) SaLSTM. (j) SSLSTMs.

4   Conclusion

In this paper, we have proposed a HSI classification method based on a LSTM network. Both the spectral feature extraction and the spatial feature extraction issues were considered as sequence learning problems, and LSTM was naturally applied to address them. Specifically, for a given pixel in HSIs, its spectral values in different channels were fed into LSTM one by one to learn spectral features. For the spatial feature extraction, a local image patch centered at the pixel was firstly selected from the first principal component of HSIs, and then the rows of the patch were fed into LSTM one by one. By conducting experiments on two HSIs collected by different instruments (AVIRIS and ROSIS), we compared the proposed method with state-of-the-art methods including CNN. The experimental results indicate that using spectral and spatial information simultaneously improves the classification performance and results in more homogeneous regions in classification maps compared to only using spectral information. We also evaluated the influences of different parameters in the network, including the patch size and the number of hidden nodes.

References

1. Bandos, T.V., Bruzzone, L., Camps-Valls, G.: Classification of hyperspectral images with regularized linear discriminant analysis. IEEE Trans. Geosci. Remote Sens. **47**(3), 862–873 (2009)
2. Chen, Y., Jiang, H., Li, C., Jia, X.: Deep feature extraction and classification of hyperspectral images based on convolutional neural networks. IEEE Trans. Geosci. Remote Sens. **54**(10), 1–20 (2016)
3. Chen, Y., Lin, Z., Zhao, X., Wang, G., Gu, Y.: Deep learning-based classification of hyperspectral data. IEEE J. Sel. Top. Appl. Earth Obs. Remote Sens. **7**(6), 2094–2107 (2014)
4. Chen, Y., Zhao, X., Jia, X.: Spectral-spatial classification of hyperspectral data based on deep belief network. IEEE J. Sel. Top. Appl. Earth Obs. Remote Sens. **8**(6), 2381–2392 (2015)
5. Cho, K., Merrienboer, B.V., Gulcehre, C., Bahdanau, D., Bougares, F., Schwenk, H., Bengio, Y.: Learning phrase representations using RNN encoder-decoder for statistical machine translation. Comput. Sci. (2014)
6. Cun, Y.L., Boser, B., Denker, J.S., Howard, R.E., Habbard, W., Jackel, L.D., Henderson, D.: Handwritten digit recognition with a back-propagation network. In: Advances in Neural Information Processing Systems, pp. 396–404 (1990)
7. Deng, L.: A tutorial survey of architectures, algorithms, and applications for deep learning. APSIPA Trans. Sig. Inf. Process. **3**, e2 (2014)
8. Friedman, J.H.: Regularized discriminant analysis. J. Am. Stat. Assoc. **84**(405), 165–175 (1989)
9. Graves, A., Jaitly, N.: Towards end-to-end speech recognition with recurrent neural networks. In: International Conference on Machine Learning, pp. 1764–1772 (2014)
10. Graves, A., Mohamed, A.R., Hinton, G.: Speech recognition with deep recurrent neural networks. In: IEEE International Conference on Acoustics, Speech and Signal Processing, pp. 6645–6649 (2013)
11. Hang, R., Liu, Q., Song, H., Sun, Y.: Matrix-based discriminant subspace ensemble for hyperspectral image spatial-spectral feature fusion. IEEE Trans. Geosci. Remote Sens. **54**(2), 783–794 (2016)
12. Hang, R., Liu, Q., Sun, Y., Yuan, X., Pei, H., Plaza, J., Plaza, A.: Robust matrix discriminative analysis for feature extraction from hyperspectral images. IEEE J. Sel. Top. Appl. Earth Obs. Remote Sens. **10**(5), 2002–2011 (2017)
13. Hochreiter, S., Schmidhuber, J.: Long short-term memory. In: Graves, A. (ed.) Supervised Sequence Labelling with Recurrent Neural Networks. SCI, vol. 385. Springer, Berlin (1997). https://doi.org/10.1007/978-3-642-24797-2_4
14. Jolliffe, I.: Principal Component Analysis. Wiley, Hoboken (2002)
15. Kingma, D., Ba, J.: Adam: a method for stochastic optimization. Comput. Sci. (2015)

16. Krizhevsky, A., Sutskever, I., Hinton, G.E.: Imagenet classification with deep convolutional neural networks. In: Advances in Neural Information Processing Systems, pp. 1097–1105 (2012)
17. Kuo, B.C., Landgrebe, D.A.: Nonparametric weighted feature extraction for classification. IEEE Trans. Geosci. Remote Sens. **42**(5), 1096–1105 (2004)
18. Lecun, Y., Boser, B., Denker, J.S., Henderson, D., Howard, R.E., Hubbard, W., Jackel, L.D.: Backpropagation applied to handwritten zip code recognition. Neural Comput. **1**(4), 541–551 (1989)
19. Lecun, Y., Bengio, Y., Hinton, G.: Deep learning. Nature **521**(7553), 436–444 (2015)
20. Melgani, F., Bruzzone, L.: Classification of hyperspectral remote sensing images with support vector machines. IEEE Trans. Geosci. Remote Sens. **42**(8), 1778–1790 (2004)
21. Palsson, F., Sveinsson, J.R., Ulfarsson, M.O., Benediktsson, J.A.: Model-based fusion of multi- and hyperspectral images using pca and wavelets. IEEE Trans. Geosci. Remote Sens. **53**(5), 2652–2663 (2015)
22. Simonyan, K., Zisserman, A.: Very deep convolutional networks for large-scale image recognition. Comput. Sci. (2014)
23. Sutskever, I., Vinyals, O., Le, Q.V.: Sequence to sequence learning with neural networks. In: Advances in Neural Information Processing Systems, vol. 4, pp. 3104–3112 (2014)
24. Tao, C., Pan, H., Li, Y., Zou, Z.: Unsupervised spectral-spatial feature learning with stacked sparse autoencoder for hyperspectral imagery classification. IEEE Geosci. Remote Sens. Lett. **12**(12), 2438–2442 (2015)
25. Wu, H., Prasad, S.: Convolutional recurrent neural networks for hyperspectral data classification. Remote Sens. **9**(3), 298 (2017)
26. Zhang, L., Zhang, L., Du, B.: Deep learning for remote sensing data: a technical tutorial on the state of the art. IEEE Geosci. Remote Sens. Mag. **4**(2), 22–40 (2016)
27. Zhao, W., Du, S.: Spectral-spatial feature extraction for hyperspectral image classification: a dimension reduction and deep learning approach. IEEE Trans. Geosci. Remote Sens. **54**(8), 4544–4554 (2016)
28. Zhou, Y., Peng, J., Chen, C.L.P.: Dimension reduction using spatial and spectral regularized local discriminant embedding for hyperspectral image classification. IEEE Trans. Geosci. Remote Sens. **53**(2), 1082–1095 (2015)

A Novel Framework for Image Description Generation

Qiang Cai[1], Ziyu Xue[1], Xiaoyu Zhang[2(✉)], Xiaobin Zhu[1],
Wei Shao[3], and Lei Wang[4]

[1] Beijing Key Laboratory of Big Data Technology for Food Safety,
School of Computer and Information Engineering, Beijing Technology and Business
University, No. 11, Fucheng Road, Haidian District, Beijing, China
caiq@th.btbu.edu.cn, xueziyucs@gmail.com, brucezhucas@gmail.com
[2] Institute of Information Engineering, Chinese Academy of Sciences, A, No. 89,
Minzhuang Road, Haidian District, Beijing, China
zhangxiaoyu@iie.ac.cn
[3] CCTV High-Tech Television Development CO., Ltd., 12B, Huabao building,
Lianhua bridge, Haidian District, Beijing, China
shaowei@cctvht.cn
[4] Information Technology InstituteAcademy of Broadcasting Science, SAPPRFT,
No. 2 Fuxingmenwai Street, Xicheng District, Beijing, China
wanglei@abs.ac.cn

Abstract. The existing *image description* generation algorithms always
fail to cover rich semantics information in natural images with single sentence or dense object annotations. In this paper, we propose a novel semi-
supervised generative visual sentence generation framework by jointly
modeling *Regions Convolutional Neural Network* (RCNN) and improved
Wasserstein Generative Adversarial Network (WGAN), for generating
diverse and semantically coherent sentence description of images. In our
algorithm, the features of candidate regions are extracted with RCNN
and the enriched words are polished by their context with an improved
WGAN. The improved WGAN consists of a structured sentence generator and a multi-level sentence discriminators. The generator produces
sentences recurrently by incorporating region-based visual and language
attention mechanisms, while the discriminator assesses the quality of generated sentences. The experimental results on publicly available dataset
show the promising performance of our work against other related works.

Keywords: WGAN · Image description · RCNN

1 Introduction

Image description can be used in many real application scenarios, e.g., video
retrieval and automatic video subtitling. It is a challenging task due to the intersection of computer vision, natural language processing and other disciplines. In
recent years, image description generation has attracted lots of focus in research

© Springer Nature Singapore Pte Ltd. 2017
J. Yang et al. (Eds.): CCCV 2017, Part I, CCIS 771, pp. 589–599, 2017.
https://doi.org/10.1007/978-981-10-7299-4_49

domain. And great progress has been made in labeling images with a pre-defined close set of visual categories, as shown in Fig. 1.

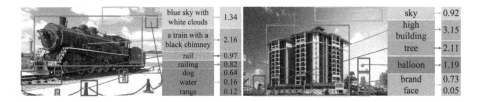

Fig. 1. Image description of dense regions.

Image description generation is thoroughly studied by the computer vision communities, where some well established models have been developed. A majority of algorithms proposed to describe image using individual semantic sentence. Everingham et al. [1] focused on labeling images with a fixed set of visual categories, while Kulkarni et al. [2] relied on hard-coded visual concepts and sentence templates to generate sentence descriptions. These two methods all ignored the location of dense objects in images. Recent works [3–5] intended to use Recurrent Neural Network to generate the dense image descriptions and corresponding location information, which were spliced according to the trained sentence model. However, the simplified sentence models always generate single smoothly sentences using fixed set of visual categories without consideration of rich vocabulary, which will fail to cover rich underlying semantics of images.

To address above-mentioned issue, we propose a novel approach for image description generation, by cooperation between modeling Convolutional Neural Network (CNN) and improved Wasserstein Generative Adversarial Network (WGAN). When using CNN in visual feature acquisition, alignment objective method will be used in order to avoid the deviation of regions from words. We have described the transformations that map every image and sentence into a set of vectors in a common h-dimensional space. After matching the dense objects and words, the higher score words are more likely to be used to generate sentence. The improved WGAN model is an adversarial training mechanism, which jointly by a structured sentence generator and multi-level sentence discriminators. The discriminator learns to distinguish between real and synthesized sentences which from generator. The visual features with higher scores are used as the input in improved WGAN to generate sentences.

The rest of the paper is organized as follows. Section 2 overviews the related works. In Sect. 3, we elaborate our method. The experimental evaluation is given in Sect. 4, and we draw conclusion in Sect. 5.

2   Related Work

Image description generation is an active area of research on its own [5]. Traditional methods concentrated on features extraction. Deep model is developing rapidly because of its accuracy and applicability, which generally divide into dense object recognition, visual sentence generation and visual paragraph generation. Generating high score sentence plays an important role in paragraph generation. In this section, we will briefly review sentence generation related works.

Kiros et al. [4] proposed a model that CNN is used to feature extraction and object detection. Borrowing the above method of feature extraction, Karpathy et al. [3] modeled Recurrent Neural Networks framework learns a joint image-label embedding to characterize the semantic label dependency as well as the image-label relevance. However, RNN failed to adapt to generate sentences, because it has no time sequence. RNN with Long Short-Term Memory (LSTM) can effectively model the long-term temporal dependency in a sequence [6]. To reason about long-term linguistic structures with multiple sentences, hierarchical recurrent network has been widely used to directly simulate the hierarchy of language. In [8,9], a framework based on CNN-LSTM was developed to image description that the generated descriptions significantly outperform baselines. However, LSTM uses fixed set of visual categories without consideration of rich vocabulary, which will fail to cover rich underlying semantics of images.

Dai et al. [15] proposed Conditional GAN (CGAN) to generate sentence, which proposed a cooperation model between native CNN and CGAN. CGAN overcomes the difficulty by Policy Gradient, a strategy stemming from Reinforcement Learning, which allows the generator to receive early feedback along the way. Liang et al. [10] proposed Recurrent Topic-Transition GAN (RTT-GAN) to generate paragraph, which composed of LSTM model and generated diverse and semantically coherent sentences by reasoning over both local semantic regions and global sentence context. The generator selectively incorporates visual and language cues of semantic regions to produce each sentence. Nevertheless, to the best of our knowledge, the previous methods failed to consider the alignment between dense objects and words, which had a great impact on the accuracy of the words.

3   Our Method

The framework of our algorithm is shown as Fig. 2. The red box part illustrates the flowchart of RCNN model. The green box part details the procedure of sentence generation. In our algorithm, we first generate descriptions for candidate image regions with RCNN. Sequentially, the sentence generator will generate meaningful sentences by incorporating the fine-grained visual and textual cues in a selective way. To ensure quality of sentences, we apply a sentence discriminator on each generated sentence to measure the plausibility and smoothness of semantic transition with preceding sentences. The generator and discriminator are learned jointly within an adversarial framework.

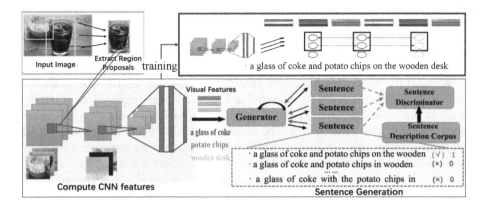

Fig. 2. The framework of our algorithm.

3.1   Learning to Align Visual and Language Data

Image-Feature Extraction. Following prior works, we observe that sentence descriptions make frequent references to objects and their attributes. Therefore, we adopt RCNN to detect object, which followed by [3]. The feature vectors for detected regions are computed as below:

$$v = W_M[CNN_{\theta c}(I_p)]. \tag{1}$$

where θ_c denotes CNN model parameter sets; the dimension of matrix W_M is $h * 4096$ (h is the size of the multimodal embedding space). Every image is transformed as a sequence of N words, which encoded in a representation vector. Every sequence will be transformed into a h-dimensional vector $\{V_i \,|\, i = 120\}$.

Objects Alignment. We formulate an image-sentence score as a function of the individual region-word scores, which used a Bidirectional Recurrent Neural Network (BRNN) to compute the word representations. Like the model of Karpathy et al. [3] interprets the dot product vi T-st between the i-th region and t-th word as a measure of similarity and use it to define the score between image k and sentence l as Eq. 2:

$$S_{kl} = \sum_{t \in g_l} \sum_{i \in gi} \max(0, v_i^T s_t). \tag{2}$$

where, g_i denotes the *i-th* image fragments. l denote the images and sentences in the training set respectively. s_t is a function of all words in the entire sentence.

Every word s_t aligns to the single best image region. As we shown in the experiments, this simplified model also leads to improvements in the final ranking performance. Assuming that $i = l$ denotes a corresponding image and sentence pair, the final max-margin, structured loss remains:

$$\partial(\theta) = \sum_i [\sum_{l(images)} \max(0, S_{il} - S_{ii} + 1) + \sum_{l(sentence)} \max(0, S_{li} - S_{ii} + 1)]. \tag{3}$$

This objective encourages aligned image-sentences pairs to have a higher score than misaligned pairs, by a margin. We get the high scores to form word vectors.

3.2   Sentence Generator

The architecture of the generator G shown in Fig. 3, which recurrently retains different levels of context states with a hierarchy constructed by sentence LSTM, word LSTM, and two attention modules. Firstly, the visual attention module selectively focuses on semantic regions, generating the visual representation of the sentence from the visual vectors. The sentence LSTM can be able to encode a topic vector for a new sentence. Secondly, the language attention module embedded in local phrases of focused semantic regions to facilitate word generation from the word LSTM, meanwhile learnt to incorporate linguistic knowledge.

LSTM for Sentence Description. Sentence LSTM is a single layer model. When we training the model, it takes the image pixels I and a sequence of input vectors from the region description $(v_1, \ldots, v_T)$. And computes a sequence of hidden states $(h_1, \ldots, h_t)$ and a sequence of outputs $(y_1, \ldots, y_t)$ by iterating the following recurrence relation from 1 to T. An attention mechanism is applied in the visual features V of all semantic regions, resulting in a visual context vector $f_v t$ which represents the next sentence at t-th step:

$$\mathrm{y}_t = soft \max\{W_{oh} \times f(W_{hx}x_t + W_{hh}h_{t-1} + b_h + W_{hi}[CNN_{\theta_c}(i)]) + b_o\} \quad (4)$$

here W_{hi}, W_{hx}, W_{hh}, W_{oh}, x_i and b_h, bo are learnable parameters. $CNN_{\theta}c(i)$ is the last layer in all net. y_t is the output vector, which holds the log probabilities of words in the dictionary, according to the highest score.

Word LSTM with Language Attention. To get plausibility and smoothness sentences, the proposed model should recognize and describe substantial details such as objects, attributes, and relationships, word replacement model is appropriate for context. We selectively incorporate the embedding of local phrases based on the topic vector and use Word LSTM to generate the better word representation. Considering that each local phrase relates to the respective visual feature, we reuse the visual attentive weights to enhance the language attention. By computing the contribution of a word to the whole sentence in hidden layer, word LSTM embeds the words which have the high contribution into sentence.

3.3   Sentence Discriminator

The sentence discriminator D_s aim to distinguish between real sentences and synthesized ones based on the linguistic characteristics of a natural sentence

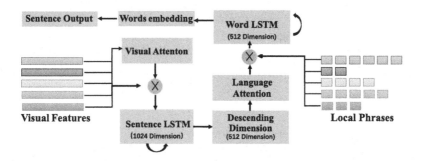

Fig. 3. The diagram of generator.

description. In our algorithm, the discriminator D_s is an LSTM that recurrently takes the input word embedding within a sentence, meanwhile produces a real value plausibility score of the synthesized sentence, which evaluates the plausibility of individual sentences.

The discrete nature of text samples disappeared gradient back-propagation from the discriminators to the generator. To overcome this problem, we use deterministic approximation. At each word generation step, we select the most probable word according to the soft-max emission distribution, and propagate gradient only through the words. More simply, we apply max-pooling operation over the soft-max to overcome the gradient disappearance in GAN model. Figure 4 shows the discriminator of the framework.

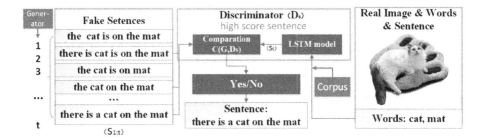

Fig. 4. The diagram of discriminator.

The Sentence Input (as shown in Fig. 4 with Fake Sentences) is the output of generator in Sect. 3.2. The objective of the adversarial framework, which followed WGAN [10] method by minimizing an approximated Wasserstein distance, is thus written as:

$$\min_{G,D^s} \max C(G, D^s) = E_{\hat{s} \sim S}[D^s(\hat{s})] - E_{\hat{s} \sim S_{1:t}}[D^s(\hat{s})] \tag{5}$$

where $\hat{s}$ is the true sentence, S and $S_{(1:t)}$ denote the true data distributions of generation sentences and true sentences, which are constructed from a sentence description corpus. sentence discriminator D^s that optimizes a critic between real/fake sentences. $\hat{s} \sim S_{(1:t)}$ distribution of generated sentences by the generator G. The objective for the generator G is:

$$G' = \arg\min_{G,D^s} \max \gamma C(G, D^s)$$ (6)

where γ is a balancing parameter fixed at 0:001 in implementation. The optimization is performed in an alternating min-max manner [10].

4   Experiments

To validate the effectiveness of our proposed algorithm, we conduct experiments on MSCOCO datasets: the dataset contains 123,000 images respectively and each is annotated with five sentences using Amazon Mechanical Turk. Throughout all the experiments, 5,000 images are used in both validation and testing.

To evaluate the performance of the LSTM model in generating sentences, we adopt VGGNet [11] in the full image experiments to calculate BLEU [12], METEOR [13] and CIDEr [14] scores. We evaluates a candidate sentence by matching five reference sentences written by humans. Word generator can optimize individual vocabulary. This step does not improve the accuracy of the sentence, because the above steps have generated vocabulary and phrases already. However, after the vocabulary of the replacement, we can get a smoother sentence.

4.1   Objective Alignment Evaluation

RCNN is pre-trained on ImageNet and fine-tuned on the 200 classes of the ImageNet Detection Challenge. In our algorithm, the top 19 ranked regions are selected. Image-Sentence alignment evaluation is used to build the Image-Sentence rank in our work, which retrieve the most compatible test sentence and visualize the highest-scoring region for each word and the associated scores, as shown in Fig. 5.

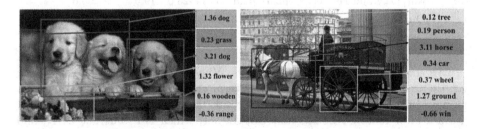

Fig. 5. Example alignments predicted by our model.

Inputting all the words into the generator in order, until the sentence generated in Sect. 3.2. The word LSTM is recurrently forwarded to optimize the words iteratively. We report the median rank of the closest ground truth result and Recall@K, which can get a correct sentence in top K results. The result of these experiments can be found in Table 1 [3].

Table 1. Image-Sentence ranking experiment results.

	R@1	R@5	R@10	Med r
Our model: 1K test images	38.4	69.9	80.5	1.0
Our model: 5K test images	16.5	39.2	52.0	9.0

4.2   The Process of Adversarial Training

The generator in WGAN can be regarded as an image captioning model due to the lack of adversarial loss, similar as Word-Concat. The discriminator identify with the closest ground truth result, comparing with corpus. It justifies that the sentence plausibility is very critical for generating long, convincing sentences.

Table 2 shows the model evaluation of full image predictions on 1,000 test images. Where B-n is BLEU-score that uses up to n-grams, its output is always a number between 0 and 1. This value indicates how similar the candidate text is to the reference texts, with values closer to 1 representing more similar texts.

Table 2. Evaluation of full image predictions on 1,000 test images.

Model	B-1	B-2	B-3	B-4	METEOR	CIDEr
Vinyals [9]	48.0	28.1	16.6	10.0	15.7	38.3
Google NIC	66.6	46.1	32.9	-	-	-
LRCN [5]	62.8	44.1	30.4	-	-	-
MS Research [7]	-	-	-	21.1	20.7	-
Chen and Zitnick [8]	-	-	-	19.0	20.4	-
Karpathy and Fei-Fei [3]	62.5	45.0	32.1	23.0	19.5	66.0
Our model	**62.7**	**45.3**	**32.5**	**23.5**	**19.7**	**66.7**

We compare our method with others. Vinyals et al. [9] use LSTM, which get the first word from through a bias term on the first step. Karpathy et al. [3] use Multimodel-RNN, the assembly result become better. Donahue et al. [5] use a 2-layer factored LSTM and GoogLeNet, which is a different CNN, and report results of a model ensemble. Others [8,12] appear to work worse than ours, but this is likely in large part due to their use of the less powerful AlexNet [7] features. Compared to these approaches, our model generates more accurate descriptions.

4.3   Personalized Sentence Generation

Different with prior works, the proposed model supports the personalized sentence generation, which will produce diverse descriptions by first words. The generator can sequentially output diverse and topic-coherent sentences for an image. If we give two different first words, two models will produce two personalized sentences for the same image, respectively. We present qualitative results of our model in Fig. 6.

Fig. 6. Example sentences generated by the proposed model for test images.

4.4   Objective Alignment Evaluation

The proposed model generates sensible descriptions of images as shown in Fig. 6. The images shown in Fig. 6 with blue background translated more accurate, the first prediction a large building with a clock tower on top of it does not appear in the training set. However, there are a large building, a clock tower, on top of are occurrence, which the model may have composed to describe the first image. However, picture of the running car, the model cannot accurately determine the state. The last two pictures, when judge the objects (especially color judgment), there have been some problems. They cannot accurately determine the color of the horse and water. Further, we will output the phrase more accurately by adjusting the recognition within the bounding boxes.

In general, we find a relatively large portion of generated sentences can be found in the training data. The proposed method can be repeated many times without any syntactic specification. Therefore, the generated sentence is more smooth and complete, and the generated words are relatively rich and appropriate.

5   Conclusion

In this paper, we propose a novel approach for cooperation between modeling CNN and RTT-GAN. When we use CNN to extract features, Image-Sentence alignment evaluation will be used to build the Image-Sentence rank, higher scores words will be selected to generate sentences. RTT-GAN uses heuristic rules to generate smoothly sentence with rich vocabularies. We evaluated the performance on full frame experiments and showed that our model outperforms retrieval baselines. As our future work, we will attempt to update RTT-GAN, improve levels and types on discriminators to enhance sentences recognition effect, and apply it to some other applications, such as paragraph generation.

References

1. Everingham, M., Van Gool, L., Williams, C.K.I., Winn, J., Zisserman, A.: The Pascal visual object classes (VOC) challenge. IJCV **88**(2), 303–338 (2010)
2. Kulkarni, G., Premraj, V., Dhar, S., Li, S., Choi, Y., Berg, A.C., Berg, T.L.: Baby talk: understanding and generating simple image descriptions. In: CVPR, pp. 1601–1608 (2011)
3. Karpathy, A., Fei-Fei, L.: Deep visual-semantic alignments for generating image descriptions. In: CVPR, pp. 3128–3137 (2015)
4. Kiros, R., Salakhutdinov, R., Zemel, R.S.: Unifying visual-semantic embeddings with multimodal neural language models. Comput. Sci. 3412–3415 (2014)
5. Zhang, C., Xue, Z., Zhu, X., Huang, Q., Tian, Q.: Boosted random contextual semantic space based representation for visual recognition. Inf. Sci. **369**, 160–170 (2016)
6. Wang, J., Yang, J., Yu, K., et al.: Locality-constrained linear coding for image classification. In: CVPR, pp. 3360–3367 (2010)
7. Fang, H., Gupta, S., Iandola, F., Srivastava, R., Deng, L., Dollar, P., Gao, J., He, X., Mitchell, M., Platt, J., et al.: From captions to visual concepts and back. arXiv:1411.4952 (2014)
8. Chen, X., Zitnick, C.L.: Learning a recurrent visual representation for image caption generation. CoRR, arXiv:1411.5654 (2014)
9. Vinyals, O., Toshev, A., Bengio, S., et al.: Show and tell: a neural image caption generator. Comput. Sci. 3156–3164 (2015)
10. Liang, X., Hu, Z., Zhang, H.: Recurrent topic-transition GAN for visual paragraph generation. arXiv preprint arXiv: 1703.07022 (2017)
11. Simonyan, K., Zisserman, A.: very deep convolutional networks for large-scale image recognition. Comput. Sci. 1543–1544 (2014)

12. Zhang, X.-Y., Wang, S., Zhu, X., Yun, X., Wu, G.: Update vs. upgrade: modeling with indeterminate multi-class active learning. Neurocomputing (NEUCOM) **162**, 163–170 (2015)

13. Denkowski, M., Lavie, A.: Meteor universal: language specific translation evaluation for any target language. In: The Workshop on Statistical Machine Translation, pp. 376–380 (2014)

14. Vedantam, R., Zitnick, C.L., Parikh, D.: Cider: consensus-based image description evaluation. CoRR, arXiv:1411.5726 (2014)

15. Zhang, X.-Y., Wang, S., Yun, X.: Bidirectional active learning: a two-way exploration into unlabeled and labeled dataset. IEEE Trans. Neural Netw. Learn. Syst. (TNNLS) **26**(12), 3034–3044 (2015)

3D Convolutional Neural Network for Action Recognition

Junhui Zhang[1,2], Li Chen[1,2(✉)], and Jing Tian[1,2]

[1] School of Computer Science and Technology,
Wuhan University of Science and Technology, Wuhan 430065, China
`mrlittlezhu@gmail.com, chenli@wust.edu.cn`
[2] Hubei Province Key Laboratory of Intelligent Information Processing
and Real-time Industrial System, Wuhan University of Science and Technology,
Wuhan 430065, China
`http://www.wust.edu.cn`

Abstract. Compared with traditional machine learning methods, deep convolutional neural network has more powerful learning ability. The convolutional neural network model of the depth learning algorithm has made remarkable achievements in the field of computer vision. As a branch of neural network, 3D Convolutional neural network (3D CNN) is a relatively new research field in the field of computer vision. To extract features that contain more information, we develop a novel 3D CNN model for action recognition instead of the traditional 2D inputs. The final feature consists spatial and temporal information from multiple channels. We demonstrate the efficacy of our method on KTH dataset.

Keywords: 3D Convolutional neural networks
Human action recognition · Surveillance video

1 Introduction

The identification of human behavior in people's daily life has a wide range of application, especially in urban life. The identification of human behavior can ensure people's daily life away from danger, this confirms its importance for people's lives. However, using surveillance video to identify and understand the human behavior has a relatively high challenge. The traditional 2D convolutional neural network (CNN) can only be used to predict the static image, and it is very weak to find dynamic images such as human behavior.

Large-pose face alignment via CNN-based dense 3D model fitting [1] propose a face alignment method for large-pose face images, by combining the cascaded CNN method and 3D Morphable Model, in which CNN is a regression network that can be learned by creating a high-level structure from a low-level structure to a Hierarchical structure of the characteristics, so as to automatically

---

View codes on GitHub: https://github.com/mrlittlepig/cnn.git.

© Springer Nature Singapore Pte Ltd. 2017
J. Yang et al. (Eds.): CCCV 2017, Part I, CCIS 771, pp. 600–607, 2017.
https://doi.org/10.1007/978-981-10-7299-4_50

complete the creation of features, such a learning process can be supervised or unsupervised learning to complete, in the field of computer vision this learning training result shows its excellent competitiveness such as Deep residual learning for image recognition [2]. CNN can achieve superior performance on visual object recognition tasks without relying on handcrafted features. In addition, CNN have been shown to be relatively insensitive to certain variations on the inputs.

At present, most of the research on deep convolutional neural network is in the field of computer vision, such as grid based object detector using convolutional neural networks (G-CNN) [3], which starts with a multi-scale grid of fixed bounding boxes, train a regressor to move and scale elements of the grid towards objects iteratively. The experiments training of the image data under deep convolutional neural network technology, are mostly two-dimensional deep convolutional neural network experiments, such as fast region-based convolutional network method (Fast R-CNN) [4], Fast R-CNN builds on previous work to efficiently classify object proposals using deep convolutional networks, compared to previous work, Fast R-CNN employs several innovations to improve training and testing speed while also increasing detection accuracy. With the rapid development of artificial intelligence, two-dimensional convolutional neural network in the field of computer vision research is a mature science technology. The 3D CNN derived from the neural network conforms to the needs of the time, and the research on the 3D CNN has made a major breakthrough. Different from the 2D convolutional neural network, the 3D CNN uses 3D data as the input data of the network. Using continuous fixed frames number of images not single image to extract temporal and spatial domain information of the video when we do experiment, thus, the understanding of the characteristics of human behavior achieves a good result. The proposed model utilizes 3D CNN network models by extracting temporal and spatial domain features in the video, shows its great development potential in action recognition.

Our experiments show that the developed 3D CNN model outperforms other baseline methods on the surveillance video data set, and the 3D CNN model is used to verify the results of the KTH data set, and the accuracy of the prediction of human behavior is 91.67%. We also observe that the performance differences between 3D CNN and other methods tend to be larger when the number of positive training samples is small.

2   Convolutional Neural Network

In 2D CNNs, 2D convolutional is performed at the convolutional layers to extract features from local neighborhood on feature maps in the previous layer. Then an additive bias is applied and the result is passed through a sigmoid function. Formally, the value of unit at position (x, y) in the jth feature map in the ith layer denoted as v_{ij}^{xy}, is given by

$$v_{ij}^{xy} = tanh(b_{ij} + \sum_{m} \sum_{p=0}^{P_i-1} \sum_{q=0}^{Q_i-1} w_{ijm}^{pq} v_{(i-1)m}^{(x+p)(y+q)}) . \tag{1}$$

where $tanh()$ is the hyperbolic tangent function, b_{ij} is the bias for this feature map, m indexes over the set of feature maps in the $(i-1)th$ layer connected to the current feature map, w_{ijk}^{pq} is the value at the position (p, q) of the kernel connected to the kth feature map, P_i and Q_i are the height and width of the kernel [5].

2.1   3D Convolutional Neural Network

In 3D convolution, the same 3D kernel is applied to overlapping 3D cubes in the input video to extract motion features. previous layer, thereby capturing motion information. Formally, the value at position (x, y, z) on the jth feature map in the ith layer is given by

$$v_{ij}^{xyz} = tanh(b_{ij} + \sum_{m} \sum_{p=0}^{P_i-1} \sum_{q=0}^{Q_i-1} \sum_{r=0}^{R_i-1} w_{ijm}^{pqr} v_{(i-1)m}^{(x+p)(y+q)(z+r)}) . \tag{2}$$

where R_i is the size of the 3D kernel along the temporal dimension, w_{ijm}^{pqr} is the $(p, q, r)th$ value of the kernel connected to the mth feature map in the previous layer [5]. The entire convolutional operation contains the input layer of the current layer and a max-pooling layer. The convolutional operation obtains the output layer of the current layer, that is, the input layer of the next layer, and the feature of the current feature map is extracted by convolutional calculation, show as Fig. 1.

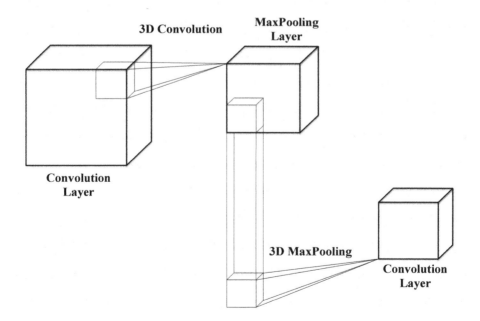

Fig. 1. The feature extracting cell as in the figure, a convolution-pool structure with a 3D convolution kernel and a 3D max-pooling kernel shown in this sample caption.

2.2 The Proposed 3D Convolutional Neural Network Model

Through multiple 3D convolutional layers connect together constitute a convolutional neural network, the network model used in this paper as shown in Fig. 2. We use 32×32 continuous 32-frame picture data as input data, all convolutional calculation stride is 1. In the first layer 3D convoluted kernel size is $7 \times 7 \times 7$, after convolutional calculation becomes $24 \times 24 \times 24$, there are 16 filters get 16 different features maps, the first 3D max-pooling kernel size of pooling layer is $2 \times 2 \times 2$, stride is 2, after max-pooling operation, 16 times $12 \times 12 \times 12$ output is obtained. In the second layer, the 3D convolutional kernel size is $5 \times 5 \times 5$, after convolutional calculation bring about $8 \times 8 \times 8$, there are 32 filters get 32 different feature maps, the second 3D max-pooling kernel size of pooling layer also $2 \times 2 \times 2$, stride is 2, after max-pooling operation, 32 times $4 \times 4 \times 4$ output is obtained. In the third layer, the 3D convolutional kernel size is $3 \times 3 \times 3$, after convolutional calculation bring about $2 \times 2 \times 2$ of the output size, there are 64 filters get 64 A different feature map, after the pooling size $2 \times 2 \times 2$ of the max-pooling operation 128 times $1 \times 1 \times 1$ output is obtained, and finally after full connection the six different output is generated; In the above convolutional layer and pool layer has a *ReLU* activation function the full architecture is

$$C(16,7,1)-R-P(2,2)-C(32,5,1)-R-P(2,2)-C(64,3,1)-R-P(2,2)-FC(64). \tag{3}$$

where $C(d,m,s)$ indicates a convolutional layer with d filters of spatial size $m \times m \times m$, applied to the input with stride s. All hidden weight layers use the *ReLU* activation function. And the max-pooling $P(m,s)$ is performed over $m \times m$ spatial windows with stride s.

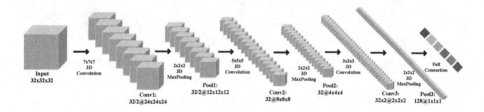

Fig. 2. The 3D convolutional network structure shown as upon, with down-sampling layers extracting features network gets six channels of output classification probabilities by a full-connection layer.

3 Experimental Result

Traditional 2D convolutional neural network with static pictures as training data has a good adaptability to the 2D matrix calculation, commonly used in the field of image processing, image clarity class recognition and evaluation, such as Learning-based Leaf Occlusion Detection In Surveillance Video [6], in this paper, the authors use 2D Convolutional Neural Network to train pictures that are blocked by leaves, and then combine the test results with Bayesian theory to optimize the final result, which is a typical example of CNNs.

3D convolutional neural network is using three-dimensional data as input, has a better adaptability to the data with continuous temporal and spatial domain characteristics of the video, so we also use the surveillance videos with jitter exception as our experiment data set. Using the jitter exception surveillance video as the experimental data on the model performance test has a very good credibility. The analysis of the video jitter exception can be viewed on the paper [7–9].

3.1 Data Collection

We trained our network in KTH data set http://www.nada.kth.se/cvap/ actions/, evaluate the developed 3D CNN model for action recognition. The KTH data set is shown as Fig. 3. We also collected about 1600 video data with jitter exception, and after multiple rounds of screening, 500 videos of the most typical video with jitter exception as the experimental data, in order to distinguish the abnormal data (here called negative samples), and added 500 Normal video as a positive sample of the experiment, and then take each 100 samples in the positive and negative samples to test the parameters after training, the rest samples used to train our designed model. In the collected data, there are roughly the following resolutions, such as 352×288, 528×284, 704×576, 1280×720, in experiment we unified images to same size.

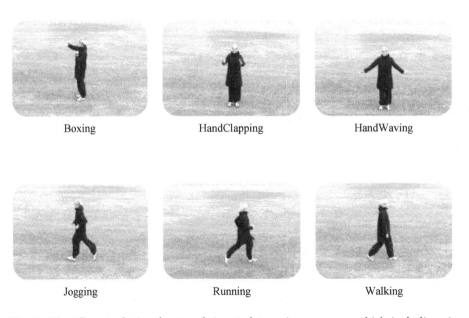

Boxing   HandClapping   HandWaving

Jogging   Running   Walking

Fig. 3. The 3D convolutional network input data using as upon, which including six classes of images, boxing, hand-clapping, hand-waving, jogging, running and walking.

3.2 Data Augmentation

The easiest and most effective way to prevent training over-fitting is to expand the data, is to artificially enlarge the data set using label-preserving transformations. [10], in which paper they employ two distinct forms of data augmentation, both of which allow transformed images to be produced from the original images with very little computation, so the transformed images do not need to be stored on disk. Different from 2DCNNs, 3DCNN architectures is more complicated, thus, we use ways as blew to data expansion, we start from the different frames of the video to capture frames as different experimental data, since the number of frames extracted at one time is fixed, so this is a valid method of data expansion from different frames. However, in order to reduce the similarity between data, the interval frames between the two start frames can not be too close.

3.3 Network Performance Evaluation Criterion

In order to make the experimental results better, we continuous try new network parameters, modify the size of the input image and extract the number of frames, optimization the size of the network layers, a number of tests was done. The classification error between the predicted class and the ground truth is used to evaluate the performance of the proposed approach [6]. The proposed network is implemented using python programming language. The algorithm is run on a PC with GPU GTX980 to speed up the process without optimization.

3.4 Input Size

Experiments show that the input data size in a certain range will affect the accuracy of the results. The input data size selection may affect the performance of the algorithm. If we choose lager size of input, the experimental network model also adapts deeper and wider, but the computational resources consumed by the experiment increased. Therefore, the maximum input size of network used in this paper is $54 \times 54 \times 54$, but the experimental results don't become better as the input increases, result shows as Table 1.

Table 1. As shown in the following table, the average accuracy of test data reach 0.9176 scores rank first with input size of $32 \times 32 \times 32$ above other input size of our model.

Input size	HandClapping	Boxing	HandWaving	Running	Walking	Jogging	Average
$16 \times 16 \times 16$	0.82	0.84	0.85	0.78	0.86	0.80	0.825
$32 \times 32 \times 32$	**0.93**	**0.91**	**0.93**	**0.86**	0.91	**0.93**	**0.9167**
$54 \times 54 \times 54$	0.87	0.88	0.87	0.82	0.91	0.89	0.8733

As shown in Table 1, we uses $32 \times 32 \times 32$ size of input network done a number of experiments on KTH data set exception surveillance video, the best result of accuracy has reached 91.67%, as shown in Fig. 4.

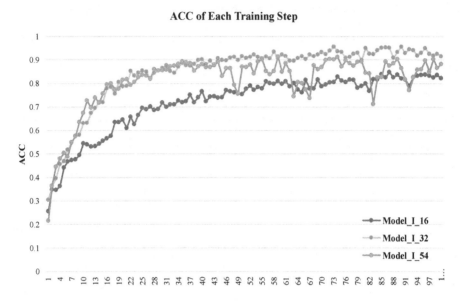

Fig. 4. Accuracy of Jitter Exception Surveillance Video.

As shown in Fig. 4, the model with input size 32 also performs best in jitter exception surveillance video data set, with the training step of output weights the model with input size 32 also performs more stable than other models in test data set.

4   Conclusion

This paper presents a 3D CNN based model for action and evaluate jitter exception surveillance video, from training step up to testing step shows that, the proposed model can extract the temporal and spatial domain information of the video by training the 3D convolutional neural networks, extract the multidimensional features, and perform convolutional and sub-sampling separately in each channel. The final feature representation is computed by combining information from all channels, we use KTH data and Jitter Exception Surveillance video on our model was evaluated, the evaluation results show that 3D CNN in the video which has a dynamic characteristics of the characteristics of temporal-spatial domain with excellent identification.

Acknowledgement. This work was supported by National Natural Science Foundation of China (61773297 and 61375017), Open Fund of Hubei Province Key Laboratory of Intelligent Information Processing and Real-Time Industrial System (2016znss01A).

References

1. Jourabloo, A., Liu, X.: Large-pose face alignment via CNN-based dense 3D model fitting. In: IEEE Computer Vision and Pattern Recognition, pp. 4188–4196 (2016)
2. He, K., Zhang, X., Ren, S., Sun, J.: Deep residual learning for image recognition. In: IEEE Conference on Computer Vision and Pattern Recognition, pp. 770–778 (2016)
3. Najibi, M., Rastegari, M., Davis, L.S.: G-CNN: an iterative grid based object detector. In: IEEE Conference on Computer Vision and Pattern Recognition, pp. 2369–2377 (2016)
4. Girshick, R.: Fast R-CNN. In: IEEE International Conference on Computer Vision, pp. 1440–1448 (2015)
5. Ji, S., Yang, M., Yu, K.: 3D convolutional neural networks for human action recognition. IEEE Trans. Pattern Anal. Mach. Intell. **35**(1), 221–231 (2013)
6. Liu, J., Chen, L., Tian, J., Zhu, D.: Learning-based leaf occlusion detection in surveillance video. In: IEEE Conference on Industrial Electronics & Applications, pp. 1000–1004 (2016)
7. Gulliver, S.R., Ghinea, G.: The perceptual influence of multimedia delay and jitter. In: IEEE International Conference on Multimedia & Expo, vol. 14, pp. 2214–2217 (2007)
8. Gulliver, S.R., Ghinea, G.: The perceptual and attentive impact of delay and jitter in multimedia delivery. IEEE Trans. Broadcast. **53**(2), 449–458 (2007)
9. Do, T.T., Hua, K.A., Tantaoui, M.A.: P2VoD: providing fault tolerant video-on-demand streaming in peer-to-peer environment. In: IEEE International Conference on Communications, vol. 3, pp. 1467–1472 (2004)
10. Krizhevsky, A., Sutskever, I., Hinton, G.E.: ImageNet classification with deep convolutional neural networks. In: Neural Information Processing Systems Conference, vol. 25, no. 2, pp. 1097–1105 (2012)

Research on Image Colorization Algorithm Based on Residual Neural Network

Pinle Qin[1(✉)], Zirui Cheng[1], Yuhao Cui[1], Jinjing Zhang[1],
and Qiguang Miao[2]

[1] School of Computer and Control Engineering,
North University of China, Taiyuan 030051, China
qpl@nuc.edu.cn
[2] School of Computer Science and Technology, Xidian University,
Xian 710075, China

Abstract. In order to colorize the grayscale images efficiently, an image colorization method based on deep residual neural network is proposed. This method combines the classified information and features of the image, uses the whole image as the input of the network and forms a non-linear mapping from grayscale images to the colorful images through the deep network. The network is trained by using the MIT Places Database and ImageNet and colorizes the grayscale images. The experiment result shows that different data sets have different colorization effects on grayscale images, and the complexity of the network determines the colorization effect of grayscale images. This method can colorize the grayscale images efficiently, which has better visual effect.

Keywords: Residual neural network · Image colorization · Non-linear map
Grayscale image

1 Introduction

Image colorization is the process of pseudo colorization for the grayscale images with black and white pixels, which has a great value of research and application. The research shows that the human eyes are more sensitive to the colorful images than the variation of gray value, and the grayscale images are bad for discriminating and recognizing the image information. We can restore, enhance and transform the images' color which improves the visual effect of human eyes to the original grayscale images, makes the observers more easily and clearly to learn the content and information of the images and improves the use value of images. The pseudo colorization method of the grayscale images is also applied in processing of the remote sensing images, satellite images and medical images to make the processed image clearer. The image pseudo colorization technology is also appropriated to the artistic fields, such as cartoon colorization, ancient painting repair etc.

The traditional image coloring methods can be broadly divided into two categories, the one is based on users who make the local colors spread to the whole image; the other is to transfer colors from another colorful image to the grayscale image that need to be colorized. Levin et al. [1] proposed that the colorization can be regarded as global

© Springer Nature Singapore Pte Ltd. 2017
J. Yang et al. (Eds.): CCCV 2017, Part I, CCIS 771, pp. 608–621, 2017.
https://doi.org/10.1007/978-981-10-7299-4_51

optimization. Due to the higher computational complexity and slower speed, there are less conducive to the images that have more complex structures and colors. So, aiming at the disadvantages of Levin algorithm, Yatziv and Sapiro [2] proposed a quick colorization algorithms using Dijistra shortest path algorithm, which improved the speed but increased the complexity of user interaction. Luan et al. [3] proposed the image colorization algorithm based on stroke-interaction segmentation, which however needs a lot of input of strokes to implement more accuracy segmentation. Qu et al. [4] combined the similarity between color and feature space, and used texture feature classification to achieve coloring of black and white comics. Different from the method based on local color transfer, Welsh et al. [5] proposed a general technique for colorizing the black-and-white images by matching brightness and context information between black-and-white images and reference images. However, the pixel matching consumed a lot of time and couldn't come into a satisfy results. Irony et al. [6] proposed an image colorization method based on texture feature matching under the hypothesis that the similar context will has similar color, where however they need to segment the reference image and target grayscale image. Liu et al. [7] proposed a colorization method based on reference image that directly searches the reference colorful image related to black-and-white image on the Internet and computed the intensity difference robustness. However, this only applies to images that can be found perfectly matched, which is a very time-consuming task.

Recently, with the development of deep learning [8], the deep learning technology was applied to image colorization continuously. Cheng et al. [9] proposed a grayscale image colorization method based on neural network, in which this method use neural network to extract features from grayscale images. However, due to the fewer dataset and the simpler network, this method greatly limits the applicable type and colorization effect of the images. Larsson et al. [10] used convolutional neural network to extract the low dimensional features and semantic information from grayscale images and used super column model to predict the color distribution of each pixel. Iizuka et al. [11] used double flow structure. On the one hand, they extracted the global features of the image, on the other hand, they extracted the low dimensional local features, and fused the local features with global features to predict color information of each pixel. Zhang et al. [12] extracted the grayscale images used VGG [13] style convolutional neural network and predicted the color distribution histogram to colorize the grayscale images. But these three methods adopt the up-sampling methods to restore the image size, which made the image lose a part of information.

This paper proposed an image colorization method based on residual neural network [14]. Firstly, the network used the VGG network to extract the classification information of the image, and used the residual neural network to extract the texture and detail features of the image, then combined the classification information with the texture detail information, and carried on the feature extraction through the residual neural network. Finally, we computed mean squared error between the network output and chromaticity of the colorful images, which will find the mapping relationship between grayscale images and chromaticity of images. The network will not modify the size of feature map, so there is no existence of up-sampling and feature scaling in the network, and the deeper residual neural network model is benefit for grayscale images in feature extraction, which make the network has better colorization ability.

2 The Residual Neural Network

The convolution neural network (Convolution Neural Network, CNN) [15] is an artificial neural network, which has become the research hotspot in recent years in some fields such as image recognition, speech analysis, natural language processing. The CNN is a weight shared network structure, which has fewer number of parameters and lower complexity of learning. The CNN generally contains data input layer, convolutional layer, pooling layer, full connection layer and output layer. In recent years, it is because of the emergence of CNN, the deep learning was able to have the development of blowout. In the large-scale image recognition challenge such as ILSVRC (Image Large Scale Visual Recognition Competition) and ImageNet [16] over the years, there have been a large number of excellent networks based on CNN, such as AlexNet, VGG, GoogleNet [17], ResNet etc., which makes the error rate of Top5 has been reduced to below 5%, which is sufficient to explain the effectiveness of convolution neural network.

The emergence of the residual neural network (ResNet) well solves the problem that the gradient of the network will disappear with the increasing number of layers in the CNN. The ResNet mainly consists of convolutional layers and residual block. Compared with the previous study is a non-reference function, the significantly improved network structure of the ResNet can study its residual function based on the input of the network, then use a multilayer network to fit a residual mapping rather than a network layer that is expected to be in a few layers of each group to fit directly into the actual mapping that the network expects. Using $H(x)$ to represent the desired actual mapping, the ResNet uses the stacked nonlinear multilayer network to fit another mapping relation $F(x) := H(x) - x$, then the actual mapping relation can be expressed as $F(x) + x$, which can be implemented via a "shortcut link" forward neural network, which refers to a connection that skips one or more layers. As shown in the Fig. 1, the shortcut link is a residual unit, which does not introduce new parameters, nor does it increase the computational complexity. Besides, He Kaiming's experiments show that such a residual structure in the extraction of image texture and detail features are significant results.

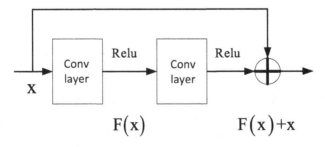

Fig. 1. Residual neural network

3 The Image Colorization Algorithm Based on Residual Neural Network

At present, in the study of grayscale image colorization based on deep learning, the colorization problem of image is regarded as a network prediction model. Iizuka et al. exploited the advantages of CNN, on the one hand, they extracted "global features" from grayscale images with 10 convolutional layers and 3 fully connected layers of VGG style; on the other hand, they extracted the "local features" of the grayscale image with 8 convolutional layers, and then merged the "global feature" and "local features" through the fusion algorithm. Iizuka used the "global features" and "local features" of the image to describe the semantics of the image. The two networks simultaneously extracted the features of the grayscale images and did not interfere with each other and the parameters of both networks were updated at the same time in back propagation. The global features vector outputted through the global feature network can use the global characteristics of the category label to get more performance. However, when Iizuka extracted the local features of the image, the use of the nearest neighbor difference algorithm on the feature map up-sampling will also lose the image information, resulting in poor coloring, and only used 10 convolution layers which is much less compared with the proposed method to extract the image features.

With the deepening of CNN layers reach a certain number, the greater the depth of the network has reached the worse effect, while the ResNet solved the gradient degradation problems in the training set and validation set and had proved that the increase of the depth of the network, the error rate is more and more small. However, in this paper, the network structure is proposed. On the one hand, we uses the CNN to obtain the classification information of the image. On the other hand, the deeper residual neural network is used to extract the texture and detail features of the image, where the classification information and texture detail information are merged. Then the ResNet further extracts the fused features so that entire network model constitutes mapping from the L channel of images to a and b channels. Compared with Iizuka, the ResNet can extract more texture and detail features of images, and in the proposed network structure, there is no up-sampling layer, so the coloring effect will tend to be the real image.

3.1 The Structure of the Network

The grayscale image colorization method based on ResNet proposed in this paper fully considers the relationship between the grayscale images and its corresponding a and b color channels, that is, the network is a kind of Non-linear mapping from grayscale image to a and b channel corresponding to the grayscale image. The network structure of this paper is shown in Fig. 2.

As shown in Table 1, the network model consists of four parts, (a) is the classification subnetwork (orange and blue part in Fig. 2), (b) and (d) are two residual sub-networks (yellow part in the Fig. 2), and (c) is the fusion layer. Each residual sub-network contains five residual cells, each residual cell consists of a three-layer convolution layer, corresponding 3 convolution operations. For example, in Fig. 2, the

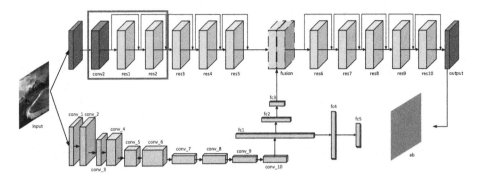

Fig. 2. Network structure diagram (Color figure online)

Table 1. The detail of network's structure

(a) Classification Features Network				(b)Residual Networks 1				(d) Residual Networks 2			
Name	Kernel	Stride	Outputs	Name	Kernel	Stride	Outputs	Name	Kernel	Stride	Outputs
conv_1.	3x3	2x2	64	conv1.	5x5	2x2	64	fusion2	3x3	1x1	256
conv_2.	3x3	1x1	128	conv2.	3x3	1x1	256	res6_1	1x1	1x1	64
conv_3.	3x3	2x2	128	res1_1	1x1	1x1	64	res6_2	3x3	1x1	64
conv_4.	3x3	1x1	256	res1_2	3x3	1x1	64	res6_3	1x1	1x1	256
conv_5.	3x3	2x2	256	res1_3	1x1	1x1	256	res7_1	1x1	1x1	64
conv_6.	3x3	1x1	512	res2_1	1x1	1x1	64	res7_2	3x3	1x1	64
conv_7.	3x3	2x2	512	res2_2	3x3	1x1	64	res7_3	1x1	1x1	256
conv_8.	3x3	1x1	512	res2_3	1x1	1x1	256	res8_1	1x1	1x1	64
conv_9.	3x3	2x2	512	res3_1	1x1	1x1	64	res8_2	3x3	1x1	64
conv_10.	3x3	1x1	512	res3_2	3x3	1x1	64	res8_3	1x1	1x1	256
fc1	-	-	4096	res3_3	1x1	1x1	256	res9_1	1x1	1x1	64
fc2	-	-	256	res4_1	1x1	1x1	64	res9_2	3x3	1x1	64
fc3	-	-	64	res4_2	3x3	1x1	64	res9_3	1x1	1x1	256
fc4	-	-	1024	res4_3	1x1	1x1	256	res10_1	1x1	1x1	64
fc5	-	-	205	res5_1	1x1	1x1	64	res10_2	3x3	1x1	64
(c) Fusion Layer				res5_2	3x3	1x1	64	res10_3	1x1	1x1	256
Name	Kernel	Stride	Outputs	res5_3	1x1	1x1	256	output	3x3	1x1	2
fusion.	1x1	1x1	128	fusion1	1x1	1x1	64				

purple marquee section contains two residual cells, which are enlarged and shown in Fig. 3, conv2 is the input of residual unit res1, res1_3 and conv2 will be "shortcut linked" ($F(x) + x$ as mentioned above) by pixel adding to a new res1_3 layer, which will continue to be propagated as the input of the second residual unit res2.

The L-channel of the image based on the CIE L * a * b color space is the input of network, which will enter the classification subnetwork (a) and the residual sub-network (b) respectively. The classification subnetwork (a) will extract the classification features of the image, and finally the 1×64 feature vector is the output of the full connection layer fc3. After entering the residual sub-network (b), the image is continually extracted from the texture and details features to the fusion1 layer, that is the structure of the pre-fusion residual sub-network. The 1×64 feature vector fc3 output by the classification network will contain a large number of global semantic information about the gray scale of the starting input. Therefore, it is merged with the fusion1 layer of the first residual subnetwork to obtain the fusion layer with

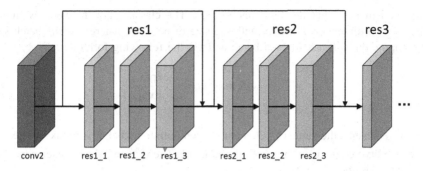

Fig. 3. The residual unit structure diagram

classification information. The concrete fusion method is to extend FC3 replication to the same dimension (H * W * 64) as fusion1, where the values of each pixel in each layer are the same, and then the extended FC3 and fusion1 are directly connected, so we can obtain 128 feature maps. The 128 feature maps contain a large number of semantic information and classification information of the image, and then these 128 feature maps will do convolution operation by using convolution kernel with size is 1 * 1, stride is 1, number is 128, so that the number of feature map is 128 in the fusion layer. And the fusion layer will be the input of the second residual subnetwork, and the feature extraction will be performed again to obtain the final output layer.

3.2 Loss and Training

After the network is built, it is necessary to train the network to optimize the weight parameters in the network, including the convolution kernel parameters and bias parameters for each convolution layer. The loss of network training consists of two parts, the one is the loss of mean square error of the residual network, and the other is the classification loss of the classification network. The residual neural network finally outputs the feature map with the size of $H * W * 2$, and make mean square errors with the supervisory signal of the network, the a and b channels of the image, and the mean square error is used as the loss function of the residual neural network, that is, the loss function mentioned above (1).

$$L_1 = \frac{1}{n} \sum_{i=1}^{n} [\frac{1}{wh} \sum_{j=1}^{w} \sum_{k=1}^{h} \|D_i(j,k) - X_i(j,k)\|^2]$$ (1)

where, n is the number of each training batch sample, w and h are the width and height of each sample, respectively, D is the network output ab channel, X is the ab channel of the original image. When the loss function propagates backwards, only the weights and bias parameters in the residual neural network are updated, while the classification network is not affected.

And the classification network will eventually output 205 categories, which are made softmax cross entropy loss with the classification label corresponding to the image, and then we get the loss function (2). The classification network eventually outputs a two-dimensional matrix with the size of n * 205, where *n* is the batch size. We compute the softmax in each batch and get the following formula:

$$y_i^{\text{logtis}} = soft \max(x)_i = \frac{\exp(x_i)}{\sum_j \exp(x_j)} \tag{2}$$

where x_i represents the value of *i* element in each output, and after softmax, we get 205 the probability vectors y_i^{logtis} of which the sum is 1. The y_i^{logtis} will be cross-entropy with y_i^{label}. The formula is as follows:

$$L_2 = \frac{1}{n} \sum_{i=1}^{n} [-\sum_i y_i^{\text{label}} \log(y_i^{\text{logtis}})] \tag{3}$$

where n is the batch size. After the cross-entropy loss of y_i^{logtis} and y_i^{label}, the loss of each batch needs to be averaged. However, if the value of y_i^{logtis} is 0, the $\log(y_i^{\text{logtis}})$ is infinite, so the design of the truncation function is as follows, the value that less than $1e - 10$ is equal to $1e - 10$.

$$f(x) = \left\{ \begin{array}{ll} 1e - 10 & y_i^{\text{logtis}} < 1e - 10 \\ 1.0 & y_i^{\text{logtis}} > 1.0 \\ y_i^{\text{logtis}} & 1e - 10 \le y_i^{\text{logtis}} \le 1.0 \end{array} \right\} \tag{4}$$

Finally, we get L_2 loss function.

4 Experiments and Analysis

4.1 The Experiment Platform and Dataset

CNN training process requires a lot of matrix operations. Besides, graphics processing (Graphics Processing Unit, GPU) is more suitable for a large number of matrix operations than the central processing unit (Central Processing Unit, CPU). So, the network achieved residual neural network colorization training on NVIDIA Tesla M40, simultaneously used TensorFlow [18] deep learning framework for experiments.

In this paper, two datasets are used to train the network model of this paper. One dataset is the MIT Places Database [19], which contains more than 2.5 million scene pictures corresponding to 205 scene categories. The other dataset is ImageNet [20], which contains more than 120 million images, corresponding to 1000 categories.

4.2 The Effect of Classification Network (Guidance Network) on Coloring Effect

The classification network is similar to the VGG classification network. The feature vector after three-layer full connection will contain a large amount of semantic information about the image that can help to classify the image and meanwhile can help to guide the residual sub-network for the process of colorization. The following Fig. 4 shows the colorization results of the colorization model obtained by the residual neural network singly and the residual neural network guided by the classification network.

It can be found that the effect of guiding the network model is better than that of the network without guidance from the aspects of image authenticity, texture detail and saturation. From the first line, it can be seen that the model of the guiding network is

Fig. 4. The colorization effect of classification guidance network. (a) Grayscale image. (b) The colorization effect without guidance network. (c) The colorization effect with guidance network (d) Ground truth. (Color figure online)

more realistic than the model that without guidance when coloring the plants; The same effect is also reflected in the second line of the image, although the color of the clothes worn by the person is not ideal, from the color of the floor, it seems to be a lot better than the network model without guidance; The color of the house in the third line of the image, the indoor color in the fourth line of the image, and the color of the light in the fifth line of the image is closer to the real color compared with the coloring effect of the model without the guidance network.

The value Peak Signal to Noise Ratio (PSNR) is directly related to the Mean Squared Error (MSE) between images that is the loss function used in this paper. Therefore, this paper randomly selects 1000 images from the Places verification set to calculate the PSNR between the colorized images through the network and the real images to evaluate the colorization effect objectively. As shown in the Table 2, when there is no guidance network, the average PSNR is 25.19 db between the network colorization image and the real image, and in the case of a guide network, the average PSNR is 25.30 db, which can objectively reflect that the guidance network is useful to the residual neural network colorization. At the same time, The RMSE (Root Mean Square Error) of with guidance network is smaller than that without guidance network. PSNR statistical frequency as shown Fig. 5, we can see that the performance of PSNR numerical distribution with guidance network is more ideal than without guidance network.

Table 2. The comparision of without guidance network and with guidance network

	RMSE	PSNR
Without guidance network	0.145	25.14
With guidance network	0.137	25.25

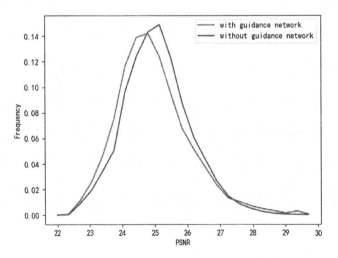

Fig. 5. PSNR histogram of without guidance network and with guidance network (Color figure online)

4.3 Comprehensive Experimental Results

In this paper, five representative images are selected to compare with Iizuka, and it is found that the colorization effect of the network model proposed by this paper is better than Iizuka in saturation and authenticity. As Fig. 6 shows, compared with the red frame in the image, the network model in this paper reflects the details that Iizuka does not show. For example, in the first pair of contrast images, whether the clouds exposed the sunlight, the color of the forest is more vivid; in the second pair of contrast images, the color of the sky and trees in colorization result is more bright than Iizuka; In contrast to the third images, Iizuka did not distinguish between the color of the sky and the beach, resulting in coloring them with the same green color, while the model in this paper not only distinguish the sky and the beach, but also restore the beach to the most true color.

a   b

Fig. 6. The colorization effect comparison of our methods and Iizuka methods. (a) The result of Iizuka methods. (b) The result of our methods. (Color figure online)

Observing the final result, that is, the *a* and *b* channel of the network output, is a fairly intuitive evaluation criterion, so we convert the image in the first column of the graph into a three-dimensional matrix based on the CIE $L * a * b$ color space, extract the matrix in the a and *b* channel, and then the value of the *l* channel is set to 100, in which we obtain a new three-dimensional matrix, and converted into RGB channel image, showed as the following Fig. 7 in the second line, so we can intuitively observe ab channel color information in the final network output. It can be seen from the Fig. 7 that the results of this paper can distinguish the sky, mountains, rivers, forests, and the boundary contour more clearly, while the Iizuka effect can't be a good distinction.

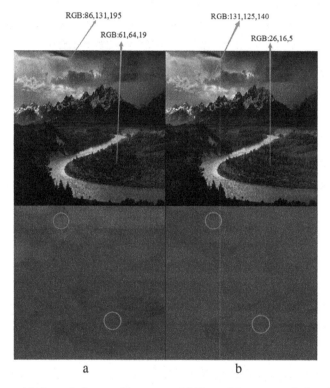

Fig. 7. The *a* and *b* channel of network's outputs. (a) Network's outputs of mine. (b) Network's outputs of Iizuka.

This paper and zhang's proposed method both used the deep learning to colorize the grayscale image. The differences are that the network structure proposed in this paper is deeper, and the algorithm proposed by zhang will make the image color distorted and unstable result in the process of image coloring. In this paper, we use the residual block to deepen the network to better extract the characteristics of the image, on the other hand, we use the image classification information to guide the ResNet for coloring, and the network did not use the up-sampling, so the image coloring effect will be better than zhang's method. Five representative images were taken from the ImageNet validation set and compared with Zhang. As shown in the following Fig. 8, the first column image is grayscale images, the second column is the colorization effect by zhang, the third column is the coloring result of the network model in this paper, and the fourth column is the true color image. From the first image, the colorization effect

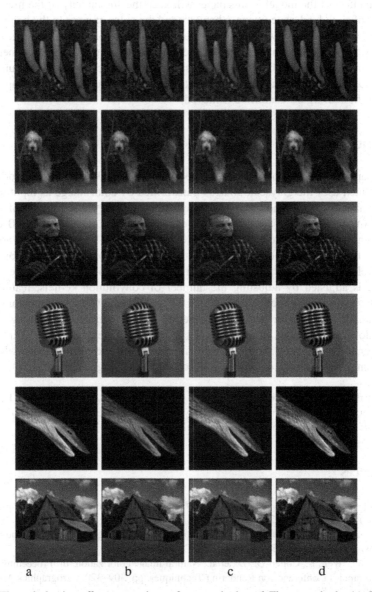

Fig. 8. The colorization effect comparison of our methods and Zhang methods. (a) Grayscale image. (b) The result of Zhang methods. (c) The result of our methods. (d) Ground truth. (Color figure online)

of this network mostly restores the true effect of the image, while the Zhang's colorization effect is like adding a yellow filter, which can't restore the real scene of the image. The same effect is also reflected in the second image, the network also restores the grass and the tree behind the dog, while the Zhang performed commonly, even in the grass appeared abnormal red; In view of the third, fourth, fifth image, Zhang will misjudge some areas and cause the colorization effect to be abnormal, while the colorization effect of the model in this paper will keep the authenticity of the image. From the view of sixth landscape image, Zhang extend the color of grass to the house and the abnormal red to the sky, while the result of the colorization of our network model is closer to the real effect than Zhang. Although the coloring effect of this network is better than Zhang's, but because of the singularity of the network structure in this paper, the coloring effect is only close to the coloring effect of the real map, and the effect of the real figure cannot be achieved.

5   Conclusion

In this paper, an image colorization method based on residual neural network is proposed, which uses the grayscale images transformed by color image as input to output the chromaticity corresponding to grayscale images, that is, the ab channel of the color image. And then splicing the grayscale L channel and the output of the AB channel, and obtain the color picture based on LAB color space, and finally formed a non-linear mapping from grayscale to the colorful image. The network consists of convolution layer, ReLU, BN layer, fully connected layer, and residual block. The network structure is adjusted by adjusting the number of convolution kernels, feature maps, hidden layers to learn more features, which will help the network output color channels of the image. This paper uses the ImageNet, Places datasets as training sets and TensorFlow to train the network model in the GPU environment. Through the experimental comparison, the deep residual neural network constructed by this paper can obtain a better image colorization effect, which has higher practical value.

Acknowledgment. This work is partially supported by Shanxi Nature Foundation (No. 2015011045). The authors also gratefully acknowledge the helpful comments and suggestions of the reviewers, which have improved the presentation.

References

1. Levin, A., Lischinski, D., Weiss, Y.: Colorization using optimization. ACM Trans. Graph. (TOG) **23**(3), 689–694 (2004). ACM
2. Yatziv, L., Sapiro, G.: Fast image and video colorization using chrominance blending. IEEE Trans. Image Process. **15**(5), 1120–1129 (2006)
3. Luan, Q., Wen, F., Cohen-Or, D., et al.: Natural image colorization. In: Proceedings of 18th Eurographics Conference on Rendering Techniques, pp. 309–320. Eurographics Association (2007)
4. Qu, Y., Wong, T.T., Heng, P.A.: Manga colorization. ACM Trans. Graph. (TOG) **25**(3), 1214–1220 (2006). ACM

5. Welsh, T., Ashikhmin, M., Mueller, K.: Transferring color to greyscale images. ACM Trans. Graph. (TOG) **21**(3), 277–280 (2002). ACM

6. Irony, R., Cohen-Or, D., Lischinski, D.: Colorization by example. In: Rendering Techniques, pp. 201–210 (2005)

7. Liu, X., Wan, L., Qu, Y., et al.: Intrinsic colorization. ACM Trans. Graph. (TOG) **27**(5), 152 (2008)

8. LeCun, Y., Bengio, Y., Hinton, G.: Deep learning. Nature **521**(7553), 436–444 (2015)

9. Cheng, Z., Yang, Q., Sheng, B.: Deep colorization. In: Proceedings of IEEE International Conference on Computer Vision, pp. 415–423 (2015)

10. Larsson, G., Maire, M., Shakhnarovich, G.: Learning representations for automatic colorization. In: Leibe, B., Matas, J., Sebe, N., Welling, M. (eds.) ECCV 2016. LNCS, vol. 9908, pp. 577–593. Springer, Cham (2016). https://doi.org/10.1007/978-3-319-46493-0_35

11. Iizuka, S., Simo-Serra, E., Ishikawa, H.: Let there be color!: joint end-to-end learning of global and local image priors for automatic image colorization with simultaneous classification. ACM Trans. Graph. (TOG) **35**(4), 110 (2016)

12. Zhang, R., Isola, P., Efros, A.A.: Colorful image colorization. In: Leibe, B., Matas, J., Sebe, N., Welling, M. (eds.) ECCV 2016. LNCS, vol. 9907, pp. 649–666. Springer, Cham (2016). https://doi.org/10.1007/978-3-319-46487-9_40

13. Simonyan, K., Zisserman, A.: Very deep convolutional networks for large-scale image recognition. arXiv preprint arXiv:1409.1556 (2014)

14. He, K., Zhang, X., Ren, S., et al.: Deep residual learning for image recognition. In: Proceedings of IEEE Conference on Computer Vision and Pattern Recognition, pp. 770–778 (2016)

15. Krizhevsky, A., Sutskever, I., Hinton, G.E.: ImageNet classification with deep convolutional neural networks. In: Advances in Neural Information Processing Systems, pp. 1097–1105 (2012)

16. Russakovsky, O., Deng, J., Su, H., et al.: ImageNet large scale visual recognition challenge. Int. J. Comput. Vis. **115**(3), 211–252 (2015)

17. Szegedy, C., et al.: Going deeper with convolutions. In: Proceedings of IEEE Conference on Computer Vision and Pattern Recognition (2015)

18. Abadi, M., Agarwal, A., Barham, P., et al.: Tensorflow: large-scale machine learning on heterogeneous distributed systems. arXiv preprint arXiv:1603.04467 (2016)

19. Zhou, B., Lapedriza, A., Xiao, J., et al.: Learning deep features for scene recognition using places database. In: Advances in Neural Information Processing Systems, pp. 487–495 (2014)

20. Deng, J., Dong, W., Socher, R., et al.: ImageNet: a large-scale hierarchical image database. In: IEEE Conference on Computer Vision and Pattern Recognition, CVPR 2009, pp. 248–255. IEEE (2009)

Image Retrieval via Balanced and Maximum Variance Deep Hashing

Shengjie Zhang, Yufei Zha$^{(\boxtimes)}$, Yunqiang Li, Huanyu Li, and Bing Chen

Air Force Engineering University, Xi'an 710051, China
`zhayufei@126.com`

Abstract. Hashing is a typical approximate nearest neighbor search approach for large-scale data sets because of its low storage space and high computational ability. The higher the variance on each projected dimension is, the more information the binary codes express. However, most existing hashing methods have neglected the variance on the projected dimensions. In this paper, a novel hashing method called balanced and maximum variance deep hashing (BMDH) is proposed to simultaneously learn the feature representation and hash functions. In this work, pairwise labels are used as the supervised information for the training images, which are fed into a convolutional neural network (CNN) architecture to obtain rich semantic features. To acquire effective and discriminative hash codes from the extracted features, an objective function with three restrictions is elaborately designed: (1) similarity-preserving mapping, (2) maximum variance on all projected dimensions, (3) balanced variance on each projected dimension. The competitive performance is acquired using the simple back-propagation algorithm with stochastic gradient descent (SGD) method despite the sophisticated objective function. Extensive experiments on two benchmarks CIFAR-10 and NUS-WIDE validate the superiority of the proposed method over the state-of-the-art methods.

Keywords: Similarity-preserving · Maximum variance
Balanced variance · Deep hashing · Image retrieval

1 Introduction

In the era of big data, Internet has been inundated with millions of images possessing sophisticated appearance variations. Traditional linear search is obviously not a satisfying choice for inquiring the relevant images from large data sets as its high-computational cost. Due to the high efficiencies in computation and storage, hashing methods [2,9,13] have become a powerful approach of approximate nearest neighbor (ANN) search in above fields. They learn stacks of projection functions to map original high-dimensional features to lower dimensional compact binary descriptors on the conditions that similar samples are mapped into nearby binary codes. By computing the Hamming distance between two binary descriptors via XOR bitwise operations, one can find the similar images of the query one.

© Springer Nature Singapore Pte Ltd. 2017
J. Yang et al. (Eds.): CCCV 2017, Part I, CCIS 771, pp. 622–637, 2017.
https://doi.org/10.1007/978-981-10-7299-4_52

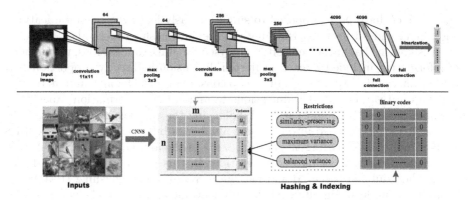

Fig. 1. Pipeline of the proposed BMDH method. When an image is fed into the pretrained deep network, it will finally be encoded into one binary vector. When a batch of m images pass through the network, the output of the full-connected layer is a $n \times m$ real-valued matrix where n is the length of generated binary codes. To learn effective yet discriminative binary codes, three restrictions are used at the top layer of the network: (1) similarity-preserving mapping, (2) maximum variance on all projected dimensions, (3) balanced variance on each projected dimension.

Generally speaking, hashing methods are divided into two categories: data dependent and data independent. Without adopting the data distribution, data independent methods generate hash functions through random projections. Nevertheless, they need relatively longer binary codes to obtain the comparable accuracy [2]. Data dependent methods utilize the data distribution and thus having more accurate results. These methods can be further divided into three streams. The first stream are the unsupervised methods which only use the unlabeled data to learn hash functions. The representative methods include [2,13]. The other two streams are supervised methods and semi-supervised methods which both incorporate label information in the process of generating hash codes. [6,9] are the typical methods of them. Compared to the unsupervised methods, the supervised methods usually need fewer bits to achieve competitive performance as a result from the using of supervised information. Here, we focus on the supervised methods.

Conventional hashing methods exploit hand-crafted features to learn hash codes which can not necessarily preserve the semantic similarities of images in the learning process. By light of the outstanding performance of convolutional neural networks (CNNs) in vision tasks, some existing hashing methods use the CNNs to capture the rich semantic information beneath images. However, these methods even some deep hashing methods [8,14] merely view the CNN architectures as feature extractors which are followed by the procedure of generating binary codes. That is to say, the image representation and hash functions learning are mutual independent two stages. It results in the feature representation may not be tailored to the hashing learning procedure and thus undermining the retrieval performance. Consequently, other hashing methods based deep learning

[4,5,7,16,17] have been proposed to simultaneously learn the semantic features and binary codes thus outperforming the previous hashing methods.

It is known that variance is often analysed for statistics. Besides, the amount of information included in each projected dimension is directly proportional to the variance of it. Thus, there can be two considerations. On one hand, maximizing the total variance on all dimensions to amplify the information capacity of the binary codes. On the other hand, balancing the variance on each dimension so that it can be encoded into the same number of bits. Whereas, most existing hashing methods including the current state-of-the-art method DPSH [5] have not taken these aspects into consideration. In this paper, a novel deep hashing method dubbed balanced and maximum variance deep hashing (BMDH) is proposed to perform simultaneous feature learning and hash-code learning. The pipeline of our method is revealed in Fig. 1. Images whose supervised information is given by pairwise labels are used as the training inputs. The similarity matrix constructed from label information indicates whether two images are similar or not. To acquire more effective hash codes, we exert three criteria on the top layer of the deep network to form a joint objective function. Stochastic gradient descent (SGD) method is utilized to update all the parameters once time. To solve the discrete optimization problem, we relax the discrete outputs to continuous real-values. Meanwhile, a regularizer is used to approach the desired discrete values. Given a query image, we propagate it through the network and quantize the deep features to desired binary codes.

The main contributions of the proposed method are as follows:

- A joint objective function is formed. We exert three restrictions on the top layer of the network to constitute a joint objective function. The first is that the learned binary codes should preserve the local data structure in the original space namely two similar samples should be encoded into nearby binary codes. The second is that the total variance on all projected dimensions is maximized so as to capture more information which is beneficial to the retrieval performance. The last is that the variance on each dimension should be as equal as possible so that each dimension can be encoded into the same number of bits.
- An end-to-end model is constructed to simultaneously perform the feature representation and hash functions learning. In the framework, the above two stages can give feedback to each other. The optimal parameters are acquired using the simple back-propagation algorithm with stochastic gradient desent (SGD) method despite the sophisticated objective function.
- The important variance information on the projected dimensions is focused on in our method. To the best of our knowledge, it is the first time that maximizing the total variance on all projected dimensions and balancing the variance on each dimension is simultaneously considered in the existing deep hashing methods.

The rest of the paper is organized as follows. In Sect. 2, we discuss the related works briefly. The proposed BMDH method is described in Sect. 3. The experimental results are presented in Sect. 4. Finally, Sect. 5 concludes the whole paper.

2   Related Work

Conventional Hashing: The earliest hashing research concentrates on data-independent methods in which the LSH methods [1] are the representatives. The basic idea of LSH methods is that two adjacent data points in the original space are projected to nearby binary codes in a large probability, while the probability of projecting two dissimilar data points to the same hash bucket is small. The hash functions are produced via random projections. However, LSH methods usually demand relatively longer codes to achieve comparable performance thus occupying larger storage space. Data-dependent methods learn more effective binary codes using training data. Among them, unsupervised methods learn hash functions using unlabeled data. The typical methods include Spectral Hashing (SH) [13], Iterative Quantization (ITQ) [2]. Semi-supervised and supervised hashing methods generate binary codes employing the label information. The representative methods include, Supervised Discrete Hashing (SDH) [10], Supervised Hashing with Kernels (KSH) [9], Fast Supervised Hashing (FastH) [6], Latent Factor Hashing (LFH) [15] and Sequential Projection Learning for Hashing (SPLH) [11].

Deep Hashing: Hashing methods based on CNN architectures are earning more and more attention especially when the outstanding performance of deep learning is demonstrated by Krizhevsky et al. on the Imagenet. Among the recent study, CNNH [14] learns the hash functions and feature representation based on binary codes obtained from the pairwise labels. However, it has not simultaneously learned the feature representation and hash functions. DLBH [7] learns binary codes by employing a hidden layer as features in a point-wise manner. [4,16,17] learn compact binary codes using triplet samples. [5,8] learn binary codes with elaborately designed objective functions.

Balance and Maximize Variance: Variance is a significant statistic value which stands for the amount of information to some degree. So, it is quite necessary for us to think about the variance on the projected dimensions. Liong et al. [8] uses the criterion of maximizing the variance of learned binary vectors at the top layer of the network. But it is not an end-to-end model. Isotropic Hashing (IsoHash) [3] learns projection functions which can produce projected dimensions with isotropic variances. Iterative Quantization (ITQ) [2] seeks an optimized rotation to the PCA-projected data to minimize the quantization error which also effectively balances the variance on each dimension.

Totally, most of the existing hashing methods try to preserve the data structure in the original space. But they seldom consider maximizing the total variance on all projected dimensions and balancing the variance on each dimension or just take one aspect into account. In this paper, we propose a novel deep architecture to learn similarity-preserving binary codes with the consideration of simultaneously maximizing the total variance on all dimensions and balancing the variance on each dimension.

3   Approach

The purpose of hashing methods is to learn a series of projection functions $h(x)$ which map u-dimensional real-valued features to v-dimensional binary codes $b \in \{-1, 1\}^v (u \gg v)$ while the distance relationship of data points in the original space is preserved. However, most existing hashing methods have not considered the variance on the projected dimensions which is an important statistic value. To learn compact and discriminative codes, we propose a novel hashing method called BMDH. Three criteria are exerted on the top layer so that a joint loss function is formed: (1) the distance relationship between data points in the Euclidean space is effectively preserved in the corresponding Hamming space, (2) the total variance on all projected dimensions is maximized, (3) the variance on each dimension is as equal as possible. In the following, we first give the proposed model and then describe how to optimize it.

3.1   Notations

Suppose there are M training data points. $\boldsymbol{x_i} \in \mathbb{R}^d (1 \leq i \leq M)$ is the ith data point, and the matrix form is $\boldsymbol{X} \in \mathbb{R}^{M \times d}$. $\boldsymbol{B}^{M \times n} = \{\boldsymbol{b}_i\}_{i=1}^M$ denotes the binary codes matrix, where $\boldsymbol{b}_i = sgn(h(\boldsymbol{x_i})) \in \{-1, 1\}^n$ denotes the n-bits binary vector of $\boldsymbol{x}_i$. $sgn(v) = 1$ if $v > 0$ and -1 otherwise. $\boldsymbol{S} = \{s_{ij}\}$ is the similarity matrix where $s_{ij} \in \{0, 1\}$ is denoted as the similarity label between pairs of points. $s_{ij} = 1$ means the two data points $\boldsymbol{x}_i$ and $\boldsymbol{x}_j$ are similar and the corresponding Hamming distance is low. $s_{ij} = 0$ means the two data points are dissimilar and the corresponding Hamming distance is high. $\boldsymbol{\Theta}$ denotes all the parameters of the feature learning part.

3.2   Proposed Model

Corresponding to the three restrictions, the overall objective function is composed of three parts. The first part aims to preserve the similarities between data pairs. That is to say, the Hamming diatance is minimized on similar pairs and maximized on dissimilar pairs simultaneously. The second part is used to maximize the total variance on all projected dimensions to boost the information capacity of the binary codes. The last part seeks to balance the variance on each dimension so that each dimension is allocated with the same number of bits.

$$
\min_{\boldsymbol{W}} J(\boldsymbol{W}) = J_1(\boldsymbol{W}) - \lambda_1 J_2(\boldsymbol{W}) + \lambda_2 J_3(\boldsymbol{W})
$$

$$
= -\sum_{s_{ij} \in \boldsymbol{S}} (s_{ij} \Phi_{ij} - \log(1 + e^{\Phi_{ij}})) + \rho \sum_{i=1}^m \|\boldsymbol{b}_i - \boldsymbol{g}_i\|_2^2
$$

$$
- \lambda 1 (\sum_{i=1}^n \mathrm{E}(\|\boldsymbol{w}_i^T \boldsymbol{H} + \boldsymbol{p}_i\|_2^2) - \sum_{i=1}^n \| \mathrm{E}(\boldsymbol{w}_i^T \boldsymbol{H} + \boldsymbol{p}_i)\|_2^2) \qquad (1)
$$

$$
+ \lambda 2 (\frac{1}{n} \sum_{i=1}^n u_i^2 - [\frac{1}{n} \sum_{i=1}^n u_i]^2)
$$

where ρ is the regularization term, λ_1 and λ_2 are used to balance different objectives. $\| \cdot \|_2$ is the L2-norm of vector.

To construct an end-to-end model, we set:

$$G = W^T H + Q \tag{2}$$

where $H = \{h_i\}_{i=1}^{m} \in \mathbb{R}^{4096 \times m}$ denotes the output matrix of the full7 layer associated with the batch of m data points, where $h_i \in \mathbb{R}^{4096 \times 1}$ denotes the output vector associated with the data point x_i. $W = \{w_i\}_{i=1}^{n} \in \mathbb{R}^{4096 \times n}$ is the projection matrix of the full8 layer, where w_i is the ith projection vector. $Q \in \mathbb{R}^{n \times m}$ is the bias matrix. It means that we incorporate the hash-code learning part with the feature learning part by a fully-connected layer. As a result, the weight W and bias Q of the hash-code layer are simultaneously updated with all the other parameters of the feature learning layers.

In order to make a better understanding of the joint objective function, we describe it in details as below.

3.3   Preserving the Similarities

The proposed model learns non-linear projections that map the input data points into binary codes and simultaneously preserve the distance relationship of points in the original space.

We define the likelihood of the similarity matrix as that of [15]:

$$p(S|B) = \prod_{s_{ij} \in S} p(s_{ij}|B)$$

while

$$p(s_{ij}|B) = \begin{cases} \frac{1}{1+e^{-\Gamma_{ij}}}, & s_{ij} = 1 \\ \frac{e^{-\Gamma_{ij}}}{1+e^{-\Gamma_{ij}}}, & s_{ij} = 0 \end{cases}$$

where

$$\Gamma_{ij} = \frac{1}{2} b_i^T b_j$$

Taking the negative log-likelihood of the observed similarity labels S as that of [5], we get a primary model:

$$\min_{W} J_1(W) = -\log p(S|B) = -\sum_{s_{ij} \in S} \log(s_{ij}|B)$$

$$= -\sum_{s_{ij} \in S} \left[s_{ij} \log \frac{1}{1+e^{-\Gamma_{ij}}} + (1-s_{ij}) \log \frac{e^{-\Gamma_{ij}}}{1+e^{-\Gamma_{ij}}} \right]$$

$$= -\sum_{s_{ij} \in S} (s_{ij} \Gamma_{ij} - \log(1+e^{\Gamma_{ij}}))$$

It is a reasonable model which makes the similar pairs have low Hamming distance and dissimilar pairs have high Hamming distance. To solve the discrete optimization problem, we set:

$$g_i = W^T h_i + q$$

where $q \in \mathbb{R}^{n \times 1}$ is the bias vector.

Then, we relax the discrete values $\{b_i\}_{i=1}^m$ to continuous real-values $\{g_i\}_{i=1}^m$ as that of [5]. The ultimate model can be reformulated as:

$$\min_{W} J_1(W) = - \sum_{s_{ij} \in S} (s_{ij} \Phi_{ij} - \log(1 + e^{\Phi_{ij}})) + \rho \sum_{i=1}^m \|b_i - g_i\|_2^2 \tag{3}$$

where $\Phi_{ij} = \frac{1}{2} g_i^T g_j$, ρ is the regularization term to approach the desired discrete values.

3.4 Maximizing the Variance

The variances on different projected dimensions are not identical. The bigger the variance on each projected dimension is, the richer the information of the binary codes convey. To expand the information capacity of all binary codes, it is a good idea to maximize the total variance on all dimensions. First, we work out the variances of different dimensions. Next, we sum the computed variances. Last, we maximize the sum so that the information contained in the binary codes is maximized.

The maximum variance theory is also the basic idea of Principal Component Analysis which can minimize the reconstruction error. Here, we set:

$$z_i = w_i^T H + p_i$$

where $p_i \in \mathbb{R}^{1 \times m}$ is a bias vector.

Then, we get the following formulation:

$$\max_{W} J_2(W) = \sum_{i=1}^n \mathrm{var}(z_i) = \sum_{i=1}^n \mathrm{E}(\|w_i^T H + p_i\|_2^2) - \sum_{i=1}^n \| \mathrm{E}(w_i^T H + p_i)\|_2^2 \tag{4}$$

3.5 Balancing the Variance

Typically, different projected dimensions have different variances and dimensions with larger variances will carry more information. So it is unreasonable to utilize the same number of bits for different dimensions. To solve the problem, we balance the variance on each projected dimension namely making them as equal as possible so that the same number of bits can be employed.

Define $u_i = \mathrm{var}(z_i)$ as the variance on each dimension and the variance vector is denoted as $u = (u_1, u_2, \ldots, u_n)$. To make $u_i(i = 1, 2, \ldots, n)$ as equal as possible, we minimize var($\boldsymbol{u}$), which denotes the variance of $\boldsymbol{u}$. We can get the following formulation:

$$\min_{\boldsymbol{W}} J_3(\boldsymbol{W}) = \text{var}(\boldsymbol{u}) = \text{E}(\boldsymbol{u}^2) - [\text{E}(\boldsymbol{u})]^2 = \frac{1}{n}\sum_{i=1}^{n} u_i^2 - [\frac{1}{n}\sum_{i=1}^{n} u_i]^2 \qquad (5)$$

Once the variance of $\boldsymbol{u}$ is minimized, the variance on each projected dimension is balanced.

3.6   Optimization

The strategy of back-propagation algorithm with stochastic gradient descent (SGD) method is leveraged to obtain the optimal parameters. First we calculate the outputs of the network with the given parameters and quantize the outputs:

$$\boldsymbol{B} = sgn(\boldsymbol{G}) = sgn(\boldsymbol{W}^T\boldsymbol{H} + \boldsymbol{Q})$$

Then, we work out the derivative of the joint objective function (1) with respect to $\boldsymbol{G}$. Since the objective function is composed of three parts, we compute the derivatives of the three parts respectively.

- The vector form of Eq. (3) can be formulated as:

$$J_1(\boldsymbol{W}) = -(\boldsymbol{S}\boldsymbol{\Phi} - \log(1 + e^{\boldsymbol{\Phi}}) + \rho\|\boldsymbol{B} - \boldsymbol{G}\|_2^2$$

where $\boldsymbol{\Phi} = \frac{1}{2}\boldsymbol{G}^T\boldsymbol{G}$.

The derivative of $J_1(\boldsymbol{W})$ with respect to $\boldsymbol{G}$ is computed as:

$$\frac{\partial J_1(\boldsymbol{W})}{\partial \boldsymbol{G}} = \boldsymbol{G}(\frac{1}{1 + e^{-\boldsymbol{\Phi}}} - \boldsymbol{S}) - 2\rho\|\boldsymbol{B} - \boldsymbol{G}\|_2$$

- The Eq. (4) is rewritten as:

$$J_2(\boldsymbol{W}) = \frac{1}{m}\sum_{i=1}^{n}\sum_{j=1}^{m}\|\boldsymbol{w}_i^T\boldsymbol{h}_j + p_{ij}'\|_2^2 - \sum_{i=1}^{n}\|\text{E}(\boldsymbol{w}_i^T\boldsymbol{H} + \boldsymbol{p}_i)\|_2^2$$

$$= \frac{1}{m}tr((\boldsymbol{W}^T\boldsymbol{H} + \boldsymbol{Q})^T(\boldsymbol{W}^T\boldsymbol{H} + \boldsymbol{Q})) - \sum_{i=1}^{n}\|\text{E}(\boldsymbol{z}_i)\|_2^2$$

where $\boldsymbol{h}_j \in \mathbb{R}^{4096 \times 1}$ denotes the outputs of the full7 layer associated with the jth data point, p_{ij}' is the bias term related to the ith weight and the jth data point.

The derivative of $J_2(\boldsymbol{W})$ with respect to $\boldsymbol{G}$ is computed as:

$$
\begin{aligned}
\frac{\partial J_2(\boldsymbol{W})}{\partial \boldsymbol{G}} &= \frac{2}{m}(\boldsymbol{W}^T\boldsymbol{H}+\boldsymbol{Q})-\frac{\partial \sum\limits_{i=1}^{n}\|\,\mathrm{E}(\boldsymbol{z}_i)\|_2^2}{\partial \boldsymbol{G}} \\
&= \frac{2}{m}\boldsymbol{G}-(\frac{\partial\|\,\mathrm{E}(\boldsymbol{z}_1)\|_2^2}{\partial \boldsymbol{z}_1},\frac{\partial\|\,\mathrm{E}(\boldsymbol{z}_2)\|_2^2}{\partial \boldsymbol{z_2}},\ldots,\frac{\partial\|\,\mathrm{E}(\boldsymbol{z}_n)\|_2^2}{\partial \boldsymbol{z}_n})^T \\
&= \frac{2}{m}\boldsymbol{G}-\frac{2}{m}\begin{bmatrix} \mathrm{E}(\boldsymbol{z}_1)\ \mathrm{E}(\boldsymbol{z}_1)\ \cdots\ \mathrm{E}(\boldsymbol{z}_1) \\ \mathrm{E}(\boldsymbol{z}_2)\ \mathrm{E}(\boldsymbol{z}_2)\ \cdots\ \mathrm{E}(\boldsymbol{z}_2) \\ \cdots\quad\ \ \cdots\quad\ \ \cdots \\ \mathrm{E}(\boldsymbol{z}_n)\ \mathrm{E}(\boldsymbol{z}_n)\ \cdots\ \mathrm{E}(\boldsymbol{z}_n) \end{bmatrix}_{n\times m} \\
&= \boldsymbol{G}'
\end{aligned}
$$

- The Eq. (5) is reformulated as:

$$
J_3(\boldsymbol{W}) = \frac{1}{n}\sum_{i=1}^{n}[\mathrm{var}(\boldsymbol{z}_i)]^2 - [\frac{1}{n}\sum_{i=1}^{n}\mathrm{var}(\boldsymbol{z}_i)]^2
$$

The derivative of $J_3(\boldsymbol{W})$ with respect to $\boldsymbol{G}$ is computed as:

$$
\begin{aligned}
\frac{\partial J_3(\boldsymbol{W})}{\partial \boldsymbol{G}} &= \frac{1}{n}\frac{\partial(\sum\limits_{i=1}^{n}[\mathrm{var}(\boldsymbol{z}_i)]^2)}{\partial \boldsymbol{G}} - \frac{\partial[\frac{1}{n}\sum\limits_{i=1}^{n}\mathrm{var}(\boldsymbol{z}_i)]^2}{\partial \boldsymbol{G}} \\
&= \frac{2}{n}(\mathrm{var}(\boldsymbol{z}_1)\frac{\partial\,\mathrm{var}(\boldsymbol{z}_1)}{\partial \boldsymbol{z}_1},\mathrm{var}(\boldsymbol{z}_2)\frac{\partial\,\mathrm{var}(\boldsymbol{z}_2)}{\partial \boldsymbol{z}_2},\ldots,\mathrm{var}(\boldsymbol{z}_n)\frac{\partial\,\mathrm{var}(\boldsymbol{z}_n)}{\partial \boldsymbol{z}_n})^T \\
&\quad -\frac{2}{n}\sum_{i=1}^{n}\mathrm{var}(\boldsymbol{z}_i)\cdot\frac{1}{n}(\frac{\partial\,\mathrm{var}(\boldsymbol{z}_1)}{\partial \boldsymbol{z}_1},\frac{\partial\,\mathrm{var}(\boldsymbol{z}_2)}{\partial \boldsymbol{z}_2},\ldots,\frac{\partial\,\mathrm{var}(\boldsymbol{z}_n)}{\partial \boldsymbol{z}_n})^T \\
&= \frac{4}{mn}(\mathrm{var}(\boldsymbol{z}_1)\cdot(\boldsymbol{z}_1-\mathrm{E}(\boldsymbol{z}_1)),\mathrm{var}(\boldsymbol{z}_2)\cdot(\boldsymbol{z}_2-\mathrm{E}(\boldsymbol{z}_2)),\ldots,\mathrm{var}(\boldsymbol{z}_n)\cdot(\boldsymbol{z}_n-\mathrm{E}(\boldsymbol{z}_n)))^T \\
&\quad -\frac{4}{mn^2}\sum_{i=1}^{n}\mathrm{var}(\boldsymbol{z}_i)\cdot(\boldsymbol{z}_1-\mathrm{E}(\boldsymbol{z}_1),\boldsymbol{z}_2-\mathrm{E}(\boldsymbol{z}_2),\ldots,\boldsymbol{z}_n-\mathrm{E}(\boldsymbol{z}_n))^T \\
&= \boldsymbol{G}''
\end{aligned}
$$

In this way, the derivative of the joint loss function (1) with respect to $\boldsymbol{G}$ is the sum of the three parts:

$$
\frac{\partial J(\boldsymbol{W})}{\partial \boldsymbol{G}} = \boldsymbol{G}(\frac{1}{1+e^{-\boldsymbol{\Phi}}}-\boldsymbol{S})-2\rho\|\boldsymbol{B}-\boldsymbol{G}\|_2-\lambda_1\boldsymbol{G}'+\lambda_2\boldsymbol{G}'' \tag{6}
$$

Last, the derivative of the parameters $\boldsymbol{W}$, $\boldsymbol{Q}$ and $\boldsymbol{H}$ can be computed as follows:

$$
\frac{\partial J(\boldsymbol{W})}{\partial \boldsymbol{W}} = \boldsymbol{H}\frac{\partial J(\boldsymbol{W})}{\partial \boldsymbol{G}} \tag{7}
$$

$$
\frac{\partial J(\boldsymbol{W})}{\partial \boldsymbol{Q}} = \frac{\partial J(\boldsymbol{W})}{\partial \boldsymbol{G}} \tag{8}
$$

$$\frac{\partial J(\boldsymbol{W})}{\partial \boldsymbol{H}} = \boldsymbol{W}\frac{\partial J(\boldsymbol{W})}{\partial \boldsymbol{G}} \tag{9}$$

All the parameters are alternatively learned. That is to say, we update one parameter with other parameters fixed.

The proposed algorithm of BMDH is summarized as follows.

Input: Original data sets: $\boldsymbol{X} = \{\boldsymbol{x}_1, \boldsymbol{x}_2, \ldots, \boldsymbol{x}_M\}$, $\boldsymbol{x}_i \in \mathbb{R}^d$; pairwise labels:$\boldsymbol{S}$; bit length: n; iterative number: R; hyper-parameter:$\rho, \lambda_1, \lambda_2$;
Initialization: Initialize $\boldsymbol{\Theta}$ with the CNN-F model. Initialize $\boldsymbol{W}$ and $\boldsymbol{Q}$ by a Gaussian distribution with mean 0 and variance 0.01.
For $i = 1{:}R$ **do**
    1. Randomly sample a batch of m data points from the original data sets$\boldsymbol{X}$.
    2. Compute the outputs of the network $\boldsymbol{H}$.
    3. Compute the binary codes of the m points, $\boldsymbol{B} = sgn(\boldsymbol{W}^T\boldsymbol{H} + \boldsymbol{Q})$.
    4. Compute the derivative of the loss function with respect to $\boldsymbol{G}$ according to (6).
    5. Compute the derivative of the loss function with respect to $\boldsymbol{W}$, $\boldsymbol{Q}$ and $\boldsymbol{H}$ according to (7)–(9).
    6. Update the parameters $\boldsymbol{W}$, $\boldsymbol{Q}$ and $\boldsymbol{\Theta}$ using back-propagation algorithm with stochastic gradient desent method.
End for
Output: The parameters $\boldsymbol{W}$, $\boldsymbol{Q}$, $\boldsymbol{\Theta}$ and the binary codes $\boldsymbol{B}$.

Algorithm 1. BMDH algorithm

---

4   Experiment

We implement our experiments on two universally used data sets: CIFAR-10 and NUS-WIDE. The mean Average Precision (mAP) as the metric is applied to evaluate the retrieval performance. Several conventional hashing methods and deep hashing methods are utilized to make comparison with the proposed method.

4.1   Datasets

- **CIFAR-10 Data set** consists of the total number of $60,000$ images each of which is a 32×32 color image. These images are categorized into 10 classes with 6000 images per class.
- **NUS-WIDE Data set** consists of nearly $27,0000$ color images collected from the web. It is a multi-label data set each image of which is annotated with one or more labels from 81 semantic concepts. Following [4,5,14], we only exploit images whose labels are from the 21 most frequent concept tags. At least $5,000$ images are associated with each label.

4.2 Evaluation Metric

We use the mean Average Precision (mAP) as the metric for evaluation which means the mean precision of all query samples. It is defined as:

$$mAP = \frac{1}{|Q|} \sum_{i=1}^{|Q|} \frac{1}{n_i} \sum_{j=1}^{n_i} Precision(R_{ij}) \tag{10}$$

where $q_i \in Q$ denotes the query sample, n_i is the number of samples similar to the query sample q_i in the dataset. The relevant samples are ordered as $\{x_1, x_2, \ldots, x_{n_i}\}$, R_{ij} is the set of ranked retrieval results from the top result until getting to point x_j.

The definition of Precision is as follows:

$$Precision = \frac{\text{the number of retrieved relevant points}}{\text{the number of all retrieved points}}.$$

4.3 Experimental Setting

For CIFAR-10, following [5], we randomly sample 1000 images (100 images per class) as the query set. The rest images are used as database images. For the unsupervised methods, the database images are utilized as the training images. For the supervised methods, the training set consists of 5000 images (500 images per class) which are randomly sampled from the rest images. We construct the pairwise similarity matrix S based on the images class labels where two images sharing the same label are considered similar. For methods using hand-crafted features, we represent each image with a 512-dimension GIST vector [9].

For NUS-WIDE, we randomly sample $2,100$ images (100 images per class) from the selected 21 classes as the query set. For the unsupervised methods, all the remaining images from the 21 classes constitute the training data set. For the supervised methods, 500 images per class are randomly sampled from the 21 classes to form the training set. The pairwise similarity matrix S is constructed based on the images class labels where two images sharing at least one common label are considered similar. When calculating the mAP values, only the top $5,000$ returned neighbors are used for evaluation. For methods using hand-crafted features, we represent each image with a 1134-dimensional feature vector, which is composed of 64-D color histogram, 144-D color correlogram, 73-D edge direction histogram, 128-D wavelet texture, 225-D block-wise color moments and 500-D SIFT features.

The first seven layers of our network are initialized with the CNN-F network that has been pre-trained on ImageNet. The size of the batch is determined as 128 which means that 128 images are fed into the network in each iteration. The hyper-parameters ρ, λ_1 and λ_2 are set to 10, 0.5 and 0.1 unless otherwise stated.

The experiments are carried out on Intel Core i5-4660 3.2 GHZ desktop computer of 8 G Memory with MATLAB2014a and MatConvNet. They are also implemented on a NVIDIA GTX 1070 GPU server.

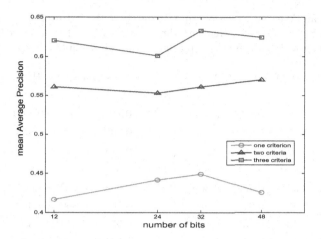

Fig. 2. mAP on different number of bits with respect to different criteria.

4.4   Retrieval Results on CIFAR-10 and NUS-WIDE

Performance Improved Step by Step: Since the loss function is composed of three parts, we validate how the performance is gradually improved with another restriction added to the objective function every time. First, we acquire the performance with the criterion of similarity-preserving. Next, the criterion of maximum variance on all projected dimensions is added. Last, we add the criterion of balanced variance on each projected dimension to get the final performance.

When the number of iterations is set to 10, the performance of different number of bits with respect to different criteria on CIFAR-10 is revealed in Fig. 2. It can be seen that the performance with the restriction of similarity-preserving is outperformed by the performance with another restriction of maximum variance added. When the last restriction of balanced variance is added, the performance is enhanced again. This can be attributed to more information captured by the binary codes and equal information contained in each dimension.

Performance Compared with Conventional Methods with Hand-crafted Features: To validate the superiority of deep features over hand-crafted features, we compare the proposed method with those conventional methods having hand-crafted features including FastH [6], SDH [10], KSH [9], LFH [15], SPLH [11], ITQ [2], SH [13]. The results are shown in Table 1. It can be found that the BMGH method exceeds the contrastive algorithms dramatically no matter it is unsupervised or supervised. It is because that deep architectures provide richer semantic features despite the tremendous appearance variations. The results of CNNH, KSH and ITQ are copied from [4,14] and the results of other contrastive methods are from [5]. Because the experimental setting and evaluation metric are all the same, so the above practice is reasonable.

Table 1. mean Average Precision (mAP) on the CIFAR-10 dataset and the NUS-WIDE dataset. The highest values are shown in boldface. The mAP for NUS-WIDE dataset is calculated based on the top 5,000 returned neighbors.

Method	CIFAR-10 (mAP)				NUS-WIDE (mAP)			
	12 bits	24 bits	32 bits	48 bits	12 bits	24 bits	32 bits	48 bits
BMDH	**0.721**	**0.746**	**0.762**	**0.775**	**0.811**	**0.841**	**0.854**	**0.865**
DPSH	0.713	0.727	0.744	0.757	0.794	0.822	0.838	0.851
NINH	0.552	0.566	0.558	0.581	0.674	0.697	0.713	0.715
CNNH	0.439	0.511	0.509	0.522	0.611	0.618	0.625	0.608
FastH	0.305	0.349	0.369	0.384	0.621	0.650	0.665	0.687
SDH	0.285	0.329	0.341	0.356	0.568	0.600	0.608	0.637
KSH	0.303	0.337	0.346	0.356	0.556	0.572	0.581	0.588
LFH	0.176	0.231	0.211	0.253	0.571	0.568	0.568	0.585
SPLH	0.171	0.173	0.178	0.184	0.568	0.589	0.597	0.601
ITQ	0.162	0.169	0.172	0.175	0.452	0.468	0.472	0.477
SH	0.127	0.128	0.126	0.129	0.454	0.406	0.405	0.400

Table 2. mean Average Precision (mAP) on the CIFAR-10 dataset and the NUS-WIDE dataset. The highest values are shown in boldface. The mAP for NUS-WIDE dataset is calculated based on the top 5,000 returned neighbors.

Method	CIFAR-10 (mAP)				NUS-WIDE (mAP)			
	12 bits	24 bits	32 bits	48 bits	12 bits	24 bits	32 bits	48 bits
BMDH	**0.721**	**0.746**	**0.762**	**0.775**	**0.811**	**0.841**	**0.854**	**0.865**
FastH+CNN	0.553	0.607	0.619	0.636	0.779	0.807	0.816	0.825
SDH+CNN	0.478	0.557	0.584	0.592	0.780	0.804	0.815	0.824
KSH+CNN	0.488	0.539	0.548	0.563	0.768	0.786	0.790	0.799
LFH+CNN	0.208	0.242	0.266	0.339	0.695	0.734	0.739	0.759
SPLH+CNN	0.299	0.330	0.335	0.330	0.753	0.775	0.783	0.786
ITQ+CNN	0.237	0.246	0.255	0.261	0.719	0.739	0.747	0.756
SH+CNN	0.183	0.164	0.161	0.161	0.621	0.616	0.615	0.612

Performance Compared with Conventional Methods with Deep Features: To verify the improvement in performance originates from our method instead of the deep network, we select some conventional methods with deep features extracted by the CNN-F network to compare with the proposed method. Table 2 shows the final results. We can see that the performance of our method outperforms that of contrastive methods on two datasets which demonstrates the progressiveness of the proposed method. In additional, the results also validate the effectiveness of the end-to-end framework. The results of the compared

Table 3. mean Average Precision (mAP) on the CIFAR-10 dataset and the NUS-WIDE dataset. The highest values are shown in boldface. The mAP for NUS-WIDE dataset is calculated based on the top 50,000 returned neighbors.

Method	CIFAR-10 (mAP)				NUS-WIDE (mAP)			
	12 bits	24 bits	32 bits	48 bits	12 bits	24 bits	32 bits	48 bits
BMDH	**0.774**	**0.796**	**0.813**	**0.822**	**0.734**	**0.743**	**0.761**	**0.765**
DPSH#	0.763	0.781	0.795	0.807	0.715	0.722	0.736	0.741
DRSCH	0.615	0.622	0.629	0.631	0.618	0.622	0.623	0.628
DSCH	0.609	0.613	0.617	0.620	0.592	0.597	0.611	0.609
DSRH	0.608	0.611	0.617	0.618	0.609	0.618	0.621	0.631

methods are from [5,12] which is reasonable as the experimental setting and evaluation metric are all the same.

Performance Compared with Deep Hashing Methods: We compare the proposed method with some deep hashing methods including DPSH [5], NINH [4], CNNH [14], DSRH [17], DSCH [16] and DRSCH [16]. These methods have not considered maximizing the total variance on all projected dimensions and balancing the variance on each dimension.

When comparing with DSRH, DSCH and DRSCH, we leverage another experimental setting the same as [16] for fair comparison. Specifically, in CIFAR-10, we randomly sample 10,000 images (1,000 images per class) as the query set. The remaining images are used to form the database set. And the database set is used as the training set. In NUS-WIDE, we randomly sample 2,100 images (100 images per class) from the 21 classes as the query set. Similarly, the remaining images are used as the database images and the training images simultaneously. The top 50,000 returned neighbors are used for evaluation when calculating the mAP values. DPSH# in Table 3 denotes the DPSH method under the new experimental setting. The results of the compared methods are directly from [5,16] which is reasonable as the experimental setting and evaluation metric are all the same. From Tables 1 and 3 we can see that the performance of our method is superior to the compared deep hashing methods including the current state-of-the-art method DPSH. It verifies the effectiveness of maximizing the total variance on all dimensions and balancing the variance on each dimension. Additionally, the superiority of our method to CNNH can also demonstrate the advantage of simultaneous feature learning and hash-code learning.

5   Conclusion

In this paper, we present a novel deep hashing algorithm, dubbed BMDH. To the best of our knowledge, it is the first time that maximizing the total variance on all projected dimensions and balancing the variance on each dimension is simultaneously considered in the existing deep hashing methods. Back-propagation algorithm with stochastic gradient desent (SGD) method and the end-to-end way ensure the optimal parameters despite the complex objective function. Experimental results on two widely used data sets demonstrate that the proposed method can outperform other state-of-the-art algorithms in image retrieval applications.

Acknowledgments. This work is supported by the National Natural Science Foundation of China (NSFC) (No. 61472442, No. 61773397, No. 61703423) and the funding (No. 2015kjxx-46).

References

1. Andoni, A., Indyk, P.: Near-optimal hashing algorithms for approximate nearest neighbor in high dimensions. In: Foundations of Computer Science Annual Symposium, vol. 51, no. 1, pp. 459–468 (2006)
2. Gong, Y., Lazebnik, S., Gordo, A., Perronnin, F.: Iterative quantization: a procrustean approach to learning binary codes for large-scale image retrieval. IEEE Trans. Pattern Anal. Mach. Intell. **35**(12), 2916–2929 (2013)
3. Kong, W., Li, W.J.: Isotropic hashing. In: Advances in Neural Information Processing Systems, vol. 2, pp. 1646–1654 (2012)
4. Lai, H., Pan, Y., Liu, Y., Yan, S.: Simultaneous feature learning and hash coding with deep neural networks. In: Computer Vision and Pattern Recognition, pp. 3270–3278 (2015)
5. Li, W.J., Wang, S., Kang, W.C.: Feature learning based deep supervised hashing with pairwise labels. Computer Science (2015)
6. Lin, G., Shen, C., Shi, Q., van der Hengel, A., Suter, D.: Fast supervised hashing with decision trees for high-dimensional data. In: Computer Vision and Pattern Recognition, pp. 1971–1978 (2014)
7. Lin, K., Yang, H.F., Hsiao, J.H., Chen, C.S.: Deep learning of binary hash codes for fast image retrieval. In: Computer Vision and Pattern Recognition Workshops, pp. 27–35 (2015)
8. Liong, V.E., Lu, J., Wang, G., Moulin, P., Zhou, J.: Deep hashing for compact binary codes learning. In: Computer Vision and Pattern Recognition, pp. 2475–2483 (2015)
9. Liu, W., Wang, J., Ji, R., Jiang, Y.G.: Supervised hashing with kernels. In: Computer Vision and Pattern Recognition, pp. 2074–2081 (2012)
10. Shen, F., Shen, C., Liu, W., Shen, H.T.: Supervised discrete hashing. In: Computer Vision and Pattern Recognition, pp. 37–45 (2015)
11. Wang, J., Kumar, S., Chang, S.F.: Sequential projection learning for hashing with compact codes. In: International Conference on Machine Learning, pp. 1127–1134 (2010)
12. Wang, X., Shi, Y., Kitani, K.M.: Deep supervised hashing with triplet labels. In: Lai, S.-H., Lepetit, V., Nishino, K., Sato, Y. (eds.) ACCV 2016. LNCS, vol. 10111, pp. 70–84. Springer, Cham (2017). https://doi.org/10.1007/978-3-319-54181-5_5
13. Weiss, Y., Torralba, A., Fergus, R.: Spectral hashing. In: Conference on Neural Information Processing Systems, Vancouver, British Columbia, Canada, December, pp. 1753–1760 (2008)
14. Xia, R., Pan, Y., Lai, H., Liu, C., Yan, S.: Supervised hashing for image retrieval via image representation learning. In: AAAI Conference on Artificial Intelligence (2012)

15. Zhang, P., Zhang, W., Li, W.J., Guo, M.: Supervised hashing with latent factor models. In: International ACM SIGIR Conference on Research and Development in Information Retrieval, pp. 173–182 (2014)
16. Zhang, R., Lin, L., Zhang, R., Zuo, W., Zhang, L.: Bit-scalable deep hashing with regularized similarity learning for image retrieval and person re-identification. IEEE Trans. Image Process. **24**(12), 4766–4779 (2015)
17. Zhao, F., Huang, Y., Wang, L., Tan, T.: Deep semantic ranking based hashing for multi-label image retrieval. In: IEEE Conference on Computer Vision and Pattern Recognition, pp. 1556–1564 (2015)

FFGS: Feature Fusion with Gating Structure for Image Caption Generation

Aihong Yuan[1,2], Xuelong Li[1], and Xiaoqiang Lu[1(✉)]

[1] Center for OPTical IMagery Analysis and Learning (OPTIMAL),
Xi'an Institute of Optics and Precision Mechanics, Chinese Academy of Sciences,
Xi'an 710119, Shaanxi, People's Republic of China
{ahyuan,xuelong_li}@opt.ac.cn, luxq666666@gmail.com
[2] University of Chinese Academy of Sciences, 19A Yuquanlu, Beijing 100049,
People's Republic of China

Abstract. Automatically generating a natural language to describe the content of the given image is a challenging task in the interdisciplinary between computer vision and natural language processing. The task is challenging because computers not only need to recognize objects, their attributions and relationships between them in an image, but also these elements should be represented into a natural language sentence. This paper proposed a feature fusion with gating structure for image caption generation. First, the pre-trained VGG-19 is used as the image feature extractor. We use the FC-7 and CONV5-4 layer's outputs as the global and local image feature, respectively. Second, the image features and the corresponding sentence are imported into LSTM to learn their relationship. The global image feature is gated at each time-step before imported into LSTM while the local image feature used the attention model. Experimental results show our method outperform the state-of-the-art methods.

Keywords: Image caption generation · Recurrent neural network
Convolutional neural network · Multi-modal embedding · Feature fusion

1   Introduction

Image caption, which automatically generates a natural language sentence to describe the content of the given image, has recently become a challenging but fundamental task of *computer vision* (CV) and *natural language processing* (NLP) [3,7,13,19]. The task is challenging because the caption generation models not only should solve the computer vision challenges of determining what objects are in an image [11,22], but also be powerful enough to describe their relationships with natural language. However, challenges and opportunities coexist. Image caption links image and natural language together, which makes it has a great potential for application in the near future. Therefore, many researchers have paid great attention to this task.

© Springer Nature Singapore Pte Ltd. 2017
J. Yang et al. (Eds.): CCCV 2017, Part I, CCIS 771, pp. 638–649, 2017.
https://doi.org/10.1007/978-981-10-7299-4_53

Approaches for image caption generation task have two main categories: (1) retrieval-based methods and (2) *multi-modal neural networks-based* (MMNN-based) methods. Before the advent *deep learning* (DL), retrieval-based methods are the most popular methods for image caption generation. These methods retrieval similar objects and then retrieval a similar sentence from the training dataset [6,16]. After that, the words are connected together according to certain grammar rules.

Although the retrieval-based methods have gained many encouraged results, many problems have not been solved. These methods need fixed visual concepts and hard-coded sentence template, which makes the sentences generated by these models are less variety.

Recently, deep learning has achieved many breakthroughs in *natural language processing* (NLP) and *computer vision* (CV), such as *machine translation* (MT) [2], image classification, object detection [15,17], *etc*. The main success of deep learning is that the ability of representation is powerful. Some image caption generating methods using deep neural networks have been proposed recently. For example, *multi-modal Recurrent networks* (m-RNN) is proposed by Mao *et al.* [12], which uses the *convolutional neural networks* (CNNs) as feature extractor and the traditional *recurrent neural networks* (RNNs) as sentence generator. From then on, the "CNN + RNN" mode becomes the most popular scheme for image caption generation.

Compared to the retrieval-based methods, the *multi-modal neural network-based* (MMNN-based) methods have shown a greater improvement on the performance of image caption generating task. Sentence generated by the MMNN-based methods is more reasonable and more changeable. However, they also have some shortages. For example, m-RNN [12], Google-NIC [18], and LRCN [4] only used the global image feature vector from the fully connected layer of the CNN, which can not dig up the subtle relationship between the image and the natural statement. On the contrary, NIC-VA [20] only uses the local image feature, which may lead to loss the global information of the image.

To overcome these shortages, we propose a *feature fusion method with gating structure* (**FFGS**). Global image features are added on the basis of the NIC-VA. The global image features are imported into the sentence generator at each time-step. Unlike the m-RNN, Google-NIC, and LRCN, which import the global image feature into RNN at the first time-step or at each time-step without any processing, we use gating mechanism for the global image features. In other words, the global image features are gated at each time-step before imported into the RNN unit.

The main contributions of the proposed algorithm are as follows:

- Feature fusion strategy is used in this work. The proposed algorithm uses the global and local image features to guarantee the information of images be more comprehensively and meticulously used.
- Gating mechanism is used for the global image feature. We use gate to control the global image feature and this mechanism can solve the argument whether should import the global image feature into the sentence generator at the

first time-step or at each time-step. Furthermore, it robustly solve how much should be imported at each time-step.
- The prosed algorithm is tested on three benchmark datasets. The experimental results show that method proposed in this paper is better than the state-of-art methods.

The rest of this paper is organized as follows. In Sect. 2, some previous works are briefly introduced. Section 3 presents our model for image caption generation. To validate the proposed method, the experimental results are shown in Sect. 4. At last, Sect. 5 makes a brief conclusion for this paper.

2   Related Work

2.1   Deep Neural Networks for MT

Recently, many works showed that deep neural network can be successfully used to solve lots of problems in NLP, such as *machine translation* (MT). In the conventional MT system, the neural network refers to as an RNN Encoder-Decoder which consists of two RNN [2]. One acts as an encoder which maps a variable-length source sentence to a fixed-length vector. And the other acts as a decoder which decodes the vector produced by the encoder into a variable-length target sentence. When RNN trained with *Backpropagation Through Time* (BPTT) [5], there exists some difficulties in learning long-term dependency due to the so-called vanishing and exploding gradient problems. To overcome these difficulties, some gated RNNs (*e.g.* LSTM and GRU) have been proposed. Some researchers treat image caption generation as a machine translation problem. However, the input is image which is not a sequence signal, so they use pre-trained CNN as encoder for image, instead encoding RNN in MT. The proposed method in this paper also follows this idea.

2.2   Generating Sentence Descriptions for Images

There are mainly two categories of methods for image caption generation task. The first category is retrieval-based methods which retrieve similar captioned images and generate new descriptions by retrieving a similar sentence from a image-description dataset. These methods project image and sentence representations into a common semantic space, which is used for ranking image captions or for image search. They retrieve similar objects and the corresponding descriptions from the training data, and then stitch these descriptions into sentences. A typical work is called BabyTalk system [9] which consists of two important steps. The first step is content planing, which is detecting the defined objects and its corresponding content words. At the second step, put these words obtained in the first step into a sentence in accordance with certain rules. Socher *et. al* propose a method named DT-RNN [16] to generate description for a given image. Dependency trees are used to embed sentences into a vector space aims to retrieve corresponding images. Another typical category is multi-modal neural network based methods. These methods based on multi-modal embedding models, generated sentence in a word-by-word manner and conditioned on image representation which is the output from a deep convolutional network. Our method falls into this category and our multi-modal embedding model is a recurrent network. Our model is trained to maximize the likelihood of the target sentence conditioned on the given training image. Some previous works seem closely related to our method such as m-RNN [12], Google-NIC [18], LRCN [4] and NIC-VA [20]. However, the proposed model in this paper uses the global and local feature fusion strategy, which is different from the aforementioned models.

3   Proposed Method

In this section, we introduce our whole model named **FFGS** (*i.e.* Feature Fusion with Gating Structure). Figure 1 shows our full model diagram and it main contains two modules: (1) image representation and (2) multi-modal embedding. VGG-19 is used for image representation. LSTM is used for multi-modal embedding. In other words, image information and the corresponding sentence are embedding in LSTM. Now, we introduce our FFGS in detail.

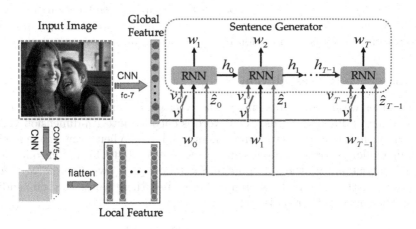

Fig. 1. Overview of our method for image caption generation. A deep CNN for image features extraction: the FC7 layer is used to extract global features and the CONV5-4 layer is used to extract local features of the given images. RNN model acts as a decoder which decodes image features into sentences. The gate for controlling the global image feature is computed with the pre-step hidden state of RNN. The local image feature are selected by the pre-step hidden state and the current word vector.

3.1   Image Feature Representation

As Fig. 1 is shown, our image feature representation concludes two parts: global feature and local feature. We use the feature vector output from FC-7 and CONV5-4 layers of VGG-19 as the global and local image feature, respectively.

Image Global Representation: The VGG-19 is pre-trained on ImageNet and used as the image encoder in our model. The global representation of image I is as follows:

$$\mathbf{v} = \mathbf{W}_I \cdot [Fc(I)] + \mathbf{b}_I, \tag{1}$$

where I donates the image I, $Fc(I) \in \mathbb{R}^{4096}$ is the output of the FC-7 layer. The matrix $\mathbf{W}_I \in \mathbb{R}^{h \times 4096}$ is a embedding matrix which projects 4096-dimension image feature vectors into the embedding space with h-dimension and $\mathbf{b}_I \in \mathbb{R}^h$ donates the bias. $\mathbf{v} \in \mathbb{R}^h$ is so-called image global feature representation because it is computed with the entire image I.

Image Local Feature Representation: The VGG-19 also be used as image local feature extractor in our model. When a raw image $I \in \mathbb{R}^{W \times H \times 3}$ is input to VGG-19, the CONV5-4 layer outputs feature map $\mathbf{v}_c \in \mathbb{R}^{W' \times H' \times D}$. Then, we flatten this feature map into $\mathbf{v}_l \in \mathbb{R}^{D \times C}$, where $C = W' \times H'$. This processing program can be written as follows:

$$\mathbf{v}_l = \{\mathbf{v}_{l1}, \mathbf{v}_{l2}, \cdots, \mathbf{v}_{lC}\} = flatten\left(Conv(I)\right), \tag{2}$$

where $\mathbf{v}_{li} \in \mathbb{R}^D$, $i \in \{0, 1, \cdots, C\}$ donates the feature of i-th location of image I. In other words, each image I is divided into C locations and every $\mathbf{v}_{li}$ represents one location. So, $\mathbf{v}_{li}$ is the location feature representation.

3.2   Sentence Representation

In our model, we encode words into one-hot vectors. For example, the benchmark dataset has N_0 different words, every word is encoded into N_0-dimension vector which only one value equals to 1 and others equal to 0. When a raw image import into our model, a corresponding sentence S is generated which is encoded as a sequence of one-hot vectors. We donate $S = (\mathbf{w}_1, \mathbf{w}_2, \cdots, \mathbf{w}_T)$, where $\mathbf{w}_i \in \mathbb{R}^{N_0}$ donates the i-th word in the sentence. We embed these words into embedding space. The concrete formula is as follows:

$$\mathbf{s}_t = \mathbf{W}_s \cdot \mathbf{w}_t, \ t \in \{1, 2, \cdots, T\}, \tag{3}$$

where $\mathbf{W}_s$ is the embedding matrix of sentences which projects the word vector into the embedding space. So the projection matrix $\mathbf{W_s}$ is a $h \times N_0$ matrix where N_0 is the size of the dictionary and h is the dimension of the embedding space.

3.3   LSTM for Sentence Generating

LSTM is used as sentence generator in our model. In other words, RNN model in Fig. 1 is LSTM. As illustrated in Fig. 2, every LSTM unit has four inputs: the local image feature $\mathbf{v_t}$, the global image feature $\hat{\mathbf{z}}_t$, the word $\mathbf{s_t}$ and the previous hidden state $\mathbf{h}_{t-1}$.

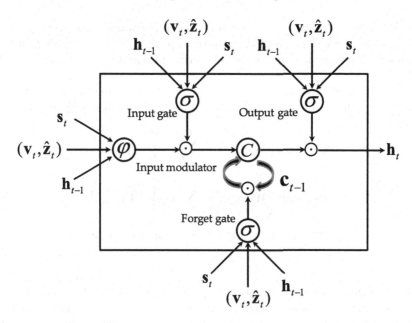

Fig. 2. The diagram of the LSTM unit used in our model. Each gate has 4 input vectors: the global feature at the time-step t $\mathbf{v}_t$, the local image feature at the time-step t $\hat{\mathbf{z}}_t$, word representation at the time-step t $\mathbf{s}_t$ and the previous hidden state $\mathbf{h}_{t-1}$.

LSTM Model: In this subsection, we introduce our formula in detail. First, three gates and updating memory content of LSTM are rewritten as follows:

$$\mathbf{i}_t = \sigma\left(W_i \mathbf{s}_t + U_i \mathbf{h}_{t-1} + V_i \mathbf{v}_t + Z_i \hat{\mathbf{z}}_t + \mathbf{b}_i\right), \tag{4}$$

$$\mathbf{f}_t = \sigma\left(W_f \mathbf{s}_t + U_f \mathbf{h}_{t-1} + V_f \mathbf{v}_t + Z_f \hat{\mathbf{z}}_t + \mathbf{b}_f\right), \tag{5}$$

$$\mathbf{o}_t = \sigma\left(W_o \mathbf{s}_t + U_o \mathbf{h}_{t-1} + V_o \mathbf{v}_t + Z_o \hat{\mathbf{z}}_t + \mathbf{b}_o\right), \tag{6}$$

$$\tilde{\mathbf{c}}_t = \tanh\left(W_c \mathbf{s}_t + U_c \mathbf{h}_{t-1} + V_c \mathbf{v}_t + Z_c \hat{\mathbf{z}}_t + \mathbf{b}_c\right), \tag{7}$$

where $W_* \in \mathbb{R}^{h \times h}$, $U_* \in \mathbb{R}^{h \times h}$, $V_* \in \mathbb{R}^{h \times h}$ and $Z_* \in \mathbb{R}^{h \times h}$ are donate weights matrixes and $\mathbf{b}_* \in \mathbb{R}^h$ donate biases. $\mathbf{v}_t \in \mathbb{R}^h$ and $\hat{\mathbf{z}}_t \in \mathbb{R}^h$ are the global and the local image features, respectively. Their calculating formulas are introduced in the next two subsections.

The current memory and hidden state are computed as follows:

$$\mathbf{c}_t = \mathbf{f}_t \odot \mathbf{c}_{t-1} + \mathbf{i}_t \odot \tilde{\mathbf{c}}_t, \tag{8}$$

$$\mathbf{h}_t = \mathbf{o}_t \odot \tanh(\mathbf{c}_t), \tag{9}$$

The LSTM module outputs a probability at each time-step. We write it as following formula:

$$\mathbf{p}_{t+1} = \mathrm{softmax}(\mathbf{y}_t) = \mathrm{softmax}(W_y \mathbf{h}_t + \mathbf{b}_y), \tag{10}$$

where $W_y \in \mathbb{R}^{N_0 \times h}$ and $\mathbf{b}_y \in \mathbb{R}^{N_0}$ donate passing forward parameters. $\mathbf{y}_t \in \mathbb{R}^{N_0}$ is a output of LSTM at the t-th time-step. $\mathbf{p}_{t+1} \in \mathbb{R}^{N_0}$ is a probability vector whose each element donates the predicting probability of the corresponding word.

Having builded the LSTM model, initializing the system is another important thing to do. The memory and the hidden state is initialized by the following formulas:

$$\mathbf{c}_0 = \tanh \left(W_{c\_init} \left(\frac{1}{C} \sum_{i=1}^{C} \mathbf{v}_{li} \right) + \mathbf{b}_{c\_init} \right), \tag{11}$$

$$\mathbf{h}_0 = \tanh \left(W_{h\_init} \left(\frac{1}{C} \sum_{i=1}^{C} \mathbf{v}_{li} \right) + \mathbf{b}_{h\_init} \right), \tag{12}$$

where $W_{c\_init} \in \mathbb{R}^{h \times h}$ and $W_{h\_init} \in \mathbb{R}^{h \times h}$ are initial weights. $\mathbf{b}_{c\_init} \in \mathbb{R}^h$ and $\mathbf{b}_{h\_init} \in \mathbb{R}^h$ are initial biases.

Gate for Global Image Feature: $\mathbf{v}_t$ occurs several times in LSTM model. And it is a output of global image feature controlled by gate. In previous works, the vast majority of them import the global feature defined in Eq. (1) at the first time-step or at each time-step into the RNN decoder, but they find that global feature imported at the first time-step is better than at every time-step. They explain that global feature imported at each time-step may bring more noise to the system. However, this reason can not convince us. So we want to design a robust algorithm that able to autonomously decide how many global feature should be imported into the decoder. Inspired by the gate technology exploited in RNN, we design a gate before the global feature imported into the system. The gate is defined as follows:

$$g_t = \sigma(\mathbf{w}_g^T \mathbf{h}_{t-1} + b_g), \tag{13}$$

where $\mathbf{w}_g \in \mathbb{R}^h$ is weight vector, b_g is bias. So the t-th gate g_t is a scaler and its value correlates with the previous time-step hidden state $\mathbf{h}_{t-1}$.

After calculating the gate, the global image feature at time-step t is computed as follows:

$$\mathbf{v}_t = g_t \mathbf{v}. \tag{14}$$

Through Eq. (14), if we set $g_t = 1$ at $t = 0, 1, \cdots, T$, $\mathbf{v}$ is imported into the decoder at each time-step. If we set $g_t = 1$ at $t = 0$ and $g_t = 0$ at $t = 1, 2, \cdots, T$, $\mathbf{v}$ is only imported into the decoder at the first time-step. So introduced gate concept, intuitively feel our algorithm is robust for global image feature and methods in previous works are special cases of our approach.

Attention Mechanism for Local Image Feature: The local image feature $\hat{\mathbf{z}}_t$ donates the local information of image. Here we use attention mechanism as introduced in references [20] for local feature. At each time-step, the attention mechanism uses the previous hidden state h_{t-1} to decide the local feature. The attention model is defined as follows:

$$\alpha_t = \text{softmax}\left(\tanh\left[\left(\mathbf{w}_a^T \mathbf{v}_l\right)^T + U_a \mathbf{h}_{t-1} + \mathbf{b}_a\right]\right) \triangleq \left[\alpha_{t1} \cdots \alpha_{tC}\right]^T, \quad (15)$$

where $\mathbf{w}_a \in \mathbb{R}^h$ and $U_a \in \mathbb{R}^{h \times h}$ are weights. $\mathbf{v}_l$ is defined in Eq. (2). $\mathbf{b}_a \in \mathbb{R}^h$ is bias. $\alpha_t \in \mathbb{R}^C$ is a probability vector whose each dimension value donates the probability of the corresponding local image feature. In our algorithm, we use the soft attention model. Therefore, $\hat{\mathbf{z}}_t$ is calculated as follows:

$$\hat{\mathbf{z}}_t = \mathbf{v}_l \alpha_t = \sum_{i=1}^{C} \alpha_{ti} \mathbf{v}_{li}. \quad (16)$$

Through Eq. (16) we know that α_t decides which locals should be used at the current time-step.

The loss function of our model can be written as the negative likelihood function, which formula is as follows:

$$L(\theta) = -\log P\left(S \mid I\right) + \lambda_\theta \|\theta\|^2 + \lambda_\alpha \left(1 - \sum_{i=1}^{C} \alpha_{ti}\right)^2, \quad (17)$$

where θ is all parameters set which concludes parameters of the LSTM, embedding matrixes in Sects. 3.1 and 3.2 and all gate models. $\lambda_\theta \cdot \|\theta\|_2^2$ is a regularization term. $\lambda_\alpha \left(1 - \sum_{i=1}^{C} \alpha_{ti}\right)^2$ is a probabilistic constraint.

The proposed model is trained with *back-propagation through time* (BPTT) algorithm to minimize the cost function $L(\theta)$.

4   Experimental Evaluation

In this section, we describe our experimental methodology and quantitative results which validate the effectiveness of our model for caption generation.

4.1   Datasets and Data Processing

Datasets. The Flickr8K [14], Flickr30K [21] and MS COCO [10] datasets are used in our experiments. Flickr8K focus on activities of people and animals (mainly dogs) and it contains almost 8,000 images and each image contain 5 corresponding descriptions. Flickr30K is a extension of Flickr8K within almost 30,000 images. Recently, MS COCO is the biggest and most challenging dataset for image caption generation. It contains 82,783 training images, 40,504 validation images and 40,775 testing images.

Data Processing. Flickr8K and Flickr30K do not have clearly training, validation and testing sets. So we choose 1,000 images for validation and 1,000 testing and the rest for training from all the datasets, which is same as reference [8]. Though MS COCO have clearly training, validation and testing sets, but the testing set does not have sentence descriptions for images, so we randomly extract both 5,000 images and their corresponding descriptions from the verification set as validation and testing data. Unlike Flickr8K and Flickr30K, some images in MS COCO having more than 5 describing sentences. To grantee each image has the same number caption, we discard data which caption in excess of 5. For all our experiments, we use a fixed vocabulary size of 10,000.

4.2   Evaluation Metrics

In order to evaluate the proposed method, three objective metrics are used in this paper. They are BLEU and METEOR. BLEU score represents the precision ratio of the generated sentence compared with the reference sentences. METEOR score reflects the precision and recall ratio of the generated sentence. It is based on the harmonic mean of uniform precision and recall.

4.3   Quantitative Evaluation and Analysis

Table 1 shows the generation results on the three standard datasets compared with the most typical and state-of-the-art models. The Results show that our model outperforms all the other models. Among them, m-RNN and LRCN use the output of the CNN FC layer as the image feature (*i.e.* image global feature in our model). The image feature is imported into the RNN unit at each time-step. The two models have a little difference which is the language model—m-RNN uses the "vanilla" RNN but LRCN uses LSTM use as the sentence generator. Different from the two aforementioned methods, DeVS [8] and Google-NIC import the whole image feature into RNN only at the first time-step. Through the results we can know that Google-NIC shows a better performance than DeVS. The main reason is that Google-NIC uses LSTM as language model which is much better than the "vanilla" RNN which DeVS used. As compared models, NIC-VA show the best performance on image caption generation task. NIC-VA is very different from other compared models, it uses the output of the CNN convolutional layer as image feature map, through the flatten operating, the feature map is changed into 196 vectors. Each vector donates a local feature of the corresponding image. Different local features are imported into LSTM unit at each time-step. At each time-step, the imported word selects local features, therefore, this model is called attention model.

Our model—**FFGS**—shows the best performance on the three datasets. The most important reason is that our **FFGS** is a general model. In other words, the compared models is one of special case of our model. For example, when $\alpha_t = 0$, $g_t = 1$ at every time-step, our model degenerates as LRCN. When only set $g_t = 0$ for all t, our model is changed as NIC-VA. When $\alpha_t = 0$ for all t, $g_t = 1$ at $t = 0$ and $g_t = 1$ for other t, our model degenerates as DeVS. So through training, our FFGS is more robustly than the compared model. The proposed model in this

Table 1. Results of image caption generation on Flickr8K, Flickr30K & MSCOCO

Model	BLEU-1	BLEU-2	BLEU-3	BLEU-4	METEOR
Flickr8K					
m-RNN	56.5	38.6	25.6	17	-
DeVS	57.9	38.3	14.5	16	16.7
LRVR [1]	-	-	-	14.1	18
Google-NIC	63	41	27	-	-
NIC-VA	67	44.8	29.9	19.5	18.9
Ours	68.2	45.2	31.2	22.5	20.1
Flickr30K					
m-RNN	60	41	28	19	-
DeVS	57.3	36.9	24	15.7	15.3
LRVR [1]	-	-	-	12.6	16.4
Google-NIC	66.3	42.3	27.7	18.3	-
LRCN	58.8	39.1	25.1	16.5	-
NIC-VA	66.7	43.4	28.8	19.1	18.4
Ours	67.9	44.0	29.2	20.9	19.7
MS COCO					
m-RNN	66.8	48.8	34.2	23.9	22.1
DeVS	62.5	45	32.1	23	19.5
LRVR [1]	-	-	-	19	20.4
Google-NIC	66.6	46.1	32.9	24.6	23.7
LRCN	62.8	44.2	30.4	21	-
NIC-VA	68.9	49.2	34.4	24.3	23.9
Ours	70.1	50.3	35.8	25.5	24.1

paper utilizes the global and local image feature which is more comprehensively using the image information.

5 Conclusion

We introduce a feature fusion with gating structure for image caption generation. Both the global image features and the local image features are used to imported into the language model. Through the gating structure, the language model robustly selects the global image features, which solves the argument that the global image feature should be imported into the language model. For the local image features, we used the attention model, which is proved very property for image caption generation. The proposed method uses the LSTM units as language model. Experimental results show that the proposed image caption generation model is better than all the compared algorithms.

References

1. Chen, X., Lawrence Zitnick, C.: Mind's eye: a recurrent visual representation for image caption generation. In: Proceedings of the IEEE Conference on Computer Vision and Pattern Recognition (CVPR), pp. 2422–2431 (2015)
2. Chung, J., Gulcehre, C., Cho, K., Bengio, Y.: Empirical evaluation of gated recurrent neural networks on sequence modeling. arXiv preprint arXiv:1412.3555 (2014)
3. Donahue, J., Hendricks, L.A., Rohrbach, M., Venugopalan, S., Guadarrama, S., Saenko, K., Darrell, T.: Long-term recurrent convolutional networks for visual recognition and description. IEEE Trans. Pattern Anal. Mach. Intell. **39**(4), 677–691 (2017)
4. Donahue, J., Anne Hendricks, L., Guadarrama, S., Rohrbach, M., Venugopalan, S., Saenko, K., Darrell, T.: Long-term recurrent convolutional networks for visual recognition and description. In: Proceedings of the IEEE Conference on Computer Vision and Pattern Recognition (CVPR), pp. 2625–2634 (2015)
5. Fairbank, M., Alonso, E., Prokhorov, D.: An equivalence between adaptive dynamic programming with a critic and backpropagation through time. IEEE Trans. Neural Netw. Learn. Syst. (TNNLS) **24**(12), 2088–2100 (2013)
6. Hodosh, M., Young, P., Hockenmaier, J.: Framing image description as a ranking task: data, models and evaluation metrics. J. Artif. Intell. Res. **47**, 853–899 (2013)
7. Karpathy, A., Fei-Fei, L.: Deep visual-semantic alignments for generating image descriptions. IEEE Trans. Pattern Anal. Mach. Intell. **39**(4), 664–676 (2017)
8. Karpathy, A., Li, F.F.: Deep visual-semantic alignments for generating image descriptions. In: Proceedings of the IEEE Conference on Computer Vision and Pattern Recognition (CVPR), pp. 3128–3137 (2015)
9. Kulkarni, G., Premraj, V., Ordonez, V., Dhar, S., Li, S., Choi, Y., Berg, A.C., Berg, T.L.: BabyTalk: understanding and generating simple image descriptions. IEEE Trans. Pattern Anal. Mach. Intell. **35**(12), 2891–2903 (2013)
10. Lin, T.-Y., Maire, M., Belongie, S., Hays, J., Perona, P., Ramanan, D., Dollár, P., Zitnick, C.L.: Microsoft COCO: common objects in context. In: Fleet, D., Pajdla, T., Schiele, B., Tuytelaars, T. (eds.) ECCV 2014. LNCS, vol. 8693, pp. 740–755. Springer, Cham (2014). https://doi.org/10.1007/978-3-319-10602-1_48
11. Lu, X., Zheng, X., Yuan, Y.: Remote sensing scene classification by unsupervised representation learning. IEEE Trans. Geosci. Remote Sens. **PP**(99), 1–10 (2017)
12. Mao, J., Xu, W., Yang, Y., Wang, J., Huang, Z., Yuille, A.: Deep captioning with multimodal recurrent neural networks (m-RNN). In: International Conference on Learning Representations (ICLR) (2015)
13. Qu, B., Li, X., Tao, D., Lu, X.: Deep semantic understanding of high resolution remote sensing image. In: 2016 International Conference on Computer, Information and Telecommunication Systems (CITS), pp. 1–5, July 2016
14. Rashtchian, C., Young, P., Hodosh, M., Hockenmaier, J.: Collecting image annotations using Amazon's mechanical turk. In: Proceedings of the NAACL HLT 2010 Workshop on Creating Speech and Language Data with Amazon's Mechanical Turk, pp. 139–147. Association for Computational Linguistics (2010)

15. Simonyan, K., Zisserman, A.: Very deep convolutional networks for large-scale image recognition. In: International Conference on Learning Representations (ICLR) (2015)
16. Socher, R., Karpathy, A., Le, Q.V., Manning, C.D., Ng, A.Y.: Grounded compositional semantics for finding and describing images with sentences. Trans. Assoc. Comput. Linguist. (TACL) **2**, 207–218 (2014)
17. Szegedy, C., Liu, W., Jia, Y., Sermanet, P., Reed, S., Anguelov, D., Erhan, D., Vanhoucke, V., Rabinovich, A.: Going deeper with convolutions. In: Proceedings of the IEEE Conference on Computer Vision and Pattern Recognition (CVPR), pp. 1–9 (2015)
18. Vinyals, O., Toshev, A., Bengio, S., Erhan, D.: Show and tell: a neural image caption generator. In: Proceedings of the IEEE Conference on Computer Vision and Pattern Recognition (CVPR), pp. 3156–3164 (2015)
19. Vinyals, O., Toshev, A., Bengio, S., Erhan, D.: Show and tell: lessons learned from the 2015 MSCOCO image captioning challenge. IEEE Trans. Pattern Anal. Mach. Intell. **39**(4), 652–663 (2017)
20. Xu, K., Ba, J., Kiros, R., Cho, K., Courville, A.C., Salakhutdinov, R., Zemel, R.S., Bengio, Y.: Show, attend and tell: neural image caption generation with visual attention. In: Proceedings of the 32nd International Conference on Machine Learning, ICML 2015, Lille, France, 6–11 July 2015, pp. 2048–2057 (2015)
21. Young, P., Lai, A., Hodosh, M., Hockenmaier, J.: From image descriptions to visual denotations: new similarity metrics for semantic inference over event descriptions. Trans. Assoc. Comput. Linguist. (TACL) **2**, 67–78 (2014)
22. Zheng, X., Yuan, Y., Lu, X.: Dimensionality reduction by spatial-spectral preservation in selected bands. IEEE Trans. Geosci. Remote Sens. **PP**(99), 1–13 (2017)

Deep Temporal Architecture for Audiovisual Speech Recognition

Chunlin Tian[1,2], Yuan Yuan[1], and Xiaoqiang Lu[1(✉)]

[1] Center for OPTical IMagery Analysis and Learning (OPTIMAL),
State Key Laboratory of Transient Optics and Photonics,
Xi'an Institute of Optics and Precision Mechanics, Chinese Academy of Sciences,
Xi'an 710119, Shaanxi, People's Republic of China
tianchunlin2015@opt.cn, {yuany,luxiaoqiang}@opt.ac.cn
[2] University of Chinese Academy of Sciences, 19A Yuquanlu, Beijing 100049,
People's Republic of China

Abstract. The Audiovisual Speech Recognition (AVSR) is one of the applications of multimodal machine learning related to speech recognition, lipreading systems and video classification. In recent and related work, increasing efforts are made in Deep Neural Network (DNN) for AVSR, moreover some DNN models including Multimodal Deep Autoencoder, Multimodal Deep Belief Network and Multimodal Deep Boltzmann Machine perform well in experiments owing to the better generalization and nonlinear transformation. However, these DNN models have several disadvantages: (1) They mainly deal with modal fusion while ignoring temporal fusion. (2) Traditional methods fail to consider the connection among frames in the modal fusion. (3) These models aren't end-to-end structure. We propose a deep temporal architecture, which has not only classical modal fusion, but temporal modal fusion and temporal fusion. Furthermore, the overfitting and learning with small size samples in the AVSR are also studied, so that we propose a set of useful training strategies. The experiments show the superiority of our model and necessity of the training strategies in three datasets: AVLetters, AVLetters2, AVDigits. In the end, we conclude the work.

Keywords: Multimodal deep learning
Audiovisual Speech Recognition

1 Introduction

Automatic Speech Recognition (ASR) has a plethora of applications, and plays a significant component in the communication between human and computers. Increasing research has been done in this area and push the accuracy up sustainedly [2,27]. But it has limitations including bad ability to fight against noise and disturbance. Furthermore, some illusion occurs when the auditory component of one sound is paired with the visual component of another sound, leading to the perception of a third sound [15] (this is called *McGurk effect* [13]).

© Springer Nature Singapore Pte Ltd. 2017
J. Yang et al. (Eds.): CCCV 2017, Part I, CCIS 771, pp. 650–661, 2017.
https://doi.org/10.1007/978-981-10-7299-4_54

Therefore, researchers have paid attention to the *Audiovisual Speech Recognition* (AVSR) utilizing both audio and visual information to tackle such problems and strengthen the ASR systems [1,17]. Besides, AVSR is, generally, an application of the multimodal machine learning. Much research in the AVSR focuses on multimodal machine learning which enhances the robustness and adaptability using multiple modalities, and particularly broadens the horizons including *Multimodal Deep Learning* [17], *Affective Computing* [20] and *Lipreading* [5].

Traditional methods on the AVSR usually use multimodal extensions of *Hidden Markov Models* (HMMs) and some statistical models like *Canonical Correlation Analysis* (CCA) [11]. Recent and related work on *Deep Neural Networks* (DNN) verified its efficiency of multimodal representation and fusion. DNN models have two main differentia and advantages with contrast to the classical methods. First, They take advantage of the superiority of feature extraction and refinement of deep learning: *Convolution Neural Network* (CNN) for extracting image features and some unsupervised deep learning algorithms for refining audio features [14,18,28]. Second, They employ DNN models to do fusion or recognition because of the better generalization and nonlinear transformation.

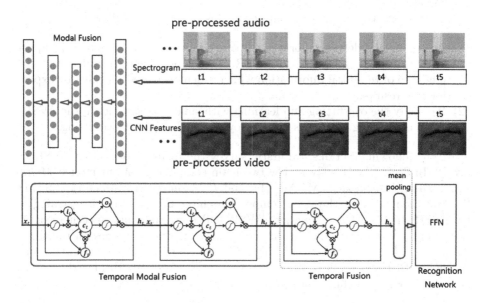

Fig. 1. The simplified architecture of our model

All these methods achieved better performance in the AVSR. However, the existed traditional methods and DNN methods cannot satisfy the demand of higher recognition accuracy and show three primary disadvantages. (1) **The first disadvantage** is that they mainly deal with modal fusion while ignoring temporal fusion. (2) **The second disadvantage** is that traditional methods fail to consider the connection among frame in the modal fusion. (3) **The third disadvantage** is that these models aren't end-to-end structure.

In order to relieve the disadvantages, we propose the deep temporal architecture for audiovisual speech recognition. The architecture of the proposed method is simply shown as Fig. 1. There are five phases in our method: **data pre-processing, modal fusion, temporal modal fusion, temporal fusion** and **recognition network**. The details of data pre-processing is not drawn in Fig. 1 as it is not the main part of the model. In the beginning, we extract the lip visual features and sound spectrogram features in pre-preprocessing phase. The modal fusion fuses jointly mouth lip features and sound spectrogram features of different frames. The temporal modal fusion considers the connection among frames in a video and further learns the joint representations. The temporal fusion encodes the modal-fused features of a video into a feature vector preparing for recognition. The proposed model takes into account the significant temporal information and helps obtain more semantic representation from both modalities. In practice, overfitting is serious as training set is insufficiently abundant, which is known as learning with small size samples, hence we adopt a set of training strategies to fight against overfitting, including visual data augmentation, aural data augmentation, multimodal data augmentation and other techniques.

In summary, the main contributions of this paper are as follows:

(1) An end-to-end deep temporal architecture mixing unsupervised with supervised learning is advanced for AVSR. The model chiefly considers the significance of temporal information which is demonstrated its value by experimental results.

(2) We study the overfitting and learning with small size samples in the AVSR. A set of training strategies are employed to fight against overfitting.

In the following sections, we firstly survey the related work about AVSR in Sect. 2. In Sect. 3, we briefly review two main components of our model: *Multimodal Deep Autoencoder* (MDAE) [17] and *Long-short Term Memory* (LSTM) [7], and then advance the deep temporal architecture and training strategies for AVSR. In Sect. 4, we conducts AVSR and cross modality speech recognition experiments for evaluating the model and training strategies on the three datasets: AVLetters (Patterson et al. 2002), AVLetters2 (Cox et al. 2008) and AVDigits (Di Hu et al. 2015), and afterwards the results are displayed and discussed. Section 5 concludes this paper.

2   Related Work

This section reviews the related multimodal models for AVSR historically.

2.1   Traditional AVSR Systems

The research on multimodal processing and interaction has long history. Humans understand the multimodal world in a seemingly effortless manner, although there are vast information processing resources dedicated to the corresponding tasks by the brain [11]. But computers are difficult to tackle this problem. When it comes to the modal fusion, which has attracted numerous investigators for a long time, it usually benefits discriminative tasks because of rich and hierarchical information.

Specifically AVSR has been studied for a few years, deal of work all focused on multimodal fusion. There are three levels of multimodal fusion: early fusion, intermediate fusion and late fusion. For early integration, it concatenates video and audio features into a single descriptor. For late fusion, fusion is done at decision part, which resembles ensemble learning. Intermediate fusion, or called hybrid fusion, lies in between late and early fusion [3,11]. Early and late fusion are uncomplicated to understand and easy to put into practice. Whereas they both are too straightforward to capture the abundant information and correlation between aural and visual features.

In the early years, investigators attempted to use probabilistic models, but they depend on strong different prior assumption. The typical models are *multistream HMMs* (mHMMs) [16] that were affirmed strong ability to model sequence data [22]. However, mHMMs don't work well, because mHMMs building a hybrid fusion structure and mapping aural and visual features jointly to a low-dimensional space are shallow models, additionally distinct modalities have various type of information representations. In this method, researchers are interested in distinct feature extraction and different feature combination [6].

2.2 Deep Learning for AVSR

Deep Learning provides strong representation ability and capacity [4], especially CNN for image feature extraction and representation [9,10,30,32]. For instance, [18] uses CNN pre-trained visual feature and aural feature refined by denoising autoencoder, and then traditional models are used for fusion and recognition. As has been argued, these models mainly make use of the rich and robust feature representation of DNN.

Some other methods are posed based on multimodal fusion by DNN. There are three levels of multimodal fusion: early fusion, intermediate fusion and late fusion. For early integration, it concatenates video and audio features into a single descriptor. For late fusion, fusion is done at decision part, which resembles ensemble learning. Intermediate fusion, or called hybrid fusion, lies in between late and early fusion [3,11]. Early and late fusion are uncomplicated to understand and easy to put into practice. Whereas they both are too straightforward to capture the abundant information and correlation between aural and visual features.

[17] used MDAE; [25] used *Multimodal Deep Network* (MDBN); [26] used *Multimodal Deep Boltzmann Machine* (MDBM); [28] uses *Deep Bottleneck Features* (DBNF); [8] used *Recurrent Temporal Multimodal Restricted Boltzmann Machines* (RTMRBM). They indeed achieve more accurate result. But the majority of them are unsupervised models, and additionally most are based on *Boltzmann Machines* that are considered difficult to train of partial function [4].

Besides, they involve several independent training process and testing process, and give rise to the loss of temporal information in the modal fusion.

3   Proposed Method

In this section, we first briefly review two main components: MDAE and LSTM of our model, and then advance the deep temporal architecture for AVSR. Finally, we propose a set of training strategies for fighting against overfitting and learning with small size samples.

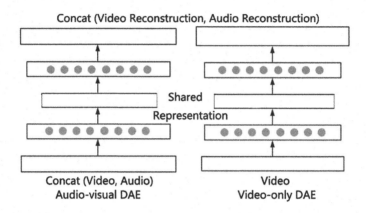

Fig. 2. Bimodal DAE and video-only DAE in our model

3.1   MDAE and LSTM

MDAE: In the multimodal fusion task, [17] advanced two types of MDAE: *Bimodal Deep Autoencoder* and *Video-Only Deep Autoencoder* (a similar model can be drawn for the audio-only setting). One reconstructs both modalities given audio and video used for AVSR, and the other reconstructs both modalities given only one modality used for cross modality speech recognition – that is, multiple modalities are available during training; during testing phase, only data from a single modality is provided [17]. The MDAE mentioned in [17] has a RBM greedy pre-training process, thereby the data ought to encode into binary.

LSTM: LSTM is the suitable algorithm to model the sequence data. There are two widely known issues with properly training vanilla RNN, the vanishing and the exploding gradient [19]. Without some training tricks in training RNN, LSTM is an off-the-shelf favorable solution. It uses gates to avoid gradient vanishing and exploding. A typical LSTM has 3 gates: input gate, output gate and forget gate which reserves the sequence information and makes *Backpropagation through Time* (BPTT) easier.

3.2   Deep Temporal Architecture for AVSR

Our DNN architecture consists of four primary components: modal fusion, temporal modal fusion, temporal fusion, recognition network.

Modal Fusion: MDAE is chosen as the main tool of the modal fusion (Fig. 2). In the model, in order to eliminates truncation error, there is no binary encoding process. We do the early fusion by concatenating video and audio features as the input and then reconstructing them. The shared representation is extracted as the modal joint fused representation. As the boost of computation power, the MDAE is easy to train. The loss function of MDAE is:

$$loss(x, y) = \frac{1}{n} \sum_{i=1}^{n} [y_i log(x_i) + (1 - y_i) log(1 - x_i)], \tag{1}$$

where n is the number features through pre-processing, x_i is the reconstruction, y_i is the original data.

Temporal Modal Fusion: LSTM is used to do the temporal modal fusion as it takes into account the temporal factors. With the decrease of attributes and deep fusion of the features after LSTM, it would be simpler for temporal fusion and recognition. Two LSTMs are taken to accomplish the temporal modal fusion (Fig. 3).

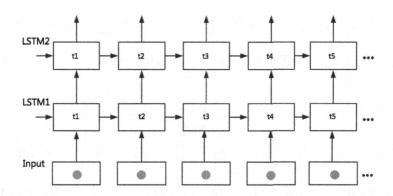

Fig. 3. Temporal modal fusion structure

The mechanism of temporal modal fusion is as follows:

$$i_t = sigmoid(W_i x_t + U_i c_{t-1} + b_i) \tag{2}$$

$$f_t = sigmoid(W_f x_t + U_f c_{t-1} + b_f) \tag{3}$$

$$o_t = sigmoid(W_o x_t + U_o c_{t-1} + b_o) \tag{4}$$

$$c_t = f_t \circ c_{t-1} + i_t \circ tanh(W_c x_t + b_c) \tag{5}$$

$$h_t = o_t \circ tanh(c_t), \tag{6}$$

In the interior of temporal modal fusion, W, U and b are the parameter matrices and vector, c_t is the cell state vector, f_t is the forget gate vector, i_t is the input gate vector, o_t is the output gate vector. The inner parts aim to preventing vanishing and the exploding gradient and passing sequence information as typical LSTM. x_t is the modal fused input, h_t is the temporal modal fused output.

Temporal Fusion: Traditional methods usually concatenate different frames of one video after modal fusion. But it needs some hyperparameters to tune, which isn't recommended by modern learning architecture. Here, another LSTM and mean pooling are used to map the audiovisual features after temporal modal fusion into a well fused feature vector (shown in Fig. 4). The mechanism of temporal fusion LSTM is roughly the same as (2)–(6), and the mean pooling works as follows:

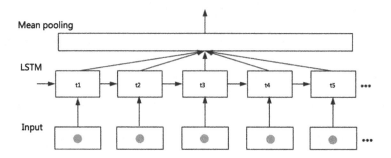

Fig. 4. Temporal fusion structure

$$z_i = \frac{1}{m} \sum_{t=1}^{m} h_t, \tag{7}$$

where m is the number of frames in a video, h_t is the temporal modal fused vector, z_i is the output of temporal fusion.

Recognition Network: Our model uses a feed forward network with batch normalization mapping the features for recognition. It has eight layers: fully connected layer, activation layer with *Rectified Linear Units* (ReLU) and batch normalization layer one by one repeatedly. The last layer is general softmax layer with squared multi-label margin loss function :

$$loss(p,q) = \frac{1}{n} \sum_{i=1}^{n} max\{0, (1 - p_i + q_i)^2\}, \tag{8}$$

where n is the number of features in a video, p_i is the output of the recognition network, q_i is the ground truth.

3.3   Training Strageties

The samples of AVSR is insufficient usually and therefore sequence learning is easy to overfit. Hence we utilize a set of training strategies for learning with small size samples and against overfitting.

Visual Data Augmentation: As is mentioned before, it is common that insufficient visual data may cause serious overfitting in the AVSR. We employ visual data augmentation with extracted lip videos. Color jittering, small angle rotation and random scaling are adopted to augment data. The augmentation will also enhance the generalization of color change, space variousness and image quality difference.

Aural Data Augmentation: To improve model's tolerance to audio noise and prevent overfitting, we apply white Gaussian noise in the training phase.

Multimodal Data Augmentation: When video and audio are both fused after the modal fusion, we make the fused features of each video simultaneously shift slight frames up and down randomly to augment data. As a benefit, the trained model will have better generalization of variance in the time domain.

Other Techniques: Common DNN training strategies including dropout [24] and early stopping [21] are also used in our model.

4   Experiments and Discussion

In this section, we show the results of the proposed model compared with some state-of-the-art methods.

4.1   Datasets

We conducted the experiments in 3 datasets: AVLetters (Patterson et al. 2002), AVLetters2 (Cox et al. 2008) and AVDigits (Di Hu et al. 2015).

AVLetters. 10 volunteers saying the letters A to Z three times each. The dataset pre-extracts the lip region of 60×80 pixels and *Mel-Frequency Cepstrum Coefficientx* (MFCC).

AVLetters2. 5 volunteers spoke the letters A to Z seven times each. The dataset provide raw video and audio in different folders.

AVDigits. 6 speakers saying digits 0 to 9, nine times each. It doesn't pre-extract the video and audio, and provides raw video of long or short time length from 1 s to 2 s.

4.2   Data Pre-processing

If the video and audio are not separated, we separate them from each other. And then truncate the video and audio into the same length.

Pre-processing of Video. Firstly, the off-the-shelf *Viola-Jones algorithm* [29] is used to extract the Region-of-Interest surrounding the mouth. The region is resized to 224 × 224 pixels, and use aforementioned visual data augmentation strategy to double the lip visual data. The features are obtained by the pre-trained VGG-16's [23] last fully connected layer. Finally, reduce features to 100 principal components with PCA whitening and center them.

Pre-processing of Audio. Double aural data with SNR of 5 dB (signal power: noise power = 5:1) white Gaussian noise as mentioned in Sect. 3.3. The features of audio signal are extracted as spectrogram with 20 ms *Hamming window* and 10 ms overlap. The spectral coefficient vector is gained with 251 points of *Fast Fourier Transform* and 50 principal components by PCA.

Visual and Aural Features Combination. When video and audio are both prepared, four contiguous audio frames correspond to one video frame for each time step.

4.3 Implementation Details

The modal fusion network has eight layers and loss is binary cross entropy. While the modal fusion finished, the shared representation use multimodal data augmentation strategy with shifting ten frames up and down randomly in Sect. 3.3, and reshape the fused data waiting for temporal modal fusion. Then the reshaped fused features are sent to the temporal modal fusion network. The temporal modal fusion network continues fusing the existed shared representation of audio-visual fusion information. The temporal fusion network maps the features of several frames in a video to a feature vector. Finally, every video has one low-dimensional well-fused vector waiting for recognition. The recognition network has two nonlinear layer to augment the nonlinearity.

The temporal modal fusion, temporal fusion and recognition network are trained by one loss function. Thus the model needs two steps of training: unsupervised training and supervised training, a reshape layer connects them at once in evaluation. One step of evaluation and testing.

4.4 Quantitative Evalution

To evaluate the proposed model, we conducted AVSR and cross modality speech recognition experiments on the multimodal data. At the same time, the experiments are also conducted to evaluate the necessity of training strategies.

Evaluation of AVSR and Training Strategies: We evaluate our methods in the AVSR task, compared with MDAE, MDBN and RTMRBM on AVLetters2 and AVDigits. Moreover, the training strategies in Sect. 3.3 are evaluated in the experiment. The quantitative results demonstrate the superiority of our model for AVSR and necessity of the training strategies. In addition, it is not uncommon that the model slowly converges without training strategies in our practice (Table 1).

Table 1. AVSR performance on AVLetters2 and AVDigits. The result indicates that our model performs better than MDBN, MDAE and RTMRBM. And the experiments show that model with training strategies in Sect. 3.3 is better than without training strategies

Datasets	Model	Mean accuracy
AVLetter2	MDBN [25]	54.10%
	MDAE [17]	67.89%
	RTMRBM [8]	74.77%
	Our method (without training strategies)	**77.71%**
	Our method (with training strategies)	**80.11%**
AVDigits	MDBN [25]	55.00%
	MDAE [17]	66.74%
	RTMRBM [8]	71.77%
	Our method (without training strategies)	**76.43%**
	Our method (with training strategies)	**78.56%**

Evaluation of Cross Modality Speech Recognition: One purpose of the multimodal deep learning is learning better single modality representations given unlabeled data from multiple modalities [17]. In cross modality learning experiments, we evaluate the accuracy of one modality (e.g. V) when given multiple modalities (e.g. V and A) during learning. We compare the model to MDAE, CRBM [1] and some single modality models including *Multiscale Spatial Analysis* [12] and *Local Binary Pattern* [31] on AVLetters. The experiments display proper cross modality speech recognition is better than single modal speech recognition, moreover our method performs better than other related multimodal models in cross modality speech recognition (Table 2).

Table 2. Cross modality speech recognition performance. The results of **V modality** suggest that cross modality speech recognition (MDAE, CRBM, our method) is better than single modality speech recognition (Multiscale Spatial Analysis, Local Binary Pattern). The **V modality and A modality** experiments show that our method performs better than other models in cross modality speech recognition.

Molidity	Model	Mean accuracy
A	MDAE [17]	61.11%
	CRBM [1]	63.42%
	Our method	**68.01%**
V	Multiscale spatial analysis [12]	44.60%
	Local binary pattern [31]	58.85%
	MDAE [17]	64.21%
	CRBM [1]	66.78%
	Our method	**70.23%**

5   Conclusion

A deep temporal architecture and a set of training strategies are proposed for AVSR task in this paper. Once the model has been trained, it's simple to do AVSR in reality. Our method considers modal fusion, temporal modal fusion and temporal fusion, such fusion enhances the robustness and the ability for sequence modeling. The experimental results suggest that the deep temporal architecture achieve better AVSR recognition and cross modality speech recognition results in three datasets. Besides our training strategies efficiently weaken the overfitting as the experiments shows.

References

1. Amer, M.R., Siddiquie, B., Khan, S., Divakaran, A., Sawhney, H.: Multimodal fusion using dynamic hybrid models, pp. 556–563 (2014)
2. Amodei, D., Anubhai, R., Battenberg, E., Case, C.J., Casper, J., Catanzaro, B., Chen, J., Chrzanowski, M., Coates, A., Diamos, G., et al.: Deep speech 2: end-to-end speech recognition in English and mandarin, pp. 173–182 (2015)
3. Atrey, P.K., Hossain, M.A., El Saddik, A., Kankanhalli, M.S.: Multimodal fusion for multimedia analysis: a survey. Multimed. Syst. 16(6), 345–379 (2010)
4. Bengio, Y., Goodfellow, I.J., Courville, A.: Deep Learning. MIT Press, Cambridge (2015). http://www.iro.umontreal.ca/bengioy/dlbook
5. Chung, J.S., Senior, A., Vinyals, O., Zisserman, A.: Lip reading sentences in the wild (2016)
6. Galatas, G., Potamianos, G., Makedon, F.: Audio-visual speech recognition incorporating facial depth information captured by the kinect. In: 2012 Proceedings of the 20th European Signal Processing Conference (EUSIPCO), pp. 2714–2717. IEEE (2012)
7. Hochreiter, S., Schmidhuber, J.: Long short-term memory. Neural Comput. 9(8), 1735–1780 (1997)
8. Hu, D., Li, X., et al.: Temporal multimodal learning in audiovisual speech recognition. In: Proceedings of the IEEE Conference on Computer Vision and Pattern Recognition, pp. 3574–3582 (2016)
9. Kruthiventi, S.S., Ayush, K., Babu, R.V.: DeepFix: a fully convolutional neural network for predicting human eye fixations. IEEE Trans. Image Process. (2017)
10. Lu, X., Zheng, X., Yuan, Y.: Remote sensing scene classification by unsupervised representation learning. IEEE Trans. Geosci. Remote Sens. (2017)
11. Maragos, P., Potamianos, A., Gros, P.: Multimodal Processing and Interaction: Audio, Video, Text, vol. 33. Springer Science & Business Media, Heidelberg (2008). https://doi.org/10.1007/978-0-387-76316-3
12. Matthews, I., Cootes, T.F., Bangham, J.A., Cox, S., Harvey, R.: Extraction of visual features for lipreading. IEEE Trans. Pattern Anal. Mach. Intell. 24(2), 198–213 (2002)
13. McGurk, H., MacDonald, J.: Hearing lips and seeing voices. Nature 264, 746–748 (1976)
14. Mroueh, Y., Marcheret, E., Goel, V.: Deep multimodal learning for audio-visual speech recognition, pp. 2130–2134 (2015)
15. Nath, A.R., Beauchamp, M.S.: A neural basis for interindividual differences in the McGurk effect, a multisensory speech illusion. NeuroImage 59(1), 781–787 (2012)

16. Nefian, A.V., Liang, L., Pi, X., Liu, X., Murphy, K.: Dynamic Bayesian networks for audio-visual speech recognition. EURASIP J. Adv. Sig. Process. **2002**(11), 1–15 (2002)
17. Ngiam, J., Khosla, A., Kim, M., Nam, J., Lee, H., Ng, A.Y.: Multimodal deep learning. In: Proceedings of the 28th International Conference on Machine Learning, ICML 2011, pp. 689–696 (2011)
18. Noda, K., Yamaguchi, Y., Nakadai, K., Okuno, H.G., Ogata, T.: Audio-visual speech recognition using deep learning. Appl. Intell. **42**(4), 722–737 (2015)
19. Pascanu, R., Mikolov, T., Bengio, Y.: On the difficulty of training recurrent neural networks. In: ICML (3), vol. 28, pp. 1310–1318 (2013)
20. Poria, S., Cambria, E., Bajpai, R., Hussain, A.: A review of affective computing: from unimodal analysis to multimodal fusion. Inf. Fusion **37**, 98–125 (2017)
21. Prechelt, L.: Early stopping—but when? In: Montavon, G., Orr, G.B., Müller, K.-R. (eds.) Neural Networks: Tricks of the Trade. LNCS, vol. 7700, 2nd edn, pp. 53–67. Springer, Heidelberg (2012). https://doi.org/10.1007/978-3-642-35289-8_5
22. Rabiner, L.R.: A tutorial on hidden Markov models and selected applications in speech recognition. Proc. IEEE **77**(2), 257–286 (1989)
23. Simonyan, K., Zisserman, A.: Very deep convolutional networks for large-scale image recognition. arXiv preprint arXiv:1409.1556 (2014)
24. Srivastava, N., Hinton, G.E., Krizhevsky, A., Sutskever, I., Salakhutdinov, R.: Dropout: a simple way to prevent neural networks from overfitting. J. Mach. Learn. Res. **15**(1), 1929–1958 (2014)
25. Srivastava, N., Salakhutdinov, R.: Learning representations for multimodal data with deep belief nets. In: International Conference on Machine Learning Workshop (2012)
26. Srivastava, N., Salakhutdinov, R.R.: Multimodal learning with deep Boltzmann machines. In: Advances in Neural Information Processing Systems, pp. 2222–2230 (2012)
27. Sutskever, I., Vinyals, O., Le, Q.: Sequence to sequence learning with neural networks, pp. 3104–3112 (2014)
28. Tamura, S., Ninomiya, H., Kitaoka, N., Osuga, S., Iribe, Y., Takeda, K., Hayamizu, S.: Audio-visual speech recognition using deep bottleneck features and high-performance lipreading. In: 2015 Asia-Pacific Signal and Information Processing Association Annual Summit and Conference (APSIPA), pp. 575–582. IEEE (2015)
29. Viola, P., Jones, M.: Rapid object detection using a boosted cascade of simple features. In: Proceedings of the 2001 IEEE Computer Society Conference on Computer Vision and Pattern Recognition, CVPR 2001, vol. 1, pp. I–511. IEEE (2001)
30. Xia, G.S., Hu, J., Hu, F., Shi, B., Bai, X., Zhong, Y., Zhang, L., Lu, X.: AID: a benchmark data set for performance evaluation of aerial scene classification. IEEE Trans. Geosci. Remote Sens. (2017)
31. Zhao, G., Barnard, M., Pietikainen, M.: Lipreading with local spatiotemporal descriptors. IEEE Trans. Multimed. **11**(7), 1254–1265 (2009)
32. Zheng, X., Yuan, Y., Lu, X.: Dimensionality reduction by spatial-spectral preservation in selected bands. IEEE Trans. Geosci. Remote Sens. (2017)

Efficient Cross-modal Retrieval via Discriminative Deep Correspondence Model

Zhikai Hu[1,2], Xin Liu[1,2(✉)], An Li[1,2], Bineng Zhong[1,2], Wentao Fan[1,2], and Jixiang Du[1,2]

[1] Department of Computer Science, Huaqiao University, Xiamen 361021, China
xliu@hqu.edu.cn

[2] Xiamen Key Laboratory of Computer Vision and Pattern Recognition, Huaqiao University, Xiamen 361021, China

Abstract. Cross-modal retrieval has recently drawn much attention due to the widespread existence of multi-modal data, and it generally involves two challenges: how to model the correlations and how to utilize the class label information to eliminate the heterogeneity between different modalities. Most previous works mainly focus on solving the first challenge and often ignore the second one. In this paper, we propose a discriminative deep correspondence model to deal with both problems. By taking the class label information into consideration, our proposed model attempts to seamlessly combine the correspondence autoencoder (Corr-AE) and supervised correspondence neural networks (Super-Corr-NN) for cross-modal matching. The former model can learn the correspondence representations of data from different modalities, while the latter model is designed to discriminatively reduce the semantic gap between the low-level features and high-level descriptions. The extensive experiments tested on three public datasets demonstrate the effectiveness of the proposed approach in comparison with the state-of-the-art competing methods.

Keywords: Cross-modal retrieval
Discriminative deep correspondence model
Correspondence autocoder · Semantic gap

1 Introduction

In recent years, cross-modal retrieval has attracted considerable attention due to the rapid growth of abundant multi-modal data. Unlike uni-modal retrieval, cross-modal retrieval searches the information of one modality by given information of another different modality, and this type of matching problem has been existing in many fields, e.g., multimedia retrieval. For instance, one often attempts to seek the picture that best illustrates a given text, or find the text that best describes a given picture. Since the samples from different modalities are always heterogeneous, it is still a non-trivial task to bridge the semantic gap, measure the similarity and establish the connections among different modalities.

© Springer Nature Singapore Pte Ltd. 2017
J. Yang et al. (Eds.): CCCV 2017, Part I, CCIS 771, pp. 662–673, 2017.
https://doi.org/10.1007/978-981-10-7299-4_55

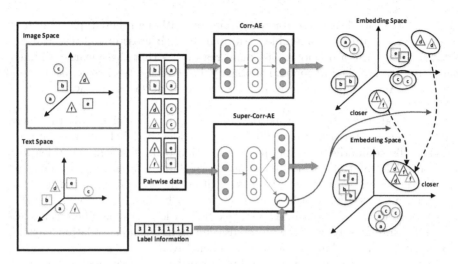

Fig. 1. The difference between the Corr-AE [5] and our proposed model.

In the past, different kinds of cross-modal retrieval approaches have been exploited, and most of them attempted to learn a common latent subspace to make all data comparable. That is, these approaches project the representations of multiple modalities into a common space with same dimension, whereby the similarity can be easily measured by counting Euclidean or other distances [13–15]. Along this line, one of the most common subspace learning methods is Canonical Correlation Analyse (CCA), which learns a pair of linear transformations to maximize the correlations between two different modalities. For instance, Rasiwasia et al. [13] proposed a semantic correlation matching scheme for cross-modal retrieval, where the CCA is applied to the maximally correlated feature representations. Later, some improved versions [12] and extensions [11] were also exploited to produce an isomorphic semantic space for cross-modal retrieval. In addition, bilinear model (BLM) [15] and partial least squares (PLS) [14], were also popularly utilized in cross-modal retrieval. Despite these contributions have been made to the solution of cross-modal retrieval, most of their performances are still far from satisfactory. A plausible reason is that these traditional approaches did not fully consider the label information or high-level featuring mapping for discriminative analysis, thereby the connections of different modalities were not well built in practice.

Recently, significant progress has been made in deep learning, and some deep models have been developed to tackle cross-modal retrieval problem [9]. For instance, Srivastava and Salakhutdinov [17] first utilized two separated Restricted Boltzmann Machines (RBM) to learn the low-level representations of different modalities, and then fused them into a joint representation by a high-level full-connection layer. Similarly, Feng et al. [5] employed a correspondence autoencoder (Corr-AE) to represent different modalities, in which an efficient loss function was designed for minimization of correlation learning error and representation learning error synchronously. Although these typical methods were able to produce considerable performance, their cross-modal retrieval performance were still a bit poor. The main reason lies that these learning models can only preserve the inter-modal similarity, but which often ignore intra-modal similarity. As a result, the corresponding retrieval performances were far from satisfactory.

In general, the more information embedded in a model for discriminative analysis, the greater performance it reaches. Evidently, label information often provides discriminative cues for the classification task, and some deep learning methods [2,19,20] thus attempted to utilize the label information in their models. For instance, Wang et al. [19] first utilized the label information to learn the discriminative representation of data from each modality individually, then investigated the highly non-linear semantic correlation between different modalities. This strategy separates the representation learning and correlation learning, which can not utilize label information sufficiently.

In this paper, we propose a discriminative deep correspondence model for cross-modal retrieval, which can learn the representation and correlation of different modalities in an integral model. As shown in Fig. 1, our proposed model inherently differs from Corr-AE model [5], and improves this method by providing the following three contributions:

– By using the label information, a supervised correspondence neural networks (Super-Corr-NN), is proposed to learn the representation and correlation of different modalities simultaneously, while preserving more intra-modal similarity.
– We propose a supervised Corr-AE (Super-Corr-AE) model and extend it into another two forms, whereby the data from different modalities can be well retrieved.
– The proposed discriminative deep correspondence model seamlessly combines the correspondence autoencoder (Corr-AE) and supervised correspondence neural networks (Super-Corr-NN) for data representation learning, which can preserve both inter-modal and intra-modal similarity for efficient cross-modal matching. The experimental results have shown its outstanding performance.

The rest of this paper is organized as follow. Section 2 presents the detail of our proposed model. In Sect. 3, we present the experimental results and related comparisons. Finally, we draw a conclusion in Sect. 4.

2   The Proposed Deep Learning Model

In this section, Corr-AE [5] is simply introduced for illustration, and then the proposed deep correspondence models are presented in details.

2.1   Correspondence AutoEncoder

As shown in Fig. 2(a), Corr-AE consists of two interdependent autoencoders, whose hidden layers are connected with each other by defining a similarity metric. Two autoencoders correspond to image and text modalities respectively,

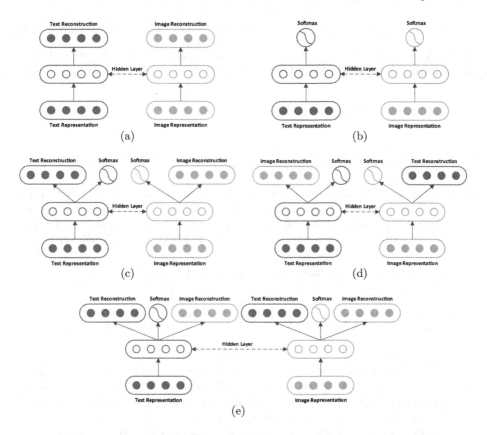

Fig. 2. (a) Correspondence autoencoder, (b) Supervised correspondence neural networks, (c) Supervised correspondence autoencoder, (d) Supervised correspondence cross-modal autoencoder, (e) Supervised correspondence full-modal autoencoder.

receive data at input layers, reconstruct them at output layers, and extract the information of hidden layers as final learned representations. Let $X_1 = \{x_1^{(i)}\}_{i=1}^n$ and $X_2 = \{x_2^{(i)}\}_{i=1}^n$ denote two different modalities, e.g., image and text, and share the same label $Y = \{y^{(i)}\}_{i=1}^n$, where n is the number of objects, $y^{(i)} \in \{1, 2, \ldots, c\}$ and c is the number of categories. The hidden layers of two autoencoders are denoted as $H_1 = f(X_1; W_f)$ and $H_2 = g(X_2; W_g)$, where W_f and W_g are parameters of two autoencoders respectively, f and g are the corresponding activation functions. The similarity metric between two modalities are defined as follow:

$$L_H(X_1, X_2; W) = ||H_1 - H_2||_F^2 = ||f(X_1; W_f) - g(X_2; W_g)||_F^2 \qquad (1)$$

where $W = [W_f, W_g]$ is the whole parameters. Accordingly, the reconstruction losses of two modalities can be described as follows:

$$L_I(X_1, X_2; W) = ||X_1 - \hat{X}_1^{(I)}||_F^2 \qquad (2)$$

$$L_T(X_1, X_2; W) = ||X_2 - \hat{X}_2^{(T)}||_F^2 \tag{3}$$

where $\hat{X}_1^{(I)}$ and $\hat{X}_2^{(T)}$ are outputs of the image and text autoencoder respectively. Consequently, the total loss can be derived as follow:

$$L_{total} = (1 - \alpha)(L_I + L_T) + \alpha L_H \tag{4}$$

where $\alpha \in [0, 1]$ is balance parameter.

2.2 Supervised Correspondence Neural Networks

In general, label information often provides discriminative cues for the vision task [18]. Inspired by the theories of Corr-AE, we proposed a Supervised Correspondence Neural Networks (Super-Corr-NN) by taking the label information into the consideration. As shown in Fig. 2(b), two autoencoders are replaced by two 3-layers multilayer perceptron in Super-Corr-NN. Given data set $X = \{x^{(i)}\}_{i=1}^n$ and its label set $L = \{l^{(i)}\}_{i=1}^n$, $l^{(i)} \in \{1, 2, \ldots, c\}$ and c is the number of categories, the softmax loss is presented as follow:

$$J(X, L; \Theta) = -\frac{1}{n} \sum_{i=1}^n \sum_{j=1}^c I\{l^{(i)} = j\} log \frac{e^{\theta_j^T x^{(i)}}}{\sum_{k=1}^c e^{\theta_k^T x^{(i)}}} \quad j = 1, 2, \ldots, c \tag{5}$$

where Θ is the parameter of network, θ_i is the i-th column of Θ, and $I\{z\}$ equals 1 if z is true, otherwise equals 0. Differing from the autoencoder model, the output of multilayer perceptron is a probability vector of a sample belonging to different classes, which can utilize label information to preserve intra-modal similarities and make results more discriminative. Accordingly, the overall softmax loss function in Super-Corr-NN can be described as:

$$L_S = J(X_1, Y; \Theta_1) + J(X_2, Y; \Theta_2) \tag{6}$$

where Θ_1 and Θ_2 are the parameters of two multilayer perceptrons respectively. Specifically, the same similarity metric is adopted to learn representation and correlation simultaneously. By taking the correspondence loss defined in Eq. (1), the total loss function can be derived by:

$$L_{total} = (1 - \alpha)L_S + \alpha L_H \tag{7}$$

where $\alpha \in [0, 1]$ is balance parameter.

2.3 Supervised Correspondence AutoEncoder

In essence, Corr-AE, as an unsupervised learning method, can learn a correspondence representation and correlation between different modalities. In contrast to this, Super-Corr-NN could learn more discriminative representation by using the label information, but which performs weak in correlation mining between different modalities. Meanwhile, Corr-AE can preserve more inter-modal similarity while Super-Corr-NN can preserve more intra-modal similarity. Inspired by this finding, As shown in Fig. 2(c), we propose a supervised correspondence autoencoder (Super-Corr-AE) by combing the Corr-AE and Super-Corr-NN, which are complementary to each other. In Super-Corr-AE, Super-Corr-NN and Corr-AE share their input layer and hidden layer, with their output layers separated independently. As a result, this model could simultaneously preserve both intermodal and intra-modal similarities and make the learned representation more discriminative. In this model, the total loss consists of three parts: reconstruction loss $L_I + L_T$, correspondence loss L_H and softmax loss L_S:

$$L_{total} = (1 - \alpha)(L_I + L_T) + \alpha L_H + \beta L_S \tag{8}$$

where $\alpha \in [0, 1]$ and $\beta \geqslant 0$ are balance parameters.

2.4 Supervised Correspondence Cross-modal AutoEncoder

As shown in Fig. 2(d), we further replace the Corr-AE in Super-Corr-AE to Corr-Cross-AE. Differing from the Corr-AE model, Corr-Cross-AE restricts the outputs of each autoencoder by the inputs of different modality. Accordingly, the reconstruction loss is given as:

$$L_{Icross}(X_1, X_2; W) = ||X_2 - \hat{X}_2^{(I)}||_F^2 \tag{9}$$

$$L_{Tcross}(X_1, X_2; W) = ||X_1 - \hat{X}_1^{(T)}||_F^2 \tag{10}$$

where $\hat{X}_2^{(I)}$ and $\hat{X}_1^{(T)}$ are outputs of image autoencoder and text autoencoder, respectively. Then, the total loss can be derived as follow:

$$L_{total} = (1 - \alpha)(L_{Icross} + L_{Tcross}) + \alpha L_H + \beta L_S \tag{11}$$

where $\alpha \in [0, 1]$ and $\beta \geqslant 0$ are balance parameters.

2.5 Supervised Correspondence Full-modal AutoEncoder

As shown in Fig. 2(e), we replace the Corr-AE in Super-Corr-AE to Corr-Full-AE. As a combination of Corr-AE and Corr-Cross-AE, Corr-Full-AE utilizes the inputs of all modalities to restrict the outputs of each autoencoder. Consequently, the reconstruction loss can be derived as follows:

$$\begin{aligned} L_{Ifull}(X_1, X_2; W) &= L_I(X_1, X_2; W) + L_{Icross}(X_1, X_2; W) \\ &= ||X_1 - \hat{X}_1^{(I)}||_F^2 + ||X_2 - \hat{X}_2^{(I)}||_F^2 \end{aligned} \tag{12}$$

$$\begin{aligned} L_{Tfull}(X_1, X_2; W) &= L_T(X_1, X_2; W) + L_{Tcross}(X_1, X_2; W) \\ &= ||X_2 - \hat{X}_2^{(T)}||_F^2 + ||X_1 - \hat{X}_1^{(T)}||_F^2 \end{aligned} \tag{13}$$

where $\hat{X}_1^{(I)}$ and $\hat{X}_2^{(I)}$ are outputs of image autoencoder, $\hat{X}_1^{(T)}$ and $\hat{X}_2^{(T)}$ are outputs of text autoencoder, respectively. Then, the total loss can be derived as follows:

$$L_{total} = (1 - \alpha)(L_{Ifull} + L_{Tfull}) + \alpha L_H + \beta L_S \tag{14}$$

where $\alpha \in [0, 1]$ and $\beta \geqslant 0$ are balance parameters.

3   Experimental Results

3.1   Datasets

For experimental evaluation, three publicly available datasets are selected:

Wikipedia[1]. This dataset [13] is collected from "Wikipedia features articles", which contains 10 categories and 2,866 image&text pairs. Images are represented by 2,296-dimensions features, which consists of three type of low-level features extracted from images, including 1,000-dimensions pyramid histogram of words (PHOW) [1], 512-dimensions gist descriptor [10], 784-dimensions MPEG-7 descriptors [8]. And texts are represented by 3,000-dimensions hign-frequency words vector. These features are all extracted by [5]. We use 2,173 pairs as training set, 231 pairs as validation set, 462 pairs as test set.

Pascal[2]. This dataset [4] contains 10 categories and 1,000 image&text pairs. Image are represented by 2,296-dimensions features, shares the same extraction with wikipedia dataset. And texts are represented by 1,000-dimensions high-frequency words vector. We use 800 pairs as training set, 100 pairs as validation set, 100 pairs as test set.

NUS-WIDE-10k. This dataset is sellected from the real-world web image dataset NUS-WIDE[3] [3], which contains 81 concepts and 269,648 images with tags. We only choose 10 largest concepts and the corresponding 10,000 image&text pairs, each concept contains 1,000 pairs. Images are represented by 1134-dimensions features, which consists of six types of low-level features extracted from images, including 64-dimensions color histogram, 144-dimensions color correlogram, 73-dimensions edge direction histogram, 128-dimensions wavelet texture, 225-dimensions block-wise color moments and 500-dimensions bag of words based on SIFT descriptions. And texts are represented by 1,000-dimensions bag-of-words. We use 8,000 pairs as training set, 1,000 pairs as validation set, and 1,000 pairs as test set.

3.2   Implementation Details

In the experiments, two RBMs were selected to pre-process the data. In the first RBM, Guassian RBM and replicated softmax RBM were adopted to process image and text data, respectively. The second RBM was a basic RBM. As a result, the pre-processed data was utilized as the input of the proposed deep correspondence models.

In particular, the settings of parameters (α, β) in different datasets were shown in Table 1, although parameters (α, β) are not sensitive to datasets (more empirical analysis will be given in the coming sections), to reach best performance, we set different (α, β) on three different datasets and different model and we fixed these values in all related experiments.

---

[1] http://www.svcl.ucsd.edu/projects/crossmodal/.

[2] http://vision.cs.uiuc.edu/pascal-sentences/.

[3] http://lms.comp.nus.edu.sg/research/NUS-WIDE.htm.

Table 1. Parameters setting on different methods and datasets

Method	Wikipedia		Pascal		NUS-WIDE-10k	
	α	β	α	β	α	β
Super-Corr-NN	0.8	-	0.9	-	0.7	-
Super-Corr -AE	0.9	11	0.9	19	0.9	1
Super-Corr-Cross-AE	0.9	50	0.8	30	0.7	3
Super-Corr-Full-AE	0.9	10	0.9	12	0.9	2

3.3 Baseline Methods

Meanwhile, three CCA basic methods: CCA-AE [6], CCA-Cross-AE [5], CCA-Full-AE [5], two multi-modal methods: Bimodal AE [9], Bimodal DBN [9,16], and three correspondence autoencoder methods [5]: Corr-AE, Corr-Cross-AE, Corr-Full-AE, were selected for comparison.

3.4 Evaluation Metric

We perform two kinds of cross-modal retrieval: retrieve text by given image query and vice verse. Mean average precision (mAP) of R top-rank retrieved data is employed in our experiments to evaluate the performance of the results:

$$mAP = \frac{1}{M} \sum_{i=1}^{M} (\frac{1}{L} \sum_{r=1}^{R} P(r) \times \delta(r)) \tag{15}$$

where M is the number of queries, L is the number of relevant data in retrieved dataset, $P(r)$ is precision of top r retrieved data, and $\delta(r)$ equals 1 if the retrieved data shares the same label with the query, otherwise equals 0. We employ $mAP@50$ $(R = 50)$ in our experiments.

3.5 Analyse of Results

Typical mAP scores obtained by different methods and respectively tested on Wikipedia, Pascal and NUS-WIDE-10k datasets, were reported in Table 2. Comparing with the baseline methods on image retrieval and text retrieval, it can be found that our proposed models have significantly improved the performances in both image and text retrieval. In particular, the retrieval performance obtained by our proposed Super-Corr-NN was competitive with the results obtained by Corr-AE. For instance, Super-Corr-NN has produced a better retrieval performance than Corr-AE on Wikipedia and NUS-WIDE-10k datasets. Meanwhile, Super-Corr-AE has achieved the better cross-modal retrieval performances than Corr-AE models on three datasets, and also produced the better result than Super-Corr-NN on Wikipedia and Pascal datasets. For instance, Super-Corr-AE improved the mAP scores, respectively 3.1% and 13.9% on image query and text

Table 2. mAP scores tested on three different datasets (I: image; T: text).

Method	Wikipedia			Pascal			NUS-WIDE-10k		
	I→T	T→I	Avg	I→T	T→I	Avg	I→T	T→I	Avg
CCA-AE [6]	0.213	0.235	0.224	0.161	0.153	0.157	0.199	0.268	0.234
CCA-Cross-AE [5]	0.197	0.230	0.214	0.137	0.182	0.159	0.199	0.344	0.272
CCA-Full-AE [5]	0.293	0.331	0.312	0.148	0.177	0.163	0.241	0.242	0.242
Bimodal AE [9]	0.295	0.307	0.302	0.253	0.278	0.266	0.251	0.295	0.273
Bimodal DBN [9,16]	0.197	0.216	0.207	0.208	0.211	0.210	0.163	0.204	0.184
Corr-AE [5]	0.326	0.361	0.344	0.290	0.279	0.285	0.319	0.375	0.347
Super-Corr-NN	0.330	0.383	0.356	0.202	0.206	0.204	0.337	**0.435**	**0.386**
Super-Corr-AE	**0.336**	**0.411**	**0.379**	**0.301**	**0.320**	**0.310**	**0.353**	0.415	0.384
Corr-Cross-AE [5]	0.336	0.341	0.338	0.271	0.280	0.276	**0.349**	0.348	0.349
Super-Corr-Cross-AE	**0.340**	**0.446**	**0.393**	0.292	**0.295**	**0.293**	0.319	**0.395**	**0.357**
Corr-Full-AE [5]	0.335	0.368	0.352	0.281	0.276	0.279	0.331	0.379	0.355
Super-Corr-Full-AE	**0.352**	0.423	**0.387**	**0.300**	**0.302**	**0.301**	**0.358**	0.415	**0.386**

query than Corr-AE tested on Wikipedia dataset, 3.8% and 14.7% on Pascal dataset, 10.7% and 10.7% on NUS-WIDE-10k. Further, Super-Corr-Cross-AE has improved the mAP scores, respectively 1.2% and 30.8% on image query and text query than Corr-Cross-AE on Wikipedia dataset, 7.7% and 5.4% on Pascal dataset. This superiority can be attributed to the physical meaning of the model illustrated in Sect. 2.3, which seamlessly combined the Corr-AE and Super-Corr-NN by using the class label information. Remarkably, Super-Corr-Full-AE has achieved the satisfactory performance, and the experimental results have shown its outstanding performance.

3.6 Visualization

Further, we utilized the t-SNE [7] to visualize the image and text representation for discrimination illustration intuitively. As shown in Fig. 3, it can be found that Super-Corr-NN was able to learn discriminative representations, and the data points of text modality can be well divided into their corresponding classes. In contrast to this, Corr-Full-AE has failed to separate the text modalities of different classes. The main reason lies that the Corr-Full-AE model focuses more on pairwise data and often fails to preserve any global information in single modality. As a result, the data points from different classes are distributed disorderly. Further, the proposed Super-Corr-Full-AE could produce discriminative separations on both image and text modalities, the data points can be well divided into different classes visually. Therefore, it can be concluded that our discriminative deep correspondence model could learn discriminative representations of different modalities for cross-modal matching and retrieval.

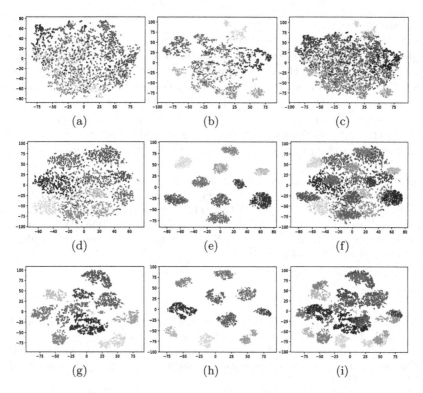

(a)   (b)   (c)

(d)   (e)   (f)

(g)   (h)   (i)

Fig. 3. Visualization of image, text and image&text pair in Wikipedia dataset, respectively shown in (a)–(c) by Corr-Full-AE, (d)–(f) by Super-Corr-NN, and (g)–(i) by Super-Corr-Full-AE.

3.7   Analyse of β

In addition, there are two parameters α and β in our models, and the sensitivity of parameter α is well analysed in work [5]. By fixing $\alpha = 0.9$, we conduct various experiments to analyse the effect of different parameter β values. The mAP values obtained by different models and tested on different datasets were shown in Fig. 4, it can be easily found that parameters β is not sensitive in our models and could produce satisfactory performance in a wide range of the values. Therefore, it is adequate to fix the β with an appropriate value in practice.

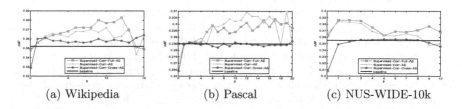

(a) Wikipedia   (b) Pascal   (c) NUS-WIDE-10k

Fig. 4. mAP values obtained by the proposed deep models with different β values.

4   Conclusion

In this paper, we have presented a discriminative deep correspondence model by seamlessly combining the correspondence autoencoder and supervised correspondence neural networks for cross-modal matching. The proposed deep model not only can learn the representation and correlation of different modalities simultaneously, but also could preserve both inter-modal and intra-modal similarity for efficient cross-modal matching. The experimental results have shown its outstanding performance in comparison with existing counterparts.

Acknowledgment. The work described in this paper was supported by the National Science Foundation of China (Nos. 61673185, 61502183, 61673186), National Science Foundation of Fujian Province (No. 2017J01112), Promotion Program for Young and Middle-aged Teacher in Science and Technology Research (No. ZQN-PY309), the Promotion Program for graduate student in Scientific research and innovation ability of Huaqiao University (No. 1611314006).

References

1. Bosch, A., Zisserman, A., Munoz, X.: Image classification using random forests and ferns. In: Proceedings of IEEE ICCV, pp. 1–8 (2007)
2. Castrejn, L., Aytar, Y., Vondrick, C., Pirsiavash, H., Torralba, A.: Learning aligned cross-modal representations from weakly aligned data. In: Proceedings of IEEE CVPR, pp. 2940–2949 (2016)
3. Chua, T.S., Tang, J., Hong, R., Li, H., Luo, Z., Zheng, Y.: Nus-wide: a real-world web image database from National University of Singapore. In: Proceedings of ACM International Conference on Image and Video Retrieval, pp. 48:1–48:9 (2009)
4. Farhadi, A., Hejrati, M., Sadeghi, M.A., Young, P., Rashtchian, C., Hockenmaier, J., Forsyth, D.: Every picture tells a story: generating sentences from images. In: Daniilidis, K., Maragos, P., Paragios, N. (eds.) ECCV 2010. LNCS, vol. 6314, pp. 15–29. Springer, Heidelberg (2010). https://doi.org/10.1007/978-3-642-15561-1_2
5. Feng, F., Wang, X., Li, R.: Cross-modal retrieval with correspondence autoencoder. In: Proceedings of ACM Multimedia, pp. 7–16 (2014)
6. Kim, J., Nam, J., Gurevych, I.: Learning semantics with deep belief network for cross-language information retrieval. In: Proceedings of IEEE International Conference on Computational Linguistics, pp. 579–588 (2012)
7. van der Maaten, L., Hinton, G.E.: Visualizing high-dimensional data using t-SNE. J. Mach. Learn. Res. **9**(2), 2579–2605 (2008)
8. Manjunath, B.S., Ohm, J.R., Vasudevan, V.V., Yamada, A.: Color and texture descriptors. IEEE Trans. Circuits Syst. Video Technol. **11**(6), 703–715 (2002)
9. Ngiam, J., Khosla, A., Kim, M., Nam, J., Lee, H., Ng, A.Y.: Multimodal deep learning. In: Proceedings of IEEE ICML, pp. 689–696 (2011)
10. Oliva, A., Torralba, A.: Modeling the shape of the scene: a holistic representation of the spatial envelope. Int. J. Comput. Vis. **42**(3), 145–175 (2001)
11. Pereira, J.C., Coviello, E., Doyle, G., Rasiwasia, N., Lanckriet, G.R.G., Levy, R., Vasconcelos, N.: On the role of correlation and abstraction in cross-modal multimedia retrieval. IEEE Trans. Pattern Anal. Mach. Intell. **36**(3), 521–535 (2013)
12. Quadrianto, N., Lampert, C.H.: Learning multi-view neighborhood preserving projections. In: Proceedings of IEEE ICML, pp. 425–432 (2011)

13. Rasiwasia, N., Costa Pereira, J., Coviello, E., Doyle, G., Lanckriet, G.R.G., Levy, R., Vasconcelos, N.: A new approach to cross-modal multimedia retrieval. In: Proceedings of IEEE International Conference on Multimedia. pp. 251–260 (2010)
14. Rosipal, R., Krmer, N.: Overview and recent advances in partial least squares. In: Proceedings of IEEE International Conference on Subspace, Latent Structure and Feature Selection, pp. 34–51 (2005)
15. Sharma, A., Kumar, A., Daume, H., Jacobs, D.W.: Generalized multiview analysis: a discriminative latent space. In: Proceedings of IEEE CVPR, pp. 2160–2167 (2012)
16. Srivastava, N., Salakhutdinov, R.: Learning representations for multimodal data with deep belief nets. In: Proceedings of IEEE ICML workshop (2012)
17. Srivastava, N., Salakhutdinov, R.: Multimodal learning with deep Boltzmann machines. J. Mach. Learn. Res. **15**(8), 1967–2006 (2012)
18. Tang, J., Wang, K., Shao, L.: Supervised matrix factorization hashing for cross-modal retrieval. IEEE Trans. Image Process. **25**(7), 3157–3166 (2016)
19. Wang, C., Yang, H., Meinel, C.: Deep semantic mapping for cross-modal retrieval. In: Proceedings of IEEE International Conference on Tools Artificial Intelligence pp. 234–241 (2016)
20. Wang, W., Yang, X., Ooi, B.C., Zhang, D., Zhuang, Y.: Effective deep learning-based multi-modal retrieval. VLDB J. **25**(1), 79–101 (2016)

Towards Deeper Insights into Deep Learning from Imbalanced Data

Jie Song[1,2], Yun Shen[1,2], Yongcheng Jing[1,2], and Mingli Song[1,2(✉)]

[1] College of Computer Science, Zhejiang University, Hangzhou, China
{sjie,yunshen,ycjing,brooksong}@zju.edu.cn
[2] Alibaba-Zhejiang University Joint Institute of Frontier Technologies,
Hangzhou, China

Abstract. Imbalanced performance usually happens to those classifiers (including deep neural networks) trained on imbalanced training data. These classifiers are more likely to make mistakes on minority class instances than on those majority class ones. Existing explanations attribute the imbalanced performance to the imbalanced training data. In this paper, using deep neural networks, we strive for deeper insights into the imbalanced performance. We find that imbalanced data is a neither sufficient nor necessary condition for imbalanced performance in deep neural networks, and another important factor for imbalanced performance is the distance between the majority class instances and the decision boundary. Based on our observations, we propose a new under-sampling method (named *Moderate Negative Mining*) which is easy to implement, state-of-the-art in performance and suitable for deep neural networks, to solve the imbalanced classification problem. Various experiments validate our insights and demonstrate the superiority of the proposed under-sampling method.

Keywords: Imbalanced classification · Imbalanced performance
Deep neural networks · Under-sampling

1 Introduction

In the binary classification problem, it is often the case that one class is much more over-represented than the other one. The more over-presented class is usually referred to as the *negative* or *majority* class, and the other one as the *positive* or *minority* class (we use these words interchangeably in this paper). Most standard algorithms (such as k-NN, SVM, decision tree and deep neural network) are designed to be trained with balanced data. If given imbalanced training data, they will have difficulty in learning proper decision rules and produce imbalanced performance on test data. In other words, they are more likely to make mistakes on minority class instances than on majority class ones.

Most existing explanations attribute the imbalanced performance phenomenon to the imbalanced training data. Therefore, many approaches which aim to balance the data distribution are proposed. For example, random over-sampling

© Springer Nature Singapore Pte Ltd. 2017
J. Yang et al. (Eds.): CCCV 2017, Part I, CCIS 771, pp. 674–684, 2017.
https://doi.org/10.1007/978-981-10-7299-4_56

method creates positives by randomly replicating the minority class instances. Although the data distribution can be made balanced in this way, it doesn't create any new information and easily leads to over-fitted classifiers. SMOTE [1] overcomes the weakness by synthesizing new non-replicated positives. However, it is still believed to be insufficient to solve the imbalanced data problem. Another direction to balance the data distribution is under-sampling the over-presented negatives. For example, random under-sampling, like random over-sampling, selects the negatives randomly to construct the balanced positives and negatives. There are also several other more advanced under-sampling methods, such as One-Sided Selection (OSS) [2] and Nearest Neighbor Cleaning Rule [3]. These methods try to remove those redundant, borderline or mislabeled samples and strive for more reliable ones. However, they are suitable to handle those low-dimensional, highly structured and size limited training data which is often in vector form. For those which are high-dimensional, unstructured and in large quantity, they behave unsatisfactorily.

In this paper, in addition to the imbalanced training data, we explore deeper and more fundamental explanations to imbalanced performance in deep neural networks. We choose deep neural network to study the imbalance problem for two reasons. Firstly, deep neural network has achieved great success in various problems in the last several years, such as image classification [4], object detection [5] and machine translation [6]. It has been the most popular model in the machine learning field. Secondly, usually used as an end-to-end model, it can handle raw input data directly without complex data preprocessing. Thus it is suitable to handle those high-dimensional and unstructured raw training data. Using deep neural networks, we find that *imbalanced data is a neither sufficient nor necessary condition for the imbalanced performance.* By saying that, we mean:

- balanced training data can also lead to imbalanced performance on test data (shown in Fig. 1b and c), and
- classifiers trained on imbalanced training data do not necessarily produce imbalanced performance on test data (shown in Fig. 1d).

Additionally, we find another important factor for imbalanced performance is the distance between the negative samples and the positives (or the latent decision hyperplane). Specifically, sampling those which are distant from the positives will push the learnt decision hyperplane towards the space of the negatives (shown in Fig. 1b). On the other hand, sampling those which are near the positives or aggressive ones (which tend to be mislabeled and invade the space of the positives) will push the learnt decision hyperplane towards the space of the positives (shown in Fig. 1c). Neither of these two sampling ways can balance the performance on the two class test data. Based on these observations, we propose to sample the negatives which are moderately distant from the positives (shown in Fig. 1e). These negatives are less likely to be mislabeled and more reliable than the nearest ones, meanwhile they are more informative than those which are distant from the positives. We call the proposed under-sampling method *Moderate Negative Mining* (MNM). Experiments conducted on various datasets validate the proposed approach.

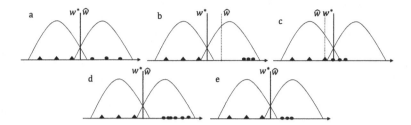

Fig. 1. Illustration of bias of a linear separator induced over training data in a one-dimensional example. Triangles and circles denote positive and negative samples, respectively. The corresponding latent distributions are shown. The dotted line ($\hat{w}$) is the hypothesis induced over training data, while the solid line (w^*) depicts the optimal separator of the underlying distributions.

2   Related Work

There is large literature on solving imbalanced data problem. Existing approaches can be roughly categorized into two groups: the internal approaches [7,8] and the external approaches [1–3,9]. The internal approaches modify existing algorithms or create new ones to handle the imbalanced classification problem. For example, cost-sensitive learning algorithms impose heavier penalty on misclassifying the minority class instances than that on misclassifying the majority class ones [8]. Although internal approaches are effective in certain cases, they have the disadvantage of being algorithm specific. That is to say, it might be quite difficult, or sometimes impossible, to transport the modification proposed for one classifier to the others [9]. The external approaches, on the other hand, leave those learning algorithms unchanged and alter the data distribution to solve the imbalanced classification problem. They are also known as data re-sampling methods and can be divided into two groups: over-sampling and under-sampling methods. Over-sampling methods increase the quantity of minority class instances by replicating them or synthesizing new ones to balance the data distribution [1]. They are criticized for the over-fitting problem [10]. Under-sampling methods, in an another direction, under-sample the majority class instances to balance the imbalanced distribution. For example, One-Sided Selection (OSS) [2] adopts the *Tomek links* [11] to eliminate borderline and noisy examples. Nearest Neighbor Cleaning Rule [3] decreases the size of the negatives by cleaning the unreliable negatives. Both these approaches are based on simple nearest neighbor assumption, thus they have limited performance when used for deep neural networks.

Albeit the large literature on imbalanced problem, few works [12–14] study the imbalanced classification problem in deep neural networks. In [12], over-sampling and under-sampling are combined by complementary neural network and SMOTE algorithm to solve the imbalanced classification problem. The effects of sampling and threshold-moving in training cost-sensitive neural networks are studied in [13]. It is demonstrated in [14] that more discriminative deep representations can be learned by enforcing a deep network to maintain both inter-cluster and inter-class margins. In this paper, we explore more insights into learning from imbalanced data by deep neural networks and propose a new under-sampling method to solve the imbalanced classification problem. We believe our approach is a good complement to existing approaches for solving the imbalanced classification problem in deep neural networks. The proposed method can be categorized as an external approach.

3   Imbalanced Data and Imbalanced Performance

In this section, we empirically validate our first observation: imbalanced data is a neither sufficient nor necessary condition for imbalanced performance in deep neural networks. The validation process consists of two stages. In the first stage, we validate that balanced training data can also lead to imbalanced performance. In the second stage, we validate that imbalanced data does not necessarily lead to imbalanced performance. The experimental settings of these two processes are largely kept the same, so we first briefly introduce them as follows.

We use the publicly available CIFAR-10 [15] dataset to construct the imbalanced datasets. CIFAR-10 has 10 classes. To construct imbalanced datasets for binary classification problem, we treat one of them as the positive class and the others as the negative one. In this way, we construct 10 imbalanced datasets (CIFAR10-AIRPLANE, CIFAR10-AUTOMOBILE, CIFAR10-BIRD, CIFAR10-CAT, CIFAR10-DEER, CIFAR10-DOG, CIFAR10-FROG, CIFAR10-HORSE, CIFAR10-SHIP, CIFAR10-TRUCK). Original training images are used for training. The test images are equally split into 2 groups. One for validation, and the other one for test.

In this paper, we study the imbalanced classification problem in deep neural networks. Our deep neural networks follow the architecture of 6 convolutional layers. Each convolutional layer is followed by a RELU [18] nonlinear activation layer. We list the detailed information of the architecture in Table 1. The number following each @ symbol represents the number of kernels in the convolutional layer.

Table 1. Architecture of deep neural networks

Input	Conv1	Relu	Conv2	Relu	Pool1	Conv3	Relu	Conv4	Relu
32×32 $\times 3$	3×3 @16		3×3 @16		MAX	3×3 @32		3×3 @32	
Pool2	Conv5	Relu	Conv6	Relu	Fc1	Relu	Fc2	Relu	Softmax
MAX	3×3 @64		3×3 @64		512		2		

These deep neural networks are optimised by the Stochastic Gradient Descent (SGD) algorithm. The base learning rate is set to be 0.01 and decreases by 50%

every 10 epoches. The maximum epoch is set to be 40. The batch size in the SGD algorithm is set to be 256.

3.1 Balanced Training Data Can Lead to Imbalanced Performance

Balanced training data can also lead to imbalanced performance, as shown in Fig. 1b and c. In Fig. 1b, the negatives which are distant from the positives are sampled. On the contrary, samples which are near or across the boundary are sampled in Fig. 1c. The intuition illustrated in Fig. 1 is simple, but it leaves the distance between two data points unsolved, especially for the high-dimensional raw data, such as images. We propose a simple method to approach this problem. Firstly, we randomly under-sample the negatives to balance the training data. We use the balanced training data to train a basic network, which we call *metric network*. The metric network follows the same architecture with the final classifier (although it can be of a different one, as shown in [16]), and is used to calculate the distance between the negatives and the decision boundary in their embedding space. Finally, we sample the same number of negatives as the positives to construct the balanced training data. To make the performance of the trained classifier imbalanced, we only sample those which are most (or least) distant from the decision boundary (or the positives).

In the trained embedding space, we assume the optimal decision hyperplane is approximately the decision hyperplane learnt by the classifier (this is a reasonable assumption because the embedding space and the classifier are optimized jointly). For softmax classifier, we usually take 0.5 as the decision threshold for a two-class classification problem. The predicted probability of a softmax classifier is

$$p_i = \frac{e^{W_i^T \cdot \mathrm{x}}}{\sum_j e^{W_j^T \cdot \mathrm{x}}}, \tag{1}$$

where i is in $\{0,1\}$, W is the learnt weights of the softmax classifier, and x is a data point in the embedding space. The decision hyperplane is thus

$$(W_0 - W_1)^T \cdot \mathrm{x} = 0. \tag{2}$$

For any negative data point x, its signed distance to the decision hyperplane is

$$
\begin{aligned}
d &= \frac{(W_0 - W_1)^T \cdot \mathrm{x}}{||W_0 - W_1||} \\
&\propto (W_0 - W_1)^T \cdot \mathrm{x} \\
&= \ln \left(p_0 \cdot \sum_j e^{W_j^T \cdot \mathrm{x}} \right) - \ln \left(p_1 \cdot \sum_j e^{W_j^T \cdot \mathrm{x}} \right) \\
&= \ln \frac{p_0}{1 - p_0} \\
&\propto p_0.
\end{aligned}
\tag{3}
$$

Thus the predicted probability is a good indicator of the distance between the data point and the decision hyperplane in the embedding space.

In our experiments, to give better understanding of the above proposed validation method, we categorize the negatives into 9 groups (group 0 to group 8) based on their predicted probabilities, as the negatives is as 9 times large in size as the positives. From group 0 to group 8, predicted probabilities of the target label (p_0) go from the largest to the smallest, thus the distance between the negatives in them and the decision hyperplane goes from the largest to smallest. Every group has the same number of negatives. Therefore, for every constructed imbalanced dataset, we can construct 9 balanced training datasets with their 9 groups of negatives and all the positives.

We use sensitivity and specificity to measure the imbalanced performance. Sensitivity, also called the true positive rate, is defined as

$$sensitivity = \frac{TP}{TP + FN}. \tag{4}$$

Specificity, also called true negative rate, is defined as

$$specificity = \frac{TN}{TN + FP}. \tag{5}$$

For a fair comparison, during the training of every deep classifier, we keep all experimental settings the same except for the different training data. In Fig. 2, we plot the sensitivity and the specificity of the classifiers trained from the training data which is comprised of the negatives in one group and all the positives.

As shown in Fig. 2, the performance of classifiers trained from the balanced datasets which are comprised of the negatives in group 0 and all the positives and is rather imbalanced, as the sensitivity is generally much larger than the specificity for group 0 in all datasets. As the group number increases within a certain range (i.e. the distance between the negative instances and the decision hyperplane increases), the difference between sensitivity and specificity becomes smaller and the imbalanced performance gradually disappears. However, when it goes further, the difference rebounds and the imbalance phenomenon arises again. From this experiment, we empirically validate that balanced training data can also lead to imbalanced performance on test data.

3.2   Imbalanced Training Data Can Lead to Balanced Performance

It is shown in Fig. 2 that the balanced training data can lead to imbalanced performance. Based on this finding, we can balance the imbalanced performance by altering the data distribution. For example, the classifiers trained from the negatives in group 0 tend to make more mistakes on negatives than on positives. If we increase the size of the negatives in training data, the imbalanced performance can be made balanced. In our experiment, we keep the positives unchanged and increase the size of the negatives in group 0 to balance the performance. Every time increasing the size of the negatives, we add to the training data the most distant negatives which are not contained in the training data. Figure 3 shows how the performance of the trained classifier changes, as the ratio between the size of

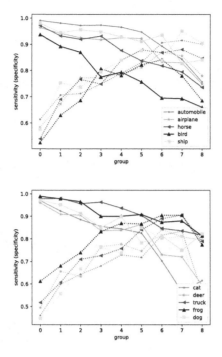

Fig. 2. Sensitivity (solid lines) and specificity (dashed lines) of trained classifiers from training data comprised of all the positives and the negatives in one group. A solid line and a dashed one of the same color correspond to the same classifier. (Color figure online)

the negatives and that of the positives in the CIFAR10-AUTOMOBILE dataset increases. As can be seen, when the ratio is about 6 (in this case, the training data is rather imbalanced), the performance becomes balanced. Although here we only plot the results on CIFAR10-AUTOMOBILE dataset, experiments on other datasets give the similar results.

4   Moderate Negative Mining

Although the method in Sect. 3.2 can balance classifiers' performance on the test data, it requires carefully regulating the size of the negatives to balance the imbalanced performance. In addition, in real life we desire for not only more balanced but also better performance. In this section, we propose a new under-sampling method named *Moderate Negative Mining* (MNM) to aid deep neural networks in learning from imbalanced data. The proposed under-sampling approach can not only solve the imbalanced problem, but also produce better overall performance in our experiments when compared with other existing under-sampling approaches.

As shown in Sect. 3, both the most and the least distant negatives will lead to imbalanced performance. Two reasons account for that. Firstly, the least

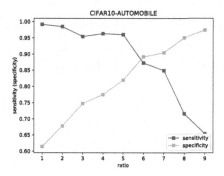

Fig. 3. Performance of the trained classifiers as we change the ratio between the size of negatives and that of the positives. We randomly choose the CIFAR10-AUTOMOBILE dataset.

distant negatives tend to be unreliable or mislabeled ones. These negatives have been shown to degrade the performance of trained classifiers. Secondly, the most distant negatives are far away from the decision hyperplane, thus they are not informative enough to represent the space of the negatives. Moderate Negative Mining exploits a simple strategy to under-sample the over-presented negatives. We hypothesize the moderately distant negatives which are between the most and the least distant ones, are ideal negatives for training the classifier. This is a reasonable assumption because the moderately distant negatives are less likely to be polluted by mislabeled ones and those unvalued redundant ones.

As described in Sect. 3, we use the predicted probability to measure the distance between a negative instance and the decision hyperplane. To give better comparison, in the same way as Sect. 3.1, we also categorize the negatives into 9 groups (from the most distant negatives in group 0 to the least distant ones in group 8) and give the performance of classifiers trained from negatives in each group. We use deep neural networks of the same architecture as described in Sect. 3. All experimental settings are kept the same during the training process for fair comparisons, except for the different training data.

As the deep classifiers often output a score for both the positive and the negative class, we must specify a score threshold (usually 0.5 for binary classification) to make a trade-off between false positives and false negatives. However, when evaluating the performance, we require a measurement which is independent of the score threshold to measure the overall performance of the trained model. We sidestep the threshold selection problem by employing AUROC (Area Under Receiver Operating Characteristic curve) and AUPR (Area Under Precision-Recall curve). Figure 4 depicts how AUROC and AUPR values change as we choose different group of negatives as training data. As can be seen, for almost all datasets, we can find a group (or several groups) which achieves better performance than random under-sampling.

The proposed Moderate Negative Mining (MNM) method selects the best classifier on the validation data. To give a better comparison with other

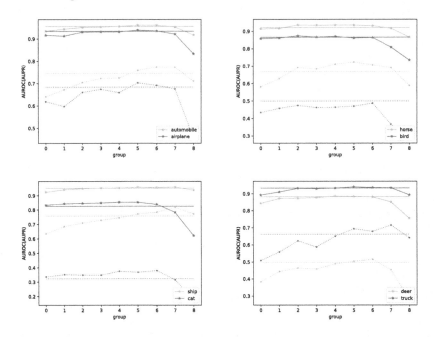

Fig. 4. AUROC (solid lines) and AUPR (dashed lines) curves of deep classifiers trained from training data comprised of all the positives and the negatives in one group. Horizontal lines denotes performance of random under-sampling.

under-sampling methods, we list the AUROC and AUPR values of the proposed method on test data in Table 2, as well as those of Random Under-sampling (RUS), One-Sided Selection (OSS) [2] and Nearest Neighbor Cleaning

Table 2. Performance on CIFAR-10

Dataset	AUROC				AUPR			
	RUS	OSS	NCL	MNM	RUS	OSS	NCL	MNM
CIFAR10-AIRPLANE	0.929	0.924	0.925	**0.938**	0.674	0.680	0.660	**0.692**
CIFAR10-AUTOMB	0.958	0.954	0.952	**0.960**	0.750	0.744	0.739	**0.774**
CIFAR10-BIRD	0.858	0.857	0.863	**0.864**	0.476	0.447	0.470	**0.487**
CIFAR10-CAT	0.831	**0.839**	0.838	0.838	0.343	0.351	0.338	**0.380**
CIFAR10-DEER	0.878	0.869	0.875	**0.881**	0.470	0.455	0.495	**0.515**
CIFAR10-DOG	0.879	0.877	0.889	**0.902**	0.497	0.501	0.515	**0.544**
CIFAR10-FROG	0.945	0.947	0.947	**0.957**	0.712	0.732	0.731	**0.763**
CIFAR10-HORSE	0.929	0.923	0.930	**0.935**	0.695	0.678	0.706	**0.724**
CIFAR10-SHIP	0.945	0.951	0.953	**0.959**	0.752	0.760	0.770	**0.817**
CIFAR10-TRUCK	0.934	0.935	0.936	**0.938**	0.685	0.687	0.668	**0.694**

Rule (NCL) [3]. These are all well known under-sampling methods to solve the imbalanced data problem. As can be seen, the proposed method achieves better performance on almost all datasets. The superiority shown in PR space is more obvious than that in ROC space. It is because the PR curves are more informative than ROC curves when the data is imbalanced [17].

5   Discussion

Although imbalanced performance is often believed to be caused by imbalanced training data, we have shown that imbalanced training data is a neither sufficient nor necessary condition in Sect. 3. Therefore, in addition to construction of balanced training data, we can also construct imbalanced training data to balance the classifier's performance on the positives and the negatives. Here we call the former method *balance* method and the latter *imbalance* method for brevity.

As deep neural networks usually get better performance given larger training data, we may expect better performance in the imbalance method. However, our experiments show the balance method achieves better performance. It is because although the imbalanced method trains the model with more examples, most of the sampled negatives are easy ones and of little information to learn the decision rule. This is validated further by our proposed Moderate Negative Mining method, in which we abandon those easy negatives and those most hardest ones. Both these two types of negatives can hinder the learning of the classifier.

6   Conclusion

In this paper we empirically prove that imbalanced data is a neither sufficient nor necessary condition for imbalanced performance on test data, although the imbalanced training data is often believed to be the chief culprit. We find another important factor for imbalanced performance is the distance between the majority class instances and the decision boundary. Based on these observations, we propose a new under-sampling, named as *Moderate Negative Mining*, to solve the imbalanced classification problem in deep neural networks. Several experiments demonstrate the superiority of the proposed method.

For future work, we aim to select the majority class instances directly rather than by boring validation. Furthermore, we desire to formulate the negative under-sampling as a learning problem, which is challenging as the sampled examples are discrete.

Acknowledgments. This work was supported in part by the National Natural Science Foundation of China (61572428, U1509206), National Key Research and Development Program (2016YFB1200203), Program of International Science and Technology Cooperation (2013DFG12840), Fundamental Research Funds for the Central Universities (2017FZA5014) and National High-tech Technology R&D Program of China (2014AA015205).

References

1. Chawla, N.V., Bowyer, K.W., Hall, L.O., Kegelmeyer, W.P.: SMOTE: synthetic minority over-sampling technique. J. Artif. Intell. Res. **16**, 321–357 (2002)
2. Kubat, M., Matwin, S.: Addressing the curse of imbalanced training sets: one-sided selection. In: ICML, vol. 97, pp. 179–186, July 1997
3. Laurikkala, J.: Improving identification of difficult small classes by balancing class distribution. In: Quaglini, S., Barahona, P., Andreassen, S. (eds.) AIME 2001. LNCS (LNAI), vol. 2101, pp. 63–66. Springer, Heidelberg (2001). https://doi.org/10.1007/3-540-48229-6_9
4. He, K., Zhang, X., Ren, S., Sun, J.: Deep residual learning for image recognition. In: Proceedings of the IEEE Conference on Computer Vision and Pattern Recognition, pp. 770–778 (2016)
5. Ren, S., He, K., Girshick, R., Sun, J.: Faster R-CNN: towards real-time object detection with region proposal networks. In: Advances in Neural Information Processing Systems, pp. 91–99 (2015)
6. Bahdanau, D., Cho, K., Bengio, Y.: Neural machine translation by jointly learning to align and translate. arXiv preprint arXiv:1409.0473 (2014)
7. Pazzani, M., Merz, C., Murphy, P., Ali, K., Hume, T., Brunk, C.: Reducing misclassification costs. In: Proceedings of the Eleventh International Conference on Machine Learning, pp. 217–225 (1994)
8. Tang, Y., Zhang, Y.Q., Chawla, N.V., Krasser, S.: SVMs modeling for highly imbalanced classification. IEEE Trans. Syst. Man Cybern. Part B (Cybern.) **39**(1), 281–288 (2009)
9. Estabrooks, A., Jo, T., Japkowicz, N.: A multiple resampling method for learning from imbalanced data sets. Comput. Intell. **20**(1), 18–36 (2004)
10. He, H., Garcia, E.A.: Learning from imbalanced data. IEEE Trans. Knowl. Data Eng. **21**(9), 1263–1284 (2009)
11. Tomek, I.: Two modifications of CNN. IEEE Trans. Syst. Man Cybern. **6**, 769–772 (1976)
12. Jeatrakul, P., Wong, K.W., Fung, C.C.: Classification of imbalanced data by combining the complementary neural network and SMOTE algorithm. In: Wong, K.W., Mendis, B.S.U., Bouzerdoum, A. (eds.) ICONIP 2010 Part II. LNCS, vol. 6444, pp. 152–159. Springer, Heidelberg (2010). https://doi.org/10.1007/978-3-642-17534-3_19
13. Zhou, Z.H., Liu, X.Y.: Training cost-sensitive neural networks with methods addressing the class imbalance problem. IEEE Trans. Knowl. Data Eng. **18**(1), 63–77 (2006)
14. Huang, C., Li, Y., Change Loy, C., Tang, X.: Learning deep representation for imbalanced classification. In: Proceedings of the IEEE Conference on Computer Vision and Pattern Recognition, pp. 5375–5384 (2016)
15. Krizhevsky, A., Hinton, G.: Learning multiple layers of features from tiny images (2009)
16. Lewis, D.D., Catlett, J.: Heterogeneous uncertainty sampling for supervised learning. In: Proceedings of the Eleventh International Conference on Machine Learning, pp. 148–156 (1994)
17. Saito, T., Rehmsmeier, M.: The precision-recall plot is more informative than the ROC plot when evaluating binary classifiers on imbalanced datasets. PLoS ONE **10**(3), e0118432 (2015)
18. Krizhevsky, A., Sutskever, I., Hinton, G.E.: Imagenet classification with deep convolutional neural networks. In: Advances in Neural Information Processing Systems, pp. 1097–1105 (2012)

Traffic Sign Recognition Based on Deep Convolutional Neural Network

Shihao Yin, Jicai Deng$^{(\boxtimes)}$, Dawei Zhang, and Jingyuan Du

School of Information Engineering, Zhengzhou University,
No. 100 Science Avenue, Zhengzhou 450001, China
iejcdeng@zzu.edu.cn

Abstract. Traffic sign recognition (TSR) is an important component of automated driving system. It is a rather challenging task to design a high-performance classifier for the TSR system. In this paper, we proposed a new method for TSR system based on deep convolutional neural network. In order to enhance the expression of the network, a novel structure (dubbed block-layer below) which combines Network-in-Network and residual connection was designed. Our network has 10 layers with parameters (block-layer be seen as a single layer); the first seven are alternate convolutional layers and block-layers, and the remaining three are fully-connected layers. We trained our TSR network on the German Traffic Sign Recognition Benchmark (GTSRB) dataset. To reduce overfitting, we did data augmentation on the training images and employed a regularization method named dropout. We also employed a mechanism called Batch Normalization which has been proved to be efficient for accelerating the training of deep neural networks. To speed up the training, we used an efficient GPU to accelerate the convolutional operation. On the test dataset of GTSRB, we achieve the accuracy rate of 98.96%, exceeding the human average raters.

Keywords: Autonomous vehicle · Traffic sign recognition
Deep learning · Convolutional neural network · GTSRB dataset

1 Introduction

With the rapid development of driver assistance systems and autonomous vehicles in recent years, traffic sign recognition (TSR) attracts attention of many researchers. Traffic signs provides lots of useful information of the traffic environment, such as speed limits, directions, danger warning, and so on. Recognizing traffic signs timely makes driving safe and convenient. However, all the traffic sign images, which were inputted to the TSR system, are photographed from real environment, and they have various lighting conditions, viewpoints, partial occlusions, and resolutions. Therefore, it is important to overcome these difficulties to ensure the robustness of the model.

Many models of TSR system have been proposed in the past decade [1–3]. Generally, the TSR process can be divided into two stages i.e. detection and

© Springer Nature Singapore Pte Ltd. 2017
J. Yang et al. (Eds.): CCCV 2017, Part I, CCIS 771, pp. 685–695, 2017.
https://doi.org/10.1007/978-981-10-7299-4_57

recognition, and we focus on the recognition stage in this paper. Lim et al. used color/shape and pictogram information to extract features and then fed them to Radial Basis Function neural network (RBFNN) to classify the traffic signs [4]. Madani and Yusof employed a pre-trained Multiclass Support Vector Machine (MCSVM) to classify the traffic signs instead of the aforementioned RBFNN [5]. Recently, convolutional neural networks (CNNs) is popular in the computer vison area, because of its excellent classification performance. Lau et al. proposed a TSR method based on CNN and achieved a good accuracy on the Malaysia traffic sign database [6]. In order to improve the accuracy of recognizing traffic signs, we proposed a deep CNN architecture and utilized it on the German Traffic Sign Recognition Benchmark (GTSRB) dataset.

The remainder of this paper is organized as follows. Section 2 presents related works. Section 3 illustrates the architecture of the deep convolutional neural network, and details of the experiment on the GTSRB datasets are showed in Sect. 4. Finally, a conclusion and future work of this study are presented in Sect. 5.

2   Related Work

2.1   Using Leaky ReLUs

Rectified Linear Units (ReLUs) is the most popular activation function used by deep convolutional neural networks in recent years. Deep CNNs with ReLUs train faster than their equivalents with tanh or sigmoid units [7]. Unfortunately, ReLU units can be fragile during training and die. Leaky ReLU was proposed to fix the dying ReLU problem [8]. Xu et al. proved that using Leaky ReLUs performs better than ReLUs in convolutional neural networks [9]. In our network, we found that using the Leaky ReLUs achieves higher accuracy than ReLUs (about 0.4%) on the GTSRB dataset. Figure 1(a) and (b) show the function of ReLU and Leaky ReLU.

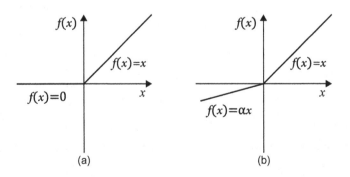

Fig. 1. ReLU vs. Leaky ReLU. (a) ReLU: $f(x) = \max(x, 0)$. (b) Leaky ReLU: $f(x) = \max(x, 0) - \alpha \max(-x, 0)$.

The function of Leaky ReLU can also be described by,

$$f(x) = \begin{cases} \alpha x, & x < 0 \\ x, & x \geq 0 \end{cases} \tag{1}$$

where α is fixed parameter ranging from 0 to 1. We set α to 0.2 in our TSR network.

2.2   Block-Layer

In traditional CNNs, layers were typically structured by stacked convolutional layers (optionally followed by normalization layer and pooling layer) and several fully-connected layers, such as LeNet-5 [10], AlexNet [11] and VGG [12]. Lin et al. proposed a novel deep neural network called Network In Network (NIN) to enhance the representation power of convolutional neural network [13]. Inspired by the NIN, Szegedy et al. designed a deep convolutional neural network architecture codenamed Inception and took the crown of the ImageNet Large-Scale Visual Recognition Challenge (ILSVRC) 2014 [14].

Residual network was firstly proposed by He et al. in [15], in which they give empirical evidence for the advantages of adding residual connection to the network. Szegedy et al. combined the Inception architecture with residual connections and got higher accuracy than without residual connections on the ILSVRC 2012 classification task [16].

In order to extract different level features by using several sizes of convolutional filter, and benefit from adding residual connection, we designed a novel architecture named block-layer as shown in Fig. 2. To simplify the representation, we omit the activation function in the figure. All the convolutional layers and max-pooling layers in the block-layer have same stride of 1 and same padding scheme of same-padding to ensure their output grids match the size of their input. In other word, the output of block-layer has same size of its input. This mechanism makes the block-layer can be added to anywhere in traditional plain convolutional neural networks. In the block-layer, 1×1 convolutions before the expensive 5×5 and 3×3 convolutions are used as dimension reduction models to limit the size of our network.

The output $\mathcal{H}(x)$ of block-layer is a function of its input x. The stacked underlying mapping of the block-layer is denoted as $\mathcal{F}(x)$. Because of adding the residual connection, the output of block-layer can be described as follows:

$$\mathcal{H}(x) = \mathcal{F}(x) + x. \tag{2}$$

3   Overall Architecture

Now we are ready to describe the overall architecture of our traffic sign recognition model. A concise representation is summarized in Fig. 3. For simplicity, the block-layer described above is represented to a single layer. The network

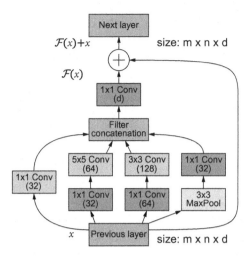

Fig. 2. The block-layer we use in our TSR network.

contains 10 layers with weights; the first seven are alternate convolutional layers and block-layers, and the remaining three are fully-connected layers. The output of last fully-connected layer is fed to a 43-way softmax which produces a distribution over the 43 class traffic sign labels. Max-pooling layers follow the layers C3, C5 and C7. The grid of pooling is 3 and the stride is set to 2. In other word, we employed overlapping pooling in our network.

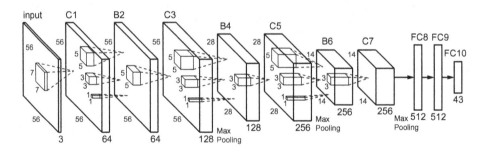

Fig. 3. A simple illustration of the architecture of our convolutional neural network for TSR.

The input of our network is a RGB image of size $56 \times 56 \times 3$. The preprocessing we do to the input image is subtracting the mean RGB value, computed on the training set, from each pixel. As depicted in Fig. 3, the layer B2, B4 and B6 are block-layers connected to previous corresponding convolutional layers (maybe pooled). The layer C1 filters the input image with 64 kernels of size $7 \times 7 \times 3$ with a stride of 1 pixel. The layer C3 has 128 kernels of size $5 \times 5 \times 64$ connected to previous block-layer. The layer C5 has 256 kernels of size $3 \times 3 \times 128$, and the layer C7 has 256 kernels of size $3 \times 3 \times 256$. The first two fully-connected layers (FC8 and FC9) have 512 neurons each. The last fully-connected layer (FC10) has 43 neurons.

A schematic view of our network in detail is also depicted in Fig. 4.

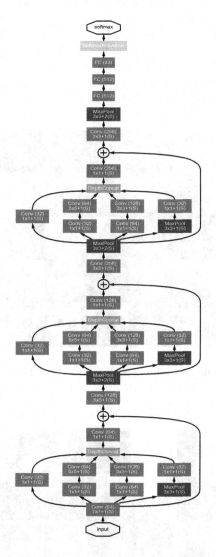

Fig. 4. A detailed representation of our TSR network.

4   Experiment

4.1   The GTSRB Dataset

The German Traffic Sign Recognition Benchmark (GTSRB) [17] dataset has 51893 images (39209 training images and 12630 testing images) of the 43 classes showing in Fig. 5. The size of the traffic sign images varies between 15×15 and 222×193 pixels. The images contain 10% margin (at least 5 pixels) around the traffic sign. All the images were collected from real environment, which have different illumination and weather conditions. Some of the traffic sign images are not easy to recognize because of low resolution, bad illumination, partial occlusion, motion blur, and so on. Figure 6 shows some example images, which are selected from the GTSRB dataset, under contaminated conditions.

Fig. 5. The 43 traffic sign classes in the GTSRB dataset.

Fig. 6. Examples which under contaminated conditions picked from the GTSRB dataset, such as low resolution, bad illumination, partial occlusion and motion blur.

4.2   Training

We trained our network using TensorFlow [18] machine learning system developed by Google Inc. Our training used Adam [19] optimizer with 0.001 learning rate. The batch size was set to 128. We used weight decay (the L2 penalty multiplier set to $5e^{-4}$) and dropout [20] regularization for the first two fully-connected layers (dropout ratio to 0.5) to reduce overfitting. We also employed batch normalization [21] between convolution operation and activation function in our network. We initialized the weights in our network from random initialization. For convolutional layers, we sampled the weights from a normal distribution with zero mean and $5e^{-2}$ variance. For fully-connected layers, we sampled the weights from a normal distribution with zero mean and $4e^{-2}$ variance. The biases of convolutional and fully-connected layers were initialized with zero and 0.1 respectively. We trained our network for roughly 200 cycles through the training set, which took about 24 h on a NVIDIA GTX 1060 6G GPU. The details of experiment is shown in Fig. 7.

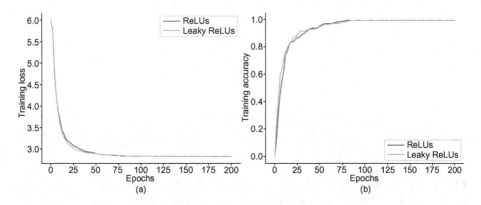

Fig. 7. Training on the GTSRB dataset of using ReLUs and Leaky ReLUs. (a) Training loss, vs. training epochs. (b) Training accuracy, vs. training epochs.

In order to reduce overfitting, we did data augmentation on training dataset. Because the images in the GTSRB dataset have different sizes, we resized all the training images to the size of 64×64 pixels first. For a resized image, we extracted random 56×56 pixels patches from it. This increased the size of our training set by a factor of 64. Another way to enlarge the training set we employed was to adjust the brightness and contrast of the image randomly. Also considering that the images of the GTSRB dataset were collected from real-world, and some of them had a bad lighting condition, we only adjusted the brightness and contrast by a factor randomly picked in a short interval.

4.3   Testing

At test time, the trained network makes a prediction by averaging the predictions made by the network's softmax layer on six patches getting from a given traffic sign image. The six patches are got as follow: As to a given image, we first resize it to two size of 64 × 64 and 56 × 56 pixels; and then extract five 56 × 56 pixels patches (the four corner patches and the center patch) from the resized 64 × 64 pixels image. The six 56 × 56 patches are the inputs which we use in per prediction. Figure 8 shows the accuracy of our trained network, as a function of the training epochs. Figure 8(a) shows that using Leaky ReLUs in our network obtains a better result on the GTSRB testing data than using ReLUs. Figure 8(b) demonstrates that average evaluation method achieves a higher accuracy (about 0.8%) on the GTSRB testing data than single evaluation (only use the resized 56 × 56 pixels image).

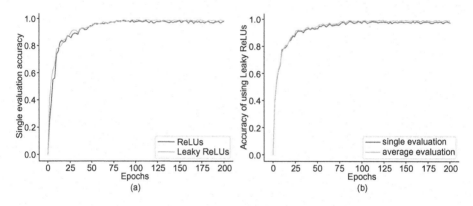

Fig. 8. The testing accuracy on the GTSRB. (a) Single evaluation accuracy of using ReLUs and Leaky ReLUs, vs. training epochs. (b) Single and average evaluation, vs. training epochs.

The final accuracy on the GTSRB of our network is 98.96%. Table 1 presents a comparison with some proposed algorithms submitted to the GTSRB and the human performance. It is worth noting that the team IDSIA only used the central region of interest (ROI) containing the traffic sign and ignored the margin. Their system may predict a error result when the input is a images containing margin region, because they employed images without margin region to train and test. Although their strategy avoids the disturbance of margin region, it may weaken the robustness of their system. In order to ensure our network can process images which contain margin noise, we directly use the traffic sign images which contain the margin to train and test.

The speed of our network on both CPU and GPU setting in test stage are summarized in Table 2. The experiment shows that using GPU is dozens of times faster than using the CPU. For average evaluation method, our network

Table 1. Comparison of results on GTSRB testing set.

Team	Method	accuracy(%)
IDSIA	Committee of CNNs	99.46
INI-RTCV	Human (best individual)	99.22
Ours	Deep CNNs	98.96
INI-RTCV	Human (average)	98.84
Sermanet	Multi-scale CNN	98.31
CAOR	Random forests	96.14

can process about 60 images per second on a NVIDIA GTX 1060 6G GPU. In the case of using GPU, our network obtains high computational efficiency in the recognition process.

Table 2. The speed of our network on Intel i7-6700K CPU and NVIDIA GTX 1060 6G GPU in test stage.

Device	Single evaluation	Average evaluation
i7-6700K	61 ms/frame	364 ms/frame
GTX 1060	5 ms/frame	16 ms/frame

Some true positive and false negative examples in test stage are shown in Fig. 9. For the images which under light contaminated conditions (Fig. 9(a)), our trained network could make true prediction. But for some images which under extreme contaminated conditions (Fig. 9(b)), our network did not work well. Because of did not use the ROI data, some images which contain a portion of another traffic sign also made our network confuse. We will also try to use the ROI data to train and test in the future work.

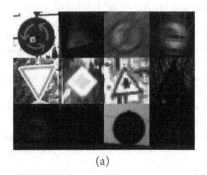 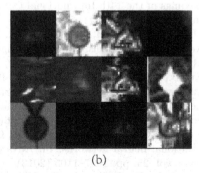

(a)   (b)

Fig. 9. Some examples in test stage. (a) True positive examples. (b) False negative examples.

5 Conclusions

This paper proposed a deep convolutional neural network used for traffic sign recognition. The network contains three block-layers except for single convolutional and fully-connected layers. The input images of our network contain 10% margin around the traffic sign; that is to say we do not use the exact region of the traffic sign. This means that our network can work well with images which have margin noise. On the test dataset of GTSRB, we achieve the accuracy of 98.96%. This result exceeds the human average rates.

In future work, we intend to improve the accuracy of our network and accelerate the calculation in prediction stage. In this regard, We are planning to consider the following issues: (1) to solve the problem of imbalanced training data (2) to estimate the impact of using the ROI data (3) to reduce the number of parameters in our network.

References

1. Nguwi, Y.Y., Kouzani, A.Z.: Detection and classification of road signs in natural environments. Neural Comput. Appl. **17**(3), 265–289 (2008)
2. Mogelmose, A., Trivedi, M.M., Moeslund, T.B.: Vision-based traffic sign detection and analysis for intelligent driver assistance systems: perspectives and survey. IEEE Trans. Intell. Transp. Syst. **13**(4), 1484–1497 (2012)
3. Lu, X., Wang, Y., Zhou, X., Zhang, Z., Ling, Z.: Traffic sign recognition via multimodal tree-structure embedded multi-task learning. IEEE Trans. Intell. Transp. Syst. **18**(4), 960–972 (2017)
4. Lim, K.H., Seng, K.P., Ang, L.M.: Intra color-shape classification for traffic sign recognition. In: 2010 International Computer Symposium, ICS 2010, pp. 642–647 (2010)
5. Madani, A., Yusof, R.: Traffic sign recognition based on color, shape, and pictogram classification using support vector machines. Neural Comput. Appl. **28**, 1–11 (2017)
6. Lau, M.M., Lim, K.H., Gopalai, A.A.: Malaysia traffic sign recognition with convolutional neural network. In: 2015 IEEE International Conference on Digital Signal Processing (DSP), pp. 1006–1010 (2015)
7. Glorot, X., Bordes, A., Bengio, Y.: Deep sparse rectifier neural networks. In: Proceedings of the 14th International Conference on Artificial Intelligence and Statistics, pp. 315–323 (2011)
8. Maas, A.L., Hannun, A.Y., Ng, A.Y.: Rectifier nonlinearities improve neural network acoustic models. In: Proceedings of the 30th International Conference on Machine Learning, ICML 2013 (2013)
9. Xu, B., Wang, N., Chen, T., Li, M.: Empirical evaluation of rectified activations in convolutional network. arXiv preprint arXiv:1505.00853 (2015)
10. LeCun, Y., Bottou, L., Bengio, Y., Haffner, P.: Gradient-based learning applied to document recognition. Proc. IEEE **86**(11), 2278–2324 (1998)
11. Krizhevsky, A., Sutskever, I., Hinton, G.E.: Imagenet classification with deep convolutional neural networks. In: Advances in Neural Information Processing Systems, vol. 25, pp. 1097–1105 (2012)
12. Simonyan, K., Zisserman, A.: Very deep convolutional networks for large-scale image recognition. In: The IEEE Conference on Computer Vision and Pattern Recognition (CVPR) (2015)

13. Lin, M., Chen, Q., Yan, S.: Network in network. In: Proceedings of ICLR (2014)
14. Szegedy, C., Liu, W., Jia, Y., Sermanet, P., Reed, S., Anguelov, D., Erhan, D., Vanhoucke, V., Rabinovich, A.: Going deeper with convolutions. In: The IEEE Conference on Computer Vision and Pattern Recognition (CVPR), pp. 1–9 (2015)
15. He, K., Zhang, X., Ren, S., Sun, J.: Deep residual learning for image recognition. In: The IEEE Conference on Computer Vision and Pattern Recognition (CVPR), pp. 770–778 (2016)
16. Szegedy, C., Ioffe, S., Vanhoucke, V., Alemi, A.A.: Inception-v4, Inception-ResNet and the impact of residual connections on learning. In: AAAI, pp. 4278–4284 (2017)
17. Stallkamp, J., Schlipsing, M., Salmen, J., Igel, C.: Man vs. computer: benchmarking machine learning algorithms for traffic sign recognition. Neural Netw. **32**, 323–332 (2012)
18. Abadi, M., Agarwal, A., Barham, P., Brevdo, E., Chen, Z., Citro, C., Corrado, G.S., Davis, A., Dean, J., Devin, M., Ghemawat, S., Goodfellow, I.J., Harp, A., Irving, G., Isard, M., Jia, Y., Józefowicz, R., Kaiser, L., Kudlur, M., Levenberg, J., Mané, D., Monga, R., Moore, S., Murray, D.G., Olah, C., Schuster, M., Shlens, J., Steiner, B., Sutskever, I., Talwar, K., Tucker, P.A., Vanhoucke, V., Vasudevan, V., Viégas, F.B., Vinyals, O., Warden, P., Wattenberg, M., Wicke, M., Yu, Y., Zheng, X.: TensorFlow: large-scale machine learning on heterogeneous distributed systems. arXiv preprint arXiv:1603.04467 (2016)
19. Kingma, D., Ba, J.: Adam: A method for stochastic optimization. arXiv preprint arXiv:1412.6980 (2014)
20. Hinton, G.E., Srivastava, N., Krizhevsky, A., Sutskever, I., Salakhutdinov, R.R.: Improving neural networks by preventing co-adaptation of feature detectors. arXiv preprint arXiv:1207.0580 (2012)
21. Ioffe, S., Szegedy, C.: Batch normalization: accelerating deep network training by reducing internal covariate shift. In: Proceedings of the 32nd International Conference on Machine Learning, vol. 37, pp. 448–456 (2015)

Deep Context Convolutional Neural Networks for Semantic Segmentation

Wenbin Yang[1,2], Quan Zhou[1,2(✉)], Yawen Fan[1], Guangwei Gao[3],
Songsong Wu[4], Weihua Ou[5], Huimin Lu[6], Jie Cheng[7], and Longin Jan Latecki[8]

[1] National Engineering Research Center of Communications and Networking,
Nanjing University of Posts and Telecommunications, Nanjing, China
quan.zhou@njupt.edu.cn
[2] Fujian Provincial Key Laboratory of Information Processing
and Intelligent Control, Minjiang University, Fuzhou 350121, China
[3] Key Laboratory of Intelligent Perception and Systems for High-Dimensional
Information of Ministry of Education, Nanjing University of Science and Technology,
Nanjing, China
[4] School of Automation, Nanjing University of Posts and Telecommunications,
Nanjing, China
[5] School of Big Data and Computer Science, Guizhou Normal University,
Guiyang, China
[6] Department of Mechanical and Control Engineering,
Kyushu Institute of Technology, Kitakyushu, Japan
[7] Huawei Technologies Co. Ltd., ShenZhen, China
[8] Department of Computer and Information Sciences, Temple University,
Philadelphia, USA

Abstract. Recent years have witnessed the great progress for semantic segmentation using deep convolutional neural networks (DCNNs). This paper presents a novel fully convolutional network for semantic segmentation using multi-scale contextual convolutional features. Since objects in natural images tend to be with various scales and aspect ratios, capturing the rich contextual information is very critical for dense pixel prediction. On the other hand, with going deeper of the convolutional layers, the convolutional feature maps of traditional DCNNs gradually become coarser, which may be harmful for semantic segmentation. According to these observations, we attempt to design a deep context convolutional network (DCCNet), which combines the feature maps from different levels of network in a holistic manner for semantic segmentation. The proposed network allows us to fully exploit local and global contextual information, ranging from an entire scene, though sub-regions, to every single pixel, to perform pixel label estimation. The experimental results demonstrate that our DCCNet (without any postprocessing) outperforms state-of-the-art methods on PASCAL VOC 2012, which is the most popular and challenging semantic segmentation dataset.

© Springer Nature Singapore Pte Ltd. 2017
J. Yang et al. (Eds.): CCCV 2017, Part I, CCIS 771, pp. 696–704, 2017.
https://doi.org/10.1007/978-981-10-7299-4_58

1 Introduction

Image semantic segmentation is a classic and challenging visual task in the field of computer vision. It aims to assign a semantic label to each image pixel to achieve object recognition and segmentation tasks synchronously. Semantic segmentation is associated with many high-level vision tasks, including image classification [1–3], object recognition [4–6] and object detection [7–9].

Typically, the previous semantic segmentation models usually first extract local appearance features, such as intensities, colors, gradients and textures, to describe different object instances. Then these hand-craft features are fed into the well trained classifiers to identify the category label for each image pixel, including regression boosting [10,11], random forests [12], or support vector machines [13]. Thereafter, great improvement have been achieved by incorporating contextual formulation [10,14] and scene global structured predictions [15,16] using conditional random field (CRF) [10,15,17] or markov random field (MRF) [16,18]. However, the performance of these systems has always been compromised by the limited expressive power of the hand-craft features.

In recent years, DCNNs have gained a lot of attention and become very popular in the computer vision community. Due to its superiority to modeling high-level visual concepts, DCNNs substantially advance the state-of-the-art results for the task of semantic segmentation [17,19,20]. As the pioneer work, LeCun et al. [21] employ the DCNNs at multiple image resolutions to compute image features, resulting in smooth predictions using a segmentation tree as postprocess. Another elegant work was proposed in [7], where the bounding box proposals and masked regions are used as inputs to train a DCNN. In the stage of classification, object shape information is taken into account within the trained DCNN. Similarly, the DCNN models can be also trained based on different image representation, such as superpixels [22]. In this work, the authors extracted the zoom-out spatial features, which are embedded into DCNNs to classify a superpixel. Although these methods can benefit from the delineated boundaries produced from a good segmentation, the object accurate shapes may be not always recovered well when there are some errors in segmentation results.

An alternative approach for dense pixel estimation relies on fully convolutional networks (FCN), where an end-to-end learning paradigm is adopted to train DCNNs [17,19,23]. Motivated from the DCNNs for image classification task, these methods directly target on estimating category-level per-pixel labels. The most representative work is [19], where the last fully connected layers of the DCNN are transformed into convolutional layers. In [17], Chen et al. proposed to learn a DeepLab model for semantic segmentation, where the receptive fields with different scales are employed using atrous convolution in deep convolutional networks. Then the final consistent segmentation results are enhanced using an independent fully connected CRF. The major limitation of FCN-based methods lies in the fact that the operation of max pooling and sub-sampling reduce feature map resolution, leading to the loss of spatial statistics for object instance. Therefore, Noh et al. [23] and Badrinarayanan et al. [24] proposed a coarse-to-fine structure with deconvolution network to learn the segmentation mask, yet with the cost of huge number of memory requirement and computing time.

In order to further advance the performance of semantic segmentation, the context cues are widely employed for image semantic segmentation [4,11,14,25]. However, we found that this major issue is not well addressed in current FCN-based models. Due to the repeated operation of max-pooling at successive layers of DCNNs, the spatial resolution is significantly reduced in feature maps when the DCNN is employed in a fully convolutional fashion [17,19]. Motivated by spatial pyramid pooling [4,25], in this paper, we present a *deep context convolutional network* (DCCNet) to explore contextual cues using the feature maps from different convolutional layers. Preciously, we exact all intermediate feature maps and fuse them together for pixel classification. Under this paradigm, the local and global contextual clues, ranging from en entire scene, through sub-regions, to every single pixel, are taken into account *jointly* to assign semantic label for each pixel. We compare the performance of our model with the mainstream models, which include contextual formulation [14], FCN-based approaches [17,19,21,22,26], and deconvolution network [24]. These are top-ranked models that previous studies have shown to significantly improve the results on PASCAL VOC 2012 dataset. In summary, the main contributions of this paper are twofolds:

- We propose a DCCNet to capture multi-scale context features in the manner of FCN for dense pixel prediction.
- We evaluate our DCCNet on PASCAL VOC 2012 dataset and achieve the state-of-the-art results for semantic segmentation.

2   The Architecture of DCCNet

In this section, we elaborate on architectural details of our DCCNet to investigate the multi-scale context information.

As shown in Fig. 1, we define our DCCNet based on VGG 16-layer network, which consists of four basic components, including convolution, rectified linear unit (ReLU), pooling (downsampling) and deconvolution (upsampling). In the main stream structure, we borrow the network architecture widely used in FCN-based model [19], where the final fully connected layers are transfered to convolution layers. It consists of the repeated convolution with 3×3 filter kernels, followed by a ReLU and a 2×2 max pooling operation with stride 2 for downsampling. The resolution of feature maps are reduced to $1/2^N$ of the original one after N max pooling operation. Here we set $N = 5$, leading to the $1/2$, $1/4$, $1/8$, $1/16$ and $1/32$ of the original resolution, and denote the final three ones as DCCNet-8s, DCCNet-16s, DCCNet-32s, respectively.

Similar to previous FCN-based approaches [17,19,23], the major limitation of our main stream structure lies in the spatial statistics of pixels is gradually lost with going deeper of our DCCNet. On the other hand, compared with shallower layers, the deeper ones have large receptive fields and are able to see more pixels. Intuitively, the feature maps with different scales can provide sufficient spatial statistics and contextual cues, which complement each other to get more reliable predictions. To this end, we integrate the main stream with two

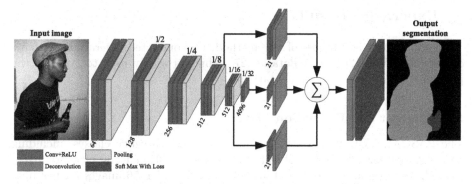

Fig. 1. Overview of our proposed DCCNet. Given an input image, we first use VGG-16 network to produce the hierarchical feature maps of intermediate pooling layers, which carry both local and global contextual cues and spatial statistics within different scales. Then the assocaited score maps are upsampled and concatenated to get the final per-pixel prediction. (best viewed in color) (Color figure online)

additional streams from the max pooling layers of DCCNet-8s and DCCNet-16s, as shown in Fig. 1. Once augmented, DCCNet allows us to fuse predictions from three streams that are learned jointly in an end-to-end architecture. More specifically, the final layers of these three streams are first convoluted with a 1×1 filter kernel to map them to three 21-component score maps, representing the confidence for each individual classes in PASCAL VOC dataset. However, these layers are with different resolution, they are thus required to be aligned by scaling and cropping. The deconvolution layers in our DCCNet utilizes the non-linear upsampling to scale score maps to the resolution with respect to DCCNet-8s, where the upsampling filter kernels are learned with the initialized weights of bilinear interpolation [19]. Subsequently, a cropping operation is performed. Cropping removes any portion of the upsampled layer which extends beyond the other layer, resulting in layers of equal dimensions for exact fusion. Finally, the three score maps are fused to a single 21-component score map, which is further upsampled to obtain the semantic output with the same resolution of original image.

At first glance of our DCCNet, it might look similar to the skip version of FCN [19], but in fact their network architectures are quite different. Essentially, FCN relies on *gradually* learning finer-scale prediction from lower layers in a stage-by-stage manner. That is, in each stage, the net is trained using the initialization of the previous stage net, where the contextual features are explored *independently*. In contrast, our DCCNet employs a all-in-once fashion to fuse the computed intermediate feature maps, where the contextual cues are investigated *jointly* to make final estimation. In addition, compared with stage-by-stage scheme used in FCN model, our all-in-once fashion results in computational efficiency and is less tedious in training process.

3   Experience Results

To validate the effectiveness of our method, we conducted several experiments on PASCAL VOC 2012 test dataset [27], which is widely used to evaluate the performance for semantic segmentation.

3.1   Dataset and Evaluation Metrics

The PASCAL VOC 2012 dataset [27] contains 21 category classes, including 20 categories for foreground object classes and additional one class for background. The original dataset contains 1464, 1449, and 1456 images for training, validation, and testing, respectively, where each image in the training and validation subset has accurate pixel-level annotated ground truth. The dataset is augmented by the extra annotations [28], resulting in 10582 training images.

In our experience, we report our results and compare with other baselines using the metrics in terms of pixel-level mean intersection-over-union (mIoU). It is commonly used to penalizes both over- and under-segmentation for semantic segmentation, which is defined as the ratio of true positives to the sum of true positive, false positive and false negative, averaged over all 21 object classes:

$$\frac{1}{\mathcal{C}} \frac{\sum_m N_{mm}}{\sum_n N_{mn} + \sum_n N_{nm} - N_{mm}} \tag{1}$$

where $\mathcal{C} = 21$ denotes the total number of different classes, N_{mn} is the number of pixels of category m labeled as class n.

3.2   Baselines

We selected 9 state-of-the-art models as baselines for comparison, including flexible segmentation graph (FSG [14]), deep hierarchical parsing (DHP [21]), fully-connected network (FCN [19]), DeepLab network (DLN [17]) and its variants (DLN[†] as DeepLab-Msc and DLN[‡] as DeepLab-Msc-CRF), deep zoom-out feature (DZF [22]), dilated convolution model (DCM [26]), and convolution and deconvolution network (CDN [23]).

3.3   Implementation Details

Our implementation is based on the public platform Caffe [29]. We minimize the soft max loss averaged over all image positions with stochastic gradient descent algorithm. The parameters of initialized network are borrowed from the pre-trained VGG-16 model using ImageNet dataset [30]. Then we fine-tune our DCCNet model using 10582 training images of PASCAL VOC 2012 dataset. Inspired by [17], we use the "poly" learning rate policy where the learning rate γ in iteration T equals to the base B multiplied by a factor:

$$\gamma = B \cdot (1 - \frac{T}{M})^{power} \tag{2}$$

where M denotes the total number of training iterations. The base learning rate is set as $B = 0.01$ and power is set as 0.9.

The performance can be further improved by increasing the iteration number. We found that our best performance is achieved when total iteration number is set to $M = 20K$, and any refinement of this parameter will result in no more significant improvement of performance.

3.4 Overall Results

We report the results in Table 1, and compare with the baselines in terms of mIoU. The results clearly demonstrate that our DCCNet outperforms prior state-of-the-art methods, including the approaches using contextual formulation [14,26], FCN-based models [17,19,21,22], and deconvolution network [23]. Trained with only PASCAL VOC 2012 data, our DCCnet achieves 71.4% mIoU among all 21 categories, and gets the highest accuracy on the classes of "bird", "boat", "bottle", "bus", "car", "cat", "chair", "cow", "dog", "person", "sheep", "train" and "TV". For fair comparison, we construct our DCCNet without any postprocessing to enhance the consistent semantic segmentation outputs. Even so, it is intriguing that our DCCNet is superior to the existing methods [17] that

Table 1. Per-class results on PASCAL VOC 2012 test set.

Method	bkg	aero	bike	bird	boat	bottle	bus	car	cat	chair	cow
DHP [21]	-	-	-	-	-	-	-	-	-	-	-
FSG [14]	-	-	-	-	-	-	-	-	-	-	-
DLN [17]	87.5	72.0	31.0	71.2	53.7	60.5	77.0	71.9	73.1	25.2	62.6
DLN[†] [17]	88.3	79.4	34.1	72.6	52.9	61.0	77.9	73.0	73.7	26.4	62.2
DZF [22]	89.8	81.9	35.1	78.2	57.4	56.5	80.5	74.0	79.8	22.4	69.6
DLN[‡] [17]	92.6	80.4	36.8	77.4	55.2	66.4	81.5	77.5	78.9	27.1	68.2
FCN [19]	91.2	76.8	34.4	68.9	49.4	60.3	75.3	74.7	77.6	21.4	62.5
DCM [26]	90.7	82.2	37.4	72.7	57.1	62.7	82.8	77.8	78.9	28.0	70.0
CDN [23]	92.7	85.9	42.6	78.9	62.6	66.6	87.4	77.8	79.5	26.3	73.4
DCCNet	93.2	84.1	39.0	82.1	67.7	78.4	87.4	83.4	85.8	38.2	77.2
Method	table	dog	horse	mbike	person	planet	sheep	sofa	train	tv	mIoU
DHP [21]	-	-	-	-	-	-	-	-	-	-	60.6
FSG [14]	-	-	-	-	-	-	-	-	-	-	64.4
DLN [17]	49.1	68.7	63.3	73.9	73.6	50.8	72.3	42.1	67.9	52.6	62.1
DLN[†] [17]	49.3	68.4	64.1	74.0	75.0	51.7	72.2	42.5	67.2	55.7	62.9
DZF [22]	53.7	74.0	76.0	76.6	68.8	44.3	70.2	40.2	68.9	55.3	64.4
DLN[‡] [17]	52.7	74.3	69.6	79.4	79.0	56.9	78.8	45.2	72.7	59.3	67.1
FCN [19]	46.8	71.8	63.9	76.5	73.9	45.2	72.4	37.4	70.9	55.1	67.2
DCM [26]	51.6	73.1	72.8	81.5	79.1	56.6	77.1	49.9	75.3	60.9	67.6
CDN [23]	60.2	70.8	76.5	79.6	77.7	58.2	77.4	52.9	75.2	59.8	69.6
DCCNet	45.6	80.9	75.2	77.3	82.8	56.7	81.5	51.8	82.4	70.5	71.4

employ CRF for further improving performance. Another interesting result is our approach outperforms deconvolution network [23]. This is probably because the proposed DCCNet has more powerful generalization than [23] due to its simple network architecture.

Some visual pleasing results of simultaneous recognition and segmentation are shown in Fig. 2. Each example shows both the original image and the color coded output. Except the boundary pixels that exhibit relative higher confusion, nearly all pixels are correctly classified. It is evident that our DCCNet can handle large appearance variations of object classes ("person", "dog", and "boat", etc.) and efficiently prohibit the clutter background (see the last four visual examples in the right column).

Image   Groundtruth   Ours   Image   Groundtruth   Ours   Image   Groundtruth   Ours

Fig. 2. Some visual examples of our semantic segmentation outputs. In each column, we show the original image, corresponding ground truth and our segmentation results. For clarity, we use different colors to denote different categories. (best viewed in color) (Color figure online)

4   Conclusion and Future Work

This paper describes a DCCNet to explore the multi-scale context information for semantic segmentation problem. Combining fine layers and coarse layers provides a more powerful representation with different receptive fields, allowing us to produce semantically accurate predictions and detailed segmentation maps. Our experimental results show that the proposed method advances the state-of-the-art results in the PASCAL VOC 2012 image semantic segmentation task.

Despite obtaining impressive results, we believe that even better results can be achieved by employing CRF models as well as [17,31] does. We also hope to demonstrate the generality of our DCCNet for other visual tasks.

Acknowledgements. The authors would like to thank all the anonymous reviewers for their valuable comments and suggestions. This work was partly supported by the National Science Foundation (Grant No. IIS-1302164), the National Natural Science Foundation of China (Grant Nos. 61401228, 61402238, 61402122, 61571240, 61501247, 61501259, 61671253, 61701252, 61762021), China Postdoctoral Science Foundation (Grant No. 2015M581841), Natural Science Foundation of Jiangsu Province (Grant Nos. BK20150849, BK20160908), Postdoctoral Science Foundation of Jiangsu Province (Grant No. 1501019A), Open Research Fund of Key Lab of Ministry of Education for Broadband Wireless Communication and Sensor Network Technology (Nanjing University of Posts and Telecommunications) (Grant No. NYKL2015012), Open Fund Project of Fujian Provincial Key Laboratory of Information Processing and Intelligent Control (Minjiang University) (Grant No. MJUKF201710), Open Fund Project of Key Laboratory of Intelligent Perception and Systems for High-Dimensional Information of Ministry of Education (Nanjing University of Science and Technology) (Grant Nos. JYB201709, JYB201710), Natural Science Foundation of Guizhou Province (Grant No. [2017]1130), and by the NUPTSF (Grant No. NY214139).

References

1. Krizhevsky, A., Sutskever, I., Hinton, G.E.: Imagenet classification with deep convolutional neural networks. In: NIPS, pp. 1097–1105 (2012)
2. He, K., Zhang, X., Ren, S., Sun, J.: Deep residual learning for image recognition. In: CVPR, pp. 770–778 (2016)
3. Szegedy, C., Liu, W., Jia, Y., Sermanet, P., Reed, S., Anguelov, D., Erhan, D., Vanhoucke, V., Rabinovich, A.: Going deeper with convolutions. In: CVPR, pp. 1–9 (2015)
4. He, K., Zhang, X., Ren, S., Sun, J.: Spatial pyramid pooling in deep convolutional networks for visual recognition. IEEE TPAMI **37**, 1904–1916 (2015)
5. Uijlings, J.R., Van De Sande, K.E., Gevers, T., Smeulders, A.W.: Selective search for object recognition. IJCV **104**, 154–171 (2013)
6. Girshick, R.: Fast R-CNN. In: ICCV, pp. 1440–1448 (2015)
7. Girshick, R., Donahue, J., Darrell, T., Malik, J.: Rich feature hierarchies for accurate object detection and semantic segmentation. In: CVPR, pp. 580–587 (2014)
8. Erhan, D., Szegedy, C., Toshev, A., Anguelov, D.: Scalable object detection using deep neural networks. In: CVPR, pp. 2147–2154 (2014)
9. Ren, S., He, K., Girshick, R., Sun, J.: Faster R-CNN: towards real-time object detection with region proposal networks. In: NIPS, pp. 91–99 (2015)
10. Shotton, J., Winn, J., Rother, C., Criminisi, A.: Textonboost for image understanding: multi-class object recognition and segmentation by jointly modeling texture, layout, and context. IJCV **81**, 2–23 (2009)
11. Tu, Z., Bai, X.: Auto-context and its application to high-level vision tasks and 3D brain image segmentation. IEEE TPAMI **32**, 1744–1757 (2010)
12. Shotton, J., Johnson, M., Cipolla, R.: Semantic texton forests for image categorization and segmentation. In: CVPR, pp. 1–8 (2008)
13. Fulkerson, B., Vedaldi, A., Soatto, S.: Class segmentation and object localization with superpixel neighborhoods. In: ICCV, pp. 670–677 (2009)
14. Zhou, Q., Zheng, B., Zhu, W., Latecki, L.J.: Multi-scale context for scene labeling via flexible segmentation graph. PR **59**, 312–324 (2016)
15. Koltun, V.: Efficient inference in fully connected CRFs with Gaussian edge potentials. In: NIPS, pp. 4–10 (2011)

16. Carreira, J., Sminchisescu, C.: CPMC: automatic object segmentation using constrained parametric min-cuts. IEEE TPAMI **34**, 1312–1328 (2012)
17. Chen, L.C., Papandreou, G., Kokkinos, I., Murphy, K., Yuille, A.L.: Semantic image segmentation with deep convolutional nets and fully connected CRFs. arXiv preprint arXiv:1412.7062 (2014)
18. Zhou, Q., Zhu, J., Liu, W.: Learning dynamic hybrid markov random field for image labeling. IEEE TIP **22**, 2219–2232 (2013)
19. Long, J., Shelhamer, E., Darrell, T.: Fully convolutional networks for semantic segmentation. IEEE TPAMI **39**, 640–651 (2017)
20. Chen, L.C., Yang, Y., Wang, J., Xu, W., Yuille, A.L.: Attention to scale: scale-aware semantic image segmentation. In: CVPR, pp. 3640–3649 (2016)
21. Farabet, C., Couprie, C., Najman, L., LeCun, Y.: Learning hierarchical features for scene labeling. IEEE TPAMI **35**, 1915–1929 (2013)
22. Mostajabi, M., Yadollahpour, P., Shakhnarovich, G.: Feedforward semantic segmentation with zoom-out features. In: CVPR, pp. 3376–3385 (2015)
23. Noh, H., Hong, S., Han, B.: Learning deconvolution network for semantic segmentation. In: ICCV, pp. 1520–1528 (2015)
24. Badrinarayanan, V., Alex, K., Roberto, C.: SegNet: a deep convolutional encoder-decoder architecture for image segmentation. arXiv preprint arXiv:1511.00561 (2015)
25. Lazebnik, S., Schmid, C., Ponce, J.: Beyond bags of features: spatial pyramid matching for recognizing natural scene categories. In: CVPR, pp. 2169–2178 (2006)
26. Yu, F., Koltun, V.: Multi-scale context aggregation by dilated convolutions. arXiv preprint arXiv:1511.07122 (2015)
27. Everingham, M., Eslami, S.A., Van Gool, L., Williams, C.K., Winn, J., Zisserman, A.: The pascal visual object classes challenge: a retrospective. IJCV **111**, 98–136 (2015)
28. Hariharan, B., Arbeláez, P., Bourdev, L., Maji, S., Malik, J.: Semantic contours from inverse detectors. In: ICCV, pp. 991–998 (2011)
29. Jia, Y., Shelhamer, E., Donahue, J., Karayev, S., Long, J., Girshick, R., Guadarrama, S., Darrell, T.: Caffe: convolutional architecture for fast feature embedding. In: ACMMM, pp. 675–678 (2014)
30. Simonyan, K., Zisserman, A.: Very deep convolutional networks for large-scale image recognition. arXiv preprint arXiv:1409.1556 (2014)
31. Zheng, S., Jayasumana, S., Romera-Paredes, B., Vineet, V., Su, Z., Du, D., Huang, C., Torr, P.H.: Conditional random fields as recurrent neural networks. In: ICCV, pp. 1529–1537 (2015)

Automatic Character Motion Style Transfer via Autoencoder Generative Model and Spatio-Temporal Correlation Mining

Dong Hu[1,2], Xin Liu[1,2(✉)], Shujuan Peng[1,2],
Bineng Zhong[1,2], and Jixiang Du[1,2]

[1] Department of Computer Science, Huaqiao University, Xiamen 361021, China
xliu@hqu.edu.cn
[2] Xiamen Key Laboratory of Computer Vision and Pattern Recognition,
Huaqiao University, Xiamen 361021, China

Abstract. The style of motion is essential for virtual characters animation, and it is significant to generate motion style efficiently in computer animation. In this paper, we present an efficient approach to automatically transfer motion style by using autoencoder generative model and spatio-temporal correlation mining, which allows users to transform an input motion into a new style while preserving its original content. To this end, we introduce a history vector of previous motion frames into autoencoder generative network, and extract the spatio-temporal feature of input motion. Accordingly, the spatio-temporal correlation within motions can be represented by the correlated hidden units in this network. Subsequently, we established the constraints of Gram matrix in such feature space to produce transferred motion by pre-trained generative model. As a result, various motions of particular semantic can be automatically transferred from one style to another one, and the extensive experiments have shown its outstanding performance.

Keywords: Deep autoencoder · Style transfer · Character animation
Spatio-temporal correlation · History vector

1 Introduction

Virtual characters are an essential constituent part of animation, and various moving styles are of crucial importance for an animator to express various postures. Different from the real world, sometimes, the storyline of animation is very exaggerated and animated character needs a dramatic gesture to perform this style. For example, an old man with a stick runs fast than a young man. Thus, this style of character suddenly changes to another distinctive style, which need animator to edit it renewedly. Evidently, motion editing by human labours often cost much time and the efficiently is lower. To tackle this problem, some motion editing techniques have been exploited for computer animation. Specifically, data-driven editing of motion capture data can produce natural

© Springer Nature Singapore Pte Ltd. 2017
J. Yang et al. (Eds.): CCCV 2017, Part I, CCIS 771, pp. 705–716, 2017.
https://doi.org/10.1007/978-981-10-7299-4_59

motion with little user involvement, and multiplying the value of these data-driven approaches by stylizing it to motion synthesis and transfer have been extensively studied. However, most of existing data-driven based transferring approaches often require the motions to be segmented, aligned and classified for each motion class. Meanwhile, these methods usually involved the human interventions to ensure the smooth and natural movements.

In this paper, we present an efficient approach to automatically transfer motion style by using autoencoder generative model and spatio-temporal correlation mining, which allows users to transform an input motion into a new style while preserving its original content. Unlike previous methods, our proposed approach does not require manual data processing or human intervention, like segmentation and labeling. Instead, we exploit an unsupervised deep learning framework to learn the style features of heterogeneous motion data for transferring. The main contributions of our proposed approach are summarized below: (1) An autoencoder generative model is proposed for motion modeling, through which different kinds of motion styles can be well characterized; (2) The deep neural network is initialized with values obtained from a restricted Boltzman Machine, which improves the generalization ability and adaptability. The extensive experiments have shown its outstanding performance.

The remaining part of this paper is structured as follows: Sect. 2 will briefly overview the related works corresponding to motion synthesis and transfer. In Sect. 3, we present the proposed model and its implementation details. Section 4 provides the experimental results and the related discussions. Finally, we draw a conclusion in Sect. 5.

2   Related Work

In the past, different kinds of motion editing approaches have been exploited for new motion generation. In this section, we mainly survey the motion synthesis and motion style transfer extensively.

Motion synthesis aims to generate new styles of motion from a series of different motion database and follow some pre-imposed constraints. Originally, optimization method is effective for synthesizing human motion while retaining control of its qualitative properties. Along this line, Arikan et al. [1] utilized the support vector machine (SVM) classifiers to generalize the user annotations to the entire database, while Mukai and Kuriyama [13] proposed to statistically optimize the interpolation kernels for given parameters at each frame. However, with the increase of user requirements, more abstract constraints and a high level should be considered, and optimization method is not suitable for reliable motion synthesis. To handle this problem, motion graphic [11] and statistical model [3] were exploited for motion synthesis, but which required the motion to be classified, segmented and aligned. During recent years, deep learning and neural network are becoming popular in data-driven based motion synthesis. For instance, Jain et al. [10] combine the power of high-level spatio-temporal graphs and sequence learning algorithm of Recurrent Neural Network (RNN) for motion synthesis. Since RNN is the long-term contextual representation of temporal sequences, this type of technology processing motion capture of continuous motion data could be significantly degraded [4]. In addition, Taylor and Hinton [15] presented three-way interactions model via conditional Restricted Boltzmann Machines (CRBM), to capture diverse styles of motion with a single set of parameters, and blend motion styles to transit among them smoothly.

Motion style transfer mainly focus on automatically transferring a motion of particular semantic from one style to another one. The pioneer works concentrated on the dominant frequency component of style transfer [2], and identification of certain emotional transforms by signal processing techniques which transfer simple style of in the same class (such as speed change). This type of approaches only replayed the captured motion data, whereby the blended motion styles were unsmooth and unnatural because of the temporal pattern not considered for continuous motion. In addition, expressive style feature of motion data by preprocessing, and utilization od linear models to transfer one style to another [8], have been developed. Although these approaches were able to transfer some motions, their performance significantly diminished when applied to another motion data of different structure. To handle this problem, non-linear models were studied. Ikemoto et al. [9] utilized Gaussian Process (GP) models of kinematics and dynamics to edit long sequences, while Xia et al. [17] introduced an efficient local regression model to predict the timings of synthesized poses in the output style.

Automatically learning feature of motion style and without manual data preprocessing is important to motion edit. Holden et al. [7] presented a framework to synthesize character movements based on convolutional autoencoder. Accordingly, the derived feedforward neural network can represent motion data in sparse components and transfer style in this space by user constraints. However, the performance obtained by this method was still a bit poor. Therefore, there is still a need to develop a practical and efficient motion style transferring method.

3   System Overview

As shown in Fig. 1 our proposed motion style transfer model consists of three steps: motion processing via pre-trained RBM, motion feature extraction and motion transfer from one style to another. We first convert the dataset into a format that is suitable for training, which retargeting them to a uniform skeleton structure with a single scale and the same bone lengths. After data preprocessing, our feature extraction model based on autoencoder is used to extract from processed motion. Specially, we add the "history" vector in autoencoder to ensure the timing can map in feature space from continuous motion and the parameters are initialized by pre-trained RBM (see Sect. 3.1). Finally, we established constraints with Gram matrix from feature space to transfer the style from the style motion to content motion (see Sect. 3.2).

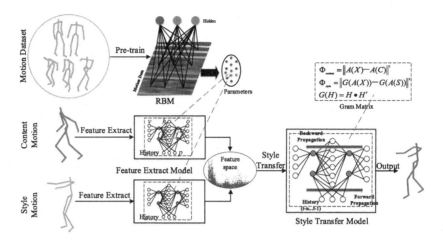

Fig. 1. Motion style transfer of our method. Our methods first utilizes the Restricted Boltzmann Machine (RBM) to pre-train large motion database to obtain the parameters. Then, these parameters are used to initialize feature extraction model, in which the content motion and style motion is mapping to style feature space with "history" vector. Finally, we construct style transfer model to transfer the style of one motion with the timing of another by Gram matrices in feature space.

3.1   Style Feature Extraction in Motion

The real character movement is natural because of a series of movements associated with all joints of the body and the previous position. Therefore, we change the autoencoder network structure with "history" vector and add the fusion factor f, decomposition factor f' to concatenated the "history" vector p (show in Fig. 1), and v, o is input and output data, h is hidden units.

The network provides feedforward propagation and backpropagation. The forward operation A has three layers that receives the input motion X in the visible unit, through the hidden unit H output the reconstruction motion $\tilde{X}$. The first forward operation as following:

$$A_2(X) = ReLU(XW^{vf}W^{fh} + PW^{pf}W^{fh} + b_1), \tag{1}$$

take visible vector $X \in \Re^{m \times d}$ and the past vector $P \in \Re^{m \times (d \times t)}$ as input, using weights matrix $W^{vf} \in \Re^{d \times k}$, $W^{fh} \in \Re^{k \times n}$ connection of visible-factor-hidden and weights matrix $W^{pf} \in \Re^{(d \times t) \times k}$, $W^{fh} \in \Re^{k \times n}$ connection of history-factor-hidden, addition of a bias $b_1 \in \Re^n$, where m is frames of continues motion to input and d is the degrees of freedom of the body model, k, n are the number of factors and hidden units in the feedforward network, which are set to 240, 73, 200 and 600. The history vector is composed of previous t frame relative to current process frame, so that has $d \times t$ lengths for every row and t is set to 12. Finally, $ReLU$ is called the activation function of rectified linear unit (ReLU).

The second forward operation is given by the following:

$$A_3(H) = ReLU(H\tilde{W}^{hf'}\tilde{W}^{f'v} + P\tilde{W}^{pf'}\tilde{W}^{f'o} + b_2), \tag{2}$$

similar to the first forward operation, using the same rectified linear operation $ReLU$ as activation function and output by feedforward neural network, where H is hidden units and P is the past vector. But we use different factor to represent weight matrix, $\tilde{W}^{hf'} \in \Re^{n \times k}$ and $\tilde{W}^{f'v} \in \Re^{k \times d}$ is connection of hidden-factor-output, $\tilde{W}^{pf'} \in \Re^{(d \times t) \times k}$ and $\tilde{W}^{sf'} \in \Re^{k \times d}$ is connection of past-factor-output, addition of a bias $b_2 \in \Re^d$.

In feedforward neural network, it has eight weight matrices and two biases of the parameters. Therefore, if it is initialized from small random values when trained, the network of convergence is slow and it is difficult to achieve the optimal value. We address this problem by construct a RBM to pre-train, and the parameters of trained RBM is used to initialize feature extraction network. The structure of RBM is similar to our extract model but adjust the input and output of motion to the same layer, and the forward propagation weight parameters initialize first layer, it transpose initialize second layer of feature extract network. The energy function of RBM is:

$$\begin{aligned} E(v,h|p,\theta) =&(1/2)\sum_i (a_i - v_i)^2 \\ &- \sum_f \sum_{ij} W_{if}^v W_{jf}^h v_i h_j - \sum_j b_j h_j, \end{aligned} \tag{3}$$

where v_i, h_j is the visible of input and hidden, a_i, b_j index the standard biases of visible and hidden, and θ is model parameters, p is conditional on the past t observations. The parameters W_{if}^v, connect input units to factors, W_{jf}^h connect output units to factors, f indexes a set of factors, i and j are index the visible units v and hidden units h. We were used contrastive divergence (CD) [16] learning algorithm to training this network. After pre-training, the parameters of RBM is used to initialize the feature model for weight matrix and bias vectors.

Finally, we train this feedforward neural network to reproduce motion input X following two forward operation Eqs. (1), (2) and output motion $\tilde{X}$. Training is, therefore, solved the following optimization problem:

$$minimze \quad Cost(X,\theta) = (1/2)\|X - \tilde{X}\|_2^2 + (\lambda/2)\phi(\theta), \tag{4}$$

In this equation, the squared-error cost function was used to input and output, $\phi(\theta)$ is a regularization term (also called a weight decay term) that tends to decrease the magnitude of the weights, and helps prevent overfitting, where λ is the weight decay parameter.

3.2 Motion Style Transfer and Generation

Style transfer by deep learning has succeeded in image synthesis. Gatys et al. [5] describe a neural network to separate and recombine the image content and style of natural images that allow user to produce new images, which encoded in the Gram matrix of the hidden layers of a neural network. But the image is a static data and the motion is associated with spatial and temporal cues. To balance this problem, we adjust the network structure with "history" vector to express the feature of continuous motion, and utilized the different styles of motion instead of input image to produce a motion that has the timing and content of one input, with the style of another. The style transfer network structure is similar to motion feature extraction but the cost function is different.

For producing new motion of high perceptual quality that combines the content of an arbitrary motion with the style of a large motion dataset, we establish content constraint:

$$\phi_{content}(X, C) = \|A(X) - A_2(C)\|_2^2, \tag{5}$$

where X, C are the input motion and content motion, $A(X)$ is hidden unit of style transfer network and $A_2(C)$ is content feature computed by Eq. (1).

Meanwhile the constraint of style motion in the Gram matrix is formulated as follows:

$$\phi_{style}(X, S) = \|G(A(X)) - G(A_2(S))\|_2^2, \tag{6}$$

where X, S is input motion and content motion, $A(X)$ is hidden units of style transfer network and $A_2(S)$ is style feature computed by Eq. (1), G is the function of Gram matrix, which expresses the correlations of expectation taken over the spatial extent of the style feature. Often, it is computed by inner product of the hidden units values:

$$G(H) = HH^T, \tag{7}$$

where H is the feature of hidden units. Finally, jointly minimise the distance of the content constraints of content motion and the style constraints of the style motion. The cost function as the following:

$$\phi_{transfer} = \alpha\phi_{content}(X, C) + \beta\phi_{style}(X, S), \tag{8}$$

where α and β dictate the relative importance given to content motion and style motion, which are set to $1.0/(1.0 + 0.01)$ and $0.01/(1.0 + 0.01)$. And we style transfer from the motion of style to another motion by minimizing Eq. (8).

4 Experimental Results

The main content of this paper is automatically transfer motion style from one style to another one. We first present a distance function to evaluate our transfer results and then test a variety of motion styles to evaluate the generalization ability. Next, we compare our proposed approach with previous methods of Shapiro et al. [14], Guo et al. [6], and Holden et al. [7]. Finally, our experiments are evaluated by transferring motion style to other different style motions, and we define a distance for describe transferred motion style as follow:

$$D = \left(\sum_{i=2}^{N}(J_i - J_o)\right)/(n - 1) \tag{9}$$

where J_i is joint 3D position of character skeleton, J_o is root 3D position of skeleton, i is index of the joints and N is the total joints of characters skeleton, D is computed to average distance from each joints of the skeleton to root joint. We use Eq. (9) to evaluate the result of transfer motion which compute the discrepancy of content distance and style content (show in Fig. 2). The result of transfer is better when the distances of transferred motion is similar to style motion, rather than the content motion.

Content \ Style	depressed	old	strong	sexy	drunk
walk	2.75	8.74	3.56	7.23	5.91
	2.14	1.79	3.01	3.78	5.27
walk-right	2.50	8.58	3.20	6.94	4.91
	1.91	1.91	2.60	3.90	4.92
stand-walk-run	4.27	8.70	4.83	7.92	7.13
	3.28	3.46	3.16	4.30	4.98
run	5.03	2.66	4.55	2.21	3.62
	3.01	3.33	3.64	4.78	5.87

Fig. 2. Distances of motion style transfer. The row of figure is content motion with walk, walk to right, stand to walk and run, run motion, and the column of figure is style motion with depressed, old, strong, sexy, drunk style motion. The top of the cell is the discrepancy for transferred motion and content motion, the below of the cell is the discrepancy for transferred motion and style motion.

We compare with Shapiro et al. [14] and Guo et al. [6] for motion style transfer. We use run as content motion, zombie and old as style motion which the result of transfer style shows in Fig. 3. And Fig. 3(A) transfer zombie motion to run motion by Shapiro et al. [14], Fig. 3(B, C) is transfer zombie and old style to run motion by Guo et al. [6]. These two method can achieve style motion transfer, but when the content motion has fast speed (like the motion of run), the irregular motion are appeared in transfer result and have big noise. In addition, the paths of transfer motion have random generation, and cannot follow the original path of movement. Comparatively speaking, the transferring results obtained by our method are smooth and natural.

For detect whether there is abnormal motion in the transferred motion sequence, we use the height (y axis) of foot joint (ltibia) to judge abnormal

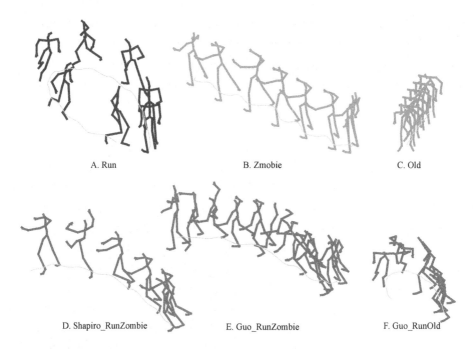

A. Run   B. Zmobie   C. Old

D. Shapiro_RunZombie   E. Guo_RunZombie   F. Guo_RunOld

Fig. 3. Motion style transfer of Shapiro et al. [14] and Guo et al. [6]. A is a content motion of run and B, C is a style motion of Zombie and Old. D is result of motion style transfer with Shapiro et al. [14] method for blend in run and zombie. E, F is style transfer with Guo et al. [6] and transfer zombie and old style to run.

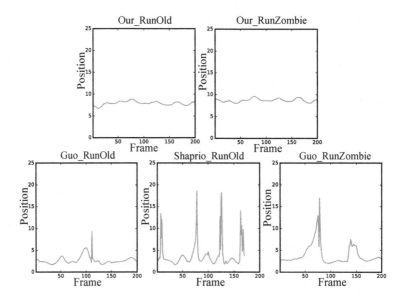

Fig. 4. Foot joint (ltibia) position with the change of frame.

A. Walk   B. Monkey

C. WalkMonkey   D. WalkMonkey

Fig. 5. Motion transfer visualization. A is orginial walk motion, B is style motion in the same spot of monkey, C is the result transfer of Holden et al. [7], D is transferred motion in our method.

posture by character skeleton data. As shown in Fig. 4, the change of hand joint obtained by Shapiro et al. [14] and Guo et al. [6] method have some a sudden change of position and this is not consistent with the reality of movement. In our method, the curve is smooth and not drastic change, which indicates that the nature of motion in our method is better than Shapiro et al. [14] and Guo et al. [6].

In recently years, the deep neural networks have a good ability to deal with non-linear problems with spatio-temporal correlation. Holden et al. [7] employed the deep learning method for automatically producing movements respected to the manifold of human motion and edit motion into the space of the motion manifold, but their method have bad style transfer for a style motion in the same spot to another motion. For example, this method failed to transfer the motion path which is short and almost not movement (Fig. 5(C)). By contrast, our method is able to deal with motion (Fig. 5(D)), and the experiments have shown its outstanding performance.

To illustrate efficiency of our method to process the static motion style, we ignore the history node and transfer motion style by antoencoder to determine the effect of the history node. The typical results were show in Fig. 6, it can be

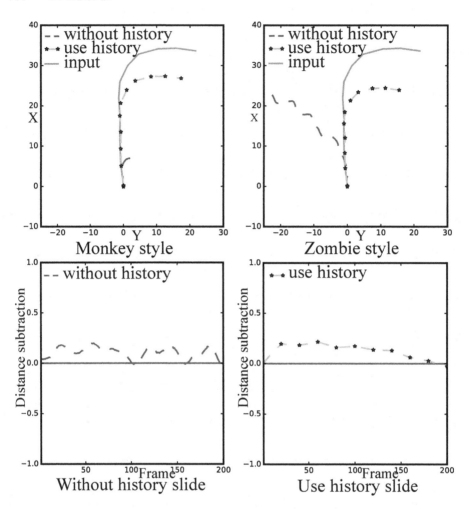

Fig. 6. History node comparison. Without history node to transfer motion of walk content and the style motion of monkey and zombie

found that the motion of transfer monkey style also movement short, and the style of zombie movement indiscriminately, if history node was not employed. Fortunately, the motion is great of transfer result, if used history node was employed. We also test a foot sliding frame of the transfer motion for the difference between the previous frame and the latter frame in foot position. The left of diagram is showing the foot-skating is reduction but there is a small amount. This problem can be addressed by using constraint-based motion editing techniques such as Lee et al. [12]. Finally, using history can reduce slide and control maintain spatio-temporal of input motion.

5   Conclusion

In this paper, we have presented an efficient approach to automatically transfer motion style by using autoencoder generative model and spatio-temporal correlation mining, which allows users to transform an input motion into a new style while preserving its original content. To this end, we introduce a history vector of previous motion frames into autoencoder generative network, and extract the spatio-temporal feature of input motion. Accordingly, the spatio-temporal correlation within motion can be represented by the correlated hidden units in this network. Meanwhile, we established the constraints of Gram matrix in such feature space to produce transferred motion by pre-trained generative model. As a result, various motions of particular semantic can be automatically transferred from one style to another one. The experiments have shown that the proposed approach can produce a variety of different movement styles, and the transferred motion styles are visually natural and vivid.

Acknowledgment. The work described in this paper was supported by the National Science Foundation of China (Nos. 61673185, 61572205, 61673186), National Science Foundation of Fujian Province (Nos. 2015J01656, 2017J01112), Promotion Program for Young and Middle-aged Teacher in Science and Technology Research (No. ZQN-PY309), the Promotion Program for graduate student in Scientific research and innovation ability of Huaqiao University (No. 1611414006).

References

1. Arikan, O., Forsyth, D.A., O'Brien, J.F.: Motion synthesis from annotations. ACM Trans. Graph. (TOG) **22**(3), 402–408 (2003)
2. Bruderlin, A., Williams, L.: Motion signal processing. In: Proceedings of the 22nd Annual Conference on Computer Graphics And Interactive Techniques, pp. 97–104 (1995)
3. Chai, J., Hodgins, J.K.: Constraint-based motion optimization using a statistical dynamic model. ACM Trans. Graph. (TOG) **26**(3), 8:1–8:12 (2007)
4. Fragkiadaki, K., Levine, S., Felsen, P., Malik, J.: Recurrent network models for human dynamics. In: Proceedings of the IEEE International Conference on Computer Vision, pp. 4346–4354 (2015)
5. Gatys, L.A., Ecker, A.S., Bethge, M.: Image style transfer using convolutional neural networks. In: Proceedings of the IEEE Conference on Computer Vision and Pattern Recognition, pp. 2414–2423 (2016)
6. Guo, X., Xu, S., Che, W., Zhang, X.: Automatic motion generation based on path editing from motion capture data. In: Pan, Z., Cheok, A.D., Müller, W., Zhang, X., Wong, K. (eds.) Transactions on Edutainment IV. LNCS, vol. 6250, pp. 91–104. Springer, Heidelberg (2010). https://doi.org/10.1007/978-3-642-14484-4_9
7. Holden, D., Saito, J., Komura, T.: A deep learning framework for character motion synthesis and editing. ACM Trans. Graph. (TOG) **35**(4), 138:1–138:11 (2016)
8. Hsu, E., Pulli, K., Popović, J.: Style translation for human motion. ACM Trans. Graph. (TOG) **24**(3), 1082–1089 (2005)
9. Ikemoto, L., Arikan, O., Forsyth, D.: Generalizing motion edits with Gaussian processes. ACM Trans. Graph. (TOG) **28**(1), 1:1–1:12 (2009)

10. Jain, A., Zamir, A.R., Savarese, S., Saxena, A.: Structural-RNN: deep learning on spatio-temporal graphs. In: Proceedings of the IEEE Conference on Computer Vision and Pattern Recognition, pp. 5308–5317 (2016)
11. Kovar, L., Gleicher, M., Pighin, F.: Motion graphs. ACM Trans. Graph. (TOG) **21**(3), 473–482 (2002)
12. Lee, J., Chai, J., Reitsma, P.S., Hodgins, J.K., Pollard, N.S.: Interactive control of avatars animated with human motion data. ACM Trans. Graph. (TOG) **21**(3), 491–500 (2002)
13. Mukai, T., Kuriyama, S.: Geostatistical motion interpolation. ACM Trans. Graph. (TOG) **24**(3), 1062–1070 (2005)
14. Shapiro, A., Cao, Y., Faloutsos, P.: Style components. In: Proceedings of Graphics Interface, pp. 33–39 (2006)
15. Taylor, G.W., Hinton, G.E.: Factored conditional restricted Boltzmann machines for modeling motion style. In: Proceedings of the 26th Annual International Conference on Machine Learning, pp. 1025–1032. ACM (2009)
16. Tieleman, T.: Training restricted Boltzmann machines using approximations to the likelihood gradient. In: Proceedings of the 25th International Conference on Machine learning, pp. 1064–1071. ACM (2008)
17. Xia, S., Wang, C., Chai, J., Hodgins, J.: Realtime style transfer for unlabeled heterogeneous human motion. ACM Trans. Graph. (TOG) **34**(4), 119–128 (2015)

Efficient Human Motion Transition via Hybrid Deep Neural Network and Reliable Motion Graph Mining

Bing Zhou[1,2], Xin Liu[1,2], Shujuan Peng[1,2(✉)], Bineng Zhong[1,2],
and Jixiang Du[1,2]

[1] Department of Computer Science, Huaqiao University, Xiamen 361021, China
pshujuan@hqu.edu.cn
[2] Xiamen Key Laboratory of Computer Vision and Pattern Recognition,
Huaqiao University, Xiamen 361021, China

Abstract. Skeletal motion transition is of crucial importance to the simulation in interactive environments. In this paper, we propose a hybrid deep learning framework that allows for flexible and efficient human motion transition from motion capture (mocap) data, which optimally satisfies the diverse user-specified paths. We integrate a convolutional restricted Boltzmann machine with deep belief network to detect appropriate transition points. Subsequently, a quadruples-like data structure is exploited for motion graph building, which significantly benefits for the motion splitting and indexing. As a result, various motion clips can be well retrieved and transited fulfilling the user inputs, while preserving the smooth quality of the original data. The experiments show that the proposed transition approach performs favorably compared to the state-of-the-art competing approaches.

Keywords: Skeletal motion transition · Hybrid deep learning
Convolutional restricted Boltzmann machine
Quadruples-like data structure

1 Introduction

Motion capture has been a successful technique for creating realistic animations, and the precise recordings of complex human movements are widely utilized in film, man-machine interaction and so forth [18]. However, the capture environment generally offers limited flexibility to produce various motion clips, and the acquisition of such motion capture (mocap) data, e.g., skeletal joints, is prohibitively expensive. Under this circumstances, the need of motion editing approaches for reusing these previously recorded data is growing. Motion transition is a technique which could produce new motions by optimally combining multiple individual motion clips in a time domain. However, effective motion transition is still a non-trivial task due to its spatio-temporal articulated complexity. Most of existing works either require much physical constraints or involve

© Springer Nature Singapore Pte Ltd. 2017
J. Yang et al. (Eds.): CCCV 2017, Part I, CCIS 771, pp. 717–728, 2017.
https://doi.org/10.1007/978-981-10-7299-4_60

large optimization procedures to achieve that, and some of which often lack of flexibilities in fulfilling the spatio-temporal consistency adaptively. Therefore, flexible skeletal motion transition is highly desirable and improving the transition performance will benefit many computer animation tasks.

The major of this paper is to create seamlessly motion transitions satisfying user demands. Following the procedures of constructing a motion graph, our proposed approach improves the state-of-the-art methods by providing the following two contributions: (1) A hybrid deep learning framework is proposed to extract the spatio-temporal features of each motion style, through which the appropriate transition points can be well obtained; (2) A novel quadruples-like data structure is exploited for motion graph building, which benefits much for the motion splitting and indexing. Without many manual pre-processing, the proposed approach can automatically detect and create transitions between different styles.

The remaining part of this paper is organized as follows: Sect. 2 will survey the related work, and Sect. 3 presents the proposed approach in detail. In Sect. 4, the experimental results are carefully introduced. Finally, we draw a conclusion in Sect. 5.

2   Related Work

Motion graph based approaches establish a connected graph for clip corpus systematically, and consider the transition problem as the specific node selection of certain properties. Along this way, Rose et al. [9] described the motions with 'verbs' and 'adverbs' to blends between different motion styles, and the complementary work has been done in [5]. Since the motion graph is constructed for efficient interpolation and prunes, many modifications have been developed to satisfy various user specifications. Lai et al. [6] found the graph nodes by clustering clip groups and generated transitions via constrained simulation. Peng et al. [8] exploited a flexible transition graph by using a two-pass clustering method. Recently, Saida and Foudil [10] imposed the inverse kinematic in constructing a motion graph to overcome the unnatural transitions. Although these approaches could produce visually pleasant performance, the global transition points searching scheme often limits their application domain in efficient motion generation.

Deep learning methods have received wide attentions in recent years, and there is a huge interest in applying deep learning techniques to synthesize novel data. A pioneer work attempted for motion transition was done by Taylor et al. [16], who proposed a conditional RBM to exploit the transition states in time series. Later, a factored conditional RBM [15] was further designed to reduce the parameter numbers in conditional RBM while appropriately preserving the stability. Gan et al. [2] suggested sampling from variational posterior in training temporal sigmoid belief networks (TSBN) which allowed for generating natural motion transitions. Song et al. [12] factorized the weight tensor of TSBN with labels to make the motion style more explicit. Experimentally, these deep learning based approaches are able to characterize the inherent motion states, but

which often limited their applications in simple motion generation. Therefore, it is imperative to develop a flexible transition approach combining the flexible motion graph and representative deep neural networks in practice. Holden et al. [3] combined the one layer convolutional autoencoder with a deep feedforward network to control and synthesize motion data. This inspired us to utilize both of the strengths of convolutional operation and RBM to extract the local spatial-temporal features of the skeletal data.

3   The Proposed Methodology

As shown in Fig. 1, our framework involves two phases: offline training and online matching. The former phase aims to extract representative features for motion splitting and construct a motion graph for transition point detection, while the latter phase supports the user interface and allows interactive operations for diverse motion transition. In the following, the skeletal data representation and the topology structure about model description are illustrated in details.

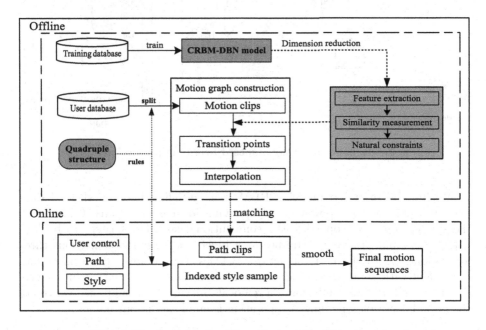

Fig. 1. The framework of the proposed transition system.

3.1   Hybrid Deep Neural Network

The popular mocap databases include CMU[1], HDM05[2] and BVH library[3], and these public available datasets almost share the similar data structure. Specifi-

[1] http://mocap.cs.cmu.edu/.

[2] http://resources.mpi-inf.mpg.de/HDM05.

[3] https://accad.osu.edu/research/mocap/.

cally, input formats of the j-th joint of i-th frame can be considered as a data point represented by a 3D position in space, $\mathbf{D_{ij}} = \{D_{ij}^x, D_{ij}^y, D_{ij}^z\}$. In this section, we provide a hybrid deep neural network to map the motion sequences into a 2D low-dimensional space, which can contribute more to the time efficiency in detecting transition nodes. Deep learning based models are of sufficient strength to extract representative features from the dataset. Since convolution operations have been making a difference in computer graphics and speech recognition, we are inspired to use the kernels to percept the local features of the skeletal data. As shown in Fig. 2, the proposed hybrid model integrates a convolutional RBM (CRBM) with deep belief network (DBN), in which the CRBM is utilized for inherent hierarchical feature learning, while the DBN is employed for low-dimensional feature space mapping.

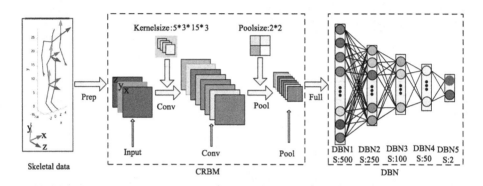

Fig. 2. The flowchart of the proposed hybrid deep neural network.

Within the proposed model, CRBM is a vital link between skeletal data and DBN for its great power in apperceiving local informations. Considering the spatio-temporal properties, the convolutional kernel size is set to be 5*3. That is, our model groups 5 frames as a block and shift the contiguous three joints adhering to the body structure. Besides, the pooling operation endues the model with some special superiorities, such as transformation invariance, denoising and irrelevant information removal. Specifically, a simplified version of the probabilistic max-pooling [7] operation is employed in our model.

Low-Level: Hierachical Real-Valued CRBM. In our model, the values of joint position are recorded in three-dimensional coordinates. Therefore, it is necessary to extend the two-dimensional convolution operation into three channels. Intuitively, the training of CRBM is similar to RBM, except for the weights $K_{c,d}$ that shared among different blocks in one 'Input' map, where subscript d corresponds to the indices of different kernels. Similar to RBM, the real-valued CRBM is also an energy-based model, and its energy function can be defined as:

$$E(V, H; \theta) = -\sum_c \sum_\alpha \sum_d \frac{V_\alpha^c}{\delta} \otimes K_{c,d} \otimes C_\alpha^d$$
$$- \sum_c \sum_\alpha \frac{(V_\alpha^c - b_c)^2}{2\delta^2} - \sum_d \sum_\alpha C_\alpha^d \cdot b_d \tag{1}$$

Physically, each part connecting 'Input' and 'Conv' layer can be treated as a tiny RBM, and the inference on the bottom-top direction is convolutional-like. Let C_α^d denote the α-th unit in d-th 'Conv' map, its value can be obtained by weighting all the blocks from the same position on different maps in 'Input' layer:

$$C_\alpha^d = \sum_{c=1}^{3} \frac{V_\alpha^c}{\delta} \otimes K_{c,d} + b_d \tag{2}$$

As an Gaussian-binary undirected graph model, each unit in hidden layers should contain two states, which are activated by random thresholds. In our work, the distribution of 'Conv' layer is calculated by dividing a sum of block related to the pooling size:

$$\tilde{C}_\alpha^c = P(C_\alpha^d = 1 | V_\alpha^c, \theta) = \frac{e^{C_\alpha^c}}{1 + \sum_{bd} e^{C_\alpha^c}} \tag{3}$$

where variable 'bd' is a 4-connected region in each 'Conv' map. The units in pooling layer won't be activated by random threshold for preserving the stabilization of our model. Subsequently, the pooling layer can be calculated by:

$$P_{xy} = \max \left(\tilde{C}_\alpha^d \right)_{2 \times 2} \tag{4}$$

the annotation $(\cdot)_{2 \times 2}$ refer to one pooling block in 'Conv' layer. To overcome the dimensional inconsistency, it is doable to fill the border of the maps in 'Conv' layer with zeros, and the bandwidth is set at 4 in our work. Different from positive phase, the visible layers can be sampled from a specified Gaussian distribution:

$$\tilde{V}_\alpha^c \sim N \left(\sum_d \tilde{C}_\alpha^d \otimes rot(K_{c,d}, 180°) + b_c, \delta^2 \right) \tag{5}$$

In practice, the original data always being normalized to have zero mean and unit variance for convenient inference. That is, the δ is set to be 1 in our work.

As the Eqs. 3 and 5 show the forward and backward inferences in Gibbs sampling method, the update of the biases and kernel can be defined as:

$$b_c = D_\alpha^c - \tilde{V}_c^{(k)}$$
$$b_d = C_{\alpha,d}^{\tilde{}} - \tilde{C}_\alpha^{d\,(k)} \tag{6}$$
$$K_{c,d} = D_\alpha^c \cdot C_{\alpha,d}^{\tilde{}} - \tilde{V}_\alpha^{c\,(k)} \cdot \tilde{C}_\alpha^{d\,(k)}$$

the annotation $\{\cdot\}^{(k)}$ means the k'th Gibbs sampling in contrastive divergence learning methods.

High-Level: Stacked DBN. Activating the hidden units into binary states by random thresholds may induce uncertainties during the forward inferences. Specifically, one same frame will take up two distinct positions in latent space while traversing the model twice. To eliminate the influence of random activation, we do not set the value of visible units as the binary hidden states in the former layer-wise RBM but the distribution of the hidden layer. In general, all units in pooling layer are fully connected to the first layer of DBN part. Each of the last four layers functions as not only binary hidden units in preceding RBM architecture but also the Gaussian visible inputs of the subsequent training. The training procedure also based on the contrastive divergence learning but for Gaussian-Bernoulli configuration [1].

3.2   Motion Graph Mining via Quadruple Structure

Motion graph is very helpful in satisfying specific paths, which consists of motion splitting, transition points detection and transition frames generation [4]. The principle of constructing motion graph is ensuring good connectedness while preserving efficient search for user constraints. To this end, we propose a quadruple data structure which is benefit to motion splitting and retrieving. Meanwhile, the proposed hybrid network aims at detecting the appropriate transition points.

Motion Splitting via Quadruple Structure. Differing from regular sampling [5] or clustering [14], we propose a quadruple-like structure to represent each node in motion graph, featuring on motion reusing and fast implementation:

$$Clip_{style} = <\text{RotAng}, \text{Start}, \text{End}, \text{Index}> \tag{7}$$

where 'RotAng' indicates the deflection which should exceed 10 degrees at the truncation frames, 'Start' and 'End' indicate the beginning and ending frame of each clip indexed from the whole motion sequence, and 'Index' records the sign of the parent motion sequence. To implement rotation on motion sequences, simple IK (Inverse Kinematics) and geometry are need to obtain a group of Euler angles which can be fast implemented by referring to the solution suggested in [11]. For details, the pseudocode for quadruple-like structure based motion splitting are summarized in Algorithm 1.

Appropriate Transition Point Detection. Motion transition often appears between two similar motion frame, and the appropriate transition points play an important role for new motion creation. Since our proposed hybrid deep model is more efficient to map the skeletal data into 2D latent spaces, the 'end to end' function for frame feature representation can be given by:

$$F_{x,y}^n = sigmoid(F_{x,y}^{n-1} \times W_l + b_l)$$

$$\text{w.r.t.} \quad F_{x,y}^0 = \left[\left(\sum_d D_{ij} \otimes K_{c,d} + b_d \right)_{mp} \right]_{fc} \tag{8}$$

Algorithm 1. Motion spilting via Quadruple Structure

Input: motion samples with style labels
Output: quadruple groups with split motions
1: **Note:**each group corresponds to one kind motion style
2: **Init:**Quad=cell(1, Num_ Of_ Styles);
3: **for** j index motion in database **do**
4:     $S_{toe} = smooth\,(O_{toe}, 30)$ (O_{toe} is the y-axis values of both feet)
5:     SamplePoint=findpeak($-S_{toe}$)
6:     SampleAngle=$Motion(SamplePoint)_{y_{Euler}}$
7:     CutPoint=find($diff(SampleAngle) > 5$)
8:     **for** i index CutPoint **do**
9:         Block(i)=Motion(CutPoint(i):CutPoint(i+1), :);
10:         **if** $Angle(i) < 10\&\ Angle(i+1) < 10$ **or**
11:             $Angle(i) > 10\&\ Angle(i+1) > 10$ **then**
12:             Clip(i)=Combine(Block(i), Block(i+1))
13:             $Quad\{label\}(i,1) = Clip(i,1)_y - Clip(i,end)_y$
14:             $Quad\{label\}(i,2) = Block(i,1)$
15:             $Quad\{label\}(i,3) = Block(i+1,end)$
16:             $Quad\{label\}(i,4) = j$
17:         **else**
18:             continue;
19: **return** Result

---

where 'mp' is max-pooling, and 'fc' represents the full connection between 'Pool' and 'DBN1' layer. Thus the similar frame can be measured by naive Euclidean metric:

$$Dis_{s,t} = \sum_{i=1}^{2} \parallel F_{s,i} - F_{t,i} \parallel^2 \qquad (9)$$

The mimimum won't be the best transition points for that the interpolated phase maybe opposed to the moving direction. It is reasonable to reset the distance $Dis_{s,t}$ to be infinite when this happened. This constraint ensures us to detect the most appropriate transition points by selected the minimum value in the updated distance matrix. Besides, the transition can also be computed like the traditional motion graphs [9].

3.3   Fulfilling User Demands

Similar to Algorithm 1, we treat the user input as combinations of moving straight and turning with different degrees. For the specified style while generating, we can turn to different groups indexed by 'Quad'. As the paths input from user interface are not irregular as the trajectories in captured motions, the deflections while moving are equal to the angles between x-axis and path segments. However, it should be noted that the lengths of this kind of path segments should not be too long for the needs of ensuring splitting the turns apart from the straights. A reasonable suggestion is to adopt the locomotive distance between less than 15 frame. The remaining operations are similar to Algorithm 1: splitting paths into small blocks and combining the continuous blocks into final clips. As a result, the clips are attached with turning degrees which can be searched in the 'Quad' spaces indexing the original motion samples. The boundary of the turning points of each movement can be noted by the x-value or z-value of each clip.

4   The Experimental Results

To evaluate the transition performance, three kind of motion styles (i.e., walk, run, jump) are picked from public available CMU dataset. Unlike discrimative deep learning models, the capabilities of the traning dataset for generative models is not so large for that the purpose of the learning procedure aims at estimating the probability distribution of the representative data. Following the size of the datasets adopted in RBM based methods [12,16], our training data is chosen from subject 2, 6, 7, 8, 9, 13, 14, 16, 37, 91, 104, 105, 111, 118, 127, 139, 141, 143, while the testing data is selected from subject 13, 16 and 35. The experiments are implemented on a desktop with a 3.30 GHz Intel Core and 8G RAM machine, implementing the coding language with Matlab 2015b.

4.1   Evaluation of Transition Quality

Without considering the flexible control of the motion path, we only evaluate the qualities of transitions between different motion styles. To this end, we compare our methods with both conventional blending methods [5] and deep learning based methods [13,16]. The linear blending is performed within two windows containing 40 frames which averagely come from two transition motions respectively. The latter two RBM based methods both generate transition frames by imposing Gaussian noises on the top hidden layer. Our methods only do simple spherical linear interpolation (Slerp) once detecting two similar frames through the hybrid CRBM-DBN model.

To analysis the naturality of the generated movements, we extract the transitional trajectories of the left toe and hand. In Fig. 3, the positions of the two end-effectors are annotated along the vertical axis while the horizontal axis shows the length of transition phase. As normal human movements always behave with slightly body shaking, both joint trajectories have noises with various degrees. Their amplitudes and frequencies are appropriate indicators for quality measurement. In view of the this, compared with the generative deep learning models, the linear blend and Slerp methods can create smoother transitions for that the noises induced by the former are of higher amplitudes. In addition to the drawbacks of intensive noises, the transitions in 'mhmublv' and 'TDBN' are out of good control, as we don't know what degree of Gaussian noises should be imposed on hidden units and the transitions may be randomly generated early or delayed. Secondly, the quality of blending within several aligned frames is problematic. If the aligned frame block is too small, the style change will be abrupt.

Relatively, the long time-series frame block will lead to foot slide, because the move speed can hardly keep pace with the blending frames. Thus, the transition length affects the blending quality greatly. This issue has also been surveyed by Wang et al. [17] who revealed that the transition length can ranges from 0.03 to 2 s where the optimal blend length is chosen by exhaustive method. In our framework, the blending length is controllable, the detected similar points can accelerate or decelerate the transitional speed. In Fig. 3, we can see our method (red line) create more gently transition than linear blending within aligned block containing 40 frames.

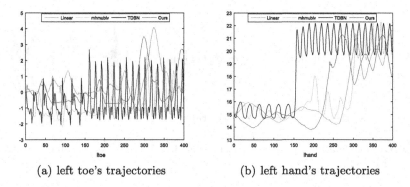

(a) left toe's trajectories   (b) left hand's trajectories

Fig. 3. Comparisons of the transition phases among conventional linear blending method, deep learning generative models and our hybrid framework. The length of the transition phase includes 400 frames and the tranjectories of the left toe and hand are both projected onto the xz-plane. The green line is drwan by the linear blending technique referring to the motion graph [5]. The blue line labelled as 'mhmublv' is generated by the conditional RBM method [16]. The black line labelled as 'TDBN' is relevant to dynamical DBN [13]. (Color figure online)

4.2   Interactive Path Synthesis with Transition

In practice, the motion transition may meet different moving path in an interactive environment. In this case, we treat the path constraints as a searching problem of turning angles in original database. Specifically, the quadruple structure matches the user specifications through the drift angles while moving, and three kinds of user inputs, linear path, complicated curves and interactive control, can be realized by our approach. As a natural operation, it is reasonable to interpolate a large number of frames (e.g., 100) between the transition points and segment them into 5 parts equally. Then, according to the transition length, down sampling each part can generate satisfactory performance.

Linear path: To clearly demonstrate that our model permits smooth transitions between arbitrary styles, we firstly picked up three kinds of movements performed by '16_22', '16_35', '16_09' corresponding to jump, walk, run along with the z-axis which are shown in Fig. 4. And then two typical transition examples between several pairs of graph nodes were shown in Fig. 5. They mainly contain transitions between walk, run and jump styles. Both simple transitions reveal that our transition system can generate compelling results with periodic motions and arbitrary length, and also could transit one motion style to another. It can be seens that the transitions between different styles are natural enough to well satisfy the visual comfort. In addition, the transition results obtained by our model can well avoid foot-slide and simultaneously prevent abrupt changes while transiting between different motion styles.

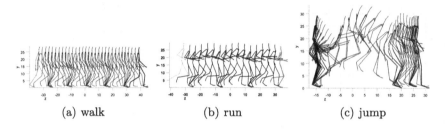

(a) walk   (b) run   (c) jump

Fig. 4. Original motion samples performed by '16_22. amc', '16_35. amc' and '16_09. amc' respectively.

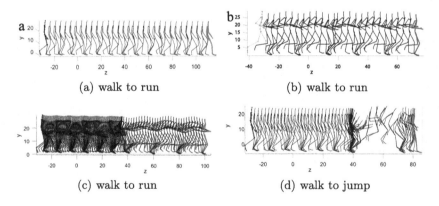

(a) walk to run   (b) walk to run

(c) walk to run   (d) walk to jump

Fig. 5. Transition performance associated with simple motion and linear path.

Complicated curves: Complex transitions consist of many simple clips, and this makes it difficult to produce smooth frames between piece-pairs while satisfying complex user inputs. In Fig. 6(a), we have extracted the trajectories of the original motion samples performed by subject 16 in three different styles. User inputs are matched with both trajectories and styles through the quadruple structure introduced in Sect. 3.2. Then, in Fig. 6(b) (d), we simulate two distinct user inputs by red line and the matched motion trajectories are displayed in blue lines. The final synthesis of the skeleton movements are shown in Fig. 6(c)

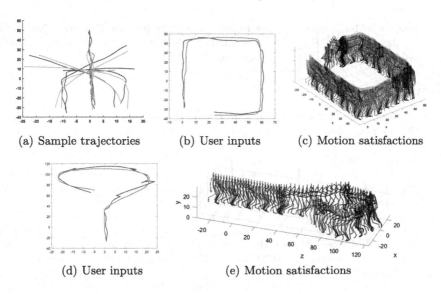

(a) Sample trajectories   (b) User inputs   (c) Motion satisfactions

(d) User inputs   (e) Motion satisfactions

Fig. 6. Complex transition performances. (a) all sample trajectories; (b) walking along a square paths and jumping at the end; (c) transition of b path; (d) path with three turning behaviors and a transition from walk to run; (d) transition of d path.

(e) respectively. Fortunately, it can found in the experiments that our proposed transition approach is able to synthesize different motion style in a user specified path, and transition motions is really human-like and natural physically. Therefore, the proposed transition framework is well suitable to the specific problem of generating different styles of locomotion along arbitrary paths.

5 Conclusion

In this paper, we have presented a flexible framework that allows efficient human motion transition from motion capture data. Within the proposed approach, a hybrid deep neural network is presented to extract the inherent skeletal feature and map the high-dimensional data into an efficient 2D latent space, through which the transition points between diverse motion styles can be well detected. Meanwhile, a quadruples-like data structure is exploited for motion graph building, which significantly benefits for the motion splitting and indexing. Matching user constraints through the move deflection can get better flexibility and efficiency than calculate the distance between trajectories directly. As a result, different kinds of diverse motions can be well transited with different user input, while preserving the smoothness and quality of the original data. The experiments have shown its outstanding performance.

Acknowledgment. The work described in this paper was supported by the National Science Foundation of China (Nos. 61673185, 61572205, 61673186), National Science Foundation of Fujian Province (Nos. 2015J01656, 2017J01112), Promotion Program for Young and Middle-aged Teacher in Science and Technology Research (No. ZQN-PY309), the Promotion Program for graduate student in Scientific research and innovation ability of Huaqiao University (No. 1511414012).

References

1. Cho, K.H., Ilin, A., Raiko, T.: Improved learning of gaussian-bernoulli restricted boltzmann machines. In: Honkela, T., Duch, W., Girolami, M., Kaski, S. (eds.) ICANN 2011. LNCS, vol. 6791, pp. 10–17. Springer, Heidelberg (2011). https://doi.org/10.1007/978-3-642-21735-7_2
2. Gan, Z., Li, C., Henao, R., Carlson, D.E., Carin, L.: Deep temporal sigmoid belief networks for sequence modeling. In: Proceedings of Advances in Neural Information Processing Systems, pp. 2467–2475 (2015)
3. Holden, D., Saito, J., Komura, T.: A deep learning framework for character motion synthesis and editing. ACM Trans. Graph. (TOG) $35(4)$, 138:1–138:11 (2016)
4. Kovar, L., Gleicher, M.: Flexible automatic motion blending with registration curves. In: Proceedings of ACM SIGGRAPH/Eurographics Symposium on Computer Animation, pp. 214–224 (2003)
5. Kovar, L., Gleicher, M., Pighin, F.: Motion graphs. ACM Trans. Graph. (TOG) 21, 473–482 (2002)
6. Lai, Y.C., Chenney, S., Fan, S.: Group motion graphs. In: Proceedings of ACM SIGGRAPH/Eurographics Symposium on Computer Animation, pp. 281–290 (2005)
7. Lee, H., Grosse, R., Ranganath, R., Ng, A.Y.: Convolutional deep belief networks for scalable unsupervised learning of hierarchical representations. In: Proceedings of 26th Annual International Conference on Machine Learning, pp. 609–616 (2009)
8. Peng, J.Y., Lin, I., Chao, J.H., Chen, Y.J., Juang, G.H., et al.: Interactive and flexible motion transition. Comput. Animat. Virtual Worlds $18(4–5)$, 549–558 (2007)
9. Rose, C., Cohen, M.F., Bodenheimer, B.: Verbs and adverbs: multidimensional motion interpolation. IEEE Comput. Graph. Appl. $18(5)$, 32–40 (1998)
10. Saida, K., Foudil, C.: Hybrid motion graphs for character animation. Int. J. Adv. Comput. Sci. Appl. $1(7)$, 45–51 (2016)
11. Slabaugh, G.G.: Computing euler angles from a rotation matrix. Retrieved on August $6(2000)$, 39–63 (1999)
12. Song, J., Gan, Z., Carin, L.: Factored temporal sigmoid belief networks for sequence learning. arXiv preprint arXiv:1605.06715 (2016)
13. Sukhbaatar, S., Makino, T., Aihara, K., Chikayama, T.: Robust generation of dynamical patterns in human motion by a deep belief nets. In: Proceedings of Asian Conference on Machine Learning, pp. 231–246 (2011)
14. Tanco, L.M., Hilton, A.: Realistic synthesis of novel human movements from a database of motion capture examples. In: Proceedings of International Workshop on Human Motion, pp. 137–142 (2000)
15. Taylor, G.W., Hinton, G.E.: Factored conditional restricted Boltzmann machines for modeling motion style. In: Proceedings of 26th Annual International Conference on Machine Learning, pp. 1025–1032 (2009)
16. Taylor, G.W., Hinton, G.E., Roweis, S.T.: Modeling human motion using binary latent variables. In: Proceedings of Advances in Neural Information Processing Systems, pp. 1345–1352 (2007)
17. Wang, J., Bodenheimer, B.: Synthesis and evaluation of linear motion transitions. ACM Trans. Graph. (TOG) $27(1)$, 1–12 (2008)
18. Yu, Q., Terzopoulos, D.: Synthetic motion capture: implementing an interactive virtual marine world. Vis. Comput. $15(7)$, 377–394 (1999)

End-to-End View-Aware Vehicle Classification via Progressive CNN Learning

Jiawei Cao, Wenzhong Wang, Xiao Wang, Chenglong Li, and Jin Tang[✉]

School of Computer Science and Technology,
Anhui University, No. 111 Jiulong Road, Hefei 230601, China
ahujwcao@foxmail.com, wenzhong@ahu.edu.cn, wangxiaocvpr@foxmail.com,
lcl1314@foxmail.com, jtang99029@foxmail.com

Abstract. This paper investigates how to perform robust vehicle classification in unconstrained environments, in which appearance of vehicles changes dramatically across different angles and the numbers of viewpoint images are not balanced among different car models. We propose a end-to-end progressive learning framework, which allows the network architecture is reconfigurable, for view-aware vehicle classification. In particular, the proposed network architecture consists of two parts: a general end-to-end progressive CNN architecture for coarse-to-fine or top-down fine-grained recognition task and an end-to-end view-aware vehicle classification framework to combine vehicle classification and viewpoints recognition. We test the technique on a large-scale car dataset, "CompCars", and experimental results show that our framework can significantly improve performance of vehicle classification.

Keywords: Deep learning · Vehicle classification
Viewpoint recognition · Convolutional neural networks

1 Introduction

Vehicle classification is one of the most important tasks for video surveillance, especially for intelligent transportation system. It has a wide range of applications for our daily life. For example, the highway toll system can charge by the vehicle types. For another example, effective vehicle classification can play a very important supplementary role in vehicle retrieval. But it's a challenging task for fine-grained vehicle classification because of viewpoint variation: cars are 3D objects whose appearances change dramatically across different angles.

Recently, Duan et al. [4] propose a method based on K-means clustering for estimate vehicle viewpoints in vehicle recognition. The authors of [2,3,17] all present that visual based vehicle type classification task should be studied from frontal view and side view. But the only two viewpoints are not enough and almost all of current methods cannot play well in practical traffic scenes because of various views as seen in Fig. 1. In fact, if our designed vehicle classification

© Springer Nature Singapore Pte Ltd. 2017
J. Yang et al. (Eds.): CCCV 2017, Part I, CCIS 771, pp. 729–737, 2017.
https://doi.org/10.1007/978-981-10-7299-4_61

Fig. 1. The vehicles in practical scenes have various viewpoints.

algorithm can be truly applied into the intelligent transportation system, we should deal with the different views together with only a united method.

To handle above issues, we propose an expressive method to address the above mentioned issues for vehicle classification in the practical complex scenes. Our method mainly contains the following two parts: a progressive CNN Architecture and an end-to-end view-aware vehicle classification framework. At the first step, we develop a progressive CNN Architecture for learning the two-level hierarchical and multi-task CNN that can be used for a coarse-to-fine or top-down recognition task. At the second step, we propose an end-to-end view-aware vehicle classification framework for combining vehicle classification with viewpoints recognition. In our hierarchical framework, we first perform viewpoints recognition in our base network and then the data will be passed into corresponding sub-networks according to the viewpoint. Then the sub-networks will classify the categories of cars. The base network and sub-networks are two configurable components in our framework. We deploy different state-of-the-art convolutional neural networks of image classification as two components and the experimental results show that our framework can significantly improve performance of vehicle classification in various viewpoints.

In summary, this paper has the following contributions. First, we introduce a general progressive CNN architecture for coarse-to-fine or top-down recognition task. Second, we develop a scheme to achieve end-to-end training and testing for above progressive CNN architecture through setting run level for layers in CNN. Third, we propose an end-to-end view-aware vehicle classification framework. We have performed evaluations on the large-scale car dataset CompCars, and our method has achieved state-of-the-art performance.

2   Related Work

Vehicle classification has achieved great development in recent years and many algorithms has been proposed for vehicle classification. These algorithms can be roughly fall into two categories: *image-based methods* and *video-based methods*. Image-based methods directly process with the handcrafted or detected vehicle images. Video-based methods will consider the corresponding relationships between the frames. The authors of [8,11,16] both use video information for vehicle type classification task. But they don't take full advantage of the corresponding relationships between the frames well. There exists several types of imaged-based methods, namely, model reconstruction method, feature representation method and convolutional neural network. Model reconstruction methods [8,14,20] estimate 3D measurement parameters for the size of a vehicle to reconstruct a three dimensional model of the vehicle. Feature representation approaches [2,4,5,15,17] extract features to represent the vehicle appearance for subsequent classification. CNN-based models [7,12,19] hold state-of-the-art performance in various computer vision tasks and the authors of [3,10,18,21] utilize CNN instead of the classical methods which depend on hand-crafted features and achieved new state-of-the-art performance in classification problem.

As for Vehicle viewpoint recognition, Chen and Lu [1] propose a discriminative viewpoint-specific component model to estimate viewpoint. Ricardo *et al.* [6] combine vehicle detection with viewpoint estimation based on the motion model. Little work has explored combining vehicle classification with viewpoints recognition and most of methods about viewpoint recognition are based on unsupervised clustering. The estimated viewpoints are rough with not high enough accuracy.

3   The Proposed Approach

Our proposed approach for vehicle classification and viewpoints recognition contains two parts: a progressive CNN architecture and an end-to-end view-aware vehicle classification framework for combining vehicle classification and viewpoints recognition. First, we develop a general and progressive CNN architecture. The general architecture is two-level and hierarchical for progressive recognition tasks. Then we design a scheme to achieve end-to-end training and testing. At last, we propose a model for end-to-end view-aware vehicle classification. We will introduce them in the following subsections, respectively.

3.1   Progressive CNN Architecture

We develop a general scheme for learning the two-level hierarchical CNN as illustrated in Fig. 2. It's designed for progressive recognition tasks. There is a top category CNN classifier for task A with K categories in base network. The next are K sub-networks for task B corresponding to the K top categories in task A. Thus, it's a two-level and hierarchical architecture. The general architecture can apply in many object recognition tasks. For example, we can use the architecture for fine-grained classification. We can take task A in base network as a coarse classifier to separate easy classes from one another. More challenging classes are routed downstream to fine category classifiers for task B in sub-networks that focus on confusing classes.

The architecture should be end-to-end in training and test phrases. Thus, we need a CNN architecture where data can flow dynamically and some layers in neural network will switch activation and deactivation accordingly. Caffe [9] is a deep learning framework made with expression, speed, and modularity in mind. Lots of researchers and engineers have made Caffe models for different tasks with all kinds of architectures and data. Thus we choose to realize our architectures based on Caffe. But Layers of CNN in the Caffe architecture can't play a selective role dynamically. Hence, if we deploy some branch networks, the data will flow through each sub-network. At the same time, the backward propagation will work in each sub-network. We need the neural network could automatically decide whether to be activated or inactive according to the classification results in base network.

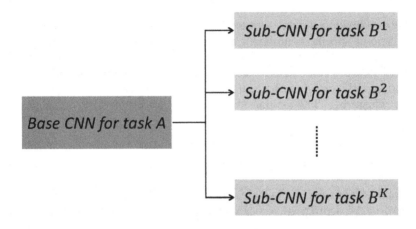

Fig. 2. The architecture of two-level hierarchical CNN.

For solving this problem, we design a novel scheme that we set different run levels for layers in CNN based on Caffe. We set run level 0 for all layers in basic network and the layers with run level 0 will always execute forward and backward propagation along with flowing data. Then we set bigger run level such as 1, 2, 3 for corresponding sub-networks and the run levels of layers in same sub-network are identical. In our new architecture, the layers whose run level isn't 0 will execute forward and backward propagation selectively according to the output of classifier in basic network. For example, in test phase, if the classifier in basic network predict that the input car image belongs to the third top category, then all layers in third sub-network (the run level is 3) will transfer data and execute forward propagation. At the same time, layers in the others sub-networks (the run level is not 3 or 0) will be closed so these layers can't perform any actions or participate in the calculation.

It is also worthy to note that the whole architecture is a general pipeline and could be trained and tested in the end to end fashion.

3.2   End-to-End View-Aware Vehicle Classification

Overall Architecture. We design an end-to-end view-aware vehicle classification CNN which combines vehicle classification with viewpoints recognition based on above architecture. As illustrated in Fig. 3, the overall architecture consists of two components: a basic network for viewpoints recognition and a series of sub-networks for vehicle classification. The two components are independent and configurable.

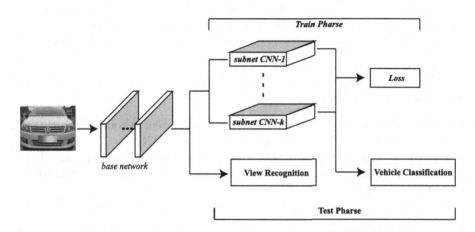

Fig. 3. Our progressive CNN framework.

Basic Network for Viewpoints Recognition. In training phrase, the data layers in this network take car images and labels as input. The labels in this network are annotations of viewpoints. The run level of this network is zero, hence it will be activated all the time. In another words, we will perfom viewpoints recognition for all car images. At last, there is a softmax layers which is not only a view classifier but also a selection switch to active corresponding branch network in sub-networks. We will switch corresponding sub-network in train phrase according to groundtruth of viewpoints. While testing, the network will active corresponding run level based on the output of view classifier. We choose the state-of-the-art CNN ResNet [7] in image classification task to build the base network.

Sub-networks for Vehicle Classification. There are a series of parallel sub-networks for vehicle classification following on the basic network. There is only one sub-network will be activated at the same time. We deploy the popular neural networks, including AlexNet, GoogleNet and ResNet-50 as sub-networks respectively. Thus, there are three CNN models and the sub-networks have same architecture in each model.

Training. The train phrase is end-to-end and we use stochastic gradient descent algorithm to train our network. The training images within the stochastic gradient descent minibatch are probabilistically routed to different sub-networks because of different viewpoints, we set minibatch size 1 and don't use batch gradient descent. The loss function we use is shown below.

$$Loss = \alpha Loss1 + \beta Loss2, \tag{1}$$

where α, β are tuning parameters. *Loss1* is the cross-entropy loss of viewpoint classifier in basic network's softmax layer and *Loss2* is the cross-entropy loss of cars classifier in each sub-network. In our network, we set α 0.3 and β 0.7.

4   Experiments

4.1   Experiment Settings

To implement our view-aware vehicle recognition system, a computer with Xeon E5-2620 v4 CPU, 64 GB memory and 8 GB memory GTX1080 GPU is employed. The program runs on a 64-bit Ubuntu LTS 14.04 operating system with CUDA-8.0, Python 2.7.6 installed. In train phrase, we set learning rate 0.0001 for the three CNN models.

Dataset Selection. We choose a large-scale public dataset CompCars [13] as our experimental data. In addition, we select 100 car models and take 61477 images as train set and 9808 images as test set. As for viewpoints of cars, we use five viewpoints, including front, rear, side, front-side and rear-side. The quantity distribution of the car images in different viewpoints is shown in Table 1.

Table 1. Quantity distribution of the car images of different viewpoints.

Viewpoint	Total images
Front	2616
Rear	2000
Side	1547
Front-side	1888
Rear-side	1757

4.2   Experimental Results

Comparison with State-of-the-art Methods. To compare our framework with state-of-the-art classification CNN models, we first train AlexNet, GoogleNet and ResNet-50 CNN models with car images in all viewpoints respectively. In these CNN models, a single category of the vehicle brand will contain different viewpoints. There are total 100 car categories according to car brand. Then we deploy basic network of viewpoints recognition with ResNet-50 and the final testing accuracy of viewpoints recognition is 0.963. For the five sub-networks in different views, the AlexNet, GoogleNet and ResNet-50 CNN models are deployed respectively compared to above models. The performances of these models are summarized in Table 2. As can be seen from the Table 2, with decomposing the classification task in all views into several sub-tasks in a series of viewpoints, our CNN models significantly outperform the baseline models.

Table 2. Vehicles classification performance.

Models	Top-1 error	Compared to baseline
AlexNet	40.15%	-
GoogleNet	10.21%	-
ResNet-50	7.81%	-
Ours based on AlexNet	16.42%	-23.73%
Ours based on GoogleNet	6.89%	-3.32%
Ours based on ResNet-50	6.35%	-1.46%

Comparison over the Components. We analyze vehicles classification performance in different views using AlexNet and our CNN model based on AlexNet. As illustrated in Table 3, we can find the recognition accuracy in rear view is obviously lower than other views because of the different appearance variation in different views. For example, many car models are very similar in their side views but rather different in the front views as shown in Fig. 4. When we divide images into different views and perform car classification individually in our progressive architecture, the inter-class difference reduces and classification performance improves significantly in each viewpoints.

Table 3. Vehicles classification performance in different views.

Viewpoint	Alexnet top-1 error	Our method top-1 error
Front	29.59%	8.83%
Rear	39.2%	9.65%
Side	50.1%	26.96%
Front-side	42.32%	20.29%
Rear-side	45.87%	20.72%

Fig. 4. The two car models are more easily distinguished in their front views than side views.

5   Conclusion

In this paper, we proposed a general end-to-end deep neural networks for fine-grained vehicle classification. Aiming at jointly training the viewpoint classification and brand recognition issue, we deploy the base-network and following sub-CNNs for these two targets, respectively. Our neural network could not only utilized in vehicle classification, but also some other tasks. Experiments on the public benchmark validated the effectiveness of the proposed network compared with existing popular classification networks.

Acknowledgement. This work was supported in part by the National Natural Science Foundation of China under Grant 61671018 and Grant 61472002, in part by the Natural Science Foundation of Anhui Higher Education Institution of China under Grant KJ2017A017 and in part by the Co-Innovation Center for Information Supply & Assurance Technology of Anhui University under Grant Y01002449.

References

1. Chen, T., Lu, S.: Robust vehicle detection and viewpoint estimation with soft discriminative mixture model. IEEE Trans. Circuits Syst. Video Technol. **27**(2), 394–403 (2017)
2. Dong, Z., Jia, Y.: Vehicle type classification using distributions of structural and appearance-based features. In: 2013 20th IEEE International Conference on Image Processing (ICIP), pp. 4321–4324. IEEE (2013)
3. Dong, Z., Wu, Y., Pei, M., Jia, Y.: Vehicle type classification using a semisupervised convolutional neural network. IEEE Trans. Intell. Transp. Syst. **16**(4), 2247–2256 (2015)
4. Duan, K., Marchesotti, L., Crandall, D.J.: Attribute-based vehicle recognition using viewpoint-aware multiple instance svms. In: 2014 IEEE Winter Conference on Applications of Computer Vision (WACV), pp. 333–338. IEEE (2014)

5. Fu, H., Ma, H., Liu, Y., Lu, D.: A vehicle classification system based on hierarchical multi-svms in crowded traffic scenes. Neurocomputing **211**, 182–190 (2016)
6. Guerrero-Gómez-Olmedo, R., López-Sastre, R.J., Maldonado-Bascón, S., Fernández-Caballero, A.: Vehicle tracking by simultaneous detection and viewpoint estimation. In: Ferrández Vicente, J.M., Álvarez Sánchez, J.R., de la Paz López, F., Toledo Moreo, F.J. (eds.) IWINAC 2013. LNCS, vol. 7931, pp. 306–316. Springer, Heidelberg (2013). https://doi.org/10.1007/978-3-642-38622-0_32
7. He, K., Zhang, X., Ren, S., Sun, J.: Deep residual learning for image recognition. In: IEEE CVPR, pp. 770–778 (2016)
8. Hsieh, J.W., Yu, S.H., Chen, Y.S., Hu, W.F.: Automatic traffic surveillance system for vehicle tracking and classification. IEEE Trans. Intell. Transp. Syst. **7**(2), 175–187 (2006)
9. Jia, Y., Shelhamer, E., Donahue, J., Karayev, S., Long, J., Girshick, R., Guadarrama, S., Darrell, T.: Caffe: Convolutional architecture for fast feature embedding. In: Proceedings of the 22nd ACM international conference on Multimedia, pp. 675–678. ACM (2014)
10. Jiang, C., Zhang, B.: Weakly-supervised vehicle detection and classification by convolutional neural network. In: International Congress on Image and Signal Processing, BioMedical Engineering and Informatics (CISP-BMEI), pp. 570–575. IEEE (2016)
11. Jiang, L., Zhuo, L., Long, H., Hu, X., Zhang, J.: Vehicle classification for traffic surveillance videos based on spatial location information and sparse representation-based classifier. In: 2016 International Conference on Progress in Informatics and Computing (PIC), pp. 279–284. IEEE (2016)
12. Krizhevsky, A., Sutskever, I., Hinton, G.E.: Imagenet classification with deep convolutional neural networks. In: Advances in Neural Information Processing Systems, pp. 1097–1105 (2012)
13. Yang, L., Luo, P., Loy, C.C., Tang, X.: A large-scale car dataset for fine-grained categorization and verification. In: IEEE CVPR, pp. 3973–3981 (2015)
14. Ma, X., Grimson, W.E.L.: Edge-based rich representation for vehicle classification. In: Tenth IEEE International Conference on Computer Vision, ICCV 2005, vol. 2, pp. 1185–1192. IEEE (2005)
15. Manzoor, M.A., Morgan, Y.: Vehicle make and model classification system using bag of sift features. In: 2017 IEEE 7th Annual Computing and Communication Workshop and Conference (CCWC), pp. 1–5. IEEE (2017)
16. Mei, X., Ling, H.: Robust visual tracking and vehicle classification via sparse representation. IEEE TPAMI **33**(11), 2259–2272 (2011)
17. Peng, Y., Jin, J.S., Luo, S., Xu, M., Cui, Y.: Vehicle type classification using pca with self-clustering. In: 2012 IEEE International Conference on Multimedia and Expo Workshops (ICMEW), pp. 384–389. IEEE (2012)
18. Yu, S., Wu, Y., Li, W., Song, Z., Zeng, W.: A model for fine-grained vehicle classification based on deep learning. Neurocomputing (2017)
19. Szegedy, C., Liu, W., Jia, Y., Sermanet, P., Reed, S., Anguelov, D., Erhan, D., Vanhoucke, V., Rabinovich, A.: Going deeper with convolutions. In: IEEE CVPR, pp. 1–9 (2015)
20. Zhang, Z., Tan, T., Huang, K., Wang, Y.: Three-dimensional deformable-model-based localization and recognition of road vehicles. IEEE Trans. Image Process. **21**(1), 1–13 (2012)
21. Zhou, Y., Cheung, N.M.: Vehicle classification using transferable deep neural network features. arXiv preprint arXiv:1601.01145 (2016)

A Fast Method for Scene Text Detection

Qing Fang$^{(\boxtimes)}$, Yanping Yang, Yali Chen, and Xiaoyu Yao

School of Electronic Engineering, University of Electronic Science
and Technology of China, Chengdu 611731, China
`fountain_uestc@outlook.com, YYanping94@outlook.com,`
`scharlie92@outlook.com, yaoxiaoyulzh@outlook.com`

Abstract. Text detection is important for many applications such as text retrieval, blind guidance, and industrial automation. Meanwhile, text detection is a challenging task due to the complexity of the background and the diversity of the font, size and color of the text. In recent years, deep learning achieves good results in image classification and detection, and provides us a new method for text detection. In this paper, a deep learning based detection method – Single Shot MultiBox Detector (SSD) is adopted. But SSD is a general object detection method, not specific for text detection and is not fast enough. Our method aims to develop a network for text detection, improve the speed and reduce the model. Therefore, we design a feature extraction network with the inception module and an additional deconvolution layer. The experiment on benchmark – ICDAR2013 demonstrates that our method is faster than other SSD-based method comparable results.

Keywords: Scene text detection · Deep network

1 Introduction

In computer vision applications, text detection from complicated backgrounds is very useful. Text detection has become a hot topic in computer vision and document analysis. However, there are still some challenges, such as noise, blur, distortion, occlusion and deformation.

The existing text detection methods are usually based on texture or component information, including Stroke Width Transform (SWT) [3] and Maximal Stable Extremal Regions (MSER) [16]. These methods are based on hand-crafted features which can not describe the semantic information well. Compared with hand-crafted features, deep learning superior performance compared with these features.

The object detection methods can be divided into two categories: region-based methods and regression-based methods. For example, region-based methods such as Faster-RCNN [18] and R-FCN [1], and regression-based methods such as SSD [14] have obtained excellent detection performance. Among these methods, SSD outperforms state-of-the-art approaches, which provides an efficient framework for text detection. However, directly using SSD to solving text

© Springer Nature Singapore Pte Ltd. 2017
J. Yang et al. (Eds.): CCCV 2017, Part I, CCIS 771, pp. 738–747, 2017.
https://doi.org/10.1007/978-981-10-7299-4_62

detection has some problems, such as the disadvantage of text feature extraction, the diversity of text aspect ratios and the unsatisfactory efficiency.

To handle these problems, we employ a model similar to GoogLeNet [20] with various specific filters and a additional deconvolution layer instead of VGGNet [19] to extract features in SSD. In this paper, we propose a fast method for text detection based on SSD. The structure of our deep network is shown in Fig. 1. Our main contribution is the replace of the feature extraction network in SSD with a designed tiny model, which can improve the speed and reduce the model. We use the inception modules with specific filters for text detection in our feature extraction network. In order to decrease the accuracy drop of reduced model, we add a deconvolution layer to concatenate multi-scale information. The experimental results prove that our method is faster in running speed with comparable results.

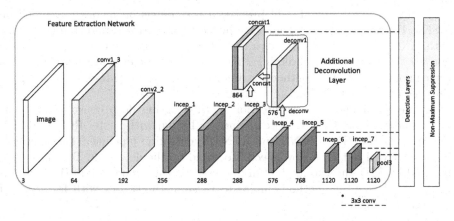

Fig. 1. Architecture of our model. Here, incep indicates inception. There are 7 inception modules in our model. A deconvolution layer is added after the $4th$ inception module. Then it is concatenated with an earlier layer. Four of the convolutional layers, including concat1, incep_5, incep_7 and pool3, are connected to detection layers. Then a detection layer predicts text scores and location offsets for 6 default boxes on every map location.

The rest of this paper is organized as follows: In Sect. 2, we briefly review the related studies. In Sect. 3, we describe the proposed method in detail, including the structure of our tiny model and additional deconvolution layer. Experimental results are presented and discussed in Sect. 4. Finally, conclusion and future work are given in Sect. 5.

2 Related Work

Deep Networks. Compared with traditional pattern recognition methods, deep neural network shows its superiority in feature extraction, classification and detection. Several well-known models such as AlexNet [12], VGG Net [19],

GoogLeNet [20] and ResNet [7] have been proposed and were widely used in many applications. In our work, we adopt the inception module of GoogLeNet by the trade-off of accuracy and computation burden.

Object Detection. Object detection methods can be divided into two main categories: region-based methods and regression-based methods. The former includes Fast-RCNN [5], Faster-RCNN [18] and R-FCN [1]. These kinds of methods obtain larger accuracy with low detection speed. The latter includes YOLO [17] and SSD [14], which purchases real-time detection but results in lower and acceptable accuracy. Our detection method is inspired by SSD by considering the fast detection speed.

Text Detection. There are several texture or component based text detection method such as SWT, MSER and SFT [8]. Yao et al. [22] propose an SWT-based algorithm. Neumann et al. [16] propose a text detection algorithms based on MSER. Huang et al. [9] propose a method combines MSER and sliding-window with CNN. However, features extracted from these conventional methods are not very satisfying. By observing that deep CNN models are more powerful than manually selected methods in image feature representation, we design a tiny deep network to extract text features.

There are a few of text detection methods using deep convolutional network. For example, CTPN [21] combines LSTM [4] with Faster-RCNN. RRPN [15] is inspired by Fast-RCNN and RPN [18] detection pipeline. These methods implement well but with slow speed. To speed up the detection, TextBoxes [13] inspired by SSD and the method in [6] based on the framework of [17] are proposed. we design our method similar to these methods, but with faster speed and improved features.

3   Proposed Method

In this section, we detail the proposed method. First, a modified tiny model is introduced to extract text features. Then, a deconvolution layer is applied on multiple maps to increase feature map resolution before detection. Finally, multi-scale features are fed into the SSD detection network for the text detection.

3.1   Feature Extraction Network

The architecture of inception is depicted in Fig. 2. We use the inception module of GoogLeNet to replace VGGNet as the base network. Network with inception module decreases the size of convolution kernel, which can increase the nonlinear expression of network and reduce the number of parameters to suppress the overfitting. At the same time, the combination of different sizes of filters is taken into account to provide more multi-scale information. Filters of different size are applied in inception modules, including 1×1, 3×3, and 7×1 which is specially designed for text. We pre-train this feature extraction network on ImageNet [2] database, and then use our own dataset to fine-tune it.

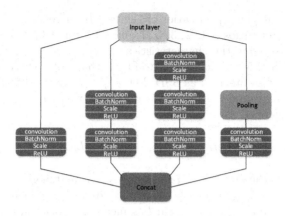

Fig. 2. Structure of the inception modules.

3.2 Additional Deconvolution Layer

In order to improve the performance of feature extraction, we add a deconvolution layer after the $4th$ inception module. Then we concatenate earlier feature maps and the deconvolution layer to integrating information from multiple maps. After weighing the time cost and accuracy improvement, we apply only one deconvolution layer. Extra layer is added before the first detection branch to enhance lower layer feature of text. Considering that lower layer contains more texture features, it is useful for small object detection. Deconvolution layer is added after the $4th$ inception module. Then feature maps of earlier module are concatenated with deconvolution output maps. We use the kernel with size of 2, stride of 2 to enlarge the feature maps.

3.3 Detection Module

After extracting text features, we next feed feature maps into the text detector. Our detector is similar to TextBoxes [13] that uses a text detector based on SSD. Detection layers simultaneously predict the bounding boxes and confidence of text. At every location on feature maps of detection branches, it outputs the offsets to correlative default boxes and the score of the text presence. On a location (i, j), the detection layer predicts the values of (dx, dy, dw, dh) to the corresponding default box $b_0 = (x_a, y_a, w_a, h_a)$. The location of predicted box $b = (x, y, w, h)$ is calculated by following formula:

$$
\begin{aligned}
x &= x_a + w_a * dx \\
y &= y_a + h_a * dy \\
w &= w_a * e^{dw} \\
h &= h_a * e^{dh}
\end{aligned}
\tag{1}
$$

Due to the different aspect ratios of the text from general objects, we adopt six designed aspect ratios – 1, 2, 3, 5, 7 and 10 – for default boxes same as in TextBoxes. Considering the characteristics of horizontal text lines, we design a kernel with width of 5 and height of 1 to fit the shape of text. At training phase, we use the same loss function as SSD. Let $x_{i,j}^p = 1$ means that the i-th default box matches j-th ground truth box of category p, while $x_{i,j}^p = 0$ otherwise. The loss function is defined as:

$$L(x, c, l, g) = \frac{1}{N}(L_{conf}(x, c) + \alpha L_{loc}(x, l, g)) \tag{2}$$

where N presents the number of default boxes that match ground truth boxes, c is the confidence, L_{conf} means the confidence loss, l is the location of predicted box, g is the location of ground truth, L_{loc} means the location loss, α is a weight coefficient between L_{conf} and L_{loc}.

4   Experiments

We pre-train the feature extraction tiny model on ImageNet database and evaluate the efficiency of our detection method on standard benchmark: ICDAR 2013.

4.1   Datasets

ICDAR2013–Focused Scene Text Dataset [11]. This dataset is introduced for Focused Scene Text Competition, which includes images containing obvious interest of text content. It's a standard dataset for text detection. Most texts of images in ICDAR2013 are horizontal and are with multiple aspects and sizes. There are 229 images for training and 233 images for testing.

HUST–TR400 [23]. This database is introduced by HUST. Some images are shot in different cities in America by several people using variant devices. Some are collected from Flickr website. And rest are from another dataset – MSRA-TD500. It's a challenging dataset as it contains images in various scenarios with different colors, fonts, sizes and orientations. We eliminate the images with incline angle bigger than 0.2 from 400 images, because we aim to detect horizontal or near horizontal text. There are 337 images remained for training.

Street View Text (SVT) [10]. It's a famous benchmark for text detection. Because of the low resolution and poor label of the street view images, it is a challenging dataset. In addition, we relabel 350 images by ourselves for training.

So far, we already have more than 8 hundred training images. But it's still too few for training. Thus we collect more image samples from the Internet. Images in our dataset are mainly pictures of medicine packages. There are 2000 images labeled for training. Some examples of our training dataset are shown in Fig. 3.

Fig. 3. The first row shows the examples of ICDAR 2013 dataset, the second row shows examples of HUST-TR400 dataset, the third row shows examples of SVT, last line is examples of our own dataset.

4.2   Implementation Details

In the proposed method, a tiny model with inception module is designed to extract text features. Input images are 300×300 and are cropped into 224×224 while training. This model is trained 8×10^5 iterations on ImageNet database. Learning rates start from 0.1, and multiplied by 0.1 after every 2×10^5 iterations. Momentum is 0.9, and weight decay is 0.0005. Batch normalization is used to accelerate convergence in our model.

Aiming at the specific detection problem, we use the text dataset to fine-tune the pre-trained network. There are 2916 images resized to 448×448 for training. For detection layers, minimum size of default boxes are set to 35, 75, 155 and 235, maximum size of default boxes are set to 75, 155, 235 and 315 respectively, aspect ratios are set to 1, 2, 3, 5, 7 and 10. Learning rate is initially set to 0.0001, and multiplied by 0.1 after every 2×10^4 iterations. In detection layers, we adopt the convolution kernel with width of 5 and height of 1.

4.3   Experiment Results

Firstly, We pre-train our feature extraction network on ImageNet. It achieves an accuracy of 67.5%. The accuracies along with iteration are shown in the Fig. 4. Then we use text dataset to train the end-to-end detection network with three branches and obtain a baseline mAP of 44%.

To find the most proper position to add deconvolution layer, we conduct several experiments on the same dataset. Results demonstrate that a deconvolution layer added after 4th inception module is the most cost-effective implementation

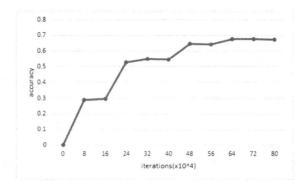

Fig. 4. This line chart shows the change of accuracy of feature extraction network on ImageNet during training phase. It tends to be stable after the iteration of 8×10^5 .

plan. It leads to 7% improvement of mAP. The improvement indicates that the fusion of multiple feature maps is helpful for the presentation of text.

After observing the test results, we find the disadvantage of the network in large object detection. Therefore another detection branch with larger default boxes is added at last additional pooling layer. It brings 2% improvement of mAP.

So far, our method achieves 0.75, 0.57, 0.6 in precision, recall and F-measure on ICDAR 2013 test set. The results are summarized and compared with other methods in Fig. 5. All these methods are trained and evaluated on the same datasets.

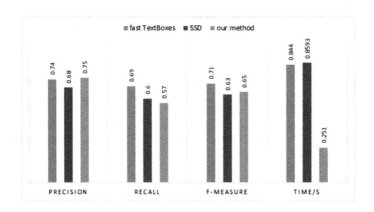

Fig. 5. This chart presents the comparison of precison, recall,F-measure and running time per image between fast TextBoxes, SSD and our method on ICDAR 2013 test dataset.

Although our method is not state-of-the-art, it has the advantage of speed. It takes only 0.251 s per image on the NVIDIA GeForce GT 640 graphics card, and

Fig. 6. The first line is the examples of correct detection, the second line is the examples of false detection, the third line is the examples of missing detection.

is 3 times faster than fast TextBoxes which takes 0.844 s per image on our platform. Convolution kernel with small size in tiny feature extraction network and less branches in detection layer are both helpful for acceleration. Experimental results indicate that our method is obviously efficient. Some experimental results are shown in Fig. 6. Some of the errors are caused by the similarity of texture and text, and some missing detections are caused by the various sizes of the text.

5   Conclusion

In this paper, we have presented a fast fully convolutional network for text detection. A tiny network is designed to extract text features and a additional deconvolution layer is applied to fuse information of multiple feature maps. Considering the character of text, we use specific filters in convolution layers. Our detection layer is based on the method of SSD. We evaluated our method on standard benchmark- ICDAR2013 and proved that it obtain comparable results with faster speed. In the future, we will improve our method with larger training dataset.

References

1. Dai, J., Li, Y., He, K., Sun, J.: R-FCN: object detection via region-based fully convolutional networks (2016)
2. Donahue, J., Jia, Y., Vinyals, O., Hoffman, J., Zhang, N., Tzeng, E., Darrell, T.: Decaf: a deep convolutional activation feature for generic visual recognition. arXiv preprint arXiv:1310.1531 (2013)
3. Epshtein, B., Ofek, E., Wexler, Y.: Detecting text in natural scenes with stroke width transform. In: Computer Vision and Pattern Recognition, pp. 2963–2970 (2010)
4. Gers, F.A., Schmidhuber, J., Cummins, F.: Learning to forget: continual prediction with LSTM. Neural Comput. **12**(10), 2451 (2000)
5. Girshick, R.: Fast R-CNN. Computer Science (2015)
6. Gupta, A., Vedaldi, A., Zisserman, A.: Synthetic data for text localisation in natural images. In: Computer Vision and Pattern Recognition, pp. 2315–2324 (2016)
7. He, K., Zhang, X., Ren, S., Sun, J.: Deep residual learning for image recognition, pp. 770–778 (2015)
8. Huang, W., Lin, Z., Yang, J., Wang, J.: Text localization in natural images using stroke feature transform and text covariance descriptors. In: IEEE International Conference on Computer Vision, pp. 1241–1248 (2013)
9. Huang, W., Qiao, Y., Tang, X.: Robust scene text detection with convolution neural network induced MSER trees. In: Fleet, D., Pajdla, T., Schiele, B., Tuytelaars, T. (eds.) ECCV 2014. LNCS, vol. 8692, pp. 497–511. Springer, Cham (2014). https://doi.org/10.1007/978-3-319-10593-2_33
10. Jaderberg, M., Simonyan, K., Vedaldi, A., Zisserman, A.: Reading text in the wild with convolutional neural networks. Int. J. Comput. Vision **116**(1), 1–20 (2016)
11. Karatzas, D., Shafait, F., Uchida, S., Iwamura, M., Bigorda, L.G.I., Mestre, S.R., Mas, J., Mota, D.F., Almazn, J.A., Heras, L.P.D.L.: ICDAR 2013 robust reading competition. In: International Conference on Document Analysis and Recognition, pp. 1484–1493 (2013)
12. Krizhevsky, A., Sutskever, I., Hinton, G.E.: Imagenet classification with deep convolutional neural networks. In: Advances in neural information processing systems, pp. 1097–1105 (2012)
13. Liao, M., Shi, B., Bai, X., Wang, X., Liu, W.: Textboxes: a fast text detector with a single deep neural network (2016)
14. Liu, W., Anguelov, D., Erhan, D., Szegedy, C., Reed, S., Fu, C.-Y., Berg, A.C.: SSD: single shot multibox detector. In: Leibe, B., Matas, J., Sebe, N., Welling, M. (eds.) ECCV 2016. LNCS, vol. 9905, pp. 21–37. Springer, Cham (2016). https://doi.org/10.1007/978-3-319-46448-0_2
15. Ma, J., Shao, W., Ye, H., Wang, L., Wang, H., Zheng, Y., Xue, X.: Arbitrary-oriented scene text detection via rotation proposals (2017)

16. Neumann, L., Matas, J.: A method for text localization and recognition in real-world images. In: Kimmel, R., Klette, R., Sugimoto, A. (eds.) ACCV 2010. LNCS, vol. 6494, pp. 770–783. Springer, Heidelberg (2011). https://doi.org/10.1007/978-3-642-19318-7_60

17. Redmon, J., Divvala, S., Girshick, R., Farhadi, A.: You only look once: unified, real-time object detection, pp. 779–788 (2015)

18. Ren, S., He, K., Girshick, R., Sun, J.: Faster R-CNN: towards real-time object detection with region proposal networks. IEEE Trans. Pattern Anal. Mach. Intell. **39**(6), 1137 (2016)

19. Simonyan, K., Zisserman, A.: Very deep convolutional networks for large-scale image recognition. arXiv preprint arXiv:1409.1556 (2014)

20. Szegedy, C., Liu, W., Jia, Y., Sermanet, P., Reed, S., Anguelov, D., Erhan, D., Vanhoucke, V., Rabinovich, A.: Going deeper with convolutions. Eprint Arxiv (2014)

21. Tian, Z., Huang, W., He, T., He, P., Qiao, Y.: Detecting text in natural image with connectionist text proposal network. In: Leibe, B., Matas, J., Sebe, N., Welling, M. (eds.) ECCV 2016. LNCS, vol. 9912, pp. 56–72. Springer, Cham (2016). https://doi.org/10.1007/978-3-319-46484-8_4

22. Yao, C.: Detecting texts of arbitrary orientations in natural images, vol. 157, no. 10, pp. 1083–1090 (2012)

23. Yao, C., Bai, X., Liu, W.: A unified framework for multioriented text detection and recognition. IEEE Trans. Image Process. **23**(11), 4737–4749 (2014)

Automatic Image Description Generation with Emotional Classifiers

Yan Sun and Bo Ren[✉]

Nankai University, Tianjin, China
rb@nankai.edu.cn

Abstract. Automatically generating a natural sentence describing the content of an image is a hot issue in artificial intelligence which links the domain of computer vision and language processing. Most of the existing works leverage large object recognition datasets and external text corpora to integrate knowledge between similar concepts. As current works aim to figure out 'what is it', 'where is it' and 'what is it dong' in images, we focus on a less considered but critical concept: 'how it feels' in the content of images. We propose to express feelings contained in images via a more direct and vivid way. To achieve this goal, we extend a pre-trained caption model with an emotion classifier to add abstract knowledge to the original caption. We practice our method on datasets originated from various domains. Especially, we examine our method on the newly constructed SentiCap dataset with multiple evaluation metrics. The results show that the newly generated descriptions can summarize the images vividly.

Keywords: Automatic image description · Emotional classifier

1 Introduction

Automatically describing the content of images in the form of a sentence is an important but challenging problem in the field of artificial intelligence. In the last few years, many works have been done and made considerable progress in the development of image caption. Generation of a standard description attach to an exact image requires comprehensive knowledge including object detection, language processing and so on.

The idea of image caption is to convey the content of images from vision to language knowledge. Particularly, much attention has been paid to object or region in the image. Referring expression and accuracy boosting are two major points in the development of image caption. As Fig. 1 shows, in the recent interest of image caption, content modeling usually results in the form of 'what is it', 'where is it' and 'what is it doing' in the image. To some extent, related captions reflect actual substances in images precisely. Either in a global or local way, details in images are expressed. Though these sentences look reasonable, it can hardly to be said as a human-like expression. There is usually some more

© Springer Nature Singapore Pte Ltd. 2017
J. Yang et al. (Eds.): CCCV 2017, Part I, CCIS 771, pp. 748–763, 2017.
https://doi.org/10.1007/978-981-10-7299-4_63

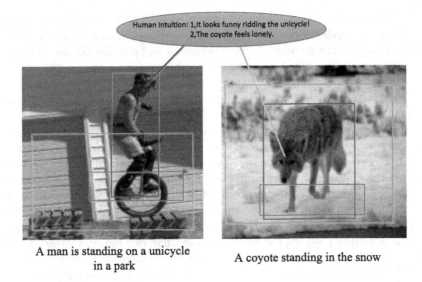

A man is standing on a unicycle in a park

A coyote standing in the snow

Fig. 1. There are two images with the corresponding caption published by [14]. The captions reflect the information: 'what is it', 'where is it' and 'what is it dong' precisely. In human intuitions, do these captions reflect all information people can capture from the above images? The answer is NO.

information extracted in human intuition. When human looks at the left one in Fig. 1, they always wonder if it is funny riding a unicycle. The man playing in the picture looks exciting on that. As to the right, there must be someone who will sigh 'What a pity coyote walking in the snow!'. Except for representational existences, sensible meanings are hidden in images which come from human culture, experiences and so on.

Concrete captions have been widely studied. The universal forms has been adopted in not only works [9,10,16,22,40–42] attempt to model the whole image, but also researches [8,11,13,21,32,37] based on local regions. Besides, there are also many works [5,6,14,28] contributing to extending the caption to hidden objects in current datasets in a various way. These work expand the range of real recognition in images. With more and more images uploaded in all kinds of social network or photo sharing sites, is it enough to mimic human intuition with simple detection on visible objects? In contrast, in another field of computer vision, researchers argue that there are multiple emotions are contained in images. Especially, they pointed out that abundant human intuition and emotion are hidden in creative applications such as art, music, but existing works lack the ability to model and utilize this information. Though the [30] modeled adjective descriptions with the aid of ANPs, sentiments haven't been presented directly or clearly as ANPs are essentially assistant tools for emotion prediction but not destinations. Strong sentiments embedded in images strength the opinion conveyed in the content, which can influence others effectively. Aware of components of emotions included in a picture will benefit the understanding of creative applications. [39] suggested *Emotional Semantic Image Retrieval* (ESIR), which is built to mimic human decisions and provide the user with tools to integrate their emotional components to their creative works. [26] classified images in affective aspect by [39] using low-level features, extracting information from texture and colors. [1] crawled large-scale images from the web to train a sentiment detector targeting at affective image classification and learning what is there in pictures. Affective information derives from the image is obtained. Images are increasing rapidly among human life, not only exist as a figural narration but also a carrier of affects. Nowadays, 'kid is not only a kid' in human intuition, the mind state of the child always being attractive, but the feeling also brought from animals in the picture usually the point. Meanwhile, vividness is a test indicator in the field of mental imagery [33]. Just like the Fig. 2 shows, traditional caption gives an illustration of what there is in the image as the left one. The description sketch can point to more than one figure. Instead, the illusion at right gives a more vivid description with affective expression 'with funny looks'. The caption with newly added phrase makes the subfigure against the orange background distinguished from others.

Prior works about captioning images have made significant progress, it is exciting and critical to further the description of images to the affective level which gives an understanding both figural and vivid. To achieve the goal of affective caption, data with emotions is needed. Constructing new datasets with affective description leads to huge consumption of time and human resources. Considering this situation, we plan a cross-domain learning process which is build on data from both image caption and affective image classification. Conclusively, in this paper, we propose to describe images with a novel concept. We mine the existent but abstract information contained in images, expressing it in a caption and convey more vivid story than speak in a flat style. Our contribution can be grown into two folds: (i) we come up with the addition of novel affective

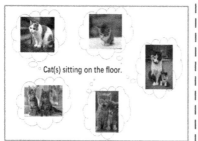

Fig. 2. The left: the traditional caption gives a description of what exists in the image and what it is doing. There are no more emotional feelings expressed. Information conveyed by traditional caption can be simply copied by a black and white picture. The right: to some extent, the addition of one more affective adverb brings colorful components, lives up to human intuition when seeing the picture.

concept in image caption to generate more vivid descriptions; (ii) To combine the caption with affective concepts, we propose to learn with cross-domain data which utilizes existing datasets designed for image caption and image sentiment. The interaction between cross domain data brings human-sources thrift to the implement of our method.

2   Related Works

Our paper draws on recent works in the image caption. The link to affective image classification tasks makes more vivid descriptions for images. Processing of recognition content and affective components allows the creation of affective caption.

2.1   Image Captioning

Describing images with natural language has become a focal task nowadays. General paradigms dealing with image caption can be split into two part: top-down and bottom-up. The top-down strategy targets at transforming images to sentences. Nevertheless, the bottom-up strategy picks out words respond to a different part of images and then combine them into sentence form. With the development of the recurrent neural network, the top-down paradigms show better performance. The models of the deep formulation are widely realized in top-down approaches. On the one hand, Incorporation of CNN and RNN is adopted in many works [9,20,38]. CNN serves as a high-level feature extractor while RNN models the sequence of words to generate captions. On the other hand, CNN and RNN models are integrated into a multimodal framework [25,28,29]. CNN and RNN interact with each other in the multimodal space and predict captions words by words. What's more, deep captioning is not addicted to merely descriptions of the whole image any longer. Goal orientation of many recent works [8,11,13,21,27,32,37] is to create exclusive descriptions in allusion to one exact region in the image. Such captioning task is named as 'referring expression'. Besides, [18] proposed a Fully Convolutional Localize Network (FCLN) for dense captioning based on object detection and language models. The FCLN model obtains a set of regions and associates them with captions. Far more abundant information remains captured with region captioning.

Apart from contributions to accurate descriptions and precise localization, the coverage of captioning is expanded as well. [14,28] succeeded in making it available to describe objects first occur. [28] constructed three novel concept datasets supplementing the information shortage of objects while [14] learned knowledge from unpaired image data and text data.

With so much contribution been devoted to the fields of image caption, we proposed to introduce another novel concept for image description. Unlike prior works focus on the coverage of real recognition, we aim to convey emotional components hidden in images for a vivid expression. Similarly, [30] proposed to add adjectives in the descriptions on the basis of sentibank [1]. According to [30], emotional information is also what they considered to capture in the generation of captions. Given ANPs they depended on, it is one assistant step for the prediction of sentiments. Images in the range of some ANPs are crawled from the Flickr without any checkout.

2.2 Emotion Modeling

Image sentiment attracts more and more attention nowadays. Many researchers attempt to model emotion existence contained in images. Particularly, affective image classification is the hottest topic in emotion modeling. Inspired by psychological and art theories [26,39] proposed to extract emotions with texture and color features. As to the *Semantic Gap* between low-level features to human intuition, mid-level representations are studied to fill the gap. [46] come up with the principle-of-arts features to predict which emotion is to one image. Meanwhile, [1] utilize the rich repository of image source on the web, generating an emotion detector. The detector is in 1,200 dimensions which are related to Adjective Nouns Pairs(ANPs). [3,7,43] tried to improve the affective image classification in various way. More recently, a deep neural network has been introduced extensively to deal with emotion modeling problems. [2] extends the sentibank to DeepSentibank based upon Caffe [17]. Besides, trial on the deep architecture [43] and reliable benchmark [44] is established as well.

3   Overview

Figure 3 gives an overview of our proposed framework on the affective caption. The process consists of two parallel parts: caption channel and affective classification channel.

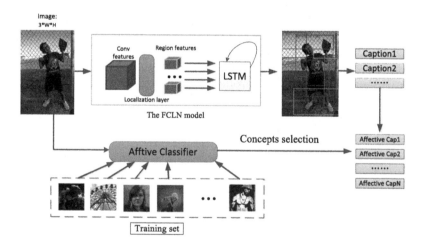

Fig. 3. Pipeline: the generation of affective captions consists of two channel. One is the extraction of traditional captions following the FCLN model. The other channel is affective concepts preparation via emotion classification.

For the caption channel, the input images are sent into the caption architecture based upon a neural network. This part is to obtain dense and traditional captions from images. The output captions serve as foundations which are to be modified with affective concepts. As for the affective classification, it manages to capture emotion attributes in images. The affective channel can be implemented with the caption channel simultaneously. Through the caption channel, images will be divided into different categories. We treat sentimental categories as different baskets, in which various affective concepts are included. Concepts will be mapped to images in the same basket as the output of the classification channel. Intermediate objects come out from different channels combined via a fusion method. Therefore, novel affective caption referring to exact images emerge. Details of channel conduction and fusion will be given in the Section Approach.

Also, there is no existing data simultaneously specialized for image captioning and affective classification to the best of our knowledge. Data collection and annotation require large demand of human sources and time consumption. Given the reality, we argue that it is reliable to train the emotion classifier with cross domain data when we evaluate our results with datasets designed for captioning. In the next, we will give a discussion on the selection of data.

3.1 Data Preparation

As mentioned in above section, we utilize emotion datasets to train the affective classification dependently. The emotion classifier will be applied to caption datasets for the extraction of emotional components.

First, we briefly browse several datasets for affective classification tasks. There are datasets collected in various types. **IAPS** is the earliest and the most common stimuli set in tasks of affective image classification, which is proposed by [23]. Pictures in IAPS are mainly depicting complex scenes. Usually, its subset **IAPSa** is often used in visual sentiment researches. Besides, [26] constructed **Artphoto** and **Abstract Paintings**. The two datasets are formed of pictures with artistic attributes. Artphotos are comprised of works of photographers. Meanwhile, Abstract Paintings are a set of works of art. Above three datasets are established on a regular scale whose magnitude is under a thousand. [44] established a larger scale dataset for image emotion recognition. For short, we call the dataset **FI**. This dataset not only expands the data to a greater magnitude but also balance in each category. The definition system *Amusement, Anger, Awe, Contentment, Disgust, Excitement, Fear, Sad* proposed by [31] is widely used in emotion models. Datasets mentioned above all followed this definition. In this paper, we will support the standard as well. Comparing above-mentioned emotion datasets, the magnitude of **FI** is large enough, and the Amazon Mechanical Turk(AMT) assures the reliability of labels. Therefore, we determine to train the emotion classifier on the basis of the dataset **FI**.

On the other hand, Flickr8k, Flickr30k, and MS COCO have become widely used datasets in the domain of image caption. **Flickr8k:** The dataset [35] consists of over 8,000 images from the largest photo sharing site: Flickr. Each image is provided with five descriptions. Expert annotation is supplied as well. **Flickr30k:** The dataset is similar to Flickr8k [45]. Five sentences are attached to each image. Both of Flickr8k and Flickr30k are designed for image-sentence retrieval process. **MS COCO:** This dataset is proposed by [24], including more than 80,000 images with five sentences annotations, whose magnitude is ten times larger than Flickr8k and Flickr30k. [30] recently annotate the **Senticap dataset** whose images are overlapped between ANPs. They designed the rewriting task upon the original captions from MS COCO. Though we argue that ANPs is just an assistant tool for emotion predict, the SentiCap annotation is closer to affective captions than traditional annotations. Thus, we evaluate our method on the **Senticap dataset**.

3.2 Affective Concepts

Affective concepts are what we will add to traditional captions. These concepts can be adverbs, phrases and so on. In the fusion with traditional captions, the affective concept is treated as an entity. It plays an important role in the generation of the affective caption. The selection of affective concepts depends on the sentimental label related to each image. Each emotion can be mapped to multiple concepts. In tasks of affective image classification, two linguistic-based language models – SentiWordNet and SentiStrength are widely utilized. They contain large quantities of sentimental vocabularies. Inspired by the two ontologies, we select some concepts referring to the eight emotion categories. Table 1 shows some examples similar to fixed categories.

3.3 Model Selection

In the early days, the model from [41] illustrated to describe images in two steps: (a) image parsing; (b) text description. Decomposition split images into parts to find out the relations between objects or part. The way [19] models captions is similar to [41]. Bidirectional model sentence mapping is proposed. Association is learned via image datasets and contexts. Fragments related to objects detection and sentences are pointed. Furthermore, as there is not always one principal object exist in pictures, referring expression started drawing attention. [20] detects regions in images and treat the associated sentence as a rich label space. From this, the model in [20] involves correspondences in the training set and then learns to generate new descriptions. Besides extension on content, novel description concepts are introduced by the model in [28]. New words are updated in original models. The datasets with novel concepts are constructed to train the model. Images with novel concepts can be described. Surprisingly, [14] models the unique concepts without the need of prepared data supplement. Simply unpaired image and text help the model recognize novel concepts. Besides, [18] propose DenseCap to capture rich information contained in images, which mines adequate sources in limited data. In that work, an end-to-end framework is constructed with localization layer. There is no need of extra bounding box preparing. Thus sentences are generated in a convenient and direct way.

In the first, captions are made on the whole image. With time goes by, captions approach to more and more correctness. Pointed objects or regions is detected. In our work, the dataset IAPS is well labeled but lack in scale. In the advantage of models strive in areas, much more referring sentences or subfigures can be cropped from original datasets. To have the privilege, we followed the model guide published by [18]. Bounding boxes are gathered to complement affective concepts.

4 Approach

The goal of this section is to generate affective captions with the aid of emotion prediction and traditional image caption. Given an instance image $\mathbf{I}$, we define the $\mathbf{X} = \{x\_1, x\_2, ..., x_d\}$ presenting the visual features, where $d \in \mathbb{R}$. A sequence of words $\mathbf{Y} = \{y_1, y_2, ..., y_n\}$ on behalf of the original caption. We target at extending the traditional caption to an affective expression, which is indicated as $\hat{\mathbf{Y}} = \{\hat{y_1}, \hat{y_2}, ..., \hat{y_n}, \hat{y_{n+1}}\}$. The extra y_{n+1} refers to the entity of affective concepts. For the generation of affective captions, we need to prepare the original caption $\mathbf{Y}$ and affective concept y_{n+1} separately.

4.1 Concrete Modelling

In this section, we aim to obtain general captions of images. We follow the FCLN model proposed by [18] which concentrate on Dense Captions. Dense Descriptions make it closer for adequate information. Also, it can be treated as a multiplier for affective data. In this model, dense information is gained based on bounding boxes which are computed in the new localization layer. Region proposals are predicted via regressing offsets from a set of translation-invariant anchors. The FCLN model adopted the parameterization in [12]. Proposals are regressed from anchors with certain scalars settings. B most confident bounding boxes are subsampled from the large quantities of proposals. As sizes and aspect ratios are varying among proposals, the FCLN model applied the bilinear interpolation for consistent dimension features. Features of chosen bounding boxes are output from the localization layer and serve as input for the generation of captions (Fig. 4).

With regions prepared, RNN language model is used to generate descriptions related to regions. The input for the RNN model is a sequence of words $\{y_0, y_1, ..., y_n, y_{n+1}\}$. Additional variances y_0 and y_{n+1} on behalf of the start code and end code. LSTM units are used to model the description sequence. A series of hidden neurons h_t are computed to formulate output generation $\hat{\mathbf{Y}} = \{\hat{y_1}, \hat{y_2}, ..., \hat{y_n}\}$. Recurrent formula function is used as Eq. 1.

$$\hat{y}_t = f(h_{t-1}, x_i); \qquad (1)$$

The subscript t means timestep in the LSTM model, which follows [15] in the FCLN model. 'Gates' are the main composition in the structure of the memory cell. Whether value from layers to be kept or not is up to 'Gates'. Gate '1'

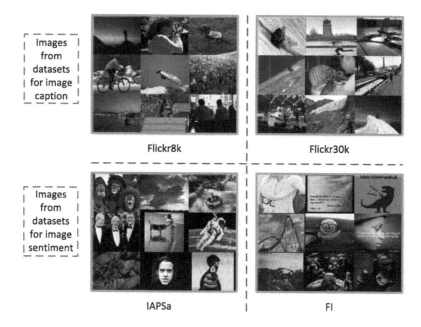

Fig. 4. Examples images: we present several images from a different domain. The top shows images from image caption. The bottom shows images for affective image classification tasks.

Table 1. Examples: affective concepts ontology according to eight emotions.

Categories	Concepts
Amuse	In funny looks; in funny pose; funnily; comic; comedy
Awe	Amazingly; marvelous; awe-inspiring; alarmingly; exclamatory
Content	Contentedly; complacently; contented; with satisfaction; happy
Excite	Excited; excitedly; feverishly; thrillingly; heatedly; wind up
Anger	Angry; angrily; annoyed; annoying; burned out
Disgust	Disgustingly; disgusted; sick; nasty
Fear	Afraid; fearful; sacred; frightened; in terror
Sad	Sad; sadly; grieved; unhappy; sorrowful; heartbrokenly

means value preserved while gate '0' stands for value to be forgotten. The input value and output value are defined as i and o separately. As h_{t-1}, c_{t-1} and x_t are already known at the time t, the hidden unit h_t can be modulated up to h_{t-1} and x_t. When meets the end code y_{n+1}, the process ends. Based on the formulation of a recurrent neural network, the original dense captions are prepared.

4.2   Affective Concepts Predicting

The prediction of affective concepts plays an important role in the generation of affective captions. An affective concept is treated as an entity to be added in a flat caption for the generation of an affective caption. Each affective concept is related to an emotion category. Therefore, we claim to select affective concepts in a hierarchical way. First of all, the emotion class of the image or region is needed. An emotion classifier satisfies the requirement inherently. Then we scan concepts refer to the predicting class and pick out a proper one to be embedded into captions. Details are given in the next. Affective concepts are denoted as $y_{\hat{n}+1}$ in the final form of affective captions. As what we mentioned before, we use the emotion dataset FI training an emotion classifier. Features $\mathbf{x}'$ will be mapped to one of eight labels l_{emo}. Corresponding to each category, labels are related to a series of affective concepts: $l_{emo} \rightarrow \{concept_1, concept_2, ..., concept_k\}$, the number of concepts selected in each category is denoted as k. A random function is used to pick out feasible concepts:

$$concept_{select} = rand(set(l_{emo})) \tag{2}$$

5   Experiments

Our experiments employ a caption model and an affective classifier in a parallel way. In the followed model supported on the project page of [18], 1,000 is set as the maximum of bounding boxes and captions. However, too much consideration on the prediction of objects is unnecessary and consume lots of time as well. It is either needless for information extraction. Concentration should be paid on the generation of vivid telling. Therefore, we changed the maximum of bounding box into 10 for convenience. In the paradigm of affective caption generation, emotional components are discriminative considering the novel role they play in the description. For the addition of emotional components, eight emotions are utilized in the paper. To satisfy the syntax structure, we convert the predicting emotion to responding adverb for the proper embedding. At the first step, the sentiment of each image is decided via the global features. Features of images are sent into the emotion classifier to obtain their referring sentiments. In the baseline experiment, simple concepts are chosen to generate comparable captions. As for the eight emotions, they are represented by adjectives. They are transformed into common forms of adverbs or phrases. Instances like : {$Amuse \rightarrow happily$, $Awe \rightarrow amazingly$, $Content \rightarrow contentedly$, $Excite \rightarrow excitedly$, $Anger \rightarrow looks\,anger$, $Disgust \rightarrow queasily$, $Fear \rightarrow in\,terror$, $Sad \rightarrow sadly$}. Thus traditional captions appended with affective concepts comprise of the novel emotional descriptions.

To have our evaluations effective, we conduct the implementation on the **Senticap dataset**. The novel annotations with ANPs gives a comparable benchmark for affective captions generated from our model. The influence of affective concepts would be better judged against emotional components.

Table 2. Evaluations on the Senticap dataset under six metrics ($BLEU\_4$, $BLEU\_3$, $BLEU\_2$, $BLEU\_1$, $CIDEr$, $METEOR$). The n in BLEU_n on behalf of the n-gram co-occurrences. $top1$ to $top5$ correspond to the confidence level referring to captions. The left-most row presents the model used in the generation of captions.

Model	rank(%)	BLEU_4	BLEU_3	BLEU_2	BLEU_1	CIDEr	METEOR
Densecap [18]	top1	0.7	2.0	5.2	13.1	4.1	4.9
	top2	0.8	1.9	4.8	12.3	3.6	4.7
	top3	0.7	1.6	4.2	11.5	2.8	4.3
	top4	0.7	1.6	4.2	11.8	2.9	4.3
	top5	0.4	1.2	4.0	11.3	2.5	4.3
$EmoCap_{vgg}$ (ours)	top1	0.9	2.4	6.6	17.7	3.3	5.2
	top2	0.9	2.3	6.2	16.8	3.1	4.9
	top3	0.7	1.9	5.5	15.9	2.4	4.6
	top4	0.8	1.9	5.5	16.1	2.4	4.5
	top5	0.5	1.5	5.1	15.7	2.1	4.6
$EmoCap_{vggft}$ (ours)	top1	0.7	2.3	6.5	17.3	3.3	5.1
	top2	0.9	2.2	6.1	16.4	3.0	4.8
	top3	0.7	1.9	5.4	15.3	2.3	4.5
	top4	0.8	1.9	5.4	15.7	2.3	4.5
	top5	0.5	1.5	5.0	15.1	1.9	4.5
$EmoCap_{bank}$ (ours)	top1	0.8	2.4	6.5	17.5	3.3	5.1
	top2	1.0	2.3	6.2	16.7	3.1	4.9
	top3	0.7	2.0	5.5	15.8	2.4	4.6
	top4	0.8	1.9	5.4	16.0	2.4	4.6
	top5	0.5	1.5	5.1	15.4	2.1	4.5

5.1   Implementation

Conduction of affective caption generation implements in two different ways. Relatively speaking, the first way is simply but rigid generation. It is feasible setting the affective adverbs appendix at the end of sentences. Emotional components are introduced forcibly.

To have the emotion classifier applicable to images may be in plain affection, we decide to adopt the VGG-16 for the extraction of features in the model $EmoCap_{vgg}$. In addition, the FI dataset is used to fine-tune the VGG net as well. Another model practices with the fine-tune features is denoted as $EmoCap_{vggft}$. The specially designed emotion detector sentibank is also applied in the classification for sentiments. The model with the aid of sentibank features is recognized as $EmoCap_{bank}$. Then each image from the field of captioning is attached with exact sentiments. With classification based on features, emotion label is fused with corresponding generation caption, and a novel description is created. Features of cropped regions are extracted via the same feature extractors mentioned before. Then they are sent into the prepared emotion classifier to tag their novel sentiments, which may be different via different features.

In the equation, I correspond to an instance image. $Score_{mother}$ represents emotion scores predicted for mother images. $Score_{r_{old}}$ on behalf of regions' emotion scores given no global thoughts which $Score_{r_{new}}$ is the final score after the combination of local and global scores. As Eq. 3 show:

$$Score_{r_{new}}^{I} = Score_{r_{old}}^{I} * \alpha + Score_{mother}^{I} * \beta \tag{3}$$

5.2 Evaluation

To evaluate our system in the generation of novel affective concepts, we adopt common automatic judgements, the $BLEU$, $METEOR$ and $CIDEr$ metrics, which derive from the Microsoft COCO evaluation software [4]. These metrics is also utilized in [30].

Table 2 shows the results of region captions from the SentiCap Dataset. Affective concepts conducted on $EmoCap_{vgg}$, $EmoCap_{vggft}$ and $EmoCap_{bank}$ are evaluated. Compared with the traditional captions generation from the $Densecap$. We extracted the top five captions which are predicted from the model. Captions in each rank are evaluated separately. The higher the rank, the preciser captions are. From the above tables, we can see that the top 1 performs well in the overall evaluation. In $BLEU$ metrics, numbers behind refer to the n in n-gram. The differences between ground truth and generation caption become stronger as the metrics go from unigram to 4-gram [34]. We use C as the candidates set and S as the reference set. The BLEU score is decided by a modification probability(P_n) and a Brevity Penalty (b). The P_n computes the counts n-grams appear in candidates set, and

$$P_n(C, S) = \frac{\sum_c \sum_{n-gram \in C} Count_{clip}(n - gram)}{\sum_{c'} \sum_{n-gram' \in C'} Count(n - gram')} \tag{4}$$

where $Count_{clip}$ means the maximum counts of words appear in annotations but not exceed the biggest counts obtained from any references of words. The subscript c correspond to $C \in Candidates$. so does c'. The Brevity Penality is determined as:

$$b(C, S) = \begin{cases} 1, & if \quad l_c > l_s \\ e^{(1-r/c)}, & if \quad l_c \le l_s, \end{cases} \tag{5}$$

where l_c refers to the length of candidate sentences while l_s refers to the length of the reference. Thus the final BLEU score is gain via the following equation:

$$BLEU\_n(C, S) = b(C, S)exp(\sum_{n=1}^{N} \omega_n log P_n(C, S)) \tag{6}$$

The evaluation of $METEOR$ is computed on the basis of uni-gram. Precision and Harmonic mean are considered. Though the relevance in segmentations is not cared while $METEOR$ gives thoughts to the coherence among uni-gram. The correlation between $METEOR$ and human judgment is higher than that between $BLEU$ and human judgment.

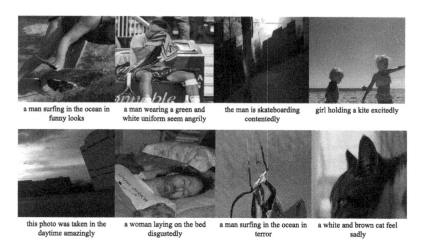

| a man surfing in the ocean in funny looks | a man wearing a green and white uniform seem angrily | the man is skateboarding contentedly | girl holding a kite excitedly |

| this photo was taken in the daytime amazingly | a woman laying on the bed disgustedly | a man surfing in the ocean in terror | a white and brown cat feel sadly |

Fig. 5. Eight selected images from the Senticap dataset. Each image represents an affective caption corresponding to one particular emotion. All of the captions are embedded with sentimental expressions.

Apart from above evaluations, [36] proposed the $CIDEr$, which is specially designed for the evaluation of image caption. The coherence in caption is evaluated via n-gram weighted via TF-IDF. The TF-IDF computes weights $g(\cdot)$ for each n-gram and take them into the final consideration of evaluations. The higher scores are, the better results are gain is applicable to all this evaluations.

5.3   Results

As the Table 2 shows, the $BLEU\_1$ to $BLEU\_4$ scores of affective captions are higher than scores of traditional captions. The differences among the three features are wispy. This condition may because evaluation metrics are up to all words in captions. The affective concepts take a sincerely small proportion. As for the $METEOR$ metrics, there is not any difference between the affective captions and traditional captions. Nevertheless, traditional captions get better scores under the $CIDEr$ standard. According to the evaluations, affective captions outperform in most evaluating metrics. It proved that affective captions give more human-like descriptions to images. The introduction of affective components makes sense against simply descriptions of objects. However, affective captions show lower performance on the $CIDEr$. As that the $CIDEr$ considers the coherence in captions, it may mean that the affective concepts are not idiomatic enough. Though there are shortages exist in affective captions, we have the confidence to say that affective concepts make senses. Emotional concepts added in captions give closer descriptions to human-intuition captions. Figure 5 presents example results with affective captions. Final emotions of each fragment are determined from themselves and weighted emotion from their mother images. In this figure, every description downward the referring fragments gives a sentimental expression of the content. Positive or negative ANPs used in [30] to present a state of objects, while affective concepts give an expression of sentiment directly. Additional adverbs in the novel caption convey moods from the principal subjects in photos. Compared to the traditional caption, conveying moods resonate human intuitions directly and make it easily associated.

6   Conclusion

From the results and example images with descriptions, it can be understood that the addition of affective concepts helps minds in photos brighter. The addition of affective concepts will make the caption of images more vivid. Affective concepts mine hidden sentiments in images and give appropriate expressions. Whats more, there are a lot of developments need to be accomplished. Out of evaluations, though emotion components are highlighted in captions, they are not smoothing enough in evaluations. There are all kinds of adverbs or phrases in the ontology of human literary describing human emotions. There can always be an elegant expression conveying various sophisticated sentiments. The prediction of emotions should be more delicate to capture complicated human feelings. Besides, the selection of position where affective concepts are embedded is to be improved. One of the expanding orientation is adaptive affective concepts addition. An end-to-end framework consists of the caption channel, and emotion classification channel would help the automatic generation of affective concepts. Moreover, the process of affective concepts could be down under various perspectives. Emotion classification gives an instruction for the selection of affective concepts. Further consideration of emotion distribution may lead to a better coverage for sentiments and preciser descriptions.

References

1. Borth, D., Ji, R., Chen, T., Breuel, T., Chang, S.F.: Large-scale visual sentiment ontology and detectors using adjective noun pairs. In: ACM MM (2013)
2. Chen, T., Borth, D., Darrell, T., Chang, S.F.: Deepsentibank: visual sentiment concept classification with deep convolutional neural networks. In: CVPR (2014)
3. Chen, T., Yu, F.X., Chen, J., Cui, Y., Chen, Y.Y., Chang, S.F.: Object-based visual sentiment concept analysis and application. In: ACM MM (2014)
4. Chen, X., Fang, H., Lin, T.Y., Vedantam, R., Gupta, S., Dollar, P., Zitnick, C.L.: Microsoft COCO captions: data collection and evaluation server. In: CVPR (2015)
5. Chen, X., Zitnick, C.L.: Learning a recurrent visual representation for image caption. In: CoRR (2014)
6. Chen, X., Zitnick, C.L.: Mind's eye: a recurrent visual representation for image caption generation. In: CVPR (2015)
7. Chen, Y.Y., Chen, T., Hsu, W.H., Liao, H.Y.M., Chang, S.F.: Predicting viewer affective comments based on image content in social media. In: ICMR (2014)
8. Deemter, K.V., Sluis, I.V.D., Gatt, A.: Building a semantically transparent corpus for the generation of referring expressions. In: INLG (2006)

9. Donahue, J., Hendricks, L.A., Guadarrama, S., Rohrbach, M., Venugopalan, S., Darrell, T., Saenko, K.: Long-term recurrent convolutional networks for visual recognition and description. Elsevier (1988)

10. Farhadi, A., Hejrati, M., Sadeghi, M.A., Young, P., Rashtchian, C., Hockenmaier, J., Forsyth, D.: Every picture tells a story: generating sentences from images. In: ECCV (2010)

11. FitzGerald, N., Artzi, Y., Zettlemoyer, L.S.: Learning distributions over logical forms for referring expression generation. In: EMNLP (2013)

12. Girshick, R.: Fast R-CNN. In: ICCV (2015)

13. Golland, D., Liang, P., Dan, K.: A game-theoretic approach to generating spatial descriptions. In: EMNLP (2010)

14. Hendricks, L.A., Venugopalan, S., Rohrbach, M., Mooney, R., Saenko, K., Darrell, T.: Deep compositional captioning: describing novel object categories without paired training data. In: CVPR (2016)

15. Hochreiter, S., Schmidhuber, J.: Long short-term memory. Neural Comput. 9(8), 1735–1780 (1997)

16. Hu, R., Xu, H., Rohrbach, M., Feng, J., Saenko, K., Darrell, T.: Natural language object retrieval. In: CVPR (2016)

17. Jia, Y., Shelhamer, E., Donahue, J., Karayev, S., Long, J.: Caffe: convolutional architecture for fast feature embedding. In: CVPR (2014)

18. Johnson, J., Karpathy, A., Li, F.F.: Densecap: fully convolutional localization networks for dense captioning. In: CVPR (2016)

19. Karpathy, A., Joulin, A., Li, F.F.: Deep fragment embeddings for bidirectional image sentence mapping. Adv. Neural Inf. Process. Syst. 3, 1889–1897 (2014)

20. Karpathy, A., Li, F.F.: Deep visual-semantic alignments for generating image descriptions. In: CVPR (2015)

21. Kazemzadeh, S., Ordonez, V., Matten, M., Berg, T.: Referitgame: referring to objects in photographs of natural scenes. In: EMNLP (2014)

22. Kuznetsova, P., Ordonez, V., Berg, A.C., Berg, T.L., Choi, Y.: Collective generation of natural image descriptions. In: ACL: Long Papers (2012)

23. Lang, P., Bradley, M., Cuthbert, B.: International affective picture system (IAPS): technical manual and affective ratings. Technical report, University of Florida, Gainesville

24. Lin, T.-Y., Maire, M., Belongie, S., Hays, J., Perona, P., Ramanan, D., Dollár, P., Zitnick, C.L.: Microsoft COCO: common objects in context. In: Fleet, D., Pajdla, T., Schiele, B., Tuytelaars, T. (eds.) ECCV 2014. LNCS, vol. 8693, pp. 740–755. Springer, Cham (2014). https://doi.org/10.1007/978-3-319-10602-1_48

25. Lynch, C., Aryafar, K., Attenberg, J.: Unifying visual-semantic embeddings with multimodal neural language models. In: TACL (2015)

26. Machajdik, J., Hanbury, A.: Affective image classification using features inspired by psychology and art theory. In: ACM MM (2010)

27. Mao, J., Huang, J., Toshev, A., Camburu, O., Yuille, A., Murphy, K.: Generation and comprehension of unambiguous object descriptions. In: CVPR (2016)

28. Mao, J., Wei, X., Yang, Y., Wang, J.: Learning like a child: fast novel visual concept learning from sentence descriptions of images. In: ICCV (2015)

29. Mao, J., Xu, W., Yang, Y., Wang, J., Huang, Z., Yuille, A.: Deep captioning with multimodal recurrent neural networks (m-RNN). In: ICLR (2015)

30. Mathews, A., Xie, L., He, X.: SentiCap: generating image descriptions with sentiments. In: AAAI (2016)

31. Mikels, J.A., Fredrickson, B.L., Larkin, G.R., Lindberg, C.M., Maglio, S.J., Reuter-Lorenz, P.A.: Emotional category data on images from the international affective picture system. Behav. Res. Methods **37**(4), 626–630 (2005)
32. Mitchell, M., Deemter, K.V., Reiter, E.: Natural reference to objects in a visual domain. In: INLG (2010)
33. Morina, N., Leibold, E., Ehring, T.: Vividness of general mental imagery is associated with the occurrence of intrusive memories. J. Behav. Ther. Exp. Psychiatry **44**(2), 221–226 (2013)
34. Papineni, K.: BLEU: a method for automatic evaluation of machine translation. Wirel. Netw. **4**(4), 307–318 (2002)
35. Rashtchian, C., Young, P., Hodosh, M., Hockenmaier, J.: Collecting image annotations using Amazon's mechanical turk. In: NAACL HLT Workshop (2010)
36. Vedantam, R., Zitnick, C.L., Parikh, D.: Cider: consensus-based image description evaluation. In: CoRR (2015)
37. Viethen, J., Dale, R.: The use of spatial relations in referring expression generation. In: INLG (2010)
38. Vinyals, O., Toshev, A., Bengio, S., Erhan, D.: Show and tell: a neural image caption generator. In: CVPR (2015)
39. Wang, W., He, Q.: A survey on emotional semantic image retrieval. In: ICIP (2008)
40. Xu, K., Ba, J., Kiros, R., Cho, K., Courville, A., Salakhutdinov, R., Zemel, R., Bengio, Y.: Show, attend and tell: neural image caption generation with visual attention. In: ICCV (2015)
41. Yao, B.Z., Yang, X., Lin, L., Lee, M.W., Zhu, S.C.: I2T: image parsing to text description. In: Proceedings of the IEEE, vol. 98, no. 8, pp. 1485–1508 (2010)
42. You, Q., Jin, H., Wang, Z., Fang, C., Luo, J.: Image captioning with semantic attention. In: CVPR (2016)
43. You, Q., Luo, J., Jin, H., Yang, J.: Robust image sentiment analysis using progressively trained and domain transferred deep networks. In: AAAI (2015)
44. You, Q., Luo, J., Jin, H., Yang, J.: Building a large scale dataset for image emotion recognition: the fine print and the benchmark. In: AAAI (2016)
45. Young, P., Lai, A., Hodosh, M., Hockenmaier, J.: From image descriptions to visual denotations: new similarity metrics for semantic inference over event descriptions. In: ACL (2014)
46. Zhao, S., Gao, Y., Jiang, X., Yao, H., Chua, T.S., Sun, X.: Exploring principles-of-art features for image emotion recognition. In: ACM MM (2014)

Author Index

Printed in the United States
By Bookmasters